WEIGHTS AND MEASURES
IN SCOTLAND:
A EUROPEAN PERSPECTIVE

Frontispiece:
Half-pint standard of the Burgh of Edinburgh, 1555, by David Rowan, master founder of the royal artillery works at Edinburgh Castle (Item **109** in the Inventory).

Weights and Measures in Scotland: A European Perspective

R. D. CONNOR

AND

A. D. C. SIMPSON

EDITED BY

A. D. MORRISON-LOW

National Museums of Scotland

TUCKWELL PRESS

Dedicated to the memory of

GORDON DONALDSON

1913-1993

The National Museums of Scotland and Tuckwell Press Limited
would like to thank the following organisations and individuals
for their assistance in the publication of this volume:

ISASC

MRS M. M. MACFARLANE

MR AND MRS ALEX R. NEISH

THE SCOTLAND INHERITANCE FUND

THE SCOTTISH HISTORICAL REVIEW TRUST

THE STRATHMARTINE TRUST

First published by
NMSE Publishing, a division of NMS Enterprises Limited
National Museums of Scotland, Chambers Street, Edinburgh EH1 1JF
and Tuckwell Press Limited
The Mill House, Phantassie, East Linton, East Lothian EH40 3DG

British Library Cataloguing-in-Publication Data
A Catalogue record for this book is available
on request from the British Library.

ISBN 1-901-663-884

The right of R. D. Connor, A. D. C. Simpson with A. D. Morrison-Low
to be identified as the authors/editor of this work has been asserted by them
in accordance with the Copyright, Design and Patent Act 1988.

Cover and internal layout design by Mark Blackadder; Illustrations by Christina Unwin;
Images managed by Richard A. Simpson; Index by Christopher Hobson.

Printed and bound in the United Kingdom by Cambridge Printing.

CONTENTS

PART I
A HISTORY OF
WEIGHTS AND MEASURES IN SCOTLAND

CHAPTER

PART II
INVENTORY OF
SURVIVING STANDARDS

PART III

APPENDICES

Acknowledgements

THIS WORK COULD NOT HAVE BEEN UNDERTAKEN WITHOUT THE HELP OF many people. We are grateful to librarians and archivists of the National Library of Scotland, the National Archives of Scotland, the Scottish university libraries, local authority libraries, and archives across the country for providing access to their material; and for their ready advice and assistance with our increasingly complex problems and with some very demanding requests. In particular, we would like to acknowledge the support of Margaret Deas, formerly of the National Library of Scotland, whose unique knowledge of the Library's source material has provided invaluable clues for us. We are also grateful to the staff of the University of Manitoba Library for their help. The National Museums of Scotland's Library has risen to the occasion when asked the impossible; and the Museums' photographers, Joyce Smith, Ken Smith and Leslie Florence, have photographed the impossible, often in seemingly impossible circumstances. In the production stage, Lesley Taylor, Director of NMS Enterprises Limited – Publishing, and her colleagues at NMS and Tuckwell Press, provided a calm, highly efficient and effective service.

Similarly, we owe a great debt of gratitude to the directors and curatorial staff of the National Museums of Scotland, the National Trust for Scotland, local authority museums, a number of independent museums, local authority weights and measures departments, and private collectors for access to relevant material in their care, and for allowing us to publish technical descriptions in this volume. We have also received valuable help from curators and collectors outside Scotland, as well as from auction houses and dealers.

We have greatly benefited from papers kindly provided by Philip Grierson at the start of our investigation, and from his help and encouragement on many occasions since then. A great many individuals working in the field of Scottish history have helped us in innumerable ways. It would be difficult to enumerate them all individually after more than fifteen years: we know who they are, and their contributions are not forgotten. All the same, we must provide honourable mention of: the late Gordon Donaldson, who encouraged us at every turn, T. C. Smout, Julian Goodare, Athol Murray, Norman Biggs, Diana Crawforth-Hitchins, David Sellar, Hector MacQueen, Alex Neish, staff at the *Dictionary of the Older Scottish Tongue* (now part of the Scottish Language

Dictionaries), the Orkney historian W. P. L. Thomson and the Shetland archivist Brian Smith (both of whom read the appendix on these islands and made valuable comments). However, any remaining errors are the responsibility of the authors.

The production of this volume has been made possible by welcome sponsorship from the International Society of Antique Scale Collectors (ISASC), American and European Chapters; Mrs M. M. Macfarlane, Edinburgh; Mr and Mrs Alex R. Neish, Barcelona; The Scottish Inheritance Fund; The Scottish Historical Review Trust; and The Strathmartine Trust.

R. D. Connor was supported by research grants from the Social Science Research Council of Canada and by the Carnegie University Trust, and acknowledges their generous contribution, as well as the extended support of the University of Manitoba. He was a Trustees' Research Fellow at the National Museums of Scotland from 1987, and thanks the Director and Trustees for this recognition. He also thanks Nancy Laxdal for typing the first draft of this work.

A. D. C. Simpson similarly acknowledges the support of the National Museums of Scotland and the NMS Charitable Trust; in 1991/92 he was a Visiting Research Fellow at the School of Scottish Studies at Edinburgh University, and is grateful for the contribution made by Professor A. Fenton, Director of the School, and his colleagues. He was helped greatly in the early years of this project by the assistance of Pat Walsh.

The support and advice of our colleague A. D. Morrison-Low over the years has been greatly appreciated, but her practical help and editorial labours in the closing stages of this enterprise have been crucial to getting it to publication.

But perhaps our greatest thanks go to our wives and families, who have lived with this book for far too long.

R. D. CONNOR, WINNIPEG
A. D. C. SIMPSON, EDINBURGH
JUNE 2004

FOREWORD

THE HISTORY OF WEIGHTS AND MEASURES IS CENTRAL TO UNDERSTANDING the history of a country. When wool is the main item of the export trade, it is crucial to know the volume of the woolsack as well as the number shipped abroad. When famine and plenty alternated, the movements of grain prices can only be understood if one is aware how the boll varied from place to place and from time to time. Consumers then, as now, had choice, but to compare their decision as to whether to buy a pound of candles or a pound of butter, we need to know which kind of pound each is measured in. If a historian does not know the true volumes of trade, the movement of prices or the worth of goods, how can the history of a nation be properly explained? Whole areas of economic experience will remain a closed book.

Unfortunately, to the non-specialist, the previous history of Scottish weights and measures looks positively chaotic, a term which has even been used by specialists who in the past have tried to unravel it. The study itself seems so esoteric, the conclusions so slippery and the opinions so various, it is hard ever to feel on safe ground. Earlier investigations with incomplete methodology and perspectives have sometimes only added to the confusion.

This volume, for the first time, brings a truer authority to the search. Not only is the study hugely erudite over many centuries, but it combines the skills of a very expert documentary historian from the University of Manitoba with the expertise of a museum curator who has examined all the surviving objects that were used as early Scottish weights and measures, scattered through the collections of burghs and museums large and small. This latter point makes it so appropriate that the National Museums of Scotland should undertake with Tuckwell Press the publication of this volume: the special strength of this investigation is its rootedness in the material objects measured and illustrated here. Only when you get hold of, for example, the medieval Inverkeithing ell of 1500 or the Craigengelt Weight (a trone stone of 1553) and subject them to properly minute investigation, can the true complexities be grasped and unravelled.

Complexities there certainly were, and complex turns out to be a better description of the situation than chaotic, for there was logic too. It is clearly established, for example, that David I based his suite of Scottish measures, the earliest we have, on English example – the same inch and the same ell, but the English changed their ell later and confused everyone. Similarly, the medieval woolsack was a standard, unvarying size: it had to be when international wool trade, in which several countries participated, was so largely based on a single centre at Bruges. Apparent variation was based on allowances for the purchaser implicit on whether the sack was weighed on a level or on a tilting beam. The pound weight would vary depending on whether a good was for internal or external trade: if the latter, it was likely to conform to an internationally recognised standard. Firlots and bolls would vary depending on whether they were heaped or levelled off, or how they were heaped. Landlords and others anxious to get hold of grain dues or purchases preferred their firlot heaped as high as possible; the sellers and producers of grain preferred it level. When the balance of agrarian power was in favour of landlords there was a tendency to inflate measures. The Union of 1707 produced its own complexities. Officially, transactions were to be in English weights and measures, uniform across the United Kingdom, but for local trade, people long preferred to use familiar local weights, and indeed it might be legally necessary under pre-existing leases and other agreements to use the older systems. When eighteenth-century mathematicians and experts then set to work to unravel and explain the variations and their English equivalents, they sometimes did as much to obfuscate the existing situation and its past history as to illuminate it.

All this Robin Connor and Allen Simpson, with the excellent assistance of Alison Morrison-Low, have explained as clearly and lucidly as anyone can. At last, in the fog of ancient measurement, we have a beacon to guide us through.

T. C. SMOUT
HISTORIOGRAPHER ROYAL IN SCOTLAND

INTRODUCTION

DIVERSE WEIGTHES, AND DIVERSE MEASURES,
BOTH THESE ARE ABHOMINATION UNTO THE LORD.

PROVERBS, chapter 20, verse 10

Quoted on the title-page of Alexander Hunter's
A Treatise of Weights, Mets and Measures of Scotland (Edinburgh, 1624).

ALEXANDER HUNTER, BURGESS OF EDINBURGH, PUT THIS APPOSITE quotation from the King James Bible on the title-page of his slender volume on land surveying practice, which was the first to discuss Scottish metrology and its equivalences. Who was Alexander Hunter? An Edinburgh merchant, wealthy enough to become a bailie on the town council in 1596-7, Hunter was named in 1592 as the King's Exchanger, responsible with the Master Coiner of the Scottish Mint for buying in sufficient bullion and old coin to supply a major recoinage which was then in progress.[1] He dedicated his Treatise to the Lord Chancellor of Scotland, George Hay of Kinfauns. His 68-page book, although designed by its author to obtain a post as a surveyor, is the starting-point for anyone thinking about Scottish weights and measures.

Hunter's previous position at the Edinburgh Mint gives his introductory comments on Scottish weights and measures added authority, but their brevity gives an indication generally of the slender nature of the literature on the subject. There are two further landmark works, one eighteenth-century, the other nineteenth-century. The first was that of John, Lord Swinton, *A Proposal for Uniformity of Weights and Measures in Scotland* (Edinburgh, 1779); the second by George Buchanan, *Tables for Converting the Weights and Measures hitherto in use in Great Britain ... also Abstracts of the Jury Verdicts throughout Scotland in regard to the Weights and Measures of Each County* (Edinburgh, 1829). The intention of both of these was to demonstrate the variety of current usage, Swinton in particular pressing for reform and standardisation; Buchanan demonstrating why it had been necessary to provide the legislation for the Imperial system in 1824. Both these works present a depressing picture of chaos and anarchy, and regional diversity which could not have assisted the economic well-being of the country.

In this, these authors have been all too successful: as a premise for Scottish metrological history, people have assumed that there really was chaos, and this is seen particularly in R. E. Zupko's 1977 article.[2] A serious attempt to unravel the problem was made by Lawrence Burrell, a senior Scottish weights and measures Inspector, in 1961.[3] However, because he was not used to using objects critically as evidence, his approach can now be seen to be incorrect, and, indeed, he made the problem worse. Several uncritical studies have appeared during the past 25 years, which have unfortunately perpetuated Burrell's misconceptions.

More recently, however, two serious studies have been published which address the problem of the evolution of Scottish weights and measures in the course of producing an economic history price series. These are the works by Elizabeth Gemmill and Nicholas Mayhew for the medieval period up to 1542; and by Alexander Gibson and T. C. Smout for the more recent period.[4] We had the pleasure of sharing information with all these authors in the course of their work.

We should perhaps explain how the present work has come about. Allen Simpson set up a small exhibition of weights and measures in 1972 in the then Royal Scottish Museum, which had been intended to set in context the introduction of metrication as then planned by the Metrication Board. This led him to an awareness that quite a number of early Scottish standards had survived, often languishing unrecognised in collections of museums and other bodies around the country. In turn, this made him realise that Burrell's analysis was deeply unsatisfactory and that it had not resolved the fundamental problem of how Scottish weights and measures had evolved. In the well-known manner of museum curators, he began to fill box files with scattered references over the years, with the intention of doing something about it eventually. That opportunity arose in 1986, when Robin Connor visited the Museum as his own volume on the *Weights and Measures of England* was drawing to completion. He had recently been presented with Philip Grierson's preliminary notes on Scottish sources, and was planning to undertake a study on the Scottish system. We agreed to collaborate, and Robin Connor joined the Museum as a Visiting Research Fellow in 1987, during which time he saw the English work through the press. The University of Manitoba was generous in its allocation of sabbatical leave, and good headway was made by Robin Connor in this period, making an independent collection of relevant texts, photocopies and reprints. Allen Simpson became more closely involved after the completion of another volume in 1989, and visits by one or other of us to Scotland and Canada continued over the following years. In 1992 Allen Simpson had nine months' sabbatical leave as a Visiting Fellow at the School of Scottish Studies, University of Edinburgh. The early 1990s was a fruitful time in which we were able to resolve most of the principal historical problems that the subject presented. However, Allen Simpson became progressively unwell during the late 1990s and finally retired in 1998. His immediate colleague, Alison Morrison-Low, was able to pick up the reins after finishing her doctorate in 1999. She has brought together and completed the text for publication.

The emphasis in this book is rather different from those made by other studies. Here, we have been concerned with how one should approach regulations and legislation in this rather distant period: to work out what can be interpreted literally, and what must have depended on unwritten measurement and trading procedures which may not square with our own modern notions of fair practice. Some aspects were meant to be followed to the letter; but others were not even mentioned, since this was not the intention of the legislation, and such silence would have been understood by contemporaries. We have been careful not to interpret the rules literally, but to recognise that these needed to be interpreted sympathetically in the light of the customary practice of the day.

We also recognise the pressing problem of looking critically at surviving objects, many of which had never been properly examined before. This, of course, is exactly the purpose of museums and the task of the museum curator, and it is an important feature of this book that we were able to bring together objects and documentary evidence to show how each can reinforce the other and inform interpretation. It would be fair to say that we could not have reached our conclusions about the customary allowances without the independent evidence of crucial surviving items. Indeed, some of the most remarkable material turned up in the most unexpected places, and perhaps the best example of how the holdings of a small burgh museum can contain items of national importance is exemplified by the Inverkeithing firlot gauge, which is also the first item to be discussed in the Inventory section.

However, this book is not just about Scotland. Weights and measures concern the market-place. The measurement practice of any nation must be influenced by the custom and practice of its trading partners, perhaps especially so in the case of a small nation like Scotland. The same principles, of course, apply to her more powerful neighbour England, and it has been by looking more critically at the trading links with other north European nations that we have been able to provide a more satisfactory context for both Scottish and English practice. Again, this view has allowed a more radical appraisal of the established view of the overall framework than we had thought possible.

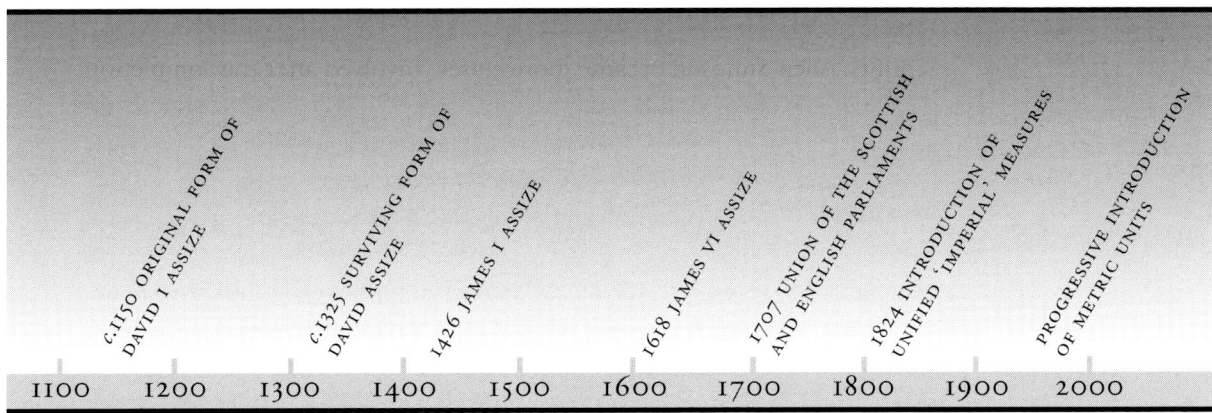

Significant dates in the evolution of the weights and measures of Scotland.

It is doubtful whether a single volume can provide a comprehensive view of weights and measures; we do not, for example, concern ourselves here with issues such as the enforcement of legislation, which provides clues as to its sources in the market-place. Our intention has been to look only at the legislation and regulations that survive, to try to discover the legislators' purpose in framing them the way they did, to see if they can be interpreted in a consistent way that may occasionally reveal the policy behind these changes. It would be expected that this would unfold without too much disruption to trading practice, but with continuity where at all possible. One suspects that legislation often followed market practice rather than leading it; indeed, legislation that attempts to introduce disruptive change has little chance of success. And the test of any interpretive framework has to be whether it has maintained market continuity and introduced the minimum amount of enforceable change.

This book provides a new framework for examining the details of Scottish metrology. Because we have started with the assumption that one must test surviving versions of legislation for subsequent changes that have been introduced, we have begun with a brief survey of the early legislation. The first myth that must be laid to rest is that the Scots inch differed from the English inch: Chapter 2 demonstrates why this cannot be the case, and provides the basis on which we can move ahead to discuss other legislation that involves length measure, for example in the definition of capacity measures. In the third chapter we look more closely at the family of length measures used in Scotland and show the type of pressures that can lead to changes in new standards developed for specific purposes – and the ease with which these can be misunderstood.

In Chapter 4, we get to the heart of the problems of methodology relating to the trading systems in which Scotland participated, looking at the structure and relationship of early weight systems and investigating the size and method of weighing of the woolsack – the unit of an essentially international trade. Chapter 5 examines the first detailed legislation to survive in Scotland, and we show how the various sources for this material have to be assessed to reconstruct an acceptable original text for interpretation. The operation of trading allowances endemic to practice in Scotland and elsewhere is a feature of early customary practice, and Chapter 6 describes how the use of 'charities' or allowances was in practice reduced to simple and repeatable operations. In particular, it allows us to appreciate how the liquid and dry capacity measures depended on each other, and in Chapter 7 this is expanded to produce a scheme which explains the progressive growth of the dry measures, and resolves what has perhaps been the most confusing and intractable problem in Scots metrology.

In Chapter 8 we return to the principal weight systems used in Scotland, and examine how the weights used in the market-place can be related to the restricted weight types described in the legislation. Finally, we investigate the historiographical problems surrounding the eighteenth- and nineteenth-century attempts to interpret these early systems, many of which had fallen out of normal trading use but were still found to have relevance because they had become embedded in legal contracts and obligations of all sorts.

For many people the least familiar aspect of all this will be the measurement standards themselves. Although a significant number survive, they often carry

SCOTLAND	ENGLAND
Malcolm III (Canmore) 1058-93	William I 1066-87
Donald Bane 1093-4	William II 1087-1100
Edgar 1097-1107	Henry I 1100-35
Alexander I 1107-24	
David I 1124-53	Stephen 1135-54
Malcolm IV 1153-65	Henry II 1154-89
William (the Lion) 1165-1214	Richard I 1189-99
Alexander II 1214-49	John 1199-1216
Alexander III 1249-86	
Margaret (of Norway) 1286-90	Henry III 1216-72
John Balliol 1292-6	Edward I 1272-1307
Robert I (the Bruce) 1306-29	Edward II 1307-27
David II 1329-71	Edward III 1327-77
Robert II (Stewart) 1371-90	Richard II 1377-99
Robert III 1390-1406	Henry IV 1399-1413
James I 1406-37	Henry V 1413-22
James II 1437-60	Henry VI 1422-61 (restored 1470-1)
James III 1460-88	Edward IV 1461-70; 1471-83
James IV 1488-1513	Richard III 1483-5
	Henry VII 1485-1509
James V 1513-42	Henry VIII 1509-47
	Edward VI 1547-53
Mary (Queen of Scots) 1542-67	Mary 1553-8
	Elizabeth I 1558-1603
James VI 1567-1625	James I (& VI) 1603-25

Kings and Queens of Scotland and England from the Conquest to the Union of the Crowns.

little information, or the inscriptions are cryptically mysterious and difficult to identify. We have, for the first time, brought together in Part II descriptions and analysis by Allen Simpson of a notable proportion of what survives. Similarly, the texts of the key parliamentary acts are reproduced in the first of a series of Appendices, principally by Robin Connor, in Part III. These Appendices cover a number of comparatively self-contained topics, which relate directly to the main text, but which would have disrupted the flow of the argument. The final directory of Scottish scale, weight and measure makers was prepared by Alison Morrison-Low. These topics are nonetheless important, and could not be omitted from any treatment of Scottish metrology: they bring together essential information not readily found elsewhere, and useful, we hope, to different sections of our readership.

NOTES AND REFERENCES

1   Joan E. L. Murray, 'The Organisation and Work of the Scottish Mint 1358-1603', in D. M. Metcalf (ed), *Coinage in Medieval Scotland (1100-1600): Second Oxford Symposium on Coinage and Monetary History* (Oxford, 1977), 155-69, p. 161.

2   R. E. Zupko, 'The Weights and Measures of Scotland before the Union', in *Scottish Historical Review*, 56 (1977), 119-45.

3   Lawrence Burrell, 'The Standards of Scotland', in *The Monthly Review: the Journal of the Institute of Weights and Measures Administration*, 69 (1961), 49-62.

4   Elizabeth Gemmill and Nicholas Mayhew, *Changing Values in Medieval Scotland: A Study of Prices, Money, Weights and Measures* (Cambridge, 1995); A. J. S. Gibson and T. C. Smout, *Prices, Food and Wages in Scotland, 1550-1780* (Cambridge, 1995).

The Craigengelt
Weight: trone stone
by Hans Cochran,
Edinburgh, 1553.
(Item 31)

Half-pint standard for the Burgh of Edinburgh, 1555, by David Rowan, master founder of the royal artillery works at Edinburgh Castle. (Item 109)

The 'Jeddart Jug': ale pint standard for Jedburgh, 1563. (Item 110)

Brewers' pint standard for St Andrews,
by David Rowan of Edinburgh, 1574. (Item 111)

Standard troye weights made in London for the Edinburgh Mint, 1687. (Item 41)

Standard weights of Stirling Guildry, probably those ordered from the Low Countries, in 1676. (Item 40)

Standard wheat firlot for the Justices of the Peace of the County of Stirling, constructed by John Miller, David Robertson and James Stark of Edinburgh, under the direction of Professor John Stewart in 1754. (Item 216)

Wheat and barley firlots, adjusted by Professor John Robison for the County Commissioners of Perthshire, 1793. (Items 223 and 224)

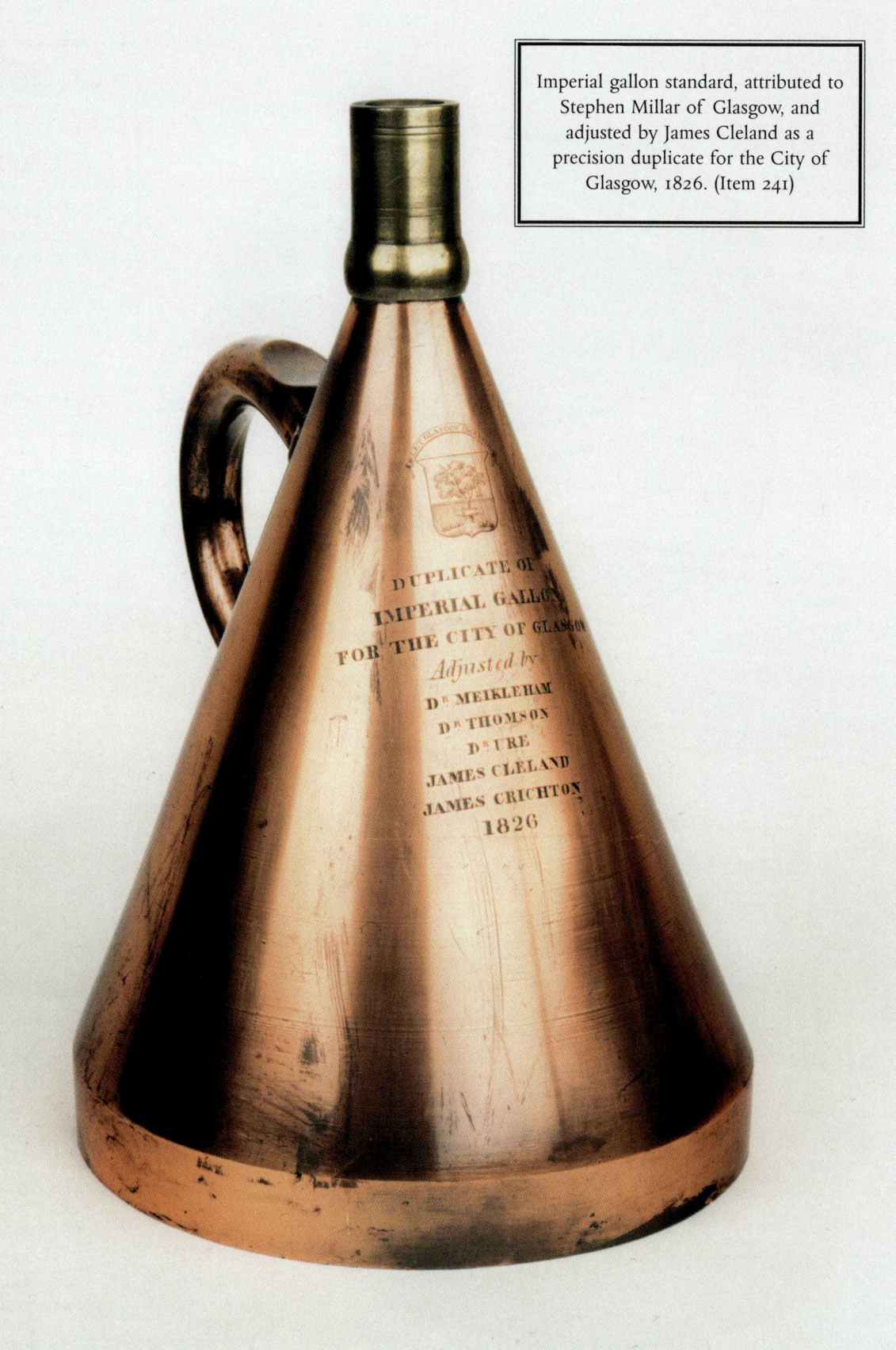

Imperial gallon standard, attributed to Stephen Millar of Glasgow, and adjusted by James Cleland as a precision duplicate for the City of Glasgow, 1826. (Item 241)

PART I

A History of Weights and Measures in Scotland

ti qm consilia dicit
el dm Regis. aut ea
rone vel assisa depo
narrat. Item. si
ius. vel redditus in
corp. vel aliorum suorum
sint sine licencia dm
si sint qm p purpre

ti inductores false
form seco Regno.
a judicia fuint de
rgo no psentata
deberent.

HISSA Regis David. facta
apud nostrum mestrum sup
tyna. de ponderibus et
mesuris. In primo.
Sterlingus debet pond
are. xxxii. grana boni 7 rotundi
frumenti.

Uncia debet pondare. xx. denar
de st Alingr. Sed modo. tempe scilt
Roberti de bruys. Uncia gtinet. xvi
denar st Alingorum. qma monetha
minuebatr in tanto. de st lingr
Libra debet pondare. xv. uncias
videlz libra debet pondare Targ
gento. xxvii. s. m d. Et hoc est
ppt minorationem denarm que
fuit modno tempe B. Roberti
de bruys. Na antiquity scilt te
pore Regis David. libra estimebat
msi. xxv. solid.

Petra debet gtinere octo libras.
scilt petra ad retã. Petra lune
debet gtinere 7 pondare. xv. libs.
Urunisa debet pondare. xii. petr
unig ponds debet gtinere octo libr
Lagena debet gtinere. xii. libras
aque. ita videlz qd quatuor libre sint
de mari. quatuor de lacu. 7 quatuor
libre sint de aqua currenti. scilt
clara. Item lagena debet esse
in pfundditate sex pollicas 7 dimm
In latitudine insioris debet esse octo
pollices. 7 dimm. in sfssitudine
earum. vtrunsq plis. Et in rotun

1

THE BEGINNINGS
OF SCOTTISH
METROLOGY

SCOTLAND IS A SMALL COUNTRY AT THE VERY EDGE OF EUROPE. ITS border with its more prosperous and powerful southern neighbour England is close to what was for a time the frontier of the old Roman Empire. Beyond Hadrian's Wall was inhospitable territory, the home of the Celtic tribes that the Romans never managed to subdue. For all its size, Scotland is a land of great geographical variety. Travelling north across the uplands of the borders, the hills gradually give way to the central valley, running from the Lothians in the east to Strathclyde in the west, and crossed by the estuaries of the Forth, Clyde and Tay. Volcanic outcrops provided naturally fortified sites for royal castles at Dumbarton, Stirling and Edinburgh, and the fertile soil lies over the coal deposits that fuelled the industrial revolution. This agricultural belt extends through Fife up the east coast of the country, forming a rich hinterland serving numerous settlements and natural harbours. The whole lowland area is separated by a geological fault line from the comparatively inaccessible highland region, which is intersected by deep valleys and glens. Historically, communications were poor and administrative control has been difficult to exercise in the highlands. It is scarcely surprising that the Scottish centres of influence, power and trade have developed in the eastern lowlands, and our concerns in this book are largely with these lowland areas.[1]

Effective administration came to Scotland in the eleventh and twelfth centuries in what has been described as 'a peaceful "Norman Conquest" of Scotland'.[2] The process can be said to have begun under Malcolm III – or Malcolm Canmore as he is often known – under whom the disparate peoples of Scotland were first fused into a single kingdom. His English wife Margaret brought with her from the Saxon court a retinue of English followers and established a pattern of marriage ties and increasing English influence. In no one is this influence more apparent than their son David, who came to the Scottish throne as David I in 1124. David was brought up in the English court where his sister Matilda was married to King Henry I, and he held estates in England. Conditioned by his Anglo-Norman upbringing he demonstrated a

Opposite:
Detail of Fig. **1.6**.

commitment to the introduction of English-style administrative structures in Scotland. This was a gradual and progressive process continued by his successors, but at the end of a period of about two hundred years to 1300, with a dozen intermarriages between the two royal houses, it could safely be claimed that the Scottish royal house was 'Norman in blood and heart'.[3]

Christopher Smout has characterised four aspects of this anglicising change.[4] The first was the growth of a type of feudalism through the granting and sub-granting of land tenure in return for obligations of loyalty and service, and this had its greatest impact in the areas settled by David's followers in the central lowlands and the east of the country. Second, reforms in the organisation and government of the church had been begun by Margaret, but were consolidated and extended to include the establishment of abbeys such as the working Cistercian orders that helped generate royal revenue by demonstrating improved sheep farming techniques. Third, and most important, was the establishment of burghs – communities with specific rights to conduct internal and foreign trade – a concept copied directly from English practice. The burghs were mainly under royal protection and were individually self-governing, although common interest led them to develop a unified code of trading practice. Fourth, and finally, effective government machinery was introduced by dividing the country into sheriffdoms, each with a royal court administered by a sheriff whose responsibilities included the collection of the king's rents. The royal burghs were the responsibility of another royal household official, the chamberlain, whose duties were similarly the upholding of the king's law and maintenance of his revenues.

In order to assess rents and taxes, and to regulate the trade which David I was so keen to foster, it was necessary to establish a centrally-controlled system of weighing and measuring and to enforce its use. The earliest assize (or legislative act) concerning weights and measures is one attributed to David, and it provided a firm basis for all the legislation that was to follow. Whatever traditional measures may have been in use in the disparate parts of his kingdom, the unified legal system introduced by David was that of England, modified to suit Scotland's needs.

To a large extent David's adoption of the best of the English systems of administration was a natural consequence of his background and upbringing; but it perhaps also reflects the history and nature of the early Scottish kingdom. After Malcolm Canmore's death his family was driven to seek refuge at the English court and it took English military intervention before David's brother Edgar was eventually put on the Scottish throne in 1097. He was succeeded first by his brother Alexander (whose army fought in support of the English in Wales) and then by David in 1124. All three had felt obliged to accept vassal status to the English king for their lands in Lothian and Cumbria.

David had spent most of his life in England at the court of his sister Matilda, and he was aged over forty when he gained the Scottish throne. Much

more than his brothers he adopted English manners and customs and granted Scottish lands to English nobility. By marriage to Matilda of Northumbria, David became earl of the shires of Huntingdon, Northampton, Bedford and Cambridge, and Geoffrey Barrow has described David by 1126 as the greatest of the English earls.[5] He had a substantial area under his jurisdiction in London, which provided an economic as well as a political power base. It was from this that he traded the animals and produce of his English lands, with additional land at Tottenham serving as a holding zone for fattening cattle before moving them and other goods into the lucrative London markets.[6]

The introduction of the English into Scotland, with the gift of lands and offices of church and state, began with Margaret and Malcolm and proceeded at an increasing pace with their children Edgar, Alexander and especially David. These territories were for the most part south of the Forth, with the royal estates beyond the Tay remaining largely intact. This provision of lands to the monasteries and to the barons continued until the death of David II in 1371, a period of almost three centuries during which Anglo-Norman penetration was pronounced. During this time there were many inter-marriages between the royal house of Scotland and that of England, and with English nobility. Add to this the repeated claim of English feudal overlord-ship of Scotland and the fact that Scottish kings and barons held lands in England for which they did homage, and we can appreciate that England was never far away from the Scottish mind.

Scottish armies were again active in England in the mid-twelfth century during the English civil wars, but the capture of the Scottish king William ('the

Lion') in 1174 led to him being obliged to swear homage to the English monarch, only being released from this some 15 years later with a payment to fund Richard I's crusade to the Holy Land.[7] A period of comparative stability, with two marriages between the royal houses, followed. But the death of the Scottish king Alexander III in 1295, without a close heir, encouraged the direct involvement of Edward I of England in determining the Scottish succession, followed by a brutal military intervention and the extended and disruptive Scottish Wars of Independence.

Throughout all this time Scotland's external trade had been largely restricted to coastal traffic which had looked south to English ports, and the principal influences on Scottish metrology continued to be English. As Alexander Stevenson has stressed, it was only in the late thirteenth century that improved merchant ship design made direct trade across the North Sea to the Low Countries and Germany more reliable.[8] It has been argued that the disruption to trade caused by the wars with England was severe, and that Scotland's overseas trade never fully recovered; and this was certainly a dangerous time for mariners, when ships' cargoes were at risk of being seized and when profits were to be made by Continental merchants running English goods to Scottish ports.[9] It is with these English influences that we are concerned initially, and with the trade conducted from Scotland's new centres of commerce – David's royal burghs.[10]

TRADE AND THE ROYAL BURGHS

The first burghs made their appearance in the Scottish records in the twelfth century with royal grants of trading privileges. A burgh's charter gave it a trading monopoly over a surrounding area of countryside (known as the 'liberty' of the burgh) so that all produce and goods from that area had to be bought or sold in the burgh's market or at periodic fairs held in the burgh. Those coming from outside the burgh to trade had to pay tolls, from which only the privileged residents in the burgh – the burgesses – were exempt. These tolls, together with other burgh rents and the money exacted as 'cain' or customs from goods entering burgh ports, provided an important source of royal revenue.[11] Many of the early burghs (and perhaps 17 had been established by the end of David's reign in 1153) will have developed from existing communities associated with fortified castles or early religious centres and located at sites such as fords, natural harbours or the intersection of trade routes.

The provision of formal burgh status indicated pressure for the monarch to modernise the Scottish economy along lines established in England and elsewhere in Europe. So too did the arrival of merchants and tradesmen from England, France and particularly from Flanders, bringing new trading and manufacturing skills to the burghs in which they settled as burgesses, and

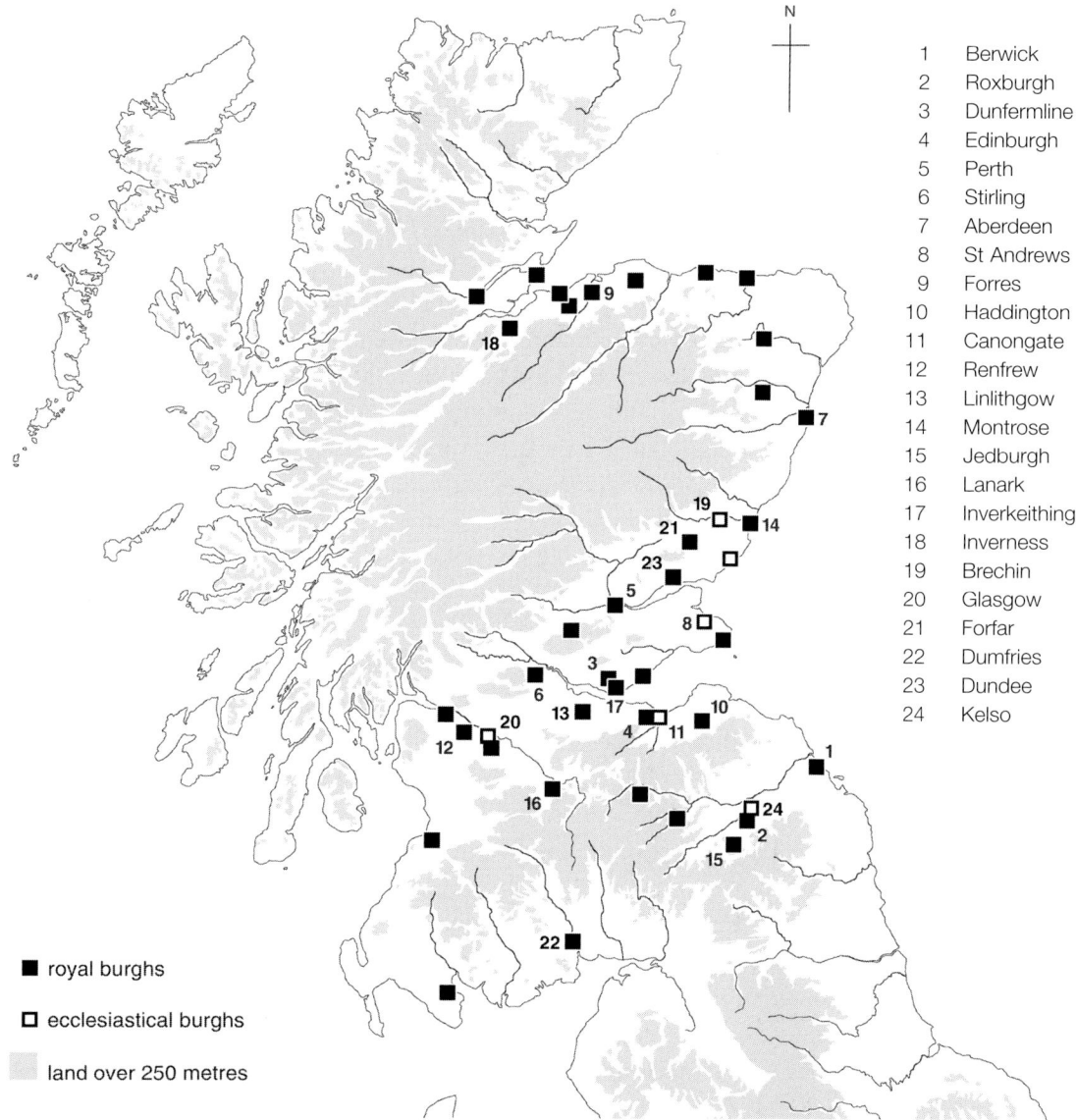

1	Berwick
2	Roxburgh
3	Dunfermline
4	Edinburgh
5	Perth
6	Stirling
7	Aberdeen
8	St Andrews
9	Forres
10	Haddington
11	Canongate
12	Renfrew
13	Linlithgow
14	Montrose
15	Jedburgh
16	Lanark
17	Inverkeithing
18	Inverness
19	Brechin
20	Glasgow
21	Forfar
22	Dumfries
23	Dundee
24	Kelso

■ royal burghs

□ ecclesiastical burghs

▨ land over 250 metres

Fig. 1.2
Royal and ecclesiastical burghs by about 1300.

drawn, as Hector MacQueen and William Windram have noted, 'by the assurance of individual rights and privileges within the burghs and in the kingdom at large'.[12] It has been commented that most of the prominent early burgesses were Flemings, and indeed the limited range of exports from both England and Scotland were principally channelled to the rapidly expanding urban markets of Flanders.[13] The close dependence of Scotland on the market at Bruges, the chief entrepôt of Flanders, is underlined by the existence of a community of Scots merchants resident in Bruges by the late thirteenth century.[14]

MacQueen and Windram have emphasised the legal nature of the concepts of 'burgh' and 'burgess', meaning that the privileges implied were enforceable by law; and they have summarised the evidence for law-making by David I in this area, for example associating the prerequisite of owning land in the burgh with feudal obligations to provide 'watch and ward' in the burgh's defence.[15]

These early burghs had little independence initially but came under the control of officers of the burgh lord, who was normally the king, but occasionally (as at St Andrews) a leading cleric, or a baron, acting with the king's authority. Very often burghs had grown from settlements around a royal castle, few of which now survive: for example, Perth's castle was destroyed in floods in 1209 and not rebuilt.[16] In these instances oversight was exercised by the king's resident sheriff – the royal officer who dealt with the business of government in his region, or sheriffdom, and who was responsible for administrative, legal and financial functions, which included the collection of the king's revenues from the burghs. The system of sheriffdoms had also been reformed by David I, and the majority of his new sheriffs were also Continental incomers.[17] The area of the sheriffdom might equate to the burgh's liberties – the hinterland from which the burgh drew produce to its single permitted market and within which only its burgess merchants could trade.

The burgh itself was normally of a characteristic and well-controlled layout. In its simplest form it comprised a single high street, with a widening at its centre to make space for the market and the erection of traders' stalls. This would also be the site of the market cross and perhaps public buildings such as a tolbooth, or townhouse, and the parish church. The main street frontages at the market would comprise the shops and houses of the more substantial merchants and craftsmen, built at the head of long narrow strips of land, radiating away from the high street and sometimes separated by closes or vennels.

These 'burgage' plots were of more or less standard widths and were laid out in a rectangular pattern, suggesting a level of centrally-controlled town-planning both in the initial layout and in the expansion of burghs. The ownership of land was central to the holding of burgess status, allocated initially for the king's settlers by occupation for a year and a day, and constrained where possible to pass to a burgess's heir or risk forfeiture; but the overlord's revenue was also dependent on the annual feudal rents from the property of his burgesses.[18] Therefore it is understandable that the officials of the king (or of the bishop or baron in the case of other burghs sanctioned by the king) would take a direct interest in controlling the layout and disposition of property. That this was a skilled procedure is shown by the invitation of Mainard the Fleming, a burgess of Berwick, and Ranoulf of Haddington to lay out the episcopal burghs of St Andrews and Glasgow respectively, Berwick having previously been laid out by Mainard and Haddington by Ranoulf.[19]

The importance placed on the burgess property qualification is emphasised by the appointment of burgh officials known as 'liners', whose responsibility was to measure land and property boundaries, especially if boundary stones were thought to have been moved.[20] This helped ensure that boundaries

remained unaltered during the long periods over which early wooden properties were progressively replaced with more permanent stone-built structures.

Cultivated land was also divided into strips, with the width of the strip
being a standard 'rod' or 'rood' (the equivalent English term being the 'rod',
'pole' or 'perch'), and this is discussed in Chapter **3**. An early but undatable
Scottish legal fragment described the width of a rood of land as a 'fall' of six ells
(about 18 feet) in the country, but noted a specific difference in burghs, where
the width of rood was to be 20 feet.[21] Although there was certainly variation in
property widths within individual burghs, and also between different burghs,
regularly-spaced 20-foot burgage widths were found convincingly in excavations in medieval Perth in the mid-1970s, with some widths of around 26 feet
perhaps also accommodating an original access vennel.[22] Rods of this length
have also been recorded for building in England, for example in late thirteenth-
century York, which may indicate a wider currency of this burgage width.[23]

Behind the burgess's house there might be other accommodation with
workshops and storage and even cultivation in the back-lands that ran down to
the edge of the burgh. The burgh's boundary might be defined by a stream, or
simply by a man-made ditch, and although the vennels between the plots
principally gave access to buildings to the rear, they could also lead across the
boundary to the country beyond the burgh. Early burghs were seldom fortified; protection for the inhabitants could, if necessary, be provided by the
castle, which would in any case stand outside the burgh. Inverness, exception-
ally, at about the extremity of the king's control, was planned to have a wooden
palisade; the walls at Berwick and Perth, on the other hand, are more properly
connected with occupying English forces.[24]

The burgh's officers included burgesses who were appointed as principal
officer, or 'provost', and 'bailies', and whose first responsibilities were to the
overlord. The bailies of royal burghs collected the king's revenues from the
burgh on behalf of the sheriff, and were answerable to the king's chief financial
officer, the chamberlain, on his annual 'ayre' or progress of inspection of the
royal burghs. The bailies also conducted the burgh court, where minor mis-
demeanours were heard and a burgess could be tried by his peers, but which
was likely to have been presided over initially by the sheriff.[25] This court, which
had only limited criminal jurisdiction, was distinct from the sheriff's own
court, held in the castle (but subsequently in the burgh tolbooth), which tried
all but the most serious crimes in the sheriffdoms – those of the 'four pleas
of the crown', of murder, arson, rape and robbery – which the sheriff had to
transmit to the king's court of the 'justiciar', who made regular progresses
around his region in a 'justice ayre'.[26]

With the passage of time the prosperity of most of the royal burghs
increased, although a few burghs failed. Administering burgh affairs became
more complex, with a greater reliance on professional clerks and paper records,
but the burghs also gained more autonomy and self-government. For example,

officials such as bailies were elected by the burgess community and were not appointed externally, and the burghs or individual burgesses were able to lease (or 'ferm') from the chamberlain the right to collect many types of burgh revenues for which only fixed annual sums were paid to the superior.[27] The independence of the burgh courts from the sheriff (in perhaps the late twelfth century) may have coincided with the institution of the chamberlain's ayre in order to secure the king's financial interests.[28]

Effective collaboration between the burghs was an essential component of their autonomous operation and this was fostered to a large extent by the acknowledgement that the burghs and their burgesses were governed by a common code of burgh laws. As early as 1211 burgesses from the four burghs of Berwick, Roxburgh, Stirling and Edinburgh were meeting to consider their collective response to the king's financial demands. What emerged from such discussions was an institution known as the Four Burghs (or the Court of the Four Burghs) in which burgess representatives from each of these burghs met together to provide interpretation of burgh law.[29] This came to act as a court of appeal from the burgh courts and the chamberlain's inquisitions at his ayre. The Court of the Four Burghs met annually under the chairmanship of the chamberlain at Haddington, and its fines were payable to the king. With the loss of Berwick and Roxburgh to the English army, the burghs of Linlithgow and Lanark were substituted in 1369.[30] By this time the burgh community was also represented as one of the three 'estates' of parliament.

The origin of the burgh laws has been the subject of much debate. They appear, as the *Leges Burgorum* and firmly ascribed to David I, in the earliest surviving compendium of Scots law, the 'Berne' manuscript, believed to have been compiled in about 1270.[31] The attribution to David I has to be treated with great caution. A similar claim is made for the first extended piece of Scots legislation about weights and measures, the so-called Assize of David I, which will be discussed below. The latter may well contain elements that echo regulations of David's time, but if so it has clearly been modified for fourteenth-century use. The attribution is more likely to imply no more than that the law is old and carries authority – and what better than the undoubted legal authority that can be ascribed to the law-maker David I.

The text of the *Leges Burgorum* has been associated with that of an early customs document of Newcastle in order to support a twelfth-century origin. But in their 1988 analysis of the Scottish burgh laws, Hector MacQueen and William Windram have exposed the circular argument of claiming that the Newcastle customal was produced at David's instigation (Newcastle was under Scottish control between 1139 and 1157), and subsequently applied elsewhere in Scotland, whilst separately basing the date of the customal on the supposed David dating of the burgh laws which it was assumed to precede.[32] They concur with A. A. M. Duncan's conclusion that the *Leges Burgorum* were probably drawn up at or for the Scottish burgh of Berwick, but they argue that there is insufficient evidence to say whether the composition has extended over a long period, perhaps beginning in the late twelfth century, or whether it is the result of a gathering together of existing regulations, with some earlier material, in the thirteenth century.[33] They conclude, however, that 'the immediate source for much burgh law was the laws of the English boroughs'; and that, aside from aspects concerning trading privileges specific to the early burghs, other material was undateable, although probably later, and revealed no substantive distinction between the effect of laws of the burghs and those of the landward areas.[34]

The Scotland of David I's time was an agrarian country in which trade between communities was conducted on a barter basis. The establishment of the royal burghs as centres for commercial trade, and particularly foreign trade, increased the long-term pressure towards a fiscal-based economy. It was a considerable time before barter began to be replaced as the principal form of transaction, but the burgh revenues were collected in cash. Circulating coinage had become well-established in England in the Saxon period and it eventually achieved reliable standards of weight and quality which were maintained in Norman times. It is no surprise therefore to find a coinage being introduced into Scotland by David, nor that it was based directly on the English example and was interchangeable with it. David's early coinage was not extensive. Its significance for us is not the effect it may have had on David's economy, but the control that was exercised on its weight and quality.

The privilege which was peculiar to merchant burgesses of the royal burghs, and some ecclesiastical burghs, was the ability to engage in foreign trade. Virtually all of this was conducted through the various ports, although the drove roads over southern Scotland saw the passage of black cattle to English markets. Major inland burghs in the lowlands had dependent ports – such as Leith for Edinburgh and Blackness for Linlithgow – whereas north of the Forth the majority of the royal burghs (characterised by Alexander Stevenson as 'foreign trading colonies') were on the coast and had their own harbours.[35]

It was largely raw materials that formed the staple of Scottish burgh exports – notably wool, woolfells (fleeces) and hides. From the outset the Scottish export economy was dominated by wool, which was in great demand for the emerging cloth trade in Flanders and those parts of southern Flanders, such as Artois, that for a time were part of France. Increasingly it was the city of Bruges, with its satellite ports of Sluys and Damme, which formed the focus of Scottish trade, whether conducted by Scots or Flemish merchants. The close interest that the Flemish had in Scottish wool certainly extends back to the twelfth century when contacts were established with the Cistercian abbey at Melrose, one of several Cistercian houses producing wool good enough to be graded in different qualities for export to Flanders.[36]

In common with other border abbeys Melrose sent its wool overland to its stores in Berwick, the principle port at the south-east tip of Scotland. Berwick, and the east-coast burghs of Aberdeen, Montrose and Perth were distinguished as the separate ports of origin for the best Scottish wool sold in Flanders. But not all wool was gathered and exported by the most obvious route: for example, a daughter-house of Melrose at Couper Angus sent its wool initially by sea to Berwick, to be exported together with Melrose's wool.[37] Although the great Scottish abbeys, such as Melrose and later Arbroath, were probably the largest and wealthiest wool producers, much of the wool from the hinterland was gathered from small holdings by middlemen acting on behalf of merchant-burgesses in the burghs. Foreign merchants could normally only purchase material for export from resident burgesses. There were, however, opportunities to bypass them and deal directly with producers: A. A. M. Duncan has shown the extent to which Scottish religious houses borrowed from the Italian banking agents who were active in Scotland, and Stevenson has stressed that such debts were often repaid in wool rather than in coin.[38]

At the point of export, merchants had to pay 'royal' or 'great' customs (as opposed to the 'petty' customs of the internal market), initially only on the staple goods of wool, weighed by the 'sack', and woolfells and hides, numbered in hundreds. These customs presumably followed a similar imposition in England, introduced in 1275, and they constituted the principal source of royal income in the fourteenth century.[39] The level of duty was significantly greater than the petty customs and amounted to half-a-merk or mark (£⅓) for a sack of wool or 360 (later 240) woolfells, and a merk (£⅔) for a 'last' (200) of hides. The collection of customs revenues, and their payment to the exchequer, was the responsibility of royal officials known as 'customars', who operated in the restricted number of burghs which were permitted to export staple goods and which had been granted a 'cocket'. This was a seal, applied in two separate parts, each held by one of the burgh's two customars, to the letters drawn up by the 'clerk of the cocket' to confirm that the merchant had paid the appropriate customs on the goods to be exported.[40]

The strict weighing (and counting) procedures used in customing goods were soon pressed down to the internal market-place. In the 1360s came confirmation that the burgh had to ensure that an official was present in the weigh-house to prevent fraud; the requirement to have a weigh-beam in the care of a 'tronar' was extended to all burghs so that wool as well as other market goods could be weighed; and the role of these beams in export is shown by the references to the tronar (paid at the rate of one penny per sack weighed) and the maintenance of the beam in customs accounts.[41] Official enumerators were required by the end of the century.[42]

With the growing success of the Flemish cloth industry, Bruges rose to be the most important trading centre in north Europe. Not only was it the main entrepôt for Venetian and Genoese trade and a major centre for Italian banking and financial services, but also an important market for Spanish trade and the principal *kantor* of the Hanseatic League.[43] Particularly in the fourteenth and

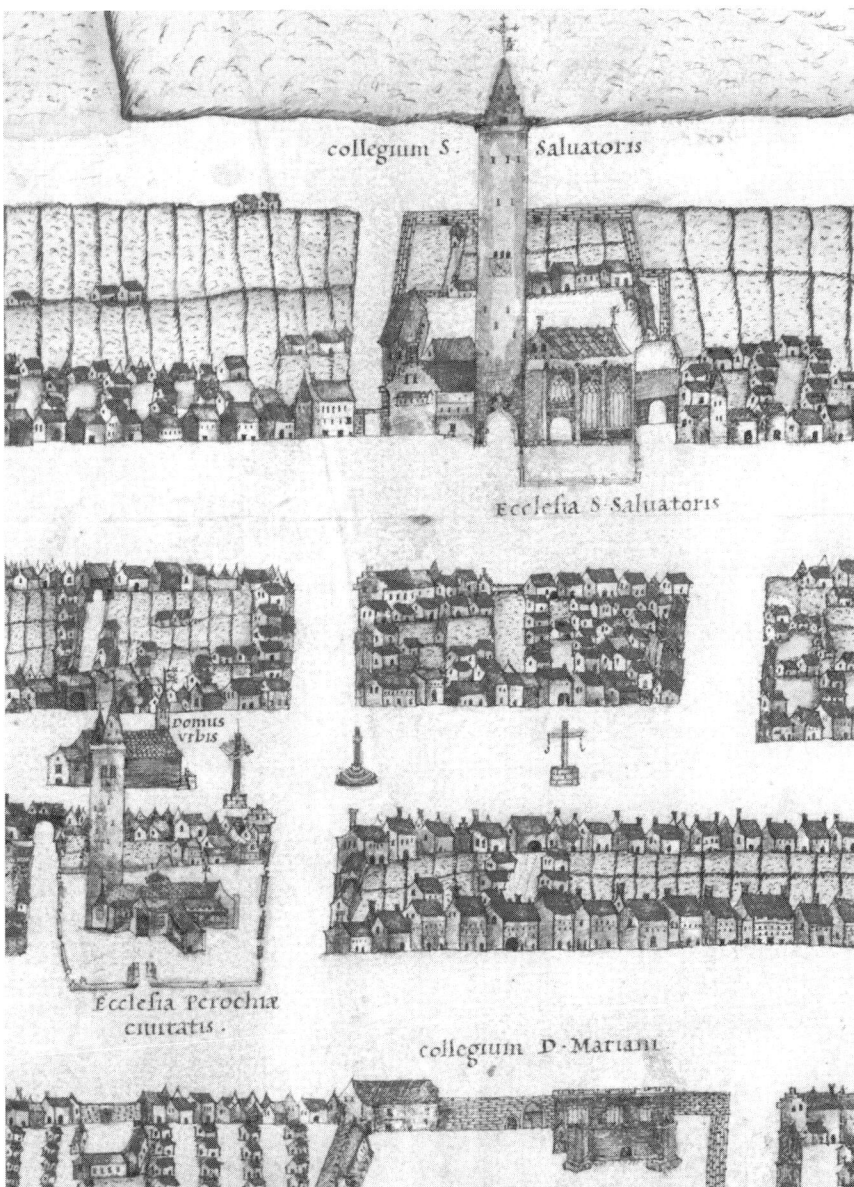

collegium S · Saluatoris

Ecclefia S·Saluatoris

domus
vrbis

Ecclefia Perochiæ
ciuitatis ·

collegium D·Mariani

Fig. 1.3
Detail from the bird's-eye view
of St Andrews, Fife, by John
Geddy, *c.*1580, showing the
tron, or public weigh scales.
(Reproduced by permission
of the Trustees of the
National Library of Scotland.)

early fifteenth centuries, Bruges dominated Scottish trade, and much of
Scotland's diplomatic, ecclesiastical and commercial correspondence also
passed through Bruges. For most of the period from the 1330s until the 1470s,
Bruges acted as Scotland's trading 'staple' – the foreign port to which Scottish
merchants were preferentially directed – except for a few brief interludes when
temporary staples were established at Middleburg in neighbouring Zeeland.[44]

The staple agreements provided various privileges for Scots merchants
taking goods to Bruges, but they did not exempt them from paying customs
and carriage duties on wool at Sluys or petty customs at Bruges, nor could they
initially avoid the storage and brokerage charges of the agents they were obliged

to use in Bruges or what could be extended delays in the official weighing of goods.[45] However, Scottish wool represented a significant proportion of the wool imported at Bruges, and the goods returned to Scotland – including fine cloth, and manufactures of all sorts – was a valuable component of the Flemish economy. Disagreements, therefore, tended to be resolved fairly quickly, albeit after some brinkmanship; and at times of difficulty for Scotland in her troubled relationship with England, the Flemish were often able to provide practical assistance.[46] Purchases for the king's household, and for those of the prelates and substantial barons, included luxury goods, exotic foodstuffs – and of course wine, much of which may have come from the French wine staple at nearby Damme.[47]

Since the customs, transport and other charges for exporting wool were not insignificant, it was in the Scottish merchant's interest to export only the best wool in order to maximise his returns. Lesser grades of wool were sold in the internal market and used to produce native Scottish cloth.[48] High-quality cloth would not have been welcomed in Flanders, where it would have been seen as in competition with local produce, but there was a market for much cheaper Scottish cloth of poorer quality. This represented almost the only aspect of manufactured goods in Scotland's exports, but Stevenson has argued that it was of considerable significance for the Scottish economy.[49]

The rate of customs duty depended on the king's financial requirements. As part of the terms for peace negotiated with England in 1328, under which Edward III unreservedly recognised Robert the Bruce as king of Scotland, the Scots had to pay an indemnity of £20,000 in three yearly instalments. The principal sources of the king's revenue were the rents of royal lands (including the royal burghs) and the much more substantial returns (at least in the fourteenth century) from the customs: occasional direct taxation in Scotland was rare.[50] The 'payment for peace' was largely recovered from a doubling of the customs duty for goods exported by foreign merchants, and also from an import duty of four pence in the pound levied on English imports.[51]

Although Berwick was briefly again in Scottish hands between its recapture from the English in 1318 and its loss in 1333, the increased customs duty was not exacted there; and this was probably a result of the lasting damage inflicted when Edward I had sacked the town in 1296, killing many of its burgesses and destroying the exporting factories of the Flemish and Cologne merchant communities.[52] Berwick had been the main burgh of the southern economic region of Scotland – the land south of the marshy upper reaches of the Forth – with Perth and Aberdeen having equivalent roles in the central and northern regions, which were separated by the Grampians.[53] These three burghs acted as the principal markets in these regions, and they also accounted for most of the coastal trade between the regions as well as that to foreign markets. The south of the country was always the most prosperous, and the earliest customs accounts show that receipts in the south region were twice those of the central and northern regions.[54] The international trade, on which customs were levied, certainly brought great wealth to many of the merchant-burgesses, and indeed to the monastic houses such as Melrose, which maintained the few major flocks of sheep. But the great proportion of wool and woolfells was gathered from ordinary peasant farmers, and Alexander Grant has stressed that 'the benefits of Scotland's international trade were not restricted to the burghs and landowners; they would have percolated through rural society as well'.[55]

The renewed outbreak of war in 1332 followed Robert I's death and represented a further determined attempt by the English king to impose a regime change on Scotland. England's candidate, Edward Balliol, was installed, and Bruce's young son and successor, David II, was rushed to safety in France. Civil war, and devastating plunder by the English army in southern Scotland, dragged on until the mid-1350s.[56] David had returned in 1341, but had been injured and captured in battle by the English in 1346. After over ten years in relatively civilised captivity in London, he was ransomed by Edward III of England for the enormous sum of 100,000 merks (about £70,000 Scots), payable in instalments. (A quarter of this ransom was still outstanding at Edward's death in 1377 and was never paid.) It has been argued that Edward recognised that the money inevitably would be raised through the wool trade. The Scottish response to this financial crisis was indeed to treble the customs rate in 1358 and then quadruple it in 1368.[57]

Berwick's former trade came to be distributed between Linlithgow (which now drew the exports from the west-coast burghs), Haddington, and more particularly Edinburgh.[58] Edinburgh's comparatively humble status in the 1320s can be appreciated from its very low level of feu-ferm of 52 merks, at the time when Berwick's feu-ferm was 400 merks.[59] By the 1420s the value of Edinburgh's annual customs had risen to 45 per cent of that for the whole of Scotland, and 67 per cent in the 1530s, so that 'by the sixteenth century it had become the economic focal point of the entire country'.[60] Berwick had earlier exported wool smuggled from south of the border which was attracted by a customs rate that was considerably lower than the English rate at Newcastle; and this traffic now mainly passed through Haddington and Edinburgh, where increased customs revenue helped compensate for monies diverted to the ransom payments.[61]

Grant has argued that the effect of draining Scotland of so much money was to lead to depreciation of the Scots currency, which had dropped in weight by 20 per cent by the mid-century, with further significant reductions by the end of the century. The old and accepted parity between the Scots and English currency, which had led to the coins being used interchangeably, was rejected by the English, and an exchange rate of 3:4 was imposed in 1373, sinking further to 1:2 in 1390.[62]

The period after the extended English wars saw the exported portion of Scottish wool rise to over 5,000 sacks per annum, peaking at an average of over

Fig. 1.4
Scotland's medieval maritime trading routes to the Low Countries and the south.

7,000 sacks per annum in the early 1370s. Grant estimates that the total clip of Scottish wool for export and domestic use was over 9,000 sacks in 1372, with an additional 2,000 sacks of smuggled English wool exported as well.[63] Thereafter, exports began a decline towards a virtual collapse of the market in about 1400 when customs was paid on under 2,000 sacks. A number of reasons have been advanced for this: long-term climate changes had probably reduced the quality of both English and Scottish wool; political and trading difficulties in Flanders had significantly reduced demand; and trade had moved into a recession which was intensified by military conflicts and naval blockades.[64] Unlike England, Scotland found itself unable to shift the emphasis of its export trade into quality cloth production, and although the foreign market for wool did improve somewhat in the 1420s, Spanish wool began to replace Scottish wool in the Flemish cloth industry.[65]

The response of David's nephew and successor, Robert II ('the Steward') was to extend customs duty to further existing exports – salmon, salt meat, suet, butter and horses – to help finance the country's imports.[66] But a marked change in emphasis in royal finances – with a greater reliance on income from crown lands rather than falling customs revenue – came in the 1420s with the forfeiture of five earldoms and a consolidation of crown estates: a policy that Grant has described as 'giving revenue precedence over patronage'.[67] We will see later (in Chapter 7) that this can be associated with increases in the size of the measures in which rental was paid in kind. However, the suggestion that the duty on wool was dropped by 19 per cent at the same time, in order to stimulate trade, is incorrect.[68] This view has arisen because inappropriate weight units have been used in calculating the size of the woolsack. In fact, during this period the sack had a constant size, which in Scotland was even given in Flemish weight units, and this is pursued in Chapter 4.

WEIGHTS AND MEASURES IN THE 'AULD LAWES'

No market, whether in the burghs or abroad, could operate successfully unless all parties to trade had a clear understanding of the rules and procedures, and unless they had reasonable confidence that these would be observed and (if necessary) enforced. Central to this was an assurance that the weights and measures used for assessing the quantities traded were just. The control and authorisation of traders' weights and measures was an early responsibility of the burgh officers, and the *Leges Burgorum* required that the weights and measures of burgesses be marked with the seal of the burgh, and that anyone found with inaccurate weights and measures be fined.[69] Transactions that used weights that were too light or measures that were too small were considered serious: offenders would similarly be fined (eight shillings for each offence) and punished by the bailies, with compensation also going to the injured party. However, a fourth offence would put the burgess 'in the kyngis mercy of lyff and of membrys [life and limb]', and would additionally incur a fine of £10.[70]

Burghs held reference weights and measures, which were used to check those of traders at markets and fairs in the burgh. Some burghs grew to have several markets specialising in particular materials: Edinburgh, for example, had as many as 15 markets by 1477.[71] Reference weights and measures for these markets were necessarily held at several locations. Those burghs involved in overseas trade often had a guild organisation, which was concerned largely with the regulation of the monopoly of trade of its burgess membership, and its senior members were frequently the same influential burgesses who were the officers of the burgh. Where appropriate, the burgh's reference weights and measures might be held by the guildry. The 'dean' or 'deacon' who headed the guild was included with the magistrates in framing burgh statutes relating to trade, and the dean of guild latterly assumed responsibility for the control of weights and measures in the burgh, prosecuting offenders in the 'dean of guild court'.[72]

Although burghs developed a degree of autonomy, they had no freedom to set the sizes of their weights and measures and it was the king's chamberlain who ensured that the royal burghs all used a uniform national system. An early but undated regulation associated with the burgh laws recorded that the chamberlain's clerk carried the chamberlain's weights and measures with him on the regular 'ayre', or tour of inspection of the burghs: during this the weights at the 'tron' or public weighbeam had to be carefully checked for accuracy against the chamberlain's weights.[73] The royal burghs could not avoid this control over their affairs; so, for example, a charter of David II assured the burgesses of Inverness that only the chamberlain was authorised to supervise the burgh's weights and measures.[74] He retained this direct involvement until the early sixteenth century, and this clearly played a significant part in ensuring that any changes in the legal size or usage of measures was enforced.[75] Implications of the chamberlain's role as holder of the national standards are considered in Chapter 7. In due course, each of the four burghs of the chamberlain's court were given delegated authority to hold one of the four types of standard, and to make and sell authorised copies to other burghs. Edinburgh was associated with the 'ell', Stirling held the 'pint', Lanark the 'stone', and Linlithgow the 'firlot' capacity measure.

Two sets of regulations for the chamberlain's ayre survive, and although they are comparatively late they presumably reflect earlier instructions. They detail the enquiries which the chamberlain had to make of the various officials (who were now mainly elected burgesses) to detect whether they were being sufficiently diligent on the king's behalf in matters such as the collection of revenue, the execution of justice, and (as a particular responsibility of the bailies) the control of quality and price of foodstuffs. The *Articuli Inquirendi*, of the late fourteenth century, include nine headings relating to weights and measures and their potential for abuse and manipulation.[76] For example, the bailies were required to state whether the burgh had resorted to holding two sets of weights, capacity measures and ells – one (with larger units) for buying goods, and another (with smaller units) for selling – rather than having equality in all transactions. Since the bailies were required to control the quality and cost of bread and ale, the chamberlain wanted to know if the bailies had examined the weights and measures of the brewers and bakers, and whether fines and punishments for infringements had been executed. Had the brewers used true measures? Had any alewives sold beer by the potful and not by sealed (or authorised) measure? Had the bakers kept the assize of bread, which set the weight of bread to be sold for a penny?

The slightly later *Iter Camerarii* is much more detailed, and just a few relevant points are extracted here.[77] The chamberlain's inquisition was to be held before the provost and bailies, and he was to call all the burgesses, indwellers and outdwellers of the burgh, all the bakers, brewers, tavern keepers, ale tasters, appraisers of flesh, liners, gaugers and tronars, and to have lists of the names of all those involved. To be presented at the same time were all the measures, balances, weights and 'ellwands' with the owners' names distinctly written on them. Ale tasters were examined about the procedures for selecting and testing ale. Bakers were quizzed about their establishments and about the various qualities of bread they baked. They were to make only penny, half-penny and farthing loaves, and enquiry was to be made to see if the weights of the loaves were less than those prescribed by the bailies. The brewers had to price their ale in relation to the cost of malt, and the prescribed vessels for the selling of ale ('quarts, pints, thirds and sixths') had to be priced. Gaugers had to confirm that each length of cloth sold was stamped with the burgh seal, signifying the correct length, breadth and weight. Tronars were to ensure that everyone's wool was weighed on the beam (and they were not to accept gifts or inducements to ignore the wool of others), and all kinds of weights were to be checked.

These rules and regulations, if they did not have the force of statute, do at least show a serious intent on the part of the administration to standardise the weights and measures across the burghs. Every official had to show that he was discharging his duties according to the law. Considerable emphasis was laid on the assizes of bread, ale and wine, all of which prevented abuses by those who controlled these particular markets, and these regulations also show marked similarities to equivalent English legal texts known from mid-thirteenth century versions.[78] For bread, the penny loaf had to be the correct weight set by the bailies according to the price of wheat, and this was only to be reduced if the price of wheat rose. Ale had to be properly brewed and approved by the tester before going on sale, where it had to be sold by good measure in authorised containers and as priced by the bailies from the cost of malt. Wine similarly had to be sold in approved vessels at the agreed price, with no old or 'corrupt' wine added to new wine. How far this ideal situation was realised is hard to assess, but at least the need for standardisation was recognised at the earliest times for which we have written records, and this concern continued to the time when printed parliamentary legislation became more widely available.

The early burgh laws, and the regulations for the chamberlain's ayre, are part of a large body of medieval Scots law known collectively as the 'Auld Lawes'.[79] These have no clear formal status, but their authority can be appreciated from their survival in a small number of manuscript collections, of which the earliest, the so-called Berne manuscript of about 1270, has already been mentioned. Other notable examples are the Ayr, Bute and Cromartie manuscripts, all from the fourteenth century, and there are more numerous later manuscript copies.[80] They cover the period up to the 1420s, when the surviving authorative parliamentary record begins.

The main text of the Auld Laws is known from its opening words as *Regiam Majestatem* ('the royal majesty …'), and is frequently coupled with the text called the *Quoniam Attachiamenta* ('with respect to procedures …'). The *Regiam* begins as a heavily-edited version of an English legal compendium, titled *Legibus et Consuentidinibus Angliae*, but normally known after its supposed author simply as 'Glanvill', which is understood to have been composed in the 1180s.[81] It continues with substantial passages from Roman and canon law texts, before reverting to Glanvill, although with some omissions. It concludes with what Hector MacQueen describes as 'a miscellany of purely native laws from the 12th and 13th centuries', some of which is commonly attributed to David I, William I and Alexander II, together with passages of early Scottish customary law.[82] The close dependence of the *Regiam* on the English text of Glanvill has been recognised from at least the early seventeenth century, and David Walker has described the main source as 'undoubtedly Glanvill … 90 per cent of which is reproduced more or less exactly'.[83]

Although the preparation of the *Regiam* is inevitably described as 'by the command of King David', this cannot be literally true, since the composition of Glanvill is some decades after David's death. The dating of the *Regiam* has been a contentious issue. It is not the *Regiam* but Glanvill that appears in the Berne manuscript, nor does the *Regiam* appear in the early fourteenth-century Ayr manuscript; however, it is present in the later fourteenth-century manuscripts. A. A. M. Duncan has concluded from a detailed study of the *Regiam* text that it dates from as late as 1320, and this has been followed by both Walker and MacQueen.[84]

It can be argued that Glanvill's text, suitably modified for Scottish needs, had been adopted as a basis for legal practice following the removal and loss

Fig. 1.5
Silver 'balance' half-merk coin of James VI, issued 1591-3, showing the sword of justice and the balance of truth. (NMS H.C3272)

of the Scottish legislative record by the English during the first War of Independence, from the 1290s. Although Glanvill was by then somewhat out-of-date, it had either been in use in the earlier period and continued to reflect conservative Scottish usage, or it was considered to be more adaptable for Scottish use than other recent texts (such as those of Bracton) which represented the more rapidly-evolving English legal system.[85]

By the time a continuous parliamentary record is again available, in the 1420s, the *Regiam Majestatem* had become firmly established as the principle authority for Scots law, and a commission was set up to 'se & examyn bukis of law of this realme that is to say *Regiam Majestatem* & *Quoniam Attachiamenta* & mend the lawis that nedis mendment'.[86] However, this and later attempts to construct an updated and authoritative version of the Auld Laws produced no results. It was not until *c*.1580 that the lawyer Sir James Balfour of Pittendreich (*c*.1525-83) finished a manuscript digest of the Auld Laws and of subsequent parliamentary legislation, arranged by subject and known simply as Balfour's 'Practicks'. It was highly regarded, although never an official compilation, and it circulated widely in manuscript form before eventually being published in 1754.[87] The first printed version of the early law books appeared in Latin and Scots in 1609, edited by Sir John Skene of Curriehill (1534?-1617), who had previously assisted Balfour. Although Skene's *Regiam Majestatem: The Auld Lawes and Constitutions of Scotland* became (and remained) the standard edition of the Auld Laws, Skene has been sharply criticised for his careless use of his source material.[88]

In the early nineteenth century a much more ambitious programme was begun of editing and publishing the text of the legislation of the Scottish parliament; and the first volume (Volume II), edited by Thomas Thomson (1768-1852) and covering the period beginning in 1424, appeared in 1814.[89] The differences between these various printed editions, and the nature of the decisions taken by Balfour, Skene, Thomson and his successor Cosmo Innes (1798-1874), has a significant bearing on how the early parliamentary record is interpreted; and this issue is discussed more carefully in Chapter 5 when we examine the crucial Weights and Measures Assize of James I, passed in 1426.

Thomson's ambition of producing a critically-edited version of the Auld Laws as Volume I of the 'Record Edition' was not fully realised before he left the Record Office, and the volume was completed by Innes for publication in 1844. In his introduction to this volume, Innes described 26 early manuscripts which contained sections of the Auld Laws, the great majority already in public collections.[90] He also produced a nearly-complete breakdown, on a clause-by-clause basis, mapping the legislative structure he and Thomson had evolved, to the content of the individual manuscripts, but without indicating to what extent they had considered or accepted variants between the manuscripts.[91] It is unfortunate for us that amongst the few aspects omitted from this analysis were the assizes for weights and measures, wine, bread and ale, all traditionally attributed to David I, but which they chose to place separate from other legislation that they could more safely assign to David.

Some of the sources used by Thomson and Innes had only recently been acquired by the Record Office in Edinburgh. The Berne manuscript had been located in the public library at Berne in Switzerland, but because of its perceived significance for Scotland the authorities there had presented it to the Record Office in 1814. The Ayr manuscript was picked up in a bookshop in Ayr by a local schoolmaster in 1824 before being offered to the Record Office. Others had been known for much longer. In particular the other two fourteenth-century documents – the Cromartie and Bute manuscripts – had both belonged to the advocate Sir George Mackenzie of Rosehaugh in the late seventeenth century. The Cromartie manuscript (Fig. **1.6**) was presented by Mackenzie to the library he had established at the Faculty of Advocates (the foundation collection of the National Library of Scotland), and Innes believed it had come originally from the priory of the Chartreux at Perth. The other was owned by the Marquess of Bute when consulted by Thomson and Innes, and it was eventually purchased by the National Library in 1987.[92]

The significance of the Bute manuscript is that it had been used by both Balfour and Skene, who considered it a valuable source. From 1564-6, Balfour was the head of the commissary court of the former diocese of Edinburgh and St Andrews.[93] The Bute manuscript has the 1565 inscription of William Skene, commissary of St Andrews; and a dictionary of legal terms in the Auld Laws

Fig. 1.6

The fourteenth-century Cromartie Manuscript, open at the David Assize of Weights and Measures. (Reproduced by permission of the Trustees of the National Library of Scotland)

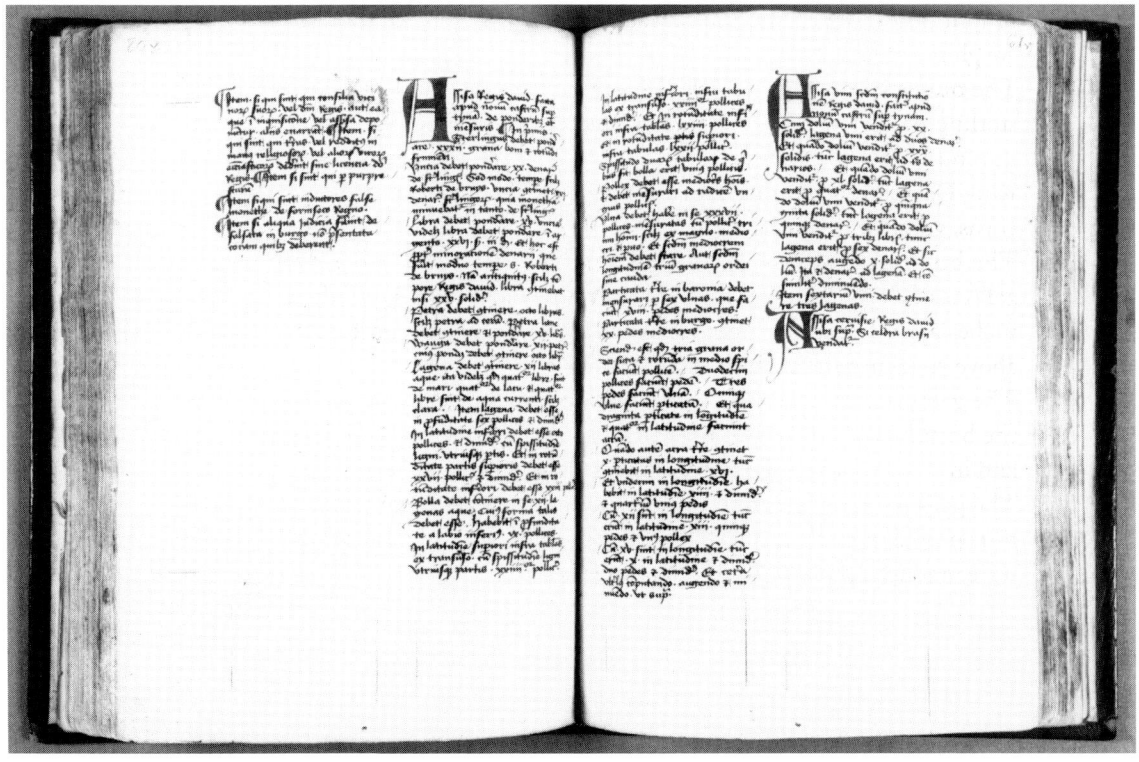

published by Sir John Skene in 1597 identified this same William Skene as his brother.[94] Balfour made specific reference to this volume, cited as 'commissarii'; and so, for example, the David Assize of Weights and Measures is correctly described by him as 'in lib. Commiss. fol.119'.[95]

The Ayr, Bute and Cromartie documents were the principal early manuscripts sources used by Thomson and Innes for the Auld Laws; and they published two slightly different versions of the assize of weights and measures, one in Scots and one in Latin, based on their own assessment of the best available sources. These two versions are printed in Appendix **A.1** and **A.1a** to this volume, and the clauses remain unnumbered. However, for convenience, the clauses have been numbered in the conflated version in modern English which follows:

1. The ell of king David ought to contain in length 37 inches with inches measured by the thumbs of three men, that is to say a large, a middle-sized, and a small man. It ought to be that of an average-sized man or else by the length of three good barley corns without tails. The thumb ought to be measured at the root of the nail.

2. (a) The stone for weighing wool and other goods ought to weigh 15 pounds.
   (b) The stone of wax ought to contain eight pounds.
   (c) The wey of wool ought to contain twelve stone.

3. The pound in David's day weighed 25 shillings. Nowadays the pound ought to weigh 26 shillings and three sterling pennies because of the diminution of the new pennies at the present time. The pound should weigh 15 ounces.

4. The ounce in king David's time contained 20 good and sufficient sterling pennies and now it shall contain 21 pennies because of the diminution of the money.

5. King David ordained that the sterling penny should weigh 32 wheat corns, good and round.

6. The boll [measure] shall contain a sexterne, that is to say twelve gallons of ale, and it should be in depth nine inches and in width 24 inches apart from the thickness of the wood, and in circumference above 72 inches and at the bottom 71 inches.

7. The gallon [measure] should be 6½ inches deep and in breadth at the bottom 8½ inches with the thickness of the wood on both sides, and in circumference above 27½ inches and below 23 inches.

8. The gallon ought to contain twelve pounds of water, that is to say four pounds of salt water of the sea, four pounds of lake [standing] water and four pounds of running water.

The form of this set of regulations is very similar to later parliamentary acts concerning weights and measures. It runs through each type of measurement unit in turn, providing definitions in terms of other units, and it gives us the first indication of the sizes of the chamberlain's standards. The assize uses terms such as the 'ell' and the 'boll', which we will later characterise among the principal Scottish measures. However, other terms such as the 'inch', 'ounce', 'pound' and 'gallon' are names more familiar from their English usage. In fact, nearly all the metrological names given in the assize are found in the early English system, and the assize gives every impression of being based directly on English standards and practice.

We should expect the two metrological systems to be firmly linked. David I, and the Anglo-Norman retinue that constituted much of his feudal administration, had grown up with the English system, and what could be more natural than to extend this into Scotland? At most we might expect some lingering use of older terms for land measure, but as these would apply in the landward areas and not in the burghs, they should not appear in these regulations. The metrologies of the two countries may have drifted apart in later and more troubled times; but in the comparatively peaceful and fruitful co-existence during the reigns of David and his immediate successors, there was every advantage to be gained by ensuring that Scotland's weights, measures and currency remained in line with those of England.

The value of such an assize to the king's officials in the control of taxable trade and the assessment of revenues is readily apparent, but it is far from clear when it was introduced. The assize no longer exists in its original form, but in a modified version which contrasts the situation 'in king David's time' with the situation 'now' when the coinage had depreciated. The assize is not in the Berne manuscript, but it is in the Ayr manuscript, which Walker dates to about 1325.[96] In this, 'now' is given as the time of *Robertus dei gratia Rex Scotorum*', without further qualification, leading Thomson and Innes to believe that the changes date from the time of Robert the Bruce, in the early fourteenth century.[97] Skene, writing before the discovery of the Ayr manuscript, printed another version which, while referring to Robert the Bruce, indicated a further level of depreciation before 1393, in the reign of Robert III.[98] However, even this late version of the assize appears to repeat detail that would have been current in David's time and reflects an earlier original text. We shall establish one such aspect in Chapter 4. In the meantime, for convenience, and with a measure of justification, it will be referred to as 'David's Assize'.

Whereas Balfour considered the assize as forming late chapters of the *Leges Burgorum* (broadly attributed to David) as well as *Iter Camerarii*, Skene entered it twice, under David and Robert III.[99] The Bute manuscript has it as a section on its own, but attributed to David, and it was this convention that Thomson and Innes adopted.[100]

Versions of the assize (corrupted to some extent by copying) are to be found in various sources and have been handled in different ways by a succession of editors. It is scarcely surprising that there are differences between these versions, and these differences often relate to numerical data giving the dimensions of particular measures or the relationships between them. However, it is precisely this numerical information that is necessary in order to understand the metrological framework, and so some care needs to be taken in deciding which variant should be preferred above others.

The provisions of the act as found in these various manuscripts, together with those given by Balfour and Skene, have been listed in Table 1.1 (pages 26-7) against the version adopted by Thomson and Innes in their 1844 volume. Every manuscript omits one or other items to be found elsewhere, but the level of agreement is remarkable, and is indicated in the table by the frequency of the check marks. Where a variant reading is found, it appears in the column corresponding to the reading accepted by Thomson and Innes. Most of the alternate readings are clearly transcription errors. For example, there have never been 34 (as opposed to 37) inches in the ell, and this may have arisen from a misreading of the numeral 'v' as 'ii'. A monetary pound of 26 shillings and three pence was given as 25s 3d in the Ayr manuscript; but 26s 3d matches our understanding of the value of the money at the time of Robert I, since it is 315 pence, or 15 ounces of 21 pennies (as the fourth clause specifies), and the 15-ounce pound is being used for currency and trade generally.

The greatest disparity appears in Balfour's 'Practicks' where the boll is 19 inches deep instead of nine inches, and contains 22 gallons instead of twelve gallons. Again, these are perhaps explicable in term of an error of inserting an 'x' before each figure, and they are consistent enough with each other to be plausible, but produce a radically different result for the size of the measure. However, it is also possible that Balfour was consciously revising the figures in order to match more closely the much larger boll in use at the time when he was writing; and, indeed, such a change would echo a similar situation found in his handling of the boll of the 1426 Assize, described in Chapter 5.[101]

More problematic are the diameters and circumferences given for the boll. Most versions give the upper and lower circumferences of the boll as 72 and 71 inches respectively, and this was adopted by Balfour, Skene and Thomson; but the Cromartie manuscript gives these as 72 and 74 inches. Thomson follows most of the manuscripts in giving a single diameter of 24 inches, but the Cromartie and Ayr manuscripts give different upper and lower diameters, with 24 and 24½ inches respectively in Cromartie (followed by Skene) and 23 and 24 inches in Ayr. This strikes a new note, and may suggest a measure of independence between the sources of Cromartie and other manuscripts.

The assumption is that the double width of the staves in the vessel is an inch at the top (and more at the bottom), with the quoted diameter measured externally and the circumference internally. This approximately matches the situation at the 1426 Assize, when the diameter given for the boll was also

	ELL		STONE				POUND		
	ell = 37 inches	inch = 3 corns	of wool = 15 pounds	of wax = 8 pounds	wey = 12 stones	of London = 12 pounds	25s in pence (David I)	26s 3d ('now', or Robert I)	pound = 15 ounces
Ayr MS (3)[e] NAS PA 5/2A ff. 13r-v	✓	✓	✓	–	–	–	–	25s 3d	–
Bute MS (4) NLS 21246 ff. 119r-v	✓	✓	✓	✓	–	✓	✓	✓	✓
Cromartie MS (5) NLS 25.5.10 ff. 80v-81r	✓	✓	✓	✓	✓	–	✓	✓	✓
Cockburn MS (13) NLS 25.4.14 f. 11r	✓	✓	✓	✓	–	✓	✓	✓	✓
f. 100r-v	✓	✓	✓	–	–	–	✓	✓	✓
MS **A.1.32** (14) NLS 25.5.7 ff. 131r	✓	✓	✓	✓	✓	–	26s	✓	✓
Foulis MS (8) NLS 25.4.10 ff. 46v	34	✓	✓	✓	✓	✓	–	✓	✓
Drummond MS (11) NAS PA 5/3 ff. 88r-v	34	–	✓	✓	–	✓	✓	✓	✓
Edin Univ. MS (7) EUL Borland 206 ff. 70v-71r	✓	✓	✓	✓	–	✓	✓	✓	✓
Balfour *Practicks* (c.1580) ff. 88-90	✓	✓	✓[c]	–	–	–	✓	26s 4d	✓
Skene *Regiam Majestatem* (1609) pt. 1, ff. 161r-v (David I)	✓	✓	✓	✓	✓	–	✓	26s 4d	–
pt. 2, ff. 68v-69r (Robert III)	✓	✓	✓	✓	✓	✓	✓	26s 4d	✓
Thomson & Innes *APS*, I (1844) pp. 673-4	✓	✓	✓	✓	✓	–	✓	✓	✓

Table 1.1 Definitions in the David Assize of Weights and Measures: comparison of the version by Thomson and Innes in the 1844 'Record Edition' of *The Acts of the Parliaments of Scotland [APS]*, with some of the principal sources for their text.

OUNCE			STER-LING	BOLL					GALLON				
20d (David I)	21d ('now', or Robert I)	32d (Robert III)	= 32 wheat grains	= 12 gallons	9" deep	24" wide	72" upper circum	71" lower circum	6½" deep	8½" wide	27½" upper circum	23" lower circum	= 12 lbs mixed waters
✓	✓	–	✓	–	✓[a]	23[b]	✓	✓	✓	✓	✓	✓	–
✓	✓	✓	✓	✓	✓	✓	✓	✓	✓	✓	✓	✓	–
✓	✓	–	✓	✓	✓	{24 / 24½}	✓	74	✓	✓	✓	✓	✓
✓	✓	–	✓	✓	✓	✓	–	–	–	–	✓	–	–
✓	✓	–	✓	✓	✓	✓	✓	✓	✓	✓	✓	✓	–
✓	✓	–	✓	✓	✓	✓	–	–	✓	✓	22	✓	✓
✓	✓	–	✓	✓	✓	✓	–	✓	✓	✓	✓	✓	–
✓	✓	✓	✓	✓	✓	✓	✓	✓	✓	✓	✓	✓	–
✓	✓	–	✓	✓	✓	✓	✓	✓	✓	✓	✓	✓	–
✓	✓	✓	✓	22	19	✓	✓	✓	✓	✓	✓	✓[d]	✓
–	✓	30	✓	✓	✓	{24 / 24½}	✓	–	✓	✓	27	✓	✓
✓	✓	✓	✓	✓	✓	–	✓	✓	✓	✓	✓	✓	✓
✓	✓	–	✓	✓	✓	✓	✓	✓	✓	✓	✓	✓	✓

a xix in the margin
b xxiiii in margin
c 'The stane ... did wey xv. pund, but now xvi. pund.'

d A number of late manuscript versions of Balfour's *Practicks* erroneously give 33 inches for the lower circumference of the gallon measure.

e Number and name in the list of manuscript sources in *APS*, I, 175-210.

external and the combined thickness of the boards was specified as 1½ inches.[102] Measuring an internal circumference is difficult, so perhaps the two circumferences were the sizes of a wooden former, around which the measure was constructed. Since the boll, and the gallon, both tapered outwards, a former could be removed after the measure was bound. However, this could not have been done with the boll described in the Cromartie manuscript, which is characterised as tapering inwards. It is far from clear how this should be explained.

The gallon is of a very similar construction, but smaller. The internal circumferences of 27½ and 23 inches give upper and lower internal diameters of about 8¾ and 7¼ inches respectively, and the external diameter at the base is stated to be 8½ inches. The depth is 6½ inches, and so the size can be calculated as 330 cubic inches (in³). The boll's volume can similarly be calculated as 3,660 in³.[103] But this cannot represent twelve gallons as the regulations claimed, since a twelfth of 3,660 in³ is 305 in³. If we calculate from the dimensions in the Cromartie manuscript, the boll comes to 3,820 in³, which is significantly larger, but the gallon as a twelfth of a boll is 318 in³. These sizes will be examined in Chapter 4 to see if they can be interpreted meaningfully.

Although the range of weights and measures covered by the assize is wide, and some clauses have presumably been added in later revisions of the original assize, we can nonetheless demonstrate that the early Scottish units normally matched equivalent English units. Thus, for example, the Scottish ale sexterne was given in the assize as twelve gallons, and the English sexterne was about twelve English gallons.[104] The 'wey' or half-sack of wool was twelve stones in both countries, and this will emerge when the units of the wool trade are examined in Chapter 4.

As with other early legislation that we have discussed, some areas of David's Assize can be directly compared with equivalent English texts, helping to build the case that they have a common source. David's weight system defined a basic monetary unit – the silver penny – in terms of 32 conventional wheat grains, as in England, and it gave the ounce as 20 pennies. Fifteen ounces made the pound, and this general pound was confirmed as 25 shillings or 300 pennies, which we can express as 6,750 grains (0·44 kilograms). The English code called the *Tractatus de Ponderibus et Mensuris* had as its basis precisely the same information, with the addition that a monetary pound in England was to be twelve ounces – namely 20 shillings, or 240 pence. The *Tractatus* is normally dated to 1302, but there are grounds for taking this back to the middle of the thirteenth century, and it is also closely associated with another legal compilation of about 1290 known as *Fleta*.[105] The links between these English texts and the David Assize are explored in Chapter 4, but in the meantime Table **1.2** on the next page draws the basic weight definitions together.

Nicholas Mayhew has reviewed the difficulties of establishing the issue weights of early Scottish silver coins, but has rejected the notion that

DAVID'S ASSIZE	TRACTATUS	FLETA
King David ordained that the sterling should weigh 32 wheat corns of good round wheat. The ounce contained, in King David's time, 20 good and sufficient pennies … The pound in King David's day weighed 25 shillings … The pound should weigh 15 ounces.	An English penny which is called a sterling round and without clipping shall weigh 32 corns of wheat in the midst of the ear; 20 pence makes an ounce and 12 ounces a pound … The pound of pence, drugs, etc, consists in weight of 20 shillings; the pound of all other things weighs 25 shillings.	… the English penny called a sterling … should weigh 32 grains of average wheat. And 20 such pennies make an ounce and 12 ounces make a pound of 20 shillings by weight and count. Again the penny sterling … weighs 32 grains of wheat, the weight of 20 pennies make an ounce and 15 ounces make a pound.

Table 1.2
Scottish and English declarations as to the penny, ounce and pound.

these coins were introduced at weights greater than those of English coin.[106] He has, however, adopted as his starting position, a troy ounce of 480 grains as the basis of David's Assize.[107] Here, our approach has a different emphasis. We have used information about the early Scottish pounds to fill gaps in the English record and to demonstrate that both countries used ounces of 450 grains (literally the weight of 20 standard sterlings of 22½ grains) before the adoption of heavier troy ounces of 480 grains, and this argument is developed in Chapter 4.[108] The use of this smaller ounce, associated with the influence of Cologne, will be tracked through thirteenth and fourteenth-century sources, which provide a new view of the way in which weight systems relate to each other.

The David Assize, in its surviving form, is a composite of some original elements and subsequent modifications that reflect early fourteenth-century adaptation. It must have remained in force for an extended period because it provided the context for the major weights and measures assize in 1426, in which comparisons were made between newly-defined measures and the 'old' measures, notably the boll 'made be [by] king Dauid'.[109] From the discussion of this later assize in Chapters 5 and 6, it will be apparent that the transition from the David's Assize was not seamless and there are a few indications of further changes in the second half of the fourteenth century.

One of these indications of change comes with a brief act entitled 'Of Weichts in Bvying and selling' published by Skene in 1609 and attributed to David II: 'It is statute be king David, that ane comon and equall weicht, quhilk is called the weicht of Cathnes in buying and selling, sal be keiped and vsed be all men within this Realm of Scotland.'[110] In his Latin text, Skene described the weight as the *pondus Cathanie*.[111] This act is not found in any of the fourteenth-century manuscripts, but only in one of the fifteenth-century manuscripts which was presumably Skene's source: this is a text in Scots that was in the Advocates Library when consulted by Thomson and Innes, where the weight is named as the 'wecht cathan'.[112] In the Record Edition the act is attributed to David I, but Skene's later dating is more plausible.[113]

In this instance, there are grounds for believing that this is not the merchant weight of the David Assize, but a slightly larger weight (with no Caithness association), which came to be the standard of the internal markets and was called trone weight.[114] This weight series, discussed in Chapter **8**, may have arisen from a form of allowance for certain coarse goods brought from the landward areas, and it is possible that what may have been a market convention beforehand only became recognised in the late fourteenth century. It may have coincided with the requirement for large burgh scales in the 1360s, and may perhaps mark the introduction of the 16-ounce pounds that had already been in use for some time before the passage of the 1426 Assize. Edward Burns deduced that the Scottish Mint had switched from a 15-ounce pound to a 12-ounce pound (as in England) in 1367, but there is inadequate information to explain how, and in what stages, 16-ounce pounds replaced the earlier 15-ounce pounds of the David Assize in the market-place.[115]

Most lists of weights and measures in legal regulations begin with the measures of length, and David's Assize was no exception. These must be the first units to be considered in detail, so that we are able, for example, to discuss the dimensioned capacity measures. Fortunately, however, the linear units once established were subject to only minor alterations as the centuries passed. The linear measures of Scotland and England show remarkable parallels, and we shall demonstrate some unsuspected links between the two countries in the two chapters that follow.

1 Peter G. B. McNeill, 'Administrative Regions', in Peter G. B. McNeill and Hector L. MacQueen (eds), *Atlas of Scottish History to 1707* (Edinburgh, 1996), 27-9.

2 William Croft Dickinson, *Scotland from the Earliest Times to 1603*, second edition (London, 1965), 83.

3 T. C. Smout, *A History of the Scottish People*, 1560-1830 (London and Glasgow, 1969), 22.

4 Ibid., 22-31.

5 G. W. S. Barrow, *The Kingdom of the Scots* (London, 1973), 173; A. A. M. Duncan, *Scotland: The Making of the Kingdom* (Edinburgh, 1975), 217.

6 Derek Keane, *Social and Economic Study of Medieval London: Interim Report on the Study of the Bank of England Area* (London, 1988).

7 A. D. M. Barrell, *Medieval Scotland* (Cambridge, 2000), 73; D. D. R. Owen, *William the Lion 1143-1214: Kingship and Culture* (East Linton, 1997), 80.

8 Alexander Stevenson, 'Trade with the South, 1070-1513', in Michael Lynch, Michael Spearman and Geoffrey Stell (eds), *The Scottish Medieval Town* (Edinburgh, 1988), 180-206, see p. 184.

9 Elizabeth Ewan, *Townlife in Fourteenth-Century Scotland* (Edinburgh, 1990), 68-9.

10 The next section is largely based on the papers in Lynch, Spearman and Stell, op. cit. (8), and on Elizabeth Ewan's more recent study, op. cit. (9).

11 Stevenson, op. cit. (8), 180.

12 Hector L. MacQueen and William J. Windram, 'Laws and Courts in the Burghs', in Lynch, Spearman and Stell, op. cit. (8), 208-27, p. 208.

13 Stevenson, op. cit. (8), 182.

14 Ewan, op. cit. (9), 68.

15 MacQueen and Windram, op. cit. (12), 208-9.

16 Derek Hall, *Burgess, Merchant and Priest: Burgh Life in the Scottish Medieval Town* (Edinburgh, 2002), 19.

17 David M. Walker, *A Legal History of Scotland*, 5 volumes (Edinburgh, 1988-98), I, 228.

18 On the alienation of heritable rights, see MacQueen and Windram, op. cit. (12), 220.

19 E. Patricia Dennison, 'Urban Settlement: 1' in Michael Lynch (ed), *The Oxford Companion to Scottish History* (Oxford, 2001), 615-16, p. 616.

20 Ewan, op. cit. (9), 13, 49.

21 'The rude off lande in baronys [the country] sal conten vj elne that is to say xviij [18] fut off a mydlyn mane, the rude off the land in the burghe mesurit off a midlying mane sal be xx [20] fut', from T. Thomson and C. Innes (eds), *The Acts of Parliament of Scotland [APS]*, 13 volumes (Edinburgh, 1814-75), I, 751: *Fragmenta ... Collecta*, no. 15.

22 R. M. Spearman, 'The Medieval Townscape of Perth', in Lynch, Spearman and Stell, op. cit. (8), 42-59, pp. 56-7.

23 R. D. Connor, *The Weights and Measures of England [WME]* (London, 1987), 44.

24 Ewan, op. cit. (9), 8.

25 MacQueen and Windram, op. cit. (12), 213.

26 Walker, op. cit. (17), I, 228-9.

27 Ewan, op. cit. (9), 64.

28 MacQueen and Windram, op. cit. (12), 213-14.

29 Ewan, op. cit. (9), 146-7.

30 Dickinson, op. cit. (2), 119. Lanark's increased prosperity depended on rights to the fair that had previously been held at Roxburgh. See Michael Lynch and A. Stevenson, 'Taxation of Burghs', in McNeill and MacQueen, op. cit. (1), 308-13, p. 308.

31 The Berne Manuscript is in the National Archives of Scotland (NAS), MS PA 1/1. See later in this chapter for its history.

32 MacQueen and Windram, op. cit. (12), 209-10.

33 Duncan, op. cit. (5), 482; MacQueen and Windram, op. cit. (12), 210-11.

34 Ibid., 222.

35 Stevenson, op. cit. (8), 182.

36 Ewan, op. cit. (9), 68; Elizabeth Ewan, 'Trade: wool producing monasteries', in McNeill and MacQueen, op. cit. (1), 237.

37 Wendy B. Stevenson, 'The Monastic Presence: Berwick in the Twelfth and Thirteenth Centuries', in Lynch, Spearman and Stell, op. cit. (8), 99-115, p. 111.

38 Duncan, op. cit. (5), 428-9, 516; Stevenson, op. cit. (8), 186.

39 Duncan, op. cit. (5), 603-4; Michael Lynch and Alexander Stevenson, 'Overseas Trade: the Middle Ages to the Sixteenth Century', in McNeill and MacQueen, op. cit. (1), 238-43, 250-60, p. 250.

40 Ewan, op. cit. (9), 75.

41 Ibid., 11, 75, 128.

42 Ibid., 75.

43 Stevenson, op. cit. (8), 189.

44 Lynch and Stevenson, op. cit. (39), 239.

45 Ewan, op. cit. (9), 79.

46 Ibid., 70, 77.

47 Stevenson, op. cit. (8), 184.

48 Ewan, op. cit. (9), 74.

49 Stevenson, op. cit. (8), 192.

50 Duncan, op. cit. (5), 162-3.

51 Stevenson, op. cit. (8), 188.

52 Ibid., 188; Ewan, op. cit. (9), 76.

53 Lynch and Stevenson, op. cit. (39), 238.

54 Ibid., 238.

55 Ibid., 251; Alexander Grant, *Independence and Nationhood, Scotland 1306-1469* (Edinburgh, 1984), 72.

56 Michael Lynch, *Scotland, A New History* (London, 1991), 130.

57 Ewan, op. cit. (9), 71; Duncan, op. cit. (5), 163.

58 Lynch and Stevenson, op. cit. (39), 250; Stevenson, op. cit. (8), 190.

59 Alexander Stevenson, 'Burgh farms', in McNeill and MacQueen, op. cit. (1), 306-7, p. 306.

60 Lynch and Stevenson, op. cit. (39), 250.

61 Ewan, op. cit. (9), 87; Stevenson, op. cit. (8), 193.

62 Grant, op. cit. (55), 80-1, 240.

63 Ibid., 79.

64 Ibid., 79-80; Stevenson, op. cit. (8), 192.

65 Ibid., 194.

66 Ewan, op. cit. (9), 73.

67 Grant, op. cit. (55), 164; Lynch, op. cit. (56), 145.

68 Stevenson, op. cit. (8), 194, where the sack weights are given as 157·46 and 187·58 kgs; Lynch and Stevenson, op. cit. (39), 250. The calculation is detailed in A. W. K. Stevenson, 'Trade between Scotland and the Low Countries in the Later Middle Ages', unpublished University of Aberdeen PhD thesis, 1982.

69 *APS*, I, 342: *Leges Quatuor Burgorum*, cap. 48.

70 Ibid., 346, cap. 68; ibid., 728: *Fragmenta Collecta*, no. 43.

71 Ewan, op. cit. (9), 66.

72 Ibid., 61; Walker, op. cit. (17), I, 200-1.

73 *APS*, I, 729: *Fragmenta Collecta*, no. 50.

74 Ewan, op. cit. (9), 157.

75 Athol L. Murray, 'The Last Chamberlain Ayre', in *Scottish Historical Review*, 39 (1960), 85.

76 *APS*, I, 680-2: *De Articulis*.

77 Ibid., I, 693-702: *Iter Camerarii*.

78 The similarities are between the Scottish documents, *De Articulis* and *Iter Camerarii*, and the equivalent English *Judicium Pillorie* ('The Judgement of the Pillory') and *Statutum de Pistoribus* ('Statute concerning Bakers') which are both attributed to 1266: *WME*, 315-19, where the text of these is printed.

79 So titled, for example, in John Skene (ed), *Regiam Majestatem: The Auld Lawes and Constitutions of Scotland* (Edinburgh, 1609).

80 The Berne and Ayr Manuscripts are located at NAS, MS PA.1/1 and PA.5/2A respectively; the Bute and Cromartie manuscripts are located at the National Library of Scotland (NLS), MS 21246 and MS 25.5.10 respectively.

81 Walker, op. cit. (17), I, 112-14 and H. MacQueen, 'Law and Lawyers, 1', in Lynch, op. cit. (19), 382-4, p. 383.

82 Ibid., 383.

83 Walker, op. cit. (17), I, 110.

84 A. A. M. Duncan, '*Regiam Majestatem:* a Reconsideration', in *Juridical Review*, new series 6 (1961), 199-217; Walker, op. cit. (17), I, 114-17; MacQueen, op. cit. (81), 383.

85 Walker, op. cit. (17), I, 110.

86 *APS*, II, 10: Act 10, 11 March 1426.

87 See P. G. B. McNeill (ed), *The Practicks of Sir James Balfour of Pittendreich*, Stair Society nos. 21 and 22, 2 volumes (Edinburgh, 1962-3).

88 Walker, op. cit. (17), I, 94, 109. The Latin text of Skene is *Regiam Majestatem: Scotiae veteres leges et constitutiones* (Edinburgh, 1609).

89 *APS*, II.

90 Ibid., I, 177-210.

91 Ibid., I, 211-73.

92 This paragraph draws on Innes's account in *APS*, I, 175-210. The Bute manuscript is now NLS MS 21246.

93 McNeill, op. cit. (87), xiv-xv.

94 Innes noted the inscription 'Liber M. Gulielmi Skeyne juris licencianti' as 'commissorii Sancti Andree, 1565': *APS*, I, 182. John Skene, *De Verborum Significatione* (Edinburgh, 1597), s.v. *canum*: 'And canage of wol or hyds is taken for the custome their-of, le.navium.fol.171 in libro M. Willielme Skene commissarii Sanctandree fratris mei germani.'

95 McNeill, op. cit. (87), 89.

96 Walker, op. cit. (17), I, 122.

97 *APS*, I, 23, 106.

98 Skene, op. cit. (79), f.56v.

99 McNeill, op. cit. (87), 88-90; Skene, op. cit. (88), has the assize at f.161 of part 1, after the burgh laws, and f.68v-69r of part 2, under Robert III. In the Scots text only the Robert III form is used: Skene, op. cit. (79), part 2, f.56v-57r.

100 NLS MS 21246, f.119r-119v; *APS*, I, 673-4.

101 Balfour's figures were presumably the source for the marginal annotations in the Ayr manuscript where the boll of nine inches deep and 23 [*sic*] inches wide was corrected to 19 and 24 inches. See Table 1.1, notes a. and b.

102 W. Robertson (ed), *The Parliamentary Records of Scotland in the General Register House, Edinburgh, Vol. 1* [Edinburgh, 1804], 63. The text is printed in Appendix A.3.

103 Altering the circumferences by half-an-inch changes the capacity by about 50 in^3, which indicates that no greater range of precision than this can be claimed.

104 For ale, the English sexterne was somewhat variable, being given as twelve to 14 gallons, and later (1421) as twelve gallons: see W. H. Prior, 'Notes on the Weights and Measures of Medieval England', in *Bulletin du Cange*, 1 (1924), 77-97 and 140-70, pp. 152-3. However, in the thirteenth and fourteenth centuries, it appears to have been reasonably steady at about twelve gallons; and so it is not unlikely that

the twelfth-century Scottish ale sexterne of twelve gallons (used for dry measure and called a boll) had an English origin in the same way as other Scottish measures.

105   *Statutes of the Realm*, I, 204-5, attributed to 1302-3. Also given in *WME*, 320-1. The *Tractatus* consists of two parts, the first giving the information in Table **1.2** as well as on the generation of capacity measures from the pound weight. The second and longer part gives much information on the units of weight for lead (fotmal), wool (sack), how many are in a dicker of gloves, and others. The second part appears in the White Book of Peterborough Abbey of about 1253, but not the first part; see Hubert Hall and Freida J. Nicholas, *Select Tracts and Table Books relating to English Weights and Measures* (London, 1929), 11. Clause 2 of the text is found in the manual of estate management called Seneschaucy of about 1276; see Dorothea Oschinsky, *Walter of Henley and other Treatises on Estate Management* (Oxford, 1971), 273. The earliest text to give both parts (Clauses 1-7), would appear to be that of *Fleta* of about 1290: H. G. Richardson and G. O. Sayles (eds and trans), *Fleta*, 3 volumes, Selden Society nos. 72, 89, 99 (London, 1955-84), II, 119. Bearing in mind these dates, it would be safe to date the *Tractatus* to the second half of the thirteenth century, but not after 1290. Material from *Seneschaucy* is understood to have been the source for *Fleta*, so named as it was written by an experienced, but unnamed, lawyer who at the time was confined in the Fleet prison in London. However, *Fleta* contains additional material not found in *Seneschaucy*.

106   [N. J. Mayhew], 'Currency', in Elizabeth Gemmill and Nicholas Mayhew, *Changing Values in Medieval Scotland: A Study of Prices, Money, and Weights and Measures* (Cambridge, 1995), 111-42, pp. 114-15, especially note 13.

107   Ibid., 114.

108   The argument is outlined in A. D. C. Simpson and R. D. Connor, 'Weighing in the Early 14th Century', in *Equilibrium*, issue no. 1 (1996), 1987-98; ibid., issue no. 2 (1996), 2015-24; and 'Fourteenth-Century Weight Systems: A Response', in ibid., issue no. 1 (1997), 2107-10.

109   *APS*, II, 12: Act 22, 1426.

110   Skene, op. cit. (79), part 2, f.39v.

111   Skene, op. cit. (88), part 2, f.51r.

112   NLS, Adv. 25.4.15 (previously W.4 ult), f.103v, ch.14.

113   *APS*, I, 324: Ch. 31. Recorded in [Elizabeth Gemmill], 'Weights and Measures', in Gemmill and Mayhew, op. cit. (106), 81-110, p. 85, note 11.

114   A. D. C. Simpson, 'Scots "Trone" Weight: Preliminary Observations on the Origins of Scotland's Early Market Weights', in *Northern Studies*, 29 (1992), 62-81, p. 76.

115   Edward Burns, *The Coinage of Scotland*, 3 volumes (Edinburgh, 1887), II, 240.

2

THE SCOTS
ELL MEASURE

THE ADMINISTRATIVE REFORMS INTRODUCED BY DAVID I AND HIS IMMEDIATE successors laid secure foundations for the medieval kingdom. Central to the economic success of Scotland were organised structures for the encouragement and regulation of trade, represented by the royal burghs and the courts, and it is largely with the market-places of Scotland's burghs that we are concerned here. The basis for the majority of transactions in the kingdom's markets, and the basis for much taxation and for transactions in land, was an accepted and centrally-controlled system of weights and measures.

Opposite:
Detail of Fig. **2.1**.

It comes as no surprise that our first detailed view of legislation or official regulations governing weights and measures is attributed to the period of David I's reign in the twelfth century when the early royal burghs were established. This is not to say that weights and measures were not used in earlier periods, but rather that there is a clear change in emphasis towards organised trade in David's reign. Thus there are many earlier descriptions of land measure (which for convenience we have treated in one of the appendices to this book), and there are glimpses of other units which are representatives of earlier systems.[1] However, the regulations attributed to David I, which we have characterised as David's Assize of Weights and Measures, provide a picture of a complete system of the weights and measures necessary for trade of all sorts, and indeed for the control of rental and other payments made in kind. From this assize developed all the subsequent legislation.

Of course, such a system is not static. Over an extended period of time there were many pressures that tended to change progressively the sizes of the units and the way that they related to each other. Each piece of new legislation was a tacit admission of the difficulty of resisting these pressures for change. Sometimes, as with the complex growth over time of the large capacity measures for grain, the legislators were faced with containing and rationalising changes that may already have taken place.

In the case of the basic Scottish length measure – the 'ell' (of 37 inches, or just under 940 millimetres) – its size was successfully controlled, and it did not change in any significant fashion over the centuries until its replacement by the yard in modern times. However, because lengths were used in the various definitions of the other standards, it is important to establish the uniformity of the early length units before considering the historical evolution of the standards which depended upon them.

Most Scottish measures and weights were originally based on those of England, although with some modification. Thus both countries had feet and inches, miles and furlongs; 'rods' in England were represented by 'falls' in Scotland, and the equivalent of the English yard was the Scottish ell. There is also an English ell, but it is a larger unit of one and a quarter yards (or 45 inches) which is almost exclusively associated with cloth measurement. Perhaps because of this it is often assumed that the Scottish ell was primarily a cloth measure. Indeed this is partly true, because cloth was one of the principal commodities reckoned in ells, and most of the surviving ell measures are either the 'elwands' used by cloth merchants or are official measures used at markets for checking these trading measures. However, the ell was certainly used as the practical linear unit in most spheres of measurement activity.

In David's Assize, the *ulna* or ell was defined as 37 inches, where the inch was the breadth of a thumb at the root of the nail of a middle-sized man or else the length of three barley corns placed end to end without 'tails'; this closely mirrored early English definitions of the inch.[2] The Saxon inch appears in the laws of Aethelbert of Kent, dated to AD 602-3, which speak of inches but leave them undefined.[3] However, Asser's *Life of Alfred*, a ninth-century biography, speaks of candles 'twelve thumb-inches' long.[4] The fourteenth-century English *Certa Mensura* gives the measure of a thumb as that at the root of the nail, while the Coventry 'Leet Book' for 1474 tells us again that three barley corns from the middle of the ear make an inch.[5] Other similar declarations reaching into early modern times confirm this basic concept of the inch.

Once standard yard measures were issued in England at the end of the twelfth century, the statute inch became regulated as one thirty-sixth of the yard instead of the rather imprecise thumb's breadth or the length of three barley corns, but some traditional measures are long lived. We shall return later in this book to the question of whether the English and Scottish inches were the same. For the present let us assume that they were.

In subsequent English law it is the yard which defines the inch, so we should pause for a moment to consider the yard, especially as it is so closely related to the Scottish ell. It should be stressed first of all that in Latin texts the measure is referred to as the *ulna*, which has always been translated as 'yard' rather than 'ell' because it normally carries with it the connotation of comprising 36 inches. The term 'ell' or 'elne', applied to the English unit of one and a quarter yards, emerged only after the name 'yard' was being used for the 36-inch unit, and the earliest surviving standards for the English ell are of the sixteenth century.[6] In fact, it will be demonstrated below that there were two versions of the English *ulna*, and the one associated with land measure (until the end of the thirteenth century) was of 37 inches. This measure is the exact analogue of, and has the same length as, the Scottish *ulna*.

In the English cloth trade the traditional practice developed of giving the purchaser an extra inch in each yard measured, by using a thumb to separate each placing of the measuring stick on the cloth.[7] This practice persisted well into the nineteenth century: for example, in 1809 the weavers in Berwickshire were selling linen giving a thumb's breadth to every yard.[8] It might be tempting to think that the use of the extra inch per yard represents the origin of the 37th inch in the Scottish ell, but in fact our knowledge of the cloth-measuring practice is gleaned from documents of a comparatively late date.[9] Instead, this larger 37-inch unit is found in unrelated areas of application. In particular, the evidence of English charters of the monastery of St Peters, Gloucester, of the twelfth and thirteenth centuries, include the use of 'yards of our lord the king with inches between'.[10] The need to distinguish between yards 'with inches' and 'without inches' arose only after the introduction of official iron yard standards (recorded in the late twelfth century) which had their principal application in the control of cloth measurement; but the larger units of 37 inches, identical to the Scottish ell, may have been in widespread or even exclusive use in England for land measurement before this.[11] The only effective way of employing yards 'with inches' for measuring plots of land is by using knotted cords, where knots of perhaps an inch in size were set at 37-inch intervals, in other words with 36 inches between the knots. Land measurement will be discussed in Chapter **3** and Appendix **B**.

It has been thought until very recently that the English *ulna* or yard was formally introduced into English metrology at the beginning of the reign of Henry I, in the early years of the twelfth century.[12] The Chronicle of Malmesbury declares that the 'measure of his [Henry I's] own arm was applied to correct the false yard of the traders and enjoined on all throughout England'.[13] However, this declaration is quite consistent with the concept of a measure close to a yard having been in use for some time, perhaps a long time, and now being established more closely. The fact that the King's arm was invoked would give royal authority to the unit. But formal introduction does not mean initial creation of a unit – indeed, recent archaeological evidence has indicated that the Saxons may have been using the yard of 36 inches and the ell of 45 inches before the Norman Conquest. Pieces of a divided wooden measuring stick have been found in excavations in the centre of Winchester, Hampshire, which was the capital of England in late Saxon and early Norman times. The pieces can be fitted together exactly and it is clear from the notches on the wood that the intended lengths are those of half a yard and half an English ell.[14] Preliminary dating based on other finds in the same archaeological strata is early to mid-eleventh century.

The suggestion that the Scottish ell might have had its origins directly in the English yard with the incorporation of a conventional allowance has some initial appeal (although we find the parallel with an early English land-measuring unit far more persuasive).[15] However, the concept of building an allowance into a measure – of recognising a traditional practice and legislating to control it – is important, and it is a feature of the Scottish legislation to which we will return on several occasions. The appearance of getting 'a little extra' was always prevalent in Scottish metrology, as it was in English custom, and on a superficial level this appears to contradict the more modern caricature of the parsimonious Scot.

On a practical level, however, the operation of a system of traditional allowances tended to favour the substantial burgh merchants. The controlling interest of the merchant class in burgh politics meant that these allowances were not only exploited to the full in the burgh markets, but also that the rights to these allowances were carefully guarded. For example, if we turn briefly from the linear measures to consider the capacity measures, large grain transactions were conducted in units of a boll, which was ostensibly four firlots (the seventeenth-century firlot being about the size of a bushel, or about 35 litres). However, in use, an extra peck (one-sixteenth of a boll) was added for each boll of grain. This addition was called the 'charity' to the boll, a term which reflects the origin of the allowance as a compensation for accidental loss and spillage in the course of the measurement. Another important feature of the capacity measures which has its origin in traditional allowances was the progressive increase in size over the centuries. The control of this increase was a major preoccupation for the Scots parliament when considering weights and measures legislation and this will be examined in later chapters.

THE 37-INCH UNIT

A common origin of the 37-inch unit in Scotland and England can be appreciated through the relationship between the English land 'rod' of 198 inches (or 16½ feet of the twentieth century), and the lengths of several early 'foot' units. From early times until the Middle Ages, wealth meant land, and it would have been of the first importance to have an invariable unit for its measurement: the 'rod, pole or perch' as the arithmetic books of an earlier generation called it. Its antiquity in England can be shown from measuring land areas in English towns for which dimensions were recorded in an early tenth-century document known as the 'Burghal Hidage'. From this the Saxon 'gyrd' emerges as equal to this land rod.[16]

Philip Grierson has suggested that the rod might have its origin as 20 'natural' or human feet, and he has discussed it in the context of a late fourteenth-century entry in the York Memorandum Book giving a customary or natural foot of about 10 inches.[17] The natural foot is understood to be in the range 246 to 252 mm (9·7-9·9 inches). The data from the Memorandum Book would indicate a figure at the upper end of this range, so if we accept 9·9 inches as a likely figure, then 20 such feet would measure exactly 198 English inches; taking the foot as 9·7 inches would yield a rod of 194 inches. The score of 20 units appears repeatedly in German and Scandinavian metrology. The Roman account *Agrimensores* discusses the foot of the Tungri tribe of the lower Rhineland, having been measured by the general Nero Claudius Drusus, in around 16-13 BC, as the 12-inch Roman *pes manualis* and 1½ inches, or 13·11 English inches. An alternative but equivalent description is that this foot is two digits longer than the Roman foot of 16 digits.[18] Fifteen 'Drusian' feet amount to 196·7 inches, very close to the length of the rod. The seventh-century Old Minster of Winchester appears to have been built using a foot of 13·11 inches.[19] Wooden and iron measures of this foot are extant in France, but none appear to have survived in England.[20] However, no less an authority than Flinders Petrie declared this foot to have been the most common English building foot of the Middle Ages: 'The Belgic foot of the Tungri is the basis of the present land measures, which we see are neither Roman nor British in origin but Belgic.'[21]

Were the rod to be divided into 16 parts rather than 15, these would each measure 12·375 inches, which is a historically-significant unit, widely used for land survey in germanic Europe, and known as the Rhineland foot. It is interesting that both the Drusian and the Rhineland foot appear to have come from the same geographical region.

In the chapters which follow, it will be apparent that frequently a unit may be defined as a multiple of 15 smaller units, but in practical usage it was invariably divided into 16 parts. The historical chronicle of England records rods of 16 feet, and presumably these are Rhineland feet, in several locations long before the seventeenth century.[22] Sometimes these have been dismissed as scribal errors for 16½ feet, but the anonymous late thirteenth century *Husbandry*, attributed to Walter of Henley, twice refers to the perch of 16 feet, which makes copyist error unlikely.[23] The *Chronicle of Battle Abbey* specifically refers to the rod as of 16 feet in the early twelfth century.[24] At Harleston, Yorkshire, the iron standard was stated to be 16 feet long in 1306, and similarly at Chertsey in 1370.[25]

However, in the same periods we find rods of 16½ feet, as in the bounds of the Manor of Godbegot, Winchester in 1012.[26] The same size is recorded at Ely cathedral in the late eleventh century.[27] The width module of the Winchester city plan of about 1148 was five metres by modern reckoning, which is 16½ feet.[28] The 16½-foot rod appears in the *Statutum de Admensuratione Terre*, attributed to 1305 in the statute book, but most probably dating from the mid-thirteenth century.[29] It was also used in a land grant to the Dean of St Paul's

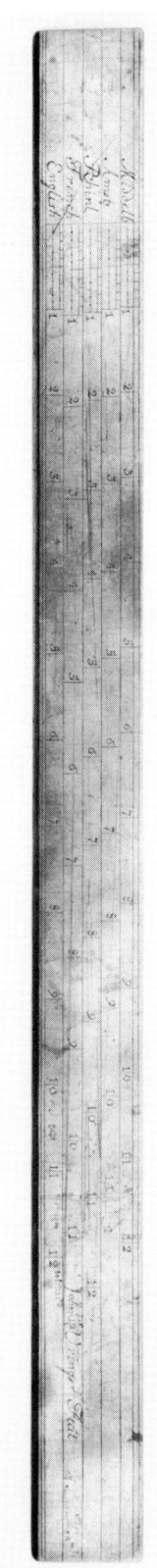

between 1229 and 1237.[30] Similarly, it appears alongside the 16½-foot rod in Walter of Henley's *Husbandry*.[31] Another specific reference is at Lawling, Essex, in 1310.[32] In general, the 16½-foot rod appears to have been in increasing use from the thirteenth and fourteenth centuries.

It would seem that rods of 16 Rhineland feet and 16½ English feet, effectively of the same overall length, were in simultaneous use in England, with the twelve-inch foot eventually becoming dominant in English metrology, and 16½ of these feet making the 'king's rod'. Only after many centuries was the rod abolished by the 1963 Weights and Measures Act.

The reason for stressing the widespread English use of Rhineland feet of about 12·36 inches is that three such feet amount to 37·08 inches, which is very close to the early 37-inch units found in both Scotland and England. One of the first commentators to draw attention to the similarity between the Scots ell and the unit of three Rhineland feet was Lawrence Burrell, who in 1960 published the first serious attempt in recent times to unravel the old Scottish metrological system.[33] Our contention is that this was the origin of the English land-measuring unit, which was then imported into Scotland by David I's administration.

There is indeed some indication that North Germanic land measurement practice was brought to England by the Angles or Saxons.[34] It has been argued that the principal English survey unit, the rod or perch, does not have its origin in the twelve-inch foot or the 36-inch yard because the statutory size confirmed in about 1300 for the 'king's rod' was the unlikely length of 5½ yards or 16½ feet. Instead, it has already been proposed by A. W. Richeson that this merely represents a compromise match between two different systems of measurement and that the survey unit was originally the North Germanic *ruthe*, of 16 Rhineland feet, which comes very close to 16½ English feet.[35] Such a suggestion would explain the references in the above early English sources to rods of 16 rather than 16½ feet, since we presume these to be references to rods of 16 Rhineland feet.[36]

The formal relationship between the English foot and the rod was first clarified in the mid-thirteenth century *Statutum de Admensuratione Terre*, when the 'Iron yard of Our Lord the King' was defined as containing three feet 'and no more', the thirty-sixth part of this yard 'rightly measured' being an inch, of which twelve made the foot, and where 'Five Yards and a half make one Perch, that is Sixteen Feet and a half, measured by the aforesaid Iron yard of Our Lord the King'.[37] Such great care has been taken to avoid possible ambiguity in these definitions that it is hard to avoid the conclusion that the intention was to differentiate between yards 'without inches', namely of the king's standard, and yards 'with inches' which had been in use for land measure. The suggestion is that yards 'with inches' were units of three Rhineland feet and the land rod was one of 16 Rhineland feet, and these were necessarily from a different measurement tradition from the yard (of 36

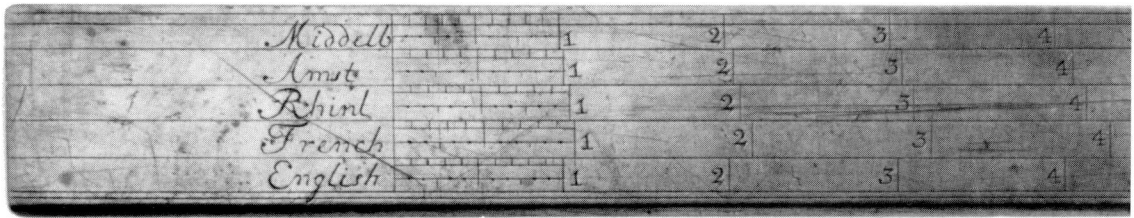

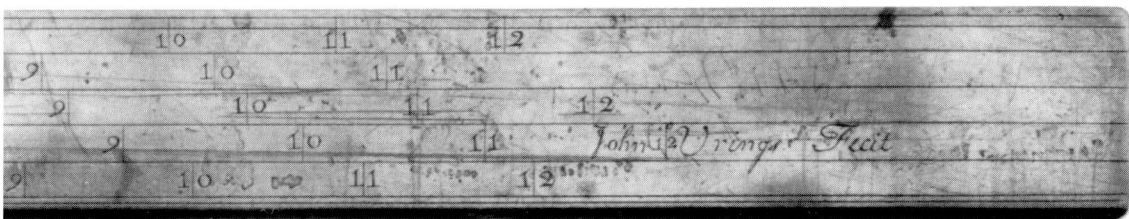

inches) and inch, which probably had specific associations with early cloth-trading practice.[38] It would inevitably be the case that there would be a small element of compromise in setting equivalents for these two Germanic units at 37 inches and 16½ feet, but it is difficult to say just how much this was without a secure knowledge of the size of the Rhineland foot. By the mid-eighteenth century the accepted value of the Rhineland foot seems to have been marginally larger than the value used by Burrell: a surviving standard foot of 1752 would have given rise to a three-foot unit of 37·09 inches, and an equivalent early nineteenth-century value was 37·10 inches.[39]

Given the uncertainty in the precise size of the Rhineland foot, Burrell has perhaps been too optimistic in finding an exact match between a unit of three Rhineland feet and the Edinburgh ell bed of 37·06 inches, particularly since we have identified the excess over the correct standard of 37 inches as a necessary measurement tolerance. It seems much more logical to conclude that the early English *ulna* (which we have already described as the 'yard with an inch') was 37 inches long because this was considered an acceptably accurate equivalence for the survey measure, and the Scottish *ulna* (or ell) was based directly on this model.

Fig. 2.1
(above and opposite)
Mid eighteenth-century brass scale, made by the London instrument maker John Urings, showing the equivalents of twelve inches in Middleburg, Amsterdam, Rhineland, French and English measures.
(NMS T.2000.90)

41

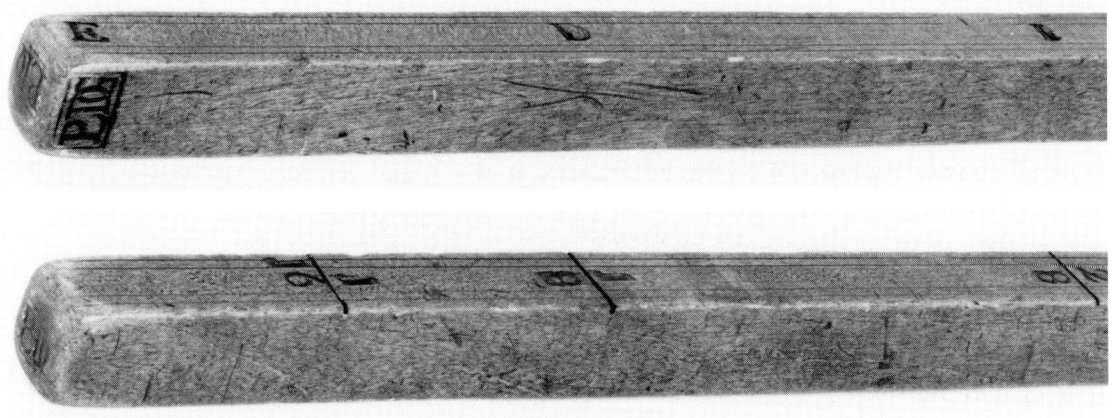

Fig. 2.2
A trader's yard measure, authorised at Edinburgh by Patrick Lindsay, Dean of Guild 1726-8, showing the usual binary division. (NMS VH.2)

THE DIVISIONS OF THE SCOTTISH ELL

Although the Scottish ell was defined as being 37 inches long, there is no evidence that it was considered for trading purposes as being composed of 37 constituent inches. Indeed this would scarcely have been practical because 37 is a prime number and therefore the ell cannot be sub-divided into factors containing whole numbers of inches. On the contrary it is clear that the ell used in trading was divided in a binary fashion (Fig. **2.2**) – into halves, quarters, eighths and so on – in exactly the same manner as early English yards (and indeed, the land-measuring rods) were divided. This was still the case in Scotland in the 1620s when Alexander Hunter, in the first book to treat Scottish metrology, described the ell as divided into quarters and every quarter into four nails.[40] The origin of the term 'nail' is in the Roman foot which was divided into 16 digits or finger breadths: since each finger carries a finger nail, a nail came to be understood as a sixteenth. Here the Scots nail is one-sixteenth of an ell (the equivalent English unit would be a sixteenth of a yard). Surviving ell measures often have the final nail divided in half to give a unit of a thirty-second, or of just over an inch.

This is not to say that the inch, the foot and the yard were not used in Scotland: indeed, it is clear from an early period that they were, although relatively rarely. An example of the yard appearing in a Scottish statute is an act of 1426 which stipulates the sizes of official batons of office:

> The kingis officiar … sal [shall] haf a horne ilk [each] ane a red wande of thre quarteris of a ȝerde [yard] lang at the lest and the officeris of the Regalite a wande of the sammyn lenth the tane [one] ends rede & the toyer [other] quhit …[41]

The foot appears in early Scottish charters concerned with measures of land, particularly with comparatively small plots of land granted in burghs.[42]

A later example of the use of the foot can be found in a parliamentary act of 1478 which relates to the specified mid-stream width of a river which had to be kept clear of obstructions such as fish-traps. This width was set at six feet, rather than the two ells that we might perhaps have expected if ells had been used to the exclusion of other units.[43] It is possible that the six-foot width replaced an earlier width which was also expressed in feet but was not a multiple of three, and therefore could not readily be given in ells since these were always used with binary division. Certainly when a new clear width of five feet was specified in a further act of 1489, this new width was not amenable to description in ells.[44] Neither of these is as colourful as the somewhat unorthodox measure of the length of a pig used in an earlier assize attributed to William I, but almost certainly dating to the second half of the thirteenth century:

> This is the kingis assise of watiris … That the mid strem aw [ought] to be fre sa mekil [so much] as a swyn [swine] of iij ȝer elde [3 years old] wel fed is of lenth sua that nother [neither] the gronȝie [snout] na the tayl may wyn [win or reach] till ony side.[45]

When a greater degree of precision was necessary, and where dimensions could not conveniently be given in binary fractions of the ell, inches were used. Thus one may turn to acts of the early fifteenth century to find the sizes of the large dry capacity measures of the 1426 Assize being given in terms of their dimensions in inches and half inches, and this is also the case in the much earlier Assize of David I. In the Assize of 1587 dimensions are given in sixths of an inch.[46] It is of course unclear whether the dimensions are defined at exact figures or whether a physical artifact has been measured with a level of accuracy reflected in these fractions, but whichever is the case these figures indicate that when precision was required accurate linear measures were available.

The second half of the sixteenth century saw the introduction of mathematical surveys and cartographic work in England, and the commercial availability of good quality divided scales. The stimulus for this activity was financial and lay in the need to measure land for valuation and revenue purposes. The manufacture of instruments for such work was the province of the mathematical instrument maker, and by the third quarter of the sixteenth century there was already a small group of Flemish craftsmen operating in London. One of the founders of the London trade was the engraver Thomas Gemini (*fl.*1524-62) who had the patronage of Edward VI.[47] The first English-born maker of distinction was Humphrey Cole (*c.*1530-91), who was an engraver and 'die-sinker of the money stamps' at the Mint in London.[48] Several of Cole's surveying instruments survive, including four fine surveyor's folding rules of 1574 and 1575 and a plane table alidade of 1582, divided in inches and sub-divided for use at a great variety of scales of reduction.[49] Two of the most splendid instruments by Cole are in Scotland (in the collection of the University of St Andrews), but their provenance is unknown so they cannot be used to demonstrate that instruments by Cole were commissioned by Scottish clients.[50] However, it would certainly be expected that instruments of the precision achieved by Cole would have been available to the officials of the Scottish Mint who conducted the work for the 1587 Assize.[51]

The London mathematical instrument makers produced specialised measuring and computing rules of increasing complexity, and it is doubtful if there was any scope for Scottish manufacturers to compete with their output.[52] Those who were required to measure or perform graphical calculation in their trade used imported English scales, a fact acknowledged by the mathematician David Gregory in the 1680s: '… other artists for the most part use the English foot, on account of the several scales marked on the English foot-measure for their use.'[53]

In Scotland the use of the ell's simple binary division seems to have become restricted to those applications where the manipulation of subdivisions was not important. In particular the ell remained the unit for

Fig. 2.3
Cloth traders at the Lawnmarket in Edinburgh, using yard sticks. 'St Giles Church …', engraved by T. Higham after a drawing by Thomas H. Shepherd. From *Modern Athens, displayed in a Series of Views, or Edinburgh in the Nineteenth Century* (London, 1829).

transactions in cloth of all sorts, and in conjunction with the larger 'fall' of six ells it was used for the delineation of plots of land.

A number of wooden ell measures, or 'elnwands', are preserved in museum collections.[54] In general they are simple wooden sticks, with a single division at the half or mid-point, another in one of these halves to indicate the quarter ell, and so on to the sixteenth (nail) or thirty-second part just over one inch. The divisions are often comparatively coarsely and inaccurately cut, although the overall length of the measure may be fairly accurate. These wooden measures are believed to have been used principally by cloth traders, and it will be appreciated that in use the accuracy of the overall length of the measure is of far greater importance than the accuracy of sub-divisions used only in the final stage of measurement of a length of cloth. In general they carry no authorisation marks, but they would have been subject to testing by guildry officials of the markets where the traders operated and to seizure if found to be of improper length.

For the burgh merchants who dealt in imported or locally-manufactured cloth, the ell measure was an important implement. The standard held by the cloth guild of a burgh (as, for example, the Stirling standard, Item **8** in the Inventory in this work) came to represent the authority of the guildry: the symbolic importance of the equivalent English measure is still acknowledged by the London guild companies.[55] The early Scottish legal compendium *Regiam Majestatem* contains a telling reference to the ell:

> The heire of ane burges, is of perfite age, quhen he is fourtene ȝears compleet, or quhen when he can number and tell silver, or measure claith (*with ane elwand*) or doe other his fathers busines and affairs.[56]

The use of the ell was therefore of no small significance if competence with it was a test of the coming-of-age of a young man.

The Scottish ell's principal application was in cloth trading, and this is clearly reflected in the form of surviving standard measures, some of which were the official standards of burghs noted for their role in the cloth trade, and a few are specifically associated with the guild organisations that controlled traders. These standards were kept by the burghs as accessible reference gauges used for checking measures. They are almost invariably of a form known as a 'bed measure', in which a direct assessment of the length of a trader's ell measure could be made by inserting it between two defining end pieces (Fig. **2.4**). Often these bed standards took the form of a divided linear scale with a projecting lug at each end of the scale.

Length standards that belonged to burghs were often physically attached to a burgh building so they would be adequately accessible. Normally this would be the 'tolbooth', which as its name suggests was the building where the tolls, customs and taxes were collected, and which might house the council chamber, court room and guardhouse. The tolbooth tended to be located at a central site, adjacent to where the markets were held and where the weighing scale or 'tron' was mounted. The earlier surviving ell standards were designed to be hung by chains; and the old Edinburgh ell, a bed measure of wrought-iron, still has chain links attached (Item **2** in the Inventory, and Fig. **2.5**). This Edinburgh ell bed, whose length is fractionally over 37 inches, has particular significance for us because at various parliamentary assizes it was decreed that the principal standard of length measure for the kingdom should be the Edinburgh ell. It is certainly an early measure, and is traditionally supposed to have been made in the sixteenth century.[57] Among the pointers we have to its age is an entry in the Edinburgh treasurer's accounts recording a payment made in December 1566 'to Nicoll Andersoun, maissoun [mason], to hing [hang] the yrne elwand in the nether Tolobuith', accompanied by an allowance of lead for securing the chain in the stonework.[58] It might be supposed that the hanging of the ell-bed followed the implementation of a major weights and measures Assize of 1563, but the measure itself may well have been constructed earlier. If this ell-bed pre-dates the 1563 Assize, then it is presumably the national standard recorded in 1552 as held by Edinburgh, in the earliest clear reference to the principal standards being in the possession of the four burghs of Edinburgh, Stirling, Lanark and Linlithgow.[59] The 1618 Act confirmed the Edinburgh ell as the national standard, and there seems no doubt that this survives as the iron Edinburgh ell bed.[60]

This wrought-iron bed measure remained the standard for a further fifty years after 1618: but in 1663 it was replaced by a new ell bed in copper (Item **4**, and Fig **2.5**), which also has a divided scale of three feet and one inch (the intervals of this scale precisely match English linear measure). However, the distance between the defining faces of the copper bed is a little greater than 37 inches and from the position of the scale divisions it is clear that an additional clearance of about 0·02 inches has been allowed at each end of the scale, so that the length of the bed is 37·04 inches. This measure similarly has an attachment chain, and the parliamentary act that required the Edinburgh magistrates to construct it stipulated that the standard foot, incorporated in this measure, should be 'preserved by the City of Edinburgh for all time comeing'.[61] Not only was every burgh to obtain a copy of the constituent foot, but the act was specific about the accessibility of these copies, which the

Fig. 2.4
The principal divisions of the ell.

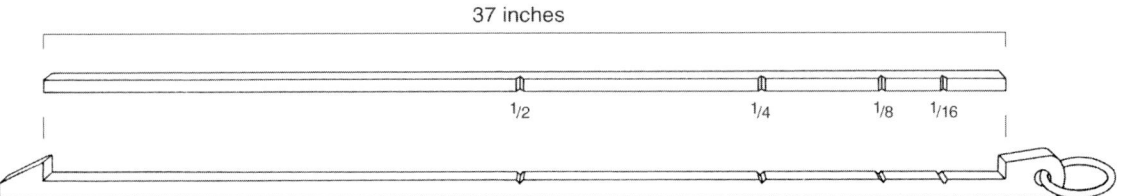

37 inches

1/2   1/4   1/8   1/16

burghs had to ensure were 'hung at their tolbuith doors or vpon their mercat croces befor the first of March 1664'.[62]

The principal purpose of the act seems to have been to introduce a statutory definition for the foot measure of twelve inches in order to control other types of measure. One of these was a peculiar short foot based on an unofficial inch:

> … notwithstanding by the antient lawes of the Kingdom, the ell is designed to be thirty seven inches, Yet many vse inches by which the ell is divyded into fourty tuo inches, and of these small inches make the foot measure of a smaller proportion than it ought to be To the great preiudice of the leidges; and that the occasion of this liberty hath been Because that hitherto ther hath no standard been appointed for foot measures alswell as other measurs; Therefor … Ordaines, that from & after the first day of Junij next 1664, no workman nor other person shall make vse of any other foot measure, then such as consists of tuelve of these inches whairof the ell containes thirty seven …[63]

The act was to apply to all wrights (joiners), glaziers, masons and other 'publict' workmen, who should use only the twelve-inch foot in their work, under penalty of a fine of £100 Scots. The small inches were therefore $^{37}/_{42}$ (or 0·88) true inches, and the reason that workmen like glaziers are specifically mentioned is that some worked to a nine-inch foot, but others to an eight-inch foot, as Alexander Hunter noted in 1624.[64] Early window glass was, of course, expensive, and only comparatively small discs could be spun; so it would not be entirely unexpected to find smaller length units being used by glaziers.

The Scottish glaziers' foot became a matter of contention soon after Hunter's volume appeared: the Convention of Royal Burghs meeting at Linlithgow later in 1624 remarked on the abuses caused by glass wrights in the measurement of the foot of glass.[65] The burghs were requested to send commissioners to the next session of the Convention fully informed about the practices in their areas. The 1625 Convention in Glasgow learned about various different values for the glaziers' foot but eventually agreed that it should properly be a quarter of an ell in length, that is 9¼ inches, rather than the nine inches that Hunter had recorded in 1624.[66] This decision was formalised at the Convention meeting in July 1626.[67]

We will return to the question of the glaziers' foot in Chapter 3 when we

Fig. 2.5
The sixteenth-century Edinburgh ell bed (above) and the 1663 Edinburgh ell bed. (Items **2** and **4** in the Inventory)

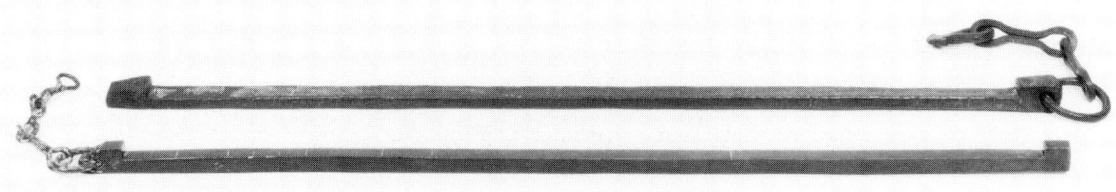

discuss the appropriate context for costing craftsmen's work. We will see that such work was assessed on area calculations using the size of the link (or hundredth part) of the land measuring chain, and in the case of glass measurement the link was termed the 'glaziers' foot'. Since the Scottish chain was longer than the English chain, we can see why Hunter recorded two sizes for the glaziers' foot – the Scottish link was a little less than nine inches long whereas the English link was a little less than eight inches long. The nine-inch glass foot was therefore the Scottish equivalent of the eight-inch English glaziers' foot, and the use of an English unit in Scotland no doubt reflects the English origin of much of the window glass in Hunter's time. Some window glass was manufactured in Scotland in the first half of the seventeenth century, notably at the glass-house established at Wemyss on the Fife coast in 1610 by Sir George Hay (1572-1634), an industrial entrepreneur and favourite of the king, who subsequently became Lord Chancellor of Scotland and first Earl of Kinnoull.[68] From 1621 Hay's glass was exempt by King James from the import restrictions placed on foreign glass entering England, even though this infringed Sir Robert Mansell's English monopoly of supply. The quality of Hay's glass was reported to the Scottish Privy Council in 1621 to be as good as that of German glass imported into Scotland through Danzig, although not as strong, but in practice English glass was probably of better quality and must have held a good proportion of the market.[69] Mansell pressed hard to enforce his monopoly against encroachment by Hay; and at the same time as the Convention of Royal Burghs was taking such an interest in the units by which glass was sold, the Scottish coal producers were countering by raising the price of the fuel that Mansell needed for his works.

Whether or not the Convention was successful in forcing an increase in the Scottish glass foot from about nine inches to 9¼ inches was soon

Fig. 2.6
Boxwood folding rule, incorporating a bubble-level, apparently divided in nominal feet of work, by Adie & Son, Edinburgh, c.1840.
(NMS T.1996.147)

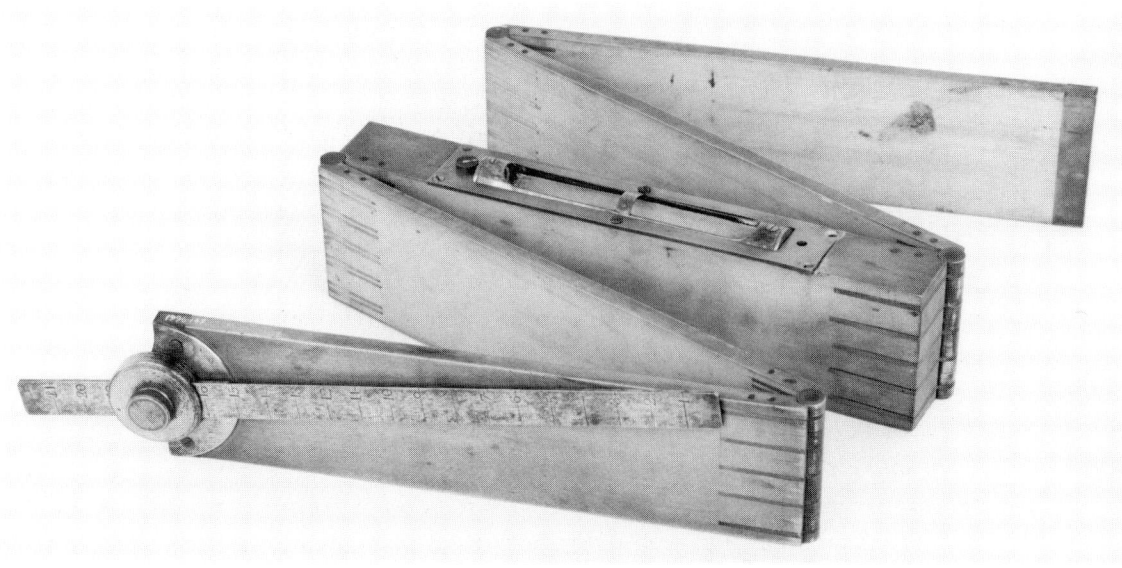

immaterial – Mansell was able to purchase control of Hay's glass-works in 1627, and they were closed not long afterwards. It is possible that the efforts of the drafters of the 1663 act to eliminate the glaziers' foot coincided with an attempt to revive Scottish window-glass manufacture at Robert Pape's important glass-works in the Citadel at Leith. But although the Privy Council provided support for his venture by forbidding merchants and others from importing glassware from abroad, window-glass was specifically excluded from the provisions of the act. This suggests that the bulk of this market was satisfied by imported English glass, and the costing of this is likely to have remained according to English practice. Thus David Gregory, writing in about 1685, recorded that an eight-inch foot was used by the glaziers, as did the garden designer John Reid in 1683.[70] John Swinton in 1779, more than a century and a half after the 1663 act, stated that the foot of the glaziers 'is commonly reckoned 8 inches', while in the same sentence referring to the 1663 act that prohibited its use.[71] Indeed, the glaziers' foot was alive and well as late as 1820, when it appeared in a government report giving the English version: 'Foot of the glaziers, commonly 8 inches.'[72]

In spite of the concern to distribute copies of the 1663 foot standard, only one of the 1663 foot measures is known to have survived (Item 5 in the Inventory). In style it resembles the Edinburgh ell, and like the standard on which it was modelled, it is in copper and carries an inscription relating it to the authorising act. It was made for the burgh of Dumfries, in south-west Scotland, a centre for the early Scottish textile trade. The use of all the later variants of linear measures is closely linked with the woollen trade and more particularly with the linen trade, and the special requirements placed on the burghs where cloth was sold in the market may have made it inevitable that Dumfries and at least a few other similar burghs should have acquired copies of the 1663 foot. The Convention of Royal Burghs took note at their Aberdeen meeting in 1665 of the diversity of lengths of the ell and foot in the various burghs and required that the measure 'be conform to the standard of Edinburgh'.[73]

In summary, it is apparent that a full range of types of linear measure was employed in Scotland, and the use of each type was appropriate to the circumstances. In normal trading use, which in practice often meant an association with textiles, the ell was applied, together with its binary divisions. However, the inch, the foot, and even the yard, were clearly understood as forming integral parts of the linear measurement system. Rope was reckoned in fathoms, as in England.[74] While this reinforces the view that the Scottish units had their origin in those of England, the adoption of a new unit such as the English glaziers' foot, and the creation of an equivalent Scottish unit, indicates not only the close commercial ties between the two economies but also a basic similarity in approach to practical computational issues.[75]

But we must not view this relationship as merely one of Scottish dependence on English measurement practice. When we come to examine the regulations that apply to textile production in the early eighteenth century in Chapter 3, we will find English acts which include the use of 37-inch units, reflecting a Scottish penetration of English practice. It is certainly hard to escape the conclusion that there was the closest equivalence between the two linear measurement systems, and that the underlying unit of the inch was identical in the two kingdoms. John Reid made the specific comment in 1683 that there was 'no distinction betwixt a Scots and English foot'.

MEASUREMENT TOLERANCE
AND THE ABSOLUTE LENGTH OF THE ELL

Because the Edinburgh ell was recognised as the national standard, we might expect to find that the length of the surviving sixteenth-century Edinburgh iron ell-bed (Item 2 in the Inventory) was precisely 37 inches. However, careful measurement of the distance between its jaws shows that it is marginally longer. The length usually quoted is 37·0598 inches (or 37·06 inches), which was obtained when the ell bed was measured in the early nineteenth century by the civil engineer, James Jardine (1776-1858), and this length was subsequently adopted as the official size of the ell in Imperial inches.[76] The Edinburgh 1663 copper ell bed (Item 4 in the Inventory) is found to have a very similar length, showing that the two measures have a close relationship.

Even such a small discrepancy between the ell's length and the length of 37 inches raised difficult issues. It opened the possibility that the origin of the ell might not be in English but in Continental practice; and it also suggested that the Scottish inch was larger than the English inch, with 37 of these larger inches to the ell. Such notions might not have emerged had there not already been other variants of the ell in use in the eighteenth century, at least one of which, an ell of 37·2 inches, was considered as comprising 37 large inches. These variant ells were all slightly longer than the basic ell, and they appear to have been related directly to regulations governing the textile trade. Initially these longer ells were applied in very specific circumstances, but by the late eighteenth century such distinctions of application had largely been lost. The problem eventually came to a head in the early years of the nineteenth century when a legal test case (which will be discussed below) required an authoritative value to be obtained for the length of the ell. It was considered prudent to go back to the venerable sixteenth-century Edinburgh ell bed; and precision measurements of this single piece made in 1811 by Patrick Copland (1748-1822), professor of mathematics at Marischal College, Aberdeen, and subsequently by Jardine, form the incorrect basis for nearly all subsequent work.

There is a range of problems in deciding, on the basis of the two surviving Edinburgh bed measures, what absolute length to assign to the Scottish ell.

**Fig. 2.7a
(opposite and next page)**
The Inverkeithing ell, 1500. The cuts on the left edge define a bed measure of one ell; those on the right edge form a gauge for the dimensions of the standard firlot. (Item 1 in the Inventory.)

For example, even from a visual inspection it is apparent that the defining faces of the two ell beds are not accurately flat or parallel to one another. Thus, in the case of the early iron ell bed, the measurements made by Copland in 1811 were taken very close to the vertices of the angles where the defining faces meet the main bar of the ell bed: he obtained two measurements of this distance as 37·06 and 37·07 inches, against our own recent measurement of 37·07 inches.[77] However, because the lower of the two faces is concave near this angle, suggesting that it has worn or eroded back, the separation of the faces over most of their area is as much as 37·09 inches, while Copland observed that the separation at the top of the faces was eight-hundredths of an inch greater than that at the bottom.[78]

In the case of the 1663 copper ell bed, the faces are inclined to one another as well as being convex: a bar of 37·05 inches can be inserted fully at the centre of the faces, although the clearance at the left edge of the measure is only 37·04 inches. The convexity of the faces may be the result of adjustment, or the cumulative effect of mechanical wear over an extended period. From the seventeenth century many trading measures were capped in metal at their ends, and repeated insertion of such measures will have led to some progressive damage of the defining faces of the standard, which is of a comparatively soft material. Whatever the cause, the result is that a bar of 37·09 inches can be at least partly inserted in the bed measure.

If we add to this uncertainty of interpreting the lengths of these measures the fact that the sixteenth-century iron ell bed is badly corroded and its surface is affected by rust pitting, it is quickly apparent that its length cannot be determined to the accuracy implied in the value of 37·0598 inches.[79] Indeed, all we can say with any confidence is that both bed measures are consistent with an original length of about 37·05 inches.[80] We can gain some support for this from the only other early ell to survive, a bed cut in one edge of a gauge measure dated 1500 (Item **1** in the Inventory, and Fig. **2.7**). Because this is cut in a sheet of brass the length can be found fairly accurately, and we have measured it as 37·05 inches.

Indeed, the various markings on the measures would enable us to come to a variety of conclusions about the 'correct' size of the ell, and so we must interpret this evidence with caution. It is certainly not adequate to obtain the simple separation of the measurement faces of the standards, as earlier commentators have done.

If we consider these measures carefully, we can see that there are several different factors coming into play. First, we must acknowledge that we do not know by what process of comparison and adjustment new standards were constructed: we know neither what level of accuracy was attainable nor what error was considered acceptable. Second, these standards were designed for practical use in testing traders' measures and may therefore incorporate different assumptions about the accuracy tolerance of a market-place measure.

In addition, however, the nature of the measure itself is a factor. Almost all the surviving ell standards are 'bed' measures, designed to test whether an inserted 'bar' measure is of adequate length. Standard measures are sometimes made as bed and bar pairs, but it is important to recognise that it is physically impossible to make these so precisely that they are both exactly the same length. In practice the bed must be fractionally longer than the bar, and it becomes a sensible question to ask which one is to the legal standard. The difference may be very small, but it is material.

There is an English example which illustrates this. New English linear standards were introduced by Elizabeth I in 1588, and the Exchequer standard took the form of bars of one yard and one English ell, together with a matching bed standard with recesses for each bar. (We have already noted that the English ell, at 1¼ yards, was larger than the Scottish ell and had particular relevance to the cloth trade. There was an analogous Scottish measure of 1¼ Scots ells, called the 'short reel' or the 'five-quarters reel', which had similar textile associations and this will be discussed in Chapter 3.) These 1588 English standards survive at the Science Museum, London, and the accepted modern measurements of their lengths are 35·99 inches for the yard bar and 45·04 inches for the 45-inch ell bar.[81] The bed measures into which they fit are longer than the bars, but not by equal amounts: the yard bed is 0·012 inches longer than its bar, but the ell bed is only 0·005 inches longer. This ell bar is in fairly good condition, whereas the yard bar has ends that are rounded by wear and at some time it has also been broken and repaired. No doubt the yard bar had been subject to heavier use in checking bed measures elsewhere, and the implication of the much larger tolerance between the yard and its bed is that the bar is now shorter than its original

Fig. 2.7b
(see also previous page)
The Inverkeithing ell, 1500.
Details of the inscription and
the reverse side at one end
of the ell bed.
(Item **1** in the Inventory)

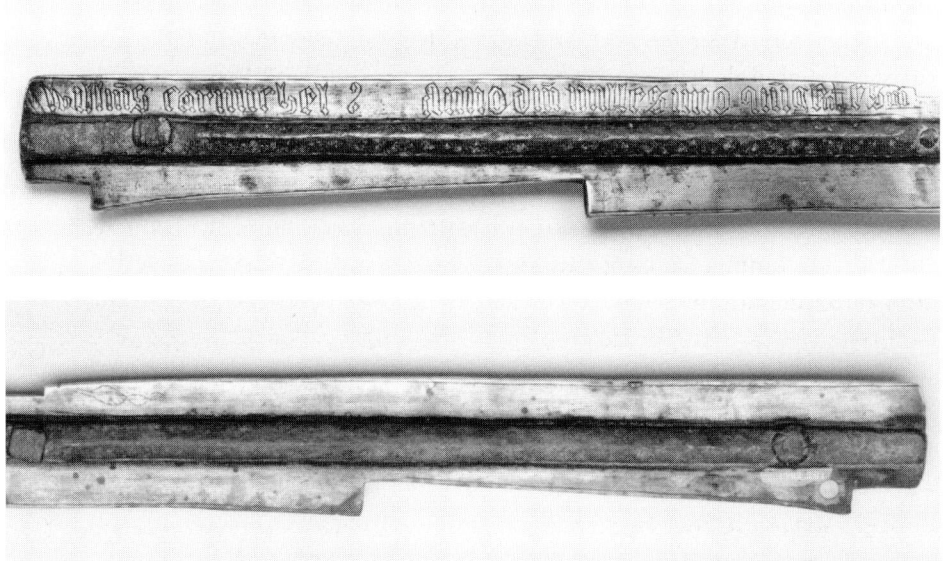

length. If the amount of clearance between each bar and its bed was initially the same, the original yard would appear to have been almost precisely the correct length of 36 inches, whereas the ell is apparently too long by 0·04 inches. This should make us cautious about accepting any single determination as reliable – aspects of construction technique and inter-comparison have clearly influenced the absolute lengths of at least one of these measures. However, the important point to emerge is that there is a length difference between a bar and its bed measure, even though either of them at the same time would have been considered as complying with the legal standard.

In Scotland the burghs held bed measures, and linear gauges for the dimensions of the capacity measures, and it was initially the responsibility of officials named gaugers to ensure that the local traders' measures conformed with the burgh measures. In turn the burgh bed measure was initially subject to checking by the royal chamberlain during his annual inspection of the burghs (the 'chamberlain ayre').[82] In a real sense, therefore, the chamberlain's ell bar performed the function of a primary standard, and it was presumably made (with the greatest degree of accuracy possible at the time) to the specified 37-inch length. However, the crucial question for us is the acceptable tolerance between this measure and the burgh bed measures. The only way we have of approaching an answer to this question is to consider the provenance of the old Edinburgh ell bed, and to examine the evidence presented by the divisions on the 1663 ell bed.

At least until the early sixteenth century the Scottish procedure for holding the primary measurement standards was analogous to that in England, where they were retained in the Exchequer (and are therefore described as 'Exchequer Standards') and authorised copies were distributed to a number of principal centres. The Scottish primary standards were held by the chamberlain.[83] Some specific burghs were issued with official copies which then served as reference standards for their localities: these burghs were listed as Edinburgh, Perth and Aberdeen at the Assize of 1458.[84] The chamberlain's standards were still in use in 1512 when Dundee was added to this list.[85]

By 1552, however, the emphasis in Scotland had to some extent shifted away from a centralised repository, although the administration continued to hold reference standards, until at least the early seventeenth century; instead the commercial opportunities to hold standards and make authorised duplicates for the burghs was ceded to Edinburgh, Stirling, Lanark and Linlithgow.[86] Each burgh had custody of one standard and was responsible for this to the Convention of Royal Burghs. It is not known when this transfer of responsibility from the chamberlain to the burghs took place. Certainly, the chamberlain's powers were curtailed in the early sixteenth century: a re-organisation of the courts led to a distribution of the functions of his Court of the Four Burghs, and he lost control over burgh accounts in 1539. The chamberlain ayre had been an infrequent affair by the turn of the century and the last was held at Edinburgh in 1517.[87] The authority of the Convention of Royal Burghs as the body which regulated the commercial and administrative interests of the burghs was established by the mid-sixteenth century, and responsibility for the quality of burgh weights and measures was presumably acquired between 1539 and 1552.

It is important to note that the chamberlain's primary standards do not seem to have been dispersed amongst Edinburgh, Stirling, Lanark and Linlithgow. Instead it appears that the four standards that were designated to be the new primary standards were merely those that the four burghs were holding at the time. It is conceivable that this radical move was made after the sacking of Edinburgh in 1544, inflicted by the Earl of Hertford as part of the English monarch's infamous 'Rough Wooing' of the child Queen Mary, before her departure for France in 1548 and eventual marriage to the Dauphin. The turbulent period that followed saw a strong English presence in eastern Scotland, with several garrisoned forts, countered by a rapidly-building French military investment. A temporary peace came in 1551 and with it a need to re-establish external trade. It is possible that the administration's primary standards may have perished during this period, and there are no definite indications that they survived. Shifting an onus of custody on-to the burghs would remedy any such loss, but in any event it would ensure the community of burghs retained greater access to reference standards.

The choice of the particular burghs of Edinburgh, Stirling, Linlithgow and Lanark, and the procedure of making each custodian of one of the standards, presumably reflects a need to demonstrate a continuity of authority by respecting the status of the burghs represented on the old chamberlain's Court of the Four Burghs.[88] The ceding of responsibility for the standards to the Convention and to the four burghs individually may represent an admission of the administration's need to establish a more effective mechanism to control trading measures, but it led to some ambiguity over the status of these standards.

Three metrological assizes were held in the second half of the sixteenth century – in 1555, 1563 and 1587 – and it seems likely that the practical work at all these was conducted by royal officials at the Mint. At the last of these assizes, it was recommended that copies of the new high-quality standards, as well as going to the burghs 'to whom they had been committed of old', should also be retained in government control in Dumbarton Castle and at the Register in Edinburgh Castle. This recommendation was made in spite of the fact that the Register had been damaged in 1573 by heavy English cannon fire that on 8 May led to the surrender of Edinburgh Castle which was being held for Mary by Sir William Kirkcaldy of Grange.[89] Although we do not have the reports of the 1555 and 1563 Assizes, there was probably some form of official repository at these earlier times too, perhaps in the Register or at the Mint. Whether Dumbarton Castle was used in 1587 is unclear, but certainly additional sets of the distributed standards were provided by the four burghs to the Privy Council for safe keeping at Edinburgh and Dumbarton Castles in connection with the 1618 Assize.[90]

This can be exemplified if we consider briefly the liquid capacity standards. We will see in Chapter **6** that the pint was defined in 1563 in terms of a water content of 55 ounces, and this represents a volume which is marginally smaller than that of the standard pint vessel for the burghs of the day, a pint standard which still survives and is now known as the Stirling Pint (Item **108** in the Inventory). The volume of this pint has been measured by us as 103·8 cubic inches, so its water content would weigh 55·6 of the then current ounces. Other examples of '55-ounce' standards with slightly smaller volumes have been identified, and it seems inescapable that official reference standards of these were also retained by the administration. The Stirling Pint may no longer have accurately represented the administration's new national standard of 1563. In spite of this it was still regarded as having some authority, because it was examined by the commissioners for the 1587 Assize, and then committed back into the care of the Stirling guildry as the standard.[91] It was not until the 1618 Assize that the Stirling Pint again became the practical as well as theoretical standard; and, even so, the '55-ounce' definition persisted.

So it appears that although the standards held by the four burghs may have been accepted nominally as the primary standards, the administration did not feel obliged to follow them and in all probability held reference standards of its own. Indeed it seems likely that the privilege extended to the four burghs was simply that of supplying (and charging for) authorised copies of their particular standard for the use of other burghs. The oldest surviving standard authorised in this way is the ell of 1500, which carries the name of William Carmichael, treasurer of Edinburgh burgh council (see Fig. 2.7, and Item **1** in the Inventory). This measure is associated with the royal burgh of Inverkeithing, on the north shore of the Forth estuary, which was from time to time an early meeting place of the Convention of Royal Burghs.[92]

It is against this background that we have to view the status of the surviving sixteenth-century Edinburgh ell, and we must recognise that it may initially have been constructed as a burgh standard and not as a national standard at all.

There were a number of assizes to reform the legal weights and measures in the early to mid-sixteenth century, and these will be discussed in Chapters **6** and **7**. Unfortunately, there is almost no record of the commissioners' deliberations or decisions, and this poses a severe problem of interpretation. However, when we get to the 1587 Assize, there is a surviving report which is preserved in the parliamentary record. The wording of this (repeated at the 1618 Assize) is specific that it was not merely the ell that was committed to Edinburgh, but the 'eln and stand thereof'.[93] The word 'stand' does not mean a support for the ell; it is being used in the sense of the components to complete a set. There are several references in burgh records to stands of weights, and this use is analogous to that of a stand of bells or of chess pieces or the stand of tools for a particular trade.[94] Capacity measures may be referred to in the form of a specified measure and its stand. Thus, for example, if measures had to conform with 'the stand and mesour' of the firlot of Linlithgow, they would have to comply with the Linlithgow firlot and all its permitted parts, namely the peck and half-peck.[95] The stand of liquid measure would be the quart, pint, 'chopin' (half-pint) and 'mutchkin' (half-chopin).[96] The 'eln and stand thereof' is likely to refer to a bar measure and its complementary bed measure, and probably additionally to sub-divisions on both measures.[97] The contemporary English Exchequer standard of 1558 exists as a combined 'yard and bed' standard.[98]

The 1587 Assize also speaks of the testing the depth and width of the firlot with 'the Elnvand'.[99] This must be a divided bar measure and not a bed measure (otherwise the depth could not be determined) and was presumably a scale that carried some authority because it has been used to obtain the dimensions (quoted to a sixth of an inch) given in the Assize. What we cannot say is whether this is the standard ell itself or an official scale related to it, but it certainly does appear to have been a 37-inch long scale divided into inches, and either sub-divided further or with sub-divisions of sixths estimated by eye. The use of sixths rather than quarters is not encountered elsewhere at this period, but it perhaps mirrors the division of the foot into twelve inches. This in turn suggests that the sub-divisions were actually marked on the scale.

Whether the Edinburgh ell standard did at this time comprise a paired bar and bed measure, with the bar being either the earlier chamberlain's measure or one that replaced it, is no longer clear. However, the existence of a bar standard in Edinburgh seems plausible, and the use of an official bar standard by the chamberlain provides a convenient mechanism for checking the local bed measures of the market-places at his inspections of the burghs. It would follow that the ell standard itself should properly be considered to be the bar measure; and therefore those who have assiduously measured the surviving Edinburgh bed measure have necessarily found a measure that is over-sized.[100]

From the sixteenth century length standards were obtained by the burghs on application to Edinburgh. The standards that survive are mainly bed measures, partly because these were more durable but mainly because the principal requirement of a burgh was for a gauge which would check the simple ellwands of the traders. (However, at least one seventeenth-century bar standard is known – the 1668 Kirkwall ell, Item **6** in the Inventory – which bears an inscription showing that it was adjusted in Edinburgh in the presence of the dean of guild: unfortunately, in its present altered form, in which one end has been damaged, it shows no obvious relationship to the Edinburgh 1663 standard.) The craftsman appointed by the Edinburgh dean of guild to supply verified ell measures presumably used an authoritative comparison piece for his bed measures and this may originally have been an official bar standard. A similar arrangement operated in Stirling in the early eighteenth century (and presumably at earlier times also) when the official standard pint was used by the craftsman who had the privilege from the burgh of making and adjusting duplicates.[101] The fact that the burghs' principal requirement was only for a bed measure was apparent even as late as the issue of the Imperial standards of 1824. The 'best set' of standards offered to the local authorities throughout Britain included paired bar and bed yards, whereas the 'second set' included merely 'the Standard Yard, being the bed only, as usual'.[102]

However, recognising that the purpose of the burgh bed measures was to check local trading measures, this comparison was undoubtedly not done with the type of rigour that would be associated with matching national standards (such as the paired English Elizabethan standards). In practice considerably greater latitude must have been allowed for these burgh standards, and an ell measure used in the market-place would not have been expected to be a push-fit in the burgh bed measure. Indeed, considering that a bed measure will more readily detect a trader's ell that is too long than one that suffers from the more serious defect of being too short, it is unlikely that burgh bed measures were made as close to the true standard as was technically possible since this would have the effect of reducing the permissible length of the trading ell. It is more logical to suppose that burgh bed measures would be made to a length that would give a recognised amount of play for a 37-inch standard, that thereby ensured that there was no downward pressure on the length of the ell of the market-place.

The combination of all these features, and the addition of some cumulative wear, suggests that we should expect all the burgh bed measures to be slightly over-standard. The importance of the Inverkeithing measure of 1500 is that it also acts as a gauge for the dimensions of the dry capacity measures, and it will be discussed in this context in Chapter 7. For the moment, we will record that its length is 37·05 inches.[103] This length, and the length of the early Edinburgh ell, seem entirely consonant with a system based on an original 37-inch bar standard. We can obtain some direct confirmation of this from the copper ell bed measure made for Edinburgh in 1663 (Item 4 in the Inventory). This also has a divided scale of three feet and one inch (with some additional sub-division) laid down on it, and the intervals of this scale precisely match English linear measure. However, the gap between the defining faces of the bed measure is a little greater than 37 inches, and from the position of the scale divisions it is clear that an additional clearance of about 0·02 inches has been allowed at each end of the scale, so that the length of the bed measure is 37·04 inches. Thus all three of the early surviving ell beds (the 1500 Inverkeithing ell, the sixteenth-century and 1663 Edinburgh ells) share a length of about 37·05 inches, and we can appreciate from the 1663 ell that these have been made to incorporate a play of about 0·05 inches (see Fig **2.8**).

The difficulty that arises if the legal ell is considered to be even fractionally longer than 37 inches is that it may be understood as comprising 37 slightly larger inches. This situation arose in the eighteenth century when a larger ell was designated for specific administrative reasons, and also in the nineteenth century when misplaced antiquarian zeal erected an apparently independent Scottish inch on a determination of the length of the old Edinburgh ell (bed) as 37·06 inches.

The fact is that the Scottish and English inches are identical. The Scottish statutes and the early law books uniformly define the ell as 37 inches without any qualification; and the early texts for both English and Scottish practice give the same length for the inch as three barley corns.[104] Some support can be obtained from an English source of 1607, in a period when it might be expected that any sensible difference between the two systems would be recorded:

> Three barley corns without tails set together in length make an inch of which corns one should be taken off the middle ridge, one off the side of the ridge and another off the furrow. Twelve inches make a foot of measure. Three feet and an inch make an ell … This is the measure of Scotland.[105]

A few years later, Alexander Hunter's quasi-official *Treatise of Weights, Mets and Measures of Scotland* of 1624 provided an equivalent definition:

> 3 Barlie cornes faire and round lying in length without the tailes maketh an inch; 12 Inches maketh a foote; 3 Foote is an English yard; 3 Foote and an inch, or 37 inches makes the Ell of Edinburgh.[106]

Hunter proceeded to give the sizes of the English mile and the larger Scottish mile (discussed in Chapter 3) in terms of yards and ells, and these sizes indicate clearly that he was using an ell of exactly 37 inches.[107] In 1685 these definitions were repeated yet again when there was an attempt (which appears to have been ineffectual) to enforce the use of the English mile so that a single mile might apply throughout Great Britain. This Scottish statute (as

Fig. 2.8
The expected dimensions of a burgh ell bed.

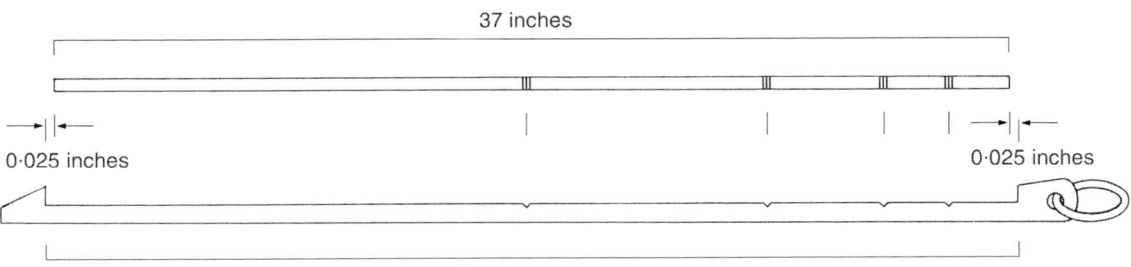

37 inches

0·025 inches   0·025 inches

37·05 inches

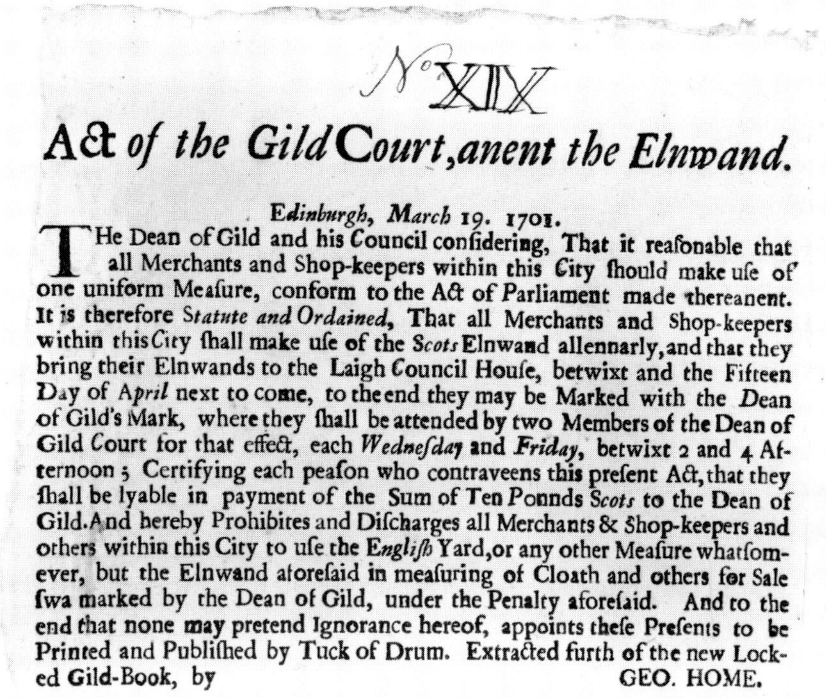

Nº XIX

Act of the Gild Court, anent the Elnwand.

Edinburgh, March 19. 1701.

THE Dean of Gild and his Council confidering, That it reafonable that all Merchants and Shop-keepers within this City fhould make ufe of one uniform Meafure, conform to the Act of Parliament made thereanent. It is therefore *Statute and Ordained,* That all Merchants and Shop-keepers within this City fhall make ufe of the *Scots* Elnwand allennarly, and that they bring their Elnwands to the Laigh Council Houfe, betwixt and the Fifteen Day of *April* next to come, to the end they may be Marked with the *Dean* of Gild's Mark, where they fhall be attended by two Members of the Dean of Gild Court for that effect, each *Wednefday* and *Friday,* betwixt 2 and 4 Afternoon; Certifying each peafon who contraveens this prefent Act, that they fhall be lyable in payment of the Sum of Ten Pounds *Scots* to the Dean of Gild. And hereby Prohibites and Difcharges all Merchants & Shop-keepers and others within this City to ufe the *Englifh* Yard, or any other Meafure whatfomever, but the Elnwand aforefaid in meafuring of Cloath and others for Sale fwa marked by the Dean of Gild, under the Penalty aforefaid. And to the end that none may pretend Ignorance hereof, appoints thefe Prefents to be Printed and Publifhed by Tuck of Drum. Extracted furth of the new Locked Gild-Book, by   GEO. HOME.

Fig. 2.9
Act of Edinburgh's dean of guild court, 1701, requiring traders to have their ellwands stamped, and prohibiting the use of the English yard. (Edinburgh University Library)

well as confirming the stubbornness of the glaziers) is the first to recite the table of length up to the mile:

> That three barly Corns set lenthways, shall make ane Inch, as it is already used, that tuelve Inches shall make a foot of measure, which is to be the only foot by which all workmen especially masons, Wrights, Glasiers, and others are ordained to measure their work in all time coming, under the pain of ane hundreth pounds toties quoties [each time], Three of these foots are to make a Yard, as three foot and one Inch makes a Scots Elne, and a thousand seven hundreth and Sixty yards are to make a myle, which is to be made the Standart of Computation from place to place in all time coming.[108]

An acceptance that Scottish and English inches are identical is an essential starting point in understanding Scottish weights and measures legislation: this allows us to interpret information about the sizes of measures given in the various acts and to relate volume and weight determinations of measures. The definition of the Scottish ell as 37 of these inches matches our understanding of the early English *ulna* of 37 inches and provides some support for the idea that the Scottish units were based directly on their counterparts at the time of David I. Although David's Assize survives in a form that has undoubtedly been altered over the centuries, it nonetheless retains elements of the earliest Scottish regulations, and these presumably include the definition of the ell as 37 inches. There are no English regulations of a comparable date – if there had been we might have found a definition of the *ulna* as 37 inches – and this gives us an indication of the potential that Scottish documents may have for illuminating gaps in the English administrative record.

However, the Scottish ell was not always taken as 37 inches. Quite separate from the issue of tolerance in the size of the burgh bed measures, longer ells were encountered in some circumstances from the late seventeenth century. Although these were probably confined initially to the cloth trade, they soon gave rise to confusion over the sizes for land measurement units. This question will be addressed in the following chapter.

1 For an earlier weight unit, see A. D. C. Simpson, 'Scots "Trone" Weight: Preliminary Observations on the Origins of Scotland's Early Market Weights', in *Northern Studies*, 29 (1992), 42-81.

2 T. Thomson and C. Innes (eds), *The Acts of the Parliaments of Scotland [APS]*, 13 volumes (Edinburgh, 1814-56), I, 673. The text is reprinted in Appendix A.1 and A.1a. For the corresponding English statements, see R. D. Connor, *The Weights and Measures of England [WME]* (London, 1987), Chapter 6.

3 F. L. Attenborough (ed), *The Laws of the Earliest English Kings* (Cambridge, 1922), 12-13.

4 H. W. Stevenson (ed), *Asser's Life of King Alfred* (Oxford, 1904), 90.

5 J. B. Sheppard (ed), 'Certa Mensura Cantuariensis: Second Report on Historical Manuscripts belonging to the Dean and Chapter of Canterbury', in *Eighth Report of the Royal Commission on Historical Manuscripts* (London, 1881), appendix 1 (part 1), 315-55; M. D. Harris (ed and trans), *The Coventry Leet Book* (London and New York, 1907), 396.

6 Paired Exchequer standards of the yard and ell survive and are attributed to the English 1588 Assize: *WME*, 240-1.

7 Ibid., 87.

8 Robert Kerr, *General View of the Agriculture of the County of Berwick* (London, 1809), 451.

9 *WME*, 87: 18 Henry VI *c*.16, 1439.

10 Ibid., 88: 'Virgas domini regis ulnaries cum pollice interposita.'

11 See, for example, Marion Gibbs (ed), *Early Charters of the Cathedral Church of St Paul, London*, Camden Society, third series, no. 58 (London, 1939), 212: no. 270: ' … 29 yards of the iron yard of our Lord King Henry, son of King John, with inches.'

12 *WME*, 83. But see also the more recent R. D. Connor, 'Perches, Pottles and Pounds', in *Physics in Canada*, 48 (1992), 63-76, especially p. 66.

13 'Mercatorum falsam ulnam castigavit; brachii sui mensura adhibita, omnibus que per Anglian proposita', William Stubbs (ed), *Willelmi Malmesbiriensis Monachi de Gestis Regum Anglorum*, 2 volumes (London, 1887), II, 489; translated in J. A. Giles (ed), *William of Malmesbury Chronicle* (London, 1847), 445.

14 We are indebted to Professor Martin Biddle and to Elizabeth Lewis of Winchester City Museums for access to this item. It is described with a drawing in Martin Biddle, *Winchester Studies 7, Part II: Object and Economy in Medieval Winchester:* *Artefacts from Medieval Winchester*, 2 volumes (Oxford, 1990), II, 925-8. For a brief notice, see Winchester Museums Service, *Newsletter*, issue 4 (June 1989), 7.

15 This suggestion was first made by Professor Philip Grierson in his 1971 Stenton Lecture, *English Linear Measures: an Essay in Origins* (Reading, 1972), 19.

16 *WME*, 39-42.

17 Grierson, op. cit. (15), 23. A. Owen, *Ancient Laws and Institutes of Wales* (London, 1841), London Record Commission, Book II, section XVII, §5. This is about one inch larger than the Welsh foot of the Venedotian Code of the tenth century. The Code declares that 'three lengths of a barley corn in the inch; three inches in the palm breadth; three palm breadths in the foot …', *ie* a nine-inch foot; Maud Sellers (ed), *York Memorandum Book, Part I, 1376-1419*, Surtees Society no. 120 (Durham, 1912), 142: a note dated 1395, after giving the customary '3 barley corns = 1 inch, 12 inches = 1 foot', then says: '3 barley corns = 1 inch, 3 inches = 1 palm, 3 palms plus 3 barley corns = 1 foot … ', *ie* a ten-inch foot.

18 For the foot of the Tungri tribe, see C. Thulin (ed), *Corpus Agrimensorum Romanorum*, 2 volumes (Leipzig, 1913), I, 86; Grierson, op. cit. (15), 35, appendix, notes 1, 2.

19 Birthe Kjølbye-Biddle, 'Interim Report: The Old Minster at Winchester in the Seventh Century' (privately circulated, 1974); Birthe Kjølbye-Biddle, 'The Seventh Century Minster Church at Winchester Interpreted', in L. A. S. Butler and R. K. Morris (eds), *The Anglo Saxon Church: Papers on History, Architecture & Archaeology in Honour of Harold Taylor* (London, 1986), 196-209; Martin Biddle and Birthe Kjølbye-Biddle, *Winchester Studies 4, Part I: The Anglo Saxon Minsters of Winchester* (Oxford, 2002).

20 Armand Machabey, *La Métrologie dans les Musées de Province* (Paris, 1962), published doctoral thesis, Paris Sorbonne University, 1959, 108-9.

21 W. Flinders Petrie, *Inductive Metrology* (London, 1877), 107. See also W. Flinders Petrie, 'Weights and Measures', in *Encyclopaedia Britannica*, eleventh edition, 29 volumes (Cambridge, 1911), XXVIII, 481.

22 The earliest reference to the Rhineland foot in English known to the editors of the *Oxford English Dictionary* is that in the *Philosophical Transactions of the Royal Society* for 1675, the data yielding a length of 12·36 inches, in [Anon], 'A Breviate of Monsieur Picart's Account of the Measure of the Earth', in *Philosophical Transactions*, 10 (1675),

261-72, p. 269: 'Suppose the Paris foot to consist of 1440 parts, the Rhynland (or Leyden) foot, contains of these 1390 [parts], the London-foot 1350 [parts], the Braccia of Florence 1686 [parts] …' For rods of 16 feet, see Grierson, op. cit. (15), 20-4.

23 Dorothea Oschinsky, *Walter of Henley and Other Treatises* (Oxford 1971), 443, 445.

24 Eleanor Searle (ed and trans), *The Chronicle of Battle Abbey* (Oxford, 1980), 50-1: 'Pertica habet longitudinis sedecim pedes.'

25 Grierson, op. cit. (15), 21.

26 *WME*, 41-2.

27 Eric Fernie, 'Observations on the Norman Plan of Ely Cathedral', in Peter Draper and Nicola Coldstream (eds), *British Archaeological Association Conference, Transactions for the Year 1976, Part II, Mediaeval Art and Architecture at Ely Cathedral* (London, 1979), 1-4.

28 Martin Biddle (ed), *Winchester Studies I: Winchester in the Early Middle Ages: an Edition and Discussion of the Winton Domesday* (Oxford, 1976), 155.

29 *WME*, 322.

30 William Hale Hale, *The Domesday of St. Paul's of the year MCCXXII* (London, 1858), 130.

31 Oschinsky, op. cit. (23), chapter 28.

32 John F. Nichols, 'The Extent of Lawling in the Custody of Essex, A.D. 1310', in *Transactions of the Essex Archaeological Society*, 20 (1933), 173-98.

33 Lawrence Burrell, 'The Standards of Scotland', in *The Monthly Review: the Journal of the Institute of Weights and Measures Administration*, 69 (1961), 49-62, see p. 54. It would be expected that there would occasionally be references to feet of a third of an ell. One such (insecure) reference in the *Fragmenta Collecta* is to the rood of land in barony as six ells or 18 feet of a middling man, but 20 feet in the burghs: *APS*, I, 752. This passage is mentioned in Chs. 1 and 2.

34 Edward Nicholson, *Men and Measures* (London, 1912), 86-7.

35 See A. W. Richeson, *English Land Measuring to 1800: Instruments and Practices* (Cambridge, Massachusetts and London, 1966), 25. Richeson used a Rhineland foot of 12·356 inches to give a ruthe of 16·475 feet.

36 See, for example, *WME*, 45. One of the examples given there is from the 1066-7 'Chronicle' of Battle Abbey, Kent, in which the perch is recorded as 16 feet in length: Searle, op. cit. (24), 50-1.

37 *WME*, 322.

38 It is possible that the rod may have pre-dated the Rhineland foot. Philip Grierson has suggested that the rod might originally have been considered as 20 'natural' or human feet, and has discussed an early fourteenth-century English source from which the natural foot emerges as 9·9 inches (the previously accepted range being 9·7 to 9·9 inches) – 20 such feet is precisely the 198-inch length of the rod: Grierson, op. cit. (15), 22-3. On the natural foot, see *WME*, chapter 2. Such a length subsequently divided into 16 (Rhineland) feet would mirror the divisions of the Roman foot into 16 digits, and would imply a foot of 12·38 inches (and a 3-foot unit of 37·13 inches). The early use of divisions of a sixteenth is a feature of Northern European metrological practice: the role of this multiple in the weight series is explored in Chapter 4.

39 A steel standard of the Rhineland foot of Leiden, 1752, in the Musée Communal de Bruxelles, measures 314·0 mm (12·36 inches): A.-M. Bonenfant-Feytmans, 'L'aune de Bruxelles', in *Cahiers bruxellois*, 12 (1967), 1-39, p. 6; David Gregory, *A Treatise of Practical Geometry … with additions [by Colin Maclaurin]* (Edinburgh, 1745), 5, gives a value of 12·362 inches quoting Picard. Patrick Kelly, *The Universal Cambist and Commercial Instructor*, 2 volumes in 1 (London, 1811), 196 (s.v. *Hamburgh*) states: 'The Rhineland foot, which is used by engineers and land surveyors, is … 12¹¹⁄₃₀ English inches' [12·37 inches], and similarly for Berlin, Stettin, Amsterdam, Rotterdam, St Petersburg and Wirtenberg. A mid eighteenth-century conversion scale of five different European units of linear measurement, including English and Rhineland inches, is in the collections of the National Museums of Scotland, and illustrated as Fig. 2.1: NMS T.2000.90. This is not a standard, but a piece of working apparatus, showing that these units were clearly in everyday currency across European political boundaries.

40 Alexander Huntar [*sic*], *A Treatise, of Weights, Mets and Measures of Scotland, with their Quantities and True Foundation …* (Edinburgh, 1624), 5.

41 For the period 1424-51, the most reliable source for the Acts is the manuscript PA.5/6/2 at the National Archives of Scotland (NAS), for Thomson's *APS* edition has been found to be defective: this issue is discussed in Chapter 5. The text here is based on William Robertson's transcript of PA.5/6/2 in the suppressed volume entitled *The Parliamentary Records of Scotland in the General Register House, Edinburgh, Vol. I* [Edinburgh, 1804], 66. For some reason Thomson ascribed a date of 1432 to this act: *APS*, II, 22.

42 Discussed in Chapter 3, in the section 'Land Survey'.

43 *APS*, II, 119: Act 6, 1478.

44 Ibid., II, 221: Act 16, 1489.

45 Ibid., I, 374. The text of the assize is erroneously dated at Perth on the 'Wednesday next before the fest of Sanct Margaret', but Queen Margaret of

Scotland (1046-93) was not canonised until 1250. Her feast day is 16 November, see A. A. M. Duncan, *Scotland, The Making of the Kingdom* (Edinburgh, 1975), 558.

46   *APS*, III, 521-2: 29 July 1587.

47   Anthony Turner, *Early Scientific Instruments: Europe 1400-1800* (London, 1987), 49-50; D. J. Bryden, 'Evidence from Advertising for Mathematical Instrument Making in London, 1556-1714', in *Annals of Science*, 49 (1992), 301-36, pp. 301-2; G. L'E. Turner, *Elizabethan Instrument Makers: The Origins of the London Trade in Precision Instrument Making* (Oxford, 2000): for Gemini, see pp. 12-20; for Cole, pp. 20-5. For the background, see Deborah E. Harkness, '"Strange" Ideas and "English" Knowledge', in Pamela H. Smith and Paula Findlen (eds), *Merchants and Marvels: Commerce, Science and Art in Early Modern Europe* (New York and London, 2002), 137-60.

48   G. L'E. Turner, 'Mathematical Instrument-making in London in the Sixteenth Century', in Sarah Tyacke (ed), *English Map-Making 1500-1650* (London, 1983), 100-1, p. 98. On Cole's position at the Mint, see M. B. Donald, *Elizabethan Monopolies: the History of the Company of Mineral and Battery Works from 1565 to 1604* (Edinburgh and London, 1961), 14. We are grateful to Dr D. J. Bryden for this reference. Also B. J. Cook, 'Humphrey Cole at the Mint', in Silke Ackermann (ed), *Humphrey Cole: Mint, Measurement and Maps in Elizabethan England* (London, 1998), 21-6.

49   On Cole's instruments see G. L'E. Turner, op. cit. (47), and Ackermann, op. cit. (48). Part of one of the Science Museum's Cole rulers is shown in *WME*, fig. 17.

50   The instruments are described in G. L'E. Turner, op. cit. (47), 149-55 and 166-8; Ackermann, op. cit. (48), 33-8; E. M. Wray, *Historical Scientific Instruments from the Collection of the Department of Physics, University of St Andrews: A Guide to Selected Exhibits* (St Andrews, 1984), and R. T. Gunther, 'The Great Astrolabe and other Scientific Instruments of Humphrey Cole', in *Archaeologia*, 26 (1927), 273. They are associated with James Gregory's attempts to establish an observatory at St Andrews University in the 1670s: see H. S. Allen, 'James Gregory, John Collins, and some Early Scientific Instruments', in *Nature*, 121 (1928), 156.

51   The work of this commission will be described in Chapters 7 and **8**.

52   Stephen Johnston, 'The Carpenter's Rule: Instruments, Practitioners, and Artisans in 16th-Century England', in Giorgio Dragoni, Anita McConnell and Gerard L'E. Turner (eds), *Proceedings of the XIth Scientific Instrument Symposium, Bologna, 9-14*

September 1991 (Bologna, 1994), 39-45; see also, S. Johnston, 'Mathematical Practitioners and Instruments in Elizabethan England', in *Annals of Science*, 48 (1991), 319-44.

53   Gregory, op. cit. (39), 3.

54   Among traders' wooden ellwands which survive in museum collections is one from Broomhills, near Keith, Banffshire (NMS H.VH.23); another from St Kilda (NMS H.VH.48); a third with no provenance, but dated 1734 (NMS H.VH.3); and a fourth, from Bathgate (NMS K.2001.522). The example illustrated, NMS VH.2, was authorised for use at Edinburgh: see *Proceedings of the Society of Antiquaries of Scotland*, 4 (1860-2), 442.

55   *WME*, 234-7.

56   John Skene, *Regiam Majestatem: The Auld Lawes and Constitution of Scotland …* (Edinburgh, 1609), 36: chapter 41. Variants are recorded by Thomas Mackay Cooper, Lord Cooper (ed), *Regiam Majestatem and Quoniam Attachiamenta*, Stair Society no. 11 (Edinburgh, 1947), 151-2. This clause was referred to by Thomson as chapters 33-5: *APS*, I, 616.

57   *Palace of History: Catalogue of the Exhibition [of the] Scottish Exhibition of National History, Art, & Industry* (Glasgow, 1911), 946, item 76.

58   R. Adam (ed), *Edinburgh Records: The Burgh Accounts,* 2 volumes (Edinburgh, 1899), II, 236.

59   J. D. Marwick, *et al.* (eds), *Records of the Convention of the Royal Burghs of Scotland (1295-1779) [RCRB]*, 7 volumes (Edinburgh, 1866-1918), I, 2: 4 April 1552.

60   *APS*, IV, 587: 19 February 1618.

61   Ibid., VII, 488: Act 57, 29 September 1663.

62   Ibid.

63   Ibid.

64   Hunter, op. cit. (40), 6.

65   *RCRB*, III, 160: meeting of 7 July 1624.

66   Ibid., III, 186: meeting of 5 July 1625.

67   Ibid., III, 221: meeting of 4 July 1626.

68   Information on Scottish glass production is taken from R. Oddy, 'Scottish Glass Houses', in *Glass Collectors Circle*, no. 151 (1966), 1-16. See also Jill Turnbull, 'The Scottish Glass Industry 1610-1750', PhD thesis, University of Edinburgh, 1999; and her *The Scottish Glass Industry 1610-1750: 'To Serve the Whole Nation with Glass'* (Edinburgh, 2001), especially pp. 58-143.

69   D. Masson (ed), *Register of the Privy Council of Scotland, 1619-1622*, first series, 14 volumes (Edinburgh, 1877-87), XII, 439-41: Report of the Commission on Scottish glass manufacture.

70   The wrights used the square foot of 144 square inches for wood, 'but the square foot which the Glaziers [use], in measuring of glass, consists only of 64 square inches': Gregory, op. cit. (39), 81. John

Reid noted that glaziers used only eight inches to the foot 'but the Act of Parliament hath reduced them to 12 as others': J. Reid, *The Scots Gard'ner* (Edinburgh, 1683), 39.

71 [John Swinton], *A Proposal for Uniformity of Weights and Measures in Scotland* (Edinburgh, 1779), 24.

72 Second Report of the Commissioners on Weights and Measures, *Parliamentary Papers* 1820, VII, appendix A, 17.

73 *RCRB*, III, 577: meeting of 6 July 1665. The 1668 Kirkwall ell (Item **6** in the Inventory) may have resulted from this directive.

74 See, for example, Sir James Balfour Paul (ed), *Accounts of the Lord High Treasurer of Scotland, 1473-1498,* 13 volumes (Edinburgh, 1877-1908), I, 291, 293, 298, 346, 347.

75 J. Reid, op. cit. (70), 38.

76 G. Buchanan, *Tables for Converting the Weights and Measures hitherto in use in Great Britain … also Abstracts of the Jury Verdicts throughout Scotland in regard to the Weights and Measures of Each County* (Edinburgh, 1829), 198-9. This figure is that obtained by James Jardine in October 1811, one month after Patrick Copland's report of 23 September on Aberdeen's measures and the Edinburgh iron ell in particular.

77 John S. Reid, 'Patrick Copland (1748-1822)', unpublished University of Aberdeen MLitt thesis, 1983, appendix 2, pp. 4-5.

78 Copland's report was not published and the manuscript has not survived, but a copy made about 1830 by his successor, William Knight, is in Aberdeen University Library, MS 167. The relevant portion is reproduced as an appendix to John S. Reid's paper, 'A New Look at Old Linear Measures', in *Annals of Science*, 46 (1989), 246-8.

79 Buchanan, op. cit. (76), 28. The only justification for specifying such a figure is to enable larger units such as the acre and mile to be given accurately. In the case of Copland's 1811 measurement, his result enabled him to have a replica constructed in copper by the Aberdeen instrument maker Charles Lunan, with the defining faces formed to more exacting standards of accuracy; but this measure does not survive: Reid, op. cit. (77), appendix 2, p. 6.

80 See the entries for Items **2** and **4** in the Inventory.

81 Science Museum, London, NMSI 1931-985 and 1931-986. The Science Museum also has a joint standard yard and ell bed measure (NMSI 1931-987). These items were presented by the Standards Department of the Board of Trade in 1931: F. G. Skinner, *Weights and Measures* (London, 1967), 104. In 1742 George Graham measured the play between the yard and its bed as 0·010 inch, see [G. Graham], 'An Account of a Comparison lately

made … of the Standard of a Yard … with the Original Standards … in the Exchequer', in *Philosophical Transactions*, 42 (1742-3), 541-56, p. 547. We are grateful to Dr Denys Vaughan for his assistance in checking these dimensions.

82 The bailies had to answer for the accuracy of the measures: *APS*, I, 702: *Iter Camerarii,* chapter 28.

83 The chamberlain's role as holder of the standards is also discussed in Chapters **1** and **7**.

84 *APS*, II, 50, 1457/8. These were the chief burghs of the three administrative regions of the kingdom.

85 The issue of an authorised set of standards is described in a remission of 1512: W. Hay (ed), *Charters, Writs and Public Documents of the Royal Burgh of Dundee, 1292-1880* (Dundee, 1880), no. 47. The addition of Dundee may reflect its rapid commercial rise and the dominance of the group of four burghs comprising Edinburgh, Perth, Dundee and Aberdeen in Scotland's external trade, and indeed in contributions to royal taxation. The Dundee weights of 1512 are discussed in Chapters **4** and **8**.

86 While the statutes are silent on this transfer of authority, it is recorded among the earliest records of the Convention of Royal Burghs, for 4 April 1552: 'Becaus of differance of mesouris within borrowis of this realme in tyme bigane [bygone], part being mair nor the rycht, and part less, It is concludit be the prouestis and commissaris [provosts and commissioners] of the borrowis foirsaidis [aforesaid], that the hale borrowis of this realme ressaue [receive] their mesouris of the burghis following, quhilkis [which] hes the iust mesouris, viz; the stane wecht of Lanark, the pynt stope of Striuiling, the ferlatt of Linlytqw, and the eluand of Edinburch, and onform to thai wechtis, mesouris, thai to vse thame selffiis within thair borrowis …': *RCRB*, I, 2.

87 A. Murray, 'The Last Chamberlain Ayre', in *Scottish Historical Review*, 39 (1960), 85.

88 The inclusion of Lanark may seem unexpected, but this burgh was at the site of an early royal stronghold: Hugh Davidson, *Lanark: A Series of Papers* (Edinburgh, 1910), 5.

89 Gordon Donaldson, *Scotland: James I to James VII* (Edinburgh, 1987), 166.

90 The provision of these standards by Edinburgh, Stirling, Lanark and Linlithgow (and additionally standards of salmon barrels of 10 gallons by Aberdeen, and of herring and white-fish barrels of 8½ gallons by Edinburgh) is recorded in J. Hill Burton, *et al.* (eds), *The Register of the Privy Council of Scotland*, three series (Edinburgh, 1877-1970), first series, XI, 325, 354, 392-4, 526-7; XII, 16-18, 27. Twenty pounds was paid to James Stewart

in Glasgow 'for transporting the measures and wechtis great and small stouppis fra Edinburgh at sundrie tymes to Dumbartane castell': from the 'Extraordiner' discharge, in Comptroller's accounts, 1617-18, NAS E24 (36), 36r. We are grateful to Dr Julian Goodare for this reference. See also Item **113** in the Inventory.

91 We cannot, of course, be sure that it was not a 1563 replacement that was submitted in 1587, but the survival of the earlier vessel suggests that this was not the case.

92 *RCRB*, I, VII: 'No record of any meeting in Inverkeithing is now extant, and the practice of holding annual Conventions there, if such a practice ever existed, seems to have been speedily discontinued …' However, this is a discussion of the period after 1552, and it is clear from other records that the Convention of the Royal Burghs met intermittently at Inverkeithing before then. Inverkeithing's status as a burgh is discussed by G. S. Pryde, *The Burghs of Scotland* (Oxford, 1965), 10-11 and 39.

93 *APS*, III, 521: Act 136, 1587. Ibid. IV, 587: 1618.

94 '… ane trone stand of bras weights to the toune …', in *Extracts from the Records of the Burgh of Glasgow 1630-1662* (Glasgow, 1881), 426; 'ane stand of bellis …', in *Accounts of the Lord High Treasurer of Scotland (1531-38),* 13 volumes (Edinburgh, 1900-78), VI, 185; 'ane stand of ches men …' in NAS, Edinburgh Testaments, X, 288b; 'ane blaksmythis stand …', in W. Cramond, *The Annals of Banff vol. I*, Spalding Club New Series no. 8 (Aberdeen, 1891), 64. We are grateful to the Editors of *The Dictionary of the Older Scottish Tongue (DOST)* for providing access to their files for these references.

95 *RCRB*, I, 476-7: meeting of 2 July 1596.

96 'Ane stand of stowpes [liquid measures] comprehending ane quart pynt chopin and mutskin stoups': Alexander Wedderburn, *The Wedderburn Book,* 2 volumes ([n.p.] 1898), II, 44. We are grateful to the Editors of *DOST* for the reference.

97 Alternatively it may simply refer to the ell bed with its marked sub-divisions, but this seems inherently less likely.

98 Described in *WME*, 240-1.

99 *APS*, III, 521: 1587.

100 As was the case with Patrick Copland and James Jardine: see Buchanan, op. cit. (76) and Reid, op. cit. (77).

101 [Robert Chambers], *A New Description of the Town and Castle of Stirling* (Stirling, 1835), 88.

102 Printed details for the supply of standard weights and measures, offered by R. B. Bate, Mathematical Instrument Maker to the Board of Excise, London, 25 February 1825: NMS T.1993.34.

103 941 millimetres.

104 See Appendix A to this volume, and the '*Compositio Ulnarum et Perticarum*' given in *WME*, 322. Although the definition of the inch might seem no more than a conventional one, it has recently been shown to be remarkable accurate and reproducible: ibid., 3. The significant feature here, however, is that the definitions in both kingdoms are the same.

105 J. Cowell, *The Interpreter* (Cambridge, 1607), s.v. *perch*.

106 Hunter, op. cit. (40), 5.

107 The English mile of 1,760 yards = 1,712 ells, making the ell 37·0093 inches; and the Scottish mile of 1,920 ells = 1,973 yards, making an ell of 36·994 inches: Hunter, op. cit. (40), 9, 10. Even if these distances were expressed to the nearest ell, an uncertainty of half an ell would only change the size of the ell by 0·01 inch.

108 *APS*, VIII, 494: Act 59, 16 June 1685.

3

The Later Units
of Length:
Cloth and Land

IF WE ACCEPT THAT THE ASSIZE OF DAVID I HAS ITS ORIGINS IN THE TWELFTH century, a period of very close relations with England, then it seems very plausible that David's ell was based directly on English practice; in all probability the ell started life as exactly 37 English inches.

The standards distributed to the burghs were bed measures designed to accept bar measures of this length, and were therefore marginally oversized. Combined with the effect of wear and tear, and acknowledging that the accuracy of division and copying could not match more modern practice, we could understand how bed standards, which have been copied from successive earlier measures, might be lengthened gradually until they reached the slightly larger sizes recorded in more recent times. However, in spite of this, we have found the lengths of the three surviving early ells beds to be remarkably constant, and we have interpreted the size of the bed measure as exactly 37 inches plus a fixed tolerance allowance. This approach is different from that of earlier commentators, who have measured these same bed standards and concluded that the legal ell was 37·06 inches (but divided into 37 Scottish inches), and this figure has unfortunately found a firm foothold in the literature.[1]

Opposite:
Detail of Fig. **3.1**.

THE 'LINEN ELL' AND THE REEL

Confusion about the true size of the ell was undoubtedly complicated by the currency of a larger ell of 37·2 inches in the eighteenth century. This larger ell has no descriptive name, but in order to emphasise the distinction between it and the normal ell of 37 inches we will term it the 'linen ell'.[2] In justification for this we will argue that it probably has its origin in textile regulations, and in particular those at the turn of the eighteenth century which governed linen production. This is one of several enhanced ells connected with textiles which we will discuss. It is not the earliest, because the 'plaiding' ells are firmly associated by their name with the woollen textile trade, which was important long before the rise of the linen industry. Nevertheless it is appropriate to discuss the linen ell first because its separate identity is more easily established and because we can relate it to the other important textile measure, the yarn 'reel'. The reel also has its origin in woollen manufactures, but it was adopted into the linen industry and it is with its use for measuring linen yarn that we shall be concerned. Although we will refer to the 37·2-inch ell as the 'linen ell', it must be emphasised that this is only a descriptive convenience: the fact that this measure had no separate name undoubtedly contributed to the confusion in the eighteenth century between it and what might be termed the 'general purpose' ell of 37 inches.

There are extant standard scales of the linen ell, which are described in the Inventory in this volume. In particular, two combined measures of an ell and a yard are associated with a group of natural philosophers active in the Philosophical Society of Edinburgh in the mid-eighteenth century, whose interest in reconstructing authoritative values for the early Scottish standards is discussed in Chapter 9. The Dalkeith standard of 1744 (Item 15 in the Inventory, and Figs 3.1 and 3.2) and the Stirling standard of 1755 (Item 16) differ from earlier measures in that they specify an exact relationship between the yard and the ell as 37·2 inches, but the ell is also divided into 37 slightly larger inches, each with decimal sub-division. The concept of a separate Scottish inch (in this case of 1·0054 English inches) and a Scots foot (of 12·065 English inches) emerges for the first time with this longer ell. The ratio between these two ells of 37 inches and 37·2 inches (and therefore also between the two feet) is 185:186. It was almost certainly this established usage that prompted early nineteenth-century commentators to propose an equivalent 'Scots inch' of 1·0016 English inches once the old Edinburgh ell bed had been measured as 37·06 inches in 1811.

Both the Dalkeith and Stirling measures are constructed as back-to-back bed measures for the ell and yard, and have divided scales on the beds.

Fig. 3.1
Combined yard and ell measure for the Burgh of Dalkeith, Midlothian, 1744. (Item 15 in the Inventory.)

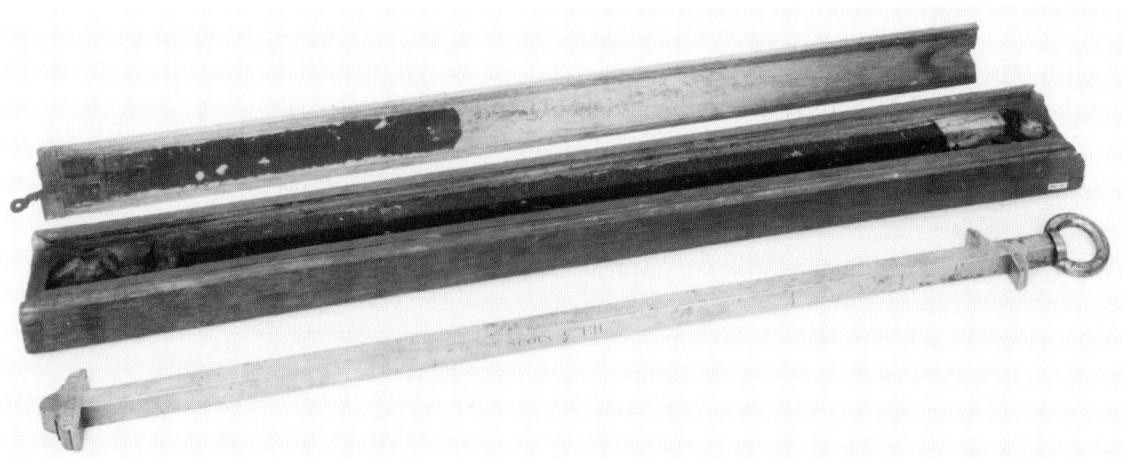

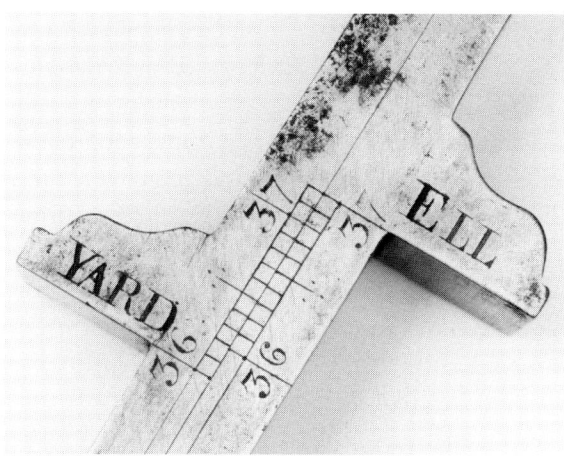

Fig. 3.2
Upper end of the Dalkeith
measure, showing the
37·2-inch length of the 'linen
ell' and the use of distinct
'Scots inches' at 37 to this ell.
(Item **15** in the Inventory.)

Neither has accompanying bar measures, and indeed it is doubtful if these
were ever made. What is clear, however, is that the beds have been accurately
constructed at 37·2 inches for the ell and 36 inches for the yard, with no
additional allowance for play between an inserted bar and the bed (Fig.
3.2). Although this is what we might expect for a mid-eighteenth century
precision measure, it represents a break with the earlier conventions of the
burgh markets and perhaps provides a warning that this measurement unit
may have emerged under different circumstances.

It is difficult to say when the linen ell of 37·2 inches was introduced. It
first seems to have been discussed in print in 1745 in a posthumous translation of a work on practical geometry and surveying technique by David
Gregory (1661-1708) who had been professor of mathematics at Edinburgh
University between 1683 and 1691, when he was appointed to the chair of
astronomy at Oxford on Isaac Newton's recommendation. Gregory returned
to Edinburgh shortly after the 'incorporating union' of Scotland and
England came into force in May 1707, remaining for several months to
supervise the re-organisation of the Scottish Mint and to put in hand a
recoinage following English Mint principles, acting on behalf of Newton,
who was then Master of the Mint in London.[3] The edition of Gregory's
Treatise was edited and revised by a successor to his Edinburgh chair, and
another protégé of Newton's, the mathematician Colin Maclaurin (1698-
1746), but in a manner that makes Maclaurin's editorial changes clear.[4]
Although it is apparent that Maclaurin considered this ell to apply universally, it was conceded that surveying operations were often carried out using
measures based on the ell of 37 English inches.[5]

Maclaurin's views became widely accepted, largely because the Gregory
Treatise was the only apparently authoritative account of Scots metrology.
Not only was it reprinted a number of times (the ninth edition appearing
in 1780), but the entire work was reprinted in the first edition of the
Encyclopaedia Britannica, published in Edinburgh in 1771.[6] The entries in

subsequent editions were based on this; so, for example, Maclaurin's values for the ell and Scots chain were unmodified in the fifth edition of *Britannica*, completed in 1817.[7]

The 1754 published description of the values of the early standards recovered by the Philosophical Society similarly included the 'Scotch ell, according to the Standard of Edinburgh' as 37²/₁₀ English inches and the 'Scotch foot' as 12¹/₁₅ English inches.[8] By the time the first survey had been made in the 1770s of the continuing use of the old unit across the Scottish counties, the advocate John Swinton was able to tabulate the larger ell of 37·2 inches as the standard for the whole country, giving its value in English inches 'because Scotch inches and feet are little used'.[9]

Much of the editorial addition made by Maclaurin to Gregory's text was drawn from the (now lost) notes of Robert Stewart (1675-1758), who had been a regent in philosophy at the University of Edinburgh since 1703. He became the first professor of natural philosophy when separate chairs were established in 1708 and held the position until 1742.[10] Stewart had described a measurement of the size of the Scots pint and of a beer cask in terms of English and Scottish gallons which had been carried out in the Edinburgh Council Chamber in October 1707 before the Excise Commissioners, the Magistrates and the Brewers' Incorporation.[11] One of the clauses of the Act of Union had been concerned with establishing an equable relationship between the excise duty on liquor in the two kingdoms, requiring assumptions about the relative sizes of the English and Scottish units; but after the passage of the act a direct measurement was presumably necessary. It is likely that both Gregory, who was already advising on government calculations, and Stewart, who had the results of the work, were at least present and may have been actively involved. Patrick Copland, professor of natural philosophy at Aberdeen, who undertook a re-measurement of the early Edinburgh ell (Item 2) in 1811, said that it was generally understood that Gregory had established the length of the ell as 37·2 inches, and he noted (possibly having been influenced by accounts of the cask measuring) that Gregory had done so in 1707.[12]

In fact Gregory's use of this unit can be taken back by perhaps a further twenty years. The text for his *Practical Geometry* specifically referred to the larger size of the Scots foot, with a ratio to the English foot of 186:185; and the ratio appears in the manuscript copies made in the 1690s for the original Latin work and in Gregory's own manuscript copy, which may date back to the claimed 1685 date of composition.[13]

This was a period in which there was an increased awareness of the size of weights and measures. Gregory's appointment to his Edinburgh chair followed closely after one of the more significant changes to affect the public: in 1679-80 the Privy Council decided that oatmeal – a staple food – must for the first time be sold by weight rather than by volume in the grain measures, a practice which had been open to widespread manipulation and abuse.[14] A parliamentary act in 1685 had sought to impose the 1593 English statute mile of 1,760 yards in place of the analogous Scottish mile, which had however no statutory basis; and a series of acts attempted to regulate the production of linen in such terms as the acceptable widths for woven cloth.[15] The interest was sufficient for the Edinburgh instrument maker, mathematician and almanack compiler, James Paterson, to produce an arithmetic textbook dedicated to Edinburgh's town council and directed to 'all those who have any Fairs in this Kingdom', who would find it particularly useful because Scots measures were used throughout.[16]

Paterson's work provides an indication that Gregory may indeed have been involved in official measurement work in the 1680s. Robert Stewart's notes describe a measurement of the Edinburgh standard pint measures, clearly made before witnesses, obtaining the rather low capacity of 102·3 cubic inches, by using an experimentally-derived value for the density of water which pre-dates a well-known determination of 1696.[17] This suggests that the measurement was made by Gregory. However, the same capacity was also given by Paterson in 1685, and it is known that the Edinburgh standards were gauged that year when an early pint (which has not survived) was found to be light compared with the 1555 chopin and 1618 pint (Items **109** and **113**) – so perhaps this measurement was conducted by Gregory.[18]

Although Paterson noted in 1685 that the Scottish and English feet were precisely the same and that the 'Standard Ell of Edinburgh' was of 37 inches, Stewart was later to insist that the ell was of 37·2 inches.[19] An associate of Stewart's in the Philosophical Society, who also used Stewart's notes, stated that the larger size was that of the old Edinburgh ell (Item **3**):

> … the standard of the Scottish ell kept in Edinburgh is an old matrix [bed measure] of rusting iron, so that the more it is consumed with time the longer the ell became and upon trial it was found 37·2 English inches although it ought to be no more than 37. But it is now fixed by a brass standard and so the Scotch foot will be 12·065 English inches.[20]

The explanation for the apparent contradiction in the length of this standard is presumably that the measurement was not made of the ell bed itself, but of an ell scale of 37·2 inches which is cut into the side of the standard. The length of this ell scale is incised round the suspension end of the measure just beyond the defining face of the conventional ell, and its binary division is cut on the side face. There is no adequate way of dating this: although the cuts are accurately spaced, they are a little uneven in depth, and so it is tentatively suggested that they were added in the late seventeenth century, and that their addition is associated with Gregory's advice. The 'fixing' of the length by a brass standard refers to the acquisition of the English yard standard of 1707 (Item **10**) to act as a companion piece.

If we look in the parliamentary acts for references to this larger ell of 37·2 inches, the evidence is at best ambiguous. The earliest statutory reference which may be to this ell is an act of 1693, and in a context which is specific to the linen industry.[21] It is our contention that this ell was introduced to meet particular requirements in regulating linen production, and that in due course, probably because it represented the only modern and accurate definition of the ell, it was applied inappropriately in other spheres of activity, including land measure.

The production of linen cloth was of major importance to the Scottish economy, and large quantities were exported, particularly to London. As an indication of the scale of growth of manufacture, 18,000 ells of Scottish linen were entered at London in 1600, but in 1700 this had risen to 650,000 ells.[22] It is estimated that about one and a half million ells of linen were exported annually to England in the decade before the 1707 Union of the parliaments of the two kingdoms. It is no surprise that in the economic arguments about the protection of Scottish exports in the run up to the Union with England, linen figured prominently.

In 1686 a somewhat unusual act was passed 'for the encouragement of linen industries', which required that bodies should be buried only in Scottish linen.[23] A similar requirement was repeated in the act of 1693. The penalties for disobedience were particularly severe: £300 for a nobleman and £200 for lower ranks. This Scottish act followed the example of English statutes of 1666 and 1678 which required burial in wool, 'for the encouragement of the woollen manufacturers of this kingdom and preventing the exportation of moneys thereof for the buying and importing of linnen'.[24] Wool was a principle English commodity just as linen was in Scotland. The Scottish position was confirmed when the 1686 act was again repeated in 1695 during a period of severe economic depression; but in the closing session of the Scottish parliament just before the Union of 1707, these two Scottish acts were repealed as a concession to the English and the burial shroud was to be made thereafter in 'plain woollen cloath or stuff'.[25]

The Scottish linen industry had assumed great importance by the mid-eighteenth century, both for the manufacture and the bleaching of cloth, but this was based on impressive growth in the seventeenth century, on vigorous centralised encouragement of the industry and on a system of tight controls on quality. The regulations of the eighteenth century (as far as they affected Scotland) were the province of the Commissioners of Manufactures, one of the government boards through which Scotland was administered in the period after the Union, and they specified the widths and lengths in which material could be sold. To confirm its quality the cloth had to be stamped before sale by the Board's officials at special 'stamp offices' around the country, and several of the unusual measures which survive are at villages and towns where linen stamping took place.

The major part of the linen output of Scotland was of cheaper types of cloth, and before the expansion of bleaching facilities in the eighteenth century a good proportion of this output was in brown (or unbleached) linen. Because linen shrank in the process of bleaching, the acceptable minimum widths of brown and white linen were different and were first specified by the statute of 1693.[26] Two sizes were permitted for cloth intended for export or for sale at markets and fairs; coarser cloth could be three-quarters of an ell broad, and two inches wider if unbleached, but finer cloth had to be an ell broad, or an ell and two inches if unbleached. However, when this ell width is given, it is defined specifically as 'a large Elne' or 'an Elne large' without further definition, but meaning, in all likelihood, that the large ell is the 37·2 inch measure.[27]

The cloth dimensions of the 1693 act differed from those of earlier acts concerning linen weaving. An act of 1661 and another 'Act for encouraging Trade and Manufacture' of 1681, both specified the width to be an ell and two inches but were unspecific about whether this was for bleached or unbleached cloth and which of the ells (large or otherwise) was to be used was not mentioned.[28] We cannot therefore say whether the large ell of 37·2 inches was introduced between 1661 and 1693.

One of the few surviving external market-place ells is the well-known standard in the village of Dunkeld in Perthshire. Dunkeld is not a royal burgh (although there was an ineffective charter obtained in 1704) but a burgh of barony under the superiority of the Duke of Atholl.[29] The existence of a standard in such a burgh would therefore be a little surprising if it were not that Dunkeld's market was a convenient centre for the many handloom linen weavers of the area, and one of the stamp offices was subsequently situated here. The standard at Dunkeld (Item 9, and Fig. 3.4) is a vertically mounted iron measure dated 1706, which incorporates two bed measures. One bed, on the left hand side of the measure, is exactly two inches longer than the Edinburgh ell bed of 37·06 inches; and the main bed on the front of the measure is the 37·2-inch 'linen' ell.[30] As with the 1663 Edinburgh ell bed, described in Chapter 2, a scale of binary divisions was laid out along the bed of the principal measure of the Dunkeld ell and this corresponds precisely to the conventional 37-inch size of the ell. This demonstrates clearly that the oversized construction of the bed is not accidental but has been designed to incorporate an allowance of 0·2 inches. The two width categories of bleached cloth in the 1693 act were to be an ell, and three-quarters of an ell. The latter can only be measured as three-quarters of a conventional 37-inch ell on the Dunkeld measure, but this appears to match an interpretation of the 1693 act as using the large ell only for bleached cloth of ell width.

There is a further oblique reference from the period just before 1650 in which the Convention of Royal Burghs agreed to petition the Privy Council for an act on cloth sizes:

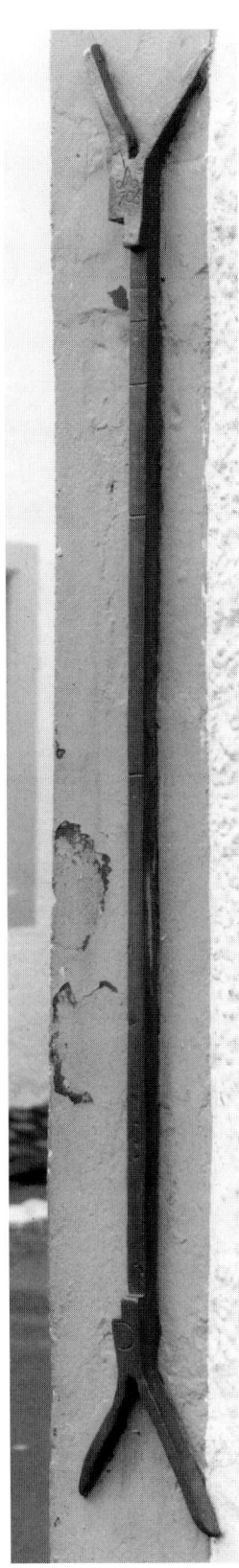

Fig. 3.4 (left) The Dunkeld ell of 1706, mounted on the left-hand corner of the frontage of St George's Hospital. (Item **9** in the Inventory.)

Fig. 3.3 The former St George's Hospital building at the centre of Dunkeld, Perthshire, now in the ownership of the National Trust for Scotland.

> Ordaines that the secret councill be supplicat for obtaining ane act ordaineing the playding to be three long quarters of ane elne and the linning [linen] sold att 10s and to be elne broad.[31]

It is presumed that the long quarter of an ell related to a longer ell that was perhaps specific to cloth manufacture and we will conclude that this is a so-called 'plaiding' ell. The only clue that might suggest that the 37·2 inch ell was in use at this date is the existence of an apparently early standard, which has been cut on the side of the Edinburgh iron ell (Item **2** in the Inventory).

THE REEL AND THE ORIGIN OF THE LINEN ELL

A further linear measure of importance to the textile industry was the 'reel', by which yarn was measured. It took the form of a large winding frame with a circumference of two and a half ells (more usually considered to be ten quarters of an ell). One turn of the reel took up a 'thread' of yarn, and 120 (or six score) threads made a 'cutt'. The linen regulations of 1693 then stipulated that a 'hesp' (hank or skein) of twelve cutts constituted the unit in which the yarn could be sold by weight.[32] A century later, the *Statistical Account* for Galashiels describes the reel also being used for wool:

> A stone of the finest of it [wool] weighed after being thus prepared will
> yield 32 slips of yarn, each containing 12 cuts, each cut being 120 rounds of
> the legal reel.[33]

Thus, by weight, a cut of wool was two-thirds of an ounce.

The reel was unusual in that it was not one of the basic metrological
standards adequately defined by statute, as for example were the ell and the
pint, and yet it was considered important enough for the burghs to be
required to hold standards. As a standardised measure defined in terms of a
primary standard (the ell), its status was somewhat similar to that of the
salmon barrel (described later in Chapter 6), specified as containing so many
gallons, of which an authoritative standard was constructed and committed
to the care of Aberdeen, the burgh that handled much of the salmon trade.
The principal standard for the reel is often described as having been similarly
committed by parliament to the burgh of Perth, but in fact this is not
recorded in the statutes.[34]

It is not known when the word 'reel' came to be applied to a device
specifically for measuring quantities of yarn: it was certainly used in this
sense in the early seventeenth century, but although isolated late sixteenth-
century references have been recorded there is no clear metrological context
to these.[35] The earliest useful references are in the records of the Convention
of Royal Burghs, and from these we can deduce something of the form of the
iron standards, although not their actual construction. The Convention
received a complaint in 1615 from the burgh of Dunfermline about 'the gritt
hurt sustenit be tredders [traders] with yairne thruch the insufficience of the
reill', and the following year's Convention stipulated the size for the reel:

> the lenth of the reill betuix heid [head] and heid sall be halfe elve [and]
> half quarter [ell] long, making the haill hesp [namely, the length of the
> hank] to be fyve quarter long of doubell yairne ... [36]

The reel therefore has to be understood as having four arms, like four
spokes of a wheel, each with a cross-bar or head at its extremity. Revolving
this about its centre will wind the yarn round the square formed by the four
heads. Another possible structure (Fig. 3.5), in the absence of an extant
example, might be two equal-armed crosses of iron mounted side-by-side on
the rotating spindle and connected at the ends of the arms by short iron rods,
one of which may have been removable. The construction of this second
form would be easier, and it would be simpler to control the significant
dimensions.

The distance between the heads, namely the length of side of the winding
square, was to be a half and a half-quarter ell (five-eighths of an ell), so that
the full circumference was 2½ ell or ten-quarters of an ell. If the thread was
to be used as double yarn, the hank could be considered as one of five-

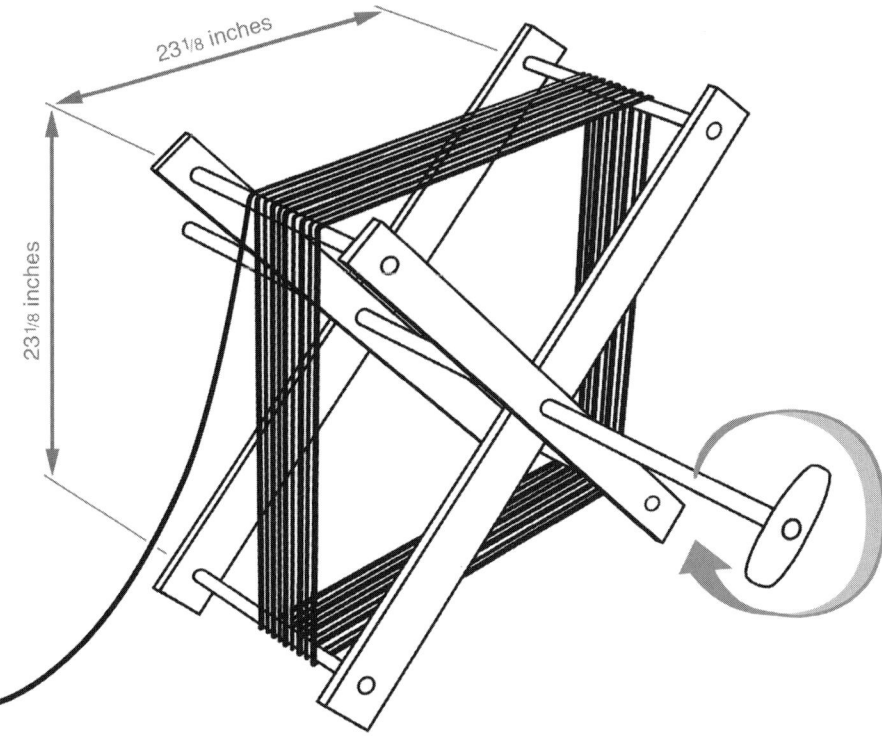

23¹/₈ inches

23¹/₈ inches

Fig. 3.5
Conjectural form of a
standard yarn reel.

quarters. The reel itself continued to be known as the 'ten-quarter reel', but
for worsted or woollen yarn the reel could also be half this size, when it would
be known as a 'short' or 'five-quarter' reel. However, it was the longer reel that
was the more significant, and in the linen acts of 1661 and 1693 parliament
decreed that 'no reill be made use of within this Kingdome under the
measure & lenth of ten quarters'.[37]

The 1616 act of the Convention of Royal Burghs does not seem to have
been put into effect – possibly as a result of the greater upheavals caused by
implementing the 1618 Assize. The issue was raised again in 1622 and 1623,
and then in 1624 it was ordered that each burgh should obtain an iron
standard made according to their 1616 act and

> hing the samin upon ther mercat croce, or in sum such uther publict
> plaice quhair the said commoditie of yairne is usuallie sauld, that the
> foresaid lenth of the said reill may be better knawin, … [and] ilk burgh to
> caus publicatioun be maid of this present act throw [through] thair burgis
> upone ane mercatt day that nane pretend ignorance …[38]

At the 1625 Convention it was reported that the burghs most directly
concerned by the measure had put it into effect.[39]

The size of the reel is yet another instance where we may draw a parallel
between Scottish and English practice. The length of this seemingly unlikely

quantity of 'ten quarters' of an ell, or 92½ inches, is only fractionally over the 90 inch length of two English ells, each of which is five-quarters of a yard.

This link is brought out in the 1721 definitions by the Convention of Royal Burghs of the permitted sizes of the two worsted reels, which uncharacteristically appear to be expressed in term of English ells:

> the short reel [to] be one quarter long, one eln about … the long reel [to] be half eln long, tuo elns about …[40]

The 'length' of the reel in this definition is the length of the side of the winding square, and the circumference of the long reel is 90 inches. Clearly this could not be two Scottish ells but must be two English ells. The Convention's committee, established in 1719, had compared local regulations for the worsted reel in particular burghs and counties, and the use of the English measure reflects the requirements of a 1719 Scottish act of the new British parliament specifying woollen cloth sizes in yards.[41] After the post-Union measures had been distributed, the burghs had standards of the English ell and would clearly appreciate this as a five-quarter measure of a yard.

The Convention of Royal Burghs had a strong interest in promoting the linen industry in the period of economic depression that followed the Union of 1707. It had a specialist committee on the linen trade from 1723, and in 1727 secured the establishment of the government Board of Trustees for Manufactures and Fisheries.[42] Also in 1727, in the Convention's draft for the parliamentary act to regulate the linen trade, we have the first independent definition of the linen reel in terms of inches:

> the uniform standart reel [to] be two Scots elns and a half in circumference or ninety three inches, commonly called the ten quarter reele.[43]

Here we should pause for reflection, because the reel should in theory be 92½ inches in circumference: a 93-inch circumference, measuring 2½ Scots ells, could arise from using the linen ell of 37·2 inches and this might imply two concurrent linear standards, one of 37 inches, the other of 37·2 inches independently generated from different sources.

There was sustained pressure from the Convention of Royal Burghs in the first half of the seventeenth century for burghs to acquire and retain standard reels for reference. If we may generalise from a single example, the presence of 'ane irone reell' amongst the standards of the small east coast fishing burgh of Dunbar in 1700 may indicate that this pressure was successful.[44] We have suggested above that the simplest construction for such standards would involve two iron crosses mounted side-by-side on a spindle and connected by four pins running between the ends of the arms (see Fig. 3.5). The holes for these pins could be positioned accurately by drilling so that the holes between adjacent arms were separated by five-eighths of an ell, the dimension stipulated in the Convention's 1616 act.[45] Our suggestion is that in the 1680s, when there was so much official activity in promoting expansion in linen manufacture, the existing reel standards were re-measured. This time the actual circumference must have been measured, and this will have been found to be greater than 2½ ells by half an inch. The difference arises because the circumference is not merely four times five-eighths of an ell – it also includes four-quarters of the circumference of the pins round which the yarn is wound, which would be about half an inch if the pins were about a sixth of an inch in diameter. It is most unlikely that this would have been considered when these standards were being made in about 1620, but it would certainly have come into play at a re-measurement in the 1680s, particularly if this task had been referred to an authority such as the professor of mathematics at Edinburgh University.

It has already been noted that David Gregory's manuscript *Treatise of Practical Geometry* of about 1685 referred to the foot (in the context of a unit of work) as enhanced in relation to the English foot, but indicated no change to the units for land survey.[46] This is certainly compatible with his possible involvement in a re-measurement of the reel. If the above is the correct explanation for the introduction of the 37·2-inch ell, then it reflects not only a new mathematical approach to technical issues, but also a practical acceptance of the need to define a new ell standard for this specific application so that the legal integrity of the existing reel standards should be preserved as 2½ ells. Further work was done after the Board of Manufactures was set up in 1727.[47] By the 1740s, this longer ell, which had now been in existence for over fifty years, was the only one that had been subject to scientifically rigorous definition; and it became natural to see its application as being as broad as Colin Maclaurin indicated in his 1745 printed edition of Gregory's *Treatise*.[48]

Unfortunately no examples of standard reels are known at present, although this may simply be because they have not been recognised as measures. However, there are several surviving examples of the wooden domestic reels on which the spinners prepared their yarn in hanks (Fig. **3.6**).[49] Museum curators have tended to associate these with woollen rather than linen yarn. In form they resemble spinning wheels, although they are hand-cranked, and they usually have five or six wooden arms, one of which hinges to release the tension of the completed cutt so that it can be removed. A simple geared counting mechanism records the necessary 120 rotations, making it a 'chakreill', a name later corrupted to 'jack reel'.[50] Circumferences of 90 inches are long reels based on the yard ($4 \times \frac{5}{8} \times 36 = 90$); whereas those of 93 inches are based on the 37·2 inch 'linen' ell.[51]

The act of 1693, which set out to control linen production, has been described as having been 'disregarded'.[52] One reason for this was that it specified that cloth exported (and especially to England) had to be made

Fig. 3.6
A domestic 'chakreill'
winding frame, for yarn.
(NMS H.RD.12)

to Scottish dimensions. The act was followed in the same parliament by another which established an existing linen works operating in Leith and Edinburgh as a 'free Incorporation' whose privileges included exemption from customs for 21 years. Within two years a further act, whilst stipulating that the dimensional requirements of the original act should be 'put to strict execution', nonetheless made the following concession:

> ... wheras the partners of the said Society [the Linen Manufactory at the Citadel of Leith] who reside ... in England have sent several patterns of Cloath not hitherto made within this Kingdom and which must be made of different lenths and breadths from these prescribed by the foresaid Act of Parliament Therefore his Majesty with advice forsaid does notwithstanding of the said Act Impower and Authorize the said Manufactory to make their Cloaths for exportation allennarly [only] of such breadths and lenths as they are or shall be advised by their said partners in England ...[53]

Another glimpse on the importance of this export trade to England can be seen in Stirling in 1704 when the Trade Incorporation of Weavers had a new 'chain' measure made. This new measure survives (Item **17**) and is made up of four hinged bars with two stretching rings for measuring the running lengths of cloth. It has clearly been made to the English standard of two English cloth ells, each of 45 inches, and it was stated to have been shorter than the old chain by the specific amount of 'one of the rings of the old chain'.[54] We can deduce that this ring was about two inches across, so that the original length of the chain would most probably have been two and a half Scots ells, the length also known as the 'long reel for yarn'. Another English cloth measure dating to before the Union is a somewhat enigmatic folding yard gauge (Item **18**) with an inscription which is only partly legible but has been identified as the standard of one of the sections of the Fife Society of Chapmen, or itinerant merchants, dated 1705.[55]

Non-compliance with the 1693 linen act continued to cause official concern. There were appeals for the question of measuring and sealing linen to be referred to the parliamentary Committee of Trade in 1698, and a further appeal in 1700 led to a draft act being read in the following year.[56] In 1702 the Convention of Royal Burghs called for the acts relating to woollen cloths to be put 'to vigorous execution'.[57]

With the passing of the Act of Union in 1707, the yard replaced the ell for official purposes, and in 1711 and 1713 two acts of the new British parliament specified new size requirements for Scottish linen in terms of the yard.[58] These were closely based on the 1693 act and specified equivalent sizes of a yard and a nail (sixteenth) or three-quarters of a yard and a nail in width for unbleached cloth, so that it might be a 'full Yard' or 'full 3 Quarters' in width once bleached.

The Scottish act of 1711 followed an English act of the same year which provides an interesting view of measurement practice in the cloth trade. The English act specified that cloth was to be measured on tables of a certain minimum size:

> … with the length of a yard nailed or marked thereupon to which shall be added one inch more which shall be used instead of that commonly called a thumb's breadth so that the same length shall contain 37 inches …[59]

This provision was repeated in an English act of 1714.[60] From these it is apparent that formalising the customary practice of giving a thumb (or an inch) to the yard meant that a unit which was equivalent to the Scottish ell was already in use in England and had been for a long time.

The Convention's concern was primarily for linen but also considered plaiding, a coarse twilled woollen cloth, often woven with a chequered or tartan pattern. A plaid was a length of this material, frequently used as an outer garment or cloak, especially in rural areas. The production of rough woollen cloth by small hand-loom weavers was a significant cottage industry in many country areas although Scottish wool and woollen cloth was seldom of adequate quality to enable weaving to develop into a vigorous export industry.[61] Plaiding also shrank and so was sold by an ell with an additional allowance of two inches, or 'two thumbsbreadths', and this was termed the 'plaiding' or 'pladden' ell.[62] The term 'plaiding ell' was certainly in use in the 1640s, and 'plaiding' ells (which we must assume incorporated the two-inch allowance) were tested and authorised, because the accounts for the bailies of Aberdeen include an entry for 1643-4 for

> ane new stamp to the playding elvandis, the stamp being losed be Robert Messour … 1 lib. 4s.[63]

We have concluded that the 37·2-inch 'linen' ell dates from at least the 1680s, but we have no definite information of its existence before then. Plausibly, therefore, the 1650 Privy Council petition, quoted above, seeking an act which would authorise the sale of plaiding 'three long quarters of ane elne' broad, properly refers to the plaiding ell rather than the 'linen' ell.

Strenuous attempts were made to organise both woollen and linen production in the seventeenth century and parliament provided some encouragement in the form of bans on the export of raw and spun materials and the import of finished goods, and exemption from customs duties. The regulations that specified the acceptable sizes of linen also referred to woollen cloths of various types. For example, the 1661 act called for no woollen cloth to be made in widths under an ell and a half, nor serge (a twilled woollen cloth similar to plaiding) under an ell and a nail (a sixteenth of an ell, slightly over 2·3 inches). The 1681 act specified that drugget (a coarse woollen mixture material) was to be sold in widths of three-quarters of an ell and a nail, but serge at an ell and two inches.

Linen took over as the dominant industry in the eighteenth century, but it was plaiding that gave its name to the large measure of two inches over the ell, and subsequently of two inches over the yard. In particular, several regional examples were recorded in the 1770s by John Swinton, for example, Aberdeen's 'plaiding yard' was 38 5/12 inches and Kincardine's 'plaiding ell' was 38½ inches, and they are variously described as for plaiding, coarse cloth or home manufacture.[64] For convenience, we have used the term 'plaiding measure' to refer to the regulation size two inches (or thereabouts) larger than the standard ell or yard, regardless of what principal type of material is being measured.

Fig. 3.7
The Mid Steeple, Dumfries, engraved from a drawing of about 1780, and published in the *Proceedings of the Society of Antiquaries of Scotland* in 1886. The bed measure, shown at the centre of the plate, is mounted on the centre of the wall of the first floor of the building, above the external staircase. By our measurement, it is 38·3 inches long. (The measure is Item **20** in the Inventory.)

A yard and a nail, as specified in the acts of 1711 and 1713, should be 38¼ inches, and two external burgh standards of this size are known. The first is an iron bed (Item **20**, and Fig. **3.7**), measured by us as 38·3 inches, mounted in the centre of Dumfries on the side of the Mid Steeple, which housed the burgh offices. Dumfries was a linen producer and had a stamp office, but was also a centre for woollen cloth. In contrast the other measure is at Dornoch, on the Moray Firth, and has no particular connection with linen production, although woollen cloth was certainly sold in its market. The Dornoch measure (Item **21**), known locally as the 'pladden ell', is unusual in that it comprises two separate end pieces set in a horizontal slab of stone. The very thin defining edges of this show signs of adjustment and are also somewhat worn so that it may originally have been somewhat shorter than its present 38·4 inches.

The only other plaiding measure for which we have some information is the old Aberdeen standard, apparently no longer extant, which was measured by Patrick Copland (1748-1822), professor of mathematics at Marischal College, Aberdeen, in 1811.[65] This was a hinged iron bed measure which he measured as 38·65 inches, rather than extending only to '38·4166 or 38 and ⁵⁄₁₂ which it ought to do'. It seems likely that Copland was merely referring here to George Skene Keith's account of the size in his 1795 report to the Aberdeen of Guild where it was set at 38⁵⁄₁₂ inches. Copland perhaps misinterpreted this as meaning that the measure had been specified at this size.[66] These figures are certainly greater than those found for the Mid Steeple ell or that at Dornoch, but a certain amount of variation is understandable, especially in the measurement of rough cloth. It might be supposed that a size of about 38·4 inches would be expected for an early plaiding yard bed, if it was a yard and a nail with a small allowance for tolerance play.

Although the old Aberdeen plaiding yard does not survive, there is a more recent combined yard and plaiding yard at Aberdeen (Item **24**, and Fig. **3.8**) which was perhaps made for Copland as a more acceptable standard for the city. This has been made to the size thought by Copland to have been the standard, namely 38⁵⁄₁₂ inches, and has been designed to mark the divisions of the yard on traders' wooden measures.

Fig. 3.8
The Aberdeen yard gauge, designed for marking binary divisions on wooden measures. The hinged end-stop in the foreground can be raised to extend the length to a plaiding yard of 38⁵⁄₁₂ inches. (Item **24** in the Inventory.)

John Swinton's 1779 account noted the use of plaiding yards in several counties, and three of these were still recorded in Buchanan's account of 1829, together with three others not mentioned by Swinton. They are given either as the conventional 38 inches (as specified in the act), or else at or just below 38½ inches, perhaps from an actual measurement.[67] By implication the ell of 37·2 inches was used elsewhere.[68] Only in Wigtownshire was there a remnant of the old plaiding ell, now described by Swinton as 40 inches, to which '41 inches are commonly allowed', presumably on the traditional basis of providing an extra thumb's width in measuring out cloth.[69] Its use was restricted to raw woollen cloth.

It might be expected that plaiding measures would have no place in the Lothians in the early nineteenth century, but their final appearance is in a composite measure made probably in the late 1820s or early 1830s and adopted as the linear standard for the County of Edinburgh in 1860 (Item 27 in the Inventory). This is a conventional divided yard in brass by the prominent Edinburgh instrument maker Alexander Adie (1775-1858), but steel extension pieces define an ell which we have measured as 37·18 inches and a plaiding yard similarly measured as 38·66 inches, although neither is identified in any way on the measure itself. It will be noted that this is a good match for the length of 38·65 inches found by Patrick Copland of the old iron plaiding yard of Aberdeen. In fact Adie had made his own ell standard, taken directly from the early Edinburgh ell-bed: Copland described this as 'a square rod of fir tipped with brass' and it was one of the Edinburgh items measured by him in 1811.[70] It would be quite in character for Adie to make himself a plaiding yard based on Copland's determination of the early Aberdeen standard, and then incorporate this size in his Edinburgh County standard.

LAND SURVEY: THE FALL, THE ACRE AND THE MILE

In Scotland, as in England, large lengths and areas were reckoned in miles and acres, and from the similarities in the definitions it again becomes apparent that Scottish practice was based on that of England.

The English mile of eight furlongs and the English acre were both generated from an intermediate unit known as the 'rod', also called the 'pole' or 'perch'. This was literally a wooden rod, of 16½ feet or 5½ yards, a length which could be carried fairly conveniently, but which was often represented by a knotted rope or cord. The rod's entry into England, as we have suggested in Chapter 2, was as 16 Rhineland feet or, in Saxon times, the 'gyrd' measure (and we have argued that this in turn was the Germanic *ruthe*). The 'furlong' and 'acre' are also both Saxon words, whose meaning associates them with ploughed and sown land. In early times the furlong of 40 rods was considered to be the length that could be ploughed by a team of oxen without a rest, and the acre was the area that could be ploughed in the animals' working day.

Since arable land was managed in strips, the acre was considered to be the area of a thin rectangle one furlong (or 40 rods) in length and four rods (or 66 feet) in width. Each furrow was about a foot in width, so the ploughing team made 33 round trips to the end of the strip and back in the process of ploughing the complete acre. The strips of land radiated out from a settlement or town, and in the parcelling out of land it was the width of the strip and therefore the length of frontage to the access road or river bank that was important. The 'acre's breadth' or four rods was therefore a significant linear measure, and the acre was counted as four 'roods' of land, each with a breadth or frontage of one rod.

The unit of length in Scotland equivalent to the rod was the 'fall' of six ells (and therefore 18½ feet), which was literally the length which would fall under the measuring rod or rope when it was laid on the ground. The jurist John Skene defined it in 1597 as follows:

> sa meikle [much] lande, as in measuring falles vnder the rod, or raip [rope], in length is called ane fall of measure, or ane lineall fall …[71]

A few years later an English commentator provided a similar definition:

> Six elnes long make one fall which is the common lineal measure and six elnes long and six broad make a square and superficial fall of measured land. And it is to be understood that one rod, one raip [rope] and one lineall fall of measure are all one for each of them contains six elnes in length.[72]

Both the acre and the rood of land are found in southern Scottish charters of the twelfth century (no doubt in an echo of the established English settlement practice), but it is unlikely that land measure was regulated by the rod until the somewhat later introduction of practical surveys.[73] Indeed, the earliest accurately dateable use of the rod or fall in Scotland that we have encountered is in a late fourteenth-century document which describes four men dividing a plot of land on the Bamff estate in Perthshire into three 'tofts', or homesteads with attached arable land:

> … the quilk foure layd the land with lyne and departit them ewynly in tria [three] …
> … the qhirylk toftis haldis in lenthe achttene [18] fal lang and thertene [13] of brede …[74]

Although this Scottish fall of six ells or 18½ feet was longer than the English rod, it must be appreciated that there were several rods in use in England other that the 'king's rod' or 'king's perch' of 16½ feet. For example, a late thirteenth-century *Husbandry* describes how rods might be of 18, 20,

22 or 24 feet, each giving rise to a different acre size; and other twelfth and thirteenth-century sources specify distinctions as fine as a six-inch difference in the rod size between different parts of the same county.[75] The longest rods were used in measuring woodland and land recently reclaimed from forest, where the productivity would be comparatively low. It can be argued that the use of a particular rod size depended on the quality of the ground and was designed to maximise rental payments, which were assessed in kind on a yield per acre. More productive land would give a higher yield, but would be measured in smaller acres; less productive land would be measured in broader acres so that the yield per acre was more comparable. It is possible then that the use in Scotland of the 18½-foot rod, a size also found in parts of England, may have its origins in an early appreciation that a slightly larger acre was appropriate to the productivity of the arable lands of Lothian.

An indication that the size of the fall was chosen to suit particular circumstances is given in an early but undateable legal fragment defining the width of the rood of land:

> The rude off lande in baronyis [the country] sal conten vj elne that is to say xviij [18] fut off a mydlyn mane, the rude off the land in the burghe mesurit off a midlyng mane sal be xx [20] fut.[76]

An apparent difference between the official definition of the fall for agricultural land as 18½ feet (six ells) and this definition as the length of the feet of 18 average-sized men is probably not significant.[77] However, the notable feature is that a larger rod of 20 foot was to be used in the burghs. And, as was noted in Chapter 1, rods of this length are also recorded for building within English cities – in York in the late thirteenth century and in Lincoln in the early fourteenth century.[78]

Two early examples from charters show that this larger unit was indeed used to specify the widths of plots of ground within burghs. In a charter of about 1160, Malcolm IV granted to 'his man Baldwin' a holding of land in Perth ten feet broad (half a rod) and 24 feet long, 'as freely as Baldwin has his own house'.[79] In another of about 1214, William the Lion granted to William, the helmet maker, a plot which was also in the burgh of Perth, 20 feet wide and 26 feet long, to be held in feu and heritage for an annual render of two iron helmets.[80] Even though virtually no medieval domestic building survives, the plots of land in the old burghs have tended to be developed individually, and therefore the widths of the plots have been preserved and the earliest large-scale Ordnance Survey maps (of the 1850s) of Edinburgh's medieval High Street still show the majority of the original building plots between the wynds as 20 feet wide.[81]

The practical measurement of land areas was transformed with the introduction of a type of land chain devised by Edmund Gunter (1581-1626), professor of astronomy at Gresham College, London, in 1620.[82] Gunter's chain, comprising 100 links, was 66 foot in length – the breadth of the acre. An area of one chain by ten, or ten square chains, was therefore an acre in extent, and the decimal division of the chain enabled the areas of rectangles of all proportions to be more readily calculated.

The chain was also applied in Scotland. Alexander Hunter wrote in 1624 of the use of chains of two, three or more falls for survey work.[83] However, it was quickly being used in the form described by Gunter, and by exact analogy with English practice it was a chain representing the acre's breadth, or four falls, and divided into 100 links; the furlong was therefore ten chains long and the acre was ten square chains. So exact was the equivalence of English and Scottish units that one of the standard early nineteenth-century textbooks on survey practice, John Ainslie's influential *Comprehensive Treatise on Land Surveying*, published in Edinburgh in 1812, could be couched almost exclusively in terms of links and chains, without the need to specify whether these were English, Scottish or indeed Irish units, all of which were tabulated.[84] No doubt this approach was important for English sales of his book – the enlarged edition was also published in London – but it also helps to emphasise that by this time a Scottish landowner was as likely to commission a survey in English as in Scottish measure.

The Scottish chain, at four falls or 24 ells, was 74 feet in length (888 inches) – eight feet longer than the English chain of 66 feet, and the size of the link was 8.88 inches. Consequently, the Scottish acre was larger than the English acre: in terms of square yards it was 6,085 square yards (0·510 hectares) as opposed to 4,840 square yards (0·405 hectares). However, the Scottish chain's relationship with the Scots acre, and its division into a hundred parts, still entitled it to be called a Gunter chain, and it is so named in John Swinton's late eighteenth-century volume on measures.[85]

Whereas the chain was immediately useful in land measure, the link was swiftly related to the output of those craftsmen, such as masons and slaters, whose work was reckoned by area and costed by the 'rood of work' – an area much smaller than the rood of land, which was a quarter of an acre. A square with the area of one rood of work would have a side of one rod and was therefore one square rod. In Scotland the rod was the fall of six ells, and so the rood of work was 36 square ells (or 342¼ square feet).[86] However, in England, using the smaller rod or perch of 5½ yards or 16½ feet, the rood of work was only 30¼ square yards (or 272¼ square feet). Because the English rod of 16½ feet (198 inches) was a comparatively coarse unit for assessing some craftsmen's work, it was further broken down into 25 short feet of 7·92 inches, each of which was a hundredth part of the acre's breadth, or chain. These short feet were therefore equal to the length of the links of the survey chain, there being 100 links to the chain, and the specific use of links in this context lasted well into the nineteenth century. John Ainslie in 1812 recommended

Fig. 3.9
A Scots chain of 74 feet with a link of about 8·9 inches (left), contrasted with an English chain of 66 feet with a link of about 7·9 inches. (NMS W.1971.321 and W.1971.322)

that land measurers should have chains, but also 'a tape … divided into links … such as carpenters, masons, and painters use for measuring their work'.[87]

As with the 'glaziers' foot', already encountered in Chapter 2, this short English foot is particularly associated with glass measuring, and quantities of window glass were normally given in square feet of this unit. Since much of the window glass used in Scotland was imported from London, the English glass foot of 7·92 inches was frequently used, although some calculation was certainly done using the equivalent Scottish unit of 8·88 inches (a hundred to the chain of 74 feet). Here we see the accurate sizes of the glass feet of 8 and 9 inches given by Alexander Hunter in his 1624 *Treatise of Weights, Mets and Measures of Scotland*. Window glass was very expensive in this period and only available in small panes, and so it is understandable that computation would be done with some care and using smaller divisions of the foot. The Scottish act of 1663 that attempted to eliminate the glaziers' foot also referred to inches, reckoned at 42 to the ell. This would make the glass inch about 0·88 inches long, and from this we can understand that the glass foot must have contained 10 'inches', each of 0·888 inches long. Since the whole calculation of area was conducted on a decimal basis, this is exactly what we would expect.[88]

It might be thought unlikely that Gunter's chain could have made suffi-cient impact for glaziers' foot sizes based on its links to have appeared in Hunter's 1624 work. However, decimally-divided surveying chains were not new, and had first appeared in the late sixteenth century.[89] Perhaps the earliest English reference to the use of chains in land measurement is in a treatise of 1582 by Edward Worsop, who refers to them as 'wires'.[90] Better known is Aaron Rathborne's highly influential work *The Surveyor*, published in London in 1616, which contained the first clear account of an English chain, with a rod or perch divided into a hundred parts.[91] The chain itself was two or three perches long, which provides an echo of Hunter's description of the surveyor's chain of two, three or more falls in length.

One of the problems raised in due course with the Scottish chain was the correct basis for its length. In terms of the ell of 37 inches it was 24 ells (74 feet), but if the longer linen ell of 37·2 inches was applied, then the chain's length should be 74·4 feet. Ground areas measured by chains of these two sizes will differ by only one per cent, but the fact that there was any difference at all might have consequences for property contracts. Unfortunately the Scots chain had no statutory definition, and so a dispute about measured areas could only be tested in law. Such a case came up in the country's highest civil court, the Court of Session in Edinburgh, in 1811, and the judge sought technical advice from Patrick Copland, professor of mathematics at Marischal College, Aberdeen.[92]

Copland believed that the 37·2-inch ell had been introduced by David Gregory in 1707, based apparently on a misreading of Gregory's post-humously published *Treatise of Practical Geometry*.[93] However, Copland was incorrect in attributing the 74·4 foot chain to Gregory – in fact Gregory did not mention the chain's length in terms of feet, and the 74·4 foot measure is first encountered in the editorial comments of Colin Maclaurin, the 1745 editor of the published edition of Gregory's work.[94] Maclaurin stressed that it was this length that generated the 'true Scots acre', rather than the 74-foot chain based on the 37-inch ell. From the context it is unclear whether the extension of use of the larger linen ell into land measure was due to Maclaurin or to Robert Stewart (1675-1758), the regent in philosophy (and subsequently professor of natural philosophy) at Edinburgh University at the time of the Union of Parliaments in 1707, whose manuscripts (now lost) were used by Maclaurin.[95]

This longer chain of 74·4 feet is certainly referred to in John Swinton's 1779 book, and by Copland's time it had become entrenched in the math-ematical literature because of a reliance on Maclaurin's version of Gregory's text.[96] This has resulted in the widespread modern assumption that 74·4 feet was the appropriate length for the chain in the eighteenth century.[97]

However, it is quite a different matter to say whether the 74·4-foot chains were actually used by practical surveyors. The scale marked on survey maps of this period is frequently given in chains of 74 feet, from which we should perhaps conclude that chains of 74·4 feet were little used, but of course the length remains ambiguous if the 74 feet could be the larger feet derived from the 37·2-inch ell. In a few instances it is possible to confirm the interpretation. For example, Thomas Milne (1768-1809), who had trained under the prominent land-surveyor Peter May (?1724-95), used a chain based on the conventional 37-inch ell whilst working for the Duke of Gordon in 1770.[98] John Ainslie's first published work on land surveying, his *Gentleman and Farmer's Pocket Companion and Assistant* of 1802, contains tables relating English and Scottish units from which the chain emerges clearly as exactly 74 feet rather than 74·4 feet.[99] The figure remained the same in Ainslie's *Comprehensive Treatise on Land Surveying* of 1812, and we must assume that this was considered to reflect the best current practice, at least in 1812.[100]

Ainslie himself was not consistent in how he described his map scales. This is neatly shown in two manuscript surveys conducted within a few years of each other in adjacent plots of land which formed part of the development area for Edinburgh's Second New Town. On one survey, conducted for the former Lord Provost David Steuart in 1787, the scale was stated to be of Scots chains of 74 feet; whereas on the other, made for the George Heriot Trust in 1797, the chain was defined exactly as 74 feet 4⁴⁄10 inches.[101] The latter size is very close to the theoretical size of the 74·4 foot chain (74 feet 4⅘ inches; 24 × 37·2 inches) and the difference may only represent a calligraphic slip. In the absence of the surveyor's working notes it cannot be deduced that he actually used the chain size specified in the scale. It may merely be that this was the size considered appropriate for a particular survey and it probably reflects the client's requirements. However, this in turn implies that the factors for the Heriot Trust (which was a major landowner in Edinburgh and further afield) used 74·4 foot chains. A few Scots chains do survive (see Fig. **3.9**), but they are of uncertain date and generally in poor condition, and so do not cast much light on this problem.[102]

Given Ainslie's use of both varieties of chains, it is clear that there was scope for confusion in land contracts and it was merely a matter of time before that issue had to be tested in law. In his report of late 1811 on the measurement of various linear measures sent from Edinburgh, Copland noted that he had become involved

> at the instance of some of the Lords of Session, from the difficulties they have experienced relating to the measure of landed property in ascertaining the proper length of the Scots chain, insomuch so as after long litigation, they have found themselves obliged to recommend settling the difference by arbitration.[103]

In the event, no formal legal decision was recorded, apparently because the case soon turned on different arguments.

The dispute can be identified as one involving an estate in Berwickshire, bought in 1809 by James Kyd, a solicitor in Fife.[104] The purchase price was to be paid in instalments, but Kyd defaulted on the final payment due in May 1811 and an action was raised against him by the seller, Nicholas Brown. The Court of Session proceedings before Lord Hermand began in June 1811, and Kyd's defence was that he had found irregularities in the title deeds and the size of the estate was smaller than he had expected. The newspaper advertisement for the sale of the estate had given its size as 'about 646 acres' and Kyd claimed he was justified in assuming that a Scottish estate advertised in a Scottish newspaper would be given in Scots acres. The areas had been taken for Brown in 1817 by the respected land surveyor John Blackadder (1793-1830), who had worked in English acres, but Kyd subsequently disputed this figure and commissioned his own survey.[105]

Patrick Copland's measurements of the sixteenth-century Edinburgh ell bed (and of other items sent from Edinburgh) in September 1811 were presumably intended to provide Lord Hermand with authoritative information from which the size of the Scots acre could be calculated. However, there was no need to invoke this information because, as a result of further memorials submitted by the parties in early 1812, Hermand concluded in June 1812 that 'the usual measurement in Berwickshire is by English acres'.[106] Although Kyd appealed unsuccessfully against other aspects of Hermand's judgement, this point was conceded, and Brown's counsel noted that:

> There is no other measure of land known in Berwickshire than the English acre. It is the common and ordinary measure of the county, and has been so, past all memory. When acres, therefore, are mentioned, it is always English acres that are understood. This is the fixed standard measure of Berwickshire, and it is believed also to be the standard measure of Roxburghshire, and the other border counties.[107]

Although the sizes of the Scots chain and acre were not embodied in a Court of Session decision, Copland's new determination of the ell as about 37·06 inches soon came to be adopted as the official figure. In 1825 the civil engineer James Jardine, who had taken the measures to Aberdeen in 1811 and had assisted Copland in this determination, conducted the official comparisons for the sheriff of the county of Edinburgh between the new Imperial standards and the old Scots standards.[108] These comparisons were adopted as the equivalents for the whole country, and the value for the ell's length was taken from the earlier measurement by Copland and Jardine and corrected to 37·0598 inches.[109]

Adopting this value for the length of the ell has the effect of increasing the length of the Scots chain by only 1·4 inches in 74 feet – an amount which is negligible in practical terms and was almost certainly widely disregarded. Nonetheless, there was at least one prominent practitioner who made use of Copland's figures, which demonstrates that they had some currency. John Wood (?-1847), was a land surveyor who trained in Yorkshire and by 1811 had settled in Edinburgh.[110] In 1818 he began a series of about fifty surveyed plans of Scottish towns, which were published separately and then gathered together in his 1828 *Town Atlas of Scotland*.[111] From 1826 until his death in 1847, most of his time was spent away from his Edinburgh home, surveying English and Welsh towns for a projected volume of plans, although he did undertake further work in Scotland. Nearly all the Scottish plans for the 1828 *Town Atlas* are dated and carry scales of chains, and most specify the lengths of the chains (Fig. **3.10**). Of these, 23 show English chains of 66 feet, seven show Scots chains of 74 feet, and ten use the slightly larger sizes of 74 feet 1¼ inches or 74 feet one inch, corresponding closely with the 1811 determination of the ell as 37·0596 inches.[112]

John Wood may have used this chain length through the influence of his close neighbour in the Edinburgh suburban estate of Canaan, Alexander Adie, the highly-regarded scientific instrument maker mentioned earlier.[113] Adie had made his own reference copy of the sixteenth-century Edinburgh

Fig. 3.10
Part of John Wood's plan of Kilmarnock, 1819, marked with a scale of Scots chains of 74 feet 1¼ inches.

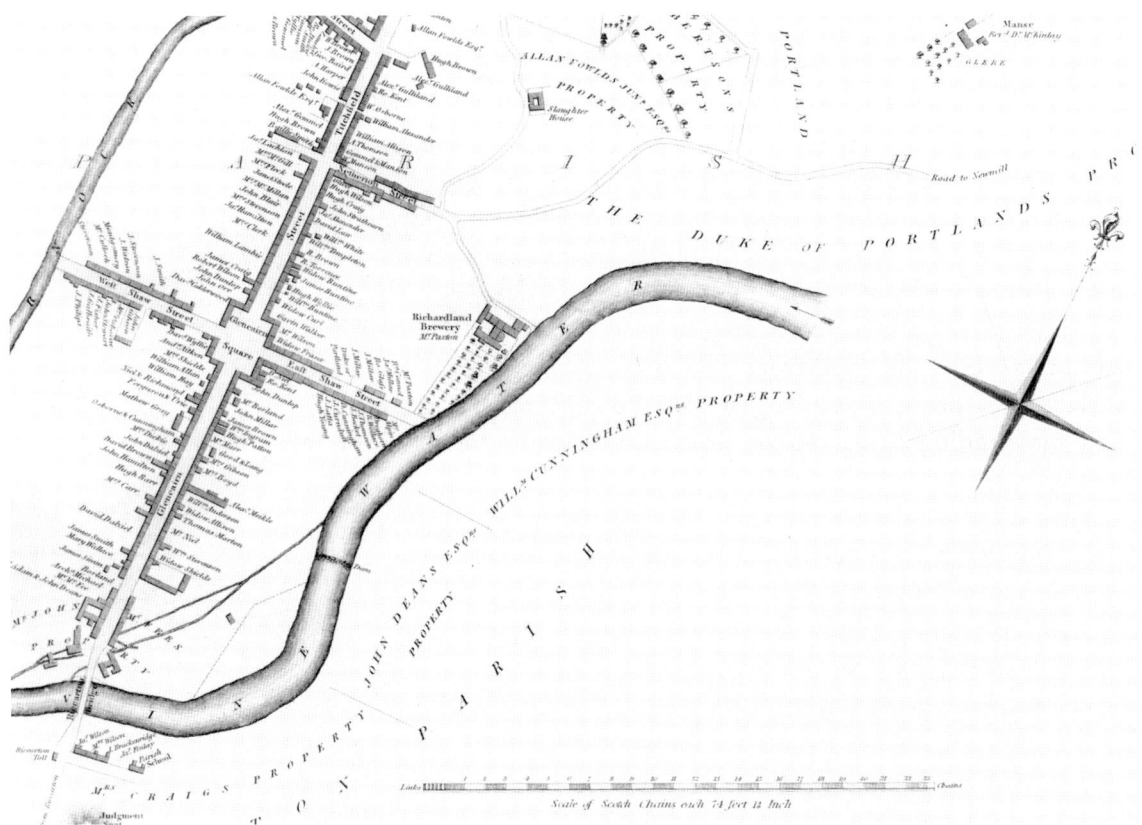

ell, in the form of a brass-tipped fir bar, and this was one of the measures taken to Copland by Jardine in 1811. Although Copland did not provide its length in his report, he noted that it was 'extremely near to the measure of the Ellbed' and that it was used by Adie to adjust the Scots chains that he made.[114] Adie was also involved in the production of another form of land measuring device: an 'odometer' or surveyor's wheel of a type described by James Hunter of Thurston, East Lothian, to the Highland and Agricultural Society in 1821. It was marketed in conjunction with another of Adie's Canaan neighbours, James Howden, and one form enabled distances to be measured directly in Scots chains.[115] Adie subsequently assisted Jardine in the 1825 measurement of the old Scots standards. In October 1826 an advertisement appeared in the local *Scotsman* newspaper, addressed to land surveyors and surveying instrument makers by John Steedman, Secretary to the Society of Scottish Land-Surveyors:

> Notice is hereby given, that a STANDARD of the IMPERIAL CHAIN of 66 Feet has lately been marked off with great accuracy on the parapet wall in front of the College Buildings, by Mr Adie, optician, under the direction of Mr Wallace, Professor of Mathematics in the University, and of the Society of Scottish Land-Surveyors. The College Janitor will give access to the area where it is situated, and show the manner of applying it, for which he is paid by the Society.
>
> Those situated at a distance may have their chains adjusted, by transmitting them to Mr Adie, who has kindly undertaken to do so at a moderate charge for each.[116]

The suggestion that the chain of 74·4 feet may not have been in practical use to any significant extent (despite its displacement of the 74-foot chain in mathematical literature), must not disguise the fact that there were undoubtedly regional and local variants of land measuring units. An example which illustrates this well is the ell used in Huntly. This Aberdeenshire burgh of barony was extended on a grid plan devised by the baronial superior, the third Duke of Gordon, in the late 1760s.[117] The land surveyor Thomas Milne was engaged in 1770 to prepare plans of the burgh and surrounding land, measured according to the town's ell of 39½ inches.[118]

It is tempting to assume that this ell of Huntly is related in some way to the plaiding ells (particularly considering the great importance of linen manufacture to Huntly) and that the use of a textile unit had therefore been extended into land measure. However, Milne noted that the Huntly convention was that the chain contained 20 ells: this allows us to see that the Huntly chain was closely following the English chain of 66 feet, and the Huntly ell was five English links.[119] It is perhaps understandable that the layout of a progressive new venture such as a rectilinear new town should reflect the use of the decimal chain so clearly, and it suggests a level of English involvement in the planning and costing of the original development. The adoption of an ell which was superficially similar to the local cloth ell can therefore be seen as a happy convenience, but one which nonetheless reinforces the eighteenth-century feeling of flexibility in adapting units to practical requirements. It would have eased Milne's task if he had appreciated this relationship; instead, he conducted his 1770 survey using a 74-foot Scots chain, and was required to convert all his measurements to conform with the Huntly ell.[120]

Like the Scots chain the Scots mile was never defined by statute. However, by direct analogy with the English mile, it was recognised as eight furlongs, each of 40 falls, and it is described as such in Alexander Hunter's 1624 *Treatise*.[121] The Scots mile therefore contained 320 falls or 1,920 ells; its length in yards depended on the size of ell assumed, but if we take the correct size of 37 inches, as discussed above, the Scots mile was equivalent to 1,973⅓ yards or about 1·12 English statute miles.[122] The use of this mile, however, seems to have been restricted to situations where measured surveys were conducted and where distances were considered as generated from smaller survey units.

Detailed survey activity in Scotland accelerated rapidly in the period of stability and economic growth after the mid-eighteenth century, as landowners enclosed and improved their agricultural estates.[123] Up to the end of the first quarter of the nineteenth century such survey work was almost exclusively conducted in Scots chains and miles, although there was always an element undertaken in English measure.[124] Before this time the county surveys were based on compass traverses in which an earlier and larger mile was used. This was a more subjective unit based on the length of the walking pace, and it reflects the fact that travel around the country that could not be conducted by sea was very largely by foot, or in easy country by horse. Even at the end of the seventeenth century very few roads were considered adequate for carts, and most inland transport was by pack-horse.

In spite of the inaccessibility of much of its land, Scotland emerged from the great Blaeu atlases of the mid-seventeenth century as one of the best mapped countries in the world.[125] Almost all the survey work used by Blaeu was undertaken by one man, Timothy Pont (*fl.*1580-*c.*1612), who travelled across the greater part of the kingdom in the interval 1583-96.[126] A great many Pont manuscripts survive, as do further maps and compilations of earlier Pont manuscripts by Robert Gordon (1580-1661) and his son James Gordon (*c.*1615-86), revised for publication by Blaeu after Pont's death. The 47 regional maps by Pont, with contributions by the Gordons, appeared in Joan Blaeu's *Atlas Novus* of 1654, and the various issues of the *Atlas Major* in the 1660s, and were the only authoritative sources for a great many years.

The majority of the Blaeu maps of Scotland were merely given a scale of miles, but the Aberdeen and Banff map, to which Robert Gordon had

considerable input, also carries the information that fifty of these miles are
equivalent to the length subtended at the Earth's surface by one degree. More
specifically, Gordon's new *Scotia Regnvm* map of Scotland, in the 1654 atlas,
has two scales – one of Scottish miles with 50 to the degree, and the other of
English miles with 60 to the degree – so that one Scottish mile was 1·2 English
miles. However, the English mile on Gordon's map scale was not the so-
called 'statute mile', defined as 1,760 yards or eight English furlongs in an
English statute of 1593, and of which there were about 69 to the degree.[127] It
was a longer mile, which was variously defined as between 9 and 9½ English
furlongs, but normally considered as 9¼ furlongs, and has romantically
been termed the 'old English mile'.[128] This English mile survived into the
eighteenth century as the 'geographical' mile, and is the mile we now call the
(British) nautical mile, of 6,082 feet.

Gordon's Scottish mile emerged in a more clearly defined form about
thirty years after the appearance of the Blaeu atlas. A substantial new atlas
of Scotland was planned in the 1680s, under the geographer Robert Sibbald
(1641-1723) and the cartographer John Adair (1647-1718), and it initially
had strong support from the Scots Privy Council.[129] Several manuscript and
printed maps prepared for this project survive, and a number have double

Fig. 3.11
John Adair's 1688 map of
the Turnings of the Forth,
with a grid of 'countrey'
miles of 1,500 paces: the
name of the common mile has
been changed to this more
accessible term by Adair's
London engraver, Herman
Moll. (Cambridge University
Library)

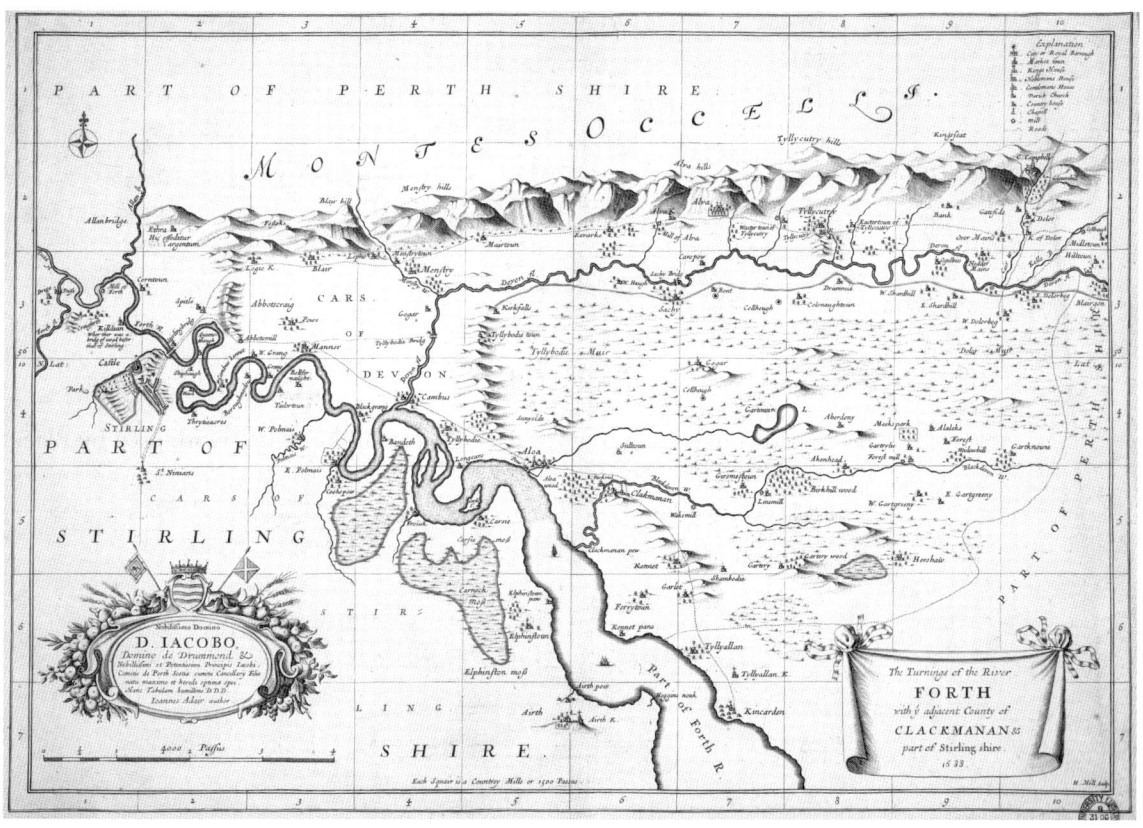

scales of thousands of paces and so-called 'common miles' of 1,500 paces (Fig. 3.11).[130] The pace was considered to be a double step of five foot (or 1⅔ yards), and was so defined in Hunter's *Treatise* in 1626.[131] A mile of 1,500 paces was therefore 2,500 yards, 1·42 English statute miles, or 1·22 of the longer 9¼-furlong English miles. Thus we can see that Adair's 'common mile' is the same as Gordon's 'Scottish mile' of 1·2 old English miles. This mile is also the unnamed mile that appears on the scales of Blaeu's Scottish maps.[132]

It is probably not coincidental that the Scots parliament tried to enforce the use of English statute miles in the 1685 Act that has already been discussed in connection with the glaziers' foot. The reason for this provision was so that the same standard of distance measure would apply to both kingdoms, and this is a concern that is understandable when one considers that the first of Adair's official maps had just appeared using the antiquated common miles. In fact, the act appears to have had little effect, and in the event Adair's county surveys came to a halt at about this time.[133] When extended commercial cartographic work resumed it was not the English statute mile but the Scots mile (of about 1,973 yards) based on survey practice that replaced the common mile.

However, the common mile remained in traditional use for the distances between separated locations, at least until the period of extended turnpike road construction in the latter part of the eighteenth century. Like its English counterpart, this Scottish mile was considered as the 'computed' mile – the mile by which distances or itineraries were computed – as opposed to the 'measured' (or surveyed) mile. One of the last instances where we can see the Scottish computed mile in use is in the well-documented tour round Scotland made in 1760 by Richard Pococke (1704-65), Bishop of Ossory and subsequently of Meath. His detailed 2,500 mile itinerary is broken down into about 250 stages which have been given in both Scottish computed miles and the equivalent number of English measured (or statute) miles.[134] Although Pococke was aware of the rough equivalence of two computed to three measured miles, the distances have not been calculated on this basis – the variation in the ratio between them shows that the figures have come from independent sources, and the average of 1·4 miles to the English statute mile shows that these Scottish 'computed' miles are the same as the old Scottish 'common' miles of Adair's survey.[135]

All the numerical information we have about the use of miles in Scotland is from the seventeenth century or later, and we have little indication of the Scottish conventions for assessing distance measure before this time. The Scottish mile of about 1,973 yards described in Hunter's 1624 volume closely mirrors the long-established English mile of 1,760 yards, which was formalised in the English 1593 statute.[136] Presumably this reflection of the definitions was intentional, and it is quite possible that the Scottish mile was not of any antiquity but had been created in conscious imitation of the

English statute mile. In contrast, the conventions of map-making up to at least 1700 show the use of a constant mile of 1,500 paces, which is a unit of considerably greater antiquity. The original Gallic league was 1,500 (Roman) paces, and this is thought to have been the origin of the English league.[137] Twice this was the *rast*, a Germanic land unit that held its value over the centuries.[138] The term 'league', although widely used in France, and to a more limited extent in England, does not appear to have been employed in Scotland. Occasional references, such as that in a poem by the prominent courtier Sir David Lindsay in 1528, may perhaps be understood in terms of known French usage.[139]

1 See, for example, W. Grant and D. D. Murison (eds), *Scottish National Dictionary [SND]*, 10 volumes (Edinburgh, 1931-76), X, Supplement, 316-7, 'Scottish Currency, Weights and Measures'. This was reprinted, but without attribution, in Ian Donachie and George Hewitt, *A Companion to Scottish History from the Reformation to the Present* (London, 1989), 244-5. For a recent corrective, see Allen Simpson, 'Weights and Measures', in Michael Lynch (ed), *The Oxford Companion to Scottish History* (Oxford, 2001), 640-2.

2 Linen cloth was made from flax and was a main component of the Scottish textile industry.

3 For Gregory's role at the Scottish Mint, see Athol L. Murray, 'Sir Isaac Newton and the Scottish Recoinage, 1707-10', in *Proceedings of the Society of Antiquaries of Scotland*, 127 (1997), 921-44.

4 D. Gregory, *A Treatise of Practical Geometry … by the Late Dr David Gregory* (Edinburgh, 1745). Colin Maclaurin noted in his 'Advertisement' to the reader that additions had been made to Gregory's text to make it more 'useful at this time'. These are 'distinguished from the Author's Text' by placing them in quotes.

5 Ibid., 86.

6 Maclaurin is named as the editor in the 1751 edition of the *Treatise*. No author was mentioned when the material was reused in the article 'Geometry, Part II, The Application of the Foregoing Principles to the Mensuration of Surfaces, Solids, etc', in *Encyclopaedia Britannica; or, a Dictionary of Arts and Sciences*, 3 volumes (Edinburgh, 1771), II, 691-710.

7 Ibid., fifth edition (Edinburgh, 1817), VIII, 21. The first five volumes of the fifth edition are reprints of the fourth (1810) edition, but subsequent volumes are revised; see R. Collison, *Encyclopaedias: their History throughout the Ages* (New York and London, 1966), 141.

8 James Gray, 'Of the Measures of Scotland, compared with those of England', in *Essays and Observations, Physical and Literary, read before a Society in Edinburgh …*, I (1754), 200-4, p. 204. Strictly, the foot is not 12¹⁄₁₅ (12·067 inches) but 12·065 inches.

9 [John Swinton], *A Proposal for Uniformity of Weights and Measures in Scotland* (Edinburgh, 1779), 24.

10 For Stewart, see Roger L. Emerson, 'The Philosophical Society of Edinburgh, 1737-1747', in *British Journal for the History of Science*, 12 (1979), 154-91. He is discussed further in Chapter **9**.

11 The measurement is referred to in Maclaurin's additions to Gregory, op. cit. (4), 114. Further details are given in a notebook of John Robison, professor of natural philosophy at Edinburgh from 1774-1805, who also had access to Robert Stewart's notes: St Andrews University Library (St. AUL), MS Q171.R8, 'Professor [John] Robison's Commonplace Book', I, p. 442.

12 Copland's report is reproduced in John S. Reid, 'Patrick Copland (1748-1822)', unpublished University of Aberdeen MLitt thesis, 1983, appendix 2.

13 Gregory, op. cit. (4), 3; MS copy of 'Geometria Practica', dated 9 March 1697 (Edinburgh University Library (EUL) MS Dc.5.57, f.5r), and Gregory's undated autograph copy (EUL MS Dc.1.75, f.1r). Maclaurin in 1745 noted that Gregory had written the work 'about sixty Years ago': Gregory, op. cit. (4), 'Advertisement'.

14 Peter Hume Brown (ed), *Register of the Privy Council of Scotland*, third series, 16 volumes (Edinburgh, 1908-70), VI, 367: 23 December 1679.

15 T. Thomson and C. Innes (eds), *The Acts of the Parliaments of Scotland [APS]*, 13 volumes (Edinburgh, 1814-75), VIII, 494: Act 59, 16 June 1685. For the acts relating to linen, see below.

16 James Paterson, *The Scots Arithmetician, or Arithmetick in all its Parts* (Edinburgh, 1685).

17 For Stewart's results, see St. AUL, MS Q171.R8, 'Professor [John] Robison's Commonplace Book', I, p. 441; and for Maclaurin's comments, see Judith V. Grabiner, 'A Mathematician among the Molasses Barrels: Maclaurin's Unpublished Memoir on Volumes', in *Proceedings of the Edinburgh Mathematical Society*, 39 (1996), 193-240, p. 236. These issues are discussed in Chapter **9**.

18 Paterson, op. cit. (16), 7; J. D. Marwick, *et al.* (eds), *Extracts from the Records of the Burgh of Edinburgh*, 14 volumes (Edinburgh, 1869-1967), XI, 158.

19 Paterson, op. cit. (16), 6 and 143, where the relationship is tabulated. Stewart's views are clearly seen in an extract from his notes used by Maclaurin in which the acre, based on the ell of 37·2 inches, is described as the 'true Scots acre': Gregory, op. cit. (4), 86.

20 Stirling of Garden Papers, National Archives of Scotland (NAS) TD 89/29/137, p. 28, quoted by permission of the Trustees, to whom we are grateful for providing the manuscripts.

21 *APS*, IX, 311-13, Act 48, 1693: 'Anent the right making and measuring of Linen Cloath.'

22 A. J. Durie, *The Scottish Linen Industry in the Eighteenth Century* (Edinburgh, 1979), 8.

23 *APS*, VIII, 598-9: Act 28, 1686. This was ratified in the 1693 Act: see ref. 21.

24 *Statutes of the Realm*, 11 volumes (London, 1810-28), V, 598, 885: 18 & 19 Charles II c. 4, 1666, 30 Charles II c. 3, 1678. The penalty for non-compliance was £5.

25 *APS*, IX, 461: Act 66, 1695; repealed in ibid., XI, 487: Act 94, 1707. The legal requirement for burial in wool remained until 1814: *Statutes at Large*, XIX, 756: 54 George III c. 108, 1814.

26 *APS*, IX, 311-13: Act 48, 1693: '… all Linnen Cloath made for Export furth of the Kingdom and Sold in ffairs and publick mercats, shall be made exact to these two Standards following, viz, either of the Breadth of three Quarters and two Inches unbleetched and of three Quarters bleetched or otherways of the breadth of ane Elne and two Inches unbleetched, or a large Elne bleetched … broad Cloath … shall be ane Elne and two Inches broad unbleetched and ane Elne large bleetched …', quotation p. 312.

27 Ibid.

28 Ibid., VII, 255-6: Act 275, 1661; VIII, 348-9: Act 78, 1681.

29 G. S. Pryde, *The Burghs of Scotland* (Oxford, 1965), 33, 57.

30 Although this is apparently a dated measure that pre-dates the 1719 (Cupar) measure (Item **19** in the Inventory), there are signs of alteration to one face of the measure. This indicates a subsequent enlargement to 37·2 inches. See discussion under Item **9** in the Inventory.

31 J. D. Marwick, *et al.* (eds), *Records of the Convention of the Royal Burghs of Scotland (1295-1779) [RCRB]*, 7 volumes (Edinburgh, 1866-1918), IV, 542: no. 146. 1649, Abstracts of Acts 1631-49.

32 *APS*, IX, 311.

33 Sir John Sinclair, *Statistical Account of Scotland*, 21 volumes (Edinburgh, 1781-99), III, 688: Galashiels.

34 George Chalmers, *Caledonia: or, a Historical and Topographical Account of North Britain*, 3 volumes (London, 1807-24), II, 40, notes the assignment of the reel to Perth (in 1810), but this does not appear in the statute book. In this he was followed by W. M. Stirling in his editorial comment to William Nimmo, *History of Stirlingshire,* second edition (Stirling, 1817), 340. It also appears to be the source for Robert Chambers's claim in 1832 that the measure for yarn had been entrusted to Perth, but he cites an Act of 1437 [*sic*] in which the standard pint at Stirling is first recorded; see R. and W. Chambers, *The Gazetteer of Scotland*, 2 volumes (Edinburgh, 1832), II, 954. This is repeated in the *New Statistical Account*, 15 volumes (Edinburgh and London, 1845), VIII, 412: Stirling. The act in question is that of James II of 1457/8 (not 1437), and provided for duplicates of the pint and newly-defined firlot to be dispatched to three burghs, including Perth, but does not mention the reel: *APS*, II, 50. Lawrence Burrell repeated this claim, but corrected the date of the act: see L. Burrell, 'The Standards of Scotland', in *The Monthly Review: the Journal of the Institute of Weights and Measures Administration*, 69 (1961), 62.

35 W. Craigie, *et al.* (eds), *Dictionary of the Older Scottish Tongue [DOST]*, 12 volumes (Chicago, Aberdeen and Oxford, 1937-2002), VII, 243, s.v. *rele*, citing St Andrews testaments, 1586, and 'Lord Fergus' Gaist' (MS of ?*c*.1560).

36 *RCRB*, III, 14, 28: meetings of 7 July 1615 and 5 July 1616.

37 *APS*, VII, 257: Act 276, 1661 and IX, 311: Act 48, 1693. In the early eighteenth century the Convention of Royal Burghs moved to standardise worsted reels: '… ordains Secret Council to be asked to obtain an Act ordaining plaiding to be three long quarters of an elne and the linning sold att 10s and to be elne broad.' *RCRB*, IV, Appendix 1, 542. The permitted sizes were given in 1721: *RCRB*, V, 226, 237, 273.

38 Ibid., III, 157: meeting of 6 July 1624.

39 Ibid., III, 187: meeting of 5 July 1625.

40 Ibid., V, 273-4: meeting of 6 July 1721.

41 The act gave breadths only in yards and inches: Owen Ruffhead and Charles Runnington (eds), *Statutes at Large*, 41 volumes (London, 1780-1865), V, 191-3: 6 George I, c. 13, 1719. The act was extended in 1723 to include lengths as well as breadths, expressed in a mix of yards, inches and Scottish ells: ibid., V, 354-6: 10 George I c. 18, 1723.

42 Durie, op. cit. (22), 15-18.

43 *RCRB*, V, 436: meeting of 4 January 1727. The 10 quarter reel yields $^{10}/_4 \times 37 = 92 \cdot 5$ inches; $^{10}/_4 \times 37 \cdot 2 = 93$ inches. The acts of the British parliament of 1711 and 1714 had been couched in terms of English measures, they referred only to finished cloth and not yarn, which was still controlled in Scotland by the Scots 1693 Act: 10 Anne, c. 16, *Statutes at Large*, IV, 509-10 and 1 George I, c. 15, ibid., V, 30-3. Both acts contains the words: '… a table with the length of a Yard nailed or marked thereupon to which shall be added one Inch more, which shall be used instead of that which is commonly called a Thumb's Breadth, so that the same Length shall contain thirty-seven inches …'

44 Copy of a manuscript 'Inventar of Working Measures and Weights and Arms belonging to ye Town of Dunbar, Delyvered to Andro Purves,

present Threasurer, 22nd Januar 1700', printed in the *Haddington Courier*, 29 November 1912. We are grateful to David Moody for the reference to this inventory. The reel has not survived.

45 *RCRB*, III, 28: meeting of 5 July 1616: '… understanding … the insufficiencie of the reill, … they have statute and ordeanit … that the lenth of the reill betuix heid and heid sall be halfe elve half quarter long, making the haill hesp to be fyve quarter long of doubell yairne …'

46 See text above, at refs. 4 and 7.

47 Durie, op. cit. (22), 41.

48 Gregory, op. cit. (4), 86.

49 Examples in the collection at the National Museums of Scotland are NMS H.RD.12, shown in Fig. 3.6, and NMS H.RD.18: both have circumferences of 92 inches. There are some 90-inch examples in the NMS Kirkness Collection.

50 *DOST*, I, 476, s.v. *chak reill*, recorded in a testament of 1635.

51 Note: 90 ÷ 2½ = 36; 93 ÷ 2½ = 37·2. An example of a 93-inch reel, acquired in 1992, is in the Highland Folk Museum, Kingussie, Inverness-shire, KIGHF. RD.0046. An Edinburgh-made 90-inch decorative reel bearing an 1829 plaque is the only datable example found (NMS K.1999.998). Examples with 92-inch winders are found, and these are probably mid-nineteenth century (see ref. 49), the equivalent short reel being of 46 inches (*eg* NMS H.RD.33).

52 Durie, op. cit. (22), 12. The act was described in Parliament in 1698 as 'never put into execution': *APS*, X, 133: 3 August 1698.

53 *APS*, IX, 430-1: 'Act in favour of the Linen Manufactory', 1695.

54 Minutes of the Trade Incorporation of Weavers of Stirling, 1700-92, entry for 6 May 1704 (Central Regional Archives, MS PD7/11/4). Also printed in D. B. Morris, *The Incorporation of Weavers of Stirling* (Stirling, 1926), 37.

55 J. L. Anderson, 'An Old Chapmen's Standard Yard-measure from Ceres, Fife', in *Proceedings of the Society of Antiquaries of Scotland*, 57 (1922-3), 167-8.

56 *APS*, X, 133: 3 August 1698; ibid., X, 228, 294: 20 December 1700 and 31 January 1701.

57 *RCRB*, IV, 329-30: meeting of 16 January 1702. Subsequently (in the 1720s), merchants in Stirling complained bitterly at being prosecuted for not selling serge according to the requirements of the 1693 Act: see Morris, op. cit. (54), 13.

58 *Statutes at Large*, IV, 714: 10 Anne, c. 21, Sess 2, 1711; ibid., IV, 627-8: 12 Anne, c. 20, Sess 2, 1713.

59 Ref. 43; and see R. D. Connor, *The Weights and Measures of England [WME]* (London, 1978), 94.

60 Ref. 43.

61 S. G. E. Lythe and J. Butt, *An Economic History of Scotland* (Glasgow and London, 1975), 48.

62 *SND*, s.v. *plaiding ell*: no source is given for the reference to the two thumbsbreadths.

63 'Extracts from the Accounts of the Burgh of Aberdeen: Accounts of the Provost and Bailies': 1643-4: in John Stuart (ed) *Miscellany of the Spalding Club, volume V*, Spalding Club no. 24 (Aberdeen, 1852), 39-181, p. 107 (cited in *DOST*, s.v. *plaiding*).

64 The increasing use of the yard is evident from Swinton's information on the length of the measure for home manufactures: Nairnshire, Wigtown, Inverness and Ross and Cromarty all give 38 inches for their ell (that is, a yard plus two inches); Aberdeen and Kincardine give 38⁵⁄₁₂ and 38½ inches respectively, somewhat larger than a yard and two inches. Brechin, Fifeshire and Langholm (Dumfries) are almost at an ell of 37.2 inches, returning 37¼, 37⅛ and 37⅛ inches respectively: see [John Swinton], *A Proposal for Uniformity of Weights and Measures in Scotland* (Edinburgh, 1779), *passim*. By implication, the ell of 37·2 inches was used elsewhere.

65 Reid, op. cit. (12), appendix 2: William Knight's transcript in Aberdeen University Library MS M167. For Copland, see Chapters 2 and 9.

66 W. Kennedy quoted extensively from an 'accurate account, which was drawn up and published in the year 1795', but did not name Keith as the author: *Annals of Aberdeen*, 2 volumes (London, 1818), II, 293-300. Writing for the Board of Agriculture in 1811, Keith noted that 'the Writer of this Report' (that is, Keith himself) had been employed by the Aberdeen Dean of Guild in 1794 to examine the Aberdeen weights and measures: see G. S. Keith, *A General View of the Agriculture of Aberdeenshire* (Aberdeen, 1811), 554. See Aberdeen Town House Archives, Guild Court Book 1793-1835, minutes of meetings on 17 January and 10 February 1795; a copy of the printed report can be found in the National Library of Scotland, 'Account of the Weights and Measures used in Aberdeen', shelf mark X.225.a.1(63). However, it is not included in James Anderson's earlier report to the Board of Agriculture, *General View of the Agriculture and Rural Economy of the County of Aberdeen* (Aberdeen, 1795).

67 Aberdeenshire 38⁵⁄₁₂, Inverness-shire 38, Kincardineshire 38½, Nairnshire 38, Ross & Cromarty 38, Wigtownshire 38 inches: [Swinton], op. cit. (64), 53, 86, 89, 101, 115, 127. Aberdeenshire 38⁵⁄₁₂, Banffshire 38⁵⁄₁₂, Elgin & Moray 38½, Inverness 38·12, Nairn 38½, Sutherland 38 inches; see G. Buchanan, *Tables for Converting the Weights and Measures hitherto in use in Great Britain … also Abstracts of the Jury*

Verdicts throughout Scotland in regard to the Weights and Measures of Each County (Edinburgh, 1829), 174, 179, 211, 220, 240, 267. The originals of the verdicts are preserved in a volume at the National Archives of Scotland (NAS) E 349/8.

68 Swinton, op. cit. (64), 79, 81, 82.

69 Ibid., 127. By way of distinction, unbleached linen was to be sold by the ell of 38 inches.

70 Reid, op. cit. (12), appendix 2, pp. 6-7. For Alexander Adie and the Adie business, see T. N. Clarke, A. D. Morrison-Low and A. D. C. Simpson, *Brass & Glass: Scientific Instrument Making Workshops in Scotland* (Edinburgh, 1989), 25-65.

71 John Skene, *De Verborum Significatione: The Exposition of the Terms and Difficill Wordes, conteined in the Foure Buikes of Regiam Majestatem, and others* …. (Edinburgh, 1597), s.v. *particata* (or *perch*).

72 Transcribed into modern English from J. Cowell, *The Interpreter* (Cambridge, 1607), s.v. *perch*. It is likely that this was based on the equivalent entry in Skene's *De Verborum Significatione* of 1597, or the second edition of 1599: Skene, op. cit. (71).

73 However, in an early but undatable legal fragment printed by Thomson it was noted that: 'In the first tyme that the law wes maid and ordanit … The aker sall contene four rude, the rude xl fallis. The fall sall hald vj ellis': *APS*, I, 751: *Fragmenta … Collecta*, no.16.

74 Sir James H. Ramsay (ed), *Bamff Charters AD 1232-1703* (Oxford, 1915), 21-2.

75 *WME*, 45.

76 *APS*, I, 751: *Fragmenta … Collecta*, no.15.

77 James Balfour attributed this clause to the *Leges Burgorum*, and apparently rationalised the 18 feet by redefining the ell as a yard in the specific context of land measurement: 'Anent the measouring of landis … the aiker sould contene iiij rudis; the rude sould contene xl fallis; the fall sould contene vj elnis; the eln sould contene iij feet, or xxxvj inches': J. Balfour, *Practicks: or, a System of the more Ancient Law of Scotland* (Edinburgh, 1754), 441-2: c. 98. (This work was collated by Walter Goodal [1706?-66], under keeper in the Advocates Library.) More accessible as Peter G. B. McNeill (ed), *The Practicks of Sir James Balfour of Pittendreich*, Stair Society nos. 21 and 22, 2 volumes (Edinburgh, 1962), I, 441-2.

78 *WME*, 44.

79 '… et concersisse homini meo Balwino quandam terram in Pert habentem decem pedes in latitudinem viginti quatuor in longitudinem juxta domum Vilchil ita liberam et quietam ut habet domum suam propriam': G. W. S. Barrow (ed), *The Acts of Malcolm IV King of Scots 1153-1165: Regesta Regum Scottorum [RRS], volume I* (Edinburgh, 1960), 216: no. 171.

80 '… concessisse et bac carta mea confirmasse Willemo Galeatori platyam illam de xx et sex pedibus in longitudime et de vigini pedibus in latitudine in burgo meo de Pert … tenendam sibi et heredibus suis de me et heredibus meis in feudo et hereditate, libere et quiete; reddendo inde singulis annis duos capellos ferri': G. W. S. Barrow (ed), *The Acts of William I King of Scots 1165-1214: RRS II* (Edinburgh, 1971), 471-2: no. 523.

81 Just as Edinburgh's ancient frontages were based on a rod (albeit one of 20 feet), we find in twelfth-century Winchester that frontages were in units of rods of 16½ feet and half rods: Martin Biddle, *Winchester Studies 7, Part II: Object and Economy in Medieval Winchester: Artefacts from Medieval Winchester*, 2 volumes (Oxford, 1990), I, 345; Derek Keene, *Winchester Studies 2: Survey of Medieval Winchester*, 2 volumes (Oxford, 1985), I, 155.

82 A. W. Richeson, *English Land Measuring to 1800: Instruments and Practices* (Cambridge, Massachusetts and London, 1966), 108-9.

83 Alexander Huntar [*sic*], *A Treatise of Weights, Mets and Measures of Scotland, with their Quantities and True Foundation …* (Edinburgh, 1624), 12.

84 John Ainslie, *A Comprehensive Treatise on Land Surveying; comprising the Theory and Practice in all its Branches* (Edinburgh, 1812).

85 Swinton, op. cit. (64), 24.

86 'But a Roode of worke, wrought by Masons or Sclaiters, contains but 36 Ells: that is, if any piece of worke bee found to bee 18 Ells in length, and 2 Ells in breadth, it makes a Roode', or similarly 12 × 3, 9 × 4, 8 × 4½, 6 × 6 ells: Hunter, op. cit. (83), 6.

87 Ainslie, op. cit. (84), 14.

88 Dr Julian Goodare has noted payments in the Master of Works Accounts for glass at Falkland Palace in the 1630s which are recorded in short rather than long hundreds of (square) feet: J. Goodare, 'The Long Hundred in Medieval and Early Modern Scotland', in *Proceedings of the Society of Antiquaries of Scotland*, 123 (1993), 395-418. For Falkland, see pp. 408-11.

89 Anthony Turner, *Early Scientific Instruments: Europe 1400-1800* (London, 1987), 81, 283.

90 Edward Worsop, *A Discoverie of Sundrie Errours and Faults daily committed by Landmeaters ignorant of Arithmetike and Geometrie* (London, 1582). The illustration in Cyprian Lucar's *A Treatise named Lucarsolace* (London, 1590) confirms that a 'wire line' was a chain. We are grateful to Dr Stephen Johnston for these references.

91 Aaron Rathborne, *The Surveyor in Foure Bookes* (London, 1616), 132.

92 See refs. 100-4. The issue is discussed in context in Chapter 9. The recommendation to consult

Copland had come from John Playfair at Edinburgh University.

93 Reid, op. cit. (12), appendix 2, p. 4.

94 Gregory, op. cit. (4), 6. For Maclaurin's role as editor, see ref. 4.

95 For Maclaurin's use of Stewart's papers, see Gregory, op. cit. (4), 86. Stewart's work is discussed in Chapter **9**.

96 Swinton, op. cit. (64), 24. For the reprinting of Maclaurin's data, see above refs. 6 and 7.

97 Thus, a tabulation of scales in chains of 74·4 feet is given in Ian H. Adams, *Descriptive List of Plans in the Scottish Record Office*, 4 volumes (Edinburgh, 1966-88), I, ix. Scots chains of both lengths are noted in Adams, *Agrarian Landscape Terms: a Glossary for Historical Geography* (London, 1976), 4.

98 Milne had to convert his use of an ell of 37 inches to the town's ell of 39¼ inches, and it is deduced below that these were conventional inches, hence Milne's ell was not 37·2 inches long: Milne to James Ross, Chief Factor at Gordon Castle, 18 November 1770: NAS, GD 44/43/19(3). We are indebted to George Dixon, Central Region Archives, for drawing this correspondence to our attention and providing a transcript. On Milne and May, see contributions by I. H. Adams to P. Eden (ed), *Dictionary of Land Surveyors and Local Cartographers of Great Britain and Ireland 1550-1850*, with *Supplement* (Folkestone, Kent, 1975-9), 447, 448; and the second edition, Sarah Bendall (ed), ibid., 2 volumes (London, 1997), II, 348, 355.

99 J. Ainslie, *The Gentleman and Farmer's Pocket Companion and Assistant; consisting of Tables for finding the contents of any piece of land by pacing, or by dimensions taken on the spot in ells …* (Edinburgh, 1802), 168, 169. It is clear that Ainslie meant 74 (English) feet and not 74 of the feet supposedly derived from the ell of 37·2 inches (equivalent to 74·4 feet) because he gave the English chain as 21 ells 1 foot 3 inches, which is correct only if the ell is 37 inches.

100 Ainslie, op. cit. (84), 164, where the chain's size can be deduced from the quoted 8·88-inch size of the link. Ainslie's text was not altered in the greatly expanded edition produced by the prominent surveyor and civil engineer William Galbraith in 1842.

101 'Plan of the Ground the property of David Steuart Esquire on which Steuartown is intended to be built, Surveyed by John Ainslie 1787', recorded by the National Monuments Record of Scotland (ref. A/75965) and reproduced in Connie Byrom, 'The Development of Edinburgh's Second New Town', in *Book of the Old Edinburgh Club*, new series, 3 (1994), 37-61, p. 42; 'Plan of the Remaining Part of Mr Wood's Farm, by John Ainslie 1797', recorded by NMRS (ref. C13230).

102 There are two examples in the National Museums of Scotland. One is an apparently early chain (NMS W.1971.321), with wire handles, perhaps of the eighteenth century. It now lacks a link, but its original length would have been 73·0 feet. It seems unlikely that fully straightening individual links would increase this to 74 feet. A second example (NMS W.1966.786, VHC.15) with cast brass handles and therefore nineteenth century, has a total length of 74·7 feet. Again, individual links are slightly distorted. Scots-manufactured chains to the English standard include a fine example by B. Reid & Co. of Aberdeen (NMS W.1959.1015, VH.74).

103 Reid, op. cit. (12), appendix 2, pp. 3-4. See Chapter **9** for a discussion of the new weights and measures acquired for Aberdeen at this time.

104 D. Hume, *Decisions of the Court of Session 1781-1822 in the Form of a Dictionary* (Edinburgh and London, 1839), 700-1: no. 533, N. Brown *vs*. J. Kyd, 1 December 1813. The case is not included in the most extensive run of printed decisions, that of the Faculty of Advocates. Papers for the case are included in the Signet Library's collection of Session Papers, volume 271 (1812-13), no. 17, and the original process papers are at the National Archives of Scotland, CS.40/13. We are grateful to the following for their advice and help in identifying the case: George Ballantyne and his colleagues, Signet Library; Margaret Deas, National Library of Scotland; Malcolm Scott, QC; David Sellar, Department of Private Law, Edinburgh University; Margaret Sturgeon and her colleagues, Edinburgh University Law Library; the late Peter Vasey, Scottish Record Office (now NAS).

105 Blackadder's 1807 survey for Brown is at NAS, RHP.1027. NAS also has a photostat of Kinghorne's undated survey for Kyd, RHP.3365. For Kinghorne, see the entry for Items **228** and **229** in the Inventory, and Bendall, op. cit. (98), II, 297.

106 *Petition of James Kyd, Writer in Cupar, against Lord Hermand's Interlocutors*, 2 December 1812, p. 3.

107 *Answers for Nicholas Brown … to the Petition of James Kyd, Writer in Cupar*, 5 January 1813, p. 24.

108 Buchanan, op. cit. (67), 198-9.

109 Ibid., 17-18. Copland's results were 37·0602 inches at 55°F, from which he obtained his value for the chain of 74·1204 feet, and 37·0707 inches for the ell at 60°F. Jardine's figure of 37·0598 inches has been calculated for 62° (the comparison temperature used in the Imperial definitions) and corrected to allow for the known relationship between Copland's 1801 scale by Troughton (obtained for

Aberdeen), and the 1760 yard by Bird adopted as the Imperial Standard in 1824.

110 For a brief account of John Wood, see D. G. Moir (ed), *The Early Maps of Scotland*, third edition, 2 volumes (Edinburgh, 1983), II, 282-3.

111 John Wood, *Town Atlas of Scotland* ([Edinburgh], 1828) and *Descriptive Account of the Principal Towns in Scotland to accompany Wood's Town Atlas* (Edinburgh, 1828).

112 Of the dated plans which specify the chain length, 74 feet was used in the only plan of the first year (1818). However, 74 feet 1¼ inches was used on all five plans of 1819 and 1820 (Irvine, Kilmarnock, Hamilton, Dumfries and Linlithgow, although the engraver has omitted the '1' inch from the Irvine plan). This was modified to 74 feet 1 inch for all four of the 1821 and 1822 plans (Stornoway, Arbroath, Brechin, and Forfar) and the single 1825 plan (Greenock), although the six plans of 1823 and 1824 use 74 feet.

113 For Adie, see Clarke, Morrison-Low and Simpson, op. cit. (70), 30-41. A possible link between Wood and Adie is provided by the site of Wood's house, Canaan Grove, which was built on a plot of land marked as in Adie's ownership on the original feu plan of 1802: NAS RPH.38141/1. We are grateful to Dr T. N. Clarke for drawing this plan to our attention.

114 Reid, op. cit. (12), appendix 2, pp. 6-7.

115 Clarke, Morrison-Low and Simpson, op. cit. (70), 34.

116 *Scotsman*, 14 October 1826.

117 D. G. Lockhart, 'The Evolution of Planned Villages of North-East Scotland: Studies in Settlement Geography, *c.*1700-*c.*1900', 2 volumes, unpublished University of Dundee PhD thesis, 1974, II, 57-8. Feus on the first of the new streets which enclosed the town square were given out in 1768: letter of William Gordon of Nethermuir to Thomas Milne, 18 October 1770, NAS GD44/43/19/20. We are grateful to George Dixon, Central Regional Archives, for drawing this Milne correspondence to our attention.

118 Thomas Milne to James Ross, Chief Factor at Gordon Castle, 18 November 1770, NAS GD44/43/19(3).

119 Thomas Milne, 'Contents of the Feus of Huntly', Central Region Archives CR8/155. Here, 20 × 39½ ÷ 12 = 65·83 feet; alternatively, the ell's length is 66 × 12 ÷ 20 = 39·6 inches. The Huntly ell of 39·5 inches is 5 × 7·9 inches.

120 Thomas Milne to James Ross, Chief Factor at Gordon Castle, 9 December 1770, NAS GD44/43/19(1).

121 Hunter, op. cit. (83), 9-10.

122 The equivalent length of the mile using the ell of 37·2 inches is 1,984 yards.

123 Ian Adams, 'Economic Process and the Scottish Surveyor', in *Imago Mundi*, 27 (1975), 13-18.

124 Adams, *Descriptive List*, op. cit. (97), *passim*.

125 J. C. Stone, 'Robert Gordon of Straloch: cartographer or chorographer?', in *Northern Studies*, 4 (1981), 7.

126 On Pont's work see J. C. Stone, *The Pont Manuscript Maps of Scotland: Sixteenth Century Origins of a Blaeu Atlas* (Tring, Hertfordshire, 1989) and Ian C. Cunningham (ed), *The Nation Survey'd: Timothy Pont's Maps of Scotland* (East Linton, 2001).

127 35 Elizabeth I, c. 6, 1593.

128 On the old English mile, see *WME*, 70-7.

129 A. D. C. Simpson, 'John Adair, Cartographer, and Sir Robert Sibbald's Scottish Atlas', in *The Map Collector*, no. 62 (Spring 1993), 32-6.

130 For example, the 1685 map of Strathearn (engraved by James Moxon), based on the 1683 manuscript (National Library of Scotland, Adair MS.2). The 1688 map 'The Turnings of the Forth' (engraved by Herman Moll) is divided by a grid, where 'Each square is a Countrey Mille or 1500 Pasus'.

131 Hunter, op. cit. (83), 8.

132 Jeffrey Stone has compared manuscripts maps of Nithsdale, Dumfriesshire, by Pont, by Gordon (based on Pont), and the published version engraved by Blaeu: J. C. Stone, 'An evaluation of the "Nidisdaile" Manuscript Map by Timothy Pont', in *Scottish Geographical Magazine*, 84 (1968), 160-71. Stone notes a linear scale for the engraved map of 1·7 miles to the inch, but assumes these are the later Scottish miles and therefore comments that the scale has been shown considerably too large. But 1·7 common miles is equivalent to 2·4 English statute miles (1·7 × 1·42), which matches the required slight enlargement from the average scale of 2·7 he derives for Pont's manuscript.

133 Simpson, op. cit. (129).

134 D. W. Kemp (ed), *Tours in Scotland 1747, 1750, 1760 by Richard Pococke* (Edinburgh, 1887), 351-5.

135 Ibid., 76. The average ratios for the 22 weekly stages vary between 1·2 and 1·6 miles to the statute mile.

136 Hunter op. cit. (83), 10.

137 *WME*, 73, 74.

138 Ibid., 74, 76.

139 'The Dreme of Schir Dauid Lyndesay of the Mont', in Douglas Hamer (ed), *The Works of Sir David Lindsay of the Mount, 1490-1555*, 4 volumes, Scottish Text Society, third series, nos. 1, 2, 6, 8 (Edinburgh, 1931-6), I, 3-38, p. 23, lines 642-4: 'The quantytie of the erth circuleir is fyftie thousand liggis … devidyng, aye, ane lig in mylis two.'

4

The Early Weights: A North European View

THE ARGUMENT FOR EARLY STRUCTURAL LINKS BETWEEN THE SCOTTISH and English systems of weights and measures is so strong that it seems very likely that when the Scottish system was formalised, perhaps in the twelfth century, it was consciously based on that of England. Thus, in terms of the linear units, the inches of the two kingdoms have been shown to be identical; the foot and yard (often thought of as exclusively English units) have been found in Scottish sources; and there is also evidence of an early English ell of 37 inches, just as that described in the Scottish Assize of David I. The similarities between the weight definitions of the David Assize and the equivalent English texts, described in Chapter I, are also striking, and it is with these evolving merchant weight systems, mainly between the twelfth and fourteenth centuries, that we will be concerned in this chapter.

The close links between the Scottish court of David I and his immediate successors, and that of his cousins in England, provided a natural administrative context for the emergence of an English-based system of weights and measures to regulate the trade of the earliest Scottish burghs. But this close relationship between the two kingdoms came to an abrupt end in the late thirteenth century when the expansionism of the English monarch Edward I was unleashed on Scotland. Although trade with England has always been important, the hostilities that began in this period have led to England being characterised as the 'auld enemy'. Scotland's economy inevitably became more heavily dependent on maritime trade with the Low Countries, and in particular on the market for Scottish produce in Bruges. Historically, the other major external influence on Scotland has been France, both through direct trade and political association, and indirectly through the Netherlands during the period of control of the Low Countries by the Dukes of Burgundy.

It is perhaps stating the obvious to say that Scottish trade and metrological practice have been strongly influenced by her dominant trading partners, to the extent that English, French and Flemish measurement units

Opposite:
Detail of Fig. 4.1.

have been incorporated into Scotland's metrology. It may be less obvious that similar external pressures should have moulded English metrology; but the trade of medieval London in particular was heavily dependent on stronger trading centres in Continental Europe such as Paris and Bruges, and this inevitably affected the units in which London's import trade was conducted. An example of this is the English trade in spices, which were largely imported into London from Italy, passing through the great markets of Antwerp. As we shall see, the wholesale trade at the London spice market was conducted using the appropriate Antwerp pound, and not the general English trade pound described in English metrological sources.

Most studies of English metrology have made use of a comparatively limited set of documentary sources that are largely concerned with operation of internal markets and have not taken account of the wider European trading context. These documents present a picture which has been possible to interpret in a largely self-consistent manner, relying almost exclusively on the English troy and trade pounds in which the texts appear to be couched. In contrast, early Scottish sources describe a metrological system which had evolved in parallel with that of England but included significant features which are not compatible with this received view of English practice. Not only do these Scottish sources provide a different perspective of the English evidence, they also help to fill gaps in the early English metrological record. The interpretation given in this chapter, which is concerned principally with north European commercial weight systems in the medieval period, attempts to redress this imbalance by providing a view of the developing metrologies of England and Scotland within a trading framework dominated by the metrologies of the Low Countries and France and influenced by those of the Italian city states of Florence and Venice. The emphasis is specifically on the second half of the thirteenth century and the first half of the fourteenth, when these metrologies were evolving into more stable and recognisably modern forms, and when we have for the first time sufficient contemporary data to assess their relationship with some confidence.

This wider European context had not been established when one of the present authors analysed the literature on English metrology in the 1987 companion to this volume, *The Weights and Measures of England*.[1] The English records have proved insufficient by themselves for a full description of the English system, and only when the present study expanded into a broader international arena was it appreciated that some interpretations of early weighing practices were erroneous, while apparently consonant with the written record. A re-evaluation was considered essential, and quite different conclusions have been now reached about English medieval weight systems, informed by connections between the metrologies of countries involved in mutual trade. This new analysis is intended to amend and amplify the account in the earlier volume, and to provide a more secure context in which to consider the Scottish weight series.

COINAGE AND WEIGHT

One of the crucial functions of an official weight series was the accurate description of transactions in precious metals, or bullion, and in particular the control of the weight and fineness of struck coinage, where the worth of a coin was the value of its bullion content. The weight of coins was regulated at the mint by the number struck to a characteristic medieval weight unit known as a 'mark' (a weight normally of eight ounces, with the pound usually 1½ marks or twelve ounces), and the gold or silver content of the coin alloy were expressed in terms of small conventional weight divisions. Such operations were undertaken with great care and precision at national mints, whose officials were amongst the most highly-skilled contemporary technicians. Since this work involved maintaining acceptable standards of accuracy whilst exploiting the permitted margins of error for the monarch's benefit (and the coiner's profit), there was presumably a strong incentive to improve the precision of measurements. Equally, merchants and money-lenders were conscious of the scope for manipulating inadequately controlled currencies: proffered under-weight or depreciated coin could be rejected, and over-weight coin was culled out for resale to the mint. On a more ambitious scale, coin could be exported in bulk to foreign mints that offered even marginally better bullion rates. Indeed, to take a Scottish example from the fifteenth century, in order to discourage merchants from depleting the circulating currency by illegally exporting coin to Dutch mints, the Scottish Parliament repeatedly insisted on merchants bringing back a set proportion of the value of their shipments in the form of silver or gold bullion.[2]

The circulating coinage of classical Rome had been in gold, silver and bronze; but after the Germanic invasions of the western provinces of the Empire the currency in these areas came to be almost entirely dependent on gold, with coin design based loosely on Roman prototypes. Silver coinage, in 'pennies' or *'deniers'* (from the Roman *denarii*), was first produced in Francia in the seventh century and grew in importance until silver deniers effectively replaced gold as the circulating currency across the west. The change to a full dependence on silver was made by Charlemagne in the late eighth century when he completed the progressive introduction of larger and thinner deniers of a recognisable medieval form and with an increased standard weight of about 1·7 grammes (g).[3] Deniers were reckoned at twelve to the 'shilling' or *'sou'* (from the Roman *solidus*) and 240 to the 'pound' or *'livre'* (from the Roman *libra*), but the shilling and monetary pound were not actually represented by coins for several hundred years – they were merely monies of account, and payments were made up with twelve silver deniers per shilling or 240 per monetary pound.

The English silver penny, or 'sterling', which had been issued at various weights before the Norman conquest, was standardised in fineness at a purity of 92½ per cent in about 1080, and this so-called 'sterling' standard has remained the accepted fineness for silver alloy. The penny's weight was also carefully controlled at 1·46 g for an extended period.[4] It was eventually reduced to 1·44 g in 1280, but as a weight unit the sterling continued to be considered as 1·46 g. Both these sizes play significant roles in the discussion to follow.

The quality and abundance of these early coins led to their wide circulation in northern Europe; and although English silver coin was subsequently produced at depreciated levels, the weight and fineness of these early sterlings became internationally accepted standards.[5] Pamela Nightingale has argued that the substantial scale of English wool exports through the Low Countries and Rhineland, much of it destined as Flemish cloth for the Italian markets, drew a ready return from Flanders of silver bullion from the German mines.[6] The importance of this trade led to Flanders, and later Cologne, adopting English weight standards in their own currencies and forming part of an emerging monetary region. In contrast, areas outside this trading region (including much of France) had to resort to progressive debasement of their denarial coins which became increasingly inconvenient in trade.

Stability was eventually provided by the issue of larger denomination silver coins of good quality, introduced initially by Venice in the early thirteenth century and generally known as *'grossi denari'* or 'large pennies'. This type of coin came to be termed the *'gros'* (or *'gros tournois'*), 'groat' or *'groschen'*, and with its incorporation into different European currencies it was issued at a variety of weights, typically of several grammes.[7]

By this time the growing success of the great medieval fairs of the Champagne region to the east of Paris had established new trade routes to Italy through France and drawn Flemish trade from the north. Europe's gold, which had earlier come from Byzantine coins, now principally flowed eastwards from Muslim Spain, much of it traded in France. French weight standards used in the gold bullion trade and later identified with the town of Troyes (and hence 'troy weight') – the commercial and administrative centre of Champagne – came to assume a significant European role. Paris in particular became associated with the finest craftsmanship in gold and by the thirteenth century was the dominant north European centre for goldsmith work.[8] When hallmarking by the Goldsmiths' Company of London was introduced in 1300 it was the 'touch of Paris' that set the fineness or purity for gold in England, and the Company's earliest weight standards are likely to have been French.[9] An indication of the importance of this bullion route for

the Low Countries was the adoption by Flanders of French troy weight
for gold imports as early as 1132 (although the troy mark was little used for
gold until the following century), whilst continuing to use English weight
standards for silver.[10]

In 1252 the first significant European gold coins were issued by Florence
and Genoa (the former being the *fiorino d'oro* or 'florin') with a weight of
about 3·5 g, followed in 1284 by the Venetian gold 'ducat'.[11] Both the florin
and ducat came to circulate across Europe and provided convenient, effective
and stable standards for weight and fineness for centuries. France and
Hungary followed suit in issuing equivalent gold coinages, as did the Low
Countries in the 1330s and England in the 1340s.

Bullion was characteristically weighed by the mark, normally of eight
ounces with 1½ marks to the pound, and coinage rates were similarly related
to a controlling mark. Although the weights and fineness of coins were prone
to adjustment as currencies were manipulated for fiscal or economic reasons,
the point to be stressed is that, at least in this later medieval period, the
mass of the controlling mark was generally unchanged. Thus, for example,
64 florins were struck to the mark of Florence until 1422 when the rate was
decreased by under half a per cent to raise the florin's weight and bring it
into coincidence with the Venetian ducat, but without altering the
Florentine mark.[12]

Frequently the divisions of the mark and ounce had conventional names
which were the same as those of current coins. But the masses of these were
chosen for the convenience of the multiples used in the weight system and
they do not necessarily match the masses of particular coins.[13] Two principal
types of mark are encountered in discussions of early English weight systems,
and these are the so-called 'tower mark' (of about 233 g), and the English 'troy
mark' (of about 249 g). The eight-ounce tower mark and the twelve-ounce
tower pound are closely associated with coining activity at the mint in the
Tower of London (hence the descriptive name), where 240 sterlings were
coined to the pound and 160 to the mark. However, the weight of the penny
was reduced so that 243 were struck to the tower pound from 1280, 266
pennies to the pound from 1344, and ultimately it sat at 540 to the pound
when tower weight was officially abolished in 1527.[14] Yet the pound and mark
did not themselves change, and a sterling (as a weight) remained at a
twentieth of a tower ounce. The larger English troy mark also has a 'pennyweight' as one-twentieth of the ounce, and this was certainly larger than the
weight of a penny. The French weight system was based on the Paris mark
(about 245 g) of eight ounces, with each ounce divided into eight *'gros'*, 24
'denier' or 20 *'esterlin'*, yet none of these weight units matched the masses of
the equivalent coins – the *gros tournois*, the *denier* and the English sterling or
penny respectively.

The smallest practical unit in a bullion weight series is usually the 'grain'.

Ostensibly this is the naturally occurring and remarkably regular grain or seed of a cereal crop – either wheat or (for more northerly climes) barley. Again, however, the mass of the weight unit depends more crucially on the conventional multiples used in the weight series. Thus, for example, the multiple for the number of grains in the ounce in the English troy series is 480 (or 20 × 24), and in the Paris system it is 576 (or 24 × 24), whereas in the troy system of the Low Countries there are 640 'asen' or grains (20 × 32) to the ounce.[15] In none of these systems do the grains accurately correspond with those in nature, nor are the three grains of the same weight.

In each of these systems the mark has a slightly different mass, although all are in a fairly narrow range of about 245 g to 249 g. The mark of Paris (245 g) was the 'marc de Troyes', and indeed the two names can be used interchangeably. Although it was also adopted as the gold mark for Bruges, it eventually became the mark standard for the whole of the Low Countries where it was later known as the 'trooise' mark. For reasons which are not well understood, Flemish weight standards subsequently diverged slightly from Paris weight, and an official standard from Brussels tested in Paris in 1529 was found to be fractionally heavier.[16] The Low Countries' mark is now considered to be about 246 g. The name of the English 'troy' mark suggests that it is also derived from the marc de Troyes, although its mass is even larger, at about 249 g. However, by the time this name was first applied (and the compilers of the Oxford English Dictionary record 'troy' only from 1390), it may well have been used generically, merely to indicate a role in bullion and monetary control.

Accurate measurements for these marks, and of numerous regional variants, were made in the late eighteenth and early nineteenth centuries for comparison with the newly introduced metric units. Although these measurements give us a snapshot of the precise differences found between particular surviving weights, they cannot tell us whether these weight systems were genuinely independent or whether some were originally linked but had diverged gradually over the course of time.

Ideally, authoritative original standards need to be measured, but by the 1790s remarkably few genuinely early standards did survive. Scant information was recorded about the weights that were examined and many were not retained after measurements were made. Little is known about how representative these (often redundant) weights were; and their status, their relationship to other contemporary standards, and their degradation through wear and damage, is a matter for speculation.

If a weight standard for a particular geographical area was to be created, it was presumably related to an earlier (and perhaps external) standard, possibly in the form of a controlled (conceivably external) currency. In any event the question arises of how accurately the standard had been constructed. In practice distributed replicas of the standard were required, each of which inevitably differed slightly from the original. But these replicas were of no use without accessible graduated weights of all the multiples and divisions of the standardised unit necessary for actual weight comparisons. With each successive copying, the scope for introducing unintentional errors increased. The magnitude of these errors can be perhaps appreciated from the performance of the best balances available in the mid-eighteenth century when some of the early scientific comparisons between English and French weights were conducted: precision balances for weights in the region of about half a mark turned on about half a grain, but the sensitivities of balances for larger weights in graduated sets was often several grains.[17]

Exhaustive comparison procedures could reduce this uncertainty, but at any given period there was a practical limit to the level of precision that could be achieved. In time, improved balance technology led to the recognition of inconsistencies within earlier sets of standards and to an understanding of the effect of wear on more frequently-used weights. Sets of weights might then be re-adjusted to improve consistency, or they might be replaced with more accurately divided weights, but implicit in each such change is a decision about the most appropriate level at which to set the precise weight level of the standard. Over a period of several hundred years a series of such changes might lead to considerable variation between different weight series which were initially considered administratively identical.

This type of inherent difficulty means that we have to be cautious not to imply greater certainty and accuracy than our limited knowledge of these early weight systems can justify. One danger area is the type of contemporary statement that says that so many pounds or ounces of one weight system equal some other number of pounds or ounces of another system. Although there is a temptation to assume that such statements are accurate, it is more likely that they were only approximations which were considered acceptable within a particular trading context, and their function was to simplify commercial calculation, perhaps also taking account of appropriate allowances or commissions. It may be that, if circumstances demanded, a different and more accurate approximation could be given.

On the other hand, certain relationships are intended to be specific and exact. Clearly an eight-ounce mark contains exactly eight ounces. Two different pounds may be based on the same ounce but use different multiples of ounces, so 15-ounce and 16-ounce pounds have masses which are in the exact ratio 15:16, and assuming competent control mechanisms are in force they will remain in this ratio. We argue below that a number of relationships of this sort exist between the metrologies of centres involved in mutual trade and these betray evolutionary features of these metrologies.

The reality of these links between certain north European metrologies is central to our argument about the processes which have helped shape the emerging metrologies of England and Scotland. However, we want to draw a clear distinction between the type of simple ratios that we believe can be justified on the basis of historical evidence, and its illogical extension. The latter is the school of thought which holds that all metrological systems – current, medieval, and ancient – can be linked through the ratios of often large integer numbers which (coincidentally) match the ratios computed from modern figures of the greatest apparent accuracy. This is an area where antiquarian zeal and numerology have combined to produce nonsensical results which would have required medieval merchants to perform miracles of calculation which were not possible with the counting boards and mathematical notation of the day.

Our purpose here is less ambitious. The period examined in this chapter is principally the second half of the thirteenth century and the first half of the fourteenth century, when the weight standards provided by coinage such as the Florentine florin were recognised and appreciated across Europe. The trading activities of the Italian merchants, and the extensive banking and credit services they provided, helped to ensure that the trading metrologies of northern Europe developed within an integrated framework which was compatible with these weight standards. What emerged was a reliance on a restricted group of closely related ounce weights, and conventions for the multiples of these ounces in pounds and larger units which reveal the nature of the relationships between the various trading systems. These relationships are implicit in a valuable contemporary commentary, the merchants' handbook of Francesco Balducci Pegolotti, which provides a European view of commercial practice for goods traded in bulk in the early fourteenth century and which is discussed later in this chapter.[18] In subsequent centuries the picture became increasingly complex as significant national and regional differences emerged, and as manipulation of trading and fiscal relationships distorted and masked what had previously been comparatively simple equivalences.

To make these equivalences more transparent in the discussion that follows, we have opted to give the majority of weight units in terms of English troy grains (15·432 Imperial troy grains are equivalent to one gramme), rather than metric units. In the case of most English and Scottish units this simplifies the numbers involved and makes the significance of weights in familiar multiples more obvious. It also allows more ready comparison with much of the English language work of earlier commentators where the convention has been to use troy grains. It must be stressed, however, that this is a mathematical convenience which is not meant to imply the contemporary use of such grains.

Before we discuss Pegolotti's fourteenth-century data, we must look at the related weight definitions in the earliest English and Scottish metrological texts and conclude that these are explicable in terms of tower rather than troy ounces. Subsequently, we will consider the large units in which bulk trade was conducted and demonstrate how the weighing practices for the larger units often differed from the more familiar procedures used for smaller quantities. This has a marked effect on how we interpret these bulk trans-actions and the weight units in which they were conducted.

THE SIZE OF THE ENGLISH TROY AND TOWER OUNCES

Both the English troy ounce (of about 31·10 g) and the tower ounce (of about 29·16 g) are bullion ounces, reckoned at twelve ounces to the pound. The troy ounce is the more familiar because it survived until very recent times as the only unit proper for the weighing of gold and silver. The troy ounce was considered to comprise 20 pennyweights each of 24 troy grains, or a total of 480 grains, so the twelve-ounce troy pound was 5,760 grains. The troy pound was the central weight unit in the English (and subsequently British) system, and in particular it was the principal legal pound of the Imperial system introduced in 1824, against which the 'avoirdupois' pound (the familiar pound of modern trade) was defined as comprising 7,000 grains.[19]

The actual physical standard of the Imperial system was a gun-metal troy pound which had been constructed in 1758 by Joseph Harris, the King's Assay Master at the Mint, for a parliamentary committee chaired by Sir John Proby, first Baron Carysfort.[20] This pound, adopted as the primary standard in 1824, was declared to contain 5,760 troy grains; and this in turn defined the avoirdupois pound which was stated to be 7,000 of the same troy grains. However, Harris's pound was destroyed in a disastrous fire at the Houses of Parliament in 1834, after which it was decided to change the emphasis of the legislation and create a new principal standard, this time of the avoir-dupois pound and constructed in more stable platinum. The new standard, completed by W. H. Miller in the 1850s, was derived from exhaustive comparisons with surviving replicas of the lost troy pound, particularly with early nineteenth-century copies in platinum whose relationships with the 1824 brass replicas of the 1758 pound were known with a high degree of precision.[21] From these detailed comparisons, and from later determinations in metric units, we can say with confidence that the weight of the lost 1758 troy pound in air under standard conditions was close to 373·24 g.[22]

The Carysfort Committee had been established at a time when there was growing unease about the legal status and sizes of the multiplicity of legal and customary measures in trading use, and its purpose was 'to inquire into the Original Standards of Weights and Measures in the Kingdom and to consider the Laws relating thereto'.[23] The earliest authoritative standards were located by Carysfort in repositories such as the Exchequer and they were measured and gauged by the most competent experts available. The work of constructing new reference standards which could be reproduced with greater inherent reliability and accuracy was begun, and reports and recommendations were produced in 1758 and 1759. However, no parliamentary bill was passed into law, and the reforms for which they pressed had to wait until British weights and measures underwent a major overhaul with the introduction of the Imperial system in the 1820s.

The principal troy weight standards that Carysfort examined were the 'pile' or set of nesting cup weights of 1588 at the Exchequer, although he was aware that these were merely part of a general issue of around 60 such sets, all of which were supposed to have been based on the earlier set of reference weights held at the Goldsmiths' Company of London, which were then considered to be the standards of authority. Indeed, the special jury which sat in London to consider the standards for Elizabeth's administration in 1574 had no hesitation in describing the Goldsmiths' 'great pile' as 'the onlie patern extant as far as we can find for the sizing of all other of the sorte'.[24] Unfortunately, the old Goldsmiths' pile was no longer in existence when Carysfort tried to borrow it in 1758.

The examination of Exchequer standards was essential because these were considered to be the standards with ultimate legal authority – a point that was clearly understood by Carysfort. This scrutiny had certainly now been extended to include troy weight. It did not follow, however, that the Exchequer weights were in the best adjustment or most accurately represented the theoretical standard, and certainly the Exchequer was not the only repository able to issue verified copies for use as reference pieces. Weights could also be issued at the London Guildhall (for avoirdupois weight) and the Goldsmiths' Company (for troy weight) – and at the Mint in the Tower of London, which was another recipient of the 1588 Elizabethan issue.[25] More than the Exchequer, the Mint had a very pressing day-to-day need to have the most accurate reference weights possible and to keep them in the best adjustment. The Mint has historically employed some of the best technical measurement experts (such as Harris), and they have often played a central role in the practical business of regulating legal metrology.

Carysfort's enquiries established that small differences existed between a range of Elizabethan and later troy pound sizes at the Exchequer and the Mint, with the pounds at the Mint being generally larger, and he concluded that the original level of the troy pound was closer to those of the Mint. This data is difficult to interpret, and Carysfort's conclusions have reinforced the conviction that the size of the troy pound has remained unchanged since its first introduction.[26] However, a more careful analysis of the eighteenth-century measurements and of surviving weight standards has demonstrated that a small increase was made in the mass of the troy pound by a switch to reliance on Mint standards.[27] The size of this increase is only a little less than a quarter of a gramme in a pound of 373 g, or less than 0.1 per cent – far too small to be noticed in the course of ordinary trade – but it has important implications for us in terms of establishing more secure values for the masses of medieval weight units and therefore in demonstrating their relationships with one another.

The situation appears to have arisen because different types of control were being exercised at official repositories after the construction of new standards by Elizabeth's administration in the sixteenth century. Standards had been issued in 1574, but had been destroyed when they were found to have been inaccurately made. Replacement standards were constructed and adjusted in 1582, the troy weights being sized against the Goldsmiths' pile, and then accessible copies of these were made and distributed in 1588.[28] However, these copies appear to have been very slightly over standard, either to allow for subsequent fine sizing that may never have taken place, or more likely to incorporate a conventional remedy appropriate for weights for trading use. Certainly at the Mint, which perhaps had a vested interest in the troy standard being higher since this was the weight used for bullion purchase, the troy weights were not adjusted down; and it was these Mint weights that formed the pattern for subsequent work.

Carysfort had relied heavily on assistance from the Mint. The standard troy pound which emerged as his recommendation for adoption as the new national standard in 1758 was one that accurately matched the standard in use at the Mint. But the pound size of the Mint's weights was about 3½ grains in the pound larger than the pound that had been embodied in the now missing, venerable old Goldsmiths' pile.

Because the troy pound was *defined* as containing 5,760 troy grains, the result of this effective increase in the pound size was to increase the grain size also. Our analysis of this episode shows that the medieval size of the English troy grain was only 64·76 milligrammes (mg) (or 15·442 grains to the gramme), against the modern accepted size of 64·80 mg (15·432 grains) for the Imperial troy grain. The grains we will use in all the discussion which follows should always be understood to be in the lighter medieval grains, unless otherwise specified.

An important consequence of accepting this shift in the size of the English troy pound is that it allows us to demonstrate the numerical relationship between the troy ounces of England and France. A 50-mark nest of standard weights of the Paris Mint, constructed in the period 1460-1510 and known as the *Pile de Charlemagne*, survives in the Musée National des Techniques in Paris and is perhaps the oldest authoritative weight standard. It has clearly been based on earlier standards, and it is thought that the level of mark which it represents can be taken back to at least 1266.[29] Paris weight came to assume great importance for the metrology of England, where we can trace its use in early metrological definitions, and also in Scotland, where 'French weight' was the troy standard for an extended period.

But the Paris ounce and mark, specifically as represented in the *Pile de*

Charlemagne, also form the fixed reference point for much metrological work from the late eighteenth century. The accepted modern value for the Paris mark, 244·7529 g, dates from this time.[30] Unfortunately, it arises from an averaging technique which is not appropriate for the medieval period, and we have proposed elsewhere that a more appropriate value, based on the way the standard was being used in the early eighteenth century, is 244·77 g.[31] This gives an ounce of 30·596 g, and in terms of Imperial grains this ounce is 472·17 grains. However, in terms of our preferred medieval troy grain, the Paris ounce is 472·47 grains, which is an excellent match for the very accurate and specific size of 472½ troy grains that emerges from the fourteenth-century data of Pegolotti which we examine below, representing a difference of only about 0·01 per cent.[32]

Besides the troy ounce, the other principal English ounce was the 'tower ounce' (of about 29·16 g) which is associated principally with the control of silver coinage at the English mints, whose operations came to be centralised at the Tower of London. Twenty silver pennies were minted to this ounce, and the tower ounce could also be considered as the monetary ounce, since the pound of twelve such ounces was the weight of the monetary pound of 20 × 12 or 240 pennies. In terms of the grain of the troy ounce, the smaller tower ounce weighs 450 grains, and so there is a simple relationship between these ounces – 16 tower ounces equal precisely 15 troy ounces. Measured in English troy grains, the sterling penny weighs 22½ grains, and when Scottish silver pennies were introduced they had the same weight.[33]

The tower pound was abolished in 1527, and it is only in the declaration of the pound's illegal status that we find clear surviving official evidence of its size in relation to the troy units: 'al maner of golde and sylver shall be wayed by the Pounde Troye which maketh 12 oz Troye which exceedith in the Pounds Towre in weight iii quarters of the oz.'[34] Since three-quarters of a troy ounce is 360 grains, the tower pound weighed 5,760 - 360 = 5,400 grains, and therefore the tower ounce was indeed 450 troy grains.

Almost the only other direct English evidence we have of the size of the tower pound comes from a single poise weight discovered in 1842 in the old Pyx Chamber at Westminster Abbey – the Trial of the Pyx was the annual testing of the quality of new coin. This weight, which was then thought to have been of the time of Henry III or Edward I (and therefore thirteenth century), was described as 'an ancient pound weight of brass, bell-shaped, less that the standard pound weight of the Exchequer by 15 dwt. 9 grs., apparently a *moneyer's pound'*.[35] The weight had been lost by 1873. In the Imperial grains of this description, it was 5,391 Imperial grains; but in terms of the older medieval grains of its construction, it was 5,395 troy grains – against a theoretical size of 5,400 grains.

Both these values encourage us to believe that the relationship between the tower and troy ounces was a simple and precise one, and that it was maintained. We shall accept that the tower ounce was exactly 450 troy grains in the discussion which follows and that the ratio of the masses of the tower and troy ounces was 15:16.

The tower ounce has been described here as a monetary ounce and this is the context in which it is normally restricted. However, for a number of reasons we believe that this is incorrect and that tower-based weights had much wider application in trade. We have been drawn to this conclusion partly by evidence from Scottish sources, and in particular by a characteristic type of Scottish merchant pound, known from a preliminary parliamentary act of 1426 as the 'Scots' pound, comprising 16 ounces of 450 troy grains, and combined in stones of 16 pounds.[36] These pounds, which are discussed in Chapter 5, were therefore distinct and different from the newer Scottish 'trois' pounds defined in the main 1426 Assize of James I in terms of 16 'trois' ounces. Internal evidence will enable us to demonstrate that these 'trois' ounces were identical with the English troy ounces of 480 grains. The descriptive name for this Scottish weight can be appreciated as a variant of 'troy' (although given here in a plural form), linking it with bullion operations.

But the monetary ounce of 450 grains has not generally been considered as playing any part in trade. Here, however, it was emerging as a mainstream trading unit in an early Scottish context that suggested that there would be a parallel application in early English trade. This view challenges the conventional modern interpretation of English commercial weights, which has been based almost exclusively on the English troy ounce of 480 grains.

It also reinforces an observation made in 1990 by Norman Biggs, who drew attention to the lack of evidence for an official adoption of the troy ounce of 480 grains into English metrology until the second half of the fourteenth century. This is not, of course, to suggest that it did not have a much longer history, but that its area of application changed at about this time. Specifically, Biggs was concerned with the use of a particular weight type controlled by the Goldsmiths' Company of London for gold and silver ware and its eventual replacement by troy weight. The earliest unambiguous reference to troy weight which he was able to cite was in a court case of 1376 where 'Troye' weight was distinguished by name from 'goldsmiths weight'.[37] If he is correct, the argument for troy weight forming the basis for weight definitions couched at least a century earlier is hardly tenable.

Our understanding of the monetary relationship between the tower and troy pounds emerges only in the late thirteenth century. In the 1270s the earliest surviving Mint coinage contracts reveal that the king's profit ('seigniorage') and minting costs together amounted to 16 pence in each pound coined from sterling-quality bullion.[38] Thus we can see that a troy pound of sterling bullion produced for the merchant a return of a tower pound of 240 pennies – the difference of 16 pence being the 360 troy grain difference between the troy and tower pounds. When 243 lighter pennies were coined to the troy pound from 1280, the seigniorage absorbed the extra three pence.

But in the late twelfth century the seigniorage and moneyer's share had totalled only twelve pence in the pound for coining sterling silver.[39] This might suggest that the particular bullion troy pound at this earlier period was the French pound, which was heavier than the tower pound by only 270 troy grains, or the weight of twelve pennies.

Pamela Nightingale has indeed made the specific claim that French troy weight was imposed in England by Henry II, the first English king of the Plantagenet house of Anjou, in the course of his major reform of Mint procedures in 1158 and as an early demonstration that he intended his English revenues to serve his Continental ambitions.[40] She describes this as the final stage in raising the sterling penny to its full weight of 1·46 g, creating the tower pound at its established size in the process, and with the penny now compatible with the French troy system as an exact fraction of the ounce. Since the tower and French twelve-ounce pounds differed by twelve pence in weight, the ounces differed by one penny. Thus a French troy ounce is readily demonstrated with 21 pennies, where 20 make the tower ounce. Whether or not this was a planned alignment between the tower and French troy systems, such a relationship was established and was maintained.

There are also other grounds for believing that Paris troy weight was used in at least some areas of London trade. We will, for example, argue later in this chapter that Goldsmiths' weight (which may have had quite wide applications) represents a residual use of French troy standards, and also that some of the earliest definitions of capacity measures appear to have been couched in terms of French weight. But if English troy is seen as replacing French-based units in some areas of application, then it does not emerge as a strong candidate to form the basis for the earliest English weight definitions. We argue instead it was an ounce of 450 grains.

DAVID'S ASSIZE AND THE TRACTATUS

The earliest record of the Scottish system of weights and measures is given in the so-called Assize of Weights and Measures of David I. This document exists in the number of variant forms in both Latin and Scots. The two versions published in the nineteenth-century 'Record Edition' of the *Acts of the Parliaments of Scotland* are given in Appendix A.[41] The form of the David Assize has already been described in Chapter 1, where it was noted that the assize cannot date from the period of David I's reign, between 1124 and 1153, although in all probability it contains elements of whatever regulations applied when a formalised system of weights and measures was first introduced. Rather, it survives as an incomplete summary of extant regulations

THE EARLY
WEIGHTS:
A NORTH
EUROPEAN VIEW
────

119

in which appeal had been made to the legislative vigour of David I in order to provide demonstrable authority. It certainly incorporates changes made after David's reign, most obviously with information on fourteenth-century coin weights, but the main text clearly echoes much earlier practice.

The assize appears in several of the collections of the Old Laws, described in Chapter I. It was, for example, included by John Skene in the first printing of the *Regiam Majestatem* in 1609, although James Balfour in his *Practicks* had attributed it to the *Leges Burgorum* and *Iter Camerarii* collections.[42] The *Regiam* is a compilation of existing practice which is believed to have been assembled in the first half of the thirteenth century and parts of it bear a close relationship to English texts, notably the early English legal compilation known as 'Glanvill', after its doubtful attribution to Ranulf de Glanville, a twelfth-century chief justiciar of England.[43] In the case of the David Assize of Weights and Measures, the comparison that can be drawn is with a portion of an English text known as the *Tractatus de Ponderibus et Mensuris*.[44]

The *Tractatus* has similarly complex origins and has been discussed by Philip Grierson in terms of its status as a collection of historical memoranda relating to official practice, having authority but not strictly the status of legislation.[45] It also exists in a number of forms. The version printed in the nineteenth-century *Statutes of the Realm* follows an earlier convention of attributing it to 1302-3.[46] However, another version (in French) appears in the 'White Book' of Peterborough Abbey, which can be dated to about 1255 (although strictly this dating may only refer to a separate portion of the Peterborough text), so its origins might be mid-thirteenth century or earlier.[47] The *Tractatus* text also appears in a series of 'husbandries' or estate management texts known as the *Seneschaucy* and the *Husbandry of Walter of Henley*, both of the second half of the thirteenth century.[48] The latter is understood to have been the source for a further version found in a valuable legal compilation of about 1290, known as *Fleta* because it was prepared by an experienced (but unnamed) lawyer who was at the time confined in the Fleet prison in London.[49] In due course we will compare evolving definitions in several of these versions, but in the meantime our concern is with the principal weight definitions that also appear in David's Assize.

The *Tractatus* falls into two distinct parts, which may have separate origins. The first provides statements (given here in modern English) about the foundation of the weights:

> By Consent of the whole Realm the King's Measure was made, so that an English Penny, which is called the Sterling, round without clipping, shall weigh Thirty-two Grains of Wheat dry in the midst of the Ear; Twenty-pence make an Ounce; and Twelve Ounces make a Pound …[50]

This is followed by a longer section which gives detail of the trading units for a wide variety of produce and includes the information that

the pound of Pence, Spices, Confections, as of Electuaries [sweet phar-
maceuticals], consisteth in weight of Twenty Shillings [namely, twelve
ounces]. But the Pound of all other Things weigheth Twenty-five
Shillings [namely, 15 ounces]. Item, of Electuaries and Confections the
Pound containeth Twelve Ounces, and an Ounce hereof is of the
Weight of Twenty-pence.[51]

For comparison, the equivalent declarations in David's Assize (again in
modern English) similarly couch the original weight definitions in terms of
the same number of standard sterling coins, each again said to weigh 32
grains of wheat:

> King David ordained that the sterling should weigh 32 corns of good and
> round wheat ... The ounce in King David's time contained 20 good and
> sufficient sterling pennies and now it shall weigh 21 pennies because of
> the diminution of the money ... The pound in King David's day weighed
> 25 shillings. Now the pound ought to weigh in silver 26 shillings and 3
> sterling pennies because of the diminution of the penny at the present
> time ... The pound should weigh 15 ounces.[52]

These two declarations are clearly closely related. Indeed, the only
apparently significant difference between them is that the David Assize only
mentions the 15-ounce pound and is silent about a twelve-ounce pound, even
as a monetary pound. However, this pound must have been known,
because it arises naturally from the common weight of the pennies, and it is
likely that this merely reflects the more universal use in Scotland of the
eight-ounce mark (which is not defined in either of these regulations).[53] The
earliest clear inference of a twelve-ounce pound in the records of the Scottish
Mint was in the 1360s.[54]

The wheat corn, the barley corn and the carob seed have conventional
weights assigned to them in metrology – four wheat corns are understood to
weigh the same as three barley corns and both are equivalent to one carob
seed (or 'carat' – a weight name still used in the diamond trade). In the
tradition of English metrology, the conventional barley 'grain' is the English
troy grain of 64·8 mg and the conventional wheat 'grain' is three-quarters of
this size, and therefore 48·6 mg. Experiment has shown that the relative
weights of large wheat and barley corns are remarkably well described by the
conventional 3:4 weight ratio, but both wheat and barley corns are very
hygroscopic and their actual weights are variable and highly dependent on
moisture content.[55] By contrast, the weight of the Mediterranean carob seed
is comparatively stable, and it is perhaps this that has encouraged the use of
seemingly exactly defined conventional weights for the barley and wheat
corns, even though such an application may be inappropriate. Clearly this is
only an accidental approximation – there is no natural evolutionary pressure
that might lead to a relationship between the masses of different grain types.

The 3:4 ratio has a long tradition, and it has been utilised with success by Philip Grierson in his examination of the silver coinage of the Carolingian Frankish kings, although he observed that 'the precise figure varied from region to region'.[56] As for the *Tractatus* statement that the penny weighed 32 wheat grains, Grierson notes that this 'was indeed never valid for England but was an expression of truth for the Carolingian penny'.[57] Charlemagne's coinage reform of 793/4 resulted in the issue of new denier coins of 1·70 g, which have been described in terms of 32 conventional French wheat grains of 53·2 mg.[58] This grain is the French or Paris grain which formed the basis of the Paris ounce (with 24 deniers, each of 24 grains, so these were 24 × 24 = 576 of these grains to the ounce). We will see shortly that this was the grain used for precise calculations by Pegolotti in the early fourteenth century and this was also the grain adopted in Scotland, but perhaps not until the late fifteenth century. However, despite the descriptive identification of the French grain with the corn of wheat, it is clearly different from the wheat grain of 48·6 mg arising from more modern English orthodoxy, and it provides a caution that these are conventional identifications that cannot necessarily be interpreted literally.

On this basis, earlier interpretations of the *Tractatus* have begun with an assumption that the 'penny' weighed the stated 32 wheat grains (4 × 8) and hence 24 barley grains (3 × 8) – and these barley grains have been conventionally understood as English troy grains. It would then follow that the 'penny' had to be the English troy 'pennyweight', of 24 troy grains (a specific weight on the troy scale, abbreviated to 'dwt') and not the sterling penny coin (which was 22½ troy grains). The ounce of 20 'dwt' would then be the troy ounce of 480 grains, and the pounds of the *Tractatus* would be twelve-ounce and 15-ounce troy pounds.

However, in addition, since the tower ounce (of 450 troy grains) was thought to have been restricted to coinage operations, with 20 pennies of 22½ grains struck to the ounce, the *Tractatus* declarations could also be seen as providing parallel definitions of the tower pounds. The twelve-ounce tower pound was the monetary pound in that it contained 240 pennies or 20 shillings, whereas the 15-ounce pound weighed the equivalent of 25 shillings – both as specified in the *Tractatus*. The apparently ambiguous term 'penny' has been interpreted as 'dwt' to give the troy weights, and as the sterling coin to give the tower weights. This dual interpretation has seemed particularly attractive in this restricted English context because of the dominant role played by the troy ounce in English metrology and trade.[59]

We shall adopt the obvious alternative view that the original *Tractatus* definitions should be interpreted in terms of tower ounces alone. Interpreting the ounce of the *Tractatus* as the tower ounce is not, of course, a new concept, but one that was widely held by influential nineteenth-century writers.[60] However, if the tower ounce (with its pound) was not simply a poise weight for checking the quality of finished coin, but was the basis for the English weight series, then we should expect that it would have pennies and grains as computational divisions as did the troy ounce. The numismatist Rogers Ruding provided some evidence for this in manuscript calculations, extracted from the papers of the London Mint by the antiquary Sir Robert Cotton (1571-1631), which show the required relationship between grains of the troy and tower ounces, and between their respective pennyweights.[61] The tower grain emerges as about 61·0 mg, and therefore still a heavy (or barley) grain, with 24 tower grains to the penny and 480 tower grains to the tower ounce. We can now appreciate that the *Tractatus* definition of the penny is of 32 light (wheat) grains, or 24 heavy (barley) grains of the *tower* and not the troy series.

But why should 'wheat' grains have been specified in the *Tractatus* in the first place? The declaration has almost certainly been established in terms of heavy grains but couched in light grains, using the long-established tradition of the 3:4 weight relationship. Perhaps initially this was done in conscious imitation of early French definitions, and the form of the declaration remained unchanged. Indeed, much later versions – which are clearly troy declarations – still use the same tried and tested formula, which must be intended to represent the authority of continuity rather than a statement of the literal truth.[62]

The *Tractatus* emerges from this present interpretation as a document in which the basic weight definition can be accepted in terms of real sterling pennies and tower ounces, with a monetary pound of twelve ounces (also used for pharmaceuticals), and a trade pound of 15 ounces. The version of these definitions in the later *Fleta* compilation of about 1290 appears to confirm this view by describing how 20 pennies make the ounce and '12 ounces make a pound of 20 shillings *by weight and count*' [our italics], so that the ounce is literally the weight of 20 pence ($20 \times 22\frac{1}{2}$ = 450 troy grains).[63] The general trade pound of 15 ounces (15×450 = 6,750 troy grains) was given the descriptive name of '*libra mercatoria*' or 'merchant pound' in *Fleta*, confusingly translated in the more accessible modern edition as a pound 'avoirdupois'.[64] But *Fleta's* pound is certainly not the trade pound which was subsequently named the avoirdupois pound: a 16-ounce pound of 7,000 troy grains, which became the Imperial pound in the nineteenth century. However, in a restricted sense, the name is appropriate, because this pound was used for heavy material and produce, which was typically sold in bulk and which might be described as 'heavy goods' or goods *avoir du pois*.

In metrologies of this period there was normally a distinction between the weighing of different classes of goods, and this can be illustrated clearly in an English ordinance of 1196 which included the stipulation (in modern English) that 'uniform weights shall be used throughout the kingdom according to the nature of the goods'.[65] There was a particular distinction between fine or costly goods on one hand, and coarser or more bulky goods, which might perhaps be termed 'heavy goods', on the other. David's Assize, the *Tractatus* and *Fleta*, all specify a goods pound of 15 ounces. In contrast, bullion was traded by the mark (normally of eight ounces), and marks came to be combined to form pounds of two marks or 16 ounces, typically for trading in higher value merchandise. In time, it was often these larger 16-ounce merchant pounds that came to dominate trade.

In the early fourteenth-century period in which Pegolotti was writing, a 16-ounce pound was used for merchant exports from Paris, and this was distinct from an older goods pound of 15 Paris ounces, both of which will be discussed later in this chapter. Thus in France, as well as in Scotland and England, fine goods and heavy goods were typically weighed in this period with pounds of 16 and 15 ounces respectively. However, different multipliers were found elsewhere. The weights of the Italian merchants of Florence and Venice, who played a substantial part in north European commerce, were termed *sottile* (or subtle) and *grosso* pounds: in Florence these pounds were of twelve and 16 ounces, but in Venice they were more unexpectedly of twelve and 19 ounces.[66] In Flanders one of the principal pounds recorded by Pegolotti, and discussed below, was a 14-ounce pound.

Here we must emphasise the difference between a formal definition of a pound and a description of its practical use in the market-place. Texts such as the *Tractatus* often define pounds as equivalent to a certain number of ounces; and we have already seen that there were several ounces, such as those of 450 grains and 480 grains, that could act as the 'definition ounces' of such pounds. However, this is not to say that these ounces necessarily formed the practical trading divisions of the pound – used when a weight that was not a whole number of pounds had to be described. Using a 15-ounce pound did not force traders to use factors of five; and worse still, the special problem of division by a factor of seven did not arise with a Flemish 14-ounce pound. Calculation that cannot be readily performed by the ordinary accounting methods and arithmetic notation of the time should be dismissed. Instead, our approach differs from other commentators in arguing simply that normal binary division was used, producing the familiar halves, quarters, eighths and sixteenths of the pound.[67] This can be neatly illustrated by a small and recently-excavated weight from south-east England of about the thirteenth century (Fig. 4.1).[68] This is unambiguously a quarter of a pound of 14 ounces of 472½ grains, but is clearly intended as a quarter-pound weight and not a weight of 3½ definition ounces. Instead it has effectively become a new four 'ounce' weight, where this 'ounce' is the sixteenth of the pound, but its use would be as a quarter-pound weight. A weight of half this size would similarly be an eighth-pound and not 1¾ ounces. It was not until 16-ounce pounds progressively replaced earlier types that the practical ounce at the retail level was the same as the definition ounce.

Fig. 4.1
Medieval quarter-pound
trading weight.
(NMS T.1993.31)

BULLION AND TRADING WEIGHTS IN PEGOLOTTI'S HANDBOOK

The major source of information about European commercial weights of
the late thirteenth and early fourteenth centuries is the writings of the Italian
mercantile agent Francesco Balducci Pegolotti. These are known from the
sole surviving fifteenth-century copy of his *'La Pratica della Mercatura'* in
the Riccardian Library, Florence, which was published in a modern edition
in 1936.[69] Pegolotti worked for one of the principal Florentine merchant and
banking houses, the Bardi company, and he was employed by them first in
Florence, and then successively as manager of the Bardi branches in Antwerp
(1315-17), London (1317-21), and Famagusta in Cyprus (1324-9), remaining
with the company until its collapse in 1346. Pegolotti's book, compiled in
the period to about 1340 but occasionally presenting late thirteenth-
century data, is not a practical merchant handbook in the conventional
sense, but rather the commercial memoranda of a merchant nearing retire-
ment. It contains, however, a wealth of detail about commercial and banking
practice which has been mined to good effect by monetary and economic
historians concerned with currency issues such as bullion rates, fineness of
coin alloy, and the operation of currencies of account.[70] Pegolotti's concerns
were largely with international trade in the various commodities, such as

pounds which survived in Britain until recent times as the 'hundredweight'.[96] The implication of these different hundreds in terms of weighing practice will be discussed later in this chapter. In his handbook, Pegolotti rarely used the word for 'hundredweight', mainly writing '100' pounds, meaning five score, as in the equivalences above.

Pegolotti stated that the Bruges pound was of 14 ounces.[97] This is a somewhat unexpected figure, although as stressed earlier its retail divisions would be sixteenths and not the 14 ounces of its definition. If commercial pounds were to be raised from 14 ounces of the silver mark (480 grains) and of the gold mark (472½ grains), they would be 6,720 grains and 6,615 grains respectively. Although 14 may seem an unlikely multiplier, Flemish 14-ounce standards certainly did exist. A surviving group of three large bronze standard weights with Bruges markings in the Gruuthuse Museum in Bruges has been shown to be based on the pound of 14 gold-mark ounces, and two associated standards are based on a pound of 16 Cologne ounces.[98] They have been tentatively dated by Gerard Houben to about 1350.[99] Both these weight types lasted for a considerable period and were amongst those recorded in about 1800 when comparisons were made with the newly-introduced metric weights.[100] Whereas we can deduce from Pegolotti's use of the 14-ounce Bruges silver-mark pound of 6,720 grains that it was appropriate for external trade, we will see later that the larger 14-ounce gold-mark pound of 6,615 grains was likely to have been a pound for the internal market. At this time we suggest that 14-ounce pounds were characteristic of early Flemish trade influence, although Machabey has also described 14-ounce pounds as being in use in northern France in the late thirteenth century.[101] However, in some applications, such as the weighing of wool, discussed later in this chapter, the 14-ounce Bruges silver-mark pound had already become obsolete by Pegolotti's time: wool had certainly been weighed by the pound of 6,720 grains in the thirteenth century, but by the fourteenth century this had been replaced by a pound of 7,000 grains.

Taking the relationship of 100 Florentine pounds to 78 Bruges pounds, and using the Bruges pound of 6,720 grains, the pound of Florence would contain $6,720 \times 78 \div 100 = 5,241 \cdot 6$ grains. This can be confirmed from the weight of the gold florin, minted at 96 to the pound (64 to the mark), which has been accepted for a number of years as 3·536 g.[102] In terms of medieval troy grains, this is 54·60 grains, implying a pound of 5,241·6 grains, which suggests that Pegolotti's equivalence, if not representing an exact match, is at least very accurate. The pound of 5,241·6 grains is the twelve-ounce *sottile* pound of Florence, for fine and high value goods, and the ounce is therefore 436·8 grains. The heavier *grosso* pound for heavy goods contained 16 of these ounces, making 6,989 grains.

From the equivalence of 100 Bruges pounds to somewhere between 88 and 89 Paris pounds, and taking the Bruges pound at 6,720 grains and the

Paris ounce at 472½ grains, the Paris pound therefore comprises between 16 and 16·2 ounces. Within the accuracy implied by Pegolotti's statement, the pound seems to be one of 16 ounces, or 7,560 grains. As a merchant pound for external trade, it should be distinguished from another pound, of 15 ounces, mentioned elsewhere by Pegolotti, which had the character of an internal pound.[103] This 15-ounce goods pound is likely to have been the earlier of the two, and apparently remained in use as the normal unit of internal trade. An ordinance of Philip IV ('*Le Bel*') in 1307 required the pound to be of 15 ounces, probably reinforcing an earlier edict, and this was reiterated in 1322.[104] We will argue at the end of this chapter that this pound formed the definition basis of the gallons of the David Assize and the *Tractatus*, as well as being the analogue of the pound of 15 Cologne (or tower) ounces described there.

From the equivalence that equates 100 Bruges pounds with 92½ London or Antwerp pounds for spices, we can get a value for the spice pound in London (and Antwerp) – or the 'London pound' as Pegolotti called it – as 7,264·9 grains.[105] In fact this is only an approximation, and we can obtain a more accurate value by using Pegolotti's information that the size of the 'load' of spices in the Paris market was 350 pounds, and that this was the same as the 364-pound spice loads of London and Antwerp.[106] Although there are some significant differences in bulk weighing practices, which we will investigate shortly, in this instance all three cities used the same method of bulk weighing for spice, and so the London spice pound is $7,560 \times 350 \div 364 = 7,269·2$ grains. Had Pegolotti chosen to equate 100 Paris pounds to 104 London or Antwerp spice pounds, the relationship would have been exact, but this presumably did not represent the real direction of actual trade.

We can use a similar argument to obtain the size of the spice pounds for the internal markets in Antwerp and Bruges. Whereas the Antwerp load of 364 pounds was for external trade, the equivalent load for internal trading was 400 pounds, as it was in Bruges, and the pound in both places was of 14 ounces.[107] If we assume that the mass of the internal and external loads at Antwerp was the same (only the pounds differing), then the Antwerp and Bruges internal pounds are of $7,560 \times 350 \div 400 = 6,615$ grains. The three 14-ounce standard weights recorded earlier at the Gruuthuse Museum, Bruges, are presumably pounds from this series.

The London and Antwerp spice pounds described by Pegolotti are unusual because a pound of 7,269 grains does not have a recognisable bullion ounce basis, although there is a link to the Paris pound. Nonetheless, the size of this pound arises clearly from Pegolotti's data, and it appears to have been the weight for wholesale distribution in the London spice market. This market was in the control of the officials of one of the London guild companies, the Pepperers' or Spicers' Company (later the Grocers' Company), and as with guild companies with similar responsibilities (such as the Goldsmiths) they

Fig. 4.2
Early London spice pound.
(Private collection)

retained their own standards, in this case the Spicers' pound and its multiples. A fine example of a one pound weight of this series has survived (Fig 4.2).[108] The current mass of this weight is only three grains less that the size deduced from Pegolotti's data.[109]

As well as giving the London spice load, Pegolotti also described the other bulk unit, the hundredweight of spice, as 104 pounds.[110] Although his interests were largely restricted to the import and wholesale markets of London, and therefore the general goods pound was of little consequence, he did record the size of its hundredweight as 112 pounds.[111] We have already identified this pound in the *Tractatus* and *Fleta* as 15 Cologne (or tower) ounces, so its hundredweight is $15 \times 450 \times 112 = 756,000$ grains. But this is precisely the same as the spice hundredweight, since $7,269 \cdot 2 \times 104 = 756,000$ grains, as well as being the mass of 100 of the Paris 16-ounce merchant pounds of 7,560 grains. There is another significant factor for this hundredweight: we will shortly see that a Flemish pound of 7,000 grains (later known as the English 'avoirdupois' or goods pound) was already in use in London by this time for particular types of material, and 108 of these pounds are also 756,000 grains.

This gives a first indication of a new and unexpected feature of London trade: that separate hundredweights were not necessarily different and there may have been a uniform hundredweight for normal London trade, defined by external reference to a hundred (five score) of Paris merchant pounds. This would bring an obvious practical advantage for shipmasters who required to distribute the burden of their ship's cargo, since the hundredweights marked on bundled goods would be standard units which were independent of the type of goods carried.

THE WEIGHING OF GOODS IN BULK

The wool industry was of great strategic importance to the economies of both Scotland and England until its progressive decline from the fourteenth century, and in both countries it was a major generator of customs revenue for the monarch. Much of this wool was exported to Flanders and in particular through the markets of Bruges. The principal accounting unit for wool was the sack, but in practice the wool was physically made up in canvas bales know as 'serplaiths' or 'sarplars' which typically weighed two to three sacks, and which would have been marked with their weights in 'sacks' and 'cloves'.

A 'clove' or 'nail' was a sixteenth part of a hundredweight, and although the wool hundredweight is not normally referred to, its cloves are nonetheless used as the divisions of the sack.[112] The modern understanding of the size of the English woolsack is that it contained 364 avoirdupois pounds (of 7,000 grains), and the hundredweight of this pound is 112 pounds. The clove would therefore be seven pounds (since $7 \times 16 = 112$), and by convention the stone is two cloves, or 14 pounds. The sack would then be 52 cloves (since $52 \times 7 = 364$). However, if we consider a long hundred of 120 (therefore forming a hundredweight which we would anticipate to be an early one), then the clove would be 7½ pounds ($7½ \times 16 = 120$) and the stone 15 pounds. Stones of this size were indeed recorded in Scotland in the David Assize described in Chapter 1.[113]

In Scotland the sack was defined indirectly in the David Assize as 360 pounds; in fact, this was done through the intermediary of the traditional unit of the half-sack, the 'wey' or 'waw', but as we will establish later the sack's weight can be securely set at 360 pounds. In the English statutes the sack was first defined in 1340, when it was given as 26 stones of 14 pounds, namely 364 pounds.[114] This is the weight given at about the same time by Pegolotti for the weight of the sack in London: it was 52 cloves, each of seven pounds and therefore 364 pounds.[115] However, this is very different from the weight given in thirteenth-century sources such as the *Tractatus* and *Fleta*, where the sack was given as 28 stones of 12½ pounds, which is 350 pounds.[116]

So, how big was the woolsack? The nineteenth-century economic historian J. E. Thorold Rogers wrestled unsuccessfully with the 350-pound size, and then opted for the 364-pound sack but without specifying the type of pound.[117] More recently, where this difference has been noted, the erroneous assumption has been that two distinct sizes of sack were being described.[118] Normally, however, the tendency has been to accept the later size of 364 pounds, although the pound is often incorrectly taken to be the current avoirdupois pound of 7,000 grains.[119] The reason for the apparent change in weight of the sack turns out to be connected with the method of weighing bulk materials at trading centres, and in fact we will see that the sack's mass remained constant and did not change at all.

This is the situation that we should expect. Because of the essentially international nature of the wool market it was imperative that the unit of that trade – the woolsack – was a constant, fixed quantity. It would scarcely be convenient for Scottish or indeed English merchants preparing wool for sale in Flanders to sort and pack it in quantities which differed from the sack as understood in the Bruges market. Equally, if the sack were to be changed in size, all the nations involved in the trade would be affected: it is not plausible that independent action to change the size of an international unit controlled by external markets would be taken by England or Scotland. T. H. Lloyd's economic history of the English wool trade describes wool movements exclusively in terms of sacks, with the clear implication that the sack's weight was uniform, although this is not stated unequivocally nor is the sack's weight given.[120]

Since very early times, all high-value materials – such as coins, precious metals and gems – have been weighed on small equal-arm balances, and the understanding has been that when equal masses (or weights) were placed in the scale pans the beam would settle in a horizontal position. Often this was shown visually by an 'indicator' pin mounted at the centre of the beam, which would align with the vertical suspension shackle when the beam was 'in balance'. With our modern concepts of fairness in measurement, we tend to think that this is the only acceptable way of conducting accurate trans-actions. However, an earlier conventional practice for weighing less valuable materials in bulk was to use a procedure which we may characterise as 'inclined-beam weighing'.

Consider the weigh-house at a civic market where dutiable goods were assessed. A large weighing beam would be supported from above, with its two pans suspended above the floor. The bundled goods to be weighed were placed on the goods pan, bringing it down to the floor. Sufficient weight was then added to the weight pan until the beam just began to lift the goods pan off the floor, but not so much as would bring the beam from this inclined position up to the horizontal. The mass of the goods purchased (or assessed) would be more than the mass indicated by the sum of the weights on the weight pan; the true mass of the goods would only be shown by adding further weights so that the weighing could be said to be by 'level beam'.

This procedure certainly stretches back to antiquity, and there is a clear description, for example, in the Babylonian *Talmud*, the comprehensive gathering of Hebrew law and regulations, which was compiled before about AD 500. This stipulates that the scales were to be suspended to give adequate clearance from the floor so that the beam could incline, and specifically that 'the shopkeeper must allow the provision [pan] to sink to a handbreadth lower than the scale [pan] of the weights, but if he gives exact weight [that is, with the beam horizontal] he must allow him ... one twentieth in the case of dry measure ...'.[121] Thus, with the beam inclined, the sum of the weights would be a twentieth (or five per cent) less than the actual weight of the goods, but this smaller 'weight' would be what the purchaser paid for, at the agreed rate. However, if the beam was to be used horizontally, the same goods would appear to weigh one-twentieth more, and so a five per cent additional payment would be demanded. To make the two weighings just, when the beam was used horizontally the seller was constrained to give the buyer an additional twentieth, or five per cent. In this way the unit cost would be the same regardless of the mode of weighing.

This distinction between inclined-beam and level-beam weighing is clarified in London in 1257, when unsuccessful attempts were made by royal officials to enforce level-beam weighing at the king's beam. The account, in the *Liber de Antiquis Legibus*, may be translated as follows:

> The usual provision is, when goods are weighed on a balance, that the beam incline towards the goods, except gold and silver, which are always weighed on a level beam, neither drawing towards the weights nor towards the gold and silver ... but from the Sabbath following the Feast of St Nicholas [6 December] in the 41st year of the reign of King Henry [III], son of King John, all goods which must be weighed by the King's beam in the City are to be weighed as gold and silver, without drawing towards the goods, in lieu of which the seller must give the buyer four pounds in every hundred.[122]

The allowance of four pounds per hundred (rather than the five pounds in the much earlier *Talmud* description) was called a 'cloffe' allowance, and its use can be illustrated for the woolsack. If a single sack was just lifted off the floor by 350 pounds, and required $4 \times 3\frac{1}{2} = 14$ pounds (4 pounds per 100) to be added to the weight pan to bring the beam level, then the total on the weight pan was 350 + 14 = 364 pounds. Thus the two 'weights' of the sack reflect different measurement practices: 350 pounds is the apparent mass by inclined beam, whereas 364 pounds is the actual mass of the sack. This is shown diagrammatically in Fig. **4.3**. The contemporary suspicion that the king's beam had been altered was countered by the statement that neither the beam nor the weights had been changed – only the form and mode of weighing had been altered.[123]

To illustrate the widespread use of this type of allowance, a French example from the late twelfth century calls for an additional four or five pounds in the hundred as cloffe for weighing by level-beam on a steelyard:

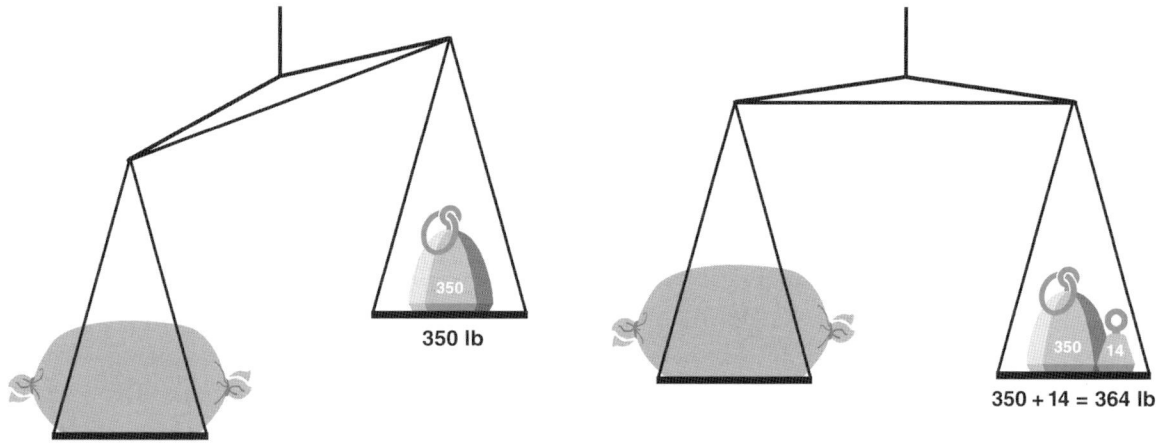

350 lb

350 + 14 = 364 lb

Fig. 4.3
Weighing the woolsack:
by inclined-beam (left),
and level-beam (right).

Le poide de la romaine [steelyard] *appellé communément 'peson'. C'est une sorte d'instrument qu'on appelle aussi balance romaine et avec quoi on pese. Sert pour les marchandises de grand poids, dans les ports il est avantageux pour le commerçant attendu qu'il vend ordinairement 104 ou 105 livres au lieu 100 …*[124]

The practice of inclined-beam weighing continued into the later middle ages, and with the figures provided by Pegolotti's handbook it can be deduced that this method of weighing was used, at least at the wholesale level and for the shipping of goods at the principal north European markets. Similar conventions applied at these centres, presumably with an acceptance that the goods pan should just clear the floor. However, the exception was provided by the weighing of wool in England: although it was initially weighed by inclined-beam, it was such an important commodity that it came to be weighed at export centres on dedicated scales by royal officials, and to ensure the greatest accuracy it was weighed by level-beam.

When Pegolotti gave the sack in London as 52 cloves, each of seven pounds, and therefore a total of 364 pounds, he was describing the official level-beam weighing of the sack – the inclined-beam weight would be 50 cloves, or 350 pounds.[125] However, at other centres where inclined-beam weighing was still in use for wool, as for example at Bruges, Pegolotti also gave the official weight as 60 cloves of six pounds, or 360 pounds, but this time in the appropriate Bruges pound and as an inclined-beam weight.[126] This provides an immediate caution about the interpretation of such figures – we must know the nature of the process before we can extract useful information or make comparisons. In a similar vein, it should be noted that we have not yet identified the size of the pound appropriate for wool weighing in England – but we will shortly find that it was a special pound of 16 tower ounces, and not the expected heavy goods pound of 15 tower ounces.

For an inclined beam to operate successfully, it must have the correct geometry, with the knife edges slightly out of line. The implication from the use of the same weighing conventions in different trading centres is that the great beams were constructed with the same characteristics. However, this would not represent a difficulty because the standard weights that were related by these different weighing procedures were well understood and readily available. It is also likely that many of these large balances would have come from the same manufacturing areas, in the same way that centres such as Nuremberg developed particular expertise in constructing and adjusting nesting weights for much of Europe.

The advantage of using inclined-beam weighing will become clearer when we consider the operation of 'cloffe' and another allowance called 'tare'. We will then be equipped to relate the woolsacks of Scotland, Flanders and England, before finally examining how the early weight units allow us to make sense of the first descriptions of the capacity measures.

To explore the practical operation of the cloffe allowance, we can turn to three important early English declarations which relate to the hundred-weight. The first is in the *Tractatus*, dating probably to the mid-thirteenth century, where the hundred of the 25-shilling pound (the internal goods pound of 15 ounces) was 100 pounds, but

> … a Hundred of Wax, Sugar, Pepper, Cinnamon, Nutmegs, and Allum, containeth Thirteen Stone and a Half, and every Stone Eight Pound. The Sum of Pounds in a Hundred, One hundred and eight Pounds …[127]

A very similar statement is made by *Fleta*, in about 1290, where

> … a hundred-weight of wax, sugar, pepper, cumin, almonds and wormwood contains 13½ stones, and each stone contains 8 pounds.[128]

Thus, the appropriate hundredweight for these comparatively costly imported goods is one of $13\frac{1}{2} \times 8 = 108$ pounds. Although the pound was not defined, it had a stone of only eight pounds, and it is therefore quite unlike the normal English goods pound of 15 tower ounces.

However, a rather different picture is presented by a slightly later English text, the *Ordinacio facta de modo ponderandi per balanciam* of 1309, which specifically mentions level-beam weighing, referring to

> … wax, almonds, rice, copper, tin and such like which are weighed by level beam …
> And that every hundred of such-like bulky heavy goods contain $v^{xx}xii$ [$5 \times 20 + 12 = 112$] pounds.
> And every hundred of small wares and spices such as ginger, saffron, sugar and such like which are sold by the pound contains $v^{xx}iiii$ [$5 \times 20 + 4 = 104$] pounds.[129]

Here the goods previously considered appropriate for the 108-pound hundredweight have been divided into two categories. Heavy goods (which were not specifically mentioned earlier) were to be weighed with a hundred-weight of 112 pounds, and this category included wax and almonds. Pegolotti later confirmed the hundredweight for heavy goods in London was 112 pounds.[130] On the other hand, higher value goods, now including spices (which may earlier have been weighed by a hundredweight of 100 pounds) as well as sugar, were to have a 104-pound hundredweight, and again this hundredweight was confirmed by Pegolotti.[131] The inference is that the 108-pound hundredweight had been displaced in London in favour of those of 104 and 112 pounds sometime between 1290 and 1309.

The *Ordinacio* has a very specific context which is relevant. The 1257 attempt to introduce level-beam weighing at the king's beam in London mentioned above was frustrated, but a cloffe allowance had been declared for moving from inclined-beam to level-beam, and this was four parts in a hundred. By the beginning of the fourteenth century there seems to have been a more general agreement about level-beam weighing for internal trade, but disagreement about the practice for imported goods. In 1303 Edward I proposed to extend to foreign merchants and strangers the privilege of weighing goods in the same manner as London merchants, but the City of London objected in 1305 that all goods coming to London had always been and were still weighed by inclined beam, and noted that the agents of important households, for example those of bishops and barons, also bought imported goods on the same basis.[132] The proposed privileges were cancelled in 1311, but perhaps only as a result of the new provisions of the *Ordinacio* of 1309, which stipulated level-beam weighing and did so by introducing a cloffe allowance. The important point, however, is that there were clearly differences between the operation of the internal market and the import market. It seems reasonable that these differences reflected the pattern in other trading centres and that broadly similar practices for weighing goods in bulk applied elsewhere.

The level beam, or balance, was required when accuracy was necessary, normally because comparatively valuable goods were being weighed in relatively small quantities. But this was also the method adopted for the bulk weighing of wool by the king's officials when the great customs on English wool was imposed in the late thirteenth century, presumably in order to ensure accuracy and acceptance. Initially many of the new regional wool beams, such as that at King's Lynn, were constructed as steelyards (which operated with the arm horizontal and therefore effectively as level beams), but by 1340 official wool beams were high-capacity balances.[133] From the early fourteenth century, equal-armed balances (which could be used level or inclined) were specified for the major market scales, and they were required by statute in 1352.[134]

Over the same period there was increasing pressure on the officials of the City of London to require level-beam weighing for more types of produce sold in bulk, but it remains unclear to what extent this pressure was successful. Many of these fourteenth-century sources indicate that a principal objection to inclined-beam weighing was that it was more prone to fraud, with the weigher able to influence the measurement by holding on to the beam. Whereas previously, the beam had to 'draw towards the better, that is towards the goods bought', now 'the beam of the balance do not bow more to one part than to the other' and the weigher should 'remove hands to show equality'.[135]

So it is likely that level-beam and inclined-beam weighing co-existed in the London wholesale markets over an extended period, and where this applied to different classes of goods it may help explain the relationship of the two types of stone weight (of twelve and 12½ pounds) defined in the *Tractatus*. In general, the implication is that heavy or bulky domestic goods, for which an inclined-beam weight might be expected at an early date, were weighed with the twelve-pound stone; but the same item on a level beam would be the same number of stones but stones of 12½ pounds. The ratio of these two stones is 96:100. In some instances both forms of stone might be permitted, and a portion of the *Tractatus* (and *Fleta*) is indeed given over to describing the equivalence of two different methods of measuring the load of lead – these were by the tron, using twelve-pound stones, and by division into weys, using 12½-pound stones. One can see how a parcel of goods could be weighed by inclined or level beam and the weight expressed correctly as the same number of stones by choosing the stone appropriate to the method of weighing. We will find that the same type of relationship is mirrored in the two principal Bruges pounds, which are also in the ratio 96:100.

The continuation of the practice of inclined-beam weighing at centres of international trade may be linked to the operation of a second type of trading allowance. Most goods had to be packaged in one form or another for shipping, for protection, and also simply to contain them in transit from market to market before reaching the retailer. For example, spices brought into London were made up in canvas-wrapped bales, and at initial purchase after import they could only be weighed conveniently when still packed. The weight of the wrapping had to be considered, that is, a 'tare' allowance had to be incorporated into the weighing process. Weighing by inclined beam provided a convenient method of assessing the contents of packed consignments, but did so on the understanding that the packing materials made up added four parts per hundred to the weight. This conventional level of tare is the one recorded, for example, in the earliest European printed arithmetic text in 1478.[136]

Thus, a bale of goods of true weight 104 pounds on a level beam would be considered (by deducting the tare) to contain 100 saleable pounds capable of sub-division when opened for sale at the retail level. But if the package was weighed by inclined-beam the apparent weight of 100 pounds (the level-

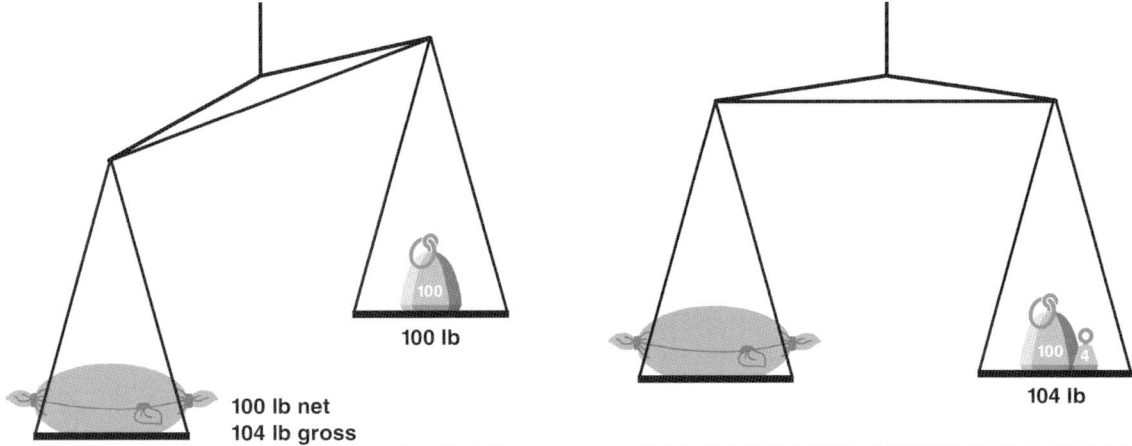

beam weight reduced by the cloffe allowance) would represent the actual net weight of the contents (Fig. 4.4). Although the tare allowance was probably generous, any excess would provide notional compensation to the purchasing merchant for loss and spillage in the course of division. In other words, it would represent additional potential profit for the importing merchant. The size of bales would be more likely to be loads rather than hundredweights, but the use of inclined-beam weighing would enable the weight of the contents to be assessed against the appropriate hundredweight for purchase.

The writer of the *Ordinacio* text was specific that the hundredweights *contained* the stated number of pounds and were therefore net weights, and it seems clear from Pegolotti's figures that he, too, was concerned very largely with the net weight of consignments. From the nature of the relationships between the various European weight systems given above, we can tell that these were direct comparisons, so that the relationships were true equivalences; and by extrapolating from certain weighings described by Pegolotti (such as that of the woolsack in Bruges, described below) that the weights given at these centres were generally by inclined beam. Our assumption, therefore, has been that there was consistent practice at these trading centres for goods at import or export, and that Pegolotti's interest related to transactions where bulk trading was done by inclined beam as a way of making tare adjustment to consignments of baled goods.

Pegolotti's data for London has shown that hundreds of 104, 108 and 112 pounds are all of the same mass, and that this uniform hundredweight was also in use in Paris. However, wool was measured by a separate hundredweight, of 112 wool pounds, each of 7,200 grains. This larger hundredweight can be seen to be 120 (a long hundred) of the 6,720-grain merchant pound of Bruges, the market through which much of the English export wool passed. It is likely that this was an earlier and independent hundredweight dictated by Flemish commercial pressure. The special status of wool in the English economy – as a principal source of royal revenue, with its measurement in the hands of royal officials – makes the use of a separate hundredweight understandable as an exception to the general hundredweight of London. It does, nonetheless, bring with it the caution that there may have been a number of earlier and displaced hundredweights using the long-hundred multiple.

In Bruges the hundredweight recorded by Pegolotti was 100 Bruges merchant pounds of 6,720 grains. This was the same as 96 of the Bruges goods pound of 7,000 grains, which was used in Pegolotti's time for wool weighing. This hundredweight is smaller than the Paris and London hundredweights discussed, but there is a simple relationship between them: the London and Paris hundredweights are 1⅛ times that of Bruges. It seems plausible to suggest the loss of an earlier Bruges hundredweight at 120 merchant pounds, which would also give rise to the English wool hundredweight and account for the origin of the seven-pound English wool clove. But this would not accommodate the 7,000 grain pound, reinforcing the feeling that the 7,000 grain pound was a later introduction. The pound of 7,000 grains is a close approximation (at least in terms of bulk trading) to the Florentine *grosso* pound of 6,989 grains, which must have been a significant influence of north European trade in the thirteenth century, and we will return to this aspect later.

It might similarly be argued that the Paris and London hundredweights, which do accommodate the 7,000-grain pound (with a hundred of 108), may perhaps post-date it. There is certainly no obvious pound with a hundred of 120 – which would have to be a pound of 6,300 grains, or 14 Cologne (or tower) ounces. Although it seems plausible to postulate an early 14-ounce pound of this type in England, there is at present no evidence to support this.

The *Ordinacio* sought to introduce level-beam bulk weighing for the more valuable spices and some other materials as 104 pounds to the hundred. It might be expected that these would previously have been classified as fine materials which would be weighed by the Paris pound. One might therefore expect a hundredweight of 100 Paris pounds, baled at 104 pounds, to be weighed by inclined beam at 100 pounds and divided for resale into 100 units of one pound. If such a consignment was now weighed by level beam at 104 pounds, but still considered as a 104-pound hundredweight, then this is tantamount to weighing in terms of a smaller spice pound whose ratio to the Paris pound is 100:104. The net weight of such a consignment could still be determined by inclined beam as 104 spice pounds. The Paris and London spice pounds have this ratio, and it follows that a consignment of (say) 100 London or Antwerp spice pounds on the inclined beam, or 104 on the level beam, would break down into 100 Paris pounds, measured for re-sale on the level beam. In this case also, the equivalence simplifies the necessary calculation.

THE WOOLSACK IN ENGLAND, FLANDERS AND SCOTLAND

We have already seen that across medieval northern Europe, one of the principal commodities to be traded was wool, much of it initially traded through Bruges to the south. The earliest English definition of the woolsack is in the *Tractatus*, perhaps of about 1250. Two of three surviving early versions give the sack as 28 stone, but perhaps the most informative version gives it as either 28 or 30 stones, where manuscript additions to the text give the two totals in pounds, and in translation the clause reads:

> A Sack of wool ought to weigh 28 stone (that is 350 pounds), and in some parts 30 stone (that is 375 pounds), and they are the same according to the greater or lesser pound.[137]

This alerts us to the fact that at least two distinct pounds were used for weighing wool in England in the thirteenth century.

The addition of the total weights confirms that both stones are of 12½ pounds, and since both totals clearly result from inclined-beam weighings, the ratio of the pound sizes is 30:28, and hence 15:14. The factor of 14 in the 'lesser pound' indicates that this is a Flemish pound, whereas the 'greater pound' must be a recognisably English pound since it is the familiar pound of which 350 give the inclined-beam weight of the sack and 364 the level-beam weight. When Pegolotti gave the level-beam weight of the sack of 'London and all of England', he described the appropriate pound as the 'English pound', clearly differentiating it from the 'London pound', which was specifically a spice pound.[138] The Flemish pound that fits this equation is the early Bruges merchant pound of $14 \times 480 = 6{,}720$ grains. The inclined-beam weight of the sack was therefore $375 \times 6{,}720$ grains, and dividing this by 350 gives the English wool pound as 7,200 grains. This is 16 tower ounces of 450 grains (thus equivalent to 15×480), where the English goods pound is 15 ounces of 450 grains, giving the wool pound a character arguably similar to that of the Paris export pound of 16 ounces.

Thus we have the interesting situation that Bruges weight was being used in at least some parts of England for the packing of wool, presumably destined for the Flemish market. One of these areas may have been the English west-country: the accounts of Beaulieu Abbey in the New Forest are available for the year 1269-70, and wool information is given only in the Flemish unit.[139]

Considering the influence that Florentine wool traders had on the establishment of the vigorous European wool trade, it might be expected that the size of the common trading unit for wool would match an early, appropriate and prominent Florentine weight unit, and indeed it has been argued that this was 500 Florentine *sottile* pounds.[140] This is the twelve-ounce pound used for fine goods (and therefore for weight definitions), which has already been identified from Pegolotti's data as of 5,241·6 grains. The official scale in Florence was the steelyard, and so in effect they were using level-beam weighing. The weight of such a sack does indeed equate to a level-beam weight of 364 English wool pounds (500 × 5,241·6 = 364 × 7,200), or 350 pounds by inclined-beam.

If the *Tractatus* dates from about the mid-thirteenth century, the weight values interlined in the text were probably added in the second half of the century, because by Pegolotti's time wool was already being weighed in Bruges by the pound of 7,000 grains, in preference to that of 6,720 grains. This follows from his statements that the sack in Bruges was the same as that in London, and that the former was of 60 cloves, each of six pounds, making a total of 360 pounds.[141] Dividing the inclined-beam weight of the sack by 360 gives 7,000 grains.

This 7,000-grain pound was also being used for a time to weigh wool in England. Its introduction can be traced through an anonymous estate remembrancer called *Seneschaucy*, which has close similarities to a better-known document attributed to Walter of Henley, although the latter does not have the *Seneschaucy*'s clause on the sack, which states that:

> He [the bailiff] ought to sell the wool by sacks or by fleeces according to what he considers to be of greater profit and advantage. If he sells by sack, each sack ought to weigh 30 stones of wool measured in cloves or twenty-eight stones measured in stones on the scale, correctly measured by the correct stone of 12½ pounds.[142]

In this form it is a modified version of the *Tractatus* definition, but it emphasises a distinction between calculation in 'cloves' (sixteenths of the hundredweight) of the Bruges pound and 'stones' of the English pound.

If the Bruges merchant weight was also in use in England, measurement in cloves would be natural. For the Bruges pound of 7,620 grains, the stone was 12½ pounds, the clove 6¼ pounds and the hundredweight was therefore 100 pounds. For the goods pound of 7,000 grains, the stone was twelve pounds, therefore the clove was six pounds and the hundredweight was 96 pounds. The hundredweights were the same (100 × 6,720 = 96 × 7,000), as were the sack definitions at 30 stones or 60 cloves (since 6¼ × 6,720 = 6 × 7,000).

There are dozens of known manuscripts of the *Seneschaucy*, and minor variations between these have been analysed to establish a copying sequence.[143] All the texts in one group of manuscripts, deduced to be amongst the latest copies, use a stone of twelve pounds for a sack defined by cloves (as 60 cloves), generating an alternative view of the sack as 360 pounds in cloves, or 336 pounds in stones.[144] The 360-pound sack is in Bruges goods pounds of 7,000 grains, as above. This indicates that the transition at Bruges from the pound of 6,720 grains to that of 7,000 grains (at least for wool), occurred between the 1270s (the likely date for these late versions of *Seneschaucy*) and about 1340 (for Pegolotti's *Pratica*). The sack of 336 pounds uses English wool pounds of 7,200 grains (since $360 \times 6,720 \div 336 = 7,200$), but it was used in a stone of twelve pounds. We shall see that the twelve-pound stone for wool is mentioned again in 1281, when it was abolished (in some areas at least) in favour of one of 13 pounds.

The fact that the *Seneschaucy* calculation for the sack had to be performed 'in stones' of the English wool pound suggests that this stone did not show the traditional relationship of two cloves equalling one stone. Because of the particular requirements imposed by the transfer from inclined-beam to level-beam weighing for wool, it is not clear how the clove of the English wool pound would have been defined. In Pegolotti's time the clove was seven pounds and the wool hundredweight was therefore 112 pounds, but the first indication that has been found of the use of a seven-pound clove was in 1288.[145] Although this reference comes from customs accounts with an implied sack of 364 pounds and therefore relates to level-beam weighing, it should not be assumed that the 14-pound stones (only formalised in the English parliamentary act of 1340) were yet in use.

It is suggested, therefore, that when the *Seneschaucy* was first written in the 1260s, the clove of the wool hundredweight was already being considered in some areas as seven pounds, so that the stone of 12½ pounds was clearly less than two cloves. A clove of seven pounds signifies a hundredweight of 112 pounds; but no hundred of 112 was recorded in the *Tractatus* or in *Fleta*, and this hundred is first noticed for traded goods in the *Ordinacio* in 1309, governing goods to be sold by level-beam. If a hundred of 112 was exclusive to the weighing of wool before this date, it provides a further indication that the wool pound was established to fit an existing (Flemish) trading unit. This unit must be a natural multiple of the Bruges merchant pound seen in the earliest English definitions of the sack and, in fact, a hundredweight of 112 English wool pounds is equivalent to 120 of the old Bruges merchant pounds ($112 \times 7,200 = 120 \times 6,720$). The early use in England of a wool hundredweight comprising a long-hundred of the dominant external pound has a reassuring ring.

A move to a 13 pounds stone is stipulated for Leicester in 1281:

> It is determined by the whole community that henceforth none may weigh wool by any 12 pound stone but it must weigh 13 pounds fully and if anyone be found hereinafter who has a 12 pound stone he shall pay the community of Leicester half a marc [six shillings and eight pence].[146]

An advantage of using a stone of 13 pounds instead of one of 12½ pounds was that this incorporated the four pounds per 100 pounds cloffe allowance for level-beam weighing (13:12½ is equivalent to 104:100), whilst retaining the weight of 28 stones to the sack specified in the earlier documents ($28 \times 12\frac{1}{2} =$

350, but 28 × 13 = 364).[147] In due course this was changed to the more familiar 26 stones of 14 pounds in 1340 (28 × 13 = 26 × 14), establishing in law for the first time the expected 2:1 relationship between the stone and the seven-pound clove.

Although the Bruges goods pound of 7,000 grains was certainly used for inclined-beam weighing of wool in England, it was not readily adaptable for level-beam weighing of the sack because then there was not a convenient or exact number of pounds in the sack's mass.[148] It was probably to prevent continued application of this type of weight for the sack (necessarily using inclined-beam weighing) that the twelve-pound stone was outlawed in the 1281 Leicester regulations. Eventually, in England in 1312, wool (together with victuals) was specifically excluded from items weighable as heavy goods (*averia ponderis*) – such as metals, dyes, wax, drugs, and spices – and was placed outside the scope of Flemish weight.[149]

The English wool pound of 7,200 grains does not seem to have had an extended history, as is probably indicated by an absence of the traditional relationship between its stone and the clove of its hundredweight. It is conceivable that this pound was defined in connection with the raising of the duty from wool at the introduction of the general wool customs of 1275. The surviving *Tractatus* documents, which date from some time after the supposed compilation in about 1250, have presumably had a domestic definition added to complement an original Flemish declaration. The use of a separate pound for wool does not appear to be specified anywhere in the surviving documents of the period. However, this was presumably because the practice was well enough understood, and because wool weighing operations were in any case conducted by separate officials and sufficiently apart from other transactions to avoid confusion. The special status of such a pound, as a unit for royal taxation, may perhaps account for the use of a 16-ounce or two-mark pound. Although no standards of this size now exist, there is some evidence that there may have been standards held at the Exchequer in London. At the time of an 'inquisition' into official weights and measures held in 1574, a small group of very early standards weights marked with a crowned 'E' were examined against the current Guildhall avoirdupois (7,000 grain) standards. It was assumed at the time that this device represented Edward III, which has subsequently been accepted as the period of the 56-pound weight of this group. These weights included standards of 28, 14 and seven pounds, each of which were respectively 201, 184 and 195 grains in the pound heavier than the equivalent London weights.[150] On the face of it, these seem very likely candidates for official wool weight standards, to a pound of 7,200 grains.

English wool was gathered from the rural producers by specialist wool-packers, middle-men acting for the major wool merchants, members of a highly-skilled profession with their own craft guilds.[151] Their responsibilities included the careful sorting and grading of wool into the price categories recognised by the international market. The sorted wool was compressed and wrapped into canvas bales in which it was transported and sold. It was often subject to further sorting and weight adjustment at later stages, with each bale being marked with suitable identifiers. The confidence of the market and the reputation of each participant in the trading process depended on the quality and honesty of these actions. These compressed bales were transported, weighed, passed from market to market and sold without unwrapping, the contents only being opened and inspected for retail and manufacture. Although we know more about the process as it operated in England, Scottish wool must have been subject to the same type of scrutiny.

Alison Hanham's analysis of the records of the Cely family of English staplers, or wool merchants, exporting through the English staple at Calais in the late fifteenth century, shows just how complex were the details and economics of this trade.[152] The wool was transported in large parcels or 'sarplars', containing usually between 2 and 2¾ sacks of wool, but reckoned in sacks and cloves, and it is still useful to think in terms of the traditional sack unit. The weighing by the customs officers was almost the last operation before shipment, since warehoused wool lost weight in drying, and therefore the bales were necessarily already packed and sealed. It follows that the recorded weight was the gross weight and considerations of cloffe and tare indicate that this should be 364 pounds, or 52 cloves each of seven 'English' pounds. In contrast the net weight of wool, indicated by inclined-beam, was 50 cloves or 350 pounds. English merchants continued to use the inclined beam, but in 1340 the woolsack was formally defined as 364 pounds, and inclined-beam weighing for wool (as for other produce in bulk) was progressively eliminated.

Although nothing in Pegolotti's account cautioned that the official weight given for the English woolsack was a gross weight, we can find support for this view in the details of the Celys' transactions, albeit for a period 150 years later. Implicit in the figures analysed for shipment weights recorded by customs officials in the 1480s is a rebate of two cloves per sack-weight, which was allowed to the merchant for the canvas covering. So, in practice, the weight of wool per woolsack liable to customs dues was only 50 cloves, or 350 pounds.[153] In line with this is the rebate of four cloves per sarpler for the canvas and turn of the scale (described as 'canvas and draught') allowed in sales between the gatherer, packer and stapler.[154] It did not matter if some definitions for the woolsack gave a gross weight – any price compensations the market considered necessary were readily made.[155] What was important was uniformity in measurement.

It must be emphasised that there was no change to the mass of the woolsack in moving from inclined-beam to level-beam measurement. However, two separate features had altered. First, a cloffe adjustment of four parts per hundred had been invoked, and this was represented in the standard characteristics of north European inclined beams. Second, a tare adjustment of the same size had to be made, since the weight was the gross and not the net weight of the goods.

From the much later time of the formal adoption of the 7,000-grain pound as the English statutory 'avoirdupois' pound in the sixteenth century, the woolsack has been recognised as 364 pounds of 7,000 grains. This does not represent a reduction in the size of the sack, even though this pound is significantly smaller than the pound of 7,200 grains – rather, we are seeing a return to a net definition of the sack's contents. The difference between 364 pounds of 7,000 grains and the gross sack weight of 364 pounds of 7,200 grains is about three per cent. This is very close to the actual tare value of the canvas recorded in a late fifteenth-century English instance as 28 pounds per sarpler (3·3 per cent at an average of 2 to 2¾ sacks to the sarpler, although this also includes a proportion for wastage), but it matches a contemporary Italian statement.[156]

Having looked at the evolution of the English woolsack and its Flemish counterpart, we must now establish that the early Scottish regulations do indeed demonstrate a sack of the same size, weighed in Bruges units. In fact there are few early Scottish definitions of the woolsack; for example, it is not specifically recorded in David's Assize, although there is a partial description in terms of the 'wey' or half-sack. The relationship between the wey and the sack emerges clearly from rules governing local tolls and customs levied by the burghs, and known collectively as the *Assisa de Tolloneis*, which is believed to be of the early fourteenth century but probably with an earlier basis.[157] The toll for wool is as follows:

> of a last of wol that is to say for ten sekkys [sacks] gaddryt togyddyr [together] aucht peniis [eight pence], of a sek of wol four peniis, of a waw [wey] of wol that is to say half a sek, twa peniis.[158]

Thus in both England and Scotland there were two weys to the sack, although a difference emerges for the last of wool, which is here given as ten sacks in Scotland against the English twelve sacks specified in the *Tractatus* and *Fleta*.[159]

The same relationship of the wey to the sack is found in another early Scottish commercial source: a rare and unusual vernacular manuscript which was apparently prepared as a ready reckoner for use by a Scots merchant trading to Flanders and dateable from internal currency references to about 1400.[160] The first commodity described is wool:

> This is the Reknyng of the Scotis woll the quhilk [which] is roght [made] be the marc … Item vj lib. of woll mak the nayle [or clove] and xxx nayle makis the walle [wey] & ij walle makis the sec.[161]

This represents the Scottish exporting merchant's view of the Bruges wool market at the very end of the fourteenth century (with a sack of 360 Bruges pounds); and the tabulated prices which follow are in terms of the clove or nail (six pounds) of the relevant Bruges hundredweight of 96 goods pounds (16×6). The reference to the wey is either to the Scottish unit (in which case it confirms the information in the *Assisa de Tolloneis*), or it indicates that the same unit name was used in Flanders. Whereas Pegolotti does not specifically use the word, his editor draws attention to 'the Flemish waghe, similar to the English unit wey'.[162] In either instance the significant feature is that the wey again emerges as the half-sack.

While the David Assize does not actually mention the sack, it gives the wey as twelve stone and each stone of 15 pounds for a total of 180 pounds. But this can only be established securely when an ambiguity in the text can be resolved. In the nineteenth-century 'Record Edition' of the Scottish statutes edited by Thomas Thomson and Cosmo Innes, the Latin and Scots versions of the text differ. The Latin clause reads (in translation):

> Of the weight of the stone. Item, the stone for wool and other heavy things ought to weigh 15 pounds. Item, the wey ought to contain 12 stones, which [stone] weight contains 8 pounds.[163]

The version in Scots reads:

> Of the wecht of the stane. Item. The stane for weying of woll and uther geir aw [ought] to wey xv pund. Item the stane of wax aw to conteyne. viij pund. Item. the vaw [wey] aw to conteyn – xij – stan.[164]

These two versions appear to present a confusing picture. In both versions the wey is given as twelve stones, and the stone of wool (and other heavy goods) as 15 pounds; but only the Scots text introduces the stone of wax as eight pounds, and only the Latin text notes that the stones of the wey should be of eight pounds.[165]

Some of this confusion is removed when we look at the sources for Thomson and Innes's text; and aspects of the relationship between texts such as these is discussed more closely in Chapter 5. The David Assize of weights and measures was earlier considered as part of the old laws associated with the *Regiam Majestatem* and it appears in two places in John Skene's first printed edition in Latin of 1609 (from which the Scots edition, also of 1609, was translated), in two slightly different forms. One is included as legislation of David I (but only in the Latin edition); and the other is placed under Robert III and dated at about 1400, no doubt because of an internal reference to Robert's reduced coinage of 1393.[166] An interesting addition found in the Robert III version is the statement that 'twelve London pounds make a stone', not present in the text attributed to David but echoing the twelve-pound stone found in some versions of the *Seneschaucy*.[167] However, the fact that three separate stones are mentioned means that there may similarly be more than one type of wey. Both versions of Skene's text have the wax stone as eight pounds and the wool stone as 15 pounds, and both say that the wey is of twelve stones of eight pounds but without specifying the material to be measured by the wey.

It seems clear that both Latin and Scots versions in Skene's volumes have to some extent been condensed, and two separate issues appear to have been conflated. First for wool, the stone should be the conventional 15-pound stone; the sack comprised two weys and therefore contained $2 \times 12 \times 15 = 360$ pounds. The half sack, or wool wey, is therefore 180 pounds. Second, wax was also sold in bulk by the wey – it was imported into Scotland for good quality candles and for writing tablets. The *Tractatus* reported that wax was sold by an eight-pound stone in London, and this is likely to reflect practice in Bruges, and therefore also have influence in Scotland. Presumably the clause originally specified the wey of wax in stones, and the stone of wax as eight pounds; alternatively it was a reminder that for goods sold in eight-pound stones rather than the 15-pound stone of the assize, the wey was still to be twelve stones. The text has then been condensed so that the eight-pound stone can be interpreted as applying to the wey for all goods, and therefore also for wool, whereas in practice we must assume that it was implicit that the wey of wool in Scotland was twelve stones of the appropriate wool stone of 15 pounds as in the assize. We might conjecture that the Scottish text should have read (in modern English):

> The stone for wool should be 15 pounds. The wey is of 12 stone [180 pounds] and the sack is of 2 weys [360 pounds]. The stone for wax should be 8 pounds, the wey of wax is of 12 stone [96 pounds].

This mass of 96 pounds suggests the anticipated link with Bruges, in that this proposed Scottish wey for wax was 100 of the Bruges thirteenth-century pounds, appearing in Scotland as 96 Bruges pounds of the fourteenth century ($100 \times 6,720 = 96 \times 7,000$ grains). The wey generally would be of twelve stone with the stone a different number of pounds according to the nature of the goods.

But by about 1400 the Scots merchant handbook was describing a very different measure for wax: '… and vj li. makis the nayle, & thretty nayle makis the walle [wey].'[168] This shows that the Bruges, and therefore presumably also the Scottish wey of wax was 180 pounds. Pegolotti also confirmed this larger size.[169] This shift from our proposed 96 to 180 pounds could easily have been effected by redefining the stone for wax from one of eight pounds to one of 15 pounds, and possibly this change has been the root of the confusion in the Scottish declarations of the early woolsack.

However, the important point about the declaration in the David Assize is that the woolsack in Scotland can be identified as weighing 360 pounds, from which we can deduce that wool was reckoned by the Bruges pound of 7,000 grains, and the measurement made on an inclined beam. This must have been a natural consequence of sending virtually all Scottish export wool for sale in Flanders. It is likely that originally the Bruges merchant pound was used for wool, as it had been in England, and that it was replaced by the goods pounds in the second half of the thirteenth century. In this sense, therefore, we can conclude that the surviving version of the David Assize has already been revised. Scottish produce will have continued to be weighed by inclined beam for as long as this was the normal practice in Flanders, and there may not have been centralised pressure to move away from the 7,000-grain pound as there was in England.

THE AVOIRDUPOIS POUND

Pegolotti's handbook has shown a number of significant links between north European weight systems; and the commercial pressures of the international market in this region, developed initially by Florentine merchants and financiers, can be recognised in a number of ways. Weight units are found to be linked in simple mathematical ratios, and units associated with dominant centres are found to have penetrated the metrologies of their trading partners. Bulk trade was conducted in each centre using large weight units (hundredweights and loads) which were characteristic of that centre but may have differed for internal and external trade. There are simple relationships between these bulk units, and some centres used the same units. A particular and important case is the woolsack, which can now be seen to be a stable international unit of constant weight. Goods in bulk were measured by inclined beam at this time, although often divided up and resold in smaller quantities by level beam, and the large beams of the north European centres were constructed to have the same controlled characteristics. The earliest Scottish burgh beams or 'trons' must have shared these characteristics and may well have been brought from the Low Countries, although there are no details of such purchases until a very much later period. These features improved the convenience of the market for merchants and simplified the processes of computation, for example allowing tare adjustments to be incorporated and relating quantities which had been weighed in different weight units.

The fact that these weight systems are intimately related may mean that key aspects were introduced at the start of this period of international trade, perhaps for the convenience of Florentine merchants. Indeed this may be the implication of the woolsack matching a conventional Florentine weight of 500 *sottile* pounds (of twelve ounces), and necessarily as a steelyard or level-beam weighing. In terms of the *grosso* pounds (of 16 ounces) in which Florentine bulk trade was conducted, the woolsack's gross weight is 375

Fig. 4.5
The relation between the
grosso pound of Florence
and the Bruges pound,
where the one is 1·04
times the other.

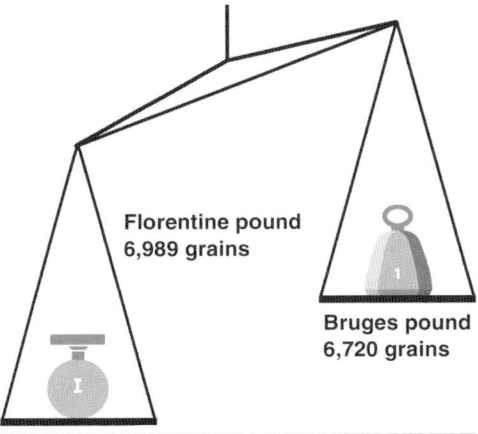

pounds (500 × 12 ÷ 16). This is exactly the same number of pounds that was recorded in the *Tractatus* for the woolsack weighed by inclined-beam in Bruges merchant pounds. This relationship arises purely from a cloffe allowance that depends on standard characteristics for the north European beams and from the application of an equal tare allowance.

In its simplest form the cloffe allowance links the Bruges and Florentine pounds directly, through inclined-beam measurement, and the link depends only on setting the amount of the cloffe and designing the beam appropriately. Thus goods weighing 100 Florentine *grosso* pounds by level beam would also weigh 100 Bruges merchant pounds by inclined beam (Fig. 4.5). So striking is this relationship that it is tempting to suggest that it is not coincidental and that the Bruges pound was initially set at this level to facilitate the strategic trade established by the Florentine adventurers. Once the procedure had been established, it would have to be adopted back along the Bruges supply routes.

If this postulated metrological link between south and north Europe is correct, it might be expected that a direct equivalence of the Florentine *grosso* pound (of 6,989 grains) would also be incorporated into Bruges market practice. It is suggested that this is the Bruges pound (later to be adopted as the English 'avoirdupois' pound) of 7,000 grains, which has been set at a value of less than 0·2% higher than the Florentine weight. In terms of bulk transactions, such a small difference would be immaterial, and it may have been much more important to establish a compromise pound that bore an exact relationship with the Bruges merchant pound. In particular, the pound of 7,000 grains fits the hundredweight of 100 Bruges merchant pounds and the clove of this hundredweight is an exact number of pounds (six pounds). It also gives a convenient fit for the inclined-beam weight of the woolsack (at 360 pounds) with binary factors (which the *grosso* pound cannot provide). It forms an exact factor of the Paris and London hundredweight, with a hundred of 108.

The pound of 7,000 grains had replaced the earlier merchant pound of 6,720 grains for weighing wool in Bruges by Pegolotti's time, and indeed we have seen that it was already being used in England for weighing wool (presumably destined for Bruges) in the 1270s. But what sort of definition basis would the English have thought appropriate for such a pound? The earliest English gallon definition, in the *Tractatus*, and the original David gallon of Scotland, both seem to be based on definitions that use the French pound of 15 Paris ounces (and this is discussed in the final section of this chapter). The English goods pound was of 15 ounces, and even the thirteenth-century English wool pound, which emerges in the *Tractatus*, could be considered as a pound of 15 ounces of the Bruges silver mark (later 15 troy ounces), and therefore compatible with the general definition in the *Tractatus*. Perhaps the 7,000-grain pound was also initially considered as a 15-ounce pound in England?[170] If so, its definition ounce would be 466⅔ grains.

The likelihood of extensive penetration of this Flemish pound into the English internal market provides a possible context for the otherwise unexpected reduction of the mass of the English penny in 1280, when the number coined to the tower pound was increased from 240 to 243.[171] Previously 21 pennies weighed a Paris ounce of 472½ grains. But now 21 of the new reduced-weight coins, minted at 243 to the pound, would weigh 466⅔ grains. Presumably the principal reason for the reduction of the coin mass was fiscal and it may therefore only be coincidence that this relationship exists. However, it is conceivable that the justification for this very specific but small reduction in the coin weight may to some extent reflect an official recognition of the increased trading importance of the Flemish goods pound, with implications for a move away from dependence on Paris standards for English definitions.

There is an indication from Pegolotti's time that an ounce of this size had a bullion association in London. The Goldsmiths' Company of London had a weight standard – known simply as 'Goldsmiths' weight' – which regulated the work of its members. Pegolotti recorded that bullion had to be traded by the tower mark, but the Goldsmiths' mark, which was heavier by 5⅓ sterlings, was used only for fabricated silverware – 'all vessels etc. of silver with which people might have dealings with the goldsmiths'.[172] Since the tower mark was 3,600 grains and the sterling 22½ grains, the Goldsmiths' mark was 3,600 + 120 = 3,720 grains, and the ounce 465 grains. However, the weight of coins was allowed to vary within well-defined margins, known as the 'remedy', which was probably about one part in 240, or nearly two grains in the ounce. An ounce standard that represented the minimum possible weight of 21 minted coins (namely about 465 grains) would ensure that no goldsmith was disadvantaged in the sale by weight of silver produce.

By 1400 the sixteenth of the pound of 7,000 grains was recognised as the operative ounce (of 437½ grains).[173] Also by this time there is reliable information about the size of the Scottish wine gallon, which is almost certainly defined from an external unit in terms of pounds of 15 of these ounces. This later phase will be discussed in Chapter 6.

THE WEIGHT BASIS OF THE EARLY CAPACITY MEASURES

Having demonstrated some of the underlying relationships between the weight systems of a number of European trading partners, it can be appreciated that the early metrologies of both England and Scotland depended to a significant degree on those of the principal markets of France and the Low Countries. By extension, we would expect to find somewhat similar associations between systems of capacity measures. These might reflect similarities in the methods of defining measures and in the custom and practice of their use in trading transactions; but since there was also a natural tendency to reckon volumes of material in terms of the weight contained in the measuring units, there are good reasons to believe that relationships between certain sorts of capacity units should exist because the weight systems were related. The specific instances that will be investigated here are the earliest definitions of the English gallon in the *Tractatus* and of the equivalent Scottish gallon in the David Assize.

Capacity measures in both England and Scotland fall into two groups. Smaller vessels, usually up to and including the gallon, were for liquid use, although there were certainly more substantial vessels (such as barrels of various sizes) for transporting liquids in bulk. Dry goods – normally wheat and other grains – were reckoned in larger units than the liquid measures, and these larger units (the bushel and quarter in England, the firlot and boll in Scotland) were themselves defined in terms of the smaller liquid units. Since the liquid measures were specified by the weight of their contents, the whole system of capacity measure has a weight basis. The nature of this weight basis will be considered partly in this chapter, but some of the discussion must wait until Chapter 6 where customary practice such as the heaping of dry measures is considered in more detail.

The essential difference that emerged between the English and Scottish metrological systems when they became more clearly codified in the fifteenth century was that the basic English capacity measures were based on the weight of their dry contents, whereas those of Scotland were based on the weight of water contents. Thus, for example, the gallon of the English assize of 1496 was defined as containing eight troy pounds of wheat, and the firlot of the Scottish assize of 1426 contained 16 times the pint's volume of wheat.

The earliest English capacity definition, in the part of the *Tractatus* dateable to the third quarter of the thirteenth century or earlier, is that the gallon contained eight pounds. The most accessible version reads (in translation):

> Eight Pounds make a Gallon of Wine; and Eight Gallons of Wine make a Bushel of London; which is the Eighth Part of a Quarter.[174]

However, there are a number of variants of this passage: some thirteenth and fourteenth-century manuscripts, for example, do not describe the gallon as a 'gallon of wine' but leave it undefined, and others are specific that the pounds are to be pounds of wheat.[175] However, the intention of the declaration appears clear: the gallon described is a gallon *for* wine, and not one containing eight pounds *of* wine. Thus, the measuring medium must be either water or grain.

The situation is complicated by the fact that the gallon from which the English bushel was raised was not defined adequately by statute until 1496, when it was defined as containing 'VIII pounds of wheat of Troy weight'.[176] The practical form of this 'statute gallon' became known as the 'corn gallon' and it was specifically a dry gallon. The volume of a given weight of wheat is known within reasonable limits, but clearly the result depends to some extent on the dryness and quality of the grain and the conditions under which it is measured, so that it can never be an exact figure. It turns out that the gallon's volume can be related to the stated weight of wheat in troy pounds (in fact it would contain nine pounds), which is considered to include an extra allowance of one-eighth (and this is discussed in Chapter 6); but since the weight unit is the troy pound, this definition presumably dates from after the official introduction of troy weight about a hundred years beforehand.[177]

Before the formal introduction of troy weight the dry gallon must have been defined in a different manner, but the similarity of the 1496 Act and the *Tractatus* declaration indicates that the formulaic description of a gallon as containing eight pounds of weight was of very long standing. The use of different measurement processes and different pounds could lead to the generation of separate gallons, and it may be that the *Tractatus* declaration allows for both wet and dry gallons: this might explain why some versions of the text relate to wine gallons and others to pounds of wheat. The type of pound is not specified and so in this respect at least the declaration is incomplete and relies on tacit knowledge of measurement practice. Certainly our experience of interpreting the *Tractatus* has been that several different pounds (such as the wool and spice pounds) are involved in the declarations but not distinguished in the text. Confirmation that separate measurement units were considered appropriate for dry goods and types of liquid can be found in *Magna Carta* (in 1215), which mentions 'one measure of wine … and one measure of ale and one measure of corn, that is the London quarter …', and implicit in this description is the understanding that the large practical units of measurement (such as the corn quarter) were multiples of different gallons appropriate to the particular material being measured.[178] These three separate gallons for wine, ale and corn remained in parallel use until at least the early nineteenth century, when they were replaced by a uniform gallon, the Imperial gallon of 1824.

The *Tractatus* text given above relates specifically to a wine gallon, and as such it would be surprising if it were not given in a liquid determination. Support for this can be found in the equivalent Scottish declaration, in David's Assize, where the gallon is clearly also a liquid measure. The clause in the nineteenth-century Record Edition of the *Acts of the Parliaments of Scotland* reads:

> Item the gallon aw to conteyn · xij · pundis of watir that is for to say iiij pundis of salt watir of the see iiij pundis of standande [standing] watir and iiij pundis of rynnand watir.[179]

The careful description of the mixture of three types of water (four pounds each of sea, loch and river water) suggests an early, practical and exact definition. Although a water determination was also used at the 1426 Assize for the pint, this measurement was performed with fresh water from the river Tay, reinforcing the feeling that the use of 'divers waters' in the David Assize characterised a much earlier determination. It seems likely that in the definition of the gallon in David's Assize we can see an echo of the twelfth-century English practice for liquid measure.

The status of the gallon definition clause in the David Assize appears problematic because Thomson and Innes included it only in the Scots and not in the Latin text published in the Record Edition (see Appendix A.1 and A.1a). However, this was probably because it does not occur in the early fourteenth-century Ayr manuscript which was their principal source for the Latin text of the parts of the 'Auld Lawes'. It has already been shown in Chapter 1 that the version of the David Assize transcribed in the Ayr manuscript is rather selective in its coverage, so perhaps in this instance the editors were being over-cautious. The clause is certainly in the late fourteenth-century Cromartie manuscript.[180] It was also accepted by James Balfour, and John Skene included it in his published edition of the *Regiam Majestatem* in 1609.[181]

The boll of the David Assize comprised twelve gallons, and here another small difference emerges between the texts. Again this declaration is missing from the Ayr manuscript, but it is present in the Cromartie manuscript (and in Skene's printed version), where the text is specific that the gallon is a liquid gallon: *'Bolla debet continere in se xii lagenas aquae.'*[182] However, in both Scots and Latin versions of Thomson's Record Edition, the boll is said to contain a *'sextarius'*, described as twelve ale gallons.[183] The distinction between the Scottish ale and wine gallons is discussed in Chapter 6, where it is suggested that this description of the David gallon as an ale gallon may reflect four-teenth-century, rather than twelfth-century, usage.

In addition to this description of the types of water necessary for defining the gallon, the David Assize also provides the dimensions of physical measures of the gallon and boll. The volumes of these vessels were calculated in Chapter **1** as about 330 cubic inches (in^3) for the gallon and about 3,660 in^3 for the boll; or, in the case of the slightly different dimensions for the boll in the Cromartic manuscript, as about 3,820 in^3. Adjusting the stated circumferences of the boll by half-an-inch (the apparent accuracy to which the dimensions are given) changes the calculated volumes by about 50 in^3 and indicates the range of precision implied by these figures. At twelve gallons to the boll the gallon would be expected to be about 305 in^3 (3,660 ÷ 12 = 305), or about 318 in^3 in the case of the Cromartie dimensions. Both these implied sizes are less than the volume given for the gallon, but they are close enough to the gallon's volume for them to be appreciated as reasonably consistent with it. The differences will be explained in terms of allowances that are specified and are seen to operate at a later date.

There is no equivalent dimensioned information for early English capacity measures. Instead, we must rely on direct measurement of the few surviving English standards, of which the earliest are the 1497 Exchequer standards of the corn gallon and bushel. (The sizes of these standards are discussed in Chapter **6**.) Thus, there is no obvious way of establishing the size of the gallon (or gallons) of the *Tractatus*. However, by taking the analogy of the Scottish water-based weight for the gallon in the David Assize, and by using the volume of an appropriately early English wine gallon, we can seek a pound such that eight pounds of water would fill the gallon with precision. Clearly this could have no historical basis if the pound was not a significant pound: it would have to be a pound of suitable status for establishing the definition of a measurement standard.

When the Imperial gallon (of 277·3 in^3) was introduced in Britain in 1824, it replaced three principal gallons which were then in current use in England. These were the corn or Winchester gallon, then operating at 272 in^3 (rather than the 268½ in^3 of the 1497 Exchequer standard), the wine gallon of 231 in^3 and the ale gallon of 282 in^3.[184] The corn gallon will be dealt with in Chapter **6** and we will concentrate here on the wine gallon.

The capacity of the English wine gallon was not formalised by statute until 1706, when it was defined as a cylinder seven inches in diameter and six inches deep, and hence with a capacity of about 231 in^3.[185] This gallon will be considered more carefully in Chapter **6**. It was acknowledged that this gallon had been in customary use for an extended period beforehand, but there had been no Exchequer standard of this size and therefore apparently no legal authority for its use.[186] This fact had enabled a London wine shipper in 1700 to challenge the right of the Excise to levy duty by this unit rather than by the larger gallons for which official reference standards existed. The inevitable result of his success in the Exchequer Court was the introduction of a formal definition of the wine gallon in the 1706 Act and the construction in 1707 of reference standards. Because this coincided with the enactment of the Act of Union, by which the parliaments of England and Scotland were combined in 1707, wine gallon standards were distributed to the Scottish burghs with the other English weights and measures that were to be brought into legal use north of the border – indeed it has been argued that the requirement to distribute standards within Scotland was the specific impetus for the production of reference gallons.[187] As a result a considerable number of these wine gallon standards survive in the collections of the Scottish burghs (and these are described as Items **146-162** in the Inventory). This wine gallon has maintained an important role to the present day because of its use in Britain's former American colonies: since these colonies had acquired independence before the introduction of the Imperial gallon in the early nineteenth century, the wine gallon of 231 in^3 became the principal liquid gallon and in due course also the American petroleum gallon.

The wine gallon of 231 in^3 was not the only wine gallon that had been in customary use: there was another gallon, of 224 in^3, which had passed out of use long before the 1706 Act but which is recorded in the seventeenth-century literature. This smaller wine gallon has no recognised descriptive name, but became known as the 'Guildhall gallon' because an ancient standard of this size was found at the Guildhall in London and was measured in 1688 by the astronomers Edmond Halley and John Flamsteed, amongst others.[188] Its earlier currency was considered of importance in the official investigations of the mid-eighteenth and early nineteenth centuries that led up to the reforms represented by the Imperial system of measures in 1824, and it has always been accepted as having been a much earlier gallon, indeed possibly even the earliest official English gallon.

At the time of the 1688 measurement it was appreciated that an earlier investigation had been carried out by John Wybard (1618-74), a physician with a strong interest in mathematics and astronomy. His *Tactometria, or the Geometry of Regulars*, published in London in 1650, discussed the mathematics of gauging capacity measures and described experiments carried out by Wybard to determine the specific gravity of various materials and the volumes of existing standard measures. One of Wybard's informants was John Reynolds (*c*.1585-*c*.1654), a member of the Goldsmiths' Company of London, whom he described as 'one of the Clerks of the Mint in the Tower of London (and sometime Assay-master at Goldsmiths-hall)'. In addition to his employment as clerk at the London Mint, Reynolds also received an annuity, first awarded to him in 1607, for assaying and testing finished coin, although his formal appointment as under-assayer at the Mint was only made in 1649.[189] Between 1619 and 1629 he had been assay master to the

Goldsmiths' Company.[190] He was perhaps related to an instrument maker and engineer of the same name, a gunner at the Tower of London from 1598, who eventually held the office of master gunner of England, and who died *c*.1636.[191]

The Tower of London was one of the locations where measurement standards were kept, although the primary standards were supposed to be those held at the Exchequer. However, by analogy with the situation in Scotland, it seems very likely that English Mint officials were closely involved in the sixteenth-century work on the revision of the standard weights and measures, and the Mint certainly maintained standard weights and had the expert technical competence to control coinage weight and fineness. For a period in the mid-seventeenth century, standard measures were certainly constructed at the Tower and a number have been recorded which carry royal authorisations and inscriptions which note their regulation by Reynolds.[192] These measures were presumably based on the standards held at the Tower rather than the commercially-accessible City of London standards at the Guildhall, because Reynolds acknowledged to Wybard that he had not made trials of the Guildhall measures,

> but only of those Measures in the Tower of London (which he pleads for to be the most ancient and true standard Measures) and at Cowpers-Hall [Coopers' Hall].[193]

Whatever the formal status of the City's reference measures at the Guildhall, Wybard was aware that they were of considerable antiquity and that they formed the effective standards for trade. His interest in density measurements mirrored that of Reynolds, but his work on the gauging of vessels was hampered by apparently contradictory information he had obtained from Reynolds and others about the sizes of the gallon units. Wybard's understanding was that the Guildhall held 'the true Standard for these Measures', and he therefore set about

> the discovering of the true contents of the Standard-measures for Wine and Ale or Beer, pertaining to the Citie of London, (which are kept at the Guild-hall) as being commonly taken for the most generally received Measures for the purpose, throughout the Kingdome, and by which I have seen the Measures which have been made for some eminent Towns farr remote from London, to be sized and sealed …[194]

The capacities of the Guildhall standard obtained by Wybard was 224 in^3. This matched the result reported by Halley to the Royal Society in July 1688, when he noted that 'the standard wine-gallon pott at Guild-hall holds 224 such [London cubic] inches, tho vulgarly reputed 231'.[195] This size was accepted by Lord Carysfort's 'Committee appointed to inquire into the Original Standards of Weights and Measures in this Kingdom' in 1758, and their report records how the Guildhall gallon was nearly recommended for authorisation for official use in taxation and how the test case in the Court of Exchequer arose when government attitudes hardened against conceding a role for gauging against this older gallon.[196]

Thus, by 1700, it had become firmly established that there had been an earlier and smaller wine gallon of 224 in³, for which there was still demand for authenticated copies in the first half of the seventeenth century, but which had been effective superseded by the 231 in³ gallon. But what was its official status? The use of such a standard by the City officials at the Guildhall indicates that this was more than a customary measure and that it must have had an extended history of authorised commercial use. Indeed Wybard recorded that copies of it were made for other English towns. Certainly it was no less official than the larger wine gallon of 231 in³ in the restricted sense that standards of neither had been retained by the Exchequer and neither was recognised as arising from a statutory definition. We will argue in Chapter **6** that the larger wine gallon of 231 in³ is closely linked with Scottish and Flemish wine units and is associated with weighing in Bruges weight: it is likely therefore to have been introduced by external commercial pressures, albeit at a relatively early date. The 'Guildhall' wine gallon of 224 in³ was clearly understood as the survivor of an obsolete gallon type whose origins were of considerable antiquity, and therefore it is a plausible candidate for the eight-pound gallon of the English *Tractutus* and related codes of practice.

Wyberd reported an apparently anomalous ale-gallon size that Reynolds had provided: 288¾ in³ value rather than the expected 282 in³. (In fact, a standard by Reynolds engraved with this capacity survived, and a description has been published.)[197] Reynolds' justification for this value brings with it a further significant indication of the earlier currency of the 'Guildhall' wine gallon. He explained that his value arose from an accepted ratio of 4:5 between the capacities of the wine and ale gallons, and that this had been described much earlier by John Godwyn. The mathematical practitioner Godwyn (*fl.*1597-1600) is believed to have been Reynolds's teacher and he and Reynolds were later coupled as two 'ancient practitioners in gauging' whose tables had remained in extended use.[198] When applied precisely to the wine gallon of 231 in³, this ratio does indeed generate a supposed ale gallon of 288¾ in³ ($231 \times \frac{5}{4} = 288.75$).[199]

In the absence of a clear understanding that there is such a simple mathematical relationship between the wine gallon of 231 in³ and the ale gallon (and we will argue later that they are based on two quite different pounds, so a simple relationship would not be expected), the ratio of 4:5 should perhaps be considered as a traditional convenience for relating two customary gallons with sufficient accuracy for acceptance in the market-place. If a factor of ⅚ is applied to the Guildhall gallon, we get a good approximation to the ale gallon, at 280 in³. Perhaps then it was the earlier currency of a wine gallon of 224 in³ that gave rise to this conventional understanding of the ratio of the volumes, and this encouraged Reynolds to apply it inappropriately to the wine gallon of 231 in^3.

We should establish whether the capacity of 224 in^3 ascribed to it in seventeenth-century sources is compatible with the *Tractatus* description. The weight of water contained in 224 in^3 is about 56,680 (early) grains, and as this is supposed to be eight pounds the pound would be about 7,085 grains. This is a close match for the pound of 15 Paris ounces ($15 \times 472\frac{1}{2} = 7,087\frac{1}{2}$ grains) which was recorded by Pegolotti.[200] It has been argued above that this pound and its hundredweight must have had considerable influence outside France, and the role of the Paris ounce as a regulator of bullion reinforces the comparative strength of the French economy at this early period. It might be expected that such a pound would form the basis for liquid measures formalised initially by an Anglo-Norman administration in England and that this is the value that should emerge in the *Tractatus* definition.

If the Scottish gallon has its origins in twelfth-century English measures, it follows that the gallon of the David Assize should be explicable using the same pounds of 15 Paris ounces. The David gallon was described as containing twelve pounds of water and this would be 85,050 grains. For fresh water this gallon would contain 336·3 in^3, but if a third of the water was sea-water this would be reduced by about one per cent to give a volume of 333·0 in^3. The dimensioned gallon described in the assize is about 331 in^3 in capacity. On its own we are not necessarily justified in interpreting this dimensioned volume as being of high precision, because the purpose of this description is to detail a vessel which will satisfy the requirements of the law, using convenient dimensions which are given (in this instance) to the nearest half-inch. Such a vessel may be an acceptable substitute for a duplicate of the standard, but its capacity may differ slightly from that of the standard in the interests of providing useful dimensional information. (In Chapter **6** we will argue that the English wine gallon of 231 in^3 has undergone just such a change, and that in this situation it is not easy to identify the 'true' gallon.) However, in the case of the gallon of the David Assize, it remained ostensibly unchanged until the 1426 Assize, when it was replaced by a much larger gallon; and details in the 1426 Assize (to be discussed in Chapter **6**) enable its size in the early fifteenth century to be set at 330·7 in^3.

The agreement between these figures is good, and so the French pound must be seen as a likely candidate for the David Assize definition. By way of contrast it might be thought that the more obvious pound to select would be the pound of 15 Cologne ounces of 450 grains, which is apparently the only pound defined in the David Assize. But twelve of these pounds of water weigh 81,000 grains and occupy 320·3 in^3 if fresh water is used, and 317·1 in^3 if a third is sea-water. These figures are distinctly different from the dimensioned gallon of 331 in^3.[201]

Another pointer to the use of French weight for the Scottish gallon is provided by the values of the sub-divisions of the gallon before 1426. In Chapter **6** it is shown that the gallon at this period comprised six pints, each of two chopins. Since the gallon weighed twelve pounds, each pint was two pounds and each chopin was one pound. The early French *pinte* is similarly claimed to have as its origin the capacity of two *livres*, and each *pinte* contained two *chopines*.[202]

Although this may have been the original basis for the gallon, it will be seen in Chapter **6** that by about the mid-fourteenth century there were closely-related Scottish ale and wine gallons of 311 in^3 and 467 in^3, used with allowances of one-sixteenth, giving rise to practical measures of 331 in^3 and 496 in^3. From the time of the 1426 Assize measurement definitions were given in terms of the basic size of the pint, rather than the size enhanced by allowance. These basic sizes can be linked to the later English wine gallon of 231 in^3 – or rather, to the volume of about 233 in^3, which we will argue was the correct volume, the evidence for this size coming from a further standard vessel by John Reynolds. These measures have their origin in external standards represented by the size of casks used for wine export from France, much of it through Flanders. Their volumes are readily compatible with definitions based on the Flemish pounds of 7,000 grains, and these will be discussed in Chapter **6** in terms of a shift away from French and towards Flemish definition types. The dimensioned gallon of 331 in^3 in the surviving fourteenth-century version of the David Assize may either represent the gallon based on Paris pounds, or it may be the measure based on the newer use of avoirdupois weight and incorporating the allowance. In any event the close similarity of the volumes arising from these two definitions suggests a possible origin of the one-sixteenth allowance in providing an effective transition from one system to the other.

1 R. D. Connor: *The Weights and Measures of England [WME]* (London, 1987). Our new approach to the description of weight-types has been tested against two collections of archaeological finds, and has enabled a significant number of weights to be classified, including several that could be identified as Flemish. These were medieval weights excavated in London, and subsequently published in G. Egan, *The Medieval Household: Daily Living c.1150-c.1450* (London, 1998), and archaeological finds at the Archaeological Resource Centre in York (a project of the York Archaeological Trust). See A. D. C. Simpson and R. D. Connor, '14th-Century Weight Systems: A Response', in *Equilibrium*, issue no. 1 (1997), 2107-10. We are grateful to Geoff Egan of the Museum of London and Dr Andrew Jones and Jim Halliday of the York Archaeological Trust for their assistance.

2 See Ranald Nicholson, *Scotland, The Later Middle Ages* (Edinburgh, 1989), 307, 433, for numerous references on this topic to T. Thomson and C. Innes (eds), *The Acts of the Parliaments of Scotland [APS]*, 13 volumes (Edinburgh, 1814-75). England's similar struggle to obtain a sufficiency of bullion for the coinage is admirably reviewed by A. E. Feavearyear, *The Pound Sterling* (Oxford, 1931), 19.

3 See in particular Philip Grierson, 'Money and Coinage under Charlemagne', in W. Braunfels (ed), *Karl der Grosse: Lebenswerk und Nachleben*, 5 volumes (Dusseldorf, 1965-8), I, 501-36.

4 *WME*, 109. Whether the full weight was achieved from about 1080 or from a later date is still uncertain because of the scarcity of early issues. For an alternative view, which proposes a final rise to 1·46 g in 1158, see Pamela Nightingale, 'The Evolution of Weight-Standards and the Creation of New Monetary and Commercial Links in Northern Europe from the Tenth Century to the Twelfth Century', in *Economic History Review*, second series, 38 (1985), 192-209.

5 Peter Spufford, *Money and its Use in Medieval Europe* (Cambridge, 1988), 141, 161-2.

6 Nightingale, op. cit. (4), 200.

7 Philip Grierson, 'The Weight of the Gold Florin in the Fifteenth Century', in *Quaderni Ticinesi di Numismatica e Antichità Classiche*, 10 (1981), 421-31; see p. 426; Peter Spufford, *Handbook of Medieval Exchange* (London, 1986), 184.

8 See C. Virginia Glenn, 'The Emergence of the Gothic Style in Thirteenth-Century Parisian Goldsmiths' Work', unpublished PhD thesis, Courtauld Institute of Art, University of London, 1990; also R. W. Lightbown, *Secular Goldsmiths' Work in Medieval France: A History* (London, 1978).

9 On the 'touch of Paris', see J. S. Forbes, *Hallmark: A History of the London Assay Office* (London, 1998), 16-18.

10 Nightingale, op. cit. (4), 202.

11 For the florin of 1252, see Spufford, op. cit. (5), 176; for the ducat of 1284, see Grierson, op. cit. (7), 426.

12 Ibid.

13 We are grateful to Professor Philip Grierson for emphasising the importance of this aspect: personal communication.

14 Charles Johnson (ed and trans), *The De Moneta of Nicholas Oresme and English Mint Documents* (London and New York, 1956), 76; on the abolition of the tower pound, see Paul L. Hughes and James F. Larkin (eds), *Tudor Royal Proclamations*, 3 volumes (New Haven, Connecticut, 1964-9), I, no. 112, 153-63, and especially pp. 160-1. For a summary of these changes, see *WME*, 113.

15 The French ounce of 30·596 grammes yields a grain of 53·118 mg in contrast to the English grain of 64·8 mg.

16 Alfred Nagl, 'Das niederländische Troyes-Gewicht und seine Verifikationen zu Paris in den Jahren 1529 und 1756', in *Numismatische Zeitschrift*, 46 (1913), 211-18.

17 The performance of balances by the London scale maker Samuel Read used in the earliest comparisons between French and English standard weights, conducted in 1742 and 1743 by George Graham for the Royal Society, are noted in [George Graham], 'An Account of a Comparison lately made by some Gentlemen of the Royal Society of the Standard of a Yard, and the Several Weights lately made for their Use; with the Original Standards of Measures and Weights in the Exchequer, and some others kept for public use at Guildhall, Founders Hall, the Tower, etc.', in *Philosophical Transactions of the Royal Society*, 42 (1742-3), 541-56, see p. 552. The comparatively poor quality of the balances at the Exchequer in 1758 is recorded in the first report of Lord Carysfort's Committee, 'Report from the Committee appointed to Inquire into the Original Standards of Weights and Measures in this Kingdom; and to consider the Laws relating thereto [Carysfort 1st Report], 26 May 1758', in *Reports from Committees of the House of Commons*, 3 volumes (1715-1801), II, 411-51, see p. 436. Improved performance was achieved by a special purpose scale

constructed by the prominent instrument maker John Bird for Joseph Harris, Assay Master at the Mint, and used by Harris in his 1758 work for Carysfort. It was claimed this would detect weight differences of 0·025 grains in the pound, but by the 1840s four surviving standard pounds adjusted by him were found to differ by 0·55 grains: 'Report from the Committee, Appointed (upon the 1st Day of December 1758) to Inquire into the Original Standards of Weights and Measures in this Kingdom; and to consider the Laws relating thereto [Carysfort 2nd Report], 11 April 1759', ibid., 453-63, see p. 456; W. H. Miller, 'The Construction of the New Imperial Standard Pound, and its Copies of Platinum', in *Philosophical Transactions*, 146 (1856), 753-946.

18   The text of the handbook is printed as (Allan Evans, ed), Francesco Balducci Pegolotti, *'La Pratica della Mercatura'* (Cambridge, Massachusetts, 1936). Our initial conclusions were published as A. D. C. Simpson and R. D. Connor, 'Weighing in the Early 14th Century', in *Equilibrium*, issue no. 1 (1996), 1987-98; ibid., issue no. 2 (1996), 2015-24; and Simpson and Connor, op. cit. (1).

19   The parliamentary act is 5 George IV c. 74. 1824.

20   Carysfort 1st Report, op. cit. (17), 437; H. W. Chisholm, *Seventh Annual Report of the Warden of the Standards … for 1872-73* (London, 1873), 39-40; see also *WME*, 246-9. For the political context see Julian Hoppit, 'Reforming Britain's Weights and Measures, 1660-1824', in *English Historical Review*, 108 (1993), 82-104; and Rebecca J. Adell, 'The English Metrological Standardisation Debate, 1758-1824', unpublished MA thesis, Carleton University, Ottawa, 2000. Our thanks to Dr Bruce Elliott for this reference.

21   Miller, op. cit. (17).

22   Ibid. See also H. W. Chisholm, 'On the Science of Weighing and Measuring and the Standards of Weights and Measures', in *Nature*, 8 (1873), 327-8.

23   Carysfort 1st Report, op. cit. (17).

24   Chisholm, op. cit. (20), 13.

25   Allen Simpson, 'English Standard Troy Weight verified at the Mint, 1588', in Silke Ackermann (ed), *Humphrey Cole: Mint, Measurement and Maps in Elizabethan England*, British Museum Occasional Papers no. 126 (London, 1998), 105-6.

26   Chisholm, op. cit. (20), 18.

27   A. D. C. Simpson and R. D. Connor, 'The Mass of the English Troy Pound in the Eighteenth Century', in *Annals of Science*, 61 (2004), 323-51.

28   For papers relating to these Elizabethan enquiries, see Chisholm, op. cit. (20), 11-15, and Carysfort 1st Report, op. cit. (17), 443-8.

29   *Collection des Poids et Mesures: Inventaire des Poids [du Musée National des Techniques, Conservatoire National des Arts et Métiers]* (Paris, 1990), 22, 20 (note 1): CNAM, Paris, Inv. 3261. For a description, see Louis Blancard, 'La Pile de Charlemagne: étude sur l'origine et les poids des deniers neufs et de la livre de Charlemagne', in *Annuaire de la Société Française de Numismatique*, 11 (1887), 595-638.

30   Armand Machabey, *La Métrologie dans les Musées de Province* (Paris, 1962), published doctoral thesis, Paris Sorbonne University, 1959, 358. The original measurements for this were made in 1797 by J. V. F. Dillon (in *grammes provisoires*) and reported in J. F. Lasparat (ed), *Métrologies Constitutionelle et Primitive, comparées entre elles et avec la Métrologie d'Ordonnances*, 2 volumes (Paris, 1801), I, 98-104.

31   Simpson and Connor, op. cit. (27).

32    $472 \cdot 174 \times 64 \cdot 80 \div 64 \cdot 76 = 472 \cdot 466$.

33   Nicholson, op. cit. (2), 23; Edward Burns, *The Coinage of Scotland*, 3 volumes (Edinburgh, 1887), I, 3; Ian Halley Stewart, *The Scottish Coinage* (London, 1955), 1-3; Elizabeth Gemmill and Nicholas Mayhew, *Changing Values in Medieval Scotland: A Study of Prices, Wages, Weights and Measures* (Cambridge, 1995), 114.

34   Hughes and Larkin, op. cit. (14), 160-1.

35   Chisholm, op. cit. (20), 16, quoting from 1864 Commons Paper, no. 115.

36   *APS*, II, 10: Act 14, 1426. See also Appendix A.2, for the second preliminary act.

37   Norman Biggs, 'English Weight Systems in the 14th Century – A New Interpretation', privately circulated paper, December 1990. The cited case of 21 October 1376 is from A. H. Thomas (ed), *Calendar of Plea and Memoranda Rolls, 1364-81* (Cambridge, 1929), 228.

38   N. J. Mayhew, 'From Regional to Central Minting, 1158-1464', in C. E. Challis (ed), *A New History of the Royal Mint* (Cambridge, 1992), 83-178, p. 108. See also *WME*, 110-12.

39   Mayhew, op. cit. (38), 102.

40   Nightingale, op. cit. (4), 205. This proposal is broadly accepted by Nicholas Mayhew: Mayhew, op. cit. (38), 89-90.

41   *APS*, I, 673-4.

42   John Skene, *Regiam Majestatem: The Auld Lawes and Constitutions of Scotland* (Edinburgh, 1609), part I, f.161; Peter G. B. McNeill (ed), *The Practicks of Sir James Balfour of Pittendreich*, Stair Society nos. 21 and 22, 2 volumes (Edinburgh, 1962), I, 90.

43   Thomas Mackay Cooper, Lord Cooper (ed), *Regiam Majestatum and Quoniam Attachiamenta*, Stair Society no. 11 (Edinburgh, 1947), Introduction, 15, 20-1, 32-40.

44   *Statutes of the Realm*, 11 volumes (London, 1810-28), I, 204-5. The text is printed in *WME*, 320-1.

45 Philip Grierson, *English Linear Measures: an Essay in Origins* (Reading, 1972), 1.

46 This dating is given for the version published in Owen Ruffhead and Charles Runnington (eds), *Statutes at Large*, 41 volumes (London, 1780-1865), I, 148.

47 Hubert Hall and Frieda J. Nicholas (eds), *Select Tracts and Table Books Relating to English Weights and Measures (1100-1742)*, Camden Society Miscellany no. 15 (London, 1929), 11-12. On the dating of this, see Grierson, op. cit. (45).

48 Dorothea Oschinsky, *Walter of Henley and other Treatises on Estate Management and Accounting* (Oxford, 1971).

49 The text of *Fleta* has been published with an English translation in G. H. Richardson and G. O. Sayles (eds and trans), *Fleta*, 3 volumes, Selden Society nos. 72, 89, 99 (London, 1955-84). The metrological section is in volume II, book II, chapter 12. For the dependence of *Fleta* on *Seneschaucy*, see Oschinsky, op. cit. (48), 81, 106-12.

50 *Statutes*, op. cit. (44), I, 204. On the divided nature of the *Tractatus* text, see Grierson, op. cit. (45).

51 *Statutes*, op. cit. (44), I, 205.

52 *APS*, I, 673-4. James Balfour in his *Practicks* gave the weight of the pound as 26 shillings and four pence: McNeill, op. cit. (42), I, 90.

53 For example, in the 1290s substantial quantities of silver goods, weighed only in marks and shillings, are listed in the inventory of the trousseau of Isabella Bruce at her marriage to the King of Norway: J. Bain (ed), *Calendar of Documents relating to Scotland,* 4 volumes (Edinburgh, 1881-8), II, 158-9. We are grateful to Dr Virginia Glenn for this reference.

54 Burns, op. cit. (33), I, 253; Stewart, op. cit. (33), 31: 'A new coinage was to be made of "a good and pure metal as the money now made in the kingdom of England …". The weight, however, was to be reduced by minting 352 pennies from a pound of silver instead of 300, the weight of the groat was reduced $61\frac{4}{11}$ grains.' Hence the penny was $61\frac{4}{11} \div 4 = 15\frac{15}{44}$ grains, and the pound weight was $352 \times 15\frac{15}{44} = 5,400$ grains, or twelve ounces of 450 grains. See also Bruce Webster (ed), *Acts of David II King of Scots 1329-71, Regesta Regum Scottorum [RRS], volume VI* (Edinburgh, 1982), 416: Act 385.

55 *WME*, 2-4.

56 Grierson, op. cit. (3), 530.

57 Ibid.

58 Ibid.

59 An assumption that the *Tractatus* was to be interpreted principally in terms of troy weight has perhaps been reinforced by an explicit reference to troy in the translated version printed in *The Statutes of the Realm*. Here the Latin text (from British Library MS Liber Horn., f.123) refers to a method of weighing lead as 'troni ponderacionem', indicating that it was 'weighed by the tron' or scales (and we are grateful to Professors Philip Grierson and Christopher Brooke for their advice on this point). However, the parallel translated text gives this as 'after Troy Weight': *Statutes*, op. cit. (44), I, 205; reprinted in *WME*, 320. See also A. D. C. Simpson, 'Scots "Trone" Weight: Preliminary observations on the Origin of Scotland's Early Market Weights', in *Northern Studies*, 29 (1992), 62-81, pp. 72, 79. In the equivalent passage in *Fleta*, the editors have translated the phrase as 'by tron weight': *Fleta*, op. cit. (49), II, 119.

60 Rogers Ruding in 1817 and the economic historian J. E. Thorold Rogers in 1866 both adopted a pound of twelve ounces of 450 grains (explicitly called tower ounces) to represent the early English 'Saxon' pound: R. Ruding, *Annals of the Coinage of Britain*, 4 volumes (London, 1817), I, 12-13; J. E. Thorold Rogers, *History of Agriculture and Prices in England* (1259-1400), 7 volumes in 8 (Oxford, 1866-1902), I, 165. Perhaps more crucially, W. H. Chisholm, in his 1873 historical analysis of early English metrology performed in his capacity as Warden of the Standards, explicitly accepted the ounce of these early regulations as one of 450 grains: Chisholm, op. cit. (22), 17.

61 Ruding, op. cit. (60), I, 13 note.

62 A version included in the *Coventry Leet Book* of 1474 specifies that the weight type is also used for bullion and metrological purposes: '… xxxii graynes of wheat take out of the mydens of the Ere makith a sterling othet-wyse called a peny and xx sterling maketh an ounce and xij ounce maketh a Pounde for syluer, golde, bred and measure': M. D. Harris (ed and trans), *Coventry Leet Book* (London, 1907), 396. However, another definition of 1496 is quite specific that operations have become the preserve of troy weight. See *WME*, 197.

63 'xij. vncie faciunt libram xx. s. in pondere et numero': *Fleta*, op. cit. (49), II, 119.

64 'quindecim vncio faciunt libram mercatoriam' and '15 ounces make a lb. Avoirdupois': ibid.

65 William Stubbs (ed), *Chronica Magistri Rogeri de Houedene*, Rolls Series, 4 volumes (London, 1868-71), IV, 33-4.

66 For the Venice 12-ounce (*peso sottile*) and 19-ounce (*peso grosso*) pounds, and the 12-ounce pound of Florence, see Patrick Kelly, *The Universal Cambist and Commercial Instructor*, 2 volumes in 1 (London, 1811; and second edition, London, 1835), s.v. *Venice* and *Florence*. By this date the 16-ounce Florentine *grosso* pound had become obsolete.

67 The argument is summarised in Simpson and Connor, op. cit. (1).

68 The mass of this weight before electrolytic treatment removed the surface layer of basic carbonate was 107·15 g (1,653·5 grains), and a quarter of a pound of 14 × 472½ grains is 1,653¾ grains: NMS T.1993.31. Excavated at Lavenham, Suffolk, and donated by the finder John Goodall; acquired through the interest of the Suffolk County Council Archaeological Service.

69 Pegolotti, op. cit. (18). The text was first published by G-F. Pagnini, *Della Decima e di varie altre gravezze imposte dal comune di Firenze, Della moneta e della mercatura de' Fiorentini fino al seco XVI ...*, 4 volumes (Lisbon and Lucca, 1765-6), III.

70 See in particular Spufford, op. cit. (5). Attention has been drawn to its potential for studies of trading weights, but this avenue has not been fully explored: see, for example, R. E. Zupko, 'Notes on Medieval English Weights and Measures in Francesco Balducci Pegolotti's "La Pratica della Mercatura"', in D. Herlihy, R. S. Lopez and V. Slessarev (eds), *Economy, Society and Government in Medieval Italy: Essays in Memory of Robert L. Reynolds* (Kent, Ohio, 1969), 153-60, p. 156.

71 Evans, op. cit. (18), xiii-xiv.

72 Joseph Ghyssens, 'Les marcs au moyen âge', in *Bulletin du Cercle d'Etudes Numismatiques*, 4 (1967), 1-6, see p. 6. 'The little mark of Flanders barely passes the beginning of the fourteenth century. It must have disappeared at the same time as the small silver deniers. From the fourteenth century, the use of the mark of Paris was imposed throughout our regions': J. Ghyssens, 'Precisions à propos des marcs', in ibid., 9 (1972), 51-7, see p. 54.

73 Pegolotti, op. cit. (18), 245: '... fa in Londra once 8 e sterlini 8 al peso della Torre di Londra.'

74 Ibid., 237: 'Lo marco dell'argento a peso di Bruggia e di tutta Fiandra si è once 6 a peso di Bruggia, e marchi 21 a peso d'argento fanno in Bruggia marchi 16 a peso d'oro'.

75 C. Wyffels, 'Note sur les marcs monétaires utilisés en Flandre et en Artois avant 1300', in *Handelingen van het Genootschaap 'Société d'Émulation' te Brugge*, 104 (1967), 66-87; see p. 71; see also C. Wyffels, 'Contribution à l'histoire monétaire de Flandre au XIIIe siècle', in *Revue Belge de Philologie et d'Histoire*, 45 (1967), 1113-41.

76 Simpson and Connor, op. cit. (27).

77 Pegolotti, op. cit. (18), 245: 'Marchio 1 d'argento di Bruggia fa in Parigi once 6 e denari 2 e grani 6⁶⁄₇, di grani 24 per 1 denaro, e di denari 24 per 1 oncia.'

78 Joseph Ghyssens, 'Quelque mesures de poids du Moyen-Âge pour l'or et l'argent', in *Revue Belge de Numismatique*, 132 (1986), 67, 68.

79 Nightingale, op. cit. (4), 206-7; Wyffels, op. cit. (75), 79.

80 Grierson, op. cit. (45), 2.

81 Ghyssens, op. cit. (78), 68.

82 Pegolotti, op. cit. (18), 255: 'In Londra si à 2 maniere di pesare argento, cioè il marchio della zecca della Torre di Londra, che è appunto col marchio di Cologna della Magna ...'

83 P. Kelly, *The Universal Cambist*, second edition (London, 1835), I, 96; Bruno Kisch, *Scales and Weights* (New Haven and London, 1966), 9; Philip Grierson, *Medieval European Coinage*, VII, section 4, Metrology, 14, forthcoming. We are grateful to Professor Grierson for giving us a preliminary text of this section of an important work.

84 Pegolotti, op. cit. (18), 244: 'con Anguersa, con Mellino, con Borsella, con Lovano, e con tutto Brabante ... E lo marchio dello argento al peso di Bruggia, ch'è once 6 in Bruggia, fa ne' detti luoghi once 6 e sterlini 8. E lo marco dell'oro al peso di Bruggia fa ne' detti luoghi once 8 e sterlini 8.'

85 Ibid.

86 Unfortunately our knowledge of the size of the Cologne ounce is largely based on eighteenth-century measurements, which give its mass as about 29·23 g, or about 451 Imperial troy grains: P. Guilhiermoz, 'Notes sur les poids du moyen âge', in *Bibliothèque de l'École des Chartes*, 67 (1906), 160-233, 402-50, p. 429. We are grateful to Professor Philip Grierson for this reference. Tillet's 1766 measurement, equivalent to 29·23 g, is reported in J. G. Berck, 'Over de standaarden van het Keulsch, Engelsch, Fransch, Hollandsch, Trooisch, Amsterdamsch en Brabantsch gewicht van de 13e tot het begin der 19eeuw', in *Natuur en Mensch*, 51 (1931), 121-7, 174-6, 194-9, 222-4, p. 122.

87 Lightbown, op. cit. (8). It is plain that towards the end of the fifteenth century considerable care was being given to the fineness of the bullion being brought to the mint. Thus in 1489 we read: '... na goldsmythe sall mak [a] mixto[u]r nor put false layis [alloy] in the said mettallis And to haue knawlege of the fynace [fineness] of the said werkes ... and that the similar werk be of fynace of the new siluer werk of bruges': *APS*, II, 221, Act 13, 1489; and again: '... and at [that] the said coiner shall give and pay for the vnce of brynt [pure] siluer 11s 6d and for paris siluer and siluer of the new werk of bruges siclik and as for siluer werk of this realme that beis [be] brocht to the mint quhilkis not so fine, the said coiner sall gif and deliure [deliver] therefore the veray avail [true worth] to avnare [value] of the said siluer and samekle [as much] as it wer na paris siluer or siluer if the new werk of bruges to be

88 Pegolotti, op. cit. (18), 257: 'Libbre 100 di Londra fanno in Parigi libbre 96 in 97.'

89 Ibid., 202: 'Libbre 100 di Londra fanno in Firenzi libbre 138 in 140.'

90 Ibid., 244: 'Con Anguersa … Libbre 100 di Bruggia fanno … libbre 92½'; ibid.: 'Libbre 100 di Bruggia fanno in Londra di cosa che non manchi libbre 92½.'

91 Ibid., 245: 'Libbre 100 di Bruggia fanno in Parigi libbre 88 in 89 …'

92 Ibid., 246: 'Libbre 100 di Firenze fanno in Bruggia libbre 78.'

93 Ibid., 256: 'Libbre 100 di Londra fanno in Anguersa libbre 100 di spezieria.'

94 Julian Goodare, 'The Long Hundred in Medieval and Early Modern Scotland', in *Proceedings of the Society of Antiquaries of Scotland,* 123 (1993), 395-418.

95 A charter granted by Alexander II in 1232 to Cambuskenneth Abbey revised the description of an earlier annual grant of cheese from neighbouring royal lands to be 20 'cowgalls' of cheese ('viginti cowgall casei'): Sir William Fraser (ed), *Registrum Monasterii S. Marie de Cambuskenneth AD 1147-1535,* Grampian Club no. 4 (Edinburgh, 1872), 316-17, no. 224. The cowgall was a weight unit of six stone, so 20 cowgalls was 120 stones or a long hundred. The issue is discussed in Simpson, op. cit. (59).

96 The hundredweight of 112 pounds was abolished by the Weights and Measures Act of 1985.

97 Pegolotti, op. cit. (18), 237: 'In Bruggia si à pure uno peso cioè libbra e oncia, e la libbra si è once 14 …' The 14-ounce pound is recorded by other commentators on Pegolotti, such as Ghyssens, op. cit. (78), 61.

98 These standards were illustrated, and their masses analysed, in D. Verlé, *Aantekeningen Betreffende de Geschiedenis van het Ijkwezen van de XVIIde tot de XXste Eeuw* (Bruges, 1971). Verlé shows that two weights of 25 pounds and one of 12 pounds are based on a 14-ounce pound of 428·8 g (equivalent to about 6,622 or 14 × 473 early English troy grains), and that the remaining 50 and 12 pound weights are based on a 16-ounce pound of 463·8 g (equivalent to about 7,163 or 16 × 447 grains). All the weights carry adjustment pieces on their suspension ring, so their original size is a little uncertain, but they have been kept in close adjustment and are excellent matches to the sizes recorded in the literature. We are grateful to Dr Jan Dieman for his help in locating Verlé's publication. The smaller of the three 14-ounce weights is also illustrated in D. A. Wittop Koning and G. M. M. Houben, *2000 Jaar Gewichten in der Nederlanden* (Lochem, 1980), 30.

99 Gerard Houben, personal communication, 27 April 1993. We are indebted to Dr Houben for drawing our attention to the survival of these weights.

100 J. Mertens, 'De Processen – verbaal van de "Commission des Poids et Mesures" van het Leidepartment 1798-1801', in *Annales de la Société D'Émulation de Bruges* (Bruges, 1967), 97-118; see in particular p. 118. Included amongst the weights of Bruges and other cities of Flanders measured in terms of the new units in about 1800 were: (1) a pound described as 'La livre de 14 onces Bruges et Courtray, 428·828 g' (equivalent to 14 × 473 early troy grains, and hence only 0·1% heavier than that derived from Pegolotti almost five centuries earlier); (2) another pound described as of 'Bruges et ci-devant [formerly] Franc pound 463·9 g [and ounce] 28·99 g' (equivalent to a pound of 16 × 448 early grains). We are grateful to Ruud de Zwarte for drawing this reference to our attention.

101 Machabey, op. cit. (30), 302: 'Ainsi, de nombreuses villes du nord (Aire-sur-la-Lys, Hesdin, Saint-Omer, etc …) ont employé une livre de 14 onces (428 g environ) comme livre de commerce courant. Il fut de même à Lyons, Tours, etc …' The only other 14-ounce pound recorded by Pegolotti was an Antwerp pound restricted to spices: Pegolotti, op. cit. (18), 250: 'Peppe, e lacca, e gengiovo, e 'ncenso, e zucchero, e indaco, e cotone mapputo, e cotone filato, e tutte spezierie grosse e sottili tutte si vendono in Anguersa a libbre d'once 14 la libbra …'

102 Grierson, op. cit. (7), 421-31, from which we see that apart from a few months in 1402, the florin did not deviate by more than a half of one per cent from its original value until the fall of the Republic.

103 Pegolotti, op. cit. (18), 245: 'Libbre 100 in Bruggia fanno in Parigi libbre 89 in 89.' He mentions 15 ounce pounds elsewhere: ibid., 148: 'Libbre 100 sottile di Vinegia fanno in Parigi libbre 62½, d'once 15 per libbra ….'; ibid., 221: 'Libbre 530 di Genova fanno in Parigi carica 1, ch'è in Parigi libbre 350 d'once 15 per libbra'; ibid., 236: 'In Parigi … la libbra si è once 15 di Parigi.'

104 A manuscript of the municipality of Carpentras of 1195 mentions the pound divided into 16 ounces, but this may reflect its division in use rather than its ounce basis; Machabey, op. cit. (30), 301, 303, 357.

105 Pegolotti, op. cit. (18), 254: 'Pepe, e gengiovo, e zucchero, e cannella, e 'ncienso, e lacca, e tutte spezierie si vendono in Londra a centinaio di libbre, e pesasi in grosso e dàssi libbre 104 per 1 centinaio.'

106 Ibid., 236: 'La carica di Parigi, ch'è in Parigi libbre 350, fa in fiera libbre 364 in 367½, e in Anguersa 364'; and ibid., 257: 'Carica 1 di spezierie di Parigi fa in Londra libbre 364.'

107 Ibid., 237: 'In Bruggia si à pure uno peso cioè libbra e oncia, e la libbra si è once 14, e le libbre 400 sono 1 carica in Bruggia'; and ibid., 250: '… tutte spezierie grosse e sottili tutte si vendono in Anguersa a libbre d'once 14 la libbra, e libbre 400 sono in Anguersa 1 carica.'

108 The weight, carrying three lions *passant*, is in a private collection. Illustrated by Norman Biggs, in *English Weights: an Illustrated Survey* (Egham, Surrey, 1992), 42.

109 The current weight is 470·6 g (7,262·5 grains, or about 7,266·5 medieval troy grains): private communication from the owner, 2 July 1989. On the mass of excavated lead weights, see Simpson and Connor, op. cit. (1).

110 Pegolotti, op. cit. (18), 254: 'Pepe, e gengiovo, e zucchero, e cannella, e 'ncienso, e lacca, e tutte spezierie si vendono in Londra a centinaio di libbre, e pesasi in grosso e dàssi libbre 104 per 1 centinaio.'

111 Ibid., 255: '… di libbre 112 per 1 centinaio.'

112 On the 'nail' and 'clove', see Grierson, op. cit. (45), 15 note 62 and 18 note 80.

113 *APS*, I, 673: 'the stane for weying of woll and uther geir aw to wey xv pund'.

114 14 Edward III, Stat. 1, c.21, 1340.

115 Pegolotti, op. cit. (18), 254: 'Lana si vende in Londra e per tutta l'isola d'Inghilterra a sacco, di chiovi 52 pesi per 1 sacco, e ogni chiovo pesa libbre 7 d'Inghilterra.'

116 The *Tractatus* provides two definitions of the sack as 28 stones or 30 stones, which are stated to be the same weight but use stones of different pounds. The manuscript version printed in the *Statutes of the Realm* (from the original in the British Library, MS Liber Horn., f.123), has an early interlined addition confirming that the first weight is 350 pounds, and so the stone is of twelve pounds: *Statutes*, op. cit. (44), I, 204. *Fleta* notes simply that '12½ such lbs. make a stone, and 28 stones make a sack of wool': *Fleta*, op. cit. (49), II, 119.

117 Rogers assumed the 15-ounce pound of *Fleta* was the 7,000-grain avoirdupois pound, but noted that the *Fleta* definition of the sack as 28 stones of 12½ pounds, or 350 pounds, represented a reduction of weight which was not compatible with the 'Saxon' (tower), troy or avoirdupois pounds. He argued that the stone size should have been 14 pounds, making a sack of $14 \times 28 = 392$ pounds of 7,000 grains, or 2,744,000 grains. Because *Fleta* also said the sack was the weight of a quarter of wheat, namely $64 \times 8 = 512$ pounds, where the pound was taken as the 'Saxon' or tower pound of 5,400 grains, the sack should be 2,764,800 grains. He felt that the difference between these of less than four pounds was sufficiently close to confirm

the interpretation, but this argument is completely erroneous: Rogers, op. cit. (60), I, 166. The sack's weight was given elsewhere as 52 cloves of 7 pounds, or 364 pounds: ibid., I, 364.

118 For example, Zupko, op. cit. (70), 156.

119 Thus, for example, it is cited as 26 stone (and hence 364 pounds) in E. E. Power, *The Wool Trade in English Medieval History* (Oxford, 1941), 23; Alison Hanham, *The Celys and their World: An English Merchant Family of the Fifteenth Century* (Cambridge, 1985), 115, 133, 307, takes the sack to be 364 avoirdupois pounds in the fifteenth century (1474-89).

120 T. H. Lloyd, *The English Wool Trade in the Middle Ages* (Cambridge, 1977).

121 I. Epstein (ed and trans), *The Babylonian Talmud, seder Nezikin, Baba Bathra* (London, 1935), 88b-89b, 361 *et seq.* It is strange that Bruno Kisch, while giving this reference for the floor clearance of the pans, made no comment on weighing by inclined-beam: Kisch, op. cit. (83), 48-51.

122 'Usualiter consuetudo est, quando pecunia, quam debet, vendi per stateram, ponderatur, quod statera debet trahere inclinando versus pecuniam, excepte auro et argento, quod semper ponderatur per medium clavum, neque trahens ad pondam neque ad aurum sive ad argentam; … Provisum fuit et statutem, die Sabbati post festum Sancti Nicholai anno regni Regis Henrici filii Regis Johannis xlj. Quod omnis pecunia, que debet vendi per stateras Regis in Civitate, ponderatur sicut aurum et argentum, nichil trahens versus pecuniam; et pro tractu predicto debet venditor dare emptori ad quemlibet centrum, quatuor libras': T. Stapleton (ed), *Liber de Antiquis Legibus* (London, 1846), 25. Our translation differs from that given by [J. A. Kingdon] in *The Strife of the Scales* (London, 1905), in that 'per medium clavum' is given as 'by level beam' and, from the sense, 'pecunia' is 'goods'.

123 Kingdon, op. cit. (122), 23.

124 Machabey, op. cit. (30), 302.

125 Pegolotti, op. cit. (18), 254, as quoted at ref. 115.

126 Ibid., 237: 'Lana si vende in Bruggia a sacco, e dàssi per 1 sacco intero 60 chiovi, e ogni chiovo si è libbre 6, e ragionasi di 60 chiovi in somma pietre 28 di Bruggia …'

127 *Statutes*, op.cit. (44), I, 204-5.

128 *Fleta*, op. cit. (49), 119.

129 'Ordinacio facta de modo ponderandi per balanciam', 1309, in Hall and Nicholas, op. cit. (47), 43. Here the editors interpolate 108 pounds for the (apparently incomplete) hundredweight of spices, but a check of the original manuscript (British Library MS Add. 37791 f.9) confirms 104 pounds: '… Auerii Ponderis, ut de cera, amigdolis, riseis,

cupro, stagno, et huiusmodi, que per balanciam, sunt ponderanda, de cetero ponderentur in equali …'; and '… merchandise averii ponderis, such as wax, almonds, rice, copper, tin, and whatever has to be weighed by balance, shall for the future be weighed level …' This translation is to be found in Kingdon, op. cit. (122), 32-3. Kingdon also gives 104 pounds for the hundredweight. The phrase about being weighed level is repeated in the text of the *Ordinacio*.

130   Pegolotti, op. cit. (18), 255: '… tutte cose grosse si vendono in Londra a centinaio, di libbre 112 per 1 centinaio …'

131   Ibid, 254: 'Pepe, e gengiovo, e zucchero, e cannella, e 'ncienso, e lacca, e tutte spezierie si vendono in Londra a centinaio di libbre, e pesasi in grosso e dàssi libbre 104 per 1 centinaio.'

132   Kingdon, op. cit. (122), 29-32.

133   Ibid., 26-7, 75-6: 14 Edward III, Stat. 1, c.21, 1340.

134   25 Edward III, Stat. 5, c.9, 1352.

135   Kingdon, op. cit. (122), 29-30, 33.

136   F. W. Swetz, *Capitalism and Arithmetic: The New Math of the 15th Century, Including the Full Text of the Treviso Arithmetic of 1478, Translated by David Eugene Smith* (La Salle, Illinois, 1987), 133-8, 232-4.

137   'Sacus lane debet ponderare xxviij petr, [hoc est ccc & l. li.] Et in aliquibus partibus xxx petr [hoc est ccclxxv. li.] Et idem sunt sedm majorem & minorem libram.': *Statutes*, op. cit. (44), I, 204 (from the original in the British Library, MS Liber Horn., f.123).

138   Pegolotti, op. cit. (18), 254: 'Lana si vende in Londra e per tutta l'isola d'Inghilterra a sacco, di chiovi 52 pesi per 1 sacco, e ogni chiovo pesa libbre 7 d'Inghilterra.'

139   S. F. Hockey (ed), *The Account Book of Beaulieu Abbey*, Camden Society, fourth series no. 16 (London, 1975), 214-21, where sacks of white and of grey wool are both defined as of 30 stone. The worth of the sack and stone (p. 219) also allows the sack to be calculated at 30 stone, and with the stone at 12½ pounds (p. 202), the sack is 375 pounds. We are grateful to Professor Norman Biggs for drawing this volume to our attention.

140   A. E. Berriman, *Historical Metrology* (London, 1953), 6; Pegolotti states that 500 pounds on the statera in Florence is equal to the Paris load, in Pegolotti, op. cit. (18), 201: 'Libbre 500 al peso dela stadera di Firenze fa in Parigi carica 1.' The Paris load is 350 pounds by inclined beam, so 364 pounds by level beam and 364 pounds was just the weight of the sack. The 'English' pound for wool is therefore $500 \times 5{,}241{\cdot}6 \div 364 = 7{,}200$ gr.

141   Pegolotti, op. cit. (18), 245: 'Sacco 1 di lana al peso di Londra, che è chiovi 52 in Londra, fa in Bruggia sacco 1 di lana'; and ibid., 237: 'Lana si vende in Bruggia a sacco, e dàssi per 1 sacco intero 60 chiovi, e ogni chiovo si è libbre 6, e ragionasi di 60 chiovi in somma pietre 28 di Bruggia …'

142   'E deit la leyne vendre par saks ou par toisons solum ceo ke il veit ke pru e la avantage seit greindre. E sy il vend par sacs, checun sak peisera xxx pieres de leyne en clovs, ou vint e ut pieres, [par pyere] a balance, bien peise par dreite piere de xii livres e demy': Oschinsky, op. cit. (48), 272-3, c.29. It is noted (p. 299) that sacks of 350 and 375 pounds differ from the conventional weight of 364 pounds, but no inferences are drawn.

143   *Seneschaucy*, in Oschinsky, op. cit. (48), 82.

144   Oschinsky, op. cit. (48), 299: 'The texts of A, except A7, reckon by the stone of 12lb, the sack of wool is here 360lb if weighed by cloves and 336lb if weighed by stones.' These are 30 and 28 stone respectively each of twelve pounds.

145   *WME*, 137, quoting from R. A. Pelham, 'Exportation of Wool from Winchelsea and Pevensey in 1288-9', in *Sussex Notes and Queries*, 5 (1935), 205-6.

146   Mary Bateson (ed), *Records of the Borough of Leicester* (London and Cambridge, 1899), 191; *WME*, 131.

147   It may have been a move from an inclined-beam measurement at 28 stones of twelve pounds to a level-beam weighing at 28 stones of 13 pounds that led Spanish wool merchants purchasing wool at Southampton in 1290 to complain that they were losing half a pound in the clove: J. Strachey (ed), *Rotuli Parliamentorum* (London, 1767-7), I, 47: 1290; *WME*, 135.

148   There were 374·4 avoirdupois pounds to the sack by level beam.

149   Eileen Power and M. M. Postan, *Studies in English Trade in the Fifteenth Century* (London, 1966), 248.

150   Chisholm, op. cit. (20), 12. These appear to differ, both in denominations represented and in the weighings recorded, from the group of weights attributed to Elizabeth's first issue of 1558 which were subsequently tested by Chisholm and which are preserved in the Science Museum, London, or at the least there may be reason to suspect damage or adjustment after 1574: see ibid., 19.

151   Power and Postan, op. cit. (149), 57.

152   Hanham, op. cit. (119).

153   Ibid., 125.

154   Ibid., 118, 119.

155   For example, Alison Hanham found no suggestion that tare allowances were made on woolsacks at Calais, and quotes G. Schanz, *Englische Handelspolitik gegen Endedes Mittelalters*, 2 volumes (Leipzig, 1881), II, no. 130, which states that 'in Calais … the canvas which is about all the sarplar is sold [as] wool': Hanham, op. cit. (119), 127 and note 92.

156 Hanham, op. cit. (119), 118. Swetz, op. cit. (136), 136: 'The hundredweight of wool …, tare being 3 per cent'; and see ibid., 232.

157 D. M. Walker, *A Legal History of Scotland*, 5 volumes (Edinburgh, 1988-98), I, 96, 106.

158 *APS*, I, 668.

159 '… et due Waye faciunt unum saccum lane & xij sacci constitutuat le last': *Statutes*, op. cit. (44), I, 204; and 'Et due wayne lane facient vnum saccum et xij. Sacca facient vnum lestum': *Fleta*, op. cit. (49), II, ch. 12.

160 National Library of Scotland (NLS), Adv. MS 34.7.6., entitled 'Treaties Scots Merchandise'. This manuscript is described and published in part by Alison Hanham, 'A Medieval Scots Merchant's Handbook', in *Scottish Historical Review*, 50 (1971), 107-20.

161 Hanham, op. cit. (160), 115.

162 Evans, op. cit. (18), 250 note 1.

163 'De pondere petre · Item lapis ad lanem et ad alias res ponderandas debet ponderare · xv · libras · Item vaga debet continere · xij · petras cuius pondus continet · viij · libras · ': *APS*, I, 673.

164 Ibid.

165 The only manuscript collection which differs from these is the Ayr manuscript, datable to *c*.1325: this omits any mention of the stone of wax, and fails to say how many stones make the wey: National Archives of Scotland (NAS), PA.5/2A (see Chapter 5). Balfour's *Practicks* only says the stone for wool is 15 pounds: McNeill, op. cit. (42), 90.

166 John Skene, *Regiam Majestatem: Scotiae veteres leges et constitutiones* (Edinburgh, 1609), part 1, f.161, and part 2, f.68v-69r, the reference to Robert III being as follows: 'Vncia continebat tempore Regis David praefati, viginti denarios. Tempore autem dicti Regis Roberti, continebat viginti vnum denarios: Nunc autem nostris temporibus; videlicet, Roberti tertii, Anno gratiæ 1393. Vncia continet de ejusdem Regis monêta, triginta duos denarios.' (The legal expression for the ounce in the time of King David was 20 pence. However, in the time of King Robert it was said to contain 21 pence. Now, however, in our time, namely Robert III, in the year 1339, the ounce similarly contained 32 pennies of the King's money.)

167 'Duodecim librae Londonienses faciunt petram': ibid., part 2, f.68v.

168 Hanham, op. cit. (160), 119.

169 Pegolotti, op. cit. (18), 238: 'Cera in pani si vende in Bruggia a peso, e dàssi per uno peso libbre 180 di Bruggia, e vendonsi a pregio di tanti marchi d'argento al peso, e di 31 soldi, denari 4 parigini il marco, ed i denari 21 di detti parigini 1 grosso tornese d'ariento.'

170 If the weights in the London Guildhall in 1372 were

171 avoirdupois (and this post-dates a surviving 1357 set of avoirdupois at Winchester), then this is specified as 'fifteen ounces to the pound': Biggs, op. cit. (37).

171 Johnson, op. cit. (14), 76.

172 Pegolotti, op. cit. (18), 255: 'In Londra si à 2 maniere di pesare argento, cioè il marchio della zecca della Torre di Londra, che è appunto col marchio di Cologna della Magna, e l'altro si è il marchio degli orfevori cioè degli orafi di Londra, ch'è più forte e più grande marco che quello della Torre sterlini 5 e ⅓, di sterlini 20 per 1 oncia e d'once 8 per 1 marco.'

173 Biggs, op. cit. (37).

174 *Statutes*, op. cit. (44), I, 204-5. Attributed to 1302/3, but more likely of the second half of the thirteenth century. See also *WME*, 320-1. This passage is not in the White Book of Peterborough Abbey of *c*.1253, which does contain much of the rest of the *Tractatus*: Hall and Nicholas, op. cit. (47), 11.

175 There are a number of variants of the description of the English gallon, for example: *Statutes of the Realm*, from Liber Horn.: 'and 8lb make a gallon of wine' [et VIII libre faciunt galonem vini]; *Statutes at Large,* from Cotton MS Claudius DII: 'and 8lb of wheat make a gallon' [et octo libre frumenti faciunt galonem]; MS Reg. 9A II, f170b, (*c*.1300), for which see Hall and Nicholas, op. cit. (47), 8, 9; 'The Record of Caernarvon', in Sir Henry Ellis, (ed), *Registrum Vulgariter Nuncupatum* (London, 1838), 242 (these two last state that '8lb wheat make a gallon of wine …', 'et octo libre frumenti faciunt galonem vini'); Harris, op. cit. (62), 396: in 1474, '… xii ounce maketh a Pound for siluer, golde, bred & Measure; wich weygth makith a pynte of Whete; and ii pyntes maketh a quart; and ii quartes maketh a Pottell and ii Pottels makith a Gallon; & VIII Gallons makith a Buysshell …'; and 12 Henry VII, c.5, 1496 '… every gallon contains 8lb of whete of troi weight'.

176 *WME*, 153; 12 Henry VII, c. 5, 1496: 'That the Measure of a Bushel contain *viii* Gallons of Wheat, and that every Gallon contain *viii* li of Wheat of *Troy* Weight, and every Pound contain *xii* Ounces of *Troy* Weight, and every Ounce contain *xx* Sterlings, and every Sterling be of the weight of *xxxii* Corns of wheat that grew in the Midst of the Ear of Wheat, according to the old Laws of this Land.' The italics are in the original.

177 Biggs, op. cit. (37).

178 *WME*, 150.

179 *APS*, I, 674.

180 'Lagena debet continere xii libras atque ita videlicet quod quatuor libre sint de mari, quotuor de lacu, et quatuor libre sint de aqua currenti sciliat clara': NLS Adv. MS 25.5.10, f.80v. We are grateful to

Professor Philip Grierson for his transcription of this manuscript.

181 For Balfour, see McNeill, op. cit. (42), I, 89, but attributed to *Iter Camerarii*, c.34; Skene, op. cit. (42), f.161r.

182 The Cromartie manuscript is discussed further in Chapter **5**: NLS Adv. MS 25.5.10, f.80v.

183 *APS*, I, 674.

184 *WME*, 162-4.

185 5 Anne, c. 27, s 17, 1706.

186 Berriman, op. cit. (140), 162-4; *WME*, 162-3.

187 Berriman, op. cit. (140), 164.

188 Carysfort 1st Report, op. cit. (17), 432.

189 Challis, op. cit. (38), 268, 325.

190 Personal communication from David Beasley, Librarian, Worshipful Company of Goldsmiths, London, December 1994.

191 E. G. R. Taylor, *The Mathematical Practitioners of Tudor and Stuart England, 1485-1714* (Cambridge, 1967), 186.

192 These were described by Maurice Stevenson, 'The Size of Liquid Measures in the 17th and 18th Centuries, Part 3', in *Libra*, 3 (1964), 9-13, p. 11; and *WME*, 158, 168 note 29.

193 John Wybard, *Tactometria, or the Geometry of Regulars* (London, 1650), 267.

194 Ibid., col. b1v-b2r, 266.

195 E. F. MacPike, *Correspondence and Papers of Edmund Halley* (Oxford, 1932), 213, listing extracts from the MS Journal Books of the Royal Society.

196 Some of the original correspondence is quoted without reference in Berriman, op. cit. (140), 163.

197 This Reynolds vessel is associated with Bridport in Somerset and is inscribed: 'A BEARE OR ALE GALLON CONTEYNING 288¾ SINGLE SQVARD INCHES AND 36 OF THESE GALLONS ARE CONTEYNED IN THE LONDON BERE BARRELL SEALED AT COOPERS HALL THIS GALLON WAS SIZED AND SEALED IN THE TOWER OF LONDON BY JOHN REYNOLDS THE 30TH OF APRIL ANO 1659.' See Stevenson, op. cit. (192), 11, which includes information supplied by C. H. Canvin. The measure was described as in Somerset County Museum, but it was unlocated in December 1994: personal communication from Mark Davis, Registrar, Somerset County Museum. The measure carries a Tower verification mark with Reynolds's initial, and the capacity given by Reynolds was confirmed in a modern measurement in 1964. This gives us confidence in his technical ability and shows at least that the accuracy of his measure-ment and calculation can be accepted.

198 Taylor, op. cit. (191), 194, citing the mathematics teacher Henry Phillippes (*fl.*1648-77). Wybard refers to the 5:4 ratio as being 'according to Mr John Goodwyn long agoe, in his little Tract entitled, A Table of gauging, published 50 years since' (that is, *c.*1600). No copy of a separate publication with this title has been found and it is not listed in A. W. Pollard and G. R. Redgrave, *A Short-Title Catalogue of Books Printed in England ...,* second edition (London, 1976-86). The Reynolds table is probably that contributed to the second edition of W[illiam] B[adcock], *A New Touchstone for Gold and Silver Ware* (London, 1679), discussed by William O'Sullivan in the introduction to the New York, 1970, reprint of Badcock's work: we are grateful to David Beasley, Librarian, Worshipful Company of Goldsmiths, London, for this reference.

199 This size of measure, however, is only known from a single standard constructed and adjusted by Rey-nolds himself, and so it provides no independent confirmation that the 4:5 ratio properly applied to the wine gallon of 231 in³. The size was inadvertently ascribed also to a gallon of 1601, from the Guildhall collection, now in the Museum of London (Inv. 13,537): *WME*, 159-60. This was apparently on the basis of its description as containing very nearly 10 wine pints (at ten pints it would exactly comply with the 4:5 ratio): 'Q. W.', 'There were Ten Pints in Elizabeth's Gallon', [*London*] *Evening News,* 10 April 1935.

200 Pegolotti, op. cit. (18), 148: 'Libbre 100 sottile di Vinegia fanno in Parigi libbre 62½, d'once 15 per libbra ...'; ibid., 221: 'Libbre 530 di Genova fanno in Parigi carica 1, ch'è in Parigi libbre 350 d'once 15 per libbra'; ibid., 236: 'In Parigi ... la libbra si è once 15 di Parigi.'

201 Note, however, that a gallon of 317·1 in³ closely matches the twelfth of the boll size calculated from the dimensions in the Cromartie Manuscript (318 in³). No obvious explanation presents itself: there is no known Scottish precedent for a large measure such as this being generated as other than a dry measure, so that a volume of twelve times the gallon's volume would not be expected because a grain compaction effect (as discussed in Chapter **6**) would be involved.

202 R. E. Zupko, *French Weights and Measures before the Revolution* (Bloomington, Indiana and London, 1978), 43, 136; Horace Doursther, *Dictionnaire Universel des Poids et Measures* (Anvers, 1840, reprint 1965), 428-9.

for ydilnes Ande yat keip pad I referrit til it be knawin quhat or yai
lieff ande at the intre be consent hit of pat Ande yairvpon ye saide
heif sall assales Ande ye saide borowris quhill ye quhilk borowris findyng ye heif
sall assay yol dave to ye ydil men to get pad maistre of to reforn pad
to lefull craftis Ande yan yol dave heand dane yat yai be fundit mad ydil
ye heif sall arreist pad apwayn ande founde pad to ye kingis ply to byde
ande be pwnyst at the kingis will Ande at yis be done als moche
I borowris as to lande thron at ye ... of

In the xviii n° 67

tem the kyng with consent of the thre estatis of the realme has ordanit yat
all statutis e ordinans of yis plament e of the tyba plamentis precedand
be kepit grat I the kingis burowis Ande ylik to ye heeff quhilk
statutis e ordinans ilk heif be haldyn to publis opinly in the ye
place of yir frandom ande in vthir notable placis Ande als to ghew
the copie of yad bath to plats baronis e borowis of yis balyery
apon ye expens of the astatis Ande at ilk heif yat be kepit
ye tenor of yis art and payn yf fynation of yis office Ande at ilk
heif salbe opinly bidung to ye pupil of yis balyery bath on lande
ande in burgh to kep e feff all statis e ordinans aland I the
said their plamentis and ye parcine to hent I artis of yad tyba
yat name yat cause to pretend na alleege any Ignorance yis

Jacobi primi   4 p

Ista hedris Jacobi de ponderibus et mensuris per totum regnum Scotis de
novaliter constituend facta apud Perth in parliamento tento Ibid yir de mensis
Marcii anno regni sui xxxi per consensum et assensum trium Statuum ibid yrfirmitum

In the xviii n° 68

In the furst thai ordanit ande statuit ye vine to reforme opinly insche as is intent
In the Statute of kyng David the furst playnly made

69

tem thai ordanit ande Statute the stane to way sex wolt and vthir marchandise
to contey yat pund tron Ilk tron punde to contey yal sex Ande that stane to
be devidit in half stane quart half quart ande punde yis

5

James the First's Assize:
Interpreting the
Parliamentary Record

OUR EARLIEST VIEW OF AN INTEGRATED SCOTTISH SYSTEM OF WEIGHTS AND measures is provided by the assize traditionally attributed to David I, who reigned from 1124 to 1153. No twelfth-century version of this document is known, and there is no clear evidence that it dates from David's time: the attribution probably rests on David's long-established reputation as the source of all early legislative wisdom. The most likely date of the version of 'David's Assize' which has entered the judicial record is the mid-fourteenth century although it may well incorporate aspects of earlier codes. Nearly a hundred years elapsed before the next important piece of metrological legislation – a weights and measures assize passed by a parliament of the young Stewart monarch James I in 1426.[1] This assize was preceded in the same year by three short but important parliamentary acts which must be considered in conjunction with the 1426 Assize.[2]

The significance of James I's 1426 Assize lies not only in the careful definitions of the permitted weights and measures, but also in the fact that some of these are given in terms of the old measures in use in the period beforehand, that is, those attributed to David I. Although this reference backwards was intended partly as an appeal to the authority of the old definitions deriving from David I, it also provides a valuable opportunity to assess the nature of the changes brought about by this impressive piece of legislation. Indeed, David's Assize can only be understood properly in the light of the content of the Assize of James I. For this reason it will be considered in Chapter **6**, once the 1426 Assize has been explored. But it is only after reviewing the printed texts of the acts of the parliaments that emerged from the hands of various editors that we will be in a position to examine the content and context of the 1426 Assize.

The 1426 Assize is the first clear piece of Scottish metrological legislation that has survived in something close to its original form. It is part of a great body of legislation passed within a very short time by a rapid succession of parliaments called by James I which opens a new chapter in Scotland's

Opposite:
Detail of Fig. **5.4**.

legislative record. Very little material from parliaments before 1424 survives, but the record of parliamentary acts from 1424 is fairly complete, and is continuous from 1466.

It is not surprising that the early official records of Scotland are fragmentary or incomplete. Many of the records were removed by Edward I of England in his punitive expedition of the late thirteenth century, and although some were returned, none of these has survived to the present day. Material has been lost in a variety of misfortunes. For example, some serious losses have been attributed to fire, and Athol Murray has pointed to the damage done to Edinburgh Castle, where the chancery Register was located, by English cannon in the siege of 1572-3.[3] The main body of records survived the siege and burning of Edinburgh by the Earl of Hertford in 1544 (although current records were lost); then during Oliver Cromwell's Scottish campaign the records were moved for safety from Edinburgh to Stirling, but confiscated by his victorious general, George Monck, at the fall of Stirling Castle, and removed to London.

In 1653 the English parliament ordered a partial return of the Scottish records, consisting in the main of land transactions, securities and contracts

Fig. 5.1

Siege of Edinburgh Castle, 1572-3, detail showing both trons on the High Street. From the *Bannatyne Miscellany; containing original papers and tracts, chiefly relating to the History and Literature of Scotland, volume II* (Edinburgh, 1886).

JAMES THE FIRST'S
ASSIZE:
INTERPRETING
THE
PARLIAMENTARY
RECORD
———

173

between private parties, and this was done in 1657. Register House in Edinburgh is now the repository of some 1600 bound volumes of this material. The public records were retained in London, but there was a subsequent decision to return them in part at the Restoration of the monarchy in 1660. The papers were sent by sea and, in the custom of the time, they were packed in large barrels and 95 of these were put aboard the frigate *Eagle*. Unhappily, a storm arose and 85 of the barrels were transferred for safety to another vessel, the *Elizabeth*. This sprang a leak shortly afterwards, but it could not be repaired because it was in a part of the ship which was blocked by the barrels themselves. The *Elizabeth* sank with the total loss of the contents.[4] No proper inventory of the returned papers had been made so the full extent of the loss cannot be assessed.[5] But of the ten barrels which remained on the *Eagle* and were subsequently delivered to Leith, a not-much-better fate was in store.[6] It is recorded that they were still unopened in 1740 and it was not until 1753 that the contents were properly sorted and examined.[7]

Where the original and authenticated record had been lost, the statutes could only be reconstructed by reference to any available copies made for legal purposes. The problem of reconstructing the record of the parliamentary acts of James I is that no original parliamentary records survive from his reign, or even from that of his successor James II, and there are remarkably few original official records before 1466. The rolls, registers (originally enrolments of grants under the Great Seal) and other archives of chancery, the exchequer and parliament were in the custody of a clerk who came to be a senior legal official, and they were retained in the Register at Edinburgh Castle. The Clerk of the Register, or Clerk Register, had responsibilities which included aspects of parliamentary draughtsmanship as well as the maintenance of the record of business. In James I's time this record of statutes and ordinances was a roll in which the proceedings were registered from working records at the end of each session.

Parliament also had judicial functions, and during the fifteenth century these were increasingly delegated to the 'council', on which the Clerk Register served. When this emerged as the highest civil court in 1504, the Clerk, now Lord Clerk Register, was one of its judges. His complex responsibilities for various categories of legal and official records changed over the centuries as the modern concept of a national record repository developed. However, the Lord Clerk Register and his deputies remained the officials responsible for the survival and quality of the legislation that is central to this study.

For the period of James I's weights and measures assize we are fortunate in that there are two important manuscript versions of the acts of parliament which are fifteenth-century contemporary copies. These are preserved in the National Archives of Scotland, the modern successor of the Register. They are now bound together as a single volume, as part of the series of Parliamentary Records, with the reference number PA.5/6.

James I recognised the deficiencies in the official record that he had inherited and in spite of various moves to improve matters the situation continued to cause concern. In 1426 he proclaimed that a committee of six wise and knowledgeable men who best knew the law should be chosen to examine the books of law, namely the *Regiam Majestatem* and *Quoniam Attachiamenta*, and 'mend the laws that needed mending'.[8] The concept of correction and revision of the laws to bring them into line with current practice is important to note. At the same time it was ordered that all statutes and ordinances of that parliament and the two preceding parliaments be registered officially and given to the sheriffs to be publicly proclaimed, with copies provided to the barons and the burghs (at their expense), so that no one could plead ignorance of the laws.[9] A similar edict was made in 1457, in the reign of James II.[10] Of course, without the technique of printing, these subsidiary copies all had to be generated by clerks in manuscript.

A new note was struck in the reign of James III, when an act of 1469 describes commissioners being empowered to put the King's laws, the *Regiam Majestatem*, acts and statutes and the contents of the other law books into one volume. This was to be authorised, with the rest being destroyed.[11] As a result of such directives, some at least of the original manuscript records were undoubtedly destroyed (although perhaps not many), but later transcripts remained; these found their way into legal compendia, but were not themselves the official records.[12]

PRINTED EDITIONS OF THE ACTS OF PARLIAMENT

Printing was introduced into Scotland only at the beginning of the sixteenth century, and in 1507 Walter Chepman and Andrew Myllar of Edinburgh were granted a royal privilege to print the books of the laws.[13] However, it was not until 1541, when a selection of the acts of James V was produced by the printer Thomas Davidson, that any legislation emerged from the presses.[14] No doubt this and subsequent publications were encouraged by an enactment of 1540 that the king's acts were to be published throughout the realm. The Clerk Register was to make an authentic extract and copy of all the acts, to be printed by a printer of his choosing.[15] The 1541 volume was followed in 1565 by the publication of the acts of one of the parliaments of Queen Mary from 1563.

It was only in 1566 that the first comprehensive or full printed edition of the acts of parliament appeared. Two editions followed in the 1680s, but considerable reliance was also placed on the semi-official editions collected by Sir James Balfour and Sir John Skene that were described in the

JAMES THE FIRST'S
ASSIZE:
INTERPRETING
THE
PARLIAMENTARY
RECORD
———

175

previous chapter. These were considered to have been very largely superseded by the early nineteenth-century 'Record Edition' begun by Thomas Thomson and completed by Cosmo Innes. The important aspect for us to decide in relation to the James I Assize is the extent to which the editors relied on the conclusions of their predecessors.

The '**Black Acts**' (**1566**) is the name given to the first extensive printing of the acts of parliament of Scotland, which appeared in two editions, published within six weeks of each other in late 1566.[16] The volumes were produced by the Edinburgh printer and binder Robert Lekpreuik, and it was his use of 'black' or Gothic type that led to the edition becoming known as the 'Black Acts'.[17] The production of two issues a bare six weeks apart arose from the fact that the first (October) version included a number of anti-Protestant statutes of the 1530s. Although it was politically insensitive to include these, they had not in fact been repealed. These Catholic acts were hastily removed for the replacement issue of November and the earlier issue was suppressed.[18]

This collection of acts was the work of seventeen commissioners appointed by Mary, Queen of Scots, in May 1566, under the Great Seal, at the initiative of John Leslie, the Catholic Bishop of Ross.[19] The commissioners' mandate empowered them not simply to print the acts as they were found in the record, but with

> full powar and autoritie, expresse command and charge in Her name and behalf, to visie, Sycht, and correct the lawis of this Realme maid be Her and Her maist nobill progenitouris … beginnand at the buikis of the law called *Regiam Majestatem* and *Quoniam Attachiamenta*, and swa consequentlie following be progress of tyme unto the dait of this commissioun.[20]

In fact, the commissioners did not do this. They began with the proceedings of the 1424 first parliament of James I and continued on to 1564 (although a number of acts were omitted). The work of examining the more ancient laws may have begun, but was not completed as 'thay requyre langar tyme to thair dew correctioun', and no material before 1424 was published by the commission. For what follows later, it must be stressed that the power to correct the laws as they were found had been given to the commissioners. We shall see that in so far as weights and measures are concerned, this may well have happened. The printed acts were to represent current law, and not to reproduce faithfully the historical sequence of parliamentary processes. In effect the volume was to be the equivalent of the present-day *Statutes in Force*.

The 'Black Acts' were published by Edward Henryson, one of the royal commission members, but the text was authenticated by Sir James Balfour (*c.*1525-83), as Clerk Register, and is therefore sometimes referred to as Balfour's 'Black Acts'.[21] Balfour's contribution is made explicit in the preface:

... and Schir James Balfour of Pittindreich, knycht, Clerk of the Register for his sincair asald and glaid concurrence to perfyte this wark and exhibitioun of the originalls out of the Register and making of thame patent at all tymes, on na wayis regardand his awin particulare outher proffeit or gloir bot onlie the commoun weill of the Realme.[22]

Balfour's *Practicks* (*c.*1580) is a manuscript compendium or digest of Scots law covering the early period as well as legislation after 1424, and appears to have been written between 1574 and 1583. The work was prepared by James Balfour, but not in his capacity as Clerk Register and it therefore cannot strictly be considered as part of the public record. Nor was it actually published until the mid-eighteenth century, although it was widely circulated in manuscript form.[23] Nevertheless, it has its origins in Balfour's service on two royal commissions for the revision of the laws and it has an authority which has given it great influence. Balfour was strongly implicated in the murder of Queen Mary's husband, Henry, Earl of Darnley, one of the events which led to her abdication in favour of her infant son James in 1567. Her half-brother the Earl of Moray secured the Regency and Balfour was ousted as Clerk Register that year in favour of his predecessor, James Makgill. By way of compensation he secured the position of Lord President of the Court of Session, but shortly afterwards lost that as well.

A subsequent regent, the Earl of Morton, established another legal commission in 1574 to create a body of Scots law, instructing that they

> sall begin and visite the bukis of the law actis of parliament and decisionis befoir the sessioun And draw the forme of the body of our lawis alsweill of that quhilk is alreddy statute as thay thingis that wer meit and convenient tobe statute ... Quhairthrow thair may be ane certain writtin law to all oure souerane lordis iugeis and ministeris of law to iuge and decyde be.[24]

Once again the task devolved on James Balfour, who was co-opted by the commission. In this labour he was assisted by John Skene (1534?-1617), an able advocate and statesman who enjoyed Morton's confidence.[25] However, unlike Balfour, Skene survived Morton's fall from grace in 1581.

Although Balfour's work continued for several years, no official publication was forthcoming, probably as a result of the political turmoil which eventually led to Morton's execution in June 1581. The effort was not lost, however, because Balfour's compilation survived in manuscript form as the *Practicks*. It has already been noted that, for the statutes passed after 1424, Balfour appears to have been content to rely on the accuracy of his 1566 publication of the acts. In spite of this, the *Practicks* was highly influential and considered a work of great authority. To judge the quality of this work, two quotations might suffice. Lord President Inglis in 1861 described it as 'of

JAMES THE FIRST'S
ASSIZE:
INTERPRETING
THE
PARLIAMENTARY
RECORD

undoubted authority', while Hector McKechnie in 1931 declared the *Practicks* to be 'the primary authority for early Scots law outside of the Register House'; indeed, as recently as 1920, Balfour's text of an Act of 1491 was preferred in the House of Lords to that of the nineteenth-century 'Record Edition'.[26]

Skene's Edition (1597) covered legislation between 1424 and 1597, and is closely related to Balfour's 'Black Acts'.[27] As noted above, Sir John Skene had previously assisted Balfour, and Skene is perhaps best remembered for his 1609 edition of the collection of the Auld Laws known as *Regiam Majestatem*, mentioned in Chapter 1. Skene's edition of the Statutes from 1424 was initiated by a parliamentary act of 1592 entitled 'For visiting and causing the laws and acts of parliament to be printed'. A new commission was to be set up of eight members including the then Chancellor (John, Lord Thirlestane), the Clerk Register (Alexander Hay), and Skene:

> to visite the laws and actis maid in this present parliament and all utheris municipall lawes and actis of parliament bygane quhairof thair is Registeris or autentik monumentis extant and to considder quhat lawis or actis necessarilie wald be knawin to the subjectis qlkis suld be kepit and obeyit be thame and to mak thame Inexcusable of ignorance To caus the samyn lawes and actis be copijt and autentik copies subscryuit to be deliuerit to his heines prentar Togidder or seuerallie as the said lord chancellair and personis Joneyd to assist him sall think expedient ...[28]

Again, it was to be the law as it stood.

Most of the work appears to have been shouldered by Skene, who replaced Hay as Clerk Register in 1594; however, Skene appears to have been guided by Balfour's printed version of 1566 rather than by the original records, which rarely seem to have been consulted.[29] Many acts were omitted or suppressed, only a few were added. However, this official compilation was well regarded, and the Privy Council directed the Lords of Session to ensure that it was purchased by all subjects of sufficient 'substance and habilitie'.[30]

The **Murray of 'Glendook' edition (1681)** was the last general collection of Scottish statutes before the 'Record Edition', and it was again restricted to the period after 1424. Sir Thomas Murray of Glendoick (1630?-84) was already a Lord of Session and had been created a baronet of Nova Scotia before his appointment as Clerk Register in 1678. The post had been vacant for some time following a period when the volume of material in the Clerk Register's care had been greatly increased with the arrival of Court of Session records and the establishment of a second building as a records repository. It is clear that in at least some respects the records were kept under damp and chaotic conditions at this time; this may go some way to explaining Murray's marked lack of diligence in checking the printed version against the original manuscripts.

Murray's authority, granted by Charles II at Whitehall in 1679, gave him

> full power and licence ... to cause the whole Acts, Lawes, Constitutiones and Ordinances of Parliament of his Majesty's ancient Kingdome of Scotland both old and new now being in force ... to be re-imprinted by whatsoever Printer within the said Kingdom of Scotland or elsewhere, it shall please him.[31]

Murray's copyright privilege, which he farmed to the Edinburgh printer David Lindsay, was ratified by the parliament in Edinburgh in 1681.[32] Two editions of Murray's work were produced. The first was a large format (folio) version in 1681; the second was a pocket-sized (duodecimo) edition which appeared in the following year, and being so small and portable it had enormous success. Although Murray was by this time out of office, Lindsay was acting on his privilege to reprint the earlier edition. (Subsequently, the pocket-sized volume was augmented by another volume which extended the work to 1685.)

Both versions carry on their title pages the declaration that the acts were 'collected and extracted from the publik records of the said [Scottish] Kingdom'. Sadly this was not the case, except for the obvious requirement to use the official statutes enacted since Skene's edition. Instead of going back to the original records for his edition, as he claimed, he 'copied implicitly from that of Skene of 1597', even to the extent of reproducing scribal or typographic errors; 'in fact even the more accurate and ample edition of 1566 does not appear to have been consulted.'[33] A dependence on Skene's edition can even be inferred from Murray's initial license, which granted him a copyright also over Skene's *De Verborum Significatione* (a glossary and digest of legal technical terms from the old books of the law), which was appended to Skene's 1597 edition and which Murray added at the end of his 1681 folio edition.

Shortly before Murray's copyright privilege was ratified by parliament in 1681, another act was passed in the same parliament which appears to set a very different context for Murray's work.[34] This act made provision for setting up a commission for the

> peruseing and considering the whole Laws Statuts and Acts of Parliament of this ancient Kingdom as weel printed as not-printed ...; to call for all the Registers and Records which containe the saids Laws or Practicks, Either from the Clerk of the Register, Justice-Clerk or their Deputs ...; to determine the true Sense meaning and Interpretation of all such Laws ... as are unclear or doubtfull in themselves ...; to digest and reduce the same into such convenient order As they shall judge fitt: Leaving out the obsolet and abrogat Acts, That the Acts in vigor in the severall Parliaments may be printed together and the rest remaine as unprinted Acts ...[35]

JAMES THE FIRST'S
ASSIZE:
INTERPRETING
THE
PARLIAMENTARY
RECORD

179

Possibly the move to set in hand such a radical review was prompted by detailed work being undertaken at the time on the Court of Session records. It is interesting to note the reference to the existence of acts that had never been printed and also the need to end up with a compilation that accurately reflected the current legal position 'leaving out the obsolete and abrogat Acts' – again, only current law was to appear. Although these requirements are very understandable, they illustrate the continued pressure to delete or modify statutes that provided administrative definitions of the evolving weights and measures. However, this machinery was not set in motion. Murray was dismissed from his post in 1681, and his edition of the acts seems to have been accepted as adequate.

Thomas Thomson's 'Record Edition' (1814-75) of the Scots statutes is undoubtedly the most extensive and detailed edition, and it has not been superseded. The advocate and legal antiquary Thomas Thomson (1768-1852) was appointed as the first holder of the new post of Deputy Clerk Register in 1806, and produced voluminous publications under the auspices of the Record Commission.[36] The most important of these was his edition of *The Acts of the Parliaments of Scotland*, covering the period from 1424 up to the union of the Scottish and English parliaments in 1707, which appeared in ten folio volumes (numbered II to XI) between 1814 and 1824. It was his intention to edit the laws prior to 1424, as far as they could be determined, as Volume I. The antiquary and professor of constitutional law at Edinburgh University, Cosmo Innes (1798-1874), joined him in 1830 to help with this volume, but it was not completed until 1844, after Thomson had left the Record Office.

The 'Record Edition' was further enlarged in the 1870s. Thomson had believed that the parliamentary records for 1639-50 (a large part of the reign of Charles I), had been irretrievably lost, and he had to rely on parliamentary warrants and similar auxiliary materials to cover these twelve years. However, in 1826, four manuscript volumes came to light containing the missing record. Innes re-edited the two relevant Thomson volumes and produced three new replacement volumes (VI, parts I and II, and VII). Innes also co-ordinated a prodigious index which was published posthumously in 1875.

We must recognise that Thomson's objectives were very different from those of the earliest compilers of the acts. Whereas they had been charged with collecting and revising the statutes to reflect current law, Thomson hoped to reconstruct a unified and definitive edition in historical sequence which adequately represented the approved text; that is, it should follow the authoritative original manuscript record from 1466, when it was first available, and should be an acceptable substitute in periods when the original was missing.[37] For earlier times he would use whatever was in existence. Where different versions of the record existed (as was certainly the case for the 1426

Assize), Thomson collated these versions and made decisions (necessarily subjective) about which parts to accept from each in order to produce a unified text.

In the context of the editing of the manuscripts of the old laws, Innes later claimed that the oldest copies were assumed to be the most correct.[38] However, Thomson made no effort to justify his choices or even to indicate which particular sources he considered to give the best reading.[39] Although Innes provided tabulations of the manuscript sources on a clause-by-clause basis for much of the old laws (unfortunately not including David's Assize), these manuscripts actually have to be examined before we can discover whether individual clauses coincide with Thomson's view. The difficulty of testing Thomson's conclusions (let alone detecting his adjustments) has meant that his edition has been widely accepted by default. However, his judgement has not passed unchallenged, and it has been stated that the 'Record Edition' is 'an edition devoid of critical apparatus'; that 'they are eclectic texts, their editors have not indicated the source of their readings'; and that it 'cannot be accepted as definite and final', at least in relation to the old laws.[40]

THE WEIGHTS AND MEASURES ASSIZE
OF KING JAMES THE FIRST OF SCOTLAND

The thirteenth century had been a period of expansion for the Scottish economy which was fuelled by the increasing success of Scottish wool exports to Flanders.[41] The central importance of Flanders as Scotland's main trading partner continued into the modern period. However, by way of contrast, the English aggression of 1296 and the forging of the 'Auld Alliance' between France and Scotland heralded a dark century in which the prosperity of the earlier age was lost. The combined destructive force of English invasions and the extended Wars of Independence dislocated the country until the 1360s and represented a serious drain on the country's resources, but it also severely damaged the burghs and their ability to provide a commercial basis for trade.

Only in the last quarter of the fourteenth century was some sort of stability achieved under the first Stewart kings – the elderly Robert II and the infirm Robert III. Comparatively few records survive of their reigns, but Michael Lynch has stressed that they were far from being ineffective and may have had more success in containing strife than their better-respected successor James I.[42]

The young James spent eighteen years as captive and pawn of successive English kings of the House of Lancaster, but returned to Scotland under a punitive ransom agreement made in 1424. A rapid succession of parliaments

JAMES THE FIRST'S
ASSIZE:
INTERPRETING
THE
PARLIAMENTARY
RECORD
————

181

held in the following years produced a stream of legislation which has given James the reputation of a progressive law-maker. Included among these was the 1426 Assize, which represented the first major revision of law relating to weights and measures since the Assize of David.

James had only been eleven when he was captured in 1406, very shortly before his father's death. Robert III had succeeded to the throne in 1390 after serving from 1384 as Guardian in place of his weak (and possibly senile) father Robert II; however, he was already an invalid by this time, having suffered an incapacitating accident in 1388. The Guardianship was passed to his brother the Earl of Fife, later created Duke of Albany. Robert's heir, the youthful and headstrong Duke of Rothesay was given responsibility as royal lieutenant in 1399, but after Rothesay's suspicious death in Albany's hands in 1402 Albany was restored as lieutenant-general. Robert sent his second son and new heir James to the comparative safety of France in 1406, and he was given a passage in a Danzig merchant vessel bound for Flanders with a cargo of Scottish hides and wool. However, the ship was intercepted by English pirates and James was sent as a captive to the Tower of London.

Robert III died very shortly afterwards and Albany was confirmed as governor in mid-1406. During his firm regency he acted as king in all but name, reflecting as much his own ambitions as the peculiar circumstance of acting as regent for an absent and uncrowned king. His half-hearted efforts to secure James's release were not effective and ceased altogether after the return from English captivity of his own son Murdoch in 1416. Albany died in 1420 and was succeeded as governor by Murdoch, now heir-presumptive to the throne. Although the second Duke of Albany lacked his father's ability, this alone does not explain Scotland's urgency to conclude a ransom agreement for James in 1423.

The reigns of Robert II and III and the governorships of the Albanys were constrained by extreme financial difficulties. The currency had been progressively debased, and from the closing decade of the fourteenth century there was a steep decline in income from the customs. The increasingly serious implications for royal finance can perhaps be illustrated by a variety of financial irregularities involving Rothesay, Albany and others. To some extent they are also complicated by the fact that the Albanys also acted as Chancellor.[43]

The events leading to James's release from captivity have been re-constructed by Alexander Stevenson in terms of the economic necessity of maintaining Scotland's external trade.[44] Although customs duty was levied on wool, woolfells (fleeces) and hides, wool was by far the most valuable generator of royal revenue. Moves were made in the 1390s to extend duty to cheap woollen cloth, which was Scotland's only manufactured export and which was being sent in quantity to Flanders, where it was not seen as competing with the much better quality Flemish cloth. But even when customs duty was levied on a wide range of exports in the later 1420s, wool still raised the great bulk of the customs duty.

An early assize of weights and measures, in which enlarged sizes were confirmed for the dry capacity measures, created the opportunity to generate additional returns from royal lands where rentals and tributes were paid in kind, and for this reason may have represented an attractive option for a cash-strapped incoming administration. Although we will demonstrate that in many respects this assize was a declaratory act, which followed the established practice of the market, it undoubtedly opened new avenues to exploit trading allowances for financial advantage. In this instance James may have been influenced by his experience in England because new features of the assize appear to mirror or emulate aspects of current English practice, including for example, the use of troy weight units.

The third parliament of James's reign was held in Perth in March 1426, although it was concluded in Edinburgh after the Lent recess.[45] The assize of weights and measures came towards the end of the session, and it was preceded by three short metrological acts, which turn out to be of crucial importance. Taken together, these four pieces of legislation allow us to appreciate some of the ways in which the somewhat artificial and stylised form of legal definitions differs from the trading and taxation context in which they were applied.

The provisions of the 1426 Assize of weights and measures are very similar to those of David's Assize. They deal in turn with the units of length, weight and capacity, but the greater part of this assize is devoted to definitions of the dry and liquid capacity units. As before, these capacity units are given both by their dimensions and by the weight of their contents. Before we can extract some general conclusions from the assize we should examine its structure. The version below has been rendered into modern English from the first (1566) printed collection of the Scottish acts, which covered the period from 1424 to 1564 (see Appendix A.3a). The text is continuous in the original, but for clarity the clauses have been numbered here:

1 On the measure of the ell:
   They [the estates of parliament] ordained and delivered that the ell shall contain 37 inches as is contained in the statute which King David the First made upon it.

2 What the stone shall contain:
   They ordained and made statute that the stone to weigh iron, wool and other merchandise shall contain 15 pound 'trois', each trois pound to contain 16 ounces and the stone is to be divided into half stones, quarters, half-quarters, pounds, half pounds and other smaller [weights].

JAMES THE FIRST'S
ASSIZE:
INTERPRETING
THE
PARLIAMENTARY
RECORD
———

183

3 Of the division of the boll and the measure of the firlot and the boll:

(a) They ordained the boll with which to measure victuals, to be divided into four parts, that is to say, four firlots should contain [or constitute] a boll, and that firlot is not to be made after the first measure nor after the measure now used but in a middle measure between the two.

(b) The boll shall contain in breadth 29 inches within the sides [at the bottom] and at top 27½ inches evenly from side to side and in depth 19 inches.

(c) The firlot shall contain in breadth evenly from side to side 16 inches under and above within the sides, the thickness of both sides shall contain 1½ inches and in depth it shall contain 9 inches, the half firlot and peck following thereafter.

(d) The firlot shall contain 2 gallons and a pint, and each pint shall contain by weight of clear water of the Tay 41 ounces, that is to say 2 pounds and 9 ounces 'trois'. So the gallon weighs 20 pounds and 8 ounces. So the firlot weighs 41 pounds, and the boll containing 4 firlots weighs 164 pounds.

(e) The old boll made by King David contained a sextern; a sextern contained 12 gallons of the old measure and each gallon weighed 10 pounds 4 ounces 'trois' of diverse waters. So the boll weighed 123 pounds; this new boll, now made, weighs more than the old boll by 41 pounds which makes 2½ gallons and a chopin of the old measure, and of the new measure [here] ordained 9 pints and 3 mutchkins.

Fig. 5.2
Balfour's 'Black Acts', showing the opening page of James I's Assize of 1426 (right), and the three preliminary acts of 1426 (left). (National Archives of Scotland)

Superficially, the assize seems straightforward, but on closer inspection it emerges as complex and apparently contradictory. Its interpretation has proved a stumbling block to earlier commentators, who have seen it as introducing sweeping change to metrological practice. In contrast, a new approach which first involves analysing the sources for the assize and assessing the internal consistency of the interpretation, allows the assize to emerge in a very different light. Now it can be seen as a declaratory act – one that enshrines current practice in new and more appropriate or secure definitions, but does not seek to introduce disruptive change.[46] In itself, this is a desirable feature for successful legislation, since otherwise it would be immensely more difficult to administer and enforce. Not only is this a valuable indicator that this interpretation of the assize is correct, it is also a testimony to the skills of James's legislators that the types of definition that they crafted remained the basis for metrological practice for over three hundred years.

There are two specific points which should be drawn out at this stage. The first concerns the weight system used to define the basic capacity unit, which is the pint, and the second involves the relationship between the capacity measures in terms of the weight of their contents.

THE STONE, POUND AND OUNCE OF 1426

The most careful of these definitions by weight is that given in clause 3(d), where the pint measure 'shall contain by weight of clear water of Tay' 41 ounces, or 2 pounds 9 ounces 'trois', whereas the larger measures merely 'weigh' a given number of pounds. The double specification of the weight in terms of ounces, and also in terms of pounds and ounces, is significant because it confirms that this 'trois' pound is one of 16 ounces. From the volume of the 1426 pint, which will be derived as 1·275 litres or 77·8 in^3 in Chapter 6, the 'trois' ounce emerges convincingly as containing 480 English troy grains, which is the weight of the English troy ounce.[47] It will be recalled that David's stone was 15 pounds, each pound being of 15 ounces; and each ounce was the weight of 20 sterling pennies, that is a monetary ounce of 450 grains. David's pound was therefore 6,750 grains and was the merchant pound at that time. The weight of the 1426 Assize trois pound was $16 \times 480 = 7,680$ grains. Earlier commentators have taken the 1426 Assize at face value and have concluded that the 15-ounce Scottish merchant pound of David's Assize was actually replaced by a 16-ounce pound in 1426 – but this is to misunderstand the nature of such definitions.

The 'trois' ounce is the same as the English troy ounce and it is unlikely to be a coincidence that the first mention in an English statute of the 'troy' pound of 12 ounces is in 1414, a mere twelve years before the 'trois' pound of the 1426 Assize appeared.[48] The earliest mention of English 'troy'

JAMES THE FIRST'S
ASSIZE:
INTERPRETING
THE
PARLIAMENTARY
RECORD
———

185

weight known to the editors of the *Oxford English Dictionary* is in 1390-1.[49] Scottish weights are given variously as pounds 'troyis' or 'trois', both meaning simply 'troy'. The Scottish administration's choice of a troy pound of 16 ounces reflected the practice of Europe, and certainly of France and the Low Countries, rather than that of England.

It must be appreciated that there is a clear distinction between the pounds appropriate for weighing different classes of goods. In England, the 15-ounce merchant pound could be used for weighing most materials, but the 12-ounce troy pound was necessary for gold, silver, pharmaceuticals, bread and 'measure'. These last two uses – for regulating the accepted weight of the staple food and the sizes of the capacity measures – were areas of important administrative concern, and they are stipulated in English regulations from at least the late fifteenth century.[50] We must, therefore, be aware of the possibility that troy units were similarly used for definition purposes in early Scottish legislation.

At this point we need the information given in the second of the three preliminary acts which were passed before the 1426 Assize of weights and measures. A principle purpose of this act was to require that weights were uniformly applied in both buying and selling. Similar sentiments were repeated at most assizes, which suggests that the fraudulent practice of unscrupulous merchants buying with heavy weight and selling with lighter weight was never eliminated and may at times have been widespread. But in addition, the preliminary act introduces a new pound, called the 'Scottis' pound. In the words of Thomson's 'Record Edition' of 1814, the act is given as:

> The king and the parliament has ordanit that thare be maid a stane for gudis [goods] saulde and bocht be wecht the quhilk [which] sall wey xv [15] lele [true] troyis pundis Ande [th]at the stane be diuidyt [divided] in xvj [16] lele scottis pundis Ande of it thare salbe ordanit half a stane a quarter ane half quarter a punde ane half punde ande vthir lese wechtis accordande therto with the quhilkis al byaris [buyers] ande sellaris of gudis within the realme sall [shall] by [buy] ande sell all gudis of wecht and with nane vther wechtis fra [from] witsonday furth nixt to cum Ande fra thin furth [then on] thir foresaide wechtis sal hafe course.[51]

Again, the stone is 15 trois pounds, and we know from the 1426 Assize that each pound is of 16 ounces of 480 grains, so the pound is 7,680 grains. The stone is thus equivalent to 15 × 16 or 240 troy ounces or 115,200 grains. But the preliminary act says that this stone is to be divided into sixteen parts, called 'Scottis' pounds, which must each be of 7,200 grains, and are clearly merchant pounds. The weights are divided in a binary fashion, down to a pound, half a pound, and 'other less weights', which implied that the binary division continues down to an unnamed ounce, or sixteenth of the pound. Each of these ounces was a monetary ounce of 450 grains, the same as in David's Assize. Confirmation of the continuing use of ounces which were of monetary weight comes in a reference in the *Iter Camerarii* of the fourteenth century which provides for the enforcement of the testing of liquid measures 'in the King's money', which should be interpreted as meaning by weight of contents, using weights based on the conventional monetary weight rather than troy weight.[52] We have to wait until 1511 for a specific reference, in Edinburgh's records, to the division of burgh weights to a sixteenth, but there is no reason to doubt that it applied at a much earlier date: the stone at one of the public scales had to weigh 'xvj lib, and euerye pund to wey xvj vnce'.[53] In fact, this reference is certainly to the heavier series of trone weights, and these will necessarily reflect the practice with the merchant weights.

If we extrapolate into the English context, we can appreciate that use of the tower or monetary ounce as the underlying unit of the London merchant pound, or *libra mercatoria*, allows for binary division and reinforces tower weight for use in commerce. The familiar, long-standing description of the merchant pound as 15 troy ounces can now be seen merely as the appropriate administrative and legal definition and not as a method of describing its use in the market-place. We have already explored this issue in Chapter 4 and shown that the 450-grain tower ounce formed the basis of a range of early English weight types. This scheme, which arises naturally from a resolution of the issues considered in Chapter 4, represents a break with the recent trend to consider English weights as based firstly on troy weight, and to see tower weight restricted to use in minting operations.[54] But the early Scottish use of monetary weight makes such a picture of English practice unlikely, and it is the second preliminary Scots act of 1426 that has proved crucial in this discussion. The nature of this break, however, is only one of perspective: we have argued that commercial weights are used in practice with binary divisions, so that the trading divisions of a 15-ounce pound are not the 15 ounces of its definition, but smaller units of a sixteenth of the pound, which are effectively 'ounces' of market-place use.

Here, however, we face the first major problem of interpretation. The text quoted above for the preliminary act was taken from Thomas Thomson's 'Record Edition' of the statutes, which is normally regarded as accurate and is in all material points the same as that in the oldest manuscripts. But another collection of statutes which is widely appealed to, and which has a much longer tradition as an authoritative work of reference, is Balfour's *Practicks*.[55] Balfour had been responsible for preparing the first printed edition of the Scottish acts; and not surprisingly, therefore, he followed the text of his 1566 edition of the acts when he extracted details of the 1426 preliminary act and the main 1426 Assize for the 'Measures and Weights' section of the *Practicks*.[56] All subsequent editors of the acts did likewise, except for Thomas Thomson, who altered the text of the assize but not the text of the preliminary act. In the 'Record Edition', Thomson used the 1566 text for the preliminary act, which we have given above, and recorded a stone

JAMES THE FIRST'S
ASSIZE:
INTERPRETING
THE
PARLIAMENTARY
RECORD

187

Fig. 5.3
Thomas Thomson (1768-1852);
engraving by Professor Carl
Schmid from a portrait by
Robert C. Bell.

weighing 15 troy pounds; but for the main 1426 Assize he gave the stone's weight as 16 troy pounds, where Balfour in the *Practicks* had given 15 pounds. The two are incompatible, but because the meaning of the preliminary act has been obscure, Thomson's alteration has passed unnoticed. Unfortunately, it has been Thomson's seemingly definitive work that has been used by commentators on the assize, and his '16-pound' version has become firmly established through repetition.[57]

In making editorial changes of this type, Thomson was conscious of discrepancies between the early manuscripts compilations he consulted, and this aspect must be explored before we can establish the true weight and the correct dimensions for the capacity measures in the assize. However, Thomson is also likely to have been influenced in this change to the stone's size, by another clause in Balfour's *Practicks*. We have already noted in Chapter 1 the group of fourteenth-century adjustments in David's Assize which take account of the reduction in weight of silver coin. Balfour gathered these together and added a further clause (which we have not found in any manuscript version of the David Assize), and, citing his sources as the *Leges Burgorum* and *Iter Camerarii* collections, stated:

> The stane, for weying of woll and uther geir, did wey xv. [15] pund, but now xvi. [16] pund.[58]

Taken together, these apparently conflicting pieces of legislation present a confusing picture which has puzzled everyone who has studied these texts. The preliminary act of 1426 which states that the stone should be 15 troy pounds or 16 Scots pounds does not mention ounces. The version of the 1426 Assize in the 'Record Edition' has only a troy pound of 16 ounces and a stone of 16 pounds, whereas earlier printed editions have 15 pounds; and finally, Balfour's *Practicks* records a change from a stone of 15 pounds to one of 16 pounds. It is certainly the case that the stone came to weigh 16 trois pounds – the 1587 Assize clearly describes such a 16-pound stone and a 16-ounce pound.[59] Unfortunately, the earliest independent statutory evidence of a 16-pound stone comes as late as 1563, and discussion of this sixteenth-century change must wait until Chapter **8**.[60] There is, however, an act of 1491 which confirms the use of a pound of 16 troy ounces in trade.[61]

As they stand these declarations are open to several different interpretations. For example, was the stone of 15 pounds to be divided into 16 smaller pounds after 1426, or was a new and heavier stone to be created by combining 16 rather than 15 pounds, and if so, then which pounds? In the late eighteenth century, the Scottish advocate John Swinton attempted to resolve the situation by arguing that the 1426 Assize increased the size of the pound from 15 to 16 ounces (no references to pounds of 15 ounces have been found after the 1426 Assize), and then by appealing to an apparent convention in Scots metrology that there should be the same number of pounds in the stone as there were ounces in the pound, settled for a 16-pound stone.[62] Thomson may have adopted a similar line, although he provided no justification for his choice of a 16-pound stone at this time. Perhaps this convention was somehow expected to exercise effective pressure on the administration to increase the stone's weight to bring it into line with the pound; but by increasing the size of the stone at the 1426 Assize, Thomson was imputing a potentially disruptive change in market practice, and, as has already been said, the evidence that the capacity measures will provide indicates that the 1426 Assize was a declaratory act which did not disrupt trade practice.

The approach that we have adopted here is fundamentally different from that of all earlier commentators. We have argued that weight definitions would necessarily be couched in terms of troy units and that this would tend to mask the way practical trading units were represented. By rejecting Thomson's version of the 1426 Assize in favour of Balfour's version (and justification for this will be given later in this chapter), we have shown that the stone of the assize is the same as the stone of the 1426 preliminary act, whose constituent parts were the 16 Scots merchant pounds whose ounces were of 450 grains. These merchant pounds were not replaced by the troy (trois) pounds of the 1426 Assize, but they continued in use; in Chapter **8**, the characteristics of an important surviving weight of 1553 will allow us to show that the Scots merchant weight survived until the mid-sixteenth century.

JAMES THE FIRST'S
ASSIZE:
INTERPRETING
THE
PARLIAMENTARY
RECORD
———

193

with the early practice of England, where water measure was used in the same manner, but in this case was a quarter larger than the statutory measure.[70]

The point to emphasise is that the size and operation of water measure is nowhere specified in the Scottish legislation. Yet, considering the importance of sea-borne trade to the economies of the royal burghs, and therefore to the economy of the country as a whole, this seems a surprising omission.[71] It serves as a further reminder of the comparatively narrow range of the provisions of a metrological assize, and of the considerable scope for mis-applying data from an assize in a trading context.

The James I Assize dates from the period when we have no original and fully authoritative versions of the statutes, and thus some aspects of the assize have to be recovered by comparing contemporary manuscript versions. There are some difficulties in interpreting the assize, and these have led commentators in the past to adopt widely differing views. However, the reading of the assize that we will develop here is that the legislators constructed the 1426 Assize as a declaratory act, in which trading practice remained largely unchanged. They retained unaltered the underlying liquid volume unit that formed the basis of the dry capacity measures: this had remained in use over an extended period and its evolution will be discussed in Chapter **6**. This demonstrates a practical stability in the liquid capacity measures, and we will see later that this continued through legislative change in the following century. In addition, the mass of the 'Scots' merchant pound remained that of 15 troy ounces (7,200 grains), and the stone of 16 of these pounds was unchanged at 115,200 grains.

The legislators also contrived to retain the existing large boll used as the basis for grain transactions. This had grown substantially (by over 50 per cent) since its definition in David's Assize, and it appears that a principle for the legislators had been the prevention of its further growth by specifying precisely (and for the first time in Scots law) the permitted excess allowed, or 'charity' in its use.

The practice of allowing a certain latitude in the use of grain measures – that is, permitting those conducting transactions to work in units which were larger than the ones legally specified in order to compensate them for potential losses such as those due to spillage – was clearly well entrenched by this time, and undoubtedly this had led to the progressive increase in the size of the boll, from the more modest measure of David's Assize to the much larger measure in widespread use in the period just before this assize.

In their attempt to stop this growth by specifying the allowance, the legis-lators defined two boll measures. We have termed these the 'legal' boll, which bears the correct proportion to the other measures, and the 'trading' boll, which had the allowance incorporated in it. Although this is the only assize which enables us to appreciate the difference between these two sizes (and in that sense it provides the legal precedent), the raising of 'trading' standards from 'legal' standards became a particular feature of the dry measures in the following two centuries, and we will use this in the chapters which follow to describe the growth of the capacity measures.

Apart from this central purpose of controlling the size of the boll, the assize enlarged several units, such as the gallon, by increasing the number of constituent parts they contained from that of David's day. However, as far as the merchant was concerned, the assize seems to have been successful in codifying the system without altering the trading sizes of the basic units of commerce – the grain boll, the stone weight, and the smaller liquid unit called the 'pint'.

Some aspects of the measurement system were certainly unchanged by the assize: the ell linear measure, for example, is stated to be as defined in David's time at 37 inches, but this is an exception, and the assize represents weights and capacity measures that have clearly changed from the earlier period. It also confronts us with significant difficulties in interpretation. Part of the problem arises because new numerical information is provided in the assize, but unfortunately there are also two significant discrepancies between various printed editions of the statutes, and this raises questions about the quality of other data. It has already been noted that the nineteenth-century 'Record Edition', edited by Thomas Thomson, differs from earlier printings of the acts (these are the printings of 1566, Murray's editions of 1681 and 1682, and the collections of Balfour and Skene) in describing the stone as containing 16 pounds. A more serious anomaly is the depth of the firlot measure, given by Thomson as 6 inches, but by all the other printed sources as 9 inches.

In order to decide on a preferred reading, we must take account of the relationship between these various printed collections and the manuscript sources on which they were based. From this, it is possible to conclude that Thomson was correct in using the 6-inch depth for the firlot; however, neither he nor the editors of other editions used the correct depth for the boll, which should be 9 inches rather than the 19 inches which we gave above when the provisions of the 1426 Assize were listed. Both of these are major differences which have a profound effect on our understanding of the capacity measures. Only when these issues have been resolved can the assize be properly assessed.

JAMES THE FIRST'S
ASSIZE:
INTERPRETING
THE
PARLIAMENTARY
RECORD
——

195

RECONCILING THE DIFFERENT VERSIONS OF THE ASSIZE

We have seen that three comprehensive editions of the Acts of Parliament of Scotland had been published before the nineteenth century, all beginning with James I's first parliament of 1424: the 'Black Acts' of Balfour (1566), Skene's edition (1597), and that of Murray of Glendoik (1681 and 1682). In addition, Balfour's *Practicks* of about 1580 provided a legal digest of recognised authority.

However, we have shown that the circumstances under which these four collections were produced have not demonstrated that improved sources were used for the later versions. In reality we find just one source – Balfour's initial compilation for the 'Black Acts' – with each successive version of the statutes being based relatively unhistorically on its predecessor. It is no surprise therefore that all four provide the same numerical information for the 1426 Assize. It does not follow that Balfour's version accurately reflected the original version of the assize. Indeed, some of its detail is not self-consistent, indicating that Balfour's source had already been subject to some alteration. Similarly, some divergence is already apparent in the earliest surviving manuscript versions, as we have seen.

These differences in the manuscripts were noted in the late eighteenth century. They were also considered during the editing of the 'Record Edition' of the statutes in the nineteenth century, and led Thomson to make some changes – unfortunately, not always the correct ones. Two significant differences emerge between Thomson's version and that of Balfour: Thomson has given a larger number of pounds in the stone and has used a much smaller depth for the firlot measure. (The figures for the assize are shown in Table **5.1**.) These are major discrepancies – particularly the reduction of the firlot's volume by one-third compared with earlier editions.

Thomson provided no justification for his changes in the assize, but fortunately his sources are known. There are two very early manuscript copies of the acts of James I in the Register Office which were used by Thomson in compiling Volume II of the 'Record Edition'.[72] An act from the second of these copies was reproduced by Thomson in plate I of his volume, but he gave no information about this manuscript save that it was 'probably written about the end of the reign of King James II'.[73] Unfortunately, these two manuscripts had been the subject of some controversy which led to Thomson advising the suppression of a printed and published version, and perhaps as a result neither he nor Innes seemed to wish to draw attention to their existence as the most important sources for the statutes of James I.

The two manuscript copies of the James I acts form part of a series of manuscripts that were considered in the late eighteenth century to form the best sources for the Scottish statutes. They were transcribed and printed with painstaking accuracy letter by letter, comma by comma, by William

Robertson (who died in 1803), one of the deputy keepers of the records under Lord Frederick Campbell, the Clerk Register. The publishing of *The Parliamentary Records of Scotland* was adopted as part of the programme of the new Record Commission of Great Britain in 1800, and the first volume covering the period 1240 to 1571 was ready for issue in 1804.[74] However, doubts had arisen about Robertson's policy of printing scrupulously exact transcriptions, allowing the inconsistencies, errors and text misplacements to be interpreted by others. The advocate Thomas Thomson was asked to report and had no difficulty in rejecting Robertson's approach in favour of the production of a clean, ordered and edited version, and the result was that this extremely valuable volume was suppressed.[75] Two years later Thomson was appointed to the newly-created post of Deputy Clerk Register and set about producing his own edition of the statutes. Cosmo Innes, in his preface on sources in the 1844 Volume I of the 'Record Edition', made it clear that the only part of Robertson's volume considered of any use was the transcription of a very important collection of fourteenth-century Scots law, known as the 'Black Book', which had been recovered from the State Papers Office in London in 1793. The only reason that Innes referred to Robertson's volume at all was because a few copies had escaped the suppression and were lodged in public repositories.[76]

The two manuscripts are now in a binding dated by Robertson to the mid-eighteenth century.[77] Originally, however, they were separate volumes. They have not acquired a provenance name, and are simply referred to as PA.5/6 in the National Archives of Scotland's collection.[78] In the first of these manuscripts (which for convenience we have called PA.5/6/1), a short miscellaneous collection of laws, is followed, in the same hand, by the statutes of 1424 to 1474. The second manuscript (PA.5/6/2) also starts in 1424, but it continues only to 1451 before breaking off abruptly. Also bound with them is a third manuscript, which is the earliest surviving authentic parliamentary record, commencing in 1466, which is early in James III's reign. Although the original record of the period 1424-66 is no longer in existence, the two early manuscript copies are nearly contemporary with the lost record and so they have considerable authority. Robertson believed the second manuscript to be earlier, on the basis of its earlier terminal date.[79] Thomson and Innes shared this view and dated the second volume to the 1450s, and the first to the 1470s, although it is not known whether they were also influenced by stylistic features of the writing.[80]

Thomson's initial reaction to these two manuscripts in 1804, when commenting on Robertson's printed text before the proposed publication, was that the first was 'in many respects inaccurate and imperfect' whereas the older second manuscript was 'still more inaccurate and imperfect', yet both would be useful 'for the purpose of collation'.[81] In fact, it is clear that Thomson came to rely on these manuscripts and altered his texts of the acts to take account of

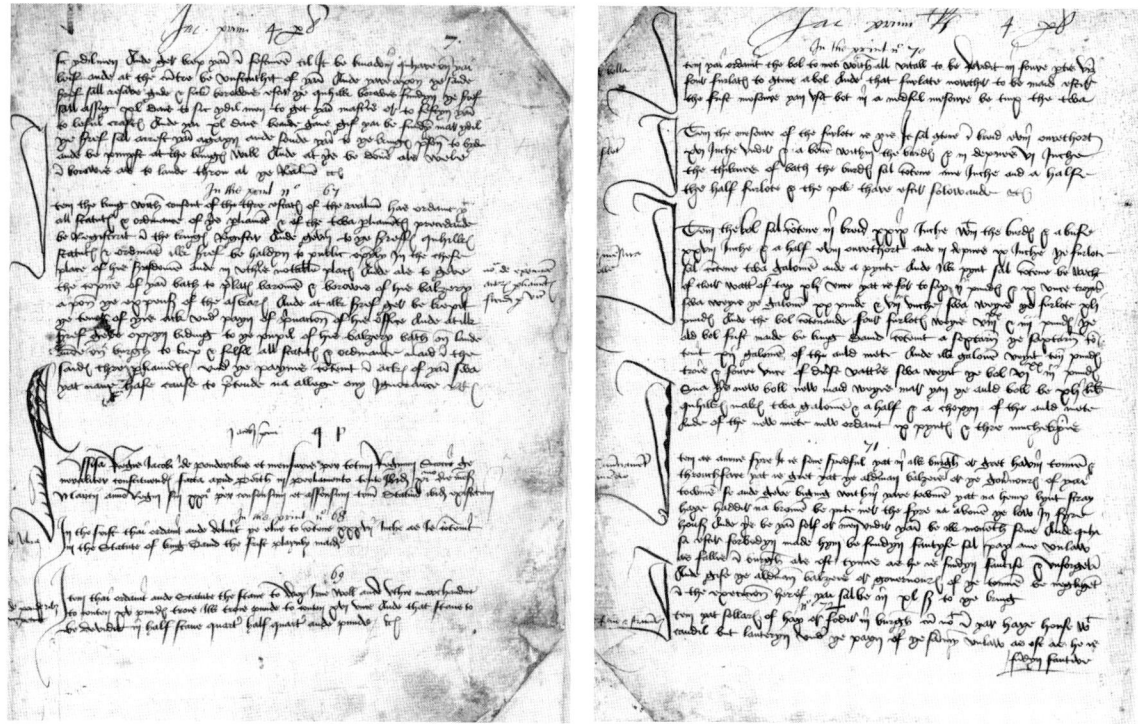

them: in his preface to his initial volume of the 'Record Edition', covering the period 1424 to 1567, Thomson noted that manuscript sources

Fig. 5.4

Manuscript copy of the 1426 Assize in PA.5/6/2. The text is given in Appendix **A.3**. (National Archives of Scotland)

> have been constantly resorted to as useful correctives [to the 1566 edition], and having furnished the means of supplying some omissions and mutilated passages, of amending many small errors, and of reducing the language and orthography of the whole to a state more nearly approaching the mode and fashion of the age to which those Statutes belong.[82]

In many respects the two versions of the James I Assize are the same. However, the differences (as shown in Table **5.1**) are in significant areas such as the size of the stone and the dimensions of the capacity measures.

The fact that there were differences between these two early manuscript copies of the assize of James I and the published versions has already been discussed in print. It was noted by the advocate John Swinton in his *Proposal for Uniformity of Weights and Measures in Scotland* of 1779, which has been largely dismissed by later commentators on the James Assize.[83] Swinton was closely identified with efforts made by the Convention of Royal Burghs of Scotland to press the case for a general reform of British weights and measures legislation.[84] Although the *Proposal* clearly stemmed from this work, it may also have been influenced by the start of a long-running legal case which began in the Court of Session in 1778 and which involved attempts to determine the true size of the large Scots grain measures that had

	PA.5/6/2 (c.1450)	PA.5/6/1 (c.1475)	BALFOUR'S BLACK ACTS (1566)	BALFOUR'S PRACTICKS	THOMSON (APS, 1814)
Ell	37 inches	38 inches	37 inches	37 inches	37 inches
Stone	15 trois pounds	16 trois pounds	15 trois pounds	–	16 trois pounds
Pound	16 ounces	16 ounces	16 ounces	–	16 ounces
Boll breadth (below)	29 inches	29 inches	29 inches	29 inches	29 inches
Boll breadth (above)	28½ inches	27½ inches	27½ inches	27½ inches	27½ inches
Boll depth	9 inches	19 inches	19 inches	19 inches	19 inches
Firlot breadth	16 inches	16 inches	16 inches	16 inches	16 inches
Firlot depth	6 inches	6 inches	9 inches	9 inches	6 inches

Table 5.1

Definitions in the 1426 Assize: comparison of the two oldest manuscript versions of the assize (taken from William Robertson's edition) with those of Balfour's 1566 printing, his *Practicks* of c.1580 and Thomson's 'Record Edition' of 1814.

last been defined in law in 1618.[85] The case will be discussed in Chapter 9. Swinton himself was later appointed a judge in the Court of Session, sitting as Lord Swinton.

Swinton's *Proposal* is usually cited to emphasis the enormous extent of regional variation in the measures, but it also included a brief section of 'Conjectures concerning the Ancient Standards of Measures and Weights in Scotland'. Having noted that the sizes given in print for the James I Assize could not be the correct ones, Swinton continued:

> This led to an inspection of the records, where it appears that the depth of the firlot, in place of 9 inches, as in the printed act, is only 6 inches; and that the depth of the boll, in place of 19 inches, as in the printed act, is only 9 inches. Farther, one of the copies in the record (for there are two) makes the stone to contain 16 lb.; though, in the other copy, and in the printed act, the stone is made to contain only 15 lb. But it is presumed that the first account is the right one; because, as all the copies agree, that an alteration was made in the pound, by adding an ounce, so from analogy it may be presumed a pound was added to the stone.[86]

Swinton found that by using these figures the relationship between the boll and the firlot was brought more nearly to the 4:1 ratio given in the assize. However, his conjecture that these sizes were close to the dimensions of the original measures, exhibiting them 'in a different light from that in which they have been hitherto understood', seems to have gone unnoticed.

Swinton appears to have opted for the manuscript PA.5/6/2 as his authority for the depths of the measures, but consciously chose the 16-pound

JAMES THE FIRST'S
ASSIZE:
INTERPRETING
THE
PARLIAMENTARY
RECORD
———

199

stone of PA.5/6/1.[87] The differences between the two manuscripts raise problems, and at least some of these may be the result of scribal error when the copy was made. There is, for example, no suggestion elsewhere that the ell has ever been defined by statute as 38 inches, as given in PA.5/6/1. Another minor difference is in the upper diameter of the boll where PA.5/6/2 gives 28½ inches and other sources have 27½ inches. There is no way of deciding between these, except perhaps by noting that because the bottom diameter is 29 inches, the importance of recording the difference in diameters is more obvious if the upper diameter is 27½ inches. Two of the late fifteenth-century compilation of the laws, the Colville and Drummond manuscripts, give 27½ and 27 inches respectively for this diameter, adding to the impression that the smaller figure may be correct.[88] We have opted for this 27½-inch diameter in the discussion that follows.

Thomson has not followed the version given in the older manuscript, which he represented as less accurate. With the exception of the incorrect ell size he appears to have followed the later PA.5/6/1. Alternatively, he may have taken the conservative view that he would not make a change in instances where both manuscripts gave the same reading. Thus the firlot depth has remained unchanged at 6 inches, by-passing the 9 inches of the 1566 edition, since both manuscripts agreed on this; but the boll depth has been retained at 19 inches even though this has the startling effect of making the boll ten times the size of the firlot.

The approximate volumes of the capacity measures can readily be obtained from the dimensions. Firlots of 6-inch or 9-inch depth will be about 1,200 in^3 or 1,800 in^3, and bolls of 9-inch or 19-inch depth will be about 5,600 in^3 or 12,000 in^3. Although the assize declared that the boll should contain four firlots, it is clear from these figures that it contained more and so it held an additional allowance or profit. These allowances are central to the use of the grain measures in trade, and they will be discussed in more detail later. In the meantime, however, it is clear that a 19-inch deep boll cannot fit into such a relationship, and a 9-inch deep boll can only relate to a 6-inch deep firlot. Both these depths are given in the earlier manuscript PA.5/6/2, but instead of taking Thomson's line and assuming that one of these is in error we will assume that these were the figures given in the lost original record. Support for this choice comes from the assize itself – one of the most difficult passages in the act deals with the permitted allowance on the boll, and by using these figures a plausible and consistent explanation can be advanced.

It is more difficult to explain why Thomson chose the stone of 16 pounds, rather than the 15 pounds used by Balfour in his printed *Acts* and by subsequent editors. This may simply reflect a preference by Thomson for the authority of PA.5/6/1 over the printed version; but it may equally have resulted from a faulty understanding of the assize and the earlier parliamentary act, taken together with a need to rationalise them with the later knowledge that, by 1563, 16 pounds were considered to be a stone.[89]

It has been argued above that the 1426 Assize retained the existing merchant stone, now defined, however, in terms of 15 of the larger troy pounds. It may have been the case that a larger stone of 16 of these pounds was used from this date for assessing royal customs on wool, necessarily leading to a reduction in revenue: Alexander Stevenson has indeed proposed that such a change was made in 1426 to help stimulate a declining market, although it seems uncharacteristic for the monarch to have been unconcerned about the short-term effect on his revenue, and we have argued in Chapter 4 for constancy in the size of the sack.[90]

A larger stone may have taken longer to enter general use. The reference to a 16-pound stone in the later manuscript PA.5/6/1 may therefore reflect a revision of the law to take account of a change in administrative practice – it is known that such a revision of the statutes took place in 1469, and PA.5/6/1 may derive from this revised source since it must have been prepared after 1474, which is the date of the last entry.[91] The most plausible explanation may simply have been an error made by a clerk who was fully aware that there were 16 (merchant) pounds to the stone and had not adequately appreciated the need to couch the definition in troy pounds. The change in the depth of the boll to 19 inches in PA.5/6/1 cannot be explained satisfactorily, and is presumed to be merely a transcription error of 'xix' for 'ix'.[92]

When James Balfour's printed edition of the *Acts* appeared in 1566, the boll depth was still given as 19 inches, but now the firlot depth had been adjusted to 9 inches (although both PA.5/6/1 and PA.5/6/2 gave 6 inches).[93] The reason for this may lie in the requirement placed on Balfour to revise rather than merely edit the laws. Balfour was undertaking his work on the statutes at a time after the legal firlot had been revised up to about 1,650 in^3 (the evolution of the capacity measures will be demonstrated in Chapter 7) and so the enhanced trading firlot would be about 1,850 in^3.[94] This is an acceptable match to the volume of a supposed 9-inch deep firlot, of about 1,800 in^3, and it seems plausible that Balfour was attempting to bring the dimension up to date.[95]

Some indication of the continuing process of revision can be gleaned from two of the late manuscript compilations of laws which contain the 1426 Assize, bearing in mind that the same authority cannot be ascribed to them as to the parliamentary copies of the record. These are the Colville Manuscript, of the last quarter of the fifteenth century, and the Drummond Manuscript, of the end of the fifteenth century, mentioned above.[96] They agree entirely with the 1566 printing of the 1426 Assize except that both omit the vital depth of the firlot. Only the Colville Manuscript includes the depth of the boll, given as 19 inches. Surprisingly, both manuscripts refer to the boll as being divided into six firlots, although they later declare it correctly as four

JAMES THE FIRST'S
ASSIZE:
INTERPRETING
THE
PARLIAMENTARY
RECORD
———

201

firlots. A new and enlarged firlot was introduced at the end of the fifteenth century, and from the arguments developed in Chapter 7, it was associated with a trading boll of about 6,600 in³. Six quarters (or one-and-a-half times) this boll would have been 9,900 in³, which is perhaps close enough to the 12,000 in³ of the supposed 19-inch deep boll to suggest that an attempt may have been made to rationalise the figures given in these manuscripts.

The depths of 9 and 19 inches, having been selected by Balfour, were then also quoted by Skene and Murray in their printed editions; and as might be expected they appear in the *Practicks*. Not all the manuscript versions of the *Practicks* include the 1426 Assize or provide numerical detail: of fifteen copies examined, only seven provided any details of the dimensions of the firlot and boll. One of these gave the depths of the vessels as 9 inches and 19 inches; four gave 9 and 20 inches; and another two were deficient but implied the firlot to be 9 inches deep.[97]

None of the manuscript versions of Balfour's *Practicks* gives the 6-inch and 9-inch depths for the firlot and boll that are found in the earliest contemporary copy of the record. By a combination of error and editorial dictat depths of 9 inches and 19 inches have been substituted, and by the mid-sixteenth century these figures had become firmly entrenched in the historical record. We have no hesitation in accepting the depths of the firlot and boll in the original legislation as 6 and 9 inches respectively. Perhaps if Thomas Thomson had approached the issue with the care of William Robertson, both figures would have been restored to the authoritative record of the 1426 Assize rather than only one. Armed with this conclusion, we can analyse the main provisions of the assize in Chapter **6**, and tackle one of the central issues of medieval metrology – the nature of allowances and heaping.

1   T. Thomson and C. Innes (eds), *The Acts of the Parliaments of Scotland [APS]*, 13 volumes (Edinburgh, 1814-75), II, 12: Act 22, 1426: Assize of James I. See Appendix A.3a of this work.

2   *APS*, II, 10; Acts 13, 14 and 15, 1426. See Appendix A.2.

3   Athol L. Murray, 'The Scottish Chancery in the Fourteenth and Fifteenth Centuries', in D. I. Guth and K. Fiann (eds), *Écrit et pouvoir dans les chancelleries médiévales: espace français, espace anglais* (Louvain-La-Neuve, Belgium, 1997), 133-51, p. 146.

4   The captain of the frigate *Eagle*, Major Fletcher, and John Weymes, captain of the *Elizabeth*, were exonerated from any responsibility in the loss by the subsequent enquiry: *APS*, VII, 11, 65: Acts 8 and 93, 1661.

5   Bruce Webster, *Scotland from the Eleventh Century to 1603* (London, 1975), 126-7.

6   *APS*, I, Preface, 26-7.

7   Ibid.

8   Ibid., II, 10: Act 10, 11 March 1426. For *Regiam Majestatem* and *Quoniam Attachiamenta*, see Chapter 1.

9   Ibid., II, 11: Act 21, 1426.

10 Ibid., II, 52: Act 39, 1457.

11 Ibid., II, 97: Act 20(e), 20 Nov. 1469: 'Item of the Reduction of the Kingis lawis Regiam maiestatem, acts, statutes and uther bukis to be put in a volume and to be authorizit the laif to be distroyit.' The Act was not included in the first printed edition of 1566, and was given the erroneous date of 10 October 1487 by John Skene in his printed edition of 1597.

12 A. A. M. Duncan, *James I, King of Scots 1424-1437* (Glasgow, 1984), 1.

13 M. Livingstone and Gordon Donaldson (eds), *Registrum Secreti Sigilli [RSS]: The Register of the Privy Seal of Scotland*, 7 volumes (Edinburgh, 1908-66), I, 223-4: no. 1546, 15 September 1507. For a recent analysis, see Alastair J. Mann, *The Scottish Book Trade 1500-1720* (East Linton, 2000).

14 *Actis (The New) … maid be … Iames the Fift … 1540* (Edinburgh, 1541). Davidson described himself as 'Prenter to the kingis nobyll grace': H. G. Aldis, *A List of Books Printed in Scotland before 1700* (Edinburgh, enlarged edition 1970), 112, and entry 21.

15 *APS*, II, 379: Act 47, 1540.

16 [Edward Henryson (ed)], *The Actis and Constitutionis of the Realme of Scotland maid in Parliamentis halden be … Kingis James the First, Secund, Thrid, Feird, Fyft, and in tyme of Marie now Quene of Scottis …* (Edinburgh, 1566).

17 The term 'Black Acts' covered all the acts of a single parliament, or a few parliaments taken together, which were printed between 1541 and 1594, but it also specifically referred to the 1566 printing which embraced the acts passed from 1424 to 1564.

18 Balfour was appointed Clerk Register by the new March 1566 administration, and he replaced James Makgill, who was associated with the hard-line Calvinist faction at court and had been implicated in the murder of David Riccio. Writing some 60 years later, Habakkuk Bisset, who had been employed by Balfour in the 1580s, reported Makgill's involvement in the suppression of the acts as follows: 'the said Acts printed by the said Lekprevick was bought unbound from him in silver by the late Mr. James Makgill of Nethir Rankeloure, clerk register for the time, and for the most part was destroyed; so that within a short space of time thereafter few or none could be found to be bought or sold by the leiges for such causes as moved the said clerk register for the time …': Habakkuk Bisset (P. J. Hamilton-Grierson, ed), *Rolement of Courtis [First Draft 1622]*, 3 volumes (Edinburgh and London, 1920-26), I, 73. When the political climate changed, Makgill was reinstated as Clerk Register in 1567. It is no longer clear whether Makgill took the initiative in suppressing the first issue and bypassed Balfour, or whether Bisset was referring to a later suppression by Makgill of the whole issue when he published for the first time the acts of the 1560 Reformation parliament in 1568. We are grateful to Dr Athol Murray and Dr Julian Goodare for their guidance here.

19 We are grateful to Dr Julian Goodare for comments on the political background to this issue.

20 *APS*, I, Preface, 29.

21 Sir James Balfour of Pittendreich was deeply involved in the turbulent intrigue between the Catholic and Calvinist factions during Mary's reign and the regencies of James VI's minority. He soon acquired a reputation for an unscrupulous attitude to politics, but as a lawyer and legal antiquary he was held in high regard. He had been trained for the legal branch of the Scottish church, becoming chief judge of the ecclesiastical court of the Archdiocese of St Andrews (which included Edinburgh and the Lothians). Although this court was abolished at the Reformation, a replacement court had to be established in 1564 to continue its

business, and Balfour was appointed to head this commissary court at Edinburgh. He became a Privy Councillor in 1565 and was a close advisor to the Queen at a time when influence was largely concentrated in Catholic hands. In 1566 he was knighted and rewarded for political services with the office of Clerk Register, which he held just long enough to complete the compilation and printing of the 'Black Acts'. Discussed in Peter G. B. McNeill, 'Balfour: Many Crimes and Little Punishment', unpublished paper given to the Scottish Legal History Group, 23 October 1999. We are grateful to Dr McNeill for giving us a copy of this paper.

22 Henryson's preface to the *Acts*, op. cit. (16). The preface also prints Henryson's license of 1 June 1566 to print and sell the volume.

23 Sir James Balfour (Walter Goodall, ed), *Practicks* (Edinburgh, 1754). This edition was reprinted by the Stair Society with an introduction by Peter McNeill: P. G. B. McNeill (ed), *The Practicks of Sir James Balfour of Pittendreich*, Stair Society nos. 21 and 22, 2 volumes (Edinburgh, 1962-3). In what follows, this later edition is the one used.

24 *APS*, III, 89: appendix, 5 March 1574.

25 Skene was a prominent lawyer, becoming King's Advocate in 1589, and was involved in foreign diplomatic missions for James VI. His appointment as Clerk Register came in 1594 and he was made a Lord of Session, sitting as Lord Curriehill. He served on a number of important commissions including that for the union of the monarchies of Scotland and England in 1604: see W. F. Skene (ed), *Memorials of the Family of Skene of Skene*, New Spalding Club (Aberdeen, 1887), 106-16.

26 McNeill, op. cit. (23), I, lxiv; Hector McKechnie, 'Balfour's Practicks', in *Juridical Review*, 43 (1931), 179-92; p. 180; ibid., 192.

27 [Sir John Skene (ed)], *The Lawes and Actes of Parliament, mayd be King Iames the First, and his successours Kinges of Scotland: visied, collected and extracted furth the Register …* (Edinburgh, 1597).

28 *APS*, III, 564: Act 45, 1592.

29 Ibid., I, Preface, 32.

30 Ibid., I, 32, note (2).

31 The Privilege is unpaginated and prefixed to the 1681 printing: Thomas Murray (ed), *The Laws and Acts of Parliament made by King James the First [and successive sovereigns to] King Charles the Second …* (Edinburgh, 1681). The duodecimo edition is *The Laws and Acts of Parliament made by King James the First, and his Royal Successors, Kings and Queens of Scotland* (Edinburgh, 1685).

32 *APS*, VIII, 388-9: Act 133, 17 September 1681. The gift was dated 9 May 1679.

33 Ibid., I, Preface, 35; William Watson, Baron Thankerton, 'Statutory Law', in *An Introductory Survey of the Sources and Literature of Scots Law*, Stair Society no. 1 (Edinburgh, 1936), 6.

34 *APS*, VIII, 388-9: Act 133, 17 September, 1681.

35 Ibid., VIII, 356: Act 94, 17 September, 1681.

36 On the nature of the post and the range of Thomson's responsibilities see M. D. Young, 'The Age of the Deputy Clerk Register, 1806-1928', in *Scottish Historical Review*, 53 (1974), 157.

37 *APS*, I, 24.

38 Ibid., I, Preface, 38.

39 Webster, op. cit. (5), 167.

40 Thomas Mackay Cooper, Lord Cooper, 'The Regiam Majestatem and the Auld Lawes', in *An Introductory Survey of the Sources and Literature of Scots Law*, Stair Society no. 1 (Edinburgh, 1936), 70-81, quote p. 74; A. A. M. Duncan, '*Regiam Majestatem* – A Reconsideration', in *Juridical Review*, new series, 6 (1961), 199-217; J. Buchanan, 'The Manuscript of Regiam Majestatem: an Experiment', in ibid., 49 (1937), 217-31, quote p. 217; George Neilson, *Trial by Combat* (Glasgow, 1890), 102-3; Duncan, op. cit. (12), 1. A new critical edition of the Scottish legislation to 1707 is currently being prepared by The Scottish Parliament Project at the Department of Scottish History, University of St Andrews.

41 For historical material following, see William Croft Dickinson, *A New History of Scotland*, second edition (London and Edinburgh, 1965); Alexander Grant, *Independence and Nationhood: Scotland 1306-1469* (Edinburgh, 1984); Ranald Nicholson, *Scotland, The Later Middle Ages* (Edinburgh, 1989) and A. D. M. Barrell, *Medieval Scotland* (Cambridge, 2000).

42 Michael Lynch, *Scotland: A New History* (London, 1991), 138-9.

43 Ibid., 135-45; Barrell, op. cit. (41), 137-55; Grant, op. cit. (41), 178-87.

44 Alexander W. K. Stevenson, 'Trade between Scotland and the Low Countries in the Later Middle Ages', unpublished PhD thesis, University of Aberdeen, 1982.

45 According to Murray (1681), the first parliament was held 26 May 1424; the second 12 March 1425; the third 11 March 1426; the fourth on the same day; the fifth 30 September 1426. In all there were 13 parliaments in this reign. The various editors do not all number the parliaments in the same way; *APS* does not number them at all.

46 It would seem that the first suggestion that the act of 1426 may have been declaratory in nature came from R. W. Cochran-Patrick, *Records of the Coinage of Scotland*, 2 volumes (Edinburgh, 1876), I,

Introduction, lxxvii-lxxviii: 'There is a probability that the Act of 1425 [1425/6] was only declaratory and that the new system of weights had been introduced even before the reign of James I. But however this may be, it certainly was in common use immediately after.'

47 Forty-one ounces each of 480 gr is 1,275·24 g, having a volume of 77·8 in^3 of water, which is just the capacity of the 1426 pint.

48 2 Henry IV c.4.

49 *Oxford English Dictionary*, s.v. *troy*. Norman Biggs has recorded an earlier published date of 1376: see Chapter 4, ref. 36.

50 R. D. Connor, *The Weights and Measures of England [WME]* (London, 1987), 127.

51 *APS*, II, 10: Act 14, 1425.

52 Ibid., I, 697, *Iter Camerarii*: ' … And at thai haif nocht thar mesuris … concordand to the kingis mone be the quhilkis mesuris the pepill may be seruit quhen mister [need] is.' Tower weight was in use in England in 1476 when measures were determined by weight: ' … XX sterlings maketh an ounce and XII ounce maketh a Pounde for sylver, golde, bred and *measure*': M. D. Harris (ed and trans), *The Coventry Leet Book* (London, 1907), 396.

53 J. D. Marwick, *et al.* (eds), *Extracts from the Records of the Burgh of Edinburgh, 1403-1718 [Edinburgh Extracts]*, 14 volumes (Edinburgh, 1869-1967), I, 133: 6 May 1511.

54 For example, see N. J. Mayhew, 'From Regional to Central Minting, 1158-1464', in C. E. Challis (ed), *A New History of the Royal Mint* (Cambridge, 1992), 83-178, pp. 90, 161.

55 McNeill, op. cit. (23).

56 Ibid., 88-90.

57 [John Swinton], *A Proposal for Uniformity of Weights and Measures in Scotland* (Edinburgh, 1779), 138, where he presumed that a 16-pound stone is the correct one. In this he was followed by L. Burrell, 'The Standards of Scotland', in *The Monthly Review: the Journal of the Institute of Weights and Measures Administration*, 69 (1961), 49-62, esp. p. 55; and by Meredyth Somerville, *The Standardization of Weights and Measures in Scotland*, Department of Geography, University of Edinburgh, Occasional Paper no. 11 (Edinburgh, 1989), 10.

58 The sections cited for the *Leges Burgorum* and the *Iter Camerarii* are Chapters 136 and 34, respectively: McNeill, op. cit. (23), I, 90. These are the chapters also shown in the manuscript copies of Balfour in the National Library of Scotland (NLS). The various versions of these early documents do not always present the same section numbering.

59 *APS*, III, 521: Act 136, 1587.

60 Ibid., II, 540: Act 14, 1563.

61 Ibid., II, 226: Act 15, 1491: The declaration that this weight is 'specialy of wechtis alswele of wax as spice [for wax as well as for spice] and xvj vnce In the pund' indicates that it is for use with heavy goods as well as for fine wares.

62 Swinton, op. cit. (57), 138.

63 McNeill, op. cit. (23), I, 90.

64 *APS*, I, 673.

65 Sir John Skene, *Regiam Majestatem: The Auld Lawes and Constitutions of Scotland* (Edinburgh, 1609), part 2, f.56v.

66 *Edinburgh Extracts*, I, 98: 2 March 1504: ' … The half watter boll of beyre and malt … cummis now to the messour of lxxvij pynttis'; ibid., II, 238, 24 February 1556: ' … the nyne furlettis [9 firlots] grundin malt of the auld mesour is to be sauld for iiij li [£4]'; see also Appendix D, where in Shetland during the 1560s the Earl Robert increased the volume of the can by one-third.

67 *APS*, II, 10: Act 13, 1426.

68 Ibid., II, 10: Act 15, 1426.

69 A. Huntar [*sic*], *A Treatise of Weights, Mets and Measures of Scotland* (Edinburgh, 1624), 5.

70 *WME*, 178-80. Water measure will be discussed further in Chapters 6 and 7.

71 For an indication of the extent of this, see Michael Lynch and Alexander Stevenson, 'Overseas Trade: The Middle Ages to the Sixteenth Century', in Peter G. B. McNeill and Hector L. MacQueen (eds), *Atlas to Scottish History to 1707* (Edinburgh, 1996), 238-43.

72 National Archives of Scotland (NAS), MS PA.5/6.

73 *APS*, II, xvii.

74 Athol L. Murray, 'The Lord Clerk Register', in *Scottish Historical Review*, 53 (1974), 154-5. William Robertson's volume of over 800 folio pages was titled *The Parliamentary Records of Scotland in the General Register House, Edinburgh, Vol. 1*, and the 'Preliminary Observations' to this work are signed merely 'WR'.

75 Murray, op. cit. (74), 154-5. For a valuable commentary on Thomson as an editor, see Marinell Ash, *The Strange Death of Scottish History* (Edinburgh, 1980).

76 *APS*, I, 29-30. There is a copy in the Library of the National Archives of Scotland. Three others are in the National Library of Scotland (all at GRO.11), one of which was presented to the Library of the Faculty of Advocates in 1922 by the Lord Lyon's Office.

77 William Robertson, 'Preliminary Observations', in op. cit. (74), 57 note.

78 We must record our thanks to Dr Athol Murray, formerly Keeper of the Records of Scotland, for his assistance and advice.

79 Robertson, op. cit. (74), 57 note.

80 Thomson's manuscript report to the Record Commissioners, November 1804, from the photocopy bound in the National Archives of Scotland copy of the suppressed Robertson volume.

81 Thomson's report, see above (ref. 80).

82 *APS*, II, xiii.

83 Swinton, op. cit. (57), 138. Comments dismissive of Swinton include: A. Stephen Wilson, *A Bushel of Corn* (Edinburgh, 1883), 10 note: 'And hence Lord Swinton in his *Conjectures* … fell into the mistake of supposing that the old money pound of Scotland was 15 oz. and 25 s. In this blunder he has been followed by numismatic and historical writers to this day'; Somerville, op. cit. (57), while mentioning Swinton twice (pp. 23, 30-2), does not discuss his 'Conjectures'; Burrell, op. cit. (57), 55: 'Lord Swinton, on the other hand, is quite explicit in his book that a pound was added to the stone and an ounce to the pound … By referring to the known capacities of the Scottish pints in existence it is reasonably certain that Swinton is correct'; on p. 56, he continues: ' … but to say as Lord Swinton does that a pint measure increased from 42 cubic inches … through various stages to 104 cubic inches by 1618 by "custom" is a bit far fetched … A misprint of 19" instead of 9" as suggested by Swinton gives an equally fantastic answer the other way'; and on p. 58: ' … Swinton and others assumed that this weight series was Troye but there is no doubt that the weights were Tron …'

84 J. D. Marwick, *et al.* (eds), *Records of the Convention of the Royal Burghs (1295-1779)*, 7 volumes (Edinburgh, 1866-1918), VII, 546, 10 July 1777.

85 Swinton's pamphlet was addressed to the sheriffs and justices of the peace of the counties and the magistrates of the royal burghs. A second edition (actually a reprinting) appeared in 1789, coinciding with the resumption of the Court of Session case dealing with grain measures which had lapsed in 1783. Final judgment was given in 1791. This is discussed in Chapter **9**.

86 Swinton, op. cit. (57), 138.

87 Ibid., 138. His comment implied that both manuscripts gave the depths as 6 and 9 inches, whereas PA.5/6/1 has 19 in place of 9 inches.

88 The Colville manuscript in Edinburgh University Library (EUL), Borland MS 208; the Drummond manuscript is NAS MS PA.5/3; and see ref. 96 below. Using 28½ inches would increase the size of the boll by 3½%.

89 As noted above, the earliest clear reference to the stone of 16 troy pounds is in an assize of 1563: *APS*, II, 540: Act 14, 1563.

90 Stevenson, op. cit. (44), 194. His figures, however, are incorrect.

91 McNeill, op. cit. (23), lviii n.; *APS*, II, 97, 1469: 'The Reductione of the kingis lawis Regiam maiestatem, acts, statutes, and uther bukes to be put in a volume and tobe autorizit and the lais [rest] to be distroyit.' The first clear reference to an English pound of 16 troy ounces is also in 1474: *WME*, 127.

92 It may be no more than coincidental, but the figure '19' is encountered after 1458: by analogy with the 1426 Assize, the permitted firlot that resulted from the 1458 Act had a capacity of 19 pints, and this may perhaps have triggered the error. The question of 9 versus 19 inches for the boll's depth appears to have spread to the text of David's Assize. The Ayr Manuscript (NAS, MS PA.5/2A), while declaring that the 'boll holds a sextern and is in depth ix inches' as the other texts do, also has a marginal note, 'xix'.

93 The first (withdrawn) printing of the *Acts* in November 1566 (copy in NAS, GA.I) reads: 'The boll sall contene in breid xxix inchis within the buirdis … and in deipnes xix inchis. The fyrlot sall contenene in breid euin ouerrthort xvi inchis under and abone within the buirdis … and in deipnes it sall contene ix inche.'

94 At the 1563 Assize the legal, trading and customary firlots can be derived (see Chapter 7) as 16⅞, 17¹⁵/₁₆ and 19¹/₁₆ pints. Using the current pint, and the grain compaction factor derived in Chapter 4, these represent volumes of about 1,650, 1,750 and 1,850 in³.

95 1,800 is ⁹/₆ × 1,200. In Chapter 7, it is shown that 16 pints of grain have a volume of 1,200 in³.

96 For these manuscripts, see ref. 88 above, and *APS*, I, xxv. See also C. R. Borland, *A Descriptive Catalogue of Western Manuscripts in Edinburgh University Library* (Edinburgh, 1916).

97 The measures in each manuscript are: 9 and 19 inches in NLS Adv. MS 24.2.4; 9 and 20 inches in NLS Adv. MS 24.2.4b; EUL MS Laing III 406 and 408; Glasgow University Library MS Murray 531; 9 inches in NLS Adv. MS 22.3.3; EUL MS Laing III 410.

6

The Abuse of Charity:
Allowances and the
Heaping of Measures

MOST OF THE SCOTTISH LEGISLATION DEALING WITH THE DEFINITIONS OF
weights and measures is concerned with the dry capacity measures – the
large measures used for dry goods such as grain. This issue was one of importance for the country's economy, not merely because barley and oats were
staple foods, but because most tenants had their rentals assessed in grain
measured in dry units. The sizes of these units helped determine the landowners' revenues, and the vested interests of the establishment were generally
served by sanctioning increases in these sizes. It is important to recognise
that some of the smaller dry units, such as the pint and gallon, are ones that
at the present day we associate only with liquid measurement. In earlier
times they played an equally important role in the dry series.

In part, this concentration of assize legislation on the dry units was because such measures are as difficult to define unambiguously as they are
to construct accurately and durably. There can be little doubt that these
difficulties were exploited, because the operation of fraudulent measures
with unofficial volumes could be highly profitable. But the principal reason
why the Scottish legislation came to be dominated by the dry capacity
measures was that official changes (always increases) were periodically made
in their sizes: with each change, greater effort had to be expended in trying to
enforce uniformity in use. Add to this the confusion caused by contracts and
rental agreements (specifying quantities in the old units) which remained in
force after the introduction of such changes, and it will be apparent that the
situation presented some administrative problems.

At times these changes came thick and fast. Indeed it is tempting to consider the sixteenth-century Scottish dry capacity measures as 'standardised'
only in the restricted sense that they were periodically defined legally. At four
of the five major assizes of weights and measures between 1490 and 1620, substantial enlargements were made in the volumes of the dry measures, although
the other measurement units remained effectively unchanged (except for one
instance) for the ordinary purposes of trade.[1] Newer and larger measures

Opposite:
Detail of Fig **6.1**.

replaced the older measures; but not in all circumstances, because the assizes incorporated a clause designed to protect the parties to agreements previously made in the old measures. Whether this protection was effective is debatable. Ideally, these existing agreements were to be re-calculated in the new measures, but there must have been a tendency to continue to use the old measures; it was therefore possible for more than one size of standard to co-exist. At times, particularly during the second half of the sixteenth century, the fixed 'standard' of dry measure must have seemed eminently adjustable.

The increases in the sizes of the dry capacity measures came about because of the operation of a system of long-standing allowances or 'charities' which were in turn directly related to the habit of heaping material in measures. If grain is to be sold by volume, it can be measured in two ways. It can be metered out in 'struck' vessels, where the act of striking involves passing a straight rule (or 'strike') over the rim of the capacity vessel and wiping off all the grain lying above the level of the rim. Alternatively, it can be 'heaped', with the grain left piled up above the rim.

A carefully constructed heap can contain a substantial quantity of grain: experiments in heaping a firlot measure in 1618 showed that the largest possible size of heap of grain comprised very nearly half the quantity held in the vessel itself.[2] For most purposes a much more modest heap of only one-sixteenth or one-eighth of the contents of the vessel was used. This was more like a gentle 'rounding' rather than a blatant 'heap', and its original justification was that it represented an appropriate allowance or 'charity' to compensate the party conducting the transaction for accidental loss or inevitable spillage in the course of the measuring operation.

In practice, the charity was taken as part of the profit in a transaction. Thus, for example, where the price of grain was fixed, the merchants who had the monopoly of the sale of surplus grain from the landward area of the burgh could insist on buying by heaped measure and raising the possibility of profiting by selling by struck measure. But it was the tenant, who generally paid his rental to the landlord in kind, who was hardest hit – he was powerless to resist the landlord's demand that the specified rental payment be received in heaped measure.

The matter came to a head in the early seventeenth century when the issue was taken up by the newly-created justices of the peace for the counties. In 1613 they disputed the rights claimed by the burghs to alter the measures decreed by parliament, and they persuaded the Privy Council of the problems arising 'betwix the maister and his tennent' because of the 'prejudice' of the measures.[3] The justices of the peace for the counties of Perth and Edinburgh urged the abolition of charities, which both groups claimed had grown to 'ane great abuse'.[4] From the county of Linlithgow report it emerges that the firlot, enhanced by charities, contained 21 pints, against the last legal definition of about 19 pints.[5]

THE ABUSE OF
CHARITY:
ALLOWANCES AND
THE HEAPING
OF MEASURES
——

209

Our first view of the system of allowances that governed the use of the dry capacity measures, and indeed our first indication of official toleration of this system, comes with the reforming assize of weights and measures of 1426. This assize also established the crucial link between the liquid and dry capacity measures: the volumes of the dry measures from this time onwards were always defined in terms of the number of fills of the pint measure they contained, and this represents the essential difference from English practice, where the dry measures were defined by the weight rather than the metered volume of contents.

David's Assize makes no mention of vessels smaller than the gallon, although fractional parts of the gallon certainly had controlled volumes by the late fourteenth century, when the *Iter Camerarii* refers to the testing of quarts, pints and smaller vessels in the burghs.[6] One of the constituent parts of the gallon was named as the 'pint' in the 1426 Assize, and as such was confirmed as the basis of the dry capacity measures. Enough can be gleaned about the liquid measures before and after the passing of the 1426 Assize to demonstrate the underlying continuity of practice, and although the descriptive names of some of the measures were changed, their volumes were unaltered by the 1426 Assize.

The two concerns of this chapter are therefore to investigate this pint measure and its role in the liquid measure series, and then to analyse the dry capacity measures based on it together with the early allowances that relate to them. However, before doing so, we must first address some specific issues that arise when grain is used as a measuring medium.

HEAPED MEASUREMENT

The heaping of measures was certainly not peculiar to Scottish metrology, and allowances of this type were also notable features of English practice. Indeed, it seems inescapable that the heaping of measures must have been a characteristic through all early European trade. Since it has not always been appreciated that heaping played as significant a role in English trade as in Scottish, we will first review some of the English evidence before looking at similar traits in the Scottish record.

The long-standing convention in England that rental payments in kind were demanded by the lords of the manor and by the king's officials in heaped measure is well illustrated in a statute of Edward III of about 1350. This records the size of the king's standard of the gallon, adding that 'every measure of corn shall be stricken, without heap, saving the Rents and Ferms of the Lords which shall be measured by such measures as they were wont in Times past'.[7] The practice of expecting the use of heaped measures certainly goes back to the thirteenth century.[8] The additional amount of material allowed is identified as one-eighth in the Customal of Sandwich, Kent, dating from 1301, which records that the gallon was to contain 8 pounds weight of wheat, but that the bushel (which was ostensibly 8 gallons) was to be 8 gallons and 8 pounds.[9]

The act of dispensing material by heaped measure could be performed in several ways. Given that a bushel comprises eight gallons, a bushel of grain could be measured using eight fills of the gallon measure, each with a conventional rounded heap of about one-eighth, or alternatively it could be measured using nine struck gallons. But the size of a heap could never be regulated accurately, and for this reason heaping as such was discouraged – innumerable statutes attempted to eliminate the practice over the centuries. With increasing pressure to use only struck measures, the bushel in practical use came to be considered as nine legal struck gallons. Although this continued the *concept* of heaped measure, the additional quantity was now controlled.

Alternatively, the measures could be enlarged so as to accommodate the heap while still functioning as struck vessels. The earliest surviving English Exchequer standards are the bronze gallon and bushel standards of Henry VII, dated 1497, the capacities of which were measured in the 1930s as 268·4 in[3] and 2,145 in[3] respectively, the bushel capacity being almost exactly eight times that of the gallon (2,145 ÷ 8 = 268·1), certainly to within the accuracy achievable at the time.[10] These standards were constructed to follow the Assize of 1496, which stipulated that:

> … the mesure of the busshell conteyn VIII gallons of whete and that every galon conteyn VIII lb of whete of troi weight and every lb conteyn XII unces of troy weight … according to the old Lawes of this Land.[11]

Nonetheless, it can be calculated that both standards are larger than the sizes given in the statute, and indeed both have been constructed to accommodate an additional eighth. Taking the density of good-quality wheat derived in 1775 by Henry Norris, who obtained a weight of 47½ avoirdupois pounds for a cubic foot of wheat, 8 troy pounds of wheat can be shown to occupy about 239 or 240 in[3], and 9 pounds about 270 in[3].[12] Thus we can appreciate that the gallon standard has been constructed to contain 9 troy pounds of wheat, whereas the bushel contains 9 struck gallons (each of 8 pounds of wheat), or 8 heaped gallons (each of 9 pounds of wheat). It can be seen that the custom and practice of the market-place exercised greater influence than the letter of the statute when it came to the construction of the standards that were intended to control commerce. There is nothing in the statute to indicate that only theoretical legal definitions were being provided, but the fact that these were Exchequer standards, and therefore measures of the highest authority, shows that there was official recognition of the central role played by traditional allowances in commercial transactions.

THE ABUSE OF
CHARITY:
ALLOWANCES AND
THE HEAPING
OF MEASURES
———

211

In spite of this, the conventional 1:8 relationship between the gallon and bushel has been maintained, and these standards have been constructed as struck vessels. This method has the added administrative advantage of controlling the size of the heap as exactly one-eighth: it represents an official toleration of the principle of the allowance whilst removing the possibility of fraudulent adjustment of the size of the heap.

The principle applied with larger measures also. Grain was sold by the 'quarter', which comprised 8 bushels, and the 1301 Sandwich Customal indicated that these 8 bushels already contained a charity of one-eighth.[13] A parliamentary act of 1391 recorded that although the quarter had been declared to be 8 struck bushels, because no penalty had been set for exceeding this amount, the widespread practice was to use 9 bushels to the quarter.[14] In spite of the later setting of a penalty, the 9-bushel quarter had become well-established, particularly in London where it was termed the *faat* or *fat*. But even this volume was exceeded: an act of 1413 lamented that the London merchants took the faat as the quarter but added an extra bushel, making ten in all, whereas the king's agents took nine heaped bushels to the quarter, enlarging the quantity by a further eighth, and so again raising it to about 10 bushels.[15] This was certainly in excess of the allowance of heaping one bushel in eight, which parliament noted in 1413 had been considered acceptable in the past.

The gallon and bushel have remained at very close to the sizes represented in the 1497 Exchequer standards, indicating that from at least the late fifteenth century the heap was formally incorporated in the volume and the enlarged measure had become the normal form. However, the mid-fourteenth century reference to measures being used struck, except for the rents and ferms of the lords of the manor, strongly indicates that the basic sizes (without the heap) were still used then, although not in all circumstances. The period between the mid-fourteenth century and the late fifteenth century apparently saw the gradual phasing out of the basic 8-pound gallon of about 239 in^3 in favour of the enhanced gallon of about 268 in^3 (in terms of the 1497 standards, $268 \cdot 4 \div 1\frac{1}{8} = 238 \cdot 6$).

At the present day, we think of the gallon solely as a liquid measure, but its legal definition as the capacity of 8 pounds weight of wheat is a reminder that it is also a fraction of the grain bushel. The 9-pound capacity of Henry VII's gallon standard, constructed to accommodate the traditional heap of one-eighth, reinforces this view of the gallon as related to the dry capacity series. The result of accommodating the heap is that a physically larger vessel is created which could be used for liquids. However, if such measures are again heaped (as in the creation of the quarter of 10 struck bushels), then by analogy an even larger liquid gallon can be envisaged.

This type of enlargement, in which a heap is incorporated within a bigger measure, is significant because it provides a mechanism for explaining the

larger sizes of some English gallons; and it is against this background that
similar enlargement of the Scottish measures will be considered. There were
indeed several English gallon sizes in current use in the second half of the
eighteenth century when pressure began to build for reform of Britain's
metrology. The reform process culminated in the imposition of a single
gallon size in 1824 – the Imperial gallon of 277·4 in^3 – superseding the old
corn, ale and wine gallons, each of which had caused confusion to successive
official enquiries because of the disputed origins of these measures, their
doubtful precise volumes, and even their questionable legal status.[16] In spite
of having been investigated on several subsequent occasions, notably in the
late nineteenth century, the relationship of these capacities has never been
satisfactorily explained and the survival of so few early reference standards
makes it unlikely that the story can ever be unravelled completely.[17]

The earliest of the English gallons for which a definition survives is
probably the so-called 'Guildhall gallon' of about 224 in^3, an ancient wine
gallon which seems to have passed out of use by the late seventeenth century.
The case for accepting the currency of this gallon was made in Chapter 4,
where it was argued that this was the gallon of the *Tractatus*, constructed to
contain 8 pounds of water using the early French pound of 15 Paris ounces.
We have no distinct knowledge of the size of the dry gallon of this period. We
know only that a new dry gallon containing eight 12-ounce troy pounds of
wheat was defined (or re-defined) in 1496, but this almost certainly replaced
an earlier type of definition (perhaps of a smaller bushel) because the English
troy pound appears to have been introduced officially only in the late
fourteenth century.[18]

Since the 1496 definition echoes the *Tractatus* so clearly, we might expect
that the earlier dry gallon contained eight 12-ounce tower pounds of wheat –
and by a happy coincidence the volume occupied by this weight of wheat is
also about 224 in^3.[19] Plausibly, therefore, the wet and dry versions of the
English gallon were both initially set as 224 in^3 (although the dry gallon came
to be used with an eighth allowance) and the *Tractatus* definition has subsequently been re-couched for the dry gallon to express the volume in terms
of seemingly appropriate dry units (although necessarily with less accuracy).
The implication of this line of argument is an increase in the dry measures of
a fifteenth, corresponding to the difference between the tower and troy
pounds. This is in addition to the effective increase of an eighth caused by the
absorption of the conventional heaping allowance.

There are, however, two further factors which must be taken into account
when considering the dry measure series. The first of these relates to the
operation of a customary measure, recognised in law in both England and
Scotland, and known as 'water measure'. Water measure was the specially
enlarged measure used at ports, literally for measuring goods that came by
water, as opposed to those that came by land. English water measure, which

THE ABUSE OF
CHARITY:
ALLOWANCES AND
THE HEAPING
OF MEASURES
——

213

applied to dry goods on board ship or in ports or maritime towns, is known from an act of 1495 to have been reckoned as 5 pecks to the bushel.[20] As the bushel normally contained 4 pecks, water measure was equivalent to 1¼ bushels, or twice the heaped grain allowance. This large rebate presumably also has its origin in losses incurred during additional acts of measurement, but it may also be an expression of the risk to the shipper of degradation of the quality of his cargo in the course of the voyage and storage.

Water measure in Scotland has already been mentioned in connection with James I's metrological legislation in 1426. One of the short preliminary acts that preceded the Scottish 1426 Assize confirmed that the existing 'water metts' would continue in use throughout the kingdom.[21] The use of water measure therefore extends back at least into the fourteenth century, and probably into the earlier period when the metrologies of Scotland and England were more closely connected. Water measure was repeatedly covered by statute in Scotland, in that it was to continue 'as before', but it is not defined in any statute that survives. For example, at the 1563 Assize it was specifically stated that water measure was exempt from the new provisions.[22] The control of water measure, that is, of the measure used for goods that came by water rather than land, seems to have been delegated by parliament as one of the privileges granted to the Scottish royal burghs, which had the monopoly of overseas trade.

Numerical information about the sizes of the water measures used at Edinburgh's port of Leith shortly after 1500 will be discussed in Chapter 7, where it will be shown that the Scottish water boll was equivalent to 1⅛ conventional bolls, or 'land' bolls. In the meantime, however, we will merely note that the use of water measures at ports was such an entrenched feature of Scottish practice that it survived the abolition of some of the 'charities' in the early seventeenth century, with the water boll remaining larger than the land boll. This traditional allowance, provided for those merchants entitled to use water measure, appears to have remained set at about one-eighth above land measure, where land measure is understood to be the permitted trading measure. However, if the size of the land measure was increased, the burghs felt free to increase water measure by the same proportion. Latterly, water measure came to be about one-eighth above the basic legal measure as shown in Alexander Hunter's semi-official *Treatise* of 1624 where he records that the 'halfe bowe [boll] mett of the water measure of Lieth [*sic*] conteines 9 peckes'.[23] Since the land boll contained 4 firlots or 16 pecks, it follows that a water boll of 18 pecks is equivalent to 1⅛ land bolls. However, as we will show in Chapter 7, for at least the century up to the time of Hunter's *Treatise* the land boll had been equivalent to about 1⅛ legal bolls. Hence the water boll in use was effectively 1¼ times the volume of the legal boll.

It will be appreciated that since the English water bushel was equivalent to 1¼ legal struck bushels, it would generate a 'quarter' of 10 struck bushels –

the size recorded in the English 1413 Act as being taken by the London merchants.[24] Perhaps, therefore, the origin of the additional bushel in the quarter (beyond the 9 bushels expected) is to be found in pressure exerted by the merchants to secure the full advantage provided by the water measure allowance.

This may provide an explanation of the otherwise puzzling size of the English ale gallon. Although this English ale gallon is understood as 282 in³, the normal standard is the quart of 70½ in³ (sometimes given, however, as 70 in³).[25] The Carysfort Committee on weights and measures tried to discover the origin of this unit in 1758 and the Commissioners of Excise provided them with a memorial of 1688 which specified the unit for gauging ale as a gallon of 282 in³. The unit is certainly quite early – there is a record of an ale standard of 282 in³ in Elizabeth's Exchequer, but unfortunately this does not survive.[26] The ale gallon remained at this size until replaced by the slightly smaller Imperial gallon in the 1824 act. If ale was reckoned by a taxable unit with allowances related to those on the grain from which the ale was produced, we might expect a London ale gallon which was 1¼ times some basic gallon. Taking the base unit as the 'Guildhall' or *Tractatus* gallon of 224 in³, this generates an enhanced gallon of 280 in³ (224 × 1¼ = 280). This is, of course, the figure we have already encountered in Chapter 4, where the traditional relationship of 4:5 between the sizes of the English wine and ale gallons was considered, but we now have a possible explanation and perhaps an indication that the ale gallon size may have crept upwards in the intervening centuries.

Although this may be a partial answer to the problem posed by the ale gallon's size, the reality was certainly more complex. An indication of this complexity is provided by John Wybard's 1647 measurements of the ale gallon standard in the City of London Guildhall, noted in Chapter 4. The capacity he obtained was 265·6 in³, corrected to 265·7 in³; and (less reliably), 267·2 in³. No reference has been found to a gallon of this size, and unfortunately Wybard gave no information about the vessel, its particular status or its application, and it is quite possible that it was no longer current. However, an indication that such a gallon size did exist comes with measurements of the 1601 standards at the Exchequer undertaken for the Carysfort Committee in the 1750s by Joseph Harris, the highly-respected king's assay master at the Mint, and the prominent instrument maker John Bird. They recorded a 1601 bushel as 2,124 in³, noting that this was 26 in³ less than the current Winchester standard of 2,150 in³; and an eighth of such a bushel is 265½ in³.[27] One possibility is that this was a size intended as a reference for standards issued to towns outside London or for some particular application; but it seems more likely that it was considered a basic size before the addition of an allowance. Increasing such a gallon by a sixteenth gives the conventional ale gallon (265½ × 1 1/16 = 282), and so perhaps this mirrors the practice

THE ABUSE OF
CHARITY:
ALLOWANCES AND
THE HEAPING
OF MEASURES
———

215

in Scotland where we will see that an allowance of a sixteenth was provided for the sale of ale and wine.

We can perhaps draw a parallel between this situation and that of the *Tractatus* wine gallon of 224 in³. If this was also used commercially with a sixteenth allowance, the enhanced capacity would be about 238 in³ (224 × 1 1/16 = 238). It has already been suggested that the English dry gallon may have been increased by a fifteenth from an early dry definition of 8 tower pounds of wheat to one of 8 troy pounds of wheat (confirmed in the 1496 act). Recognising the symbolic but imprecise nature of such grain declarations, it is tempting to suggest that the dry gallon was simply increased to match the enhanced form of the liquid gallon (and we will argue that the same happened in Scotland), and the new definition in troy pounds remained appropriate and was sufficiently accurate (expressed as a fifteenth rather than as a sixteenth) to carry authority.

Whereas the English water measure was maintained at one-quarter above the legal measure, the situation was more complex in Scotland because the legal size of the dry capacity measures was regularly increased. In practice, the water measure sizes were also increased to keep pace with the enlargement of the land measure so that they too were maintained at one-quarter above the legal level, at least until the seventeenth century. Hunter's reference shows that the use of enlarged water measures was still a feature of Scottish trade well into the seventeenth century. Although the concept of a separate water measure series which was to some extent outside the scope of normal metrology statutes helped perpetuate the practice of heaping, it was gradually modified so that it came to be applied only to certain classes of produce. Water measure was eventually abolished in the 1824 act which established the Imperial system of weights and measures, although it re-emerged as a permitted heaped measure for grain with a carefully specified conical heap comprising one-quarter of the contents.[28] A Glasgow-made set of Imperial heaped measures survives at Lanark (Item **261** in the Inventory). This final permitted form of heaped measure was abolished in the Weights and Measures Acts of 1834 and 1835.[29]

The final complication for the Scottish dry measures is that whilst most commodities, such as wheat, were supposed to be ascertained by struck measure, some other goods, including barley, were permitted to be measured by heaped measure. The size of this heap was particularly large, representing half as much material again as was contained in the firlot or peck measure (although usually referred to as 'the just third' of the total quantity), and this was about the maximum size of heap that could be achieved (Fig. **6.1**).[30] Because of this, the measure was not heaped as such, but a conventional equivalence of three struck measures for two heaped measures was used (and this in turn enabled an element of conventional heaping to be added as thought necessary).

Fig. 6.1

A late eighteenth-century
wheat firlot (Item **223** in
the Inventory), showing the
comparative sizes of con-
ventional heaps. At the top
is the struck vessel, with
the support bar visible at the
grain surface. At the centre
the firlot has a heap of one-
sixteenth of its contents,
representing the basic stage
of enlargement of the
Scottish measures.

This is contrasted at the
bottom with the maximum
heap, theoretically applied
for barley, but in practice
replaced by using three
struck firlots in place of two
heaped firlots. At the 1618
Assize a larger barley firlot
was introduced to accom-
modate the heap within a
struck measure.

The aperture and wall
thickness of this late eighteenth century firlot match
those of the 1618 wheat
firlot (although the capacity
is a little larger), and so the
measure will support the
same size of heap as in
1618.

THE ABUSE OF
CHARITY:
ALLOWANCES AND
THE HEAPING
OF MEASURES
———

217

This procedure almost certainly had a long history, but the '3 for 2' or '6 for 4' rule was first specified in the statutes in the 1587 Assize, when the commissioners unsuccessfully recommended to the Privy Council that new and larger measures for malt, barley and oats should be introduced so as to accommodate the heap within the firlot measure.[31] At the 1618 Assize the logic of this approach was accepted and two firlots were defined. One firlot of 21¼ pints was for 'Wheat Rye Beines Peas Meal Whyt Salt and such other stuff and Victuall as before this tyme hath beine in use to bee measured by straik Mett within this Kingdome'.[32] The other was for malt, barley and oats, which 'haue euer beene used to bee measured by heape'; but because the commissioners found by experiment that the maximum heap was less than a third of the total and therefore concluded that too large an allowance was being taken, they set the size of this larger firlot at 31 pints rather than the 31⅞ pints that would have been expected (since $21¼ \times ⅜ = 31⅞$).[33] Thus after 1618 the old heaping equivalence of '3 for 2' or '6 for 4' gave way to separate wheat and barley firlots.

The practice of heaping by various traditional allowances played a central part in the evolution of the later Scottish dry capacity measures, and this will be examined in Chapter 7. But the implication that grain was used in the definition of these measures means that we must be satisfied about the characteristics of grain as a measuring medium. It transpires that there are significant differences between the behaviour of grain and water, and we must take adequate account of these before looking at the earliest descriptions of trading allowances in the 1426 Assize.

MEASUREMENT IN GRAIN

In Scottish metrology, at least from the beginning of the fifteenth century, the dry capacity measures were defined in terms of the number of pints they contained. The firlot could and did grow with successive assizes, and we shall see that the mechanism involved accommodating the conventional allowances, in much the same way as has already been described for the relationship between the various English gallons. But this growth was defined in terms of an increasing number of pints contained in the firlot – the pint and liquid gallon did not grow as a consequence of the growth of the firlot.[34] In England the measures were also allowed to grow, but there was a more subtle relationship between the standards and the statutes authorising them. As said before, a principal difference in England is that it was not a volume but a weighed quantity of material that defined the dry measures.

The last major Scottish assize of weights and measures before the Act of Union was held in 1618, and we are fortunate that the detail given about the basis of the dry measures is comparatively good and complete. (The full text of the 1618 Assize is given in Appendix A.9) The 1618 firlot definitions are certainly adequate to allow us to conclude that grain, and not water, was used as the measuring medium. This can then be used as the starting point for working back to the earlier assizes for which the detail is less complete, and in particular to the pivotal 1426 Assize.

The problems inherent in dispensing dry material by volume and not by weight are well illustrated in the declarations given for the wheat firlot at the 1618 Assize (Fig. **6.2**). The wheat firlot was stated to contain 21¼ pints, and because we can be certain that the volume of the pint was 103·7 in³ at this time, it follows that this firlot should have a capacity of 2,205 in³.[35] But the assize also described the vessel as having a cylindrical interior of diameter 19⅙ inches and depth 7⅓ inches. It was to have a triangular bar across the opening of the measure and a pillar connecting this to the centre of the base, giving a volume for these metal supports of about 9 in³. These dimensions enable the internal capacity of the 1618 wheat firlot to be calculated as 2,107 in³.[36] However, we must recognise that the diameter is given to no better than one-sixth of an inch, and so there is a small intrinsic uncertainty in the measurement of these dimensions or of the construction of a vessel to match this description. In providing the dimensions for legally-acceptable vessels, the sizes had to be given precisely in inches and convenient fractions, even though it was inevitable that these would generate a volume that differed slightly from the desired theoretical volume. A well-documented example of this is the definition of the English Winchester bushel in 1696, where the use of convenient dimensions generated a volume which was known to be 4·8 in³ larger than the legal standard. However, these dimensions were adopted, 'there being no other convenient Dimensions (without counting to the hundredth part of an Inch) that would come so near as these'.[37] A reasonable measure of the potential imprecision in the match between a physical standard of the 1618 firlot and the independent definition in terms of the vessel's capacity in pints is about ± 20 in³, and so the volume of the wheat firlot obtained from the dimensions stated in the 1618 Assize can best be given in round numbers as 2,110 ± 20 in³.[38] However, this appears to create a problem because it is incompatible with the capacity of 2,205 in³ obtained by taking 21¼ times the volume of the pint measure.

There is a similar situation with the 1618 barley firlot of 31 pints. Multiplying up from the known volume of the pint gives a capacity of 3,215 in³, whereas the volume obtained from the dimensions given in the assize is 3,020 ± 25 in³.[39] The figure derived from 31 fills of the pint is clearly outside the range of capacity calculated from the dimensions.

This difficulty was recognised by a number of commentators in the mid-eighteenth century, when they tried to reconstruct the correct sizes of the lost 1618 firlot standards. They resolved the paradox by claiming that the dimensions given in the assize were erroneous (and this work will be

THE ABUSE OF
CHARITY:
ALLOWANCES AND
THE HEAPING
OF MEASURES

219

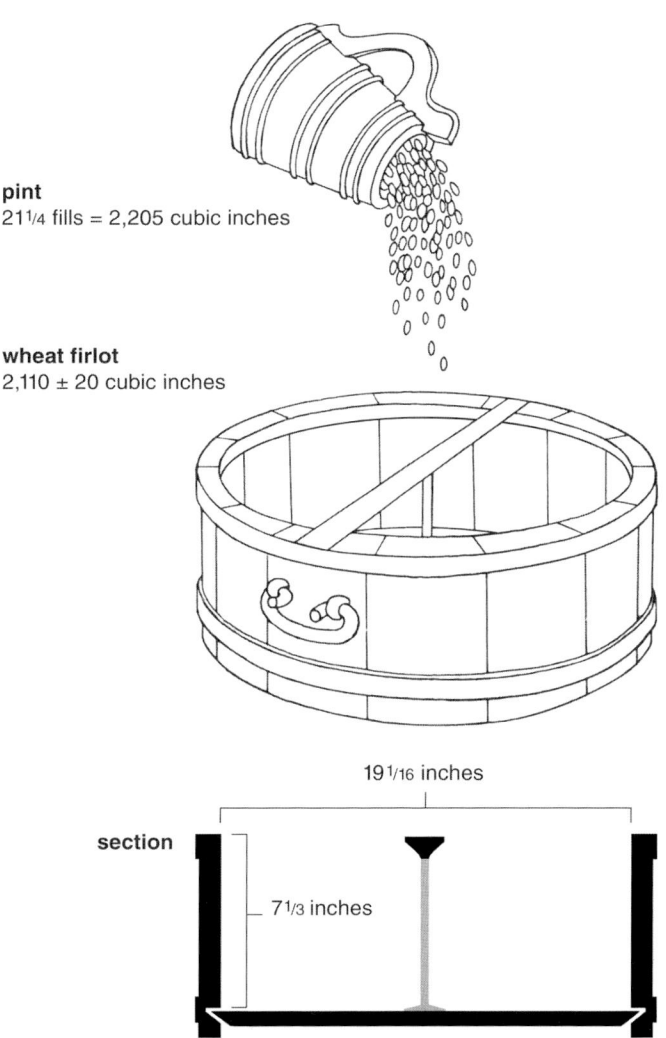

pint
21¼ fills = 2,205 cubic inches

wheat firlot
2,110 ± 20 cubic inches

19 1/16 inches

section

7 1/3 inches

Fig. 6.2
Capacity of the 1618
wheat firlot: the apparently
incompatible definitions by
volume and dimensions in
the 1618 Assize.

discussed in Chapter **9**). However, an examination of the measurement techniques used in the creation of these standards indicates that the dimensions
are sound, as we will demonstrate.

In the eighteenth century, scientific precision was increasingly being
applied in measurement activities. In the assessment of the 1618 firlot sizes,
conducted in the 1750s by the professor of natural philosophy (which we
would now term physics) at the University of Edinburgh with colleagues
in the Philosophical Society (precursor of the Royal Society of Edinburgh),
the requirements of accuracy made it self-evident that the raising of a firlot
from the pint measure should be undertaken using water as the measurement
medium. Indeed, the most obvious precedent for the precision measurement
of a standard dry capacity measure, and specifically one using water, was
Thomas Everard's official determination of the volume of the English
Winchester bushel for the setting of the malt duty in 1696.[40] However, at a

much earlier date – and certainly at the time of the 1426 Assize – these large
dry capacity measures were undoubtedly raised using the material they were
designed to measure, namely grain.

The English definitions, such as those of the 1490s, actually specify the
use of weighed quantities of wheat. In Scotland, the use of the contents of the
small capacity measure to define the large dry measures masks the nature of
the measuring medium, although by analogy with English practice we must
assume that it was grain also. The weight of the contents of the Scottish pint
was determined in a water measurement, and this can be associated with the
gallon definition of David's Assize which is certainly water-based. But
although the larger measures of the 1426 Assize also have their capacities
defined in terms of the weight of their water contents, it is clear that these
weights are simply multiples of a basic unit which itself has been determined
by a weight of water. They are given merely to establish the conventional
hierarchy of sizes that relates the measures to each other. It will be seen later
that this theoretical water-based definition of the boll measure of the 1426
Assize bears little relationship to the boll that was actually sanctioned for
use. In practice, then, the dry measures of 1426 were defined on the basis of
measurement with grain.

Superficially, grain behaves very much like a liquid, in that it flows and
it does not appear to be compressible. However, the individual grains are of
a finite size and when they lie together there are voids between them: the
precise volume that they occupy depends on the extent to which they pack
together to minimise this lost space. Their surfaces are also rough, and at
a microscopic level this makes them behave in a 'sticky' fashion.[41] Once the
grains have assumed adequate contact with their neighbours it is unlikely
that they can be compressed much further and they behave like a slightly
elastic lattice under vertical compression. However, the closeness of the
contact is established to some extent by the method of pouring the grain into
the container.

Disturbing grain within a container, for example by stirring or pressing
hard into it, has the effect of dislocating the tightness of packing between
individual grains and therefore actually increasing the volume occupied; and
the adhesion between the individual grains means that they cannot readily
be agitated into a more tightly packed configuration. There has been a long-
standing traditional awareness of this amongst vendors who sold grain by
volume and who stood to benefit by exploiting this phenomenon.[42] However, the physics of what is now described as 'dilatation', particularly as it
applies to granular solids such as sand, was first investigated mathematically
only about a hundred years ago.[43]

If the grain is poured from a reasonable height and in thin layers, as would
be the case with a large dry capacity vessel being filled repeatedly from a
smaller vessel such as a pint measure, then the grains are packed at a density

THE ABUSE OF
CHARITY:
ALLOWANCES AND
THE HEAPING
OF MEASURES
———

221

near the maximum. In contrast, practical experience has shown that the most likely method of filling the smaller vessel (by pouring grain from just above) gives rise to a slightly lower density in the pint, or in a larger vessel, because the grains are less tightly packed.[44]

These densities differ by only a few per cent. Not only is this difference unexpected, it is also difficult to detect. Even if detected, it is difficult to correct for it: attempts at packing the pint to the higher density typically found in the firlot is a laborious procedure because the neck of the pint vessel is very narrow. There is no obvious or intuitive reason to suspect that the density of material in the pint should be any lower than in the firlot, and this feeling would be reinforced by the discovery that the weights of successive fills of the pint standard, if filled in a consistent fashion, are constant to a high degree of accuracy.[45]

However, the result of this small percentage difference in density is that the volume generated by taking a set number of fills with the smaller vessel will be reduced by the same percentage. Recent experiments attempted to reconstruct the firlot of the 1618 Assize using a pint standard of the period, and obtained almost identical results to those of the 1618 commissioners.[46] In the particular case of the 1618 wheat firlot the density difference was found to be about 4½ per cent. Reducing the capacity found from 21¼ fills of the pint of 103·7 in³, namely 2,205 in³, by this amount gives a volume of 2,110 in³, which now accurately matches the capacity obtained from the dimensions of the vessel. For the deeper barley firlot of 31 pints it was appreciated that a difference of about 6½ per cent would apply: this increase is compatible with early nineteenth-century work, and it can be used as a guide for the density to be expected in deeper measures such as the boll.[47]

Although variation in grain type and quality has a bearing on the precise way in which the medium behaves, the differences introduced are very small compared with the general effects observed, which depends merely on the fact that the medium is composed of small elongated grains which are not perfectly smooth.[48]

The apparent incompatibility of the twin definitions of the firlots in the 1618 Assize has proved a serious problem of interpretation since at least the mid-eighteenth century and has led to the sweeping assumption that dimensional figures given even in this act are in error and should be disregarded.[49] The use of grain as the measuring medium allows the two volumes of the 1618 firlot to be brought into coincidence, and removes any need to invoke claims of error in the parliamentary record or to make arbitrary choices about which figures to accept. This seems altogether more satisfactory, and the success of invoking grain measurement for the 1618 firlots has led us to apply similar corrections to earlier measurements, notably those made by the 1426 commissioners.

The definition of national standards based on actual measurement with grain as late as 1618 should not be taken as implying a lack of sophistication in metrology: on the contrary, it indicates that the commissioners had an explicit appreciation of the tacit knowledge of practitioners and a pragmatic understanding of the practical requirements of the market-place. However, the physical standard created at the 1587 Assize, generated under the control of the officials of the Mint, was unambiguously based on a water determination: this only exacerbated the confusion caused by the enlargement of the measures at the 1587 Assize, and it subsequently became clear that the burghs were using measures based on the correct number of pints (and accommodating the accepted allowance) but using grain.[50] In the knowledge of this and to avoid potential confusion, the commissioners at the next assize, in 1618, provided dimensions that related to practical grain measures that would have market-place relevance.

We can tell that water was not used in the measuring of the dimensioned firlots of 1618, yet the report of the 1618 commissioners refers to water as the measurement medium. An existing measure had been presented to the 1618 commissioners as the current standard by the provost and bailies of Linlithgow, together with an authorised gauge: once the measure's dimensions had been tested against the gauge:

> the saids Commissioners caused praesentlie fill the same with water which being full they fand the same conteined Twentie ane pincts and ane mutchkin [21¼ pints] of just Sterline Jug and measure and that the foresaid jug containes within the same Thrie punds and seaven unces [55 ounces] of frensh Troys weght of clear running water of the water of Leith.[51]

A precise definition has been provided of the pint's volume, reflecting that of the 1426 Assize, and of necessity this is in terms of its water content. The clerk has apparently conflated this with the parallel definition of the firlot in terms of the pint. The result is a statement which seems internally self-consistent, but which reveals either that the clerk has not appreciated the significance of distinguishing between the media, or that a simplified description of this sort was considered preferable for the legislative record. It certainly cannot be argued that grain was not used in the commissioners' measurements, because careful assessments of the size of the maximum heap on the firlot were made by them.

It should be emphasised that the differential packing effect is quite small and requires rapid and repeated checks on quantities. It has, however, been analysed before. An awareness of the phenomenon of increased packing density in large measures is not recent. For example, an account of grain measuring procedures used in Aberdeenshire in the 1790s described a number of sharp practices. These included the measurements taking place in a grain loft where the buyer 'takes care to have several persons walking about during the whole time of the operations' and where 'the grain is usually

THE ABUSE OF
CHARITY:
ALLOWANCES AND
THE HEAPING
OF MEASURES

———

223

thrown into the firlots by shovelfuls from a considerable height, so as to make it fall with force': both of these procedures, together with the dubious methods of striking described in the account, would result in more grain being contained in each fill of the firlot and therefore a reduction in what otherwise have been the number of fills of the (pint) measurement unit, for which the buyer paid the farmer at the agreed rate.[52] Some years later, Adam Anderson (*c.*1780-1846), the rector of Perth Academy, conducted a series of experiments on the filling of bushel measures, and stressed the implications for the pricing of grain to the 1834 Select Committee on the Sale of Corn.[53] Similarly, when the Board of Trade permitted weights and measures Inspectors to test capacity measures using rape seed towards the end of the nineteenth century, very specific regulations were enforced to ensure that vessels were always filled under identical conditions to prevent density variations of the type reported by Anderson.[54]

The type of grain density effect discussed above may account for a puzzling feature of the English dry measures. In 1696 the bushel for grain (then known as the 'Winchester' bushel) was re-defined as a dimensioned vessel of 18½ inches internal diameter and 8 inches depth for the purposes of exacting a new duty on malt.[55] This generated a bushel of 2,150·4 in³. Although the gallon was not described in the same act, it would be expected to be an eighth of this, namely 268·8 in³. However, the accepted size for the corn gallon was a little larger, and subsequent legislation stated that the Winchester gallon was 272¼ in³, a discrepancy noted by the Parliamentary Committee on Weights and Measures in 1819, but never explained.[56] This larger size was of long standing – the Carysfort Commission had measured a 1601 dry gallon as 271 in³, and the three early standards located in the Exchequer in 1688 were all found to have capacities of 272 in³.[57] The answer may simply lie in the adaptation of the dry series to practical measurement in grain: given that the volume of the bushel was carefully controlled against earlier standards, the practical dry gallon probably had to be increased in size so that eight gallon-fills of grain adequately filled the bushel. The amount of this increase (about one per cent) is in line with Anderson's later findings.

The definition of the malt tax bushel in 1696 is the earliest occasion when specific information is given about a significant type of compromise accepted in the setting of a capacity standard. The existing Exchequer standard bushel was carefully gauged using water by the Excise gauger Thomas Everard and three assistants, in the presence of members of parliament, using a specially constructed vessel of 224 in³ capacity, and it was found to contain 2,145·6 in³.[58] However, since a large number of working standards would have to be produced in order to assess and collect the duty, the legislative requirement was to provide clear dimensions for a simple cylindrical vessel that would be considered as complying with the Exchequer standard.

In effect this was a manufacturer's specification. These dimensions were fixed at 18½ inches diameter and 8 inches depth:

> for a Cylindrical Vessel of these Dimensions will contain 2150·42 solid Inches, which exceeding the Content of the Standard Bushel but 4·82 [solid] Inches, and there being no other convenient Dimensions (without counting to the hundredth part of an Inch) that would come so near as these: It was enacted, in the Act for laying a Duty upon *Malt, That every Round Bushel, with a plain and even Bottom, being eighteen Inches and a half Diameter throughout, and eight Inches deep, shall be esteemed a legal* Winchester *Bushel, according to the Standard in his Majesty's Exchequer.*[59]

Although there is only a small difference (0·2%) between the dimensioned vessel and the standard of which it is a legal duplicate, this case provides a timely caution against interpreting dimensions too literally and this has been an important consideration in treating the dry measures defined in Scottish assize legislation.

Another English example is the wine gallon, belatedly legalised in 1706, and described then as a cylindrical vessel 7 inches in diameter and 6 inches deep. Although these figures give an acceptable capacity of 231 in³ (which, according to the act, 'shall be deemed and taken to be a lawful wine gallon'), we can appreciate that the original standard may have had a different volume which was not amenable to such a simple dimensioned description. In fact, it will be argued later in this chapter, that the true volume was about one per cent larger at about 233 in³, and this size is found to be compatible with the equivalent Scottish gallon. By analogy, it is possible that the English ale gallon came to be defined in a similar dimensioned form, and it should be noted that a gallon of 6 inches aperture and 10 inches depth has a capacity of 282·7 in³. Perhaps in this case, the use of a convenient dimensioned form sanctioned an effective increase from the early value of 280 to 282 in³?[60]

To sum up this discussion about grain measure, we can now clarify the difference that has emerged between English and Scottish practice, which we infer had been the same at the time of David's Assize. Thirteenth and fourteenth-century sources defined the English dry gallon in terms of 8 pounds of wheat, and it is clear that capacity standards such as the bushel (of 8 gallons) could be raised by dispensing the appropriate number of weighed quantities of wheat. In Scottish practice from at least the fifteenth century, it was not a particular weight of grain but the grain capacity of a small vessel that formed the basis of the measurement. Although a weight was associated with each pint-fill, it was the weight of the water content not the weight of the grain. Therefore the weight given for a firlot in the 1426 Assize is purely a conventional weight, derived from the number of pint-fills, multiplied by the known weight of water required to fill the pint, and does not imply an actual weighing and certainly not a weighing of water. Thus, in the English case, the only grain compaction effect relates to the small variation of density

THE ABUSE OF
CHARITY:
ALLOWANCES AND
THE HEAPING
OF MEASURES
———

225

in the filled bushel (as investigated by Anderson); whereas in Scotland, in addition, we have to contend with the effect of the rather lower density of grain in the pint measure. The reduction in volume of the ordinary firlot measures is about 4½ per cent, and this is a feature of Scottish measurement practice which we would not expect to find mirrored in English practice.

THE SIZE OF THE PINT AND
THE DIVISIONS OF THE OLD GALLON

One of the most interesting features of the reforming 1426 Assize of James I is the care taken to relate the definitions of the new measures to those of the measures that were in use at the time the assize was introduced. In two places there are references to David's Assize – first in confirming that the ell was still defined as 37 inches as in 'the Statute of king Dauid the first', and second to describe the boll 'made be [by] king Dauid' as containing a sextern, where a sextern was 12 gallons of ale 'of the ald met'.[61] Although these statements are clearly intended to establish a secure authority for the system about to be enshrined in law, it is apparent that there had already been significant departures from the metrological system ascribed to David I. These had probably occurred in the hundred years since the early fourteenth-century date of the administrative document that has been termed David's Assize. In particular, the firlot had grown substantially, and the new firlot measure defined in 1426 was not to be made 'eftir the first mesoure' (that is, the quarter of David's boll), nor 'eftir the mesoure now vsit' (which must have been considerably larger), but 'betuix the twa'.[62] The definitions in the 1426 Assize provide enough information about the 'old system' to clarify the type of change represented in the assize. In doing so, it will become apparent that the assize preserved valuable elements of continuity.

The 1426 Assize discussed the liquid measures of the old system in terms of a gallon standard, and made the inference that this was the gallon defined in the David Assize. The David definitions had been solely of the gallon and boll measures, but the Exchequer rolls for 1264 mention a 'firthelote' – a firlot (or quarter boll).[63] Although smaller units certainly came into use, they only appear in the literature at later dates. By the late fourteenth century, official control was being exercised over the 'quart' measures of ale traders; that is, over measures of a quarter of a gallon.[64] There were other divisions also, because a 'chopyn' measure is specifically mentioned in the 1426 Assize in connection with a quantity given in old gallons, and this is the earliest occasion that the term 'chopin' has been encountered.[65]

In an echo of the David Assize, the capacity of the old gallon was given in the 1426 Assize in terms of its weights of 'divers' or mixed waters. However, the 'new' pint ('new' in the sense of being defined for the first time in this statute) was defined in terms of fresh water of the river Tay, rather than the marginally denser mixture of fresh and sea waters of the David Assize.[66] The weight given for the capacity of the old gallon was not the 12 pounds stipulated in David's Assize, but 10 pounds 4 ounces of the newly-defined pound of 16 trois ounces, namely 164 ounces. Unfortunately, we cannot infer from this that any actual measurement has been performed: in 1426 the 'new' gallon was given as holding twice the weight of David's gallon, so the stated weight (of the 'old' gallon) is merely four times the weight contents of the newly-defined pint (4×41 ounces = 164 ounces). However, it does allow us to understand that the gallon of the old system, containing four quarts, was to be considered as equivalent to four of the 1426 pints. Thus, there are now eight 'new' pints to the 1426 gallon, the same multiplier as in England. Taking the trois ounce as 480 grains, as was established in Chapter 5, this weight of fresh water would occupy 77·8 in³.[67]

Although the Assize indicated that the new pint of 77·8 in³ was equivalent to the quart (or quarter) of the David gallon, the full significance of the relationship between the two is not immediately obvious. The David Assize provided the dimensions for the physical gallon measure, and as we have seen in Chapters 1 and 4, these dimensions give a comparatively large volume of approximately 330 in³, compared with the volume of about 311 in³ generated by four quarts of 77·8 in³.[68] We shall discuss further the significance of these various values for the volumes of the gallon.

At this stage, we cannot adequately account for the small differences between these volumes for David's gallons although they are similar enough for it to be plausible that the original twelfth-century gallon should have evolved into the gallon of the 1426 Assize over the intervening two-and-a-half centuries. After all, we know nothing about the copying and control procedures at this early period, and very little of the commercial and administrative pressures which might have led to adjustment. For example, there is no obvious reason why the pint should have been equated to exactly 41 ounces in the 1426 Assize – perhaps the use of this whole number reflects a compromise of some sort in setting the size of the pint? Equally, the value may have been adjusted to match external standards: in this instance, there is good reason to believe that the 1426 pint is the pint of an earlier Scottish wine gallon (and this is discussed below) which seems to be precisely double the early form of the English wine gallon. It will be argued later that a compromise was intended, and that the 1426 pint should, strictly speaking, have been set at 41¹⁄₆₄ ounces. In the meantime it is merely necessary to note that the 1426 Assize specified 41 ounces as the contents and that this gives rise to a pint of 77·8 in³.

We are, however, able to confirm from an independent source the 330 in³ size of the dimensioned gallon in the received text of David's Assize. A surviving Scots merchant's handbook, which has been tentatively dated from

THE ABUSE OF
CHARITY:
ALLOWANCES AND
THE HEAPING
OF MEASURES
————

227

internal evidence to about 1400, provides information on weights, measures and currencies encountered at Bruges.[69] French wine was exported to Flanders, and onwards, through the French wine staple at Damme in large barrels known as 'tuns', whose size was strictly controlled, and which originally must have been defined in terms of French standards. We will examine the evidence for the tun in the next section, where we will find that the effective volume for the tun at this time was about 58,200 in³ (or about 955 litres). From the definitions given in this handbook in terms of Flemish and Scots units, we can deduce that the tun contained 352 stoops, or 704 pints (at two pints to the stoop), and the size of this wine pint was therefore 82·7 in³.[70] We will return to this later, but in the meantime we merely record that such a pint is a quarter of 331 in³, which provides a good match for the dimensioned gallon of approximately 330 in³ in the David Assize.

The measures of 82·7 and 77·8 in³ are closely related: the first is a sixteenth larger than the second ($77·8 \times 1\frac{1}{16} = 82·7$). There is evidence that over an extended period both wine and ale were measured in 'customary' pints which were a sixteenth larger than the statute pints. For example, an authoritative late eighteenth-century source stated that 'the customary gallon used by the brewers, as also the pint, chopin and other ale-measures generally hold $\frac{1}{16}$ part above standard'.[71] There are indeed two extant sixteenth-century standards of this type (Items **110** and **111** in the Inventory), which will be discussed in Chapter **7**, and there are early nineteenth-century references to others.[72] An official determination of the capacity of a half-barrel cask in Scots and English measure (18 English ale gallons, or about 83 litres) was carried out in 1707 for the new Excise Commissioners for Scotland, and this also was conducted in pints which were larger than the statute pint.[73] For our purposes, we will refer to these 82·7-in³ pints as customary, brewers', or tavern pints, and the 77·8-in³ pints as 'basic' or 'legal' pints since these are the ones defined in the statutes. We must recognise that standards of this larger size must have been in regular use in the burghs, in contrast to the statute pints which seem to have been largely restricted to providing the basis for the definition of the dry measures.

In this light, we can perhaps appreciate the dimensioned gallon vessel in the received text of David Assize as a customary standard for liquid, which should more correctly be considered as about 331 in³ capacity, containing four quarts of 82·7 in³ (or 330·8 in³ to be exact). It is therefore equivalent to a gallon of about 311 in³, comprising four quarts of 77·8 in³, but with each unit enhanced by the conventional allowance for wine and ale of one-sixteenth.

The 1426 Assize defined a new gallon with a weight equivalent of 20 pounds 8 ounces, numerically twice the size of the old gallon, and comprising 8 pints. Although the first preliminary act of 1426 (see Appendix **A.2**) stated that physical standards of the gallon would be given out from Edinburgh, that reference (together with the 1426 Assize itself) is the last time that the statutes refer to a gallon as a unit of liquid measure – from now on the current pint is used exclusively.[74] (However, it remained in use for non-liquid measures; see, for example, the salmon barrel, below.) The point of interest is not that the term 'gallon' is now being applied to a measure of twice its original size, but the fact that there is an underlying unit of capacity which remains constant.

We should pause to consider the size of this new 1426 pint of 77·8 in³. No standards of this capacity (nor indeed standards or ordinary drinking vessels of 82·7 in³) have yet been located. It is of course possible that some may survive unrecognised, because this pint size remained in use until as late as the early sixteenth century. The reason why we can be certain that there was a pint of this volume, even though no survivals have come to light, is because assize legislation provides details of the number of pints in the dry measures for the remainder of the fifteenth century and this information is only compatible with the pint of 77·8 in³. This issue will be examined in Chapter 7, which addresses the growth of the dry measures. In particular, we have confirmation of the size of the firlot in 1500 from an existing standard gauge (Item 1 in the Inventory); and from contemporary figures in the Edinburgh records we know that the operating sizes of the dry measures were reckoned in terms of a pint of this size.

From about 1500, the volume of the pint was set (for administrative purposes at least) at the larger value of 103·7 in³. A considerable number of standard vessels of this size survive, and its capacity bears an exact relationship to the pint of the 1426 Assize – it is precisely one-third larger (77·8 × 1⅓ = 103·7), and we will see the significance of this shortly. In the seventeenth century, the larger pint was described as holding 55 ounces of water (although this is a slightly different size of ounce).

The Edinburgh advocate John Swinton, in a late eighteenth-century work on Scots metrology, also concluded that the pint of James I was of about 77·8 in³.[75] However, this view was rejected more recently by Lawrence Burrell, whose work was based on the unsupported supposition that the oldest surviving standard pint of the 103·7 in³ size, namely the undated Stirling Jug (Item **108** in the Inventory), was the actual standard of Stirling referred to in the Assize of 1458 and therefore represented the pint of the James I Assize.[76] Since the James I pint contained 41 (and not 55) ounces of water, Burrell had to postulate that the weights of James I's Assize were one-third larger than English troy weight (of 480 grains to the ounce) in order to generate a sufficiently large volume. He equated this heavier weight with the Scots trone weight series which, as we will see in Chapter **8**, was not one-third but one-quarter greater than the 480-grain standard. Inevitably, this approach breaks down in due course because these two factors are different. We might broadly characterise Burrell's theory as a 'constant pint' inter-pretation. In contrast, our approach may be described as one of a 'constant

THE ABUSE OF
CHARITY:
ALLOWANCES AND
THE HEAPING
OF MEASURES
——

229

ounce', in that we have accepted a broad continuity in the weight series, but we have also accepted that (at least for the limited administrative purpose of defining the sizes of the dry measures) the pint did change in size.

But if the pint standard was 77·8 in³ from 1426, and one-third larger from about 1500, we should consider carefully whether it can be identified in the period before 1426. The late fourteenth-century reference to the control of quart measures, cited earlier, also records the pint measure; and from this we can glean further information about the sub-division of the quarts of the 'old system' in use before 1426. We have previously noted an equivalence between the old quart and the new pint, but we can now appreciate even closer similarities in the underlying units of the old and new systems.

This reference is included in the *Iter Camerarii*, a set of regulations for enquiries to be conducted by the chamberlain in the burghs, and datable to the end of the fourteenth century. One of these governs the liquid measures by which brewers sold ale, and it required the brewers to present their measures so that it could be established that they were made to the correct standards. The prescribed standards were the 'quart pynt thrid pert and sext pert'.[77] The most significant feature here is the use of factors of one-third, rather than the halves and quarters that we might otherwise have expected. Recalling that the gallon was defined in David's Assize in terms of the volume of twelve pounds of water, it is plausible that the quart was divided into three units, each of which was equivalent to one pound. Although it is not specified of which unit the 'third part' and 'sixth part' are fractions, the most likely explanation is that they are divisions of the quart.

If David's gallon contained 12 pounds, his quart is equivalent to three pounds, the 'third' of the quart is the one-pound unit, and the 'sixth part' of the quart would be a half-pound unit. The capacities appear to have been given in descending order of size, from which we can perhaps deduce that the 'pint' was equivalent to two pounds, and thus was two-thirds of a quart. We therefore have the somewhat unexpected situation that before the 1426 Assize the gallon comprised six and not eight pints. The old pint would then be one-sixth of 311 in³, namely 51·9 in³. (Since we have seen that the customary gallon was used with an allowance of one-sixteenth as 331 in³, we would also anticipate a customary old pint of 55·1 in³.)

We will therefore suggest that the 'pint' was the name given in early times to the usual measure of two-thirds of a quart or one-sixth of a gallon. This would mean that the *Iter Camerarii* regulation was referring to measures of a quart, ⅔ quart, ⅓ quart and ⅙ quart, being the volumes of 3, 2, 1 and ½ pounds of water respectively (Fig. **6.3**). The terms 'chopin' and 'mutchkin' for the smaller divisions of the measures are used in the 1426 Assize and we will further propose that these were the names given to the 'thirds' and 'sixths'.

We do not know when the words 'pint' and 'chopin' were introduced, but it is likely that they came from the French *pinte* and *chopine* and therefore date back to the period of strong Norman influence in Scotland. Some support for the proposal that the early Scottish pint was two-thirds of a quart can be drawn from the early French measures, where the *pinte* is claimed to have its origin as the capacity of two *livres*, and where from at least the thirteenth century the *pinte* contained two *chopines* (each therefore equivalent to one *livre*).[78] Our proposal for the early Scottish pint is similarly that it should be equivalent to two pounds and contain two chopins.

We can now appreciate that it is not the quart but the chopin measure, representing the original capacity pound (twelve of which made the gallon), that formed the basis of the capacity series. We have noted that the old quart and the new pint seem to be closely equivalent in size, and we will now propose that they were based on the same chopin measure and that this chopin was unchanged by the 1426 legislation. The pint and the gallon were merely re-defined in terms of changed multiples of the old chopin; but as far as the basic unit of trade in liquids was concerned, the 1426 Assize introduced no disruptive change whatsoever.

An essential feature of successful legislation had to be the ease with which it could be enforced, and this type of underlying continuity would therefore be expected. It is clear that the names of some of the measures did change – whatever the pint had been before 1426, it was subsequently the name given to the old quart. If the old pint was indeed two-thirds of the old quart, then it was two chopins (51·9 in³) before 1426. It was re-defined as a three-chopin measure (77·8 in³) in the 1426 Assize. But, because the pint was re-defined yet again about 1500, we should perhaps appreciate the name as applying to a recognised type of measure rather than to a specific defined capacity: in particular, the name may refer in assizes to the current administrative measure used in the definition of the dry capacity measures. In Chapter 7 we will discuss this final increase in the size of the pint to one-and-a-third times its

Fig. 6.3

Relationship of the statute pints (as opposed to the customary pints) before and after the 1426 Assize.

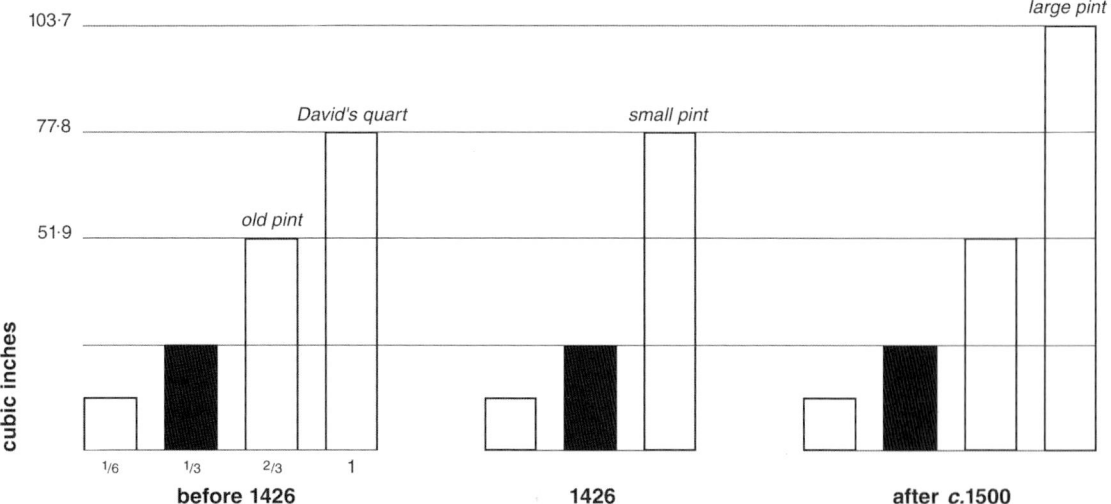

THE ABUSE OF
CHARITY:
ALLOWANCES AND
THE HEAPING
OF MEASURES
———

231

fifteenth-century level, namely from 77·8 to 103·7 in³. We can already appreciate that this was achieved by retaining the chopin again and re-defining the pint as four rather than three chopins (Fig. **6.3**). In the process the less convenient factor of three was finally lost and a system was achieved which would allow successive binary division down to the smallest unit of capacity.

The chopin and mutchkin had originally been the half-pint and quarter-pint respectively. By the time the large pint of 103·7 in³ had become well-established in the sixteenth century, the terms 'chopin' and 'mutchkin' were being applied to the half and quarter of the large pint, so that the sixteenth-century chopin was the equivalent to two chopins of the fifteenth-century statute pint.

THE SCOTTISH WINE AND ALE GALLONS

It has already been noted that the early Scottish merchant's handbook, published by Hanham, enables a Scots wine pint of 82·7 in³ to be inferred, with 704 pints to the tun.[79] Six of these pints would make the customary wine gallon of 496 in³. If, as seems very likely, the document dates from before the 1426 Assize, then the use of this enhanced pint of 82·7 in³ (and therefore of a base pint of 77·8 in³), demonstrates that this size of pint must have been introduced at an even earlier date. From this consideration, the 1426 Assize emerges yet more clearly as a declaratory act, namely one that confirms current practice rather than introducing change. It has already been indicated in Chapter **5** that the stone appears to have been increased in mass from 15 to 16 'Scots' pounds before the 1426 Assize, and so perhaps both this change and the introduction of the larger pint occurred towards the end of the fourteenth century.

However, we must be careful not to infer that the new statute pint of 77·8 in³ merely replaced the smaller pint of 51·9 in³, since they may have been appropriate in different contexts, and may therefore have remained in use concurrently. So far, we have been able to make the simplistic assumption that there was only a single current gallon in Scotland. But our experience of the situation in England is that there were several gallons in use, and in particular that there were separate gallons for ale and wine, and that these might differ from the gallon of dry measure. It would be surprising if this situation was not mirrored in Scotland, and this turns out to have been the case, at least in the fourteenth century.

There are two revealing instances where the gallon type is specified in the early Scottish statutes. The first is in the definition of the boll measure in the David Assize, where the boll is stated (in Thomson's 'Record Edition') to contain 'a sexterne viz. xij gallonis of aile'.[80] Although Thomson was here following the Ayr manuscript (attributed to the early fourteenth century), it has already been noted in Chapter 4 that the Cromartie manuscript (attributed to the second half of the fourteenth century) merely describes the gallon as of water.[81] It is therefore not clear if this was initially described as an ale gallon or simply as a liquid gallon until it became necessary to distinguish it from other types of gallon, and such a distinction may characterise fourteenth-century usage. However, at this time, we are able to demonstrate that the ale gallon was about 330 in³, which is the size of the dimensioned gallon of the David Assize.

The second instance is in a brief and undated *Assisa de Vino* (Assize of Wine), attributed by Thomson to David I, or shortly thereafter.[82] This relates the price for the *dolium* or barrel of wine to the price for individual gallons. However, the gallon type is specified, and the context clearly differs from that of the ale gallon, because (in translation) it reads:

> When the *dolium* of wine is sold for 20 shillings, the gallon of wine would be 2 pence … [and] the sextern must contain three gallons. When sold for 30 shillings … the gallon of wine would be 3 pence. When sold for 40 shillings … the gallon of wine would be 4 pence. When sold for 50 shillings … the gallon of wine would be 5 pence. When sold for 60 shillings … the gallon of wine would be 6 pence. And for every 10 shilling increase in the *dolium* the gallon should increase by one penny.[83]

The most striking difference between these is in the size of the sexterns (for ale, twelve gallons – for wine, three gallons), providing an echo of the contemporary English situation where the ale sextern was considerably larger than the wine sextern.[84]

The particular barrel size that is being discussed is the large 'tun' used by the French for wine exports; and since this was an international unit, the size of the tun traded to Scotland was very probably the same as the vessels known in England, Flanders and France. Its size was presumably originally defined in French units, and it was the largest of a whole family of barrel sizes whose volumes were mathematical fractions of the tun's volume. In the English tradition, the tun was equivalent to two pipes or butts, three firkins or puncheons, four hogsheads, and so on.[85]

From the five price levels quoted in the David Assize of Wine, ranging from two pence to six pence per gallon, the ratio of these prices for the barrel and gallon is 120:1. It might be concluded that the tun contained exactly 120 Scots gallons, but we must acknowledge that the principle virtue of the multiplier of 120 is that it facilitated price calculation, since there were 240 pence to the financial pound. In practice this may only be an approximation to the actual volume. Similarly, the construction of such a very large and elongated coopered vessel was clearly a difficult feat, so it is unlikely that the volume of the tun can be known to a very high degree of accuracy and we must expect some variation. Strangely, the full volume of the tun in English units was never adequately defined. However, there is enough evidence to demonstrate that it initially contained about 59,800 in³ (or about 980 litres), but that at some stage, and certainly before 1400, it was considered to be only about 58,200 in³ (about 955 litres). It is not possible to say whether this change represented a physical reduction in the capacity or the acceptance of a smaller volume for administrative purposes. The question of the tun's volume will be assessed shortly, once we have clearer information about the size of the wine gallons.

The dimensioned gallon of David's Assize is about 330 in³. But it cannot be the appropriate gallon for the assize of wine attributed to David, since this would imply that the tun of 59,800 in³ contained about 180 gallons, which is clearly not an acceptable approximation to 120. However, we have just seen that the customary or enhanced pint of 82·7 in³ was the appropriate pint unit for the wine tun in about 1400, when the tun's volume was considered to be the smaller volume of 58,200 in³. There were 704 of these wine pints to the tun, so with six pints to a gallon of 496 in³, there would indeed be about 120 of these gallons to the tun.[86] Although this has been deduced for a tun of a slightly later date and smaller size, it is clear that this customary wine gallon is half as big again as the David ale gallon, and that it would approximately comply with the gallon of the David Assize of Wine.

Thus, taking these two examples of Scots gallons specifically identified for ale and wine, we can now appreciate that in Scotland, as well as in England, there were indeed separate and distinct ale and wine gallons.

The basic ale gallon was nominally 311 in³ and comprised six pints of 51·9 in³, although in practice these units were used with a customary allowance of one-sixteenth, giving an ale pint of 55·1 in³ and an ale gallon of 331 in³ (which we have concluded matches the volume of the dimensioned vessel in the David Assize). The wine gallon was one-and-a-half times this size, and by analogy with the ale gallon, also contained six pints. These wine pints were nominally 77·8 in³; but, as has been seen, they were used in the customary form enhanced by one-sixteenth to 82·7 in³. The basic volume of the wine gallon was therefore 467 in³, and with the one-sixteenth allowance, the customary or tavern value was 496 in³. Thus, the sizes of the liquid capacity measures were as follows:

	ALE		WINE	
	LEGAL	CUSTOMARY	LEGAL	CUSTOMARY
PINT	51·9 in³	55·1 in³	77·8 in³	82·7 in³
GALLON	311 in³	331 in³	467 in³	496 in³

Table 6.1
Scottish liquid capacity measures, pre-1426.

The use of a capacity measure which incorporates a fixed allowance also acknowledges the acceptance of the basic legal unit, even though we have no clear evidence of the separate existence of these basic units before the 1426 Assize. The dimensioned gallon of the surviving versions of the David Assize indicates that the size of the allowance had remained unchanged from about the mid-fourteenth century, and so perhaps the concept of a uniform allowance (and therefore of the underlying basic units) may date from this period also. The suggestion made in Chapter 4 that the David gallon was originally defined as the volume of 12 French pounds of water, directly generating a unit very close in size to what we later describe as the customary gallon of 331 in^3, implies that the allowance formula was introduced subsequent to the original form of the David Assize.

Taking the basic Scottish wine gallon as 467 in^3, it will be noticed that this is almost precisely twice the English wine gallon of 231 in^3. In fact, the early size of this English wine gallon is not accurately known, and it was argued in Chapter 4 that the capacity finally attributed to it by statute in 1706 arose from the conventional dimensions of 7 inches diameter and 6 inches depth and such dimensions for the construction of replicas of the standard could only be given in whole numbers of inches and simple fractions. A similar situation with the English Winchester bushel led to the conventional dimensioned form differing in size from the historical size of the bushel obtained by direct comparison with earlier standards. In effect, two equally legal but slightly different forms of the bushel had been created. Measurement by this bushel was required for the operation and collection of the new malt tax, so it was essential that suitable standards and trading vessels were in place as soon as possible. By announcing the acceptable dimensions of a cylindrical vessel, trading standards could be manufactured at will, and officials had enough information to test measures in the market-place. The simple dimensions for the wine gallon satisfied the same type of requirement, and almost certainly the volume of this working measure differed slightly from the intended volume, which is likely to have matched a French prototype. The dimensions of 6 inches deep and 7 inches in diameter, generated exactly 231 in^3 (if π is approximated at the conventional value of $^{22}\!/_7$), which is presumably close enough to the true volume, using only convenient English figures for the dimensions. In this case the diameter has to be rather larger than might be expected for a liquid measure in order to obtain this particular approximate volume. It is presumed that there would be some retained standards of the correct size, but it is apparent that the convenient dimensioned size was the one almost universally used in working practice. The true volume had thus become only theoretically correct, and it was the dimensions of the working measure that were formally adopted in 1706.

The difficulty in establishing the original size of the English wine gallon has always been the absence of authoritative standards in the Exchequer.

THE ABUSE OF
CHARITY:
ALLOWANCES AND
THE HEAPING
OF MEASURES

———

235

However, another repository for official standards was the Tower of London, and these standards were used as references by John Reynolds (*c*.1585–*c*.1654), whose work was discussed in Chapter 4. According to John Wybard, Reynolds described the Tower standards as 'the most ancient and true standards Measures', pleading their authority in preference to those at the London Guildhall.[87] Official standards were certainly being verified at the Tower in the late sixteenth century, and in all probability had been for some considerable time beforehand.[88] Two standard wine measures, constructed by Reynolds in 1641 at the Tower, and perhaps from the same set, have been recorded. Only one is now located, and this is a wine pottle or half-gallon (Fig. **6.4**), carrying the royal arms and cipher, and stamped 'WINE POTLLE TRYED BY IOHN RENALDS AT THE TOWER'.[89]

A significant feature of this fine bronze standard is that it is not constructed to the conventional dimensions, but is considerably narrower and taller. Its capacity is also larger: in spite of the fact that Reynolds quoted the conventional volume of 231 in³ to Wybard, the capacity of this pottle is half of a gallon of 233 in³.[90] The quality of Reynolds's work can be confirmed in the Bridport ale gallon described in Chapter 4, so it is apparent that Reynolds meant the gallon of the wine pottle to be of this larger size.[91] It is proposed that

Fig. 6.4
English half-gallon standard wine measure by John Reynolds, London, 1641, representing the projected original size of the standard at the Tower of London. (Reproduced by permission of the Museum of Archaeology and Anthropology, University of Cambridge.)

this Reynolds measure was a surviving representation of the early English wine gallon copied by direct comparison from an authoritative standard which was unaffected by a need to conform to conventional dimensions.

It will be noted that twice the volume of this projected English wine gallon is 466 in³, virtually identical with the Scottish wine gallon of 467 in³. This gives the strongest indication that the ratio of the sizes of the Scottish and English wine gallons was intended to be 2:1. Indeed, considering that the French tun was an imposed external standard, this is exactly the sort of simple market relationship that would be expected.

If we take the view that these Scottish and English wine gallons are directly related in size, then we would expect to find that they had the same type of definition basis and that this was couched in terms of conventional weight definitions. By analogy with the wine and dry gallons of the *Tractatus*, it might be expected that another English wine gallon would be defined in terms of 8 or perhaps 9 pounds of water content. An early gallon of the projected volume of about 233 in³ would contain about 59,000 grains, recognising that a possible error of about 0·5 in³ represents a little over 100 grains. If this weight was equivalent to 8 pounds then the pound would be about 7,380 grains, which does not match any known standard. However, if the gallon contained 9 pounds, the pound would be about 6,560 grains, which is a very close match for a pound of 15 avoirdupois ounces. (The equivalent 16-ounce pound was used as a Flemish heavy-goods pound, where the ounce was again 437½ grains and the pound was 7,000 grains.) If such a pound also formed the basis of the definition, then an English gallon of 9 pounds would have an accurate volume of 233½ in³, which is exactly the size predicted from the equivalent Scottish gallon, and only fractionally larger than the gallon volume of 233·1 in³ obtained by direct measurement of the 1641 Reynolds pottle standard.[92]

It would follow that the basic Scottish wine gallon of 467 in³ (twice the size of the English gallon) would contain precisely 18 pounds of water, and since this gallon comprised 6 pints, each wine pint would contain 3 pounds of water. More significantly, in terms of the Scottish ale units, the ale gallon would have been precisely 12 pounds and the pint 2 pounds. Recalling the suggestion made in Chapter 4, the capacity of the original form of the David Assize gallon was that of 12 pounds of water, each of 15 Paris ounces, in a definition that related to that of the *Tractatus* gallon. We now find a parallel situation in which the basic ale gallon (or the 'old' gallon of the 1426 Assize) is again equivalent to 12 pounds, this time of 15 Flemish or avoirdupois ounces. Thus, the 12-pound descriptive definition of the David Assize has been retained, although it now relates to a vessel which is about a sixteenth smaller (311 in³), with the enhanced version of this gallon (331 in³) very close in size to the projected original gallon of David's time. The dimensions given for the gallon given in the David Assize generate a volume of about 330 in³.

THE ABUSE OF
CHARITY:
ALLOWANCES AND
THE HEAPING
OF MEASURES
——

237

In a similar manner, the early English wine gallon of 224 in³ could still be considered as compatible with the *Tractatus* definition of the gallon as an 8-pound unit. But by direct analogy with the example of the 1497 dry gallon standard (discussed at the beginning of this chapter) which had an in-built allowance of an eighth, the new wine gallon of 233½ in³ similarly operated as a 9-pound unit. It may be this type of flexibility that allowed the *Tractatus* and the David Assize to remain relevant as metrological codes for such extended periods.

The Scottish 1426 Assize defined the pint as containing 41 trois ounces of water. The new trois ounce could be interpreted in this instance as equal to the English troy ounce of 480 grains for two reasons, one of which was discussed in Chapter 5. First, one of the preliminary acts passed just before the Assize of 1426 provided the relationship between trois weight units and the merchant weights, which were clearly based on the 450-grain ounce.[93] And, second, only an ounce of 480 grains could preserve the mathematical ratios that precisely linked the capacities that could be established for the various types of pints: in particular, the Paris ounce of 472½ grains could not provide the necessary ratios. But the definition of the pint as '41' ounces is the one element in the 1426 Assize that seems superficially improbable: this number has no factors, and so it does not appear to support the finding that the pint contained three constituent parts, of which twelve had comprised the old gallon. At the very least, a factor of three would be expected. An equally serious problem was how to relate this intractable quantity to recognisable weights in the Flemish, French or English systems. The most likely explanation is that this weight had originally been given in different units which did have the appropriate factors, and had then been converted into 480-grain troy ounces (which were in any case a relatively new introduction in official definitions in England), but rounded up or down in an acceptable compromise. Giving the weight as 45 avoirdupois ounces (or 3 pounds of 15 ounces), does display the expected factor, and this converts to 41¹⁄₆₄ troy ounces. It would be understandable if such a figure was to be rounded down to 41 ounces in an assize definition: the difference between these two figures (0·03 in³ of water, or 0·04 per cent of the total) is so small as to be inconsequential. In fact, it is at about the limit of what we could achieve in measuring the water contents of sixteenth-century pint capacity standards by eye.[94]

The evidence of the wool sack being calculated as 360 pounds in English records of the late thirteenth century, and in the Scottish Assize of David I, has been discussed in Chapter 4, and from these references it is apparent that the avoirdupois pound of 7,000 grains was well-known in both countries from an early date. Pegolotti did not describe the ounce basis for the avoirdupois pound, although we can perhaps deduce from its later operation that it was considered as a 16-ounce pound in Pegolotti's time, and we can assume that avoirdupois ounces of 437½ grains formed the natural binary divisions of the pound in practical use.

Early pounds for internal use were normally defined as 15-ounce pounds, although this did not prevent their use with divisions of sixteenths. Thus the *Tractatus* and the David Assize both defined pounds of 15 ounces of 450 grains. The French internal pound of 15 ounces of 472½ grains was recorded by Pegolotti (operating in conjunction with a 16-ounce trade pound) and the likelihood of its early use in England is strengthened by the fact that it matches the definition pound for the English liquid measures. The English wool pound can be described as a pound of 16 ounces of 450 grains, but equally it can be (and probably was) considered as another 15-ounce pound, with troy ounces of 480 grains, and as such it still complied with the formulae of the *Tractatus* and later documents. The Scottish merchant pound at the time of the 1426 Assize could similarly be considered as a pound of 15 troy ounces of 480 grains, and a more striking example of a 15-ounce Scottish pound is a version which survived into the sixteenth century (and which will be discussed in Chapter **8**), which was equivalent to a pound of 15 Paris ounces of 472½ grains.

Thus a pound of 15 avoirdupois ounces is certainly compatible with this view of early definitions of pounds as multiples of 15 base ounces, and our argument for its existence is that it accounts successfully for the volumes of the wine gallons. Twelve of these pounds defined the Scottish gallon of the David Assize, if not in David I's day, then at least by the mid-thirteenth century.

Fig. 6.5

The Scottish ale and wine pints and their relationship to the statute pint.

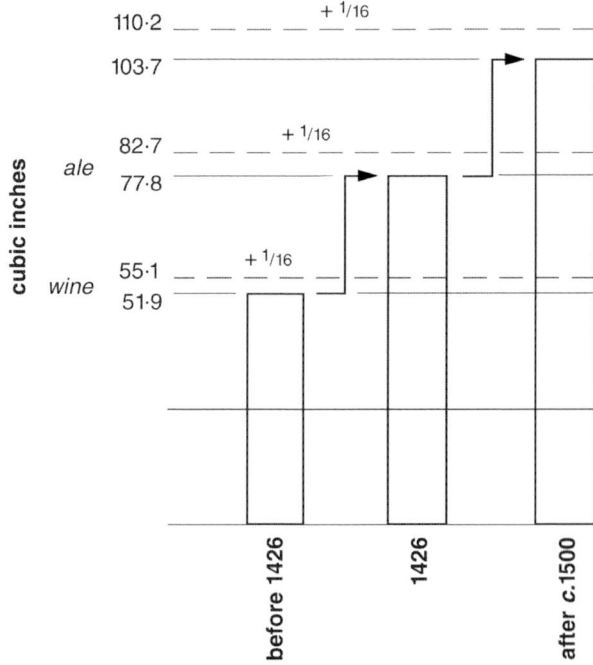

THE ABUSE OF
CHARITY:
ALLOWANCES AND
THE HEAPING
OF MEASURES
———

239

We have established a clear link between the sizes of the wine gallons of England and Scotland, and we have been able to interpret both in terms of a pound that we would associate with the markets of the Low Countries. Although wine was not produced in Flanders, Bruges acted as one of the most important trading ports for wine, particularly through its satellite port of Damme, where the French wine staple was located. As far as England and Scotland were concerned, the 'tun' barrel was the unit of the external market, and we would anticipate that the gallons in which wine was traded in both countries would have been introduced because they were intimately related to an appropriate gallon of this dominant external economy. We would expect, therefore, to find these gallons to be defined in terms of external pounds. (Considering the English wine gallon of 224 in³ in Chapter 4 has already indicated that such a use of pounds represents a general method of defining liquid units.) But if the English wine gallon was tied in this way to the gallon used in the definition of the tun, then we would similarly expect that the number of English gallons in the tun would be some metrologically significant number, preferably only with factors of 2 or perhaps 3, but almost certainly not of larger integers.

The first time that the tun is defined in terms of English gallons is in a parliamentary act of 1423, when it is stated to be 252 gallons, in what appears to be a confirmation of an earlier restriction in the tun's size. But 252, with a factor of 7, is almost certainly not significant, whereas it is just short of 256, which only has factors of 2. If the tun of an original definition was 256 gallons (expressed here in the English size of the external unit), then we can gain some support from the English dry measure series, where the grain 'quarter' is a quarter of a large, but unnamed, unit of 256 gallons. It is a feature of English practice which we now recognise that the dry and wet definitions mirror each other. Indeed this is the type of flexible symmetry that seems to be a characteristic of the English *Tractatus*. Thus the English 'hogshead' (or quarter tun) barrel, at 64 liquid gallons, was the analogue of the 'quarter' of 64 dry gallons. Such considerations might allow us to set the tun provisionally at 256 wine gallons of 233½ in³: 256 × 233½ = 59,776 in³, or perhaps 59,800 in³.

This appears to be confirmed by Pegolotti, writing before 1340, who recorded that the tun in Paris contained 96 *cesters*, each of 8 pints, so that the tun was 768 (which is 3 × 256) Paris pints, and presumably this was a commercial wine pint.[95] Taking the tun as 59,800 in³, allows us to calculate this pint as 77·8 in³. This is, of course, exactly the same size as the legal pint of the Scottish 1426 Assize. A unit of six such Paris wine pints would be 467 in³ – matching the basic Scots wine gallon and being twice the English wine gallon. Since this Paris unit is also explicable in avoirdupois or Flemish weight (45 avoirdupois ounces), it appears that the French wine measure has been determined by the principal external market to which much of its wine exports were consigned.

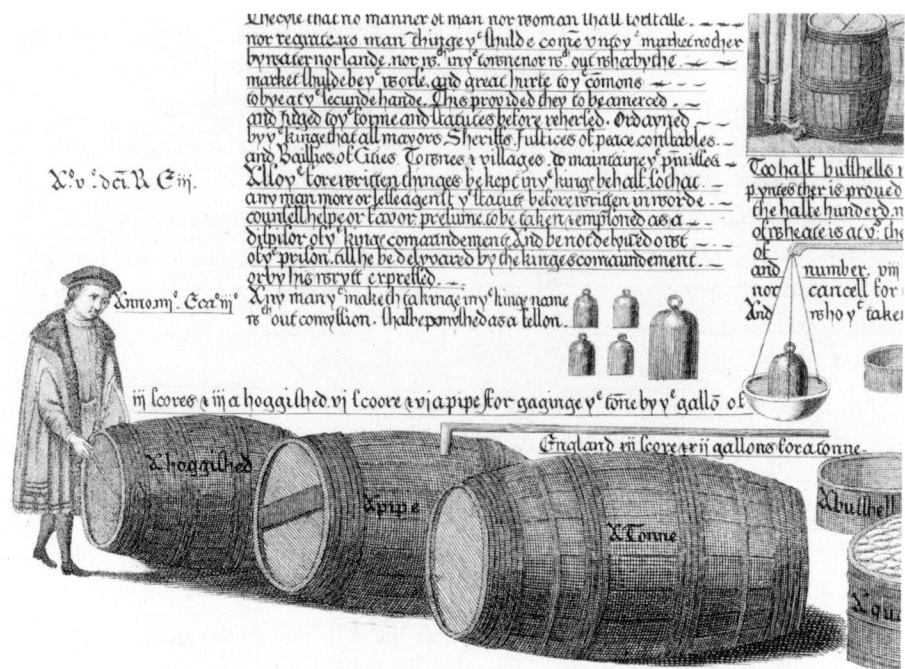

Fig. 6.6

Quayside examination of hogs-
head, pipe and tun barrels.
From the 1746 re-engraved
print of 'The Standard of
Weights and Measures in the
Exchequer. Anno. 12 Henrici
Septimi. [1497]'.
(NMS T.1992.46)

However, later (certainly by 1423) the situation had changed. The English act of 1423 set the tun at 252 English (wine) gallons, although it did not specify the gallon's size. The multiplier does not have the expected convenient factors, so the act appears to indicate that a change of some sort has taken place. This change could be applicable to English import trade only, or it might reflect a changing understanding of the tun's volume in all markets.

Was there a physical change in the tun's size? As we have already observed, the construction of these large, elongated, coopered vessels must have been very difficult. Their capacities cannot have been controllable within very tight limits, and there must inevitably have been some variation in size. Commercial incentive might have led producers to press the volume downwards, so the need might eventually arise to set a minimum permissible size. (A Scottish example, where a minimum size was set for the salmon barrel in the late sixteenth century, is cited in the next section.) Such a restriction is likely to have been imposed through Damme and therefore have affected all markets, and it would suggest a physical reduction of the barrel's size.

Another possibility is that the barrel was not reduced in size, but that a growing pressure from purchasers led to a restriction in the officially accepted volume of contents. The most likely cause for this would be an acceptance that a small proportion of the contents settled out as lees or residue, and that this had to be discarded. Again this might have been controlled by Flemish officials at Damme or in regional markets, and it could have been limited to assessing import or excise duty (and hence merchant liabilities)

THE ABUSE OF
CHARITY:
ALLOWANCES AND
THE HEAPING
OF MEASURES
———

241

rather than sale quantities.[96] This would not necessarily imply a reduction in the barrel's size.

If the English act of 1423 was confirming an aspect of market practice, we should perhaps consider this accepted volume as made up of gallons of 231 in³, which were the working standards of the market, and of the Excise. Thus the tun might be considered as $252 \times 231 = 58,212$ in³, or perhaps 58,200 in³, representing an effective decrease in the tun of nearly 3 per cent. With the statutory fixing of the wine gallon's size in 1706, this remained the tun's declared English volume until recent times. An administrative adjustment of this sort in England would perhaps also help address any merchant concerns about over-charging Excise duty through the use of the 231 in³ gallon, as it may well have been appreciated that the true standard was 233½ in³. The reduction of the multiplier (from 256 to 252) more than compensates for the reduction of the gallon size.

The English figures have been chosen as round numbers, so we cannot learn from these what was the contemporary Flemish view. However, if we use the reduced size of 58,200 in³ for the tun calculation in the Scots merchant handbook of about 1400, and take this together with the supplied information that the tun was 22 'cesters' each of 16 'stoops' and hence was 704 pints, then the pint emerges as the enhanced Scots pint of 82·7 in³. This multiplier of 704 (which again lacks the special characteristics of a number such as 256) has presumably been chosen to give the enhanced version of the gallon that the merchant used in the Scottish home market. It is unclear whether this *cester* (of 32 Scots pints) was a Scottish or Flemish unit and whether it is the analogue of the *sextarius* of 3 gallons (each of 6 pints) in the David Assize of Wine of the previous century. If so, it must represent 4 gallons of 8 pints, which also matches the English definition of the *sester*.[97] Either this demonstrates that the Scottish wine gallon was already, by 1400, composed of eight rather than six pints (and the evidence of the salmon barrel, to be discussed shortly, does not bear this out), or more likely that it provides external pressure for the move to the 8-pint definition found in the 1426 Assize. The important point to note, however, is that it was the enhanced rather than the legal Scots units that appear to match the Flemish units.

This seems to have been so at an earlier time as well. Before the reduction of the tun's effective volume, it contained 256 English wine gallons of 233½ in³, or 128 Scots legal gallons of 467 in³. But in terms of the customary or enhanced Scots wine gallon of 496 in³, it contained 120½ gallons, which is clearly compatible with the David Assize of Wine with its implication of 120 gallon to the tun. We have discussed the apparent Flemish association with legal size of the gallon, but Pegolotti's data also allows us to appreciate that the customary or enhanced size (which emerged from the merchant handbook) matches to actual units of the Bruges market, since he described the tun at Bruges as 360 Flemish *lotti* (each, therefore, the same as the *stoop* of the handbook), a simple multiple of 120.[98] So it appears that the customary form of the Scots gallon follows the Flemish unit directly, and (as we concluded for the David gallon at the end of Chapter 4) the use of an enhancement in Scotland can be seen as a method of bridging from French-based to Flemish-based standards. But, the multiple also suggests a justification of the tun in Flemish units as a long hundred, in an encouraging parallel with the early bulk weight terms explored in Chapter 4.

To summarise, there is clear evidence for the existence of separate Scottish gallons for wine and ale in the fourteenth century, and therefore presumably for other periods. The ale gallon, of 6 ale pints, is the only unit described in the surviving versions of David's Assize of Weights and Measures. The larger wine gallon, again with 6 pints to the gallon, appears in the David Assize of Wine. At the Assize of 1426 it was the wine pint rather than the smaller ale pint that was established by the administration as the statute pint, perhaps indicating a shift in emphasis towards legislating primarily for merchant trade, but not representing a disruption in metrological practice (see Fig. 6.5). Thus, in a further respect, the 1426 Assize is revealed as a declaratory act. A principal novelty may have been the redefinition of the gallon as a unit of 8 (rather than 6) pints. The 8-pint gallon and the new 16-pint firlot, which is discussed below, reflect strong features of English metrology, and it is suggested that they were introduced in 1426 in conscious imitation of the English system with which James I and many of his advisers would have been familiar. However, since this equivalence is with English dry measure, it is possible that the Scottish 8-pint gallon was initially introduced for use in dry measure definitions – certainly the only example of its use in the 1426 Assize was in defining the permitted trading firlot, which was to contain 17 pints, or rather 'twa gallonis ande a pynte'.[99]

THE SALMON BARREL AS AN INDICATOR OF CHANGES IN THE PINT AND GALLON

Four Scottish statutes of the late fifteenth century provide details of the barrel 'bind' for salmon, namely the coopered standard of iron-bound staves (or 'trees') appropriate for salmon, and they link the centre of export to the port of Aberdeen. An act of 1478 complained:

> the Realme [is] gretly sklanderit be [slandered by] strangeris & vtheris that byis salmond of the mynising [reducing] of vesschiall & barellis that the salmond Is pakit In [Therefore] It is statut & ordanit that in time to cum all salmonde be pakit in barellis of the mesur of hamburgh efter the ald assise [added: 'of Abirdene'] And na smallar barell nor veschell And that na couper within the Realme mak smallar barell to pak fische in than the said mesur of hamburgh & ald assise [added: 'of aberdene'][100]

THE ABUSE OF
CHARITY:
ALLOWANCES AND
THE HEAPING
OF MEASURES

———

243

An act of 1487 described the contents of the salmon barrel as 14 gallons, and
also went on to specify the operation of the 'bind' standards in testing the
barrel:

> the barell bind of Salmond suld kepe & contene the assise & mesour of
> xiiij gallonis & not to be mynyst [reduced] vnder the pain of eschete
> [forfeit] of the salmond … And that Ilk burgh haue thre hupe Irnis
> convenient herefore ane at Ilk end of the barell & ane in the middis for
> the mesuring of the barell & a birnyng Irne to mark the samy vnder the
> pain of eschete of the barell vnmarkit.[101]

The other two acts repeated the 14 gallons or said the measure was to be the
old bind of Aberdeen.[102]

These late fifteenth-century references are all to the barrel of 14 gallons.
No further definitions are found until the 1570s: in 1570 the Convention of
Royal Burghs asked for the barrel to be redefined as 12 gallons, and in 1573
parliament authorised the change, specifying that the measure was to be 'by
the Stirling jug'.[103] The Convention confirmed this size in 1581, noting that it
was by 'Flemish bind', and in 1584 parliament reaffirmed that the barrel was
to conform to the old acts.[104] The implication is that the barrel size had not
changed in this period, but is merely redefined in terms of different gallon
units. (For comparison, no change is recorded in the English barrel.) Further
changes in the specification were signalled after the 1618 Assize, with two
Privy Council declarations in 1619 that the 'old measure' of Aberdeen was 10
gallons, and parliament's 1641 declaration that the barrel was to be 10 gallons
'by Stirling pint'.[105] Constant reference to the 'auld gedge of Aberdeine'
would seem to indicate that no change to the barrel's volume had, in fact,
occurred.

The re-defining of the barrel in terms of smaller numbers of gallons –
initially 14, then reduced to 12, and finally 10 gallons – might imply a progressive reduction in the volume of the barrel, but we must take account of the
changes in the size and number of pints in the Scottish gallon as well as
understanding that there would be pressures for continuity imposed by the
external markets to which Scottish salmon were exported. There was
undoubtedly a tendency on the part of exporters to press barrels sizes
downwards, and in their analysis of Scottish prices Elizabeth Gemmill and
Nicholas Mayhew produce ample evidence of tension over the control of
barrel sizes from the Aberdeen records.[106] However, it does not follow from
this that the volume of the barrel was actually reduced, and the changing
official definitions can indeed be interpreted plausibly in terms of a constant
barrel size.

We have argued that the basic size of the Scottish wine gallon in the
early fifteenth century was twice that of the English wine gallon. The English
salmon barrel is uniformly described during this whole period as 42 wine
gallons.[107] Recognising that 42 gallons may be an approximate or to some

extent a conventional description of the barrel's volume, we will set the
Scottish definition of the same barrel initially at 21 Scottish gallons without
pretending to achieve greater accuracy than is warranted. However, we used
the basic version of the gallon in demonstrating this 2:1 ratio, and the
example of the wine tun calculation in the Scottish merchant handbook of
about 1400 shows that enhanced units which were one-sixteenth larger were
used where wine was concerned. The 21 gallons would reduce to about 20
enhanced gallons ($21 \div 1\frac{1}{16} = 19 \cdot 8 \approx 20$). These gallons are 6-pint gallons, but
the definition gallon was increased to 8 pints in the 1426 Assize, and so we
would expect the 8-pint gallon to be used in parliamentary definitions in the
1480s. The barrel would contain a smaller number of these 8-pint gallons,
namely $20 \times \frac{6}{8} = 15$.

It must be appreciated that salmon were very large fish, and therefore the
concept of the 'volume' of the fish was difficult to define and certainly could
not be given with any great precision. This is illustrated by a Convention of
Royal Burghs petition of 1595, which recorded that a volume for particular
fish containers should be made by the coopers at 'fyftene gallouns, or at the
leist xiiij gallouns and ane halff', demonstrating the burghs acknowledged
that there was a problem in policing the regulations, and considered that
latitude of half a gallon (in fifteen) was tolerable.[108] Against such a background, the '14-gallon' value of the 1470s may be seen as acceptable given an
understanding of re-calculation as about 15 enhanced gallons.

By the 1570s, when the capacity was redefined in Stirling measure as 12
gallons, such a calculation would have been performed without considering
an enhancement stage: the value in 8-pint gallons would have been taken as
16 gallons ($15 \times 1\frac{1}{16} = 15 \cdot 9 \approx 16$), and translating this into Stirling measure
would result in 12 gallons ($16 \times \frac{3}{4} = 12$). It is suggested that this is the figure
that has been given in the regulations of the 1570s as '12 gallons'. It seems that
a different situation applied after the 1618 Assize, when the salmon barrel
capacity was redefined as 10 gallons: here the emphasis has been on citing the
earliest statutory definition, namely the 14 gallons of the 1470s, taking this as
exact and converting to Stirling measure to get 10½ gallons ($14 \times \frac{3}{4} = 10\frac{1}{2}$),
which has been rounded to '10 gallons'.

In this interpretation, each of the three phases of definition of the same
barrel volume has been informed by the changing definitional practice of the
day. Arguing for constancy of the barrel size allows the effect of a changing
pint (and therefore the corresponding gallon) to be illustrated.

THE SIZES OF THE NEW DRY MEASURES OF THE 1426 ASSIZE

In the 1426 Assize the two principal dry measures – the firlot and the boll –
were defined in two ways. The first was in terms of their constituent parts and

THE ABUSE OF
CHARITY:
ALLOWANCES AND
THE HEAPING
OF MEASURES
———

245

in particular their relationship to the pint; the second was in terms of their physical dimensions. We have examined similar definitions which were given in the 1618 Assize and discovered that these must be interpreted with care, remembering that grain was used as the measuring medium. In the case of the 1426 Assize there is the added complication, which was examined in Chapter 5, that the physical dimensions given in the sixteenth-century printed versions of the Assize have been altered as part of a statute revision process, but the original depth dimensions for the firlot and boll can be reconstructed with confidence. Now that these issues have been resolved, we can examine the 1426 Assize itself, the full text of which is given in Appendix A.3.

The boll of the 1426 Assize is explicitly stated to contain four firlots, and this relationship is confirmed by providing conventional weights of their contents – 164 pounds for the boll and 41 pounds for the firlot. These are given against a carefully defined unit of 41 ounces, representing the weight of the water content of the pint standard. Since this weight is also given as 2 pounds and 9 ounces it is clear that there are 16 ounces to the pound, so the firlot's weight equivalent of 41 pounds shows that it contains 16 pints. We will describe this statutory firlot as the 'legal' firlot. In subsequent assizes the number of pint-fills in the firlot is explicitly stated, but in the 1426 Assize it has to be inferred from the weight equivalents. The capacity of the pint is 77·8 in^3, and therefore one might expect the firlot's volume to be $16 \times 77\cdot8 = 1{,}245$ in^3.

The firlot standard is described as internally cylindrical with a diameter of 16 inches and we have concluded that its depth was 6 inches, which indicates a volume of about 1,205 in^3. This is another instance of a formula for constructing physical standards which provide an acceptable match with the definition of the firlot's contents. In this instance the firlot dimensions are given in round numbers, but the boll's dimensions are given to half an inch, so it can be assumed that this is also the accuracy of the firlot dimensions. A reasonable measure of the intrinsic uncertainty in the match between the firlot's volume from dimensions, and from an independent definition in terms of pints, might therefore be about ±40 in^3.[109] Expressing the volume of the firlot from its stated dimensions as 1,205 ± 40 in^3 provides a range which just accommodates the combined capacities of 16 pints.

However, we must again recognise that in filling the firlot with grain, the density of grain in the firlot will be greater than that in the pint measure used for filling it, and the difference in density will be about the 4½ per cent found in the case of the 1618 wheat firlot. This implies that the volume of the firlot which contains 16 pint-fills of grain would be reduced from 1,245 in^3 to only about 1,190 in^3, which is comfortably compatible with the volume from dimensions.[110] In practice, the compaction may be a little less because the 1426 firlot is less deep than that of 1618; but in the absence, for example, of information about the shape of the early pint standards, a volume of 1,200 in^3 will be adopted as the best estimate of the size of the 1426 firlot.

The boll can be treated in a similar fashion. The dimensions given for standards are an internal diameter of 29 inches at the base, tapering in to 27 inches at the mouth, and we have demonstrated in Chapter 5 that the depth is 9 inches. These sizes generate a volume of 5,640 in^3, but because of the imprecision of the dimensions, a reasonable measure of the intrinsic uncertainty in the volume might be ±120 in^3.[111] The boll's volume from dimensions should more correctly be given as 5,640 ± 120 in^3. We will later adopt a value of 5,600 in^3 as the best estimate of the boll's size.

It can be appreciated that the physical boll standard described in the 1426 Assize is not four times the volume of the firlot, as the assize claimed. We would expect a boll of four firlots to have a volume of about $4 \times 1,200 = 4,800$ in^3, whereas the volume calculated from the dimensions is about 4¾ times that of the firlot of 1,200 in^3 (Fig. 6.7). However, there can be no doubt that the dimensions given are for the construction of actual physical measures authorised for use. It is apparent, therefore, that the boll has a built-in extra allowance or 'charity' and that it is a genuine trading standard. No dimensions for a 'legal' boll (of four 'legal' firlots) were given in the assize, presumably because it was only an administrative fiction. Indeed, the standard described has an air of market-place reality, and we shall describe it as the 'trading' boll. A clear parallel can be seen with the English Exchequer standard bushel of 1497, discussed earlier in this chapter, which was also constructed to accommodate an extra allowance.

We can obtain the approximate capacity of the boll by considering the introduction of individual pint-fills of grain. The boll is deeper than the firlot so the compaction of the grain in the boll will be slightly greater. By analogy with the 1618 barley firlot, which was of approximately the same depth, the difference in density between the grain in the boll and in the pint measure with which it is being filled is about 6½ per cent: this indicates that the boll of 5,640 in^3 would contain about 77 grain-fills of a pint of 77·8 in^3.[112] A boll of 77 pints is considerably larger than the 'legal' boll of 64 pints, or 4 firlots each of 16 pints (Fig 6.7). For one thing, it implies the use of a customary firlot of a quarter of the trading boll, namely 19¼ pints ($4 \times 19¼ = 77$). The explanation of this disparity between the boll and firlot lies in an attempt to compromise between an existing situation and an ideal of establishing the firlot at 16 pints.

Although the dimensioned firlot vessel of the 1426 Assize corresponds to the 'legal' measure of 16 pints, the act also gives a rare glimpse of the operation of an official allowance by describing how the firlot (presumably the firlot of the market-place) was to 'contene twa gallownis & a pynt' (see Appendix A.3a). In other words, the legislature was prepared to recognise a 16-pint 'legal' firlot vessel struck, plus a separate measure of one pint. In

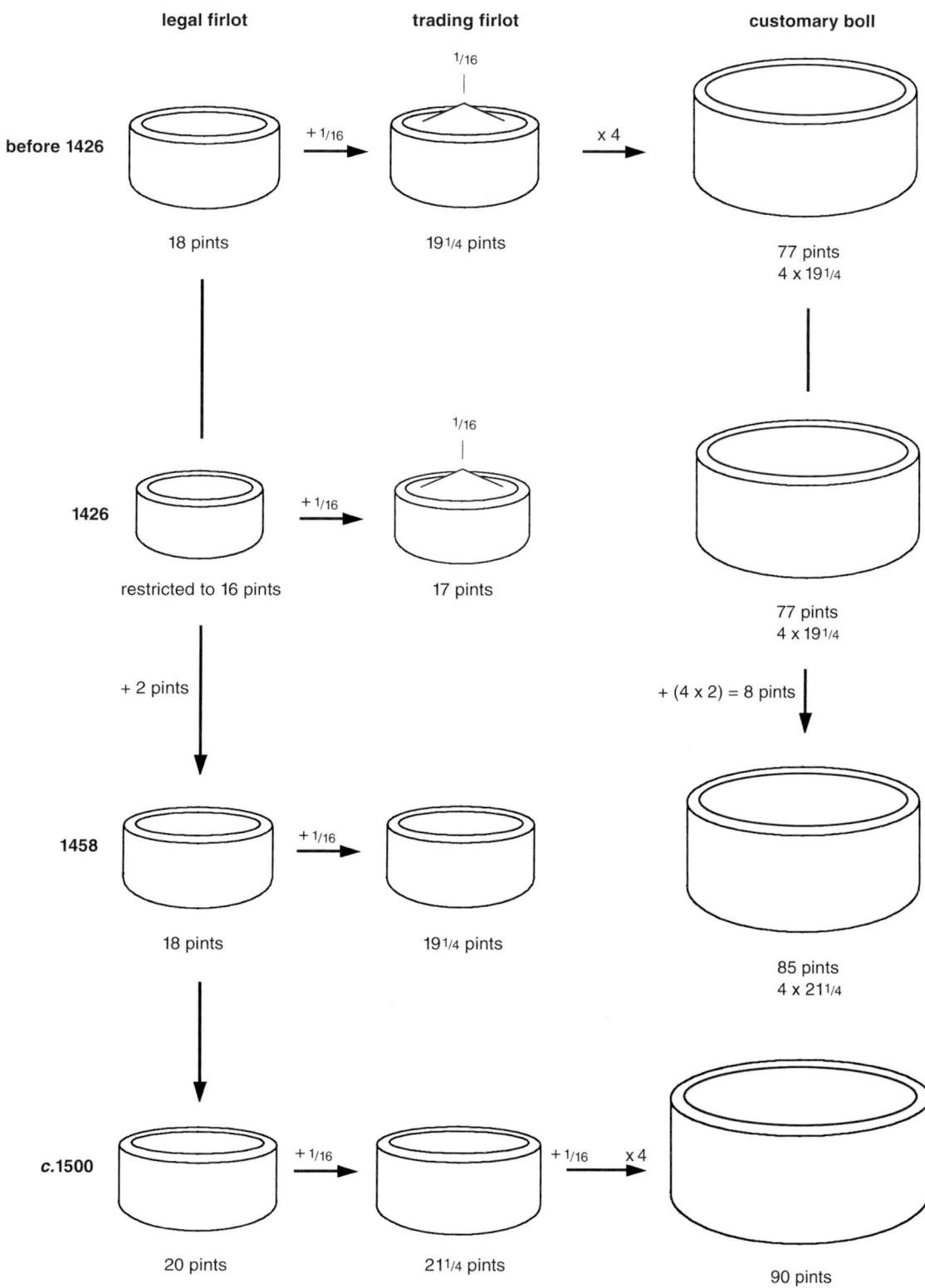

legal firlot

trading firlot

customary boll

before 1426

18 pints

+ 1/16

1/16

19 1/4 pints

x 4

77 pints
4 x 19 1/4

1426

restricted to 16 pints

+ 1/16

1/16

17 pints

77 pints
4 x 19 1/4

+ 2 pints

+ (4 x 2) = 8 pints

1458

18 pints

+ 1/16

19 1/4 pints

85 pints
4 x 21 1/4

*c.*1500

20 pints

+ 1/16

21 1/4 pints

+ 1/16   x 4

90 pints
4 x 22 1/2

more clearly in the assizes of the sixteenth century, but it appears to have originated in these two assizes of the fifteenth century.

However, the consequence of adopting the previous customary firlot as the new legal one, was that a new customary firlot emerged in the market place, enlarged by a further sixteenth. Thus, as a result of the changes in the assize of the late 1490s, a new boll of 90 pints came into use, in turn giving rise to a firlot of 22½ pints, since 4 × 22½ = 90 pints (Fig **6.9**).

Although the customary bolls maintained by the burghs and the land owners were one-eighth larger than the legal measures, this was made up of two distinct additions of one-sixteenth. The first stage initially represented the customary allowance on the liquid measures. It was couched as a permitted allowance on the firlot and generated the trading measure. The second stage was an allowance on the use of the boll which originated in the 1426 Assize. It came to have a specific name, the 'charity to the boll' and was often described as 'a peck to the boll' – and since the peck was a quarter firlot, this was one-sixteenth of a boll.

It remains to examine the last puzzle in the 1426 Assize. This is a confused statement, the first part of which we have already addressed, describing the amount of the additional allowance on the boll:

> this new bol new maid weyis mar than the auld boll be xli *lib*, quhilkis [which] makis twa galounis & a half & a chopyn of the auld mete ande of the new mete new ordanit ix pyntis & thre muchekynis. [129]

The first part tells us that the weight of the new legal boll was 41 pounds greater than that of the David boll. This weight is not the same as 2½ old gallons or 9½ new pints, given in the second part of the statement, because these volumes both have a weight equivalent of only about 25 pounds.

The answer here is probably that two clauses have been run together. One relates to the 41 pound difference between two ideal sizes – those of the 'legal' boll and of David's boll – and we have demonstrated above that this is correct. The second describes how much the new boll 'makis', and this is a practical expression of the difference between the newly-legalised customary boll and the boll of four permitted firlots.[130] The trading or customary boll has been found to have a capacity of about 77 pints and the boll of four permitted firlots is 4 × 17 = 68 pints. The difference between these is about 9 pints.

In terms of the old measures, we are told that the difference is 2½ old gallons and a chopin, which is equivalent to 1¼ new gallons and a chopin or 10⅓ new pints, on the assumption that the unit named the 'chopin' had remained a third of the old pint and was unchanged. In terms of the new measures, the same difference is 9 pints and 3 mutchkins, or 9½ pints, for the mutchkin was one-sixth of an old quart, now one-sixth of a new pint (see Fig. **6.3**). The fact that these are different suggests that two separate determi-

THE ABUSE OF
CHARITY:
ALLOWANCES AND
THE HEAPING
OF MEASURES

255

nations of the allowance have been made, one in the old and one in the new measure. Considering that they represent differences between two very large quantities of grain it is quite remarkable that the two determinations come so close to each other and also to the 9 pints that have been deduced from the independent calculation of the boll size. The stated profit of 9½ pints in the new measure is one-eighth of 76, reinforcing the message that an allowance of an eighth had been established for the boll.

To recapitulate, therefore, the agreement between the various measurements is probably as good as the figures will permit, and the boll emerges as approximately 77 pints in capacity in 1426, with the customary firlot being 19 pints or a little over. In the light of these uncertainties, we shall round off our estimate of the 1426 boll's capacity and give it as 5,600 in^3. The official view of the 1426 Assize appears to have been that the firlot contained (when struck) exactly 16 fills of grain from the pint measure. In trading use, an extra pint-fill was added for every firlot as the permitted charity. The boll contained four of these enhanced firlots, and also the permitted charity for the boll, which was defined as 9½ pints. This was equivalent to giving charities of 1 in 16 for the firlot and an additional 1 in 16 for the boll created from these enhanced firlots.

The situation may appear complicated because we have considered the boll's size from all avenues open to us from the 1426 Assize. However, from the merchants' standpoint in the 1420s it would have been more straightforward, involving only customary charities and allowances. Specific information on these is sparse at this early period, but there is enough to construct a plausible explanation for the evolution of the dry measures to bridge the gap between the David Assize to the end of the fifteenth century. The picture becomes clearer in the following period, examined in Chapter 7; some charities, such as the 'peck to the boll' are seen to have an extended life.

1   An enlargement of the market weight series was made at the 1563 Assize, and this will be discussed in Chapter **8**.

2   T. Thomson and C. Innes (eds), *The Acts of the Parliaments of Scotland [APS]*, 13 volumes (Edinburgh, 1814-75), IV, 587: 19 February 1618. See Appendix A.**9**.

3   J. Hill Burton, *et al.* (eds), *The Register of the Privy Council of Scotland [RPCS]*, three series, (Edinburgh, 1877-1970), second series, VIII, 333: letter of the Privy Council to the justices of the peace, 6 January 1613. It is plain that Edinburgh had been making standards on their own: 'The quhilk day the provest, baillies and counsale sittand in jugement ordanis Patrik Irland, baillie of his consent, to caus mak certain stane wechtis of leid and merk thame with the townis merk and that thai wey stane, half stane and sua; and ordanis the samyn to be deliverit to the personis that kepis the buithis at the ouir [over] tron and utheris that weyis woll, butter, cheis and sic stuf and that on thair expens that sall ressaif the samyn wechtis; and thairfor dischargis all wechtis that the saidis personis hes now presentlie and gif thai be fundin weyand [weighing] thairwith to be punist.'

4   Ibid., second series, VIII, 333-5, 335-6: letters of the justices of the peace for the sheriffdoms of Perth and Edinburgh, 2 and 24 February 1613.

5   Ibid., second series, VIII, 338-8: letter of the justices of the peace of the sheriffdom of Linlithgow, 25 February 1613.

6   *APS*, I, 697: *Iter Camerarii*, chapter 10.

7   *Statutes of the Realm [Statutes]*, 11 volumes (London, 1810-28), I, 321: 25 Edward III Stat. 5 c.10, 1351-2.

8   R. D. Connor, *The Weights and Measures of England [WME]* (London, 1987), 156.

9   *WME*, 156; W. Boys, *Collections for an History of Sandwich in Kent,* 2 volumes (Canterbury, 1892), II, 542.

10 The bushel is at the Science Museum, London, NMSI 1931-1011; the gallon is 1931-1013. A second bushel was acquired more recently, NMSI 1993-779; see also *WME*, 155.

11 *Statutes*, II, 637-8: 12 Henry VII c.5, 1496, 'An Act for Wayghtes and Measures'.

12 Henry Norris, 'Ancient Weights and Measures prior to Henry VII', in *Philosophical Transactions*, 65 (1775), 48. Norris's value of 47½ avoir pounds to 1 ft³ is equivalent to 192·5 grains to one cubic inch.

13 '8 gallons and 8 lbs make a bushel and 8 bushels make a London quarter': *WME*, 156.

14 *Statutes*, II, 78: 15 Richard II c.4, 1391; *WME*, 157: 'Divers statutes [decreed] … that eight bushels striked make the quarter of corn; nevertheless, because that no pain [penalty] is thereupon ordained in the said statutes, divers People … will not take … but [at] nine bushels for the quarter … It is ordained and assented … that none from henceforth do buy in the City of London, nor in any other place any manner of corn … but after eight bushels to the Quarter.'

15 *Statutes*, II, 174: 1 Henry V c.10, 1413: '… Purveyors … have taken … continually Nine bushels of Wheat and other Corn for the Quarter and that many times by Measures not sealed and also not stricken … and the Merchants and Citizens of London do use to take of every Seller for the Quarter of Wheat Nine Bushels, by the measure used within the City called the Vat with the bushel set upon the said Vat [*faat*] … and take for a Quarter of Oats Ten Bushels … It is ordained that the said good Ordinances be firmly holden and kept … and that no Purveyor of our Lord The King nor any other shall use hereafter to buy nor to take any corn by other Measure, but Eight Bushels striked for the Quarter …'

16 Report from the [Carysfort] Committee appointed to inquire into the Original Standards of Weights and Measures in this Kingdom; and to consider the Laws relating thereto, 26 May 1758 [Carysfort Report], *Reports from Committees of the House of Commons*, 3 volumes (1715-1801), II, 411-51, see esp. p. 432.

17 H. W. Chisholm, *Seventh Annual Report of the Warden of the Standards, on the Proceedings and Business of the Standard Weights and Measures Department of the Board of Trade for 1872-73* (London, 1873), 27-34.

18 Norman Biggs, 'English Weight Systems in the 14th Century – A New Interpretation', privately circulated paper, December 1990. This aspect is discussed in Chapter 4.

19 Using Norris's value for the density of wheat, the volume can be calculated as 224·4 in³. See above, ref. 12.

20 *Statutes*, II, 570-1: 11 Henry VII c.4, 1495; and see *WME*, 178: 'Water measure within the ship board shall only contain V pecks after the said standard rasen and stricken'.

21 *APS*, II, 10: Act 14, 1426.

22 Ibid., II, 540: Act 14, 1563.

23 A. Huntar [*sic*], *A Treatise of Weights, Mets and Measures of Scotland* (Edinburgh, 1624), 5.

24 1 Henry V c.10, 1413. For further on the *faat* of 9 bushels and the additional bushel making 10, see *WME*, 157.

25 *WME*, 159; Carysfort Report, op. cit. (16).

26 Chisholm, op. cit. (17), 24, mentions the gallon of 282 in[3] without indicating it was missing. Presumably this occurred between that date and 1967 when F. G. Skinner (former Deputy Keeper in the Science Museum, London) published his *Weights and Measures* in which he does not mention this gallon. See *WME*, 159-64.

27 Carysfort Report, op. cit. (16), 433.

28 *WME*, 179.

29 O. Ruffhead and C. Runnington (eds), *Statutes at Large*, 41 volumes (London, 1780-1865), XXVII, 630, 977-86: 4 & 5 William IV c.49, session 6, 1834, and 5 & 6 William IV c.63, session 7, 1835.

30 *APS*, III, 521: 29 July 1587. See also Appendix A.**8**. We are grateful to Richard Morrison-Low for the grain provided for this experiment.

31 Ibid.

32 Ibid., IV, 586: 19 February 1618. See also Appendix A.**9**.

33 Ibid., 587. It is exceptional to find evidence of experimental work at such an early date.

34 The last mention of the gallon in the Scottish statutes was at the 1426 Assize (although a reference to this clause is also included in the 1587 Assize), and after this the dry and liquid measures separated. The liquid gallon remained at 8 pints (each of 77·8 in[3], then 103·7 in[3] after *c.*1500), whereas the equivalent dry unit (the half-firlot) changed every time the firlot changed. This is apparent in the changes in the salmon barrel.

35 21¼ × 103·7 = 2,203·6 in[3]. The capacity of the pint is not known to better than about 0·05 in[3], or 1 part in 2,000, so the figure is rounded up to 2,205 in[3].

36 $\pi \times 7\frac{1}{3} \times (\frac{1}{2} \times 19\frac{1}{8})^2 = 2{,}115{\cdot}8$ in[3]; 2,115·8 - 8·88 = 2,107·0 in[3]. The act gives the sides of the triangular-section bar across the aperture as one inch, and the circumference of the support as one inch (see Appendix A.**9**), and the small volumes of the two features are 8·30 in[3] and 0·58 in[3] respectively, making 8·88 in[3] in total.

37 Thomas Everard, *Stereometry, or the Art of Gauging made Easy, by the help of a Sliding-Rule,* tenth edition (London, 1738), 192.

38 Altering the diameter alone by ¹⁄₁₂ inch changes the volume by 20 in[3], and altering the depth alone by the same amount changes the volume by 25 in[3].

39 31 × 103·7 = 3,214·7 in[3], which will be given as 3,215 in[3]. From the dimensions: $\pi \times 10\frac{1}{2} \times (\frac{1}{2} \times 19\frac{1}{8})^2 =$ 3,029·5 in[3] and 3,029·5 - 9·71 = 3,019·8 in[3], where 9·71 in[3] is the volume of the cross bar and support. Altering the dimensions individually by ¹⁄₁₂ inch changes the volume by 25 in[3], so the volume is given in round figures as 3,020 ± 25 in[3].

40 Everard, op. cit. (37), 192.

41 This microscopic 'stickiness' apparently involves some interlocking of fibrous external material, and in this respect it may differ from that between the particles of granular minerals such as sand, which will more readily distribute themselves to pack initially at a maximum density.

42 See, for example, J. Walker, *The Flying Circus of Physics* (New York and London, 1977), 172. We are grateful to Dr Allan Mills for this reference.

43 Osbourne Reynolds, 'On the Dilatancy of Media Composed of Rigid Particles in Contact', in Osbourne Reynolds, *Papers on Mechanical and Physical Subjects,* 3 volumes (Cambridge, 1900-1903), II, 203-16; [William Thomson] Lord Kelvin, *Baltimore Lectures on Molecular Dynamics and the Wave Theory of Light* (London, 1904), 624-6.

44 A. D. C. Simpson, 'Grain packing in early standard capacity measures: evidence from the Scottish dry capacity standards', in *Annals of Science*, 49 (1992), 337-50.

45 Ibid., 345, reporting variations of ± 0·4% and under.

46 Ibid., 346. The conclusion was that the firlot would have contained 21·3 ± 0·2 pints of grain, which compares favourably with the 21¼ pints found by the 1618 commissioners. This is notably different from the 20¼ pints that the physical firlot would have contained if differential grain packing had not been a factor, that is if water had been used to fill both the pint and the firlot.

47 Ibid., 347.

48 Ibid., 343-4.

49 See for example, Lawrence Burrell, 'The Standards of Scotland', in *The Monthly Review: the Journal of the Institute of Weights and Measures Administration,* 69 (1961), 49-62, *passim*.

50 This will be discussed in Chapter 7.

51 *APS*, IV, 586: 19 February 1618.

52 James Anderson, *General View of the Agriculture and Rural Economy of the County of Aberdeen* (London, 1795), 42-3.

53 'Letter from Dr Adam Anderson … communicating the Result of some Experiments made by him relative to the Condensation of Grain by different modes of Measuring', in Report from the Select Committee on the Sale of Corn, 25 July 1834, *Parliamentary Papers* 1834, VII, 517, appendix 12.

54 *Weights and Measures: Inspectors and Inspection: Model Regulations* (London, 1890), number 36. We are grateful to the late Maurice Stevenson,

Honorary Librarian of the Trading Standards Association, for his assistance with this point.

55 8 & 9 William III, c.22, s. xlv.: '… it is hereby declared that every round bushel with a plain and even bottom being eighteen inches and a halfe wide throughout and eight inches deep shall be esteemed a legal Winchester bushel according to the Standard in His Majesty's Exchequer.'

56 First Report of the Commissioners to Consider the Subject of Weights and Measures, 7 July 1819 (HC 565), *Parliamentary Papers* 1819, XI, 307-23. The Report from the Select Committee on Weights and Measures, 15 July 1862, *Parliamentary Papers* 1862, VII, 187-478, recorded that they had found 'ten different bushels in use.' Although the Winchester bushel was declared illegal in trade in 1824, it is recorded in Lincolnshire in 1885 that justices used it to fix corn rents and the average price of the Winchester bushel of 'good marketable wheat for the past 21 years formed the basis for assessing corn rents for the next 21 years': J. A. O'Keefe, *The Law of Weights and Measures*, second edition (London, 1978), appendix I, 10.

57 Carysfort Report, op. cit. (16), 433; *WME*, 162.

58 The work on the bushel is described in Thomas Everard, *Stereometry: or the Art of Gauging* (first published in 1684), but apparently only from the ninth edition (London, 1727), in a new section on malt gauging contributed by the Excise gauger and mathematician Charles Leadbetter, and cited here in the tenth edition, 1738: Everard, op. cit. (37), 192-3. Although written in the first person ('In *February* 1696 … I did … make an Experiment in order to find the true Content of the said Standard Bushel …'), the author of the account must be Everard and not Leadbetter, whose birth was in 1695. On Leadbetter see R. V. and P. J. Wallis,

Applications, Part II, 1701-1760 (Letchworth, Herts, 1986), 85. For Everard, see E. G. R. Taylor, *The Mathematical Practitioners of Tudor and Stuart England* (Cambridge, 1967), 279, 399.

59 Everard, op. cit. (37), 192-3. It must be remembered that the measuring techniques of the early eighteenth century do not match those of today. The volume of the Elizabethan bushel, which was the standard till 1824, was found to be 2,148·28 in³ in 1931, the difference with the Winchester bushel as defined being only 2·14 in³ rather than 4·82 in³.

60 See above, for the gallon of 280 in³, and *WME*, 160 for the gallon of 282 in³.

61 *APS*, II, 12: Act 22, 1426.

62 Ibid.

63 J. Stuart, *et al.* (eds), *Exchequer Rolls of Scotland*, 23 volumes (Edinburgh, 1878-1908), I, 1.

64 This reference is from *APS*, I, *Iter Camerarii*, 697, and is discussed below.

65 *Dictionary of the Older Scottish Tongue [DOST]*, 12 volumes (Chicago, Aberdeen and Oxford, 1931-2002), I, 524, s.v. *chopin*.

66 Lawrence Burrell noted, somewhat flippantly, that the 1426 parliament was held at Scone, just north of Perth, and 'the Tay happened to be handy': Burrell, op. cit. (49), 56. There is no particular significance in the use of water from the Tay (water from the Water of Leith was used in 1618 Assize); however, it is significant that fresh water was stipulated. Using the 'divers waters' specified in the David Assize would have had the effect of reducing the volume of the pint by 1%.

67 The accepted modern density of fresh water is 252·89 grains per cubic inch. The volume of 41 ounces is therefore $41 \times 480 \div 252·89 = 77·8$ in³.

68 The David Assize described a tapered cylindrical vessel of 6½ inches depth, base diameter 8½ inches, and upper and lower circumferences of 27½ and 23 inches: *APS*, I, 674. These circumferences imply diameters of 8·75 and 7·32 inches respectively, and because the latter is less than the quoted base diameter of 8½ inches, it is assumed that the circumferences are internal. Within the accuracy of the stated dimensions, the volume can be calculated as 330 in³.

69 Alison Hanham, 'A Medieval Scots Merchant's Handbook', in *Scottish Historical Review*, 50 (1971), 107-20.

70 Taking the English wine gallon as 231 in³, the capacity of the 252-gallon tun was $252 \times 231 = 58,212$ in³ $= 704 \times 82·69$ in³.

71 [John Swinton], *A Proposal for Uniformity of Weights and Measures in Scotland* (Edinburgh, 1779), 29.

72 For example, the Glasgow ale pint, described in George Buchanan, *Tables for Converting the Weights and Measures hitherto in use in Great Britain into those of the Imperial Standards …* (Edinburgh, 1829), 232-3.

73 David Gregory, *A Treatise of Practical Geometry … with Additions* [*by Colin Maclaurin*] (Edinburgh, 1745), 86: 'The cask measured by Edinburgh brewers before the Commissioners of Excise in 1707 was found to contain 46⅞ Scots pints and 18¹/₁₆ English ale gallons [282 in³] giving a Scots pint of 108·664 in³ but it is suspected on several grounds that the experiment was not made with sufficient care and exactness.' At the time the Scottish pint augmented by one-sixteenth would have been 110·18 in³ instead of the 108·664 in³ given in the text. For the context, see Chapter **9**.

74 Apart from an isolated reference in 1587 to earlier Acts: *APS*, III, 521: 29 July 1587, and two others, of

which one dealing with the Excise on wine stated that 'the tun held sixty [four] gallons': *APS*, VI, 149, 1644. These 'gallons' are of 909 in³ reflecting the increase of the firlot to that date, the gallon being taken as a half-firlot. The second is in the exemplification of the Act of Union, which stated that 24 gallons of English beer equals 12 gallons in Scotland: *APS*, XI, 448: 1707.

75 Swinton, op. cit. (72), 139. This is contained in an appendix entitled 'Conjectures concerning the Ancient Standards of Measures and Weights in Scotland'. Swinton derived the volume as 77·695 in³, but the small difference between this value and 77·8 in³ arises principally from the use of an early value for the weight of a cubic inch of water. Swinton gave 1 in³ = 0·578532 ounces avoirdupois (of 437½ gr), or 253·11 gr: ibid., 48. Using the modern value of 252·89 gr raises his volume for the pint's capacity to 77·76 in³.

76 Burrell, op. cit. (49), 49, 56, 57; *APS*, II, 50: 6 March 1458, see Appendix A.4.

77 *APS*, I, 697: *Iter Camerarii*, ch. 10.

78 R. E. Zupko, *French Weights and Measures before the Revolution* (Bloomington, Indiana and London, 1978), 43, 136; Horace Doursther, *Dictionnaire Universel des Poids et Measures* (Anvers, 1840, reprint 1965), 428-9.

79 Hanham, op. cit. (69), 119. The dating of this medieval handbook is considered on pp. 111-13.

80 *APS*, I, 674.

81 The Ayr manuscript is located at the National Archives of Scotland (NAS), PA.5/2A; the Cromartie manuscript is located at the National Library of Scotland (NLS), Adv. MS 25.5.10, and the relevant clause is given in Chapter 4, ref. 180. Sir John Skene's Latin edition of the Auld Laws, J. Skene (ed), *Regiam Majestatem: Scotiae Veteres Leges et Constitutiones* (Edinburgh, 1609), f.161r of part I, follows Cromartie with: 'Bolla debet continere in se xii lagenas aquae.' It was shown in Chapter 4 that the size of the gallon deduced for the boll in the Cromartie manuscript does indeed match a definition in 'mixed waters'.

82 *APS*, I, Preface, 38: 'In the existing charters of his own time, and of the reigns immediately succeeding … we find numerous references to the laws of David [and] they serve to establish beyond question the authenticity of many of those which are ascribed to him.' Ibid., I, Preface, 50: 'In many instances, the transcriber [of the Assize of Wine, etc.] copied … from his older authority, and thus the ascertained age of the manuscript furnishes only the limit below which the calculations cannot come.'

83 'Cum dolium vini suerit ad xx s lagena vini erit pro

duobus d … Item sextarium debet continere tres lagenas': *APS*, I, 676: *Assisa de Vino*.

84 *Fleta* (of the late thirteenth century) and later sources give wine sextern as 4 gallons ('et quodlibet sextarium quotuor ialones'): H. G. Richardson and G. O. Sayles (eds and trans), *Fleta*, volume II, Selden Society no. 72 (London, 1955), 120. The ale sextern is less definite but was certainly larger than that for wine and it is recorded in the fifteenth century as 12 gallons, yet in 1486 only 4 to 6 gallons (an error for wine sextern?), and 13 to 19 gallons in the sixteenth century: see J. E. Thorold Rogers, *History of Agriculture and Prices in England,* 7 volumes in 8 (Oxford, 1866-1902), III, 278-9; M. D. Harris (ed and trans), *Coventry Leet Book* (London, 1907), 531, 678, 696; and W. H. Prior, 'Notes on the Weights and Measures of Medieval England', in *Bulletin du Cange*, I (1924), 77-97 and 140-70; see pp. 151-4.

85 See 2 Henry VI, c.14 (formerly c.11): *WME*, 171.

86 Hanham, op. cit. (69), 119.

87 J[ohn] W[ybard], *Tachometria, seu Tetagmenometria, or the Geometry of Regulars* (London, 1650), 267.

88 Allen Simpson, 'English Standard Troy Weight Verified at the Mint, 1588', in Silke Ackermann (ed), *Humphrey Cole: Mint, Measurement and Maps in Elizabethan England,* British Museum Occasional Paper no. 126 (London, 1997), 105-6.

89 Cambridge University Museum of Archaeology and Anthropology, Inv. Z.2531. A wine pint at the Courage 'World of Beer', London, measured in 1980 but not accurately gauged in water, is no longer located: *WME*, 153.

90 The measure was gauged in 1994 and found to contain 1,909·6 g of water, equivalent to the volume of the gallon being 233·06 in³: personal communication from Roger Mason, Head of Metrology, Cambridgeshire County Council Trading Standards Department. We are grateful to Dr David Phillipson, Curator of the Museum of Archaeology and Anthropology for permitting this work to be carried out.

91 The Bridport ale gallon, discussed in Chapter 4, has the size inscribed on the side as '… 288¾ SINGLE SQVARD INCHES …' Although now unlocated, the measurement was confirmed in 1964: see Maurice Stevenson, 'The Size of Liquid measures in the 17th and 18th centuries, Part 3', in *Libra*, 3 (1964), 9-13, esp. pp. 11-13.

92 See above, ref. 90.

93 *APS*, I, 10: Act 14, 1426. See Appendix A.2.

94 See the Introduction to the Inventory.

95 (Allan Evans, ed), Francesco Balducci Pegolotti, *'La Pratica della Mercatura'* (Cambridge, Massachusetts,

1936), 201: 'Tinello 1 di vino di Parigi, ch'è in Parigi cesteri 96, e ciascuno si è 8 pinte, …'

96 An indication that the physical size of the tun remained unchanged comes in a petition of 1410 to Henry VI complaining about wine shipped from Gascony in vessels which were 'not filled by viij or ix ynches', holding wine that was not yet fully fermented and containing excessive residue: W. Herbert, *The History of the Twelve Great Livery Companies of London*, 2 volumes (London, 1837), II, 631.

97 *Fleta*, op. cit. (84), 120.

98 360 Flemish lotti or 288 Antwerp lotti; thus 5 Bruges lotti equals 4 Antwerp lotti: Pegolotti, op. cit. (95), 238: 'Olio, vino, mele si vendono in Bruggia a tinello a pregio di tanti reali d'oro il tinello, e di soldi 2 di grosso tornesi per 1 reale d'oro, e ogni tinello si è a misura 360 lotti di Bruggia e di tutta Fiandra'; and ibid., 251: 'Olio, vino, mele si vendono in Anguersa e per tutta Brabante a tinello, a pregio di tanti soldi [di] grossi tornesi il tinello, e ogni tinello si è a misura 288 lotti d'Anguersa.'

99 [W. Robertson (ed)], *The Parliamentary Records of Scotland in the General Register House, Edinburgh, Volume I* (Edinburgh, 1804), 16.

100 *APS*, II, 119: Act 9, 1 June 1478.

101 Ibid., II, 178-9: Act 16, 1 October 1487.

102 Ibid., II, 213: Act 3, 14 January 1489; ibid, 237: Act 23, 1493.

103 J. D. Marwick, *et al.* (eds), *Records of the Convention of the Royal Burghs of Scotland (1295-1779) [RCRB]*, 7 volumes (Edinburgh, 1866-1918), I, 23, 5 January 1570; *APS*, III, 82: Act 4, 1573.

104 *RCRB*, I, 114-15: 20 April 1581; *APS*, III, 302: Act 19, 1582.

105 *RPCS*, first series, XI, 525-7: 2 March 1619; ibid., first series, XII, 16-18: 15 July 1619; *APS*, V, 417: Act 116, 1641. The Privy Council declared the salmon barrel to contain 10 gallons. The provost and bailies of Aberdeen produced two standards of their salmon measure each of 10 gallons, according to the 'auld gedge and standart of Aberdeine'. Duplicates were to be sent to the castles of Edinburgh and Dumbarton for safe keeping. Act 116 reads in part: '… the barrelles conteyne no les then ten gallones of the stirvling pinte conforme to ane act of his Majesty's consell of the dait at Halyrudhouse the fyfteene day of Julii 1619 …'

106 Elizabeth Gemmill and Nicholas Mayhew, *Changing Values in Medieval Scotland: A Study of Prices, Money, Weights and Measures* (Cambridge, 1995), 103-7.

107 2 Henry VI c. 11, 1423; 22 Edward IV, c. 2, 1482; 13 Elizabeth I, c. 11 s. 4, 1571 (states that the barrel sizes are given in wine gallons); 5 George I, c. 18,

ss.6, 12, 13, 15, 1718; 43 George III, c. 69, Schedule (c), 862, 1803; Second Report of Commissioners to consider the Subject of Weights and Measures; 13 July 1820 (HC 314), *Parliamentary Papers*, 1820, VII, appendix A., s.v. *barrel*. The wine gallon was swept away by the 1824 Act (5 George IV, c. 74.), but a Notice given by the Commissioners of the Herring Fishery, 15 May 1852 stated that 45 wine gallons were equal to 37·5 Imperial gallons, so 42 wine gallons were exactly 35 Imperial gallons, a very convenient relation: see J. A. O'Keefe, *The Law of Weights and Measures* (London, 1966), 629.

108 *RCRB*, II, 12-14: 25 April 1595.

109 Changing the diameter alone by ¼ inch alters the volume by 35 in³; changing the depth alone by ¼ inch alters the volume by 50 in³.

110 $16 \times 77 \cdot 8 \div 1 \cdot 045 = 1,191$ in³.

111 Reducing the diameter alone by ¼ inch reduces the volume by 100 in³; reducing the depth alone by ¼ inch reduces the volume by 150 in³.

112 The compaction figure that gave the best fit for the 1618 barley firlot was 6½%, which gives a boll of $1 \cdot 065 \times 5,640 \div 77 \cdot 8 = 77 \cdot 2$ pints. If we take 77 pints exactly, this would give a volume of 5,625 in³, which is an excellent agreement with the dimensional volume of 5,646 ± 120 in³.

113 *Actis and Constitutionis of the Realm of Scotland …* (Edinburgh, 1566), James I, fol. xii. The nearcontemporary manuscript copy of the assize, discussed in Chapter 5, has a wording which is more clipped than in Balfour's 1566 printed version but carries the same sense. In Robertson's transcription it reads: 'that furlate nowther [neither] to be maid eftir the frist mesoure than vsit bott in a medful mesoure betwix the twa': Robertson, op. cit. (99), 63. The word 'than' is used here with the meaning of 'nor' (as in *Oxford English Dictionary*, s.v. *than*).

114 Swinton, op. cit. (71), 139.

115 Robertson, op. cit. (99), 63; text printed in Appendix A.3; *DOST*, s.v. *medful*.

116 Robertson, op. cit. (99), 63.

117 Strictly 18 pints with an allowance of one-sixteenth would generate 19⅛ pints, giving a boll of 76½ pints. However, we do not have a sufficiently accurate knowledge of these sizes to be able to discriminate between these options, other than to suggest that 19¼ pints may have been taken, since 77 pints fits the progression of sizes for the boll. Another possibility, bearing in mind that the new pint (and therefore the wine pint also) may have been used with divisions of a third and a sixth, is that the customary firlot was considered as 19⅙ pints, giving a boll of 76⅔ pints. This appears to fit the boll progression equally well.

118   *APS*, IV, 586-7: 19 February 1618.

119   Ibid., III, 521: 29 July 1587.

120   Ibid., IV, 587: 19 February 1618.

121   Ranald Nicholson, *Scotland: The Later Middle Ages* (Edinburgh, 1974), 303.

122   *APS*, II, 50: Act 18, 6 March 1457/8.

123   Ibid., II, 51: Act 28 [1458]; and see Nicholson, op. cit. (121), 382.

124   This item, and the assize to which it relates, are discussed in Chapter 7.

125   1,200 x $^{18}/_{16}$ = 1,350 in³. The internal diameter was to be 16½ inches, and $\pi \times 8\frac{1}{4}^2 \times 6\frac{1}{3}$ = 1,354 in³.

126   Peter G. B. McNeill (ed), *The Practicks of Sir James Balfour of Pittendreich,* Stair Society nos. 21 and 22, 2 volumes (Edinburgh, 1962-3); in his 1566 edition of the acts, Balfour states that the quart was also distributed. However, the quart is not mentioned in the near-contemporary manuscript compilation of the acts of this period, NAS PA.5/6/1, which covers the acts of 1424 to 1474: the text of this is printed in Robertson, op. cit. (99), 42. Thomas Thomson's *APS* text follows Robertson rather than Balfour in this respect. The manuscript PA.5/6/2, which has been preferred for acts in the early part of this period, breaks off in 1451. These manuscripts are described in Chapter 5.

127   *APS*, II, 50: 6 March 1457/8. Thomas Thomson was incorrect when he stated that the pint was given to the burgh of Stirling – the pint was sent to the chamberlain as a national standard, and he happened at the time to be on circuit and sitting at Stirling. This is discussed more fully in Chapter 7.

128   Ibid.

129   Robertson, op. cit. (99), 63; see Appendix A.3.

130   For this sense of 'mak', see *DOST*, s.v. *mak*.

7

THE RIDDLE
OF THE FIRLOT

Opposite:
Detail of Fig. **7.7**.

SOME METROLOGIES CENTRE ON THE UNIT OF LENGTH, OTHERS ON THE unit of weight. But in the Scottish system of weights and measures it was certainly the unit of capacity which principally occupied the legislators; and it is the evolution of the capacity units that is addressed in this chapter. After the Assize of 1457 four major revisions of the metrological system can be identified.[1] These occurred in about 1500, in 1555 (with modifications in 1563), in 1587, and in 1618. Each involved a substantial increase in the volume of the grain firlot, the basic measure of capacity for dry goods. After 1618, different circumstances prevailed, and apart from the progressive spread of English measures into Scotland, there was no further significant alteration in the Scottish system.

Previous commentators have been well aware of the increase in the capacity of the firlot with time, but the complexity of the legislation has tended to mask the nature of this growth, which has not been well understood. Normally it is characterised as chaotic, and it has been assumed that the size of grain measures increased in a steady and unchecked fashion, with the administration unable to establish effective control.[2] However, the evidence of the statutes and of surviving measures shows that this was not the case: the growth seems to occur in discrete steps and to have been brought about by a determination on the part of the burghs and merchants to retain their traditional allowances on the measures.

Enlargement of the capacity measures represented the principal business of each metrological assize, and on each occasion the size of the legal firlot was increased to match the larger firlot that was in current and widespread trading use. On two occasions the legislators said that they called for the existing measures, found them larger than specified by the current legislation, and then legalised a firlot that corresponded to the size of the one that had been produced for them. Their efforts were rapidly frustrated because the burghs and the merchants were fully aware of the official allowances to which they had always been entitled and they merely increased the sizes of

263

the new measures by the traditional proportion so that measures in use were again larger than those specified in the legislation. The cycle of official increase could then commence all over again at the next assize.

Clearly this is a simplification of what must have been a more complex situation, and the specific circumstances at the various assizes certainly differed. Equally, the changes proposed cannot have been instantaneous, and in practice, once the administration had conceded the use of customary measures which had been larger than the legal sizes, there must have been some delay before burghs ordered and obtained measures which incorporated further allowances on the newly-legalised sizes and before the effect of these larger measures was felt. Nonetheless, in broad terms, the process outlined above appears to describe the mechanism of enlargement of the dry measures and it can be tested satisfactorily against documentary sources and surviving capacity measures. As before, the upward pressure on these sizes can be explained in terms of the vested interests of the landowners, who gathered their rentals in kind, and the profit margins of the major merchants who came to dominate burgh government.

These changes came at an economically and politically troubled period for Scotland, notably during the mid-century reign of Mary, Queen of Scots. There seems little doubt that allowing the capacities of the measures to increase at assizes was financially advantageous for the monarch and for major landowners, because it could be used unscrupulously to increase the return from rentals and other victual-based revenues. But it is difficult to say whether it was the hope of such advantage that initiated the assize process. As it happens, the Scottish assizes of this period all occurred within a year or so of equivalent English assizes, which may perhaps indicate that these revisions were prompted by wider external considerations, even though the decisions that enlarged the firlot may have been opportunistic.

The purpose of this chapter is to examine the information for the increase in size of the dry measures and to demonstrate how the system of allowances outlined in Chapter **6** led to controlled increases in capacity. Far from being chaotic, this progressive enlargement can be calculated accurately, and a precise and well-developed metrological framework emerges from the legislative record. The extent to which the actual measures used in trade complied with this framework is more debatable, but at least the administrative structure can be seen to be compatible with that of a sophisticated trading nation.

THE ASSIZE OF JAMES IV

In the period after 1426, the Scottish dry and liquid capacity measures were regulated in terms of a pint of 77·8 in³ capacity and an enhanced version, one-sixteenth larger, of 82·7 in³. The role of these pints was explored in Chapter **6**, where it was noted that no surviving vessels with these capacities have yet been located.

In the sixteenth century a larger pint, of 103·7 in³ capacity, was introduced for metrological definitions, and dated standards of this general size (or its enhanced version) are known for the 1560s and 1570s. It has already been established in Chapter **6** that the underlying unit of liquid measure (the pre-1426 chopin, or the chopin of the ale gallon) had remained constant, although the number of these contained in the statutory pint had increased. So, the old pint in use before 1426 contained two such units, the pint of the 1426 Assize comprised three units, and the new 'large' pint of the sixteenth century contained four units.

It is more of a problem to discover just when and how this larger administrative pint was introduced, but the indications are that it happened shortly after 1500. The trigger for this was an assize of weights and measures (perhaps one of a group of separate enactments) in which the grain firlot was again increased, from the 18 pints of the 1458 Assize to 20 pints. There is at least one other change that can be associated with this period, and this involved a minor reduction in the ounce size, which will be discussed in Chapter **8**. Whether this was covered in the same assize or in a related act cannot be said, but there is at least an indication of metrological activity on more than one front within a period of a few years around 1500.

Although the date of this assize cannot be established from the parliamentary record – which is remarkably incomplete for the reign of James IV – it will be seen that the new firlot size was in use by 1500. The justification for inferring that the assize took place at all has to be assembled from scattered references, but the evidence for change in the firlot size is clear. Provisionally, the assize could perhaps be dated to a parliament of 1496, of which the legislative record has largely been expunged, or to within a year or so of this date. It can scarcely be a coincidence that this dating corresponds with that of Henry VII's English assize of 1496, which gave rise to the English Exchequer standards of 1497 which were discussed in Chapter **6**.

The Scottish assizes of weights and measures that have attracted attention in the past are those for which an extended account survives in the parliamentary record. However, the legislative procedure varied at different periods: for example, in 1555 and 1563 we have only the parliamentary acts that established commissions for revising the weights and measures, but not their reports. It is not clear whether it was always considered necessary to have such reports confirmed by parliament, but for three successive assize periods in the reigns of James IV and Mary, the conclusions of the commissions were either not laid before parliament or else were subsequently removed from the record.

The only surviving parliamentary reference to the James IV Assize is a brief act of early 1504. This act called for the burghs to fetch their measures from Edinburgh, and provided a stipulation, similar to that used in 1458, for rendering in the new measures payments which had been agreed in the old measure:

> And quhar ther Is ony fermez aucht [owed] in heritage of the auld mete that the said fermez be parportionat to the quantite of the auld mete & pait with the new mete to the avale [amount] of the auld mete per-portionaly And gif [if] ony persones vse ony other messouris or wechtis in tyme tocum bot the messouris & wechtis now to be maid … thai [they shall] be Inditit …[3]

The lack of an extant statute describing the nature of the changes made at this time to both the measures and the weights was to prove a problem at the Assize of 1587, when the legislators could only appeal to the 1458 Assize. This difficulty may also have been behind the otherwise inexplicable references, in the parliamentary acts which established the 1555 and 1563 commissions, to earlier statutes ascribed to James III.[4] However, the only surviving candidate act of James III is one of 1467 which merely called for the 1458 Assize to be observed, and was presumably prompted by the increase of the customary firlot beyond the intended level of about 19¼ pints to a capacity of around 21¼ pints, as described in Chapter **6**.[5] However, we can be fairly certain that the changes did not take place in James III's reign, but in that of his son, because at the parliament of 1491 an act called for the 'auld statutes and ordinances' relating to weights and measures to be observed.[6]

Parliaments and councils were not policy forming bodies under either James III or his son James IV, but rather the venues where policy formulated at court was publicised.[7] James III continued the established practice of calling parliaments on an almost annual basis in the twenty years of his adult rule, from 1469 to 1488, but during this period the weights and measures were not revised. Although regular parliaments were held in the years of James IV's minority, they were associated with the suppression of political dissent and unrest, and after his assumption of full power in 1495 he largely dispensed with parliament and ruled through a variety of forms of council, of which there is now an inadequate record. His reign, which was characterised by vigorous personal government, was marred by repeated insurrections in the Isles. Norman Macdougall has argued that the overriding reason for calling his parliaments in 1504, 1506 and 1509 was to obtain the sentences of for-feiture against the Highland rebels which only parliament could pronounce.[8]

An important mechanism used by James IV was his wide-reaching act of revocation of 1504, which gave him arbitrary power to cancel any previous grants, acts or statutes, of parliament or general council. His instruction that these be 'put furtht of the bukis and writingis' led to the removal from the original record of all of the proceedings of the parliaments of 1493 (during his

minority) and 1496: these records could only be partially assembled for the
first (1566) printing of the parliamentary acts.[9] No proceedings at all survive
for the parliaments of 1492 and 1494.

Between 1496 and 1504 James did not summon parliament, although
there is some evidence for a number of general councils being held in this
period, in addition to his regular 'daily' inner council of advisors, but there
is scant detail of their business.[10] It was only with the eventual calling of
parliament in 1504 that James's professional civil servants had real opportunity to legislate for improvements in civil and criminal justice.[11] It is against
this background that the act quoted above for the use of the new capacity
measures should be understood.

It is suggested that a principal motive for increasing the firlot's size in
the 1490s was the need to generate increased income to fund James's expensive court style, his progressive palace building programme and his military
expansion. In particular, large tracts of land were annexed by the crown during
James's reign, and revenues from crown lands provided the comptroller with
more than three times the returns from the customs.[12] These lands were made
to generate valuable income in cash and victuals, and rental agreements were
increasingly replaced with feu charters, with the new feu-duty often exceeding
the old rent. The ability to increase returns by the additional expedient of
enlarging the size of the measures must have seemed tempting; and it perhaps
became irresistible later in the century once it was apparent that the fixed
nature of the new feu-duties represented a financial disadvantage in times of
inflation and rising rents. Given the existing pressures for allowances on
bulk goods measured by capacity, James IV could have followed the English
practice of retaining the measures size. Instead, there seems to have been a
conscious (and perhaps opportunistic) decision to incorporate the existing
allowances in the measure, opening the way for additional allowances to be
taken subsequently.

Whether or not the lost assize formed part of the business of one of the
parliaments whose proceedings were purged cannot be established. It seems
more likely, however, that the matter was handled by James's efficient household administration and the outcome authorised by council. Indeed, the
1504 act implies as much in its opening sentence:

> It is statute & ordanit that all mesouris & wechtis baith pynt quart ferlot
> pec elwand stane & pund be of ane quantite & mesour quhilk [which]
> salbe ordanit in Edinburgh be our souerane & his chavmerlane and
> consale …[13]

Conceivably, the earliest indication of some administrative change can be
deduced from the significant number of indictments for selling ale by short
measure recorded in the Edinburgh minutes in early 1499, although this may
merely reflect increased vigilance on the part of the burgh's officials.[14]

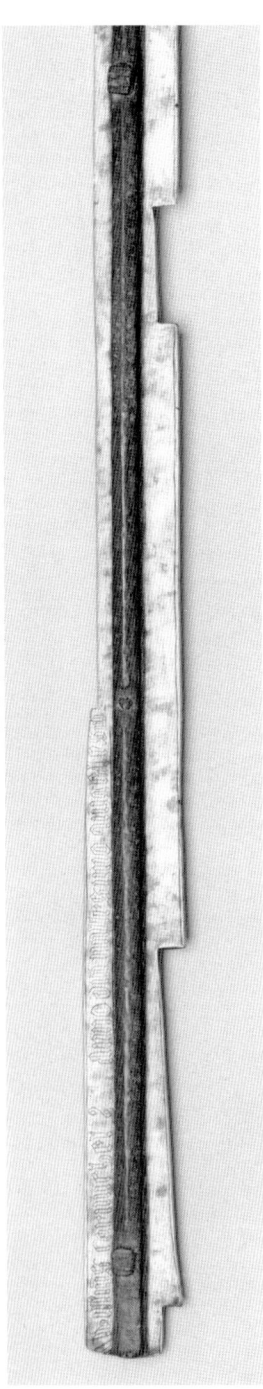

Fig. 7.1
Inscription with William
Carmichael's name and the
date 1500 on the Inverkeithing
firlot gauge and ell bed.
A photograph of the complete
measure is shown in Fig. **2.5**.
(Item **1** in the Inventory.)

The first positive evidence of the James IV Assize is to be found embodied in what is probably the oldest surviving Scottish standard measure, a remarkable brass gauge, dated 1500, which was the length standard of the royal burgh of Inverkeithing on the Fife coast (Fig. **7.1** and Item **1** in the Inventory). Although this is clearly an ell bed-measure, its original use, which has passed unrecognised, was as a gauge for testing the internal depth and diameter of firlot measures using the distances between notches cut in the opposing edge. This was the only practical way in which rapid spot checks could be made by burgh officials on the accuracy and legality of the traders' measures in the burgh markets. It also appears to carry more rudimentary markings which may be for the customary boll measure.

The use of such gauges must have been very widespread, but they were presumably replaced each time the official firlot size was revised and so perhaps no other early examples are extant. In this instance, the firlot gauge has probably survived only because it is combined with an ell gauge, for which there would have been a continuing requirement in the burgh market.

The gauge is an impressive device. It is formed from a broad strip of brass with stepped cuts in both edges, held between two profiled iron bars by bolts with decorative nuts, and is engraved in gothic lettering with the name of William Carmichael and the date 1500 (Fig. **7.1**). Almost certainly this is the William Carmichael who was a prominent and well-connected bailie on Edinburgh's town council, and who was burgh treasurer in 1500.[15] The inscription, therefore, probably represents the mark of authority of issue by the principal burgh as required by the 1458 Assize. It has already been noted in Chapter **2** that the length of the ell bed incorporates the same amount of play as found in the early Edinburgh standard.

The inscription does not relate in any way to Inverkeithing itself, and it remains supposition that this was originally Inverkeithing's standard.[16] However, there is no reason why Inverkeithing's measure should differ in its operation from that of any other prosperous burgh – in 1500 it was a cocket port and one of the principal entry points for travellers on the busy pilgrim routes to Dunfermline and St Andrews.[17] It is even possible that the measure may have had a reasonably high status because at this period Inverkeithing was a meeting place for the regular conventions of the royal burghs: the Edinburgh parliament of 1487 ratified an act of 'the haill commissaris of burrowis' calling for an annual meeting of the commissioners of the royal burghs both north and south of the Forth to be held at Inverkeithing,

> to comoune and trete apoune the welfare of merchandis, the gude Rewll [rule] and statutis for the commoune proffit of borowis and to provide for Remede apoune the scaith and iniuris sustenit within burowis.[18]

The dimensions on the Inverkeithing gauge define a cylindrical measure with a volume of 1,580 in³ (Fig. **7.2**). From the known capacity of the 1618 firlot, the volume of which was 2,110 in³, as we saw in Chapter **6**, the capacity of the firlot measure derived from the dimensions on the Inverkeithing gauge can be calculated to be about 21·2 small pints of 1426.[19] The size of the firlot of James IV's Assize can be set with some confidence at 20 pints and allowance of a sixteenth part would generate a permitted firlot of 21¼ small pints. It is our contention that it is this permitted firlot that is represented on the Inverkeithing gauge. Of course, the size computed from the dimensions of the gauge is approximate, to the extent that no account has been taken of play in the use of the gauge (which would reduce the capacity slightly) or of possible small differences in grain density from those in the larger vessels used in 1618; but in spite of these unknowns, the match between the two capacities is remarkable.

If this interpretation of the operation of the Inverkeithing gauge is correct, then the important feature of administrative practice to emerge is that a specific allowance of one-sixteenth, first encountered in the Assize of 1426, was still sanctioned in 1500. Our ability to see the allowance in 1500 as a sixteenth rather than merely as two additional pints, gives us confidence that the allowance permitted at the 1458 Assize was also a sixteenth, giving a trading firlot of about 19¼ pints. This was the size of the quarter of the

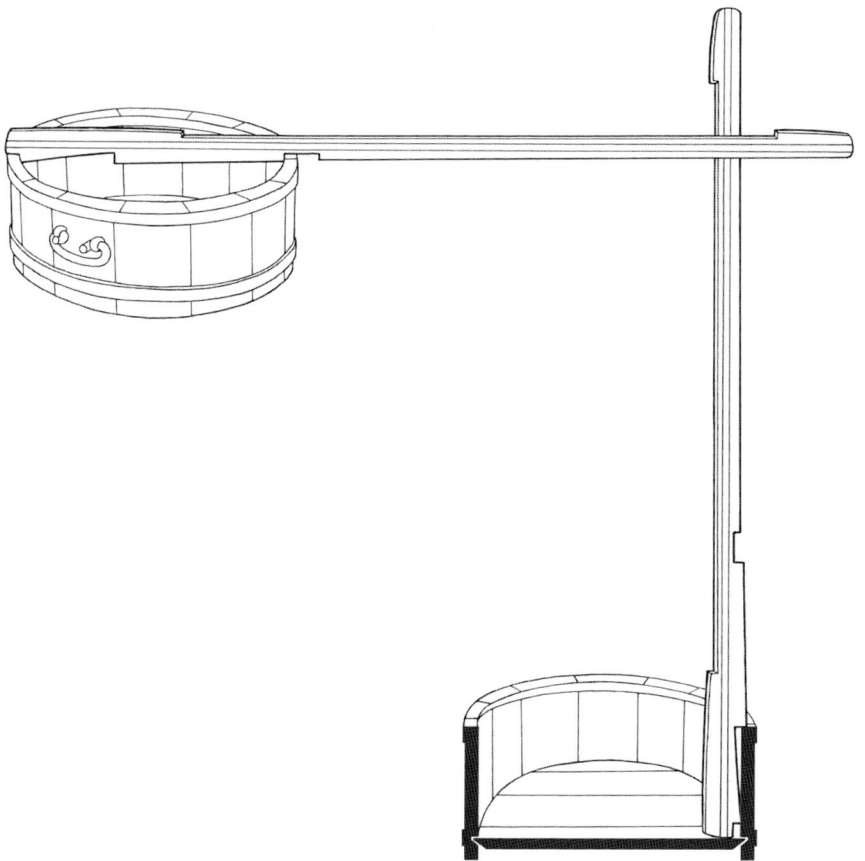

Fig. 7.2
Use of the Inverkeithing gauge to check the depth and diameter of a firlot measure.

trading boll of about 77 pints of the 1426 Assize, and it strongly suggests that the increase in 1458 was designed to allow the permitted trading firlot to 'catch up' with the measure that was already in practical use.

The newly-increased sizes of the capacity measures can be seen in two extracts from the minutes of Edinburgh council in 1504 and 1508. These relate specifically to the large 'water metts' for measuring material shipped through Edinburgh's port at Leith. The first is dated in March 1504, preceding by only a few days the parliamentary act that required observance of the new measures – a strong indication that the town was putting its house in order before the passage of the act. It takes the form of a report from the official measure maker of the capacities to which the water measure 'now comes':

> it was declarit be Willie Couper, maker of the mettis, that be auld consuetude [custom] the half watter bol of quheit [wheat] and ry contenis xlviij [48] pynttis; sua cummis [comes] the watter chalder of homyll [unbearded] corne to xviij [18] bollis … the half watter boll of beyre [barley] and malt sould contene [*blank*] sua cummis it now to the messour of lxxvij [77] pynttis.[20]

The second extract comes four and a half years later, in November 1508, when further measures were to be made to the same standard:

> the provest ballies and hale [whole] counsale ordanis the thesaurare to mak certane watter mesouris to serve our Soverane Lordis liegis and all vtheris equalie, and that the wattir hale boll to contene lxxxxvj [96] pyntis, the half boll xlviij [48] pynttis and the firlott xxiiij [24] pyntis, and that thir mesouris serve till [for] quheit beir and ry and all vther victuallis, … swa that the chalder be the watter met contene xviij [18] bollis land mett.[21]

The relationship between water and land measure is brought out clearly in this definition of the chalder measure. This was too large to be represented by a physical standard and was defined purely as a multiple of the boll. But, whereas the chalder was normally understood to be 16 bolls, the enlarged water chalder was 18 land bolls – an allowance of one-eighth. A very similar definition, this time of the boll, was given towards the end of the century at one of the annual conventions of the royal burghs – whereas the land boll comprised 16 pecks (the peck being a quarter firlot), the water boll was equivalent to 18 pecks, which again represented an allowance of one-eighth.[22]

If the water boll is 96 pints and the water firlot is a quarter of this, namely 24 pints, then to preserve the relationship between land and water measure the land firlot should be 21⅓ pints, which is a good match for the trading firlot of 21¼ pints that we have already postulated in Chapter **6** for the increase of *c*.1500 (see Fig. **6.9**). This size was a quarter of the earlier customary boll, and it fits the dimensions found in the Inverkeithing gauge.

The various sizes for the Leith measures have been drawn together in Fig.

7.3 to indicate the projected relationship between the legal sizes of the measures and their usual trading forms. The basic legal size of the firlot is believed to have been 20 pints, with a permitted trading allowance of one-sixteenth, giving the trading (or land) firlot of about 21¼ pints. This would imply a legal boll of 80 pints and a permitted trading boll (the land boll) of 85 pints. However, it has been pointed out in Chapter **6**, that as a result of the increases of the 1458 Assize a further allowance of a sixteenth was taken on the boll (the 'charity to the boll' of one peck for each boll), so that in practice the customary boll of the market-place was 90 pints. The quarter of this was the customary firlot of 22½ pints. Increasing this by a further sixteenth gives the water measure, with a water firlot of 24 pints and a water boll of 96 pints, as specified in the Edinburgh records. Water measure is therefore two factors of about one-sixteenth, or one of about one-eighth, greater than land measure. Whilst some of these figures are not exact (for example, the customary boll should perhaps be given as 90¼ pints), they represent a reasonable compromise that may reflect the accuracy that was considered necessary at the time. From Fig. **7.3** it will be seen that similar schemes can handle the further increases introduced at subsequent assizes in the sixteenth and early seventeenth century. Taken together, this indicates that the approach is sound, even though the evidence for individual assizes (such as the James IV Assize) is incomplete.

Fig. 7.3
Projected relationship between the assumed legal definitions of dry measure and the land and water measure series, using information of *c*.1500-10.

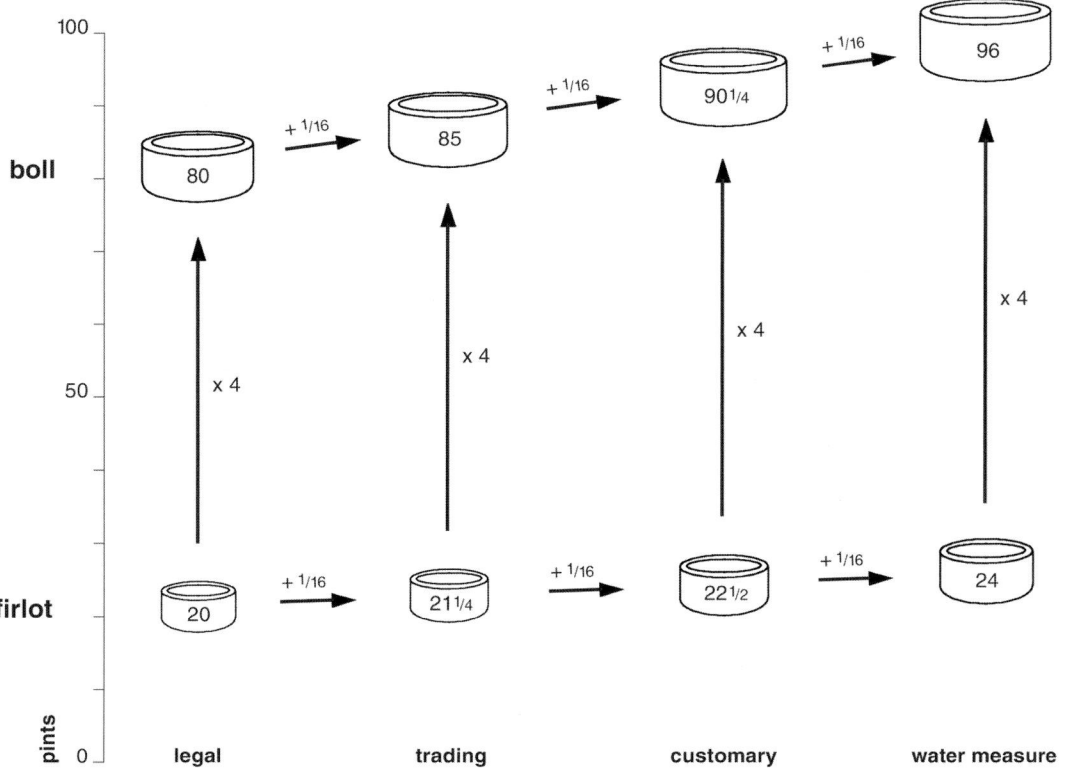

One aspect of the references of 1504 and 1508 to the size of water measure at Leith remains a puzzle.[23] The only significant difference between the two extracts is in the handling of barley and malt: this difference may only be procedural but there is inadequate information to resolve it. In the 1508 extract the same half-boll measure is to be used for all materials, including barley. Malt and barley were traditionally measured as notionally 'heaped' materials, counting two fills for every three taken. There were attempts to eliminate this '3 for 2' or '6 for 4' practice at the 1587 Assize, and it was eventually outlawed at the 1618 Assize when specially large firlot measures were constructed for barley. If barley and malt were to be measured on a '3 for 2' basis in 1508, we would expect a water half-boll for barley to be equivalent to 1½ x 48 = 72 pints. However, the 1504 extract, whilst leaving blank the description of the water half-boll vessel for barley and malt, nonetheless gives its capacity 'now' as 77 pints. This may represent a simple scribal error, but equally it may indicate that a different and provisional arrangement was in use between 1504 and 1508. Whether or not the figure has been entered in error, it represents an echo of a familiar size – the permitted land boll (four times the trading firlot of 19¼ pints) in force up to the time of the Edinburgh town council minute of 1504.

THE CUSTODY OF THE STANDARDS

The evidence of the Inverkeithing gauge, together with the evidence of the 1504 and 1508 extracts from the Edinburgh records, concern the specific issue of the enlargement of the firlot in the James IV Assize, but these provide only negative evidence about the introduction of the large pint. However, the fact that the water measure definitions given in 1508 were still in terms of the small pint of 77·8 in³ does tend to suggest that the larger pint of 103·7 in³ had not yet been introduced as an administrative unit: to take this further we will have to consider more carefully the construction and issue of standards. Nor does it throw any light on a related metrological change, which will be examined in Chapter **8**, namely a slight reduction in the mass of the burgh weights. Both these changes appear to have been put into effect about fifteen years after the assize that changed the capacity of the firlot.

In March 1512 the burgh of Dundee was granted a general remission by James IV preventing prosecution of the council or of any individuals at chamberlain or justice ayres for financial arrears or transgressions arising from use of the burgh's standard weights and measures. Of these standards, which had recently been issued to the burgh by the chamberlain, the act recorded:

> We gif licence to the saidis provest, ballies, counsale and communite and inhabitaris of our said burgh, to vse and hald the mettis, wechtis and

meseuris siclike as wes deliuerit to thame in our last chaumerlane air
haldin in our said burgh vnto the tyme that new mettis, wechtis and
mesouris be deuisit and gevin to thame be ws and our grete chaumerlane,
and will that thai bena in na wise accusit nor trubillit thairfore in thare
personis, landis, or gudis in tyme cuming: Gevin vndir our prive sele at
Edinburgh the xx day of Marche in the yere of God, j[m] v[c] and ellevin
yeris, and of our regne the xxiiij yere.[24]

The chamberlain ayre was held very infrequently in this period, and
indeed the last recorded ayre has been identified as that held at Edinburgh in
1517.[25] Formerly it was assumed that the Dundee, Perth and Cupar ayre of
February 1512, to which James IV's remission refers, had been the last to take
place.[26] The purpose of the remission, issued at Edinburgh shortly after the
chamberlain's return, was to ensure that the burgh possessed an adequate
legal authority to use these standards, until such time in the future as a new
general issue was made by the crown. However, the necessity of issuing such
an authority indicates that it was appreciated that these standards differed
from those previously in force. The most likely difference, and the one likely
to lead to legal difficulties, was the small reduction in the ounce size.

As well as helping to confirm the timing of the change in the weight
system, Dundee's remission tells us that the Great Chamberlain of Scotland
still had responsibilities for the burghs' standards, and that the standards of
the principal burghs were constructed and issued by the crown.

At the 1426 Assize, the standards were available for the burghs to collect
from Edinburgh, but there was no indication of whether any particular
standards were considered of greater authority than others. However, the
arrangements at the 1458 Assize were more specific: three sets of reference
standards of the capacity measures were to be made, each comprising a pint
and a firlot, and these were to be sent to Edinburgh, Perth and Aberdeen,
where they were to remain, perhaps in the hands of the burgh officials.[27]
Copies could then be obtained by those that needed to use measures, but
these copies had to be marked with the seals of the issuing burgh.[28] The
implication is that each administrative region was being provided with a
reference set of equal status. These standards were however based directly
on a particular pint vessel namely:

> ye pynt quhilk was giffen by ye ordinance of ye thre estatis [that is, parliament] Sir Johne Fostar yet [that] tyme beande [being] chawmerlane to
> the ayr of Strivilling as for standart …[29]

Sir John Forrester of Corstorphine (d?1448) was a relatively minor noble
of burgess origin who was a favourite of James I and served as Master of the
Household, in about 1424. He was Chamberlain from 1424 to 1448, having
earlier served as depute chamberlain south of the Forth.[30] Although as
Chamberlain he was the crown's chief financial officer, the position was

downgraded in 1428 when fiscal affairs were transferred to the new offices of comptroller and treasurer.[31] However, the chamberlain continued to supervise burgh administration and burgh finances through the chamberlain ayre and through the Court of the Four Burghs (the court of appeal from the burgh courts) which sat before him: it was undoubtedly through this latter formal link that custody of the standards came to be vested in the four principal burghs in the sixteenth century. Forrester was chamberlain until 1448, and so the pint standard must have been made before then; however, the association with a parliamentary instruction suggests that this pint may have been constructed at about the time of the 1426 Assize or shortly afterwards.

The extract from the 1458 Assize quoted above is taken from the surviving near-contemporary manuscript copy of the statutes, and not from James Balfour's more accessible *Practicks* of the late sixteenth century which has been altered in two respects. The first alteration is Balfour's introduction of a 'quart' standard as one of the measures provided by Forrester and as one of the distributed standards, but in this instance Thomas Thomson corrected Balfour's text so that his nineteenth-century edition of the *Acts* followed the near-contemporary manuscript.[32] The second alteration is Balfour's adjustment of Forrester's provision (this time retained by Thomson), so that the pint 'was gevin … into the burgh of Striviling', whereas the original refers to the 'ayre' and not the 'burgh' of Stirling.[33] Therefore, it is by no means clear that a standard left the chamberlain's hands and passed into the burgh's possession. The ready assumption has been made that it was presented to the burgh by Forrester, but this has been conditioned by the sixteenth-century understanding that the standard pint was held by the burgh of Stirling. It is not even clear whether the 1458 Assize refers to a physical artifact: what was 'given' may have been parliament's stipulation that the new system of 1426 was to be based on the unit of the pint as standard. In practice, we have no reliable information about the existence or custody of primary national standards until the end of the sixteenth century, and if any standards had this status it would be more logical to expect that they would remain in the control of the crown rather than the burghs. This has led some commentators to the erroneous assumption that the Stirling jug is the artifact 'given' by Forrester in 1458 to Stirling.

After the Assize of James IV, the Act of 1504 specified the authority for legalising particular weights and measures and called on the burghs to obtain standards from Edinburgh:

> It is statute and ordainit, that all mesouris and wechtes baith pynt, quart, fyrlot, pec, elwand, stane & pund, be of ane quantite & mesour, quhilk salbe ordainit in Edinburgh, be our Soverane his Chamerlane and Counsale and that ever ilk Burgh cum and fech the mesouris furthout of Edinburgh, selit [sealed] and maid, and keip the samyn.[34]

Thus, the act specified that the king, with his chamberlain and council, could define the permitted weights and measures of trade. But, strictly, it was a single measure that each burgh was to collect, indicating that it was the enlarged firlot that was being made available – and certainly the remainder of the act was restricted to the impact of this new firlot on existing contracts.[35] Authorised copies of the firlot were presumably provided by the chamberlain to some other principal burghs, as had happened in 1458 when copies of the firlot had been sent to Perth and Aberdeen. We have an indication of the enlargement of this small group of centrally-supplied burghs with the royal remission to Dundee covering the standards provided by the chamberlain in 1512.

The supply of standards to Dundee occurs at the start of a period when the administration was well-provided with expertise and facilities for casting weights and measures, and this seems the most likely time for the creation of the first of the large pint standards, of which only the Stirling example survives (Item 108 in the Inventory). The construction of modern artillery weapons was introduced to Scotland by French technicians imported by James III; and it was greatly accelerated by James IV in a remarkable arms race with Henry VIII of England from about 1505. Perhaps the most conspicuous aspect of this was the rebuilding of the Scottish navy, whose flagship was the vast *Great Michael*, completed in 1511.[36] Casting of bronze cannon was begun by James III in Edinburgh in the 1470s, and continued by James IV at Stirling Castle from about 1507, probably under the charge of the French gun-smith John Veilnaif.[37] David Caldwell, who has examined this activity, has noted that there is some evidence of gun casting for James continuing in Edinburgh, and that the centre of royal ordnance operations was transferred to a major foundry in Edinburgh Castle by 1511, the year in which a group of French gunners was brought to work in the castle.[38]

In early 1513 the Scot, Robert Borthwick, was described in the royal accounts as gunner and 'maister meltare of the Kingis gunnis', a position he may have held since late 1511 when the issue of his livery gown was recorded.[39] He retained his office as master founder until 1532 when he was succeeded as 'principale maister maker and meltar of our soverane lordis [James V's] gunnys and artilyeary' by his former assistant, the Frenchman 'Peris Rowane' (or Piers de Rouen). He in turn was succeeded in 1548, now in the service of Mary Queen of Scots, by his son David as 'principale maister makar and meltar of oure soverane ladyis gunnis and artaillierie'.[40] Throughout the sixteenth century a substantial complement of gunners was maintained on the royal payroll at Edinburgh Castle, and it is a measure of the importance of their work that many of the individuals recorded in the Treasurer's Accounts and the Exchequer Rolls were French, Flemish or Dutch specialists retained by the king.

The accounts also give an indication of the scale of the casting operations, recording the purchase of raw materials and construction of casting moulds. An example of this is the casting of cannon undertaken in the castle in late 1539 by a Dutch founder, Hans Cochran, who was listed as a master gunner from 1538, and appears under Rowan's command in the late 1550s.[41] Cochran was paid for 'necessaris apoun the making of vj gun muldis', which included clay, horsehair, wire, wax, steel and coal; Andrew Masterton was paid for carving wooden patterns of details 'witht lyoun heidis and flour de lices' of the six guns, and six men worked 'in the gun hous upoun the making of the samin muldis, and wirking of the clay'; casting sand was obtained, and finally Cochran mixed the bronze alloy and cast the cannon.[42]

With this type of expertise available, we would expect the administration would turn to the ordnance workshops in Edinburgh Castle for the casting of the standard weights and measures provided to the principal burghs. Indeed, two prominent salaried craftsmen from the royal workshops can be identified with this type of work in the mid-sixteenth century. Unfortunately, no separate charges for metrological work can be found in the treasurer's accounts, but presumably these were masked by the very much larger ordnance undertakings.

The first of these specialists is David Rowan (*fl.*1543-93), whose distinctive 'DR' monogram appears on the Edinburgh half-pint standard of 1555 (Fig. 7.4, and Item 109 in the Inventory).[43] The status of this vessel is not entirely clear, but the assumption is that it is the sole surviving standard resulting from the work of the parliamentary commission established in Edinburgh in 1555 to revise the weights and measures. No report of this commission is known, although it can be established that the assize took effect because there is a reference in the Edinburgh council minutes in early 1556 to firlots 'of the auld mesour'.[44] Rowan's half-pint standard is a very competent single-piece casting of high quality. It carries a crowned Scottish shield with a lion *rampant* as well as a representation of the three-towered castle of Edinburgh: the first of these devices must denote its status as part of a royal issue, whereas the second is the burgh seal which provided its authority as a standard which could be copied.[45]

The second craftsman is the gun-founder Hans Cochran, who can be identified as the Johannes Coquhren whose name appears on a rare surviving standard weight of 1553 (Item 31 in the Inventory, and see Fig. 7.3).[46] The mass of this impressive cast-brass weight is a stone in the 'trone' series, which will be described in Chapter 8, and is the only weight standard which we have located for this early period. Gothic lettering cast in relief round the weight records that it was made at the instruction of John Craigengelt when provost of the burgh of Stirling. However, this does not necessarily imply the standard belonged to Stirling, merely that it was produced as a result of the meeting of the Convention of royal burghs held at Stirling in 1553 and convened by Craigengelt to discuss arrangements for equalising the weights

Fig. 7.4
The mark of David Rowan,
master founder at the royal
ordnance workshops at Edinburgh Castle, on a half-pint
standard of 1555. (Item **109**
in the Inventory.)

and measures used in the various burghs. Its construction by one of the royal
founders at Edinburgh, its basis on the official 'French' weight standard and
not the merchant equivalent (the significance of this will be discussed in
Chapter **8**), and Cochran's use of the *fleur-de-lys* motif, all suggest that this
weight forms part of an officially-sanctioned central issue of standards to the
principal burghs.

Although the office of chamberlain continued for a considerable period,
the effective suppression of the chamberlain's powers came in 1535 when the
supervision of the finances of the royal burghs was vested in the Lords
Auditors of the Exchequer.[47] In the process, his control over the burghs'
privileges and trading affairs passed to the Convention of Royal Burghs, a
body which acted effectively as a mercantile parliament. Conventions of
commissioners representing the royal burghs generally, or of specific
groups of burghs, had certainly been held since the late fifteenth century;
and although we have no minutes of these early meetings, they undoubtedly
noted changes in metrological legislation. We might speculate, for example,
that the Convention meeting in 1497, to which Aberdeen sent a representative 'for the gude of merchandice', may have been concerned in part with
the implications of the lost James IV Assize.[48]

The principal royal burghs already had a common code of burgh laws,
and appeals from the burgh head courts, whose business was legislative,
administrative and judicial, were held in the chamberlain's Court of the
Royal Burghs. Changes in the structure and organisation of the courts in the
early sixteenth century enabled appeal procedures to be handled by the
ordinary courts of law, and remaining functions of the chamberlain and the
old Court of the Four Burghs were merged into the court of the Convention
of Royal Burghs.[49] The Convention emerged in the mid-sixteenth century,
when surviving records of its proceedings begin, as the body which co-

ordinated the interests of the burghs and was empowered to regulate matters of common welfare and to resolve disputes between its members.

One of the functions passed to the Convention was the responsibility for ensuring a uniform administration of weights and measures control. The earliest surviving Convention minutes, for a meeting held in Edinburgh in April 1552, recorded agreement that there was a

> differance of mesouris within borrowis of this realme in tyme bigane [bygone], part being mair nor [more than] the rycht, and part less …[50]

This was to be addressed by having authorised copies available from the those 'burghis following, quhilkis hes the iust mesouris, viz; the stane wecht of Lanark, the pynt stope of Striuiling, the ferlatt of Linlytqw, and the eluand of Edinburch …'.[51] This is the first indication that custody of the standards had been given to four individual burghs, and although there have been attempts to justify the choice of these burghs on the basis of their particular trading strengths, the likelihood is that the distribution merely reflects the fifteenth-century membership of the Court of the Four Burghs.[52]

It is not known when this change in practice was made, although it possibly dates from the 1535 reduction of the chamberlain's authority. The earliest statutory reference is in the 1587 Assize when it was noted that copies of the four standards were to be held by the burghs 'to quhome thay haif bene committit of auld'.[53] Nor is it known whether physical standards held at Edinburgh were distributed, or whether it was merely stipulated that the four relevant standards already held by these burghs were considered to be the just standards. It may have been the latter, because there is some indication from the late sixteenth century that the Lanark standards were adjusted to the merchant variety of ounce, which was marginally heavier than the legal 'French' ounce.[54]

The status awarded to these four standards was presumably responsible for the survival of the Stirling Jug itself. There must have been several copies of the pint standard made initially by Borthwick, or whoever was responsible for the work; but these were probably replaced when a small change was made to the legal basis of the pint at the 1563 Assize, which will be discussed below. However, the Stirling pint continued to be appreciated as *the* nominated standard, even though it did not quite comply with the new definition. There would therefore be no question of the Stirling authorities replacing it, and indeed it was later re-introduced as the 'national' standard at the 1618 Assize.

Once custodial rights to the primary standards had been established, the four burghs retained this privilege – there was after all some profit in making copies for sale to the other burghs. The privileges were even revived in the early eighteenth century when these four burghs became the issuing centres for the newly-introduced English weights and measures supplied for the

burghs as a result of the legal abolition of the Scottish system under the Act
of Union of 1707, and this will be discussed in Chapter **9**.

THE INTRODUCTION OF THE LARGE PINT

The best-known standard of the 'large pint' is also probably the earliest. It is
an undated vessel, at Stirling (Fig. **7.5**, and Item **108** in the Inventory), often
described simply as the 'Stirling Jug', which came to be regarded as the
national standard. The body of the Stirling Jug is cast in a bronze or bell-
metal alloy in two main pieces (the base has been made separately) and its
handle is riveted top and bottom. It carries no markings except for two
shields in relief, one with a lion *rampant* (the Arms of Scotland) and the
other with a crouching child which has defied adequate description.[55] This
is almost certainly an early version of the seal of Stirling, which in the
seventeenth century is represented as a lamb or wolf *couchant*.[56] Plausibly
this is a pre-Reformation version of the seal, depicting the Christ child, an
image likely to have been changed after the Reformation to the alternative
representation of the paschal lamb and subsequently adjusted to remove

Fig. 7.5
The Stirling Jug, a standard
of the 'large' pint; probably
constructed in Edinburgh,
c.1510. (Item **108** in the
Inventory.)

any religious connotations.[57] These two devices are also found as the marks of authority on the various seventeenth-century local standards based directly on the Stirling Jug.

It has often been supposed, apparently on the basis of its demonstrably 'ancient' character, that the Stirling Jug is the very vessel referred to in the Assize of 1458 as having been 'gifted' earlier to Stirling, and indeed Lawrence Burrell explicitly accepted this claim.[58] However, there is no real justification for such extravagantly early dating, and instead it seems more plausible that it dates only from the early sixteenth century when we deduce that measures of this size were first introduced.[59]

We have determined the capacity of the Stirling Jug as 103·8 in[3]. This was done by weighing the pint empty, and then carefully filling it with water and weighing it again full. The water surface was viewed obliquely to judge when it most accurately coincided with the vessel's rim. (The procedure for gauging these measures is described more fully in the introduction to the Inventory section of this volume.) Dividing the difference between these two weights by the density of water gives the volume of the vessel. This is substantially the technique that has been used by all those who have investigated the Jug since the eighteenth century, but there is a considerable spread in the calculated capacities. In part this is because different values have been accepted from time to time for the density of water, but more crucial has been the difficulty of deciding the precise level to which the Jug should be filled, given that its rim is uneven.[60] These measurements are discussed in more detail in the entry for this standard in the Inventory (Item **108**). There is an added complication in that the vessel has clearly been subject to some alteration and repair at unknown dates, and indeed currently has a slow leak at one of the rivets for the handle: it is therefore not entirely clear whether the best measurement of its volume that can now be obtained accurately represents its original capacity.

The most careful modern determination is Adam Anderson's measurement of 1827 in which the weighed water contents were 26,286 grains, equivalent to a capacity of 103·9 in[3].[61] His results are compatible with the recent measurements, although he adopted a different and somewhat implausible water level which was not flat. The capacity that has come to be accepted in the literature, however, is the one obtained by James Jardine in Edinburgh in 1825, which used an inappropriate method. Jardine placed a flat glass 'strike' on the rim of the Stirling Jug and the final addition of water was introduced through a small central hole in the glass plate until there was no remaining air space. This method necessarily gave a larger volume because the water surface was level with the highest points on the rim of the pint. Jardine's result for the water contents was 26,307 grains, and this was officially adopted as the correct capacity. It was given in contemporary tables as 104·2 in[3], but would be derived as 104·0 in[3] using the modern density.[62] Again, this is compatible with Anderson's result and the recent measurements, making allowances for the larger value gauged.

A much lower value of 26,180 grains for the capacity was obtained in the earliest of the modern determinations of which we have records. Alexander Bryce's measurement of the Stirling Jug in about 1752 was given in the published account as 103·4 in³.[63] In part the difference between the capacities obtained by Anderson and Bryce may result from Bryce's method of determining the water level by contact established with a fine wire stretched over the rim of the vessel. His experimental technique, which is inherently likely to give a low result, is referred to in Chapter 9.

Once differences between the water levels and water densities adopted by the various investigators of the Stirling Jug's capacity are taken into account, their results are all found to be coincident within reasonable limits of observational error. Clearly it is important to establish that the visual technique used in the most recent measurements is actually appropriate, and this was done by repeated checking of the capacity of a fine brewers' standard pint of 1574 by David Rowan for St Andrews (Item 111 in the Inventory). This was designed as an overflow vessel regulated by an accurately cut notch in the rim. When the vessel is filled to nearly overflowing, the liquid settles with the meniscus observably above the rim due to surface tension. The water is held back at the notch until flow suddenly commences and the level of the water in the jug drops to that of the rim. This was confirmed by masking off the notch and refilling the pint, to record the same weight of water content. From this it was apparent that our criteria used to judge whether the water surface was visually level with the rim or not must have been close to those used by Rowan in the original adjustment of the vessel (even if they did not coincide with those of Bryce, Anderson or Jardine).

Although the Stirling Jug has been measured as 103·8 in³, there is no really adequate way of resolving the question of its original capacity because we know too little, first about the procedures used for adjusting one measure against another, and second about the history of adjustment and repair to the Stirling Jug. However, we can obtain some help from the high quality copies of the Stirling Jug made for the other principal burghs at the time of the 1618 Assize. The Edinburgh standard of 1618 (Item 113 in the Inventory) has a rim which carries 1618 adjustment markings and which is accurately flat: it is therefore intrinsically easier to gauge and its volume has been measured as 103·7 in³.[64] The care with which the Edinburgh 1618 pint has been constructed and adjusted indicates that it was seen as a high-precision standard, and the inference from the 1618 Assize is that it was adjusted against the Stirling Jug. It may therefore be that this 1618 comparison provides at least as good an expression of the Stirling Jug's capacity as the 103·8 in³ obtained by direct experiment.

The dimensions of all the liquid capacity vessels discussed so far have been given to the nearest 0·1 in³ (which for a large pint represents an accuracy of ± 0·05%). In practice this is about the limit of accuracy that was achievable before the availability of recognisably modern apparatus in the second half of the eighteenth century, and it therefore does not make sense to try to specify these capacities to a further place of decimals. It is clearly preposterous to give the Stirling Jug's capacity, as Jardine did, to eight significant figures. Our experience of gauging a number of these pint standards was that, providing the vessels had an adequately flat rim, the water surface could be judged to match the vessel's rim with an uncertainty in the water contents of about ± 0·5 g; but normally some unevenness in the rim increased this to about ± 1 g. This is equivalent to an intrinsic uncertainty of about ± 0·05 in³ in the volume, arising solely from the method of observing the water level.

In the light of this, the difference between considering the Stirling Jug as 103·7 or 103·8 in³ seems almost insignificant. The point has been laboured, however, because establishing a working value for the volume of the large pint is of central importance. In fact, for the purposes of calculation, we will assume that the theoretical size of the large pint, based on contemporary administrative regulations, was 103·7 in³. This figure arises as the best compromise between the rounded figures adopted for the smaller pints that still preserves their numerical relationship and it is compatible with the size of pint assumed to have been adopted at the 1618 Assize.

We proposed in Chapter 6 that the 'pint' was the particular administrative unit used in the definition of the dry capacity measures, and that for convenience it was increased in size as the capacity of the dry measures grew. It can be shown to have been equivalent to two (old) chopin units in the

Fig. 7.6
The comparative capacities of the three principal statute pints.

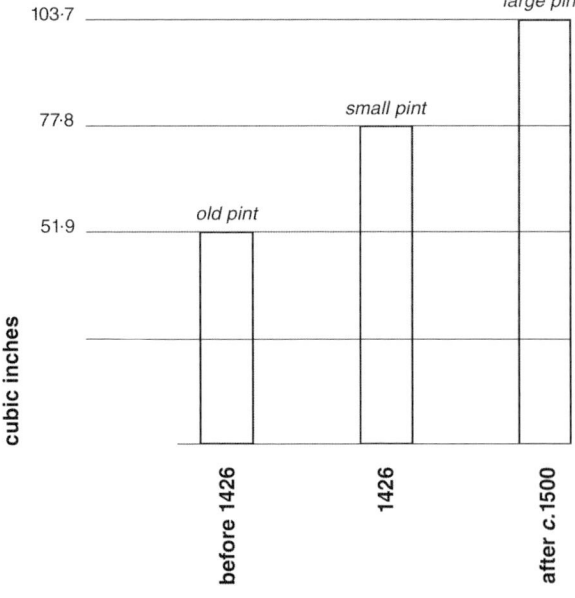

period before 1426, three units after 1426, and then four units after about 1500. This relationship is indicated in Fig. **7.6**. We have taken the volume of the old (pre-1426) pint as 51·9 in³, the pint of 1426 as 77·8 in³ and the new large pint is 103·7 in³. However, equating the 1426 pint to precisely the specified 41 ounces of water gives the latter volume as 103·76 in³.[65] Whatever the precise original size of the Stirling Jug, there is an excellent match between this projected size and the measured 103·8 in³ capacity, considering that we know nothing about the vessels used as intermediaries in adjusting the Stirling Jug.

The reason for adopting the weight of the pint as exactly 41 trois ounces in 1426 was explored in Chapter **6**, and it was noted that there were grounds for believing that the standard might correctly have been 41 ¹⁄₆₄ ounces, and that it was rounded down to 41 ounces in the assize. It is anticipated that at least some standards created as a result of the assize (including, presumably, standards for individual burghs) will have followed the assize definition and been adjusted to hold 41 ounces of water. However, it is plausible that the administration's own reference standards (such as the Forrester pint) will have been to the marginally larger size of 41¹⁄₆₄ ounces. This small difference might still be reflected in the size of the first large pints, which we believe were made in the royal ordnance workshops in Edinburgh in about 1510 and adjusted by officials at the Mint. If so, the volume of a large pint could be expressed more accurately as 103·80 in³.

In addition to the Stirling Jug, there will almost certainly have been another version constructed at the same time and retained in Edinburgh. This standard would initially have had a comparable status to that of the Stirling Jug, but given the comparatively poor quality of the rim of the Stirling Jug it is almost certain that some variation between the capacities of these vessels would have been detectable in Rowan's time. If this Edinburgh pint was still available in 1555, it is intrinsically more likely that Rowan would have based his half-pint standard on it. (Such a pint would not have survived the reassessment of the pint's basis conducted by the 1563 commission, which will be described below.)

Rowan's Edinburgh half-pint of 1555 has been measured as half of a pint of 103·9 in³, and we can appreciate therefore that it has the same basis as the Stirling Jug. The slight discrepancy in the measured volumes may arise from differences between the Edinburgh and Stirling pints or from alteration to the vessels. There is in fact a possibility that the Stirling pint was once marginally larger than at present: the rivets that attach the handle have large heads inside the vessel and may perhaps be replacements for flush rivets. If so, this would imply a volume of about 103·85 in³ before the alteration, and if this took place between 1555 and 1618 it would help bridge the small gap between the volume of the Stirling pint and the reference pint used by Rowan. This slightly increased volume is a better match for the supposed volume of 41¹⁄₆₄ ounces of water.

However, the argument that these early pints might preserve a possible pre-1426 volume unit is tenuous and inconclusive. It is more likely that individual early standards differed by small quantities (of perhaps about 0·1 or 0·2 in³) representing the tolerance within which such vessels could be adjusted in the period before about the mid-sixteenth century.

Two other dated pint standards survive from the sixteenth century, as does another which is undated (Items **110-112** in the Inventory). All these standards, and the more numerous seventeenth-century surviving standards, are based on pints which are larger than 102 in³. The undated sixteenth-century standard, and a dated pint standard of 1574, are both by Rowan, and the quality of his work is so good that we are able to compare these Rowan vessels with each other and with his 1555 half-pint and determine that they are all based on slightly different principles. They show capacity differences of a few cubic inches – large differences compared with the degree of un-certainty in the precise size of the Stirling Jug – and we will return to the significance of these differences later.

The administrative vessels may have changed in 1426 and about 1500, but our contention is that the usual vessels of trade were not directly affected, and certainly this approach preserves continuity in the under-lying old chopin unit, which was not altered by the 1426 legislation. We would expect, therefore, that the 1426 pint continued in use for some time after 1500, before eventually being superseded by the large pint. Equally, if we are to maintain this feeling of practical continuity, we should propose that an ale vessel of the same size as the old pint continued in use after 1426. In fact, we have no clear understanding of the extent to which separate ale and wine units continued to be used in parallel during the remainder of the fifteenth century, as happened in England. From the perspective of the burgh officials the ale context may have been the more significant: the post-1426 administrative pint may still have been seen as a wine pint, amounting to three ale chopins in size, although of course comprising two larger wine chopins. Indeed, Elizabeth Gemmill has noted an entry covering an import of Genoan wine at Aberdeen in 1458/9, the price detail of which shows that the gallon contained 16 chopins, and hence was 8 pints, each of 2 (wine) chopins.[66] But this is in a context which is specific to wine, and may therefore have no bearing on separate ale units. Thus although it may be excusable to think of the post-1426 pint in terms of three of the usual ale chopins, perhaps even into the sixteenth century the pint properly constituted two wine chopins or four wine mutchkins.

The possibility that a usual measure of two old chopin units survived into the sixteenth century raises a difficulty in deciding the identity of the 1555 standard vessel by David Rowan. This vessel carries no descriptive name and we have previously called it a half of the large pint; but we are not justified in

using the existence of this vessel in 1555 to demonstrate that the large pint was
in official use, because it could equally be described as a standard for the old
two-chopin measure which was the statutory pint before 1426 and apparently had a continuing role in its enhanced form as the ale pint. Ultimately,
the term 'chopin' came to be applied to half of the large pint, but whether
this change in the use of the word had taken place by 1555 is a moot point, and
we have therefore avoided calling the 1555 vessel a standard chopin and refer
to it merely as a half-pint. Some support for this tentative suggestion that
the large pint was considered to be twice the size of the (old) ale pint may
perhaps be found in the alternative name of 'Stirling Stoup' sometimes given
to the Stirling Jug. This tag may have been adopted simply for its alliterative
quality, but it will be recalled from Chapter 6 that in the Scots merchant's
handbook of around 1400 the 'stoup', or 'stope', was used as a specific capacity
of two pints (in this case wine pints) rather than merely to describe a liquid
capacity vessel.[67]

The first clear evidence that the large pint had passed into widespread use
comes in the 1560s. There was an assize of weights and measures in 1563, for
which no report survives; however, there is a standard pint vessel at Jedburgh
dated 1563, which was presumably produced as a result of this assize,
inscribed 'THIS IS YE COMVN MVSVR OF IEDBVRGHT' (Fig. 7.7, and Item 110 in
the Inventory). This is slightly more than a large pint in capacity – we have
measured it as 108·7 in³ – and the vessel has an unusual form with a spout that
allows excess water to run off in a controlled fashion so that it automatically
adjusts itself to the correct level. The only other dated sixteenth-century
standard pint measure which we have located is of a similar type. The St
Andrews standard – the first vessel which describes itself specifically as a
'pint' – is dated 1574 and was constructed by David Rowan (Item 111 in the
Inventory). The capacity of this latter pint is 109·1 in³, and it also is a self-
adjusting measure, although in this instance Rowan has cut a complex notch
in the rim rather than forming a spout.

The likely reason for constructing these two pints as self-adjusting
measures is to enable them to be used repeatedly for testing small measures of
trade. But this was a situation in which a traditional allowance was given, and
John Swinton described the continued operation of this allowance in the late
eighteenth century:

> the customary gallon used by the brewers, as also the pint, chopin, and
> other ale-measures generally hold 1/16 part above standard … [68]

The Jedburgh and St Andrews pints are surviving examples of standards
of the brewers' pint (hence, presumably, the description of the Jedburgh pint
as the 'common measure'), and from their volumes of 108·7 and 109·1 in³ we
can calculate that they are based on pints of 102·3 and 102·7 in³ respectively,
assuming the allowance is accurately one-sixteenth. These capacities are

Fig. 7.7
The 'Jeddart Jug' or
Jedburgh Jug, a standard
of the customary or brewers'
pint, dated 1563. (Item **110**
in the Inventory.)

sufficiently different from the expected 103·7 in³ pint to require some explanation. Fortunately, the smaller of the two sizes can be confirmed from an
independent measurement: an undated standard pint at Dundee, which
bears no inscription but carries David Rowan's mark, also has a capacity of
102·3 in³ (Item **112** in the Inventory). All three variant sizes (102·3, 102·7,
103·9 in³) are therefore represented by standards made by Rowan; and,
because the quality of his technical workmanship is so good, we can interpret the differences in capacity in terms of changing definitions of the
pint's volume.

The small reduction in the basis of the Scottish weight system in or
before 1511 in which the ounce mass was reduced from that of the English
troy ounce of 480 grains to match the Paris ounce of 472½ grains has
already been mentioned. In Chapter **8**, the one surviving standard weight
of this period – a trone stone of 1553 (Item **31** in the Inventory) – will be
used to demonstrate that the Paris ounce did indeed form the basis of
officially-sanctioned weights. The Scots pint was defined in terms of the
weight of its water contents, and it might therefore be expected that the

pint's size would be reduced when the ounce basis was changed, or else that it would be re-defined to contain a greater number of Paris ounces so that its volume remained constant. The evidence of Rowan's 1555 half-pint standard is that no such reduction in volume was made initially or at the 1555 Assize. However, the dated Jedburgh pint and the Dundee pint indicate that the smaller size must have come into use as a result of the 1563 Assize.

For burghs that traded abroad, it was not Paris weight but the marginally heavier weight of the former Burgundian Netherlands that was commercially significant for the merchant community. It is clear that many burgh standards were constructed accordingly to the Flemish or Dutch trooise version of troy weight, with an ounce of about 474·4 grains. The St Andrews pint is one of these standards – constructed to contain the correct number of ounces, but Flemish ounces rather than Paris ounces. The confusion caused by this situation eventually forced the administration to adopt Flemish ounces at the 1587 Assize. This is not known from the assize report, which is entirely silent on this matter, but is inferred from comments at the Privy Council made some twenty-five years later.[69]

The three basic pint sizes obtained from the Rowan standards, namely 102·3, 102·7 and 103·9 in³, bear the same ratio to one another as the three ounces on which they were based – 472½, 474·4 and 480 grains – and each therefore contains the same number of ounces. (The changes in the weight system in the second half of the sixteenth century will be treated in Chapter 8.)

To derive these figures we have needed to consider the role of the brewers', tavern or customary pint, incorporating a specific allowance, which was first introduced in Chapter 6. The importance of such standards is that they are of immediate practical relevance to burgh trade, but they have tended to be eclipsed in metrological studies by a preoccupation with the capacity of the Stirling pint. In many instances it is not apparent which type of pint is being described, but there are enough important references which relate to the brewers' pint to suggest that the two measures have frequently been confused. For example, it was presumably the brewers' pint that was referred to by David Gregory in the 1680s when he noted the capacity of the Scots pint as 109 in³.[70] Similarly, his successor in the mathematics chair at Edinburgh, Colin Maclaurin, described the measurement of a cask performed by the brewers of Edinburgh in 1707 for the newly-established Commissioners of Excise, on which the setting of the Excise rates after the Union depended, using a pint of 108·66 in³.[71]

From at least the early eighteenth century, a distinctive type of pewter tankard known as a 'tappit hen' is associated with Scots tavern measures. The name is thought to derive from a similarity between the crest on a hen and the thumb-finial on the lid of the measure. Strictly the name applies only to

Fig. 7.8
A 'tappit hen' (the handle
probably restored), made by
A. N. of Edinburgh, 1733.
(Reproduced by permission
of the Trustees of the
Shakespeare Birthplace Trust.)

the pint measure, although the same construction is found for the many smaller sizes. The correct capacity is found by filling the measure up to a small pewter stud, or 'plowk', attached inside the neck of the measure, so in effect it is used as a modern 'line' measure in which the level mark is engraved on the glass. Many of the surviving measures of this type are of the late eighteenth or early nineteenth century, but one of the oldest known dated tappit hens is a tankard bearing a 1733 Edinburgh touch-mark (Fig. 7.8).[72] The capacity of this, measured to the plowk, was 104·0 in³, but measured up to the rim of the tankard it was found to be 111·4 in³. Increasing 104·0 in³ by one-sixteenth gives 110·5 in³. This rim measure is, therefore, very close to the brewers' pint, and it is suggested that the two capacities were originally embodied in the measures in this way.

The brewers' pint continued in use well into the nineteenth century. The series of county reports on weights and measures produced in the 1820s to show how local standards and practice differed from the newly-introduced Imperial measure, recorded the use of the brewers' pint in two Scottish counties. In Glasgow the pint standard was found to contain 105·1 in³ in 1821 (with a sixteenth allowance this would generate 111·7 in³), and there was another measure, also 'called a Scotch Pint', of 111·6 in³, which 'has been long in use in Glasgow, for the sale of Ale, Beer, Porter, and Butter Milk'.[73]

Standards of this size were in use in Dunbartonshire as well as Lanarkshire, and in Renfrewshire there was a similar ale pint of 112·8 in³.[74]

Another possible instance of the use of the brewers' pint was in the generation of permitted firlots. At the Assize of 1426 an allowance of only one-sixteenth was granted for the firlot, even though a rather larger measure was already in practical use. This attempt to hold the size of the firlot down was ultimately unsuccessful, and its size was revised upwards to 18 pints in 1458. Both at the 1458 Assize and at another in the 1490s, when we will argue that the firlot became 20 pints, allowances again appear to have been permitted; and from the scheme (which we develop below) for the progression of the firlot's capacity, it seems that these allowances were both of one-sixteenth. However, adding this further quantity in the process of checking trading measures may not have been a simple matter, because the number of pints was not a multiple of 16 and because the early pint, initially at least, was considered to be of six parts and was therefore not divisible by the required factors. A more straightforward procedure for the burgh officials might be to measure out the specified number of pints, but using the permitted brewers' enhanced pint rather than the legal pint. This difficulty was removed with the introduction of the large pint because it was certainly based on four rather than three old chopins and so allowances could be calculated and measured with greater ease.

From the 1563 dated standard of the customary pint for Jedburgh, it is clear that the large pint (and not the smaller 1426 pint) was regulating the ale trade from at least the 1563 Assize. The evidence for the 1555 Assize is inconclusive, but we can be sure that the use of the large pint was enforced from 1563.

Some independent confirmation of this comes from changes observed in the measures used for the assessment of rental in the northern isles. In both the Orkney and the Shetland islands, extended Norse control led to the development of land-holding and metrological practice which was separate from that of mainland Scotland. With the effective annexation of the northern isles by Scotland in the fifteenth century, the pressures of assessing and collecting rentals brought these practices progressively into line with those of the rest of Scotland. These issues are discussed in more detail in Appendix D. In sixteenth-century Shetland, however, the principal unit of capacity was the 'can', of which there were 48 to the barrel, and which was equivalent to two pints. Robert Stewart (the illegitimate son of James V), who held the earldom of Orkney in feu from 1564, increased the size of the Shetland can by a third. The barrel itself was unchanged, and it therefore held 36 cans of the increased size. This change is first noted in the 1560s, when tenants objected to the increase by one-third of rentals which were assessed by the can.[75] The enlarged can was equivalent to two of the large Scots pints of 103·7 in³, and this increase presumably reflects

Stewart's self-interest in raising rentals within the framework provided by the 1563 Assize, but may also represent the end of a period of grace in which the older pint unit was still permitted.

We have assumed that the large pint would be introduced as an administrative vessel some time before it superseded the 1426 pint in trading, and we might expect this to have occurred in the course of the James IV Assize of the late 1490s. However, even in the restricted definition of the large pint as the vessel used to define the dry capacity measures, it is by no means clear whether it was yet available to the burghs in 1508: in this year the revised sizes of the operating dry measures were given in the Edinburgh council minutes in terms of the smaller 1426 pints.[76]

The likely explanation for this is that the large pint was not introduced in the course of this legislative change, but as a result of it. The evidence of the Inverkeithing gauge and the Leith water measure sizes indicates with reasonable certainty that the firlot was increased to a capacity of 20 pints in the James IV Assize; that is, it was increased by the same amount as in 1458, but not by the same proportion of the total. An increase of two pints would have been administratively simple to achieve, but such an increase would not match the growth of an eighth (or a little over two pints) that would have taken place in the customary measure by 1500. It is suggested that the change to a larger pint was opportunistic, and was only possible because the firlot had been set at exactly 20 pints. By taking a larger pint of four rather than three old chopins, 15 large pints would be equivalent to 20 of the smaller pints, but because the large pint would permit binary division, the allowances could from now on be calculated exactly. The awkward factor of three, which had its origins in the original definition of the gallon as containing twelve pounds of water, had finally been eliminated and had been replaced by a system which was more readily adapted for the types of calculations required.

The introduction of the large pint will therefore be set at about 1510 and it will be noticed that this coincides with the reduction of the ounce basis of the weight system. It is possible, therefore, that these episodes follow the transfer of the royal ordnance operations to the major new foundry at Edinburgh Castle in 1511. On this basis, the casting of the Stirling Jug is tentatively attributed to Robert Borthwick, the 'maister meltare of the Kingis gunnis'.[77]

THE GROWTH OF THE FIRLOT IN THE SIXTEENTH CENTURY

The increase in volume of the firlot in the James IV Assize set a new precedent in Scottish metrology. It can be argued that the enlargement of 1458 was designed to bring administrative procedures into line with entrenched practice, whereas it seems to have been different economic arguments that prompted the increase in the late 1490s. This in turn was followed by similar changes in 1555 (modified in another Assize of 1563) and in 1587, and led up to a final enlargement in 1618.

There are adequate surviving records of only two of these assizes, namely those of 1587 and 1618, yet even for these the information provided is incomplete. In both instances the major part of the assize was devoted to the dry measures, which were carefully defined in terms of their physical construction, dimensions and capacity in pints. Both sets of commissioners called for the current measures, found them larger than specified in the existing statutes, and legalised the larger size presented to them. Fig. 7.9 indicates the increasing capacity of the firlot over the sixteenth century, rising from 15 pints (that is, 15 large pints or 20 of the 1426 pints) in the 1490s to just over 19 pints in 1587 and 21¼ pints in 1618. On the assumption that similar forces prompted the increases at each assize period and that the growth factor was therefore likely to be roughly the same, we can interpolate a firlot size of about 16⅞ pints at the 1555/1563 Assize, with individual increases of about one-eighth at each assize.

We can already anticipate that the same mechanism that was seen to operate at the late fifteenth century assizes in Chapter 6 was also coming into play here. This involved two stages of allowance of one-sixteenth being applied to the basic legal firlot to generate a larger customary firlot for use in the burgh markets. The first stage gave an authorised trading firlot, and the second stage incorporated the further sixteenth allowance for the customary boll. From the mid-sixteenth century the simple trading firlot probably had no real existence in the market-place and the full allowance was applied straight away. However, the two stages of allowance were presumably considered administratively separate because one was partly eliminated at the 1618 Assize.

Only the parliamentary act appointing the commissioners for the 1555 Assize survives, and this is given in Appendix A.6; we therefore have no direct knowledge of the purpose of the commission or its findings. However, it is noticeable that the bishops of Orkney and Dunblane had prominent roles and so it may be that the bolstering of ecclesiastical revenues was a deciding factor: certainly this was a period when there was great activity in feuing off church lands and when the political influence of the church was perhaps at its peak.[78]

A similar Act in 1563 (reproduced in Appendix A.7) established largely the same group of commissioners (a notable addition being James Makgill of Rankeillor, the Deputy Clerk Register) to undertake the same work, and did so in terms of the unfulfilled remit of the 1555 commission: the previous act was said to have taken no effect and the commission members not to have performed their tasks:

And because it is understand [sic] be the Quenis grace and [the] thre Estatis in this present Parliament that the said act [of 1555] tuke nane effect nor the foirsaides persounis contenit thairin performit not the contentis of the said act saw [so] that the samin mycht haue takin full effect.[79]

The most pressing problem seems to have been the sanctioning of an enlarged series of grain measures, and there is no doubt that this at least was achieved, because the Edinburgh town council minutes were already speaking of the 'old measures' of grain in 1556.[80] It would be understandable if, in such very troubled times, the distribution and enforcement of new standards around the country had not been completed, and this would represent a financial obstacle for Mary's new administration given the urgency of maximising royal revenues.[81]

However, if we are correct in detecting that the 1563 commissioners made a change in the method of defining the pint, and therefore the dry capacity measures which depended on the pint, then it is clear that the commission must have concluded that there were some basic errors in the assumptions made by their predecessors. Possibly for this reason, any report that the 1555 commission may have presented to parliament is likely to have been suppressed from the first printed version of the collected parliamentary acts, which was published in 1566. Equally, it would scarcely be necessary to print a report of the 1563 commission if the error was considered to have been rectified.

At roughly the same time as these Scottish assizes were taking place, similar activity was being conducted in England. Elizabeth I's initial issue of standard weights was made in 1558, the first year of her reign. However, these were judged to be too heavy against the earlier standards, and problems with establishing the 'correct' basis for the avoirdupois weights led to the establishment of specialist juries in 1574 and 1582, issuing new standards in 1574 (condemned by the 1582 jury) and 1588.[82] The precise nature of the difficulties experienced during this period has remained unclear.

Some of these problems are likely to have been familiar to the Scottish official most closely involved with precision metrology, Sir Archibald Napier (1534-1608), General of the Scottish Mint, and father of the more famous John Napier, inventor of logarithms. Archibald Napier was appointed in 1576, and made at least one official visit to England (in 1580) before embarking on a major weights and measures assize in 1587.[83] Although he is recorded as one of eight commissioners for the assize, there seems little doubt that Napier was intimately involved in the operation and that the work devolved on Mint officials in Edinburgh. Their detailed report, which the Clerk Register was instructed to insert in the parliamentary register, contrasts sharply with the paucity of information about the work of the previous assize commissions. It also provides some insights into the problems created by the lack of earlier formal evidence, because the commissioners, under the guidance of the Lord Advocate, were specifically required to consider the existing parliamentary act relating to weights and measures.

The standard firlot produced for the 1587 assize commissioners was found to contain 19 pints 'and a jowcat', where the jowcat is presumed to be a sixteenth of the pint, or a gill. Whereas it is clear that this measurement was made with a large pint, the assize gave the capacity of 'the pint of Stirling' as that of 2 pound and 9 ounces (or 41 ounces) trois weight of clear water.[84] But this is simply a repetition of the definition in the 1426 Assize, which related to the smaller pint of 77·8 in³ and was given in old 480-grain ounces.[85] Confusingly, these 1426 ounces were also called trois ounces, but the term is merely descriptive of its use – as a troy ounce – and does not define its mass basis.

It is difficult to say why such an obviously incorrect size as 41 ounces should have been given in the 1587 Assize, except to suggest that the commissioners were more concerned with the problem of bridging the difficult (and embarrassing) gap in the legislative record authorising earlier increases in the size of the dry measures. They felt it necessary to invoke an 'errour of the prentair' to cover the increase in the firlot's capacity from its last definition at 18 pints in 1458 ('in an Act of James II') to the 19¹⁄₁₆ pints found by measurement in 1587.[86] This increase may seem small – and this was probably the intention – but in fact it was substantial because the 18 pints were the small 1426 pints, whereas the 19¹⁄₁₆ pints were the large pints. Although the commissioners confirmed the enlarged size of the firlot, they did not refer to the altered pint size, and therefore merely repeated the last legalised definition, even though the pint had increased in the intervening period by a third.

In spite of this, repeating the 1426 definition of the pint was a reasonable option. This small pint size had probably continued in use well into the second half of the sixteenth century: in the 1560s, and presumably also in the 1580s, it would be a commonplace that the administrative pint had one-and-a-third times the capacity of the earlier pint. It followed that the large pint contained 1⅓ x 41 = 54⅔ ounces; and it has already been noted that David Rowan's three pint sizes appear to have been adjusted on this basis. It is possible that the commissioners considered they had covered themselves by noting that the measurement had been performed with the 'stoup of Stirling', which we have suggested implied a large pint based on the earlier vessel. In this case it would presumably be a large pint of Paris ounces, as introduced at the 1563 Assize and which had a capacity of 102·3 in³.[87]

Two particular features of the 1587 Assize stand out. The first is a concern for engineering quality that is new and probably reflects Napier's thinking, influenced perhaps by the best English practice. The firlot standards were to be made in brass and were to be strengthened by having a triangular section cross-bar over the opening of the measure, connected by a pillar to the centre of the base. This structure became a feature of Scottish capacity measures. Because the dimensions of the firlot standard are also given in the assize, its capacity can be calculated as about 1,940 in³ and this enables it to be seen as a water-based standard (on the English pattern) and not a grain-based measure.[88] In spite of this, the firlot capacity definition was interpreted by the burghs as a conventional grain-based definition, and when the Linlithgow standard was returned to Edinburgh in 1618 for assessment at the next assize it was clearly a grain measure that was produced.

Second, two sets of principal reference standards were to be made in brass, in addition to the standards that were to be consigned to the care of the principal burghs, and these were to be held in the Register at Edinburgh Castle, and at Dumbarton Castle in the west. Not only does this reinforce the idea that the administration had retained reference standards at previous assizes, it also acknowledges the greater safety of dispersed standards. This policy may have been influenced by the likely loss of the Mint's own weight standards in the destruction of the Mint's buildings during the siege of Edinburgh Castle in 1573.[89]

Before these various capacities can be compared, allowance must be made for the slight differences in the volume of the pints in which the firlot's capacity was defined. It will be assumed, for example, that the firlot defined in 1555 will have been redefined in 1563 as containing the same number of pints, but using the pint based on the Paris ounce instead of the English troy ounce. Equally, the firlot of 1587 (or rather, the customary measure based on it) was re-measured in 1618 using the larger Stirling pint, and so will have been found to contain a slightly smaller number of pints.

In Fig. **7.9** the progression of sizes of the firlot has been tabulated more exactly. The first figure against the date for each change represents the legal firlot in (large) pints. The second is the permitted or trading firlot, namely the legal firlot enhanced by one-sixteenth. The third is the quarter of the trading boll with the charity included, which came to be the customary firlot in practical use in the burghs. It is proposed that it was this customary firlot which was presented to the assize legislators and formed the basis of the next legal firlot.

A number of assumptions have been made in generating Fig. **7.9** which shows our projected scheme for the evolution of increases in the firlot in the sixteenth century. The starting point is the group of three sizes for the firlot in about 1500, which was generated in Fig. **7.3**. This shows the customary firlot of 22½ small pints, or 16⅞ large pints, obtained from considering the Inverkeithing gauge and the Leith water measure figures. The suggested sequence of events is that this firlot was submitted to the commissioners for the next assize – in 1555 – and subsequently engrossing the allowances of sixteenths would allow it to rise to a customary firlot of 19¹⁄₁₆ pints, where

Fig. 7.9

Projected scheme of increases in the firlot in the sixteenth century. The horizontal lines indicate the allowances of a sixteenth to generate the customary firlot; the diagonal lines indicate the presentation of a standard to the commissioners for the following assize. The firlot sizes in solid boxes are those for which there is some separate evidence.

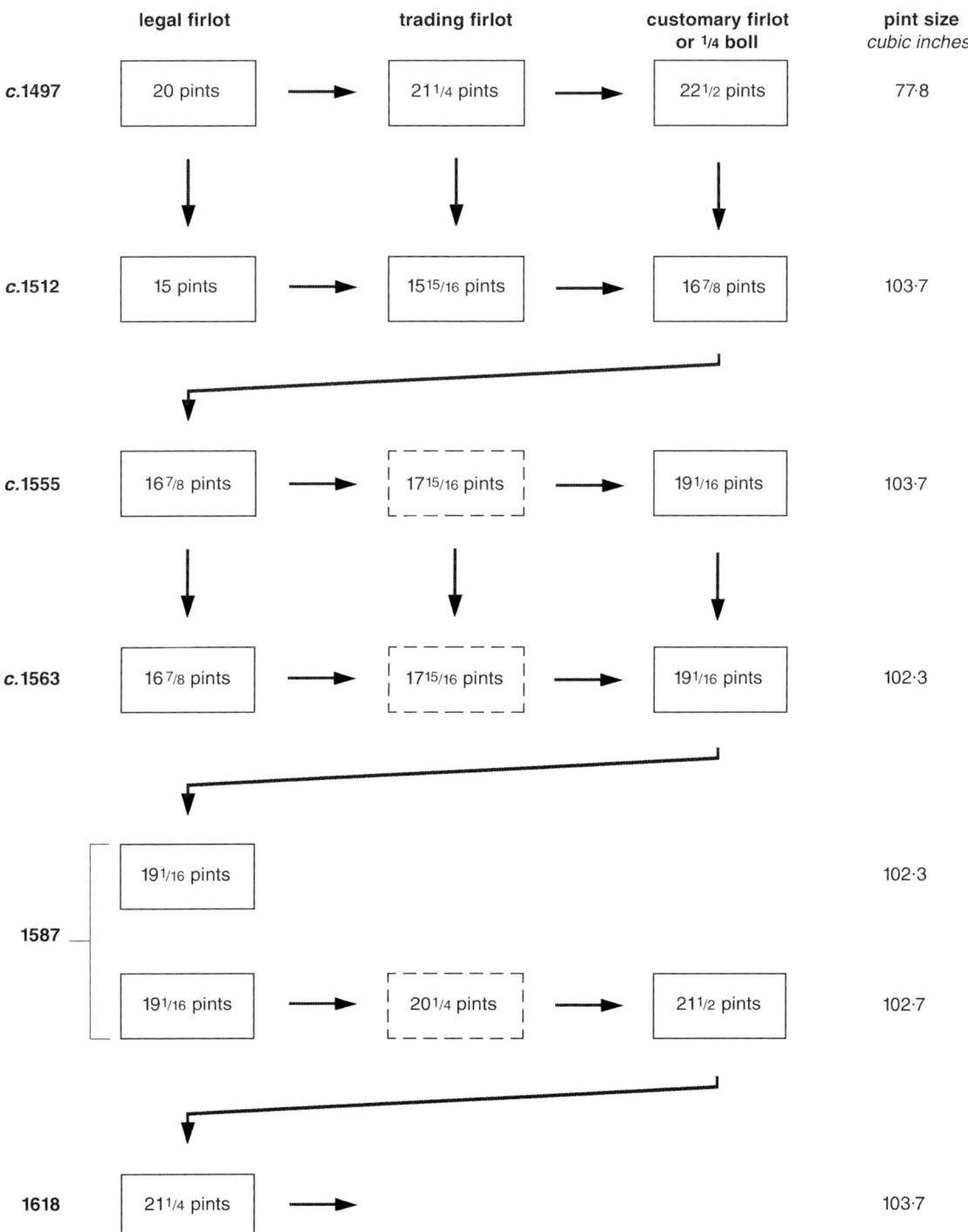

	legal firlot	trading firlot	customary firlot or ¼ boll	pint size cubic inches
*c.*1497	20 pints	21¼ pints	22½ pints	77·8
*c.*1512	15 pints	15¹⁵/₁₆ pints	16⁷/₈ pints	103·7
*c.*1555	16⁷/₈ pints	17¹⁵/₁₆ pints	19¹/₁₆ pints	103·7
*c.*1563	16⁷/₈ pints	17¹⁵/₁₆ pints	19¹/₁₆ pints	102·3
1587	19¹/₁₆ pints			102·3
	19¹/₁₆ pints	20¼ pints	21½ pints	102·7
1618	21¼ pints			103·7

these sizes are given only to the nearest sixteenth of a pint. By this time, it is suggested that the intermediate step of a permitted trading firlot with one-sixteenth allowance had ceased to have any real market-place relevance, and both stages of allowance were taken from the outset. Settling on the smaller pint of 102·3 in³ in 1563 led to the procedure being repeated, resulting in a smaller customary firlot of 19¹⁄₁₆ pints of 102·3 in³.

The procedure at the 1587 Assize is far from clear, in spite of the apparent wealth of detail in the parliamentary record. This mainly arises from un-certainty about the size of pint used. It is assumed, however, that the initial measurement of the firlot which established its size as '19 pints and 1 jowcat' was made in the pint of the 1563 Assize. It was stated that this was 'aug-mented' to 19 pints and 2 jowcats, or 19⅛ pints. This revised size was perhaps still in 1563 pints, because it equates to almost exactly 19¹⁄₁₆ of the pints of 102·7 in³, which we argue was the variety selected at the 1587 Assize.[90]

Adding the allowances or charities to a firlot of 19¹⁄₁₆ pints generates a customary firlot of 21½ pints. The commissioners at the 1618 Assize reverted to the use of the Stirling pint of 103·7 in³, and this was the pint size used in Chapter **6**, in assessing the effect of differential grain packing when replicating the 1618 measurement. When the customary firlot of 21½ pints of 102·7 in³ was measured with the pint of 103·7 in³, its volume was found to be 21¼ pints.[91]

Again, the enlarged volume was accepted, even though there was no statutory authority. The attitude of the 1618 commissioners to the inadequacy of the legislative record is revealing. Although accepting that the last assize definition of the firlot had been one of 19¹⁄₁₆ pints in 1587, they insisted on transferring responsibility for the subsequent increase onto the officials from Linlithgow by getting them to swear that it was the original and legal dry capacity measures that had been brought before them. In practice, it is very hard to believe that either the 1587 or 1618 commissioners were unaware of the nature of the previous changes: the wording of their reports merely incorporates an element of self-protection, as well as conventional reassurances about the stability of the system.

But the difficulty with the 1587 Assize rests not only with the pint size. The Commissioners are said to have thought it '… expedient to have double sets of the aforesaid weights, metts and measures made of brass', with the implication that these are the sets to be deposited in Dumbarton and Edinburgh castles; but then the text goes on to describe the nailing of the crossbar and ringing the firlot with iron, surely referring to a wooden firlot since one in brass would have no need of being ringed or nailed. A cooper is invoked later in the text and brass firlots are referred to a second time, to be made by the burgh of Linlithgow. If brass firlots were made, none have survived. It is presumed that wooden firlots are the reality being described, especially when the 1587 Act speaks of the thickness of the boards of the firlot

being three-quarter of an inch: 'in thiknes of bayth the burdis ane insch and
ane half.'[92]

Although it must be stressed that this development sequence is conjectural, it does have the advantage of accommodating the available information about the sizes of the firlots and it preserves the numerical multipliers
in both 1563 and 1587 when adjustments were made to the pint size for technical reasons. These adjustments match changes we can detect in existing
liquid capacity standards and changes in the weights system that will be
examined in Chapter **8**.

THE FIRLOT AFTER 1618

The 1618 Assize marked the start of a new period in which some control of the
growth of the dry measures was at last achieved; but, unfortunately, it is also
a period in which our direct knowledge of the units is very slight. The trigger
for legislative action in 1618 appears to have been a dispute over the implementation of the 1587 Assize between on the one hand the justices of the
peace, responsible for the administration of justice in the sheriffdoms since
their very recent creation in imitation of English practice by James VI, and on
the other hand the vested interests of the royal burghs. In trying to promote
discussion between the justices of the peace, the commissioners of the burghs
and the Privy Council, the views of the justices of the peace were sought in
late 1612.

The most indignant surviving reply was from Edinburgh, where the
justices of the peace recalled that the earlier assize had been supposed to
create a universal measure, but this purpose had been obstructed by the
burghs, causing great problems between masters and tenants and in other
contracts where dues were paid in kind. They were clear that charity on
measures, 'quhilk is growin to ane extraordinar abuse', should be abolished:
not only were charities applied 'without ony warrand of law', the burghs
'have not the libertie nor oversicht to eik or alter this measur' and recourse
should be had to the reference measure ('the just double') which had been
retained in the Register in Edinburgh Castle after the last assize.[93] Both the
Edinburgh and Perth justices of the peace noted that the burghs wanted to
secure a new firlot rather than put the last one into effect, and the justices of
Linlithgow sheriffdom concluded that a new firlot of 21 pints was necessary.[94]
In the event, the 1618 Assize commissioners measured (and accepted) the
Linlithgow burgh firlot standard at 21¼ pints.

This growing concern to control charities was reflected in the 1618 Assize:
there was now more specific insistence that the newly-defined firlots were to
be those used in the markets; a new injunction was that four firlots were to
make the boll 'allenerlie' (presumably here in the legal sense of solely or

exclusively, in other words without augmentation); and demands were made of officials in five sheriffdoms in the borders and south-west, where different local measurement conventions prevailed, that they were to ensure the use of the new firlots.[95] To a limited extent this seems to have been successful, but we have to look to eighteenth-century sources for evidence of its effects. The result appears to have been a separation of the two stages of allowance that applied at the earlier assizes, with the 'peck to the boll' applied only at the level of the boll and only when the measurement was specifically to be 'charitied'.[96] However, the divergent practice of the five uncooperative sheriffdoms was not successfully brought into line.

There was no subsequent general assize of weights and measures in the seventeenth century at which the dry measures could have been altered again, and from which the size of any allowance that might have been incorporated for the burghs at 1618 could be inferred. All dry measures were, in theory at least, replaced at the Union in 1707; however, in practice, Scots measure was such an integral part of so many practices and legal procedures that it remained in almost universal use throughout the eighteenth century, although now operating effectively outside central legal control.

Although our knowledge of the way the dry capacity measures were used in the seventeenth century is so limited, we can make some inferences from the more copious details available from the second half of the eighteenth century. In particular, a group of mathematicians and practical scientists associated with the Philosophical Society of Edinburgh attempted in the 1750s to re-establish a basis for the old system of measures using scientific precision. Their work will be discussed and set in context in Chapter **9**; but it has already been established in Chapter **6** that the firlot volumes that they calculated were too large, because they made what seemed a natural assumption that the specified 21¼ and 31 pints of the wheat and barley firlots of 1618 were measured in water. In fact the volume and dimensioned information about the dry measures in the assize is only consistent with the traditional use of grain as the appropriate measuring medium for a dry capacity standard.[97] Their conclusions were that the correct volumes of the 1618 firlots should be 2,197·3 and 3,205·6 in³, and the impressive prototype reference standards of these sizes made by them for the justices of the peace of Stirlingshire in 1754 still survive in Stirling and are described in the Inventory as Items **216** and **217**.

It was these values that entered the literature and which were promoted as the standards for dry measure in the tables published by the advocate John Swinton in 1779 as part of a long-running campaign for the legal reform of weights and measures discussed in Chapter **9**.[98] Swinton also published details of the great variety of measure sizes used in counties across Scotland. His purpose was to demonstrate great diversity, particularly in the dry measure size, and in this he certainly succeeded. But, if we discount, for

30 Sir James Balfour Paul, *The Scots Peerage*, 9 volumes (Edinburgh, 1907-14), IV, 82.

31 Nicholson, op. cit. (7), 313.

32 The near-contemporary manuscript is National Archives of Scotland (NAS), PA.5/6/1, and the relevant act is printed in Robertson, op. cit. (28), 42. Sir James Balfour's addition of a quart standard to the pint provided by Forrester appears to have arisen from misreading the abbreviation 'q^{lk}' (quhilk, which) as 'qt' (quart), but there is no obvious basis to his addition of quarts to the distributed pints and firlots: *APS*, II, 50; Peter G. B. McNeill (ed), *The Practicks of Sir James Balfour of Pittendreich*, Stair Society nos. 21 and 22, 2 volumes (Edinburgh, 1962-3), I, 89-90. Burrell, therefore, was incorrect in drawing attention to this as the first use of the term 'quart' in the statutes: Burrell, op. cit. (2), 57.

33 McNeill, op. cit. (32), 89; *APS*, II, 50; Robertson, op. cit. (28), 42.

34 *APS*, II, 254: Act 42, 1504.

35 The singular is given in Balfour's versions, but Thomson in *APS* changed this to the plural 'mesouris': *APS*, II, 246.

36 Michael Lynch, *Scotland: A New History* (Edinburgh, 1991), 160; Norman Macdougall, '"The greatest scheip that ewer saillit in Ingland or France": James IV's "Great Michael"', in Norman Macdougall (ed), *Scotland and War* (Edinburgh, 1991), 36-60; and Macdougall, op. cit. (8), chapter 9, 'Royal Obsession: The Navy', 223-46.

37 D. H. Caldwell, 'Royal Patronage of Arms and Armour Making in Fifteenth and Sixteenth-Century Scotland', in D. H. Caldwell (ed), *Scottish Weapons and Fortifications 1100-1800* (Edinburgh, 1981), 73-93, p. 75.

38 Ibid., 76.

39 *Treasurer's Accounts,* IV, 261: '1511: Item, to Robin of Borthuik, gunnar, for his 3ule leveray goune, tua stekis tanne chamlot; price xj li [£20] Item, to lyne the samyn … price x li [£10]. Item, to bordoure and begary the said goune … summa v li xij s. vj d [£5 12s 6d]. Item, for his doublet, iij elnis tanne welvot … summa viij li v s [£8 5s]'; ibid., 442: '1512-13: Item, the x day of Januar, to Robert Borthwik, gunnar, maister meltare of the Kingis gunnis, for him self and vj men his servandis [servants], of the quhilk he hes for himself vij li. x s. [£7 10s 0d] in the moneth, and for v servandis ilk ane iiij li. iiij s. [£4 4s 0d] and for ane uthir xlij s. [42s], for the moneth of December; summa xxx li xij s. [£30-12-0.]'. See also C. E. Whitelaw, *Scottish Arms Makers* (London, 1977), 136. This work was edited by Sarah Barter from Whitelaw's 1939 manuscript.

40 Whitelaw, op. cit. (39), 137-8.

41 *Treasurer's Accounts*, VII, 199; VIII, 105; X, 355, 402.

42 Ibid., VII, 343-50.

43 David Rowan had a pension of £4 per month for life, to come from the Edinburgh customs, his father Peris having died in August 1545. David was then described as 'Engineer to the Queen': *ER*, XVIII, 50: 1543. In 1550 his pension was increased to £5 per month. He is last mentioned on 1 November 1593 when he was paid for 7 months at £5.

44 This is a response to instructions from the Queen Dowager, Mary of Guise-Lorraine, and the Privy Council, to fix bread and ale prices: *Edinburgh Extracts*, II, 238: 24 February 1555/6.

45 The subsequent use of the royal arms on pint standards apparently manufactured for the burgh of Stirling after the 1618 Assize (see Items **109-126** in the Inventory) probably only reflects a copying of the arms on the Stirling Jug.

46 The name is given as 'm:hanes coqhren'.

47 *APS*, II, 349: Act 35, 1535.

48 *RCRB*, I, 504.

49 J. Colson, *The Guildry of Edinburgh* (Edinburgh, 1887), 30-1. The Convention of Royal Burghs continued till 1973 when it was replaced by the Convention of Scottish Local Authorities, in accordance with 22 Elizabeth II, Local Government (Scotland) Act, c.65.

50 *RCRB*, I, 2; *Dictionary of the Older Scottish Tongue [DOST]*, 12 volumes (Chicago, Aberdeen and Oxford, 1931-2002), s.v. *mare*.

51 *RCRB*, I, 2.

52 For example, Sir George Mackenzie advanced the spurious argument that the 'standard of stone weight remains at Lanark because the chief commodity of old that was weighed by the stone weight was wool, which was paid in to the King at Lanark and was therefore called *Lanae-Arca*': G. Mackenzie, *Observations on the Acts of Parliament* (Edinburgh, 1687), 118. The correct derivation is Cumbric, from the Old Welsh *llanerch*, a clear space or glade: W. J. Watson, *Celtic Place-Names of Scotland* (Edinburgh, 1926), 356. We are grateful to Ian Fraser, formerly of the Place-Name Survey, School of Scottish Studies, for his advice.

53 *APS*, III, 522: Act 136, 1587.

54 This will be discussed in the section on trone weight in Chapter **8**.

55 Nineteenth-century sources frequently refer to the figure on the Stirling pint as an ape. See the description in the Inventory, Item **109**. For example, Robert Chambers (ed), *A Biographical Dictionary of Eminent Scotsmen,* 9 volumes (Edinburgh, 1854-5), I, 394: 'Near the bottom another shield, and an ape, passant gardant, with

the letter S below, supposed to have been intended as the arms of Stirling.'

56 The lamb was recorded on the smaller burgh seal by Stevenson: J. H. Stevenson and M. Wood, *Scottish Heraldic Seals … Public Seals* (Edinburgh, 1940), 81.

57 We are grateful to Charles Burnett, Ross Herald, for his guidance on this question.

58 Burrell, op. cit. (2), 49, 52.

59 Doubts were first expressed to us about the putative date of the Stirling Jug by N. J. Mayhew and E. Gemmill: personal communication.

60 The issue of deciding on the appropriate figure to use for the density of water used for gauging is discussed in the entry for Item **108** in the Inventory.

61 A. Anderson and T. Bruce, *Report on the Stirling-shire Weights and Measures* ([Stirling], 1827). Anderson used the density of 252·458 grains per in³ specified in the 1824 Act.

62 George Buchanan, *Tables for Converting the Weights and Measures hitherto in Use in Great Britain …* (Edinburgh, 1829), 200-1.

63 James Gray, 'Of the Measures of Scotland, compared with those of England', in *Essays and Observations, Physical and Literary*, I (1754), 200-4.

64 The Linlithgow copy, cast at the same time, does not appear to have been fully adjusted to standard: see entry for Item **114** in the Inventory.

65 41 ounces of water multiplied by 480 grains to the ounce, divided by the accepted modern density value for water density of 252·89 grains/in³.

66 Gemmill and Mayhew, op. cit. (14), 394, citing *ER*, VI, 501-2.

67 Alison Hanham, 'A Medieval Scots Merchant's Handbook', in *Scottish Historical Review*, 50 (1971), 107-20; A. Stephen Wilson, *The Botany of Three Historical Records: Pharaoh's Dream, the Sower, and the King's Measure* (Edinburgh, 1878), 113, writes: 'The stoup of Stirling is the oldest standard of length, capacity and weight, now existing in Great Britain.' R. W. Cochran-Patrick, *Mediaeval Scotland* (Glasgow, 1892), has a frontispiece entitled 'The Stirling Standard Stoup in the Custody of the Burgh of Stirling', and states (p. 164) that 'in 1457 the firlot was ordered to contain 18 pints of the Stirling Stoup'. In fact, the statute uses the word 'pint', and 'stoup' never appears in official documents. The Gaelic *stòp* is still used to describe a measure of liquids in the Western Isles: we are grateful to Dr John MacInnes for this information.

68 [John Swinton], *A Proposal for Uniformity of Weights and Measures in Scotland* (Edinburgh, 1779), 29.

69 J. Hill Burton, *et al.* (eds), *The Register of the Privy Council of Scotland [RPCS]*, three series

(Edinburgh, 1877-1970), second series, VIII, 333: 6 January 1613. Here the Council minute refers to the diversity of trois weight, French, Flemish (Dutch), and English with no mention of alteration of tron weight in the record: 'whilk is the ordinair and proper weght of the kingdome', but not enough J.P.s were present at the meeting to resolve the matter. See also Chapter **8**.

70 David Gregory, *A Treatise of Practical Geometry … by the Late Dr David Gregory* (Edinburgh, 1745), 110.

71 Ibid., 112. See also the discussion in Chapter **8**.

72 This 'tappit hen', or pewter tankard, is now in the Shakespeare's Birthplace Trust, Stratford-upon-Avon, where it has the inventory number 1996-44/860/472. Formerly in the collection of Alex Neish, who kindly allowed us to measure it. He had acquired it from the collection of Richard Mundey, who in turn had acquired it at Sotheby's, 21 May 1971, Lot 105a. George Dalgleish of the National Museums of Scotland has provisionally identified the maker as the pewter smith, Archibald Napier of Edinburgh.

73 Buchanan, op. cit. (62), 232-3.

74 Ibid., 250-1. Basic pints of about 105-6½ in³ may have become established as a result of the distribution of poorly-adjusted standards in 1622: see Item **116** in the Inventory.

75 A. W. Johnston, 'Notes on the Fiscal Antiquities of Orkney and Shetland', in *Old-lore Miscellany of Orkney, Shetland, Caithness and Sutherland*, 9 (1933), 53-72; 133-9 and 234-52, esp. pp. 247-8. See also Appendix **D**.

76 *Edinburgh Extracts*, I, 118: 29 November 1508.

77 *Treasurer's Accounts*, IV, 442.

78 Lynch, op. cit. (36), 181-3.

79 *APS*, II, 540: Act 14, 1563.

80 The entry for 24 February 1555/6 reads in part: 'nyne furllettis grundin malt of the auld mesour is to be sauld for iiij li [£4]': *Edinburgh Extracts*, II, 237.

81 Lynch, op. cit. (36), 210.

82 The work of the Elizabethan juries is discussed in H. W. Chisholm, *Seventh Annual Report of the Warden of the Standards* (London, 1873): R. D. Connor, *The Weights and Measures of England* (London, 1987), 240-2.

83 *Treasurer's Accounts*, XII, 311.

84 *APS*, III, 521: Act 136, 1587.

85 Ibid., II, 12: Act 22, 11 March 1426.

86 Ibid., III, 521: Act 136, 1587. See Appendix **A.8**.

87 Ibid., II, 540-1: Act 14, 1563. The size is represented by the smallest of the three Rowan pints.

88 The firlot's diameter was given as 18⅛ inches and the depth as 7½ inches, implying a volume of

1,944 in³. The intrinsic uncertainty in the match between this volume and the ideal volume, arising from the precision with which the measurements were quoted is estimated as ± 20 in³. The volume occupied by the triangular bar across the opening of the measure and the pillar connecting this with the base of the measure is 8 in³, giving an internal capacity of 1,936 in³, which we can express as 1,940 ± 20 in³. Taking 19 1/16 pints of 102·7 in³ (the relevant ounce here being 474·4 gr) gives 1,958 in³, which represents an increase of only 1% over the figure from dimensions instead of the 4 to 5% expected had grain been used.

89 Ian Stewart, 'Scottish Mints', in R. A. G. Carson (ed), *Mints Dies and Currency* (London, 1971), 165-290: 'In 1572 Edinburgh Castle was under siege, and the mint destroyed.'

90 A possible alternative to this is that the additional sixteenth of a pint was intended to compensate for about the same volume being occupied by the newly-required crossbar and pillar (about 8 in³) – but this seems unlikely because the assize specified that it was the enlarged volume that was to be the capacity unit.

91 21·25 × 103·7 = 21·46 × 102·7.

92 *APS*, III, 522: Act 136, 1587; see Appendix A.8. It is possible, therefore, that wooden standards were sent to Dumbarton and Edinburgh castles. Standards of some sort may have been sent for secure custody there, because this provision was repeated in at the 1618 Assize and there is a confirmation of the deposit in September 1619: see Item **113** in the Inventory; and also Chapter **2**, ref. 90.

93 *RCPS*, second series, VIII, 335-6.

94 Ibid., 334, 336, 338.

95 *APS*, IV, 585-9, esp. pp. 587, 588. The text is printed in Appendix A.**5**.

96 The recording of the charity level after 1618 is seen, for example, in the price of produce from the Scottish estates of Charles, Prince of Wales, where bere barley and oatmeal from the 1619 crop, at £3 per boll, have been recorded as 88 chalders 13 bolls (or 88·8 chalders) if charitied, but also at the numerically larger 96 chalders 11 bolls 3 firlots (or 94·7 chalders) if uncharitied: NAS E42/1, accounts of the earldom of Ross, lordships of Ardneanach and Ettrick Forrest. Strictly, the additional sixteenth allowance would generate 94·4 chalders uncharitied measure from 88·8 chalders charitied measure. We are grateful to Dr Julian Goodare for drawing this reference to our attention.

97 This is discussed further in A. D. C. Simpson, 'Grain packing in early standard capacity measures: evidence from the Scottish dry capacity standards', in *Annals of Science*, 49 (1992), 337-50.

98 Swinton, op. cit. (68).

99 Ibid., 75, 100.

100 Ibid., 75-6.

101 Ibid., 91, 100, 110.

102 Ibid., 84-5, 80, 97-8.

103 Ibid., 82-3.

104 Ibid., 122-3.

105 Simpson, op. cit. (97).

106 Swinton, op. cit. (68), 75.

107 Ibid., 123.

108 Ibid., 110-11.

109 *APS*, IV, 587.

110 This assumes a grain compaction similar to that found in the wheat firlot.

111 Buchanan, op. cit. (62), 40.

8

'ONE JUST WEIGHT
THROUGHOUT
THE KINGDOM'

Opposite:
Detail of Fig. **8.8**.

A ROYAL COMMISSION WAS SET UP IN 1617 TO REVISE THE WEIGHTS AND measures of Scotland – part of a programme of reforms to be administered by justices of the peace, recently introduced by James VI and I on the English pattern.[1] In its report of 1618 the commission required that

> there shall bee only one Just Weght through all the parts of this Kingdome which shall Universallie serue all his Majesties Lieges (by the which and no other) they shall buy and sell all and whatsomever Wares [are] accustomed to be boght and sauld by Weght als weell Foraine as Cuntrie-Wares; in all tyme hereafter …[2]

This is an injunction, apparently echoing those of earlier legislators, that there should be a uniform standard of weight used across the country, and specifically that no one should use different weights for buying and for selling. But this clause goes much further: it seeks to enforce a single type of weight for all classes of goods, applicable equally to 'foreign' and 'country-wares'.

Foreign goods were weighed using the merchant pound, since it was the merchant burgesses in the royal burghs who imported them or purchased them from Continental merchants. However, transactions in 'country' goods – meaning, for example, meat, cheese and other farm produce sold at the burgh markets and fairs, as well as raw materials such as lead – were conducted in a heavier weight series known as 'trone' weight. This is invariably spelt 'trone' in early records, although the form that has survived to the present day in place-names is 'tron', which may perhaps reflect the original pronunciation of the word.

We should be in no doubt that trone weight was the dominant type of weight used in the burgh markets: indeed the very name it had acquired by the mid-sixteenth century suggests that this was the weight system encountered at the burgh weigh-beam or 'trone'. In 1613 the Privy Council described trone weight for the benefit of the newly-created justices of the peace as 'the ordinair and proper weght of the kingdome'.[3] In spite of this, trone weight is not mentioned in the statutes – except for the surprising announcement of its formal abolition five years later in the Assize of 1618.[4] The ban was ineffective: trone weight was so deeply entrenched in market practice that it survived well into the nineteenth century.

The fact that trone weight does not figure more prominently in the statutes is not unexpected. We have already concluded, from examining the 1426 Assize in Chapter 5, that assize legislation was couched in somewhat stylised definitions given in terms of troy weight. The assizes, therefore, tend not to give us explicit information about the weights actually used in the burghs for trading. This has not been adequately appreciated by earlier commentators, who have assumed that the assize definitions represented the trading units. The correct interpretation is that there were two or three series of weights operating in parallel, but applying to different types of transaction. The fact that there is no mention of trone weight in 1426 is merely a reflection of the limited scope of the particular legislation that is available to us: in practice, merchant weights and market weights must have been used in tandem for the whole of the period that concerns us here.

Neither the existence nor the operation of this merchant weight was apparent from the 1426 Assize. Had it not been for the glimpse of the 'Scots' weights afforded by one of the preliminary parliamentary acts to the Assize of 1426 (discussed in Chapter 5 and reproduced in Appendix A.2), we might have come to very different conclusions about the practical weights of merchant trade. Similarly, another of the preliminary acts of 1426 demonstrated that at least one of type of customary measure which was not mentioned in the assize (the so-called 'water measure') was officially recognised and in long-standing use in the royal burghs. A much later assize, in 1587, reveals that water measure still had a legal existence, although again its use was not specified. It is clear, therefore, that the assizes do not provide a complete picture of practice. In addition, we learn from another preliminary act of 1426 that physical standards were distributed to burghs; and we can deduce from the figures given in the 1426 Assize itself that the standards of the large capacity vessels were 'trading' measures which incorporated customary allowances. Although this preliminary act does not speak of distributing weight standards, the permitted denominations which were listed at the time were certainly of actual trading weights and not troy weights. Indeed, the oldest standard weights to survive are also practical weights – trone standards of the sixteenth century.[5]

The purpose of the present chapter is to look beyond the restrictions of the assize definitions to examine the operation of the separate trading weights for 'foreign' and 'country' goods and the relationship between them. These weights were not constant, and a number of small changes in the ounce size can be detected in the sixteenth and early seventeenth centuries.

The most pronounced was the move from the English troy ounce to a French ounce shortly after 1500, and this has already been discussed in Chapter 7.

At least some of these changes also affected the official troy weights, but just how many is uncertain: it is a surprising fact that we have a more secure knowledge of the trading weights than we do of Scotland's troy weights. This is in marked contrast to the situation in England where the English troy ounce of 480 grains had a very prominent role and retained a constant fixed value through the whole period from its introduction, probably in the late fourteenth century. (Nonetheless, a small discontinuity in the English weight series emerged in the mid-eighteenth century which has already allowed us to derive an improved value for earlier English troy weight in Chapter 4.)

In Scotland the position is complicated by two main factors. First, for the latter part of the period under review, the mass of the Scottish troy weight was clearly very similar to (if not identical to) that of the official merchant weight. This often resulted in the merchant weight (whether official or unofficial) being described as 'troy', making it difficult to identify references to the actual troy weight. For example, the unified weight introduced at the 1618 Assize was described as 'troys' and was apparently intended for all categories of weighing – but was it also intended for weighing precious metals (the category of weighing still associated with troy weight)? It is possible that by this time the term 'troy' was being used merely to describe the weight appropriate for fine or precious goods, as opposed to that for heavy or 'trone' goods. This probably reflects a similar broad distinction which had clearly emerged by the mid-sixteenth century in England between troy and avoirdupois weight.

The second area of confusion is the difficulty in defining the nature of the official Scots troy ounce, even as late as 1600. There are two possible interpretations, depending on whether the official troy ounce is taken as the Mint's weight basis for the coinage or as the marginally lighter ounce used to control the purchase of silver bullion by the Mint. The assumption we make here is that the official weight should be defined as the former, although it is acknowledged that the Mint's bullion rate must have been widely used for troy transactions. The difference between the two ounces is small, reflecting the range of permitted weight of minted coin (the 'remedy' of the coinage, discussed in Chapter 4), but it is nonetheless significant. However, the difference is also small enough to make it difficult to say with any great confidence which ounce is being referred to in individual cases, although taken together the evidence points in one direction.

Part of the reason for this confusion between types of ounce must be the dependence of Scotland's economy on those of her dominant trading partners. The focus of this trade changed over the centuries and was subject to political as well as economic constraints, leading to domestic tensions between the use of English, Low Countries and French weight standards. There can be little doubt that the importance of the trade with the Low Countries resulted in the widespread use of Flemish merchant weights in the burghs in the sixteenth century, and then at a later stage to the adoption of 'Dutch' (or Amsterdam) weight (normally understood as about 476·8 grains to the ounce). The Flemish troy weight, known in the Low Countries as 'troois' weight, was ostensibly to the French standard (472½ grains), but was in fact slightly heavier at about 474½ grains to the ounce when in use in Scotland. Its name amongst the Scots merchants was 'trose' weight, and sometimes *fleur-de-lys* weight – from the mark stamped on sets of weights by the Nuremberg weight makers apparently to indicate that they were to a 'French' standard.

The problems posed by such a variety of weight types were tackled with varying degrees of success at the assizes of the sixteenth and early seventeenth centuries. This is perhaps best appreciated from comments made in 1613, as pressure built up for the reforms that were to come in the 1618 Assize, when the Privy Council complained that the 1587 weights and measures commissioners had 'proceidit upoun the diversitie of the trois [here 'merchant'] weght allanerlie [only], being than [then] thrie in nomber, Franshe, Flemis and Englische', but had not addressed the question of trone weight.[6] We will begin by establishing the nature of the main changes in the weight series during the fifteenth and sixteenth centuries, before addressing the trone, merchant and troy weights individually in more detail.

THE PRINCIPAL CHANGES IN THE WEIGHTS

Two hundred years elapsed between the accession of the young James I of Scotland, and the eventual union of the crowns of Scotland and England in 1603 when his descendant James VI of Scotland also became James I of England. But although the monarchies of the two countries were united in a single sovereign, the administrations were separate, and another century was to pass before the incorporating union of 1707 abolished the independent parliament in Edinburgh. Between the weights and measures assize of 1426, and the time when James VI left in triumph for London, the weights of Scotland's markets evolved from an essentially medieval system to a modern structure which remained much the same up to (and indeed beyond) the Union of 1707.

In 1426 the units of merchant weight were a stone of 16 pounds, and a pound of 16 ounces. Each ounce was the 450-grain monetary ounce, although the weight system was defined in terms of the 480-grain troy ounce, with the mass of the merchant pound equivalent to that of 15 troy ounces. By 1600, the situation – shown diagrammatically in Fig. **8.1** – was very different.

The merchant weight, which had been for a time based on the French ounce of 472½ (English troy) grains, was now based on a Flemish ounce of about 474½ grains. As in 1426, the merchant pound of 1600 comprised 16 ounces and the stone 16 pounds. The difference was that the ounce of the merchant weight had risen substantially from the 450 grains of 1426, so that the ounce, pound and stone were now all about five per cent larger. It is the nature of this rise that we must now address.

The other apparent difference was that in 1600 there were certainly two distinct types of weight in use in the burghs: the merchant weight and the heavier market or trone weight. The earliest mention of trone as a weight type in official records is as late as 1565.[7] However, we can readily deduce that this type of weight was in operation before 1500 and we will argue that it has much earlier origins.[8] The fact that there is no mention of trone weight in 1426 is merely a reflection of the limited scope of the particular legislation that is available to us: in practice, trone weight must have been in use over the whole period.[9]

Although the mass of the trone or market weight is not given in the statutes, there are enough official and semi-official definitions in the seventeenth century for it to be quite clear that the market stone weighed 20 merchant pounds, and was therefore 1¼ merchant stones; and the market pound was similarly 20 merchant ounces or 1¼ merchant pounds (Fig. **8.1**).[10] We will see that this relationship held good in 1553 and survived changes in the sizes of the market and merchant units. There is every reason to believe, therefore, that this numerical relationship between the market and merchant weight was consciously preserved by officials over an extended period. There are apparent similarities between this and the parallel numerical relationship maintained between the water and land measures, although in practice this may turn out to be no more than coincidental.

To bridge the gap between 1426 and 1600, the introduction of a number of distinct changes have to be detected. Thus, if we consider the merchant pound of 1426 as equivalent to 15 troy ounces, the size of the pound has to be increased to 16 ounces. Similarly, the stone in 1426 was equivalent to 15 troy pounds, and this has to become 16 pounds by 1600. In addition, the ounce, pound and stone must all be reduced slightly when the French and Flemish troy ounces replaced the English troy ounce of 480 grains. The conventional view of these changes is based on a faulty understanding of the 1426 Assize – namely that a 16-ounce (troy) pound actually replaced the earlier pound of 15 ounces, and that (following Thomas Thomson's incorrect version of the 1426 Assize) the stone was raised to 16 of these larger pounds. In fact the changes are more complex and are spread over an extended period.

An understanding of the progressive nature of these changes has been obscured by a growing (but mistaken) acceptance that the French ounce was used as the weight basis of the Scottish coinage from at least the 1430s. This

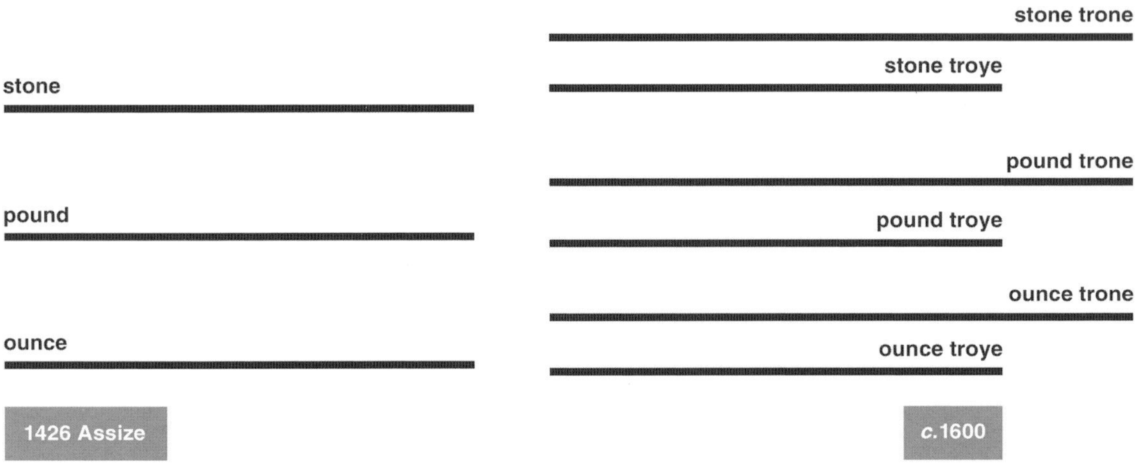

stone		stone trone
		stone troye
pound		pound trone
		pound troye
ounce		ounce trone
		ounce troye

1426 Assize   *c.*1600

Fig 8.1

The masses of the trading weights in 1426 compared with those of 1600. The height of the horizontal lines represent (on a logarithmic scale) the mass of each weight unit.

has arisen because Edward Burns in his authoritative three-volume *Coinage of Scotland* described the pound of the Scottish Mint as 'conforme to the French wecht' in a discussion of Mint accounts of 1436-8 where the coinage pound is first identified as containing 16 ounces: in fact, the source of this quotation is a 1578 edict of the Convention of Royal Burghs of Scotland.[11] There are in any case problems in trying to extrapolate from the specialised weights of the Mint into a more general situation, and so we cannot be sure, for example, whether a general troy weight (for bullion weighing) might be affected at the same time as a coinage weight. Some have tried to claim that the troy ounce was transferred to a French basis in 1426, and have used this to signal a general change in the weights at the 1426 Assize.[12] From such a perspective, all the changes that the weights had to undergo before 1600 (excepting the final adjustment to a Flemish weight standard) would have been introduced at the 1426 Assize: we shall see that this was not so.

The statutes are totally silent about the introduction of the French ounce size, and it is not known accurately when, or at precisely what level, French weight became the official troy weight of the kingdom. When it came, it involved a small reduction in the ounce size from the 480-grain trois ounce to the mass of the French ounce of 472½ grains. This change was not as drastic as might be supposed, and it represented a reduction of between one and two per cent. To some extent it may reflect James VI's inclinations from about 1510 to realign his foreign policy towards France.[13] It also represented a practical recognition that the variety of French ounce used in the Burgundian Netherlands had already become the commonly used weight of the Scottish importing merchants, although the Flemish troois ounce was already operating in Scotland at the level of about 474½ grains in the 1520s. The reduction had apparently been made by May 1511, the date at which Edinburgh's burgh council upheld a complaint by the collectors of the petty customs (who had purchased the tack, or lease, from the council) that they were losing revenue because the town's official weights at the Over Tron had not been reduced to comply with the law, and so the quantities being weighed (on which duty was charged) appeared too light. With a heavier weight there were fewer stones in the consignment on which to charge duty at the agreed rates. The council had previously agreed that 'thair commoun wechtis wer mynnist [or 'reduced'], becaus thay war to lairge agane the law'.[14] Here the word 'minis' was being used in the sense of lessening, diminishing or reducing by a small amount, which we take to refer to the small percentage reduction in weight to reflect the change in the ounce basis.[15] The council now required that the holders of the weights (and indeed, all others in the burgh) ensure that their weights did match the standard, with the stone weighing 16 pounds and the pound 16 ounces.

This entry is contemporary with the March 1512 Dundee charter which was mentioned in Chapter 7. In this charter, James VI authorised the council and merchants of Dundee to continue *pro tem* to trade by the new standard weights and measures which had been provided by the Chamberlain during his 1512 circuit, without risk of prosecution:

> … we gif licence to the saidis provest, ballies, counsale and communite and inhabitaris of our said burgh, to vse and hald the mettis, wechtis, and mesouris siclike as wes deliuerit to thame in our last chaumerlane air haldin in our said burgh vnto the tyme that new mettis, wechtis, and mesouris be deuisit and gevin to thame be ws [by us] and our grete chaumerlane …[16]

The provision of a set of standards to Dundee may be seen as part of the process of raising the burgh to a status comparable with that of Perth, and the action may indicate that Dundee did not previously possess standards of such authority. It has already been noted that the issue of these weights and measures followed so soon after the establishment of the royal artillery workshops at Edinburgh Castle that it is tempting to suggest that they were cast there.

It is unclear, however, at what date it was decided to authorise the change in the basis of the weights. A parliamentary act of 1504 related to the availability of new weights and measures from Edinburgh and indicated that the sizes of the measures had been changed.[17] We have argued in Chapter 6 that a dated measure of 1500 (the Inverkeithing gauge, Item 1 in the Inventory) follows a re-definition of the firlot. If these changes formed part of a single assize, then it seems reasonable to assume that it took place in the late 1490s; but in fact there may have been a number of separate ordinances, made over several years.

It has already been shown in Chapter 4 that the 'Scots' merchant pound (with a mass equal to 15 ounces of 480 grains) continued after the 1426 Assize, and that the larger pound of 16 ounces (of 480 grains) defined in the assize was a true 'troy' pound and not a replacement for the merchant pound.

However, the situation is not as clear-cut as this statement might suggest, because there was a Scottish pound of 16 troy ounces mentioned in an act of 1491, which was clearly a trading pound.[18] But its use was not universal: it appears to have been restricted to certain types of costly imported goods, and the act specifically mentions spices and candle wax. But the act failed to specify what type of ounce was to be used, leaving it open to speculation whether this was the Paris or even the Flemish ounce. This omission may have encouraged the more general adoption of the Flemish ounce. Certainly, after the reduction in around 1511 of the ounce size for the *official* trading weights from 480 grains to 472½ grains, the merchant weights would have been a better match for the Low Country troois weight, differing by less than half a per cent, and it is anticipated that by 1500 there must have been a sizeable import trade in Scotland in goods reckoned by this type of Continental weight unit. The Scots 1491 pound would be a Scottish analogue of the English 'haberty-poie' pound, of 16 English troy ounces, which is known from fifteenth-century English documents: it is specified in a source of 1496 as the appropriate pound for spice, and another fifteenth-century document has a more extended list of goods including metals, fine manufactured materials, spices and wax.[19] The importation of spices and wax into London from Antwerp in the fourteenth century can be deduced from Pegolotti's account, and the pound used in London for spices was the Antwerp spice pound. But such a pound would have become irrelevant in Antwerp when level-beam weighing was introduced for bulk goods, and this is presumably reflected in the use of a pound of 16 troy ounces (of 480 grains) in London in the fifteenth century.

Scotland's principal trading partner in the thirteenth and fourteenth centuries was Flanders, where Scottish wool and rough cloth found a ready market in Bruges, which also was the source of much of Scotland's imports of finished goods. However, the Scots staple – the port which offered preferential conditions to Scots merchants and commercial agents – moved to Veere in neighbouring Zeeland in the mid-fifteenth century; and in due course it was the Amsterdam commercial market that came to exert the greatest influence on Scottish trading.[20] By the late fifteenth century the various weight systems used in the ports of the Low Countries operated exclusively with 16-ounce pounds, rather than the 14-ounce pounds which at an earlier date were still being referred to in Pegolotti's account.

In the second half of the sixteenth century there is evidence that sets of weights were being imported by Scottish burghs from the Low Countries, and particularly through Antwerp, to act as local standards. The oldest surviving weights of this type are an incomplete set of nested weights of 1572 for the burgh of Perth (Item **33**, and Fig. **8.2**). Such sets were manufactured in Nuremberg and were to the troois rather than the current Paris standard, although they were ostensibly French weights. But the pounds of these sets

Fig. 8.2

Incomplete set of Flemish
troois weights by G. Weinmann
of Nuremberg, and stamped
at Antwerp, acquired as standards by the burgh of Perth in
1572. (Item **33** in the Inventory.)

of weights were combined in stones of 16 pounds, and so we must suppose that 16-pound stones were also now in widespread use in Scotland.

The increasing dominance of trading weights of 16 troy ounces represented a pressure for change in the statutory weights to which the administration was slow to respond. Definitive action appears to have been taken only at the 1563 Assize, when the requirements laid down for the commissioners for weights and measures included the specific instruction that the stone should be of 16 pounds 'trois' weight. In a move that may be seen as a device to demonstrate historical authority for the new stone, the commissioners were to have the old stone and pound brought from Lanark. Although we can appreciate that these two weights were burgh market or trone weights and therefore strictly irrelevant, the commissioners were instructed 'be [by] the samin as thay find to mak ane vniuersall wecht of the stane of the wecht xvj pund trois wecht'.[21]

Rather than claiming that the old 'Scots' merchant pound was replaced by the larger trois pound, we suggest merely that changing patterns of trade led to the trois pound becoming dominant and the old merchant pound slipping progressively out of use. The changed situation was eventually recognised by the administration in the adjustments made in the 1563 Assize. However, by the mid-sixteenth century there must also have been significant use of the English avoirdupois pound by merchants in Lothian at least, reflecting a growth in trade with England that was to accelerate as the century progressed. The mass of the 'Scots' merchant pound after the 1511 reduction was about 7,090 grains – only about one per cent larger than the avoirdupois pound of 7,000 grains. This commercial pound was also already a long-standing feature of Scottish merchant trade with Flanders (its use in wool export was discussed in Chapter 4), and it formed an increasing component of English trade. The demise of the old Scots merchant pound was therefore probably caused by a combination of the penetration of the English commercial pound and changing patterns of trade in the Low Countries.

Although we do not have documentary confirmation of the exact nature of the changes in the Scots trading weights made in 1511 and 1563, this version of events is supported by the mass of an important and official weight standard of 1553. The so-called 'Craigengelt Weight' (Item **31**, and Fig. **8.3**) was named in the 1880s after John Craigengelt, Provost of the burgh of Stirling, whose name is recorded in the weight's Gothic-lettered inscription as the authorising official. It was an official weight constructed by Hans Cochran, a master gunner at the royal artillery works in Edinburgh Castle, and it was apparently made for one of the burghs represented at the Convention of Royal Burghs meeting convened by Craigengelt at Stirling in 1553.

The Craigengelt Weight carries no denomination because it is the basic unit of the weight system – the stone. It is certainly larger than a merchant stone, and so it is plausible that it is a market or trone stone. It has already been noted that the later relationship between the merchant and trone stones was that the trone stone was one-and-a-quarter merchant stones, or 20 merchant

Fig. 8.3
The Craigengelt Weight: a trone stone of 1553, by Hans Cochran. (Item **31** in the Inventory.)

pounds. The mass of the Craigengelt Weight is 141,560 grains (9·17 kg), and if this does correspond to 20 merchant pounds, then these pounds would have a mass of 7,078 grains and the ounce of such a pound would be 442·4 grains. The practical ounce of the fifteenth-century Scots merchant pound was 450 grains, with 16 ounces to the pound and 16 pounds to the stone. Since there were only 15 'trois' pounds to the same stone and each was 16 'trois' ounces, it followed that the merchant pound could be defined as one of 15 'trois' ounces of 480 grains. We have argued that after about 1510 all the trade weights were reduced to take account of the replacement of the troy ounce of 480 grains by the French troy ounce of 472½ grains. This would have resulted in a pound of 7,087½ grains – the familiar pound of 15 Paris ounces (15 × 472½ = 7,087½), which in Pegolotti's day was used as the normal internal pound in France. The theoretical trone stone (of 20 such pounds) would be 141,750 grains. But in giving this equivalence in terms of English troy grains, we must be conscious that we have established in Chapter 4 that the English troy pound was marginally increased in size in the 1580s. Expressed in modern troy grains, the stone is 141,660 grains (9·18 kg), with the pound at 7,083 grains, which is a close match for the Craigengelt Weight.[22]

If we have found the correct interpretation for this weight, then it is the earliest surviving standard trone stone, and has suffered only about 100 grains of mass loss through wear. The comparatively slight amount of wear in use can perhaps be explained by the fact that this size of stone was superseded at the 1563 Assize. It is hard to believe that this virtual coincidence of the Craigengelt Weight with the stone generated from the pound of 15 Paris ounces is purely accidental. Equally, the possibility of a deception (in altering the Craigengelt Weight at a later date to match a known mass standard) can be eliminated, because all previous commentators have concluded that the trade stone had already been raised to 16 troy pounds in the fifteenth century.

If we assume that the weight has correctly been identified as a trone stone, then there are a number of important consequences. First, we can see that the relationship between the trone and the merchant weights (1¼:1), which was only made explicit after 1600, applied at a much earlier date. The fact that we find it expressed for weights that had been reduced to take account of the troy ounce change in about 1510 implies that the relationship previously held good in the fifteenth century, before this adjustment, and therefore in all probability it dates back to before the 1426 Assize. Second, it is clear that the ordinary stone still comprised 16 merchant pounds (equivalent in mass to 15 troy pounds) and had not yet been raised to 16 troy pounds. It is quite possibly the case that this merchant pound and stone were little used by this time, but they must still have been officially recognised for the trone stone to have been set at 20 merchant pounds. The specific requirement for the 1563 Assize commissioners that the stone should be 16 troy pounds indicates that the change was made only ten years after the date of the Craigengelt Weight.

There is a further important detail we can extract. The mass of the particular French-derived version of the troy ounce used in the Scottish Mint for bullion receipt is accurately known as 471·15 grains from comparisons with English troy weight at the Mint in 1604, and again, applying an appropriate correction, this is about 471·0 modern troy grains.[23] If we were to consider that this was the size of the French ounce applied to the trade weights from about 1510, then the equivalent trone stone would be only 141,300 modern grains, and so any surviving weight would inevitably be worn to a slightly lower mass. However, the mass of the Craigengelt Weight is about 250 grains *greater* than this, and so it follows that the French troy ounce cannot have been introduced at about 471 grains, but must have been at 472½ grains. This is an important distinction to make, because monetary historians have tended to apply this smaller ounce universally, whereas the evidence of the Craigengelt Weight is that the coinage ounce is likely to have been the Paris ounce of 472½ grains.[24]

It is largely because of the survival of the Craigengelt trone stone in such good condition that we can determine the evolution of the trade weights between the assizes of 1426 and 1618. This scheme of evolution was represented in Fig. 8.4, with the mass of the Craigengelt stone represented on the top line, after the ounce reduction of about 1510 and before the stone increase of 1563. Such a diagram raises a number of questions – particularly how the trone stone emerged and how the relationship between the trone and merchant stones became established. These issues have already been identified at the end of Chapter 1, where a preliminary analysis is referred to; and we will offer further support for this in the section on the 'merchant Weights' later in this chapter. In the meantime, we should note merely that the Edinburgh minutes cited for the reduction of the trone weights at the Over Trone in 1511 contain the reminder that the stone should comprise 16 pounds, and the pound 16 ounces. It follows that these factors must have applied before the reduction in the ounce size, and therefore that trone weight has its origins in the fifteenth century or earlier.

TRONE WEIGHTS IN THE BURGHS

It has already been noted that the trone weight makes no appearance in the statutes until its brief entry in the reforming Assize of 1618, when it was declared illegal. It is not particularly surprising that the ban on the use of the trone weight was not effective – it had become entrenched in market practice and survived well into the nineteenth century. For a time after the 1618 Assize had been confirmed by parliament in 1624, little was heard of trone weight in official papers, but there were repeated references in the second half of the century. An example is the Privy Council's confirmation in 1682 that meat

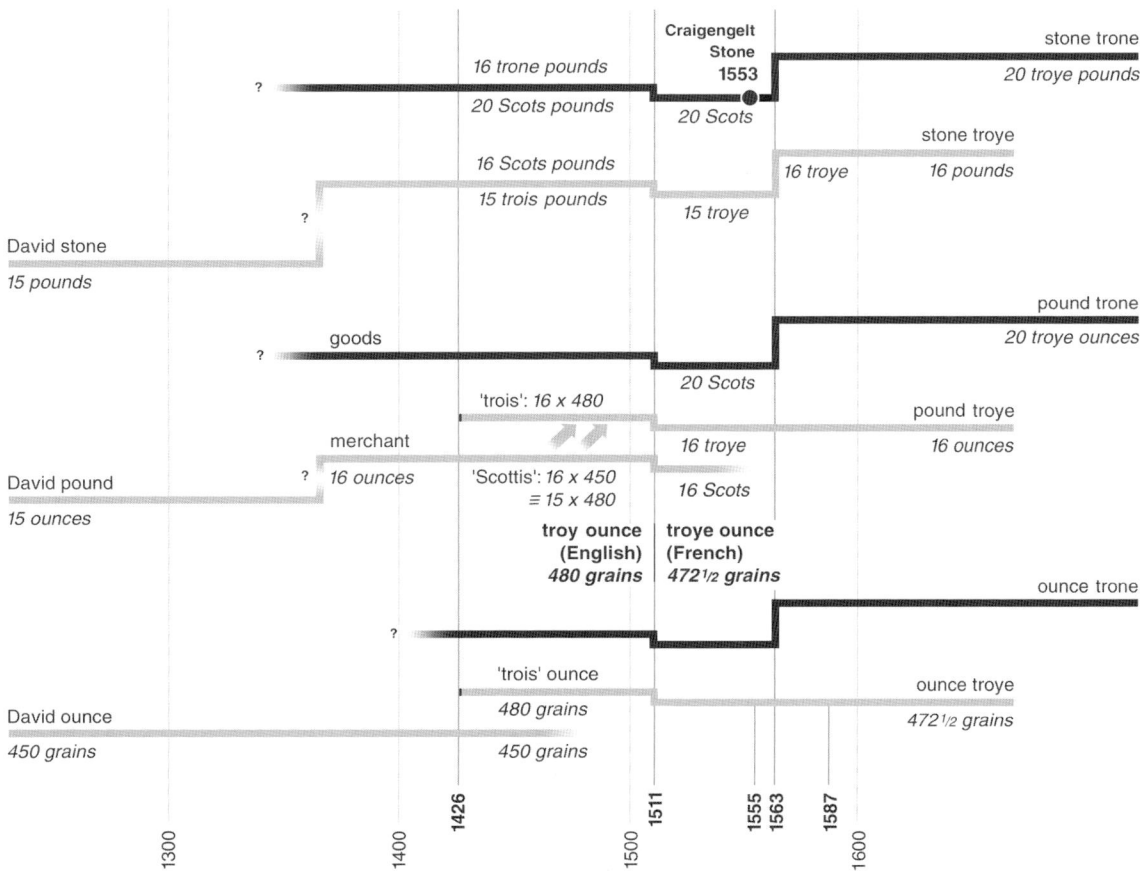

Craigengelt
Stone
1553

16 trone pounds   stone trone
20 Scots   20 troye pounds

20 Scots pounds

16 Scots pounds   16 troye   stone troye
15 trois pounds   15 troye   16 pounds

David stone
15 pounds

goods   pound trone
20 troye ounces

20 Scots

'trois': 16 x 480   pound troye
merchant   16 troye   16 ounces

David pound   16 ounces
15 ounces   'Scottis': 16 x 450
≡ 15 x 480   16 Scots

troy ounce   troye ounce
(English)   (French)
480 grains   **472½ grains**

ounce trone

'trois' ounce   ounce troye
David ounce   480 grains   472½ grains
450 grains   450 grains

1300   1400   1426   1500   1511   1555 1563   1587   1600

should be sold in burgh markets 'at tuenty trois ounces, being a tron pound weight'.[25] In Chapter **9** we point to the widespread use of trone weight long after the Act of Union, together with its eventual adaptation to the English avoirdupois system.

    Alexander Hunter provided an important clue in 1624 to the role of the recently-abolished trone weight by specifying that it was used in the weighing of 'Butter, Cheese, Wooll, Tallow, and such other Countrie commodities as carryeth refuse'.[26] In other words, it represents another example of an allowance to compensate the buyer for aspects such as wastage in the rind of cheese, extraneous matter included with farm produce, for spillage and for similar losses. Here the allowance is simply a conventional quarter of the weight, since the trone stone is 20 pounds, of which 16 pounds make an ordinary stone: therefore, in effect, five quarters must be given to each stone. As with the operation of similar allowances in the capacity measures, it can be seen that the allowance disadvantages the peasant producer both in the market and in the payment of rentals rendered in kind to the landowner.

    When responsibility for the issue of burgh standards was ceded to the

Fig. 8.4
The projected evolution of the official Scottish trading weights between the fourteenth and sixteenth centuries. The height of the horizontal lines represent (on a logarithmic scale) the mass of each weight unit, and the steps indicate increases and decreases in mass. The Craigengelt trone stone is indicated on the top line, between the changes of about 1510 and 1563.

four burghs of the Chamberlain's court in the first half of the sixteenth century, Lanark became custodian of the stone, with the pint to Stirling, the firlot to Linlithgow and the ell to Edinburgh. Our earliest account of this comes in 1552, at the start of the surviving records of the Convention of the Royal Burghs, the body which emerged as effective successor to the Court of the Four Burghs and acted as a form of commercial parliament for the burghs.[27] In April 1552 representatives of the burghs met in Edinburgh and agreed that because of differences between their various weights and measures, every burgh was to receive standards from the four burghs 'quhilkis [which] hes the iust mesouris'.[28] This was claimed to be the result of parliamentary pressure, but no relevant parliamentary act has been preserved.[29]

It was agreed that a further meeting would be held in July 1552 at which the standards would be produced. John Craigengelt, provost of Stirling, duly brought the pint measure, and his colleague the provost of Edinburgh produced the ell. Unfortunately, no representatives from either Lanark or Linlithgow were present and the Convention members sent a written protest to these two burghs. The matter was deferred to a meeting that Craigengelt was asked to convene in Stirling in July 1553. No minutes of the 1553 meeting have survived, and it is not known whether Lanark's standard was produced. However, the existence of the Craigengelt Weight (Item **31**) indicates that some action was taken, with or without Lanark's cooperation. With the contractions expanded, the attribution on the weight reads:

> iohn craigingelt of that ilk me conding [conforming]
> maid [,] quhen he ves prouest of striviling [,]
> anno domini mdliii [1553] : magister : hanes coqhren[30]

Hans Cochran, the maker of the weight, can be identified as the master gunner in the royal artillery works at Edinburgh Castle, responsible for major gun-founding operations at the time and recorded in royal accounts between 1539 and 1558.[31] Craigengelt did demit office during 1553, apparently in the autumn.[32] So, following the inscription on the weight, we would expect that it was commissioned as a result of the Convention meeting in July, and constructed a little later that year.

Although this is a trone stone that clearly takes account of the adoption of a French troy ounce of 472½ grains, it could have been created independently from the administration's own standards, and we have no information about how it related to the particular stone standard held by Lanark. The inscription on the weight associates it with Craigengelt (but almost certainly in his capacity as chairman of that year's convention, held in Stirling), but nothing further is known of its history other than its emergence in Glasgow from a private collection in Renfrewshire in 1889.[33] It seems, therefore, that it is one of a group of accurate standards, the construction of which was instigated by Craigengelt and carried out in the royal workshops, but it does not follow that it was the standard made for Stirling. It is even conceivable that it may have been a copy made for Lanark.

The diversity in the weights of the various burghs that had caused concern for members of the Convention in 1552 may indicate that some burghs had not yet implemented the weight reduction of about 1510 which was the consequence of the adoption of the French ounce. Although it was certainly appreciated in Edinburgh that the trone weights had to be reduced as well as the merchant weights, the requirement may not have been sufficiently clear to small or landward burghs, and there may have been no effective way to enforce the change. Another likely cause of confusion would be the adoption by some burghs of the Flemish form of the French ounce, with a mass of about 474½ grains instead of 472½ grains: it has already been noted that the administration moved to the Flemish standard in 1587, which suggests that it was previously in extensive use in the burghs.

We have to turn to later sources for confirmation of the type of weight represented by the Lanark standard. In 1578 the Convention had stipulated that 'trois wecht keip the iust wecht of xvj vnces for the pund, conforme to the Frenche wecht', and required all burghs to have brass standards of stone, half-stone, pound and half-pound, 'trone wecht, according to the stane of Laneark', which certainly implies that the Lanark standard was a trone stone.[34] The Convention was more specific in 1596 when it stipulated the type of trade to be conducted in each weight series:

> … all Scottis wayris [to] be ressauet [received], sauld, and delyuerit with the iust wecht, according to the stane of Lanerik, pund and half pund, and vtheris effeirand [corresponding] thairto, conforme to the actis of parliament maid thairanent; and the wecht of all forane wayris to be correspondent to the siller wecht, quhilk is saxtene vnce for the pund …[35]

Glasgow did not comply adequately with this instruction, and in 1607 the Convention required the burgh 'to conforme thair trone weyt to the weytis of Lanerik, and thair tros wecht according to the Frenche weycht'.[36]

These instructions emphasise a distinction in the custodial status of the standards of trone weight and the merchant weights. The implication is, that, while Lanark held physical standards of the trone weights, mercantile trade was conducted in the trose or 'French' weights, which did not rely on official retained standards (although the merchants and the exclusive merchant guilds kept standards for their own purposes). This might be because the quality of control exercised by the Continental weight-makers and their guilds was considered adequate; but it seems more likely that Lanark's role had always been restricted to the issue of the weights for ordinary landward trade. In contrast to 'country wares', imported goods were to be sold by the 16-pound 'silver' or troy weight, that is by the weights of the merchants. At the 1618 Assize this distinction was abolished – the new 'troys' weights of 1618 were to be used 'als weell [for] Foraine as Cuntrie-Wares'.[37]

The 1563 Assize commissioners were required to establish a merchant stone at 16 troy pounds, and we have already noted that the (trone) stone and pound were to be brought from Lanark and

> … be the samin as thay find to mak ane vniuersall wecht of the stane of the wecht xvj pund trois wecht allanerlie [only] the pund and uther smaller wechtis equivalent thairto …[38]

Since no report of this commission survives, we are left to consider how to interpret this instruction. It seems clear that the intention was to increase the size of the stone so that it was now based on 16 troy pounds (totalling 16 × 16 × 472½ = 120,960 grains) – rather than 16 of the old merchant pounds or 15 troy pounds (113,400 grains), no doubt reflecting an existing preference for the Flemish weights, with 16 pounds to the stone. However, the stipulation that the stone should be a 'universal' stone comprising 'only' 16 pounds, may suggest an attempt to eliminate the use of the large trone stones, which had been set at 20 merchant pounds. A similar intention is implicit in the 1587 Assize, which speaks of the stone of 16 troy pounds as 'proportional' to the troy pound, and stresses the need to retain the same proportion for all time coming.[39] This intention was only made explicit in the 1618 Assize, when the trone stone was named for the first time, but only so that it could be abolished.[40] If the administration's intention in 1563 was to eliminate trone weight, then this was certainly unsuccessful. The trone stone was central to the regulation of the markets and the payment of rentals, and it was tied to the size of the merchant stone – presumably through a multitude of legal and technical agreements which took account of the long-standing relationship between the masses of the two stones and which could not lightly be laid aside. The trone stone simply re-emerged as one-and-a-quarter times the enlarged merchant stone, and although there is no statutory basis for this change, the new trone stones were certainly acknowledged by the Convention of Royal Burghs.[41]

There is some evidence of the construction and distribution of new trone weights from Lanark in the ten years following the 1563 Assize. In 1567 Lanark council ruled that trone weights had to be of brass or lead, and that they should be to a set pattern.[42] Also in the records of Lanark, there is a surviving draft letter which has been dated to '*circa* 1570', declining a request from another burgh to borrow the standard stone weight so that a copy could be made, because lending the stone was thought not to be lawful: instead, the unnamed burgh was to send two commissioners to adjust a replica in Lanark, as had been done by Ayr, Irvine, Glasgow, Linlithgow, Stirling, Peebles 'and wther tounes'.[43] The change in the trone weights may have taken some time for the Convention of Royal Burghs to implement amongst its members, but it did eventually take place. The occasional infringement is recorded in the Convention's minutes; such as Haddington's complaint against St Andrews in 1607 'for haueing thair stane weight augmentit to tuenty pundis weight and ane half' instead of the correct 20 pounds.[44]

The clearest confirmation that the trone stone increased in mass after 1563 to take account of the increase by a fifteenth in the size of the merchant stone (from 15 to 16 trois pounds) can be found in a small group of early trone weights (Items **32**, **34** and **42**) preserved in a collection of weights and measures belonging to South Lanarkshire Council, successor to the Royal Burgh of Lanark. The mass of the trone stone before 1563, represented by the Craigengelt Weight, has been deduced to be 141,660 grains (9·18 kg). Increasing this by a fifteenth to reflect the parallel increase in the merchant stone in 1563 would give a trone stone of about 151,100 grains (9·79 kg), and a trone pound of 9,445 grains, and indeed this is very close to what we find when we examine the trone weights at Lanark.

The Lanark weights have tapering square-section shapes and ring-handles, and a group of three are clearly companions because each has an identical double-headed eagle, which is the crest of Lanark, cast in relief on one face (Item **32** and Fig. **8.5**). The denominations are not indicated on the weights – except that a further small weight of matching shape and apparently from the same series (also described under Item **32**) is marked as being of 4 ounces trone ('4 V TROYN') – but it is clear from the proportions of the measured masses that the stone was divided into 16 pounds, each of 16 ounces.

The three matching weights have masses of 1 stone, 4 pounds and 2 pounds and all three are worn, but the stone and 2-pound weights are worn to a greater extent than the 4-pound weight. If the mass of this last weight is divided by four it can be seen that the pound (reduced somewhat by wear) is 9,465 grains. Since this pound was 20 merchant ounces, the mass indicated for the merchant ounce is at least 473·3 grains. It is clear, therefore, that these weights have been made after the 1563 increase, because otherwise this ounce size would have been about 443 grains. However, it is equally clear that the Paris ounce of 472½ grains has not been used, because the ounce size found is larger. Instead, the ounce basis appears to be the Flemish ounce, which operated at about 474½ grains. This provides a valuable indication that the Flemish standard was adopted at the 1587 Assize – an issue that will be addressed shortly.

There are two definitions of the mass of the trone stone at this period: probably both were obtained from an official source, but they are difficult to interpret. Sir John Skene, the Clerk Register, provided a size of 19½ 'trois' pounds for the trone stone in his glossary of technical legal terms *De verborum significatione*, published in 1597.[45] This value was repeated by Alexander Hunter (presumably having been obtained through his contacts at

Fig. 8.5
Group of three trone weights,
from 1 stone, with the arms
of the burgh of Lanark, late
sixteenth century.
(Item **32** in the Inventory.)

the Mint) in his 1624 volume as 19 pounds and 8 ounces 'Paris weight'.[46] These two definitions are apparently inconsistent with an understanding of the trone stone as 20 pounds, but the more precise information given by Hunter may provide the answer. The Paris weight referred to by Hunter is possibly the 'kings Paris weight' for which Sir James Hope (1614-1661), who had been appointed General of the Scottish Mint in 1642, gave comparison figures in 1647.[47] The ounce of this pound was 489 grains, so a stone of 19½ such pounds would be 152,570 grains. After 1618 the Scots merchant ounce was raised marginally to 475·6 grains (this will be established in the next section), so that a 20-pound stone would be 152,190 grains: this is well within the range of accuracy of the definition as 19½ Paris pounds. However, it is more likely that a specific commercial context for Skene's statement has to be understood.

Lanark council does not appear to have been required to send its weights to Edinburgh for inspection by the commissioners for the 1618 Assize. Although the Lanark standards must have been primarily (or even exclusively) market or trone weights, the weights and measures commissioners of 1555 and 1563 had been required to call the pound and stone from Lanark, presumably to demonstrate a continuity of authority. The 1618 Assize commissioners concluded that trone weight should be abolished, but they also confirmed a continuing role for Lanark as guardian of the weight standards and the surviving 1618 weights (Items **35-37**) are all stamped with the double-headed eagle of Lanark.

THE MERCHANT WEIGHTS

One of the inherent difficulties in understanding any of these early weight types is recognising its area of application. It is only very occasionally that contemporary sources indicate these categories of use – as, for example, when the Convention of Royal Burghs described the use of trone weight for Scottish or 'country' goods, and 'silver' (or troy) weight for 'foreign goods'.[48] Two broad categories are seen to emerge as time progresses. One encompasses goods of comparatively low unit value, and in general these are bulky or heavy materials: at a later period this category was weighed using the 'avoirdupois' system (literally for stuff 'having weight'). Such material was likely to be traded at burgh markets, and although it might be exported (as was lead ore) it would nonetheless be considered domestic (or country) produce. The weight system for this domestic market was initially a general weight but was later characterised as an internal or heavy-goods weight.

The other category is for fine or costly materials, which (in the Scottish economy) generally meant imported goods. Because trading in material of this sort was the particular province of the merchant guilds, we have tended to describe the weight type appropriate for this category as 'merchant weight'. It would be understandable if the weight systems used for foreign trade were influenced by the merchant weights of dominant external trading partners: Pegolotti's account (see Chapter 4) has illustrated how London's import trade used foreign weight systems for aspects of the wholesale trade, and we have argued that Paris weight played a part in the retail sale of goods such as spices. This Paris pound, and the Flemish version subsequently used in Scotland, were both 16-ounce pounds, and the European trend for trading pounds was towards pounds of 2 marks or 16 ounces. By contrast, the early domestic (or heavy-goods) pounds of England, Scotland and France were 15-ounce pounds.

The Scottish fine goods category included precious materials such as gold and silver bullion, and the weighing of bullion was latterly considered a specific 'troy' activity (the name 'troy' is not found in Scottish sources before 1426). In England a separate 12-ounce pound was used for bullion, initially because this was also the weight of the silver monetary pound of 240 pennies. With the progressive reduction of the weight of the penny from about 1300, this special relevance was lost, but the 12-ounce pound for troy weighing was retained in England. Scotland's original silver coinage was the same weight as that of England, and the relative weight of Scottish and English silver coins was always an issue of economic importance, particularly during periods of coinage depreciation. Although the monetary pound of 12 ounces of 450 grains must inevitably have been well known in Scotland, bullion seems to have been weighed initially by the general pound of 15 ounces, and then by the merchant pound of 16 ounces once the latter had been introduced.[49]

(There is evidence for a 12-ounce pound having been used in the second half of the fourteenth century as a coinage pound in the Scottish Mint, as will be discussed below, but there is nothing definite to indicate that this was also the size of the bullion or troy pound.)

Both the English *Tractatus* and the Scottish Assize of David described weights of 15 ounces of 450 grains and by implication they were general goods weights applicable to domestic trade. A 16-ounce version of the English weight does emerge, but its use in England seems to be restricted to the special case of wool weighing (which was officially controlled because it was a valuable source of customs revenue). The Scottish merchant weight of 16 ounces of 450 grains is the 'Scots' weight recorded in one of the preliminary parliamentary acts of 1426, and discussed in Chapter 5. The stone was given as 16 'Scots' pounds, and alternatively as 15 'trois' pounds, where the 'trois' pounds were each of 16 ounces of 480 grains. The novel aspect of this statutory definition was apparently the use of 'trois' units, defined in terms of the pre-existing 'Scots' pound and stone.

There is no reason to doubt that the new trois weight of the 1426 Assize, which we can identify as having the same ounce as the English troy ounce of 480 grains, was intended for weighing proper troy materials such as gold and silver. There is no earlier statutory use of 'trois' in surviving Scottish legislation, although it is conceivable that this weight type might have been introduced before this time, perhaps closely following its official introduction in English metrology, which is believed may have been in the third quarter of the fourteenth century. Before the Scottish introduction of 'trois', our contention is that troy weighing was performed with the 'Scots' merchant weight, so that the Scottish troy pound was a 16-ounce (or 2-mark) pound, whereas the English troy pound was a 12-ounce pound.

This distinction between Scottish and English practice was reinforced when the troy ounce of 480 grains came into official use. The Scottish troy pound (the 'trois' pound of the 1426 Assize) was 16 ounces of 480 grains, and the use of a pound of 16 ounces at the Scottish Mint was also noted in accounts for 1436-8.[50] The English troy pound was 12 ounces of 480 grains, and it is plausible that this also represented a 2-mark pound, using the Bruges silver mark of 6 ounces of 480 grains. The Scottish troy weight is seen to be operating as a definitional ounce for liquid measure in the 1426 Assize; and the area of application of English troy weight includes the specification of 'measure' in documents later in that century.[51]

Returning to the period before the official introduction of the 480-grain troy ounce, the English goods weight of 15 ounces (of 450 grains) was larger than the 12-ounce troy weight by a factor of 1¼. However, in Scotland, the equivalent 15-ounce weight was less than the 16-ounce Scottish weight for troy and fine goods. At a later date we do encounter a Scottish goods pound which is one and a quarter times the 16-ounce troy pound, and this will be recognised as the 20-ounce trone pound. Considering that trone weight is clearly identified with the internal market, and is thus the direct analogue of the English 15-ounce weight, the fact that each bears the same relation to the appropriate troy pound is perhaps significant. We will suggest, therefore, that when the 'Scots' pound emerged as a complement to the 15-ounce pound of the David Assize, to satisfy troy and merchant requirements, the trone pound was created as a heavy-goods pound by direct analogy with the situation in England. Presumably this happened before 1426 (when the troy function was separated from the fine goods function and became the province of the larger troy ounce of 480 grains) and so the implication is that trone weight may date back to the late fourteenth century.

In the early manuscripts of the Scottish 1426 Assize, the word troy is used in a plural form, agreeing with the noun, as for example in 'ii pundis … troyis', although to some extent this is disguised by using the conventional contraction 'punde', where the final hook stands for the plural 'is'.[52] However, it must be appreciated that in the fifteenth century there was little need for consistency in spelling, and certain letters (in this case 'y' and 'i') are free variants. It follows that the spelling 'troyis' and 'trois' are simply spelling variants, and indeed the earlier of the two manuscript versions of the assize also uses the form 'trois', and this is the spelling used in the 1804 printed edition by William Robertson. (As we have seen in Chapter 5, Thomas Thomson also consulted the manuscripts used by Robertson, but seemed to rely more heavily on the later manuscript, and spelt the word consistently as 'troyis'.)

By the time the first printed editions of the parliamentary acts appeared, in the second half of the sixteenth century, the adjective used to describe the character of these weights had taken on the set form 'trois', which was no longer considered to be either singular or plural, but merely descriptive. Not only was this spelling now used by the editors of the published acts for the troy weight of the 1426 Assize, it was also applied somewhat confusingly to the new merchant pounds of 16 troy ounces.

However, here we have to make an important distinction with the earlier usage. At the time of the 1426 Assize there is no particular reason to doubt that 'trois' ounces were used for troy purposes (including the weighing of precious metals), and this appears to be the case until at least the 1555 Assize. But from at least the 1587 Assize it is clear that the ordinary commercial 'trois' pound was slightly heavier than the administration's troy standard. By this time the word 'trois' was being used to describe the ordinary commercial weights of the kingdom, without the implication of an official troy function: in fact we can conclude that the adjectives 'trois' and 'trone' merely differentiated between the two classes of weighing in trade. For the rest of this period (to 1618) we will use 'trois weight' to describe the weights we have previously termed the merchant weights, and this distinguishes them from the market or trone weights.

It is necessary to note that these names did not imply a fixed mass standard (in the same sense that the English troy ounce had a fixed mass of 480 grains) – both trois and trone weight underwent a number of small changes in mass in the sixteenth and seventeenth century, but the descriptive names for the weights remained the same until well into the seventeenth century. There are, however, some exceptions to this. The most significant is at the Assize of 1618 where the weights are referred to as 'troys' throughout, and where the standard weights issued as a result of the Assize are also stamped 'TROYS' (Items **35-37**). It is conceivable that this represents little more than a studied anglicisation of 'trois', of a type that is not uncommon in the period that followed the union of the crowns and the departure of the court to London.[53] At first sight it appears as a more archaic form (the use of 'y' as a free variant of 'i' is associated with earlier usage), but here the spelling may perhaps represent a conscious echo of the English 'troy'.

English transactions in bulky or heavy merchandise were conducted at this period using the commercial avoirdupois system, with a stone of 14 pounds, each of 7,000 grains. This division into two classes of weight was also mirrored on the Continent. Scotland's dominant trading partner was Flanders, a component of the French Burgundian Netherlands. Here the Dutch troy or 'troois' weight, which was used throughout the Netherlands, had an ounce of about 474½ grains.[54] There is no clear understanding of the origins of troois weight in the earlier troy systems of the Low Countries, although the discussion in Chapter 4 indicated how commercial weights based on the troois ounce fitted conveniently into the existing goods hundredweights using ounce multiples of 16 rather than 14. By the time historical analyses of the early Dutch standards were undertaken in the late eighteenth and early nineteenth centuries, very few examples had survived of the earlier use of 14-ounce multiples, and the troy ounce types encountered by Gerrit Moll (1785-1838) and others were almost exclusively troois.[55]

The merchant weight, based on a pound of 16 troois ounces, was a significant influence on Scots trade and had clearly penetrated through much of the domestic market by the late sixteenth century. Sir James Hope, General of the Scots Mint in the mid-seventeenth century, prepared a report for parliament in 1647 when a question arose of accepting certain under-weight coins as currency. In this he referred to a type of weight in widespread use which he termed 'ordinarie French Trose weight used be [by] our Marchands in Scotland which is ordinarie marked with a fleure de lyce'.[56] From the comparisons Hope provided, based on a 'new pyle 4 lb marked with a fleure de lyce boght from Jᵒ Falconar Warden [of the Mint] from Holland', this weight series had an ounce of 474·2 grains; but again, taking account of the small changes in the size of the English troy weight, this is 474·1 modern grains, but would have been recognised as 474·4 grains in the sixteenth century.[57] The word 'trose' may represent a corruption of the Dutch 'troois', although the close similarity to the Scots 'trois' probably means that no distinction is meaningful and the words should perhaps be considered as interchangeable. It is more likely that the qualification to be stressed here is that this was the *ordinary* French trose weight, meaning perhaps the Flemish or Dutch form as distinct from the original French form. It is not clear why the mass of Hope's trose ounce is slightly less than the modern understanding of the mass of the trooise ounce, which is 30·76 g (or 474·7 modern grains). Clearly the size that can be derived from Hope's figures depends on the accuracy of his 4-pound weight, but in fact this same ounce size emerges from several independent sources, so we must assume that it is correct.

There are a few surviving sets of imported Continental weights used as standards by the merchant guilds of Scottish burghs, and these all appear to be adjusted to this ounce size. All that remains of the standard weights of the Dundee Guildry is an incomplete 8-pound set (or 'pyle') of nested cup weights (Item **39**) characteristic of the type described as 'Nuremberg' weights, named for the German city which was particularly known for the manufacture and adjustment of weights. The Nuremberg weight makers served an international market, and manufactured to many accepted weight standards. In such sets the weights form a binary sequence, with the outer lidded cup weighing half the total of the set, and the mass of each cup is the same as that of the group of cups it contains. An 8-pound set would therefore typically comprise an outer cup of 4 pounds, and inner cups of 2 and 1 pound, 8, 4, 2, and 1 ounce, with a central plug of 1 ounce.

The outer cup of the Dundee set has a lid stamped with the date 1672 and with the master-sign of G. B. Weinmann of Nuremberg (a bow and arrow), and there are three remaining inner cups. All four cups are stamped with a *fleur-de-lys* mark (see Fig. **8.6**). The outer cup shows some signs of alteration, but the present masses of the inner cups are consistent with an original standard of over 474·0 grains (474·2 old grains) to the ounce. The Jedburgh 16-pound set (Item **38**) is also late seventeenth-century and is by a member of the Nuremberg Schirmer family. The outer cup is damaged but the larger two of the three inner cups are original, and their present masses indicate that they were originally adjusted to a standard of at least 473·9 grains.

All these sets are stamped with the *fleur-de-lys* mark, which indicated that they were a troy weight made to a 'French' standard, rather than that they were made for use in France. However, it is easy to understand how these weights came to be considered as 'French', although in fact they were Flemish weights and made in Nuremberg.

If these nested sets are representative of the early merchant weight that the burgh guildries found necessary, then we can perhaps draw the conclusion that early merchant trade was based on an external Flemish 'trose' standard of about 474½ grains to the ounce. We have already noted that the earliest surviving burgh standard is the incomplete 8-pound nest carrying the name

Fig. 8.6
Stamps on Nuremberg
nested weights at Dundee,
1672. (Item **39** in the
Inventory.)

of the Perth dean of guild and the date 1572 (Item **33**, and Fig. **8.2**). These are by an earlier Nuremberg master from the same family as the maker of the Dundee set of a hundred years later, G. Weinmann: they carry his master-sign (a Turk's head), but they also have the 'hand' stamp, indicating export from Antwerp, and the *fleur-de-lys*, for 'French' troy weight. As with the Dundee Guildry set, the outer cup of the Perth set has been modified, but the masses of the two original inner cups show that they were made to an ounce of at least 473 grains. However, this value is more uncertain than those of Dundee and Jedburgh because the bases of the weights show clear signs of two stages of alteration at minimum. Thus, at least in their present state, they are trose weights, and too heavy to have been adjusted to the true French standard of 472½ grains (472·2 modern grains) to the ounce.

The evidence of the Craigengelt Weight of 1553 is that at least some classes of standards were adjusted to the French ounce of 472½ grains, although the merchant guilds of even substantial burghs were regulating trading standards according to the heavier *fleur-de-lys* ounce of about 474½ grains. The difference between these two systems was small (less than 0·5 per cent) but detectable, and it represented a problem that had become very necessary to address by the late 1580s. A contemporary indication of this confusion comes in an act of the Convention of Royal Burghs in 1578 when the burghs were instructed to keep their trois weights 'conforme to the Frenche wecht', although as we see, the interpretation of such statements in terms of specific weight systems remains ambiguous.[58]

The other feature that is readily apparent from such surviving *fleur-de-lys* weights is that most are made up in sets of 8 pounds, indicating that a stone of 16 troois pounds was the unit of trade. An increasing use of imported sets of *fleur-de-lys* weights by merchants, and presumably also by royal burghs of the second rank which were not provided with weights by the Chamberlain, must in turn have provided some pressure for the adoption of a troy stone of 16 trois pounds. Indeed, it is possible there may have been some provision for such a troy stone in the 'missing' James IV Assize of the late 1490s.

A resolution of the vexed question of the correct ounce basis for weight definition was tackled by the commissioners for the 1587 Assize. In the extract from the 1613 Privy Council minutes already quoted, the complaint was made that the commissioners had 'proceidit upoun the diversitie of the trois weght allanerlie [only], being than [then] thrie in nomber, Franshe, Flemis and Englische', but had not addressed the issue of trone weight.[59]

The work for the 1587 Assize appears to have been entrusted to Sir Archibald Napier (1534-1608), the General of the Mint and a member of the commission, and it was presumably his influence that led to the rigorous new definitions provided for the dry capacity measures. In spite of a clear concern for precision in measurement and in the physical specification of reference standards, the language of the report is vague in some key respects – intentionally so, it appears, in relation to the authority for changes in the dry measure sizes, and this aspect was discussed in Chapter 7. Although the clear implication of the Privy Council's comment in 1613 was that a conclusion had been reached by the 1587 commissioners, their report provides no clues – the weight system is merely defined as having a stone containing 'sextene pund trois', where each pound comprised 16 ounces, with the Stirling pint containing 'tua pund and nyne vnce trois wecht of cleir watter'.[60]

In fact, the pound and ounce were not given any other descriptive name in the statutes until the 1618 Assize, when they were named as 'frensh Troys' weight.[61] It will be shown in the next section that the basis of the 1618 weight was certainly not the Paris ounce, but a heavier ounce associated with the Low Countries. The interpretation of the 1618 description of 'French' weight must therefore be similar to that of Sir James Hope in his 1647 comment quoted earlier, when he described as 'French' an ounce which can be identified as being Flemish or Dutch.

Fortunately a few standards for this period do survive, and there are references to others which are specific enough to allow their weight basis to be deduced. On balance, the evidence from this material is that the administration conceded in 1587 that the Flemish (or *fleur-de-lys*) version of French troy weight was now almost universally applied in the burghs. Our supposition is that the 1587 commissioners adopted the Flemish weight formally – accepting in law the *de facto* practice of the market. This assumption has already been used in developing the succession of firlot sizes discussed in Chapter 7. A consequence of this is that references in the statutes to 'trois' weight in this period have to be interpreted as descriptive references to an external system, rather than specific references to a retained standard. This is in line with sixteenth-century edicts of the Convention of the Royal Burghs which imply that the retained burgh standards were restricted to Lanark trone weight and that by extension there was adequate confidence in Continental manufacturing controls for the quality of troy weight.

The surviving standards of the period provide somewhat ambiguous evidence about a change in the weight basis at the 1587 Assize. The first is the 1574 pint of the burgh of St Andrews made by David Rowan, which is clearly based on the Flemish ounce (Item **111**). Although this pint dates from well before the 1587 Assize, it nonetheless indicates the extent of penetration of Flemish weight into the market-place. Presumably Rowan supplied what the client required, but it is perhaps significant that (in spite of his status as the monarch's master founder) Rowan was not constrained to adjust the vessel to the Paris standard used in earlier decades. The pint has not been adjusted by physical comparison against the Stirling pint (or a duplicate) as had been the earlier convention, but rather against a pint constructed on a new ounce basis; and yet the conscious choice has been made to follow the Flemish ounce rather than the Paris ounce (as used for the Dundee pint, Item **112**). At the least this represents a level of official toleration for the Flemish ounce that implies a general acceptance of its near-universal use. The other surviving standards are the three trone weights carrying the Lanark double-headed eagle (Item **32**, and Fig. **8.5**), which are certainly based on the Flemish ounce. Although these are undated, they can be assigned with some confidence to the late sixteenth century, and they are presumed to be Lanark's standards for the burgh weights or an unissued set based on Lanark's standards. Again, these provide a clear illustration of the currency of Flemish weight in the burghs, but we are still left with some doubt about whether the administration's policy had actually been brought into line with the burghs' practice.

It is not at all clear whether Edinburgh's existing standard weights might have matched the administration's before the 1587 Assize and therefore perhaps they might have been based on the true French ounce. Certainly, the capital's standards were renewed after the assize; and although this was presented as merely taking account of wear and tear, it may equally indicate an official move to a trose standard. In October 1589 the council noted that 'the awld braysin wechts are worn and brokken and thairby become unjust' and the dean of guild was instructed to replace them with new adjusted standards.[62] The work was undertaken by David Rowan, who made a trois set and also a trone set.[63] One weight at least – the trois pound – still survived in 1647 when the Edinburgh dean of guild's standards were weighed by Sir James Hope, General of the Mint, against the new four-pound pile of *fleur-de-lys* trose weights which his warden had obtained from Holland.

Hope's principal concern was to provide equivalences for the stone and the 4-pound weights of the Edinburgh standards, but he gave no identification details for these weights; in spite of this, he did describe the pound weight as marked with a castle, 'I·P' (for one pound) and the familiar David Rowan monogram, of which he provided a sketch (Fig. 8.7).[64] However, we can be sure that all the 1589 weights were marked because the Edinburgh goldsmith John Bartane was specially commissioned to make 'the forme of thrie castellis to put upone thir wechtis'.[65] The stone and 4-pound weight both emerge from Hope's calculations as accurately to the 475·6 grain (475·5 modern grains) ounce which we will associate with the 1618 Assize; but, the one-pound weight was considerably lighter than the others, and Hope's comparisons against the *fleur-de-lys* weights show it was to an ounce of only 473·5 grains.[66] Hope accounted for this difference by saying that the weight 'was worne-like', but it seems more plausible that he selected Rowan's pound weight because it carried authorisation marks, and that it was not the pound weight of the dean of guild's current standards. We suggest that it was a survival of the 1589 set and that, although now worn, it was originally made to the smaller *fleur-de-lys* standard of about 474½ grains to the ounce.

Fig. 8.7
Illustration based on James Hope's drawing of the markings on a 1589 standard weight by David Rowan: from Cochran-Patrick's *Records of the Coinage of Scotland*.

'TROYS' WEIGHT OF 1618 AND DUTCH WEIGHT

When Sir Archibald Napier and his fellow commissioners undertook their revision of the weights and measures in 1587 they made it very clear that they saw the definitions of the various standards as inter-dependent.[67] The dry capacity measures are described in terms of their linear dimensions, and are defined as containing a specific number of pints. The pint is defined as containing a stated weight of water, and so the weight and capacity systems are tied to each other. We have seen how this led to variant pint sizes in the 1560s and 1570s, and we will now examine the implications for the weight system of the reversion to the earlier pint size in 1618.

The major part of the 1618 Assize was concerned with the reform of the large grain measures, and especially with provisions for eliminating the use of heaping. But these capacity measures were themselves raised from the pint, and this time the legislators were quite specific about which pint was to be accepted as the standard. They relied on historical precedent and stated explicitly that the primary standard pint was to be the existing Stirling Jug, but also that it was to contain 55 ounces of water. Because of the difficulties inherent in measuring the Stirling pint, which has an uneven rim (see the entry for Item **108**), we have adopted the 103·7 in³ capacity of the dated 1618 Edinburgh standard pint (Item **113**) as representing the findings of the 1618 commission. If this is indeed to be exactly 55 ounces, then it follows that the ounce would be about 477 troy grains. This is the weight type later known as the Dutch or Amsterdam commercial weight, but we cannot be sure that ounces of this larger size appeared in Scotland until after 1700. For the century after the passing of the 1618 Act a more modest ounce of about 475·6 grains was in legal use (the ounce of the Edinburgh dean of guild's weights measured by Hope in 1647), although undoubtedly merchants and many burghs continued to use weights based on the old *fleur-de-lys* ounce (of about 474½ grains).

We have called these weights 'troys' because this is the spelling used in the 1618 Assize and on the weights themselves, and because it helps distinguish them from the earlier and lighter trois weights. However, it will be appreciated that these spellings are interchangeable, so that this form is employed only as a convenience.

It is unclear what prompted this small shift in the weight basis in 1618. The change from about 474½ to about 475·6 grains to the ounce does not seem to be enough to enable 55 ounces to generate one pint, or if this was the intention then there has been a slight error in the work. Instead we must conclude that the 55 ounces were supposed to be descriptive rather than prescriptive, and that, for all its apparent accuracy in definition, the 1618 Assize did not succeed in constructing mutually dependent weight and capacity standards.

Copies of the standards created by the 1618 commission were again distributed to the burghs – or rather, the burghs were encouraged to purchase them. The arrangements were handled through the Convention of Royal Burghs, where the prices to be charged by the four distributing burghs were agreed in July 1618. Edinburgh was to charge £10 Scots (at £12 Scots to £1 Sterling) for the ell, Stirling £24 for the pint and Linlithgow £60 for the set of six dry capacity measures (firlot, peck and half-peck combined, and quarter peck, both for wheat and for barley measure). The weights were to be available from Lanark, 'keipar of the wechtis', either at £33 6s 8d (or 50 merks) for the stone or £66 13s 4d (100 merks) for a full set from a stone to an ounce:

> … for the single stane wecht to the soume of threttie thrie pundis sax s. viij d.; and in caice the saids borrowis salbe vrgit to ressaive from the said burgh [of Lanark] the haill number of wechtis following, viz., stane, half stane, quarter stane, tua pund, pund, half pund, quarter pund, tua vnce, and ane vnce, they modifie the haill pryces thairof to be the soume of ane hundreth merkis …[68]

A few of these weights survive, but not in such numbers as to suggest that many burghs felt it necessary to replace their existing standards or were successfully 'urged' to indulge in a complete set. There is a single extant stone weight at Culross, a half-stone at Lanark itself, and part of a full set (lacking the stone and pound) at Edinburgh. They are cylindrical weights of brass, the larger ones with bar handles, stamped with the double-headed eagle which represents the arms of Lanark, with the date 1618 and with their denomination in 'Lanerk Troys'. Nothing is known about the making of these weights: however, they are of such good quality that it seems unlikely they were made in Lanark, and it is suggested that they were made, and adjusted in Edinburgh, possibly by the Mint officials.

Unfortunately, the few weights that exist now show considerable variation and their value as evidence is limited. For example, the Culross stone weight (Item **35**) has a damaged base and the handle has been repaired with extra weight added around the neck. It has been adjusted and is now to an ounce of 476·7 grains, but this adjustment has undoubtedly been done at a late stage, probably in the eighteenth century. This weight was described in the Perthshire report on weights and measures compiled in the late 1820s to show the relationship of local units to the newly introduced Imperial system.[69]

The 8-pound weight at Lanark (Item **36**) is in reasonably good condition, and now weighs 60,697 grains, representing a present ounce of 474·2 grains. This is the sole survivor of what must once have been at least one full set of weights at Lanark, and which should have been considered as forming the nation's principal reference after 1618. Seven of these had survived to be weighed in 1827, when the ounce of the 8-pound weight was 474·25 grains.[70] The 4-pound and 1-pound weights were in better condition, and were to ounces of 475·1 and 475·5 grains respectively, but the 8, 4, 2 and 1-ounce weights were all to ounce sizes of less than 474 grains. The writer of the 1827 Lanarkshire report had taken a weighted mean of these, which produces an ounce of 474·5 grains, but it is surely more realistic to assume that the standard is more accurately reflected in the weights with the largest ounce sizes, given only that these weights are substantial enough. In this case the pound and 4-pound weights both give ounces of over 475 grains, and so we deduce that the weight basis is an ounce of about 475·5 grains.

Luckily, we can get some independent confirmation of this from a stone weight of the same series and with the correct markings, which the writers of the Edinburgh report of 1825, James Jardine, Alexander Adie and David Murray, understood to have been the missing stone from Lanark's set. This appears to have been borrowed from Lanark in the eighteenth century and it was found as part of a suite of 1618 weights by John Robison in 1778 'in the possession of Mr Thomas Simpson, pewterer, in Edinburgh and used by him to adjust the weights in the burgh, made of brass inscribed "Lanark Troys 1618"'.[71] Robison found that the stone was 'good' but that the smaller weights were 'worn and readjusted'.[72] The stone remained for a time in the Edinburgh dean of guild's care, marked with the weight that Robison had measured, and as having been examined by him on 24 February 1800, and it was recorded in the dean of guild's inventory as 'The original Lanark Troy

Stone'.[73] The stone may subsequently have been returned to Lanark, but is not known to survive. The mass of the stone was found by Jardine and his colleagues as 121,743·2 grains, which yields an ounce of 475·56 grains. Not only does this correspond with the best results from the 1827 measurements of the weights at Lanark, it is also an excellent match for the mass of the dean of guild's stone weight obtained by Hope in 1647, which was to an ounce of 475·5 (modern) grains.

It is not known why there should have been such variation in the weights at Lanark tested in 1827, but one possibility is that they may have been a mixed set. The likelihood is that the 1618 standards made for distribution were constructed in Edinburgh and then dispatched to Lanark. It would be readily apparent to officials at Lanark that an alteration had been made to the ounce standard. It is suggested that at least some weights (including at least one reference set retained by Lanark) were adjusted to match the previously accepted burgh standard – the Flemish ounce of about 474·4 grains – and the 8-pound weight (and the fractions of the pound) are from this adjusted group. The adjustment was presumably made to the bases of the weights, and the 8-pound weight now has Lanark's stamp impressed into its base.

Another indication of the slight increase of the ounce size in 1618 is seen in an 8-pound trone weight in the collection of early standards and unissued weights at Lanark. This weight has a tapering square section, very similar to the group of trone standards described earlier but lacking the double-headed eagle (Item 34). Considering the pound as composed of 20 merchant ounces, as before, gives an ounce of 475·6 grains – exactly matching the new 1618 weight basis. But the weight is not in its original form – some additional metal has been added carefully at the top of the weight around the attachment for the suspension ring. It is suggested therefore that it was originally to the lighter Flemish standard and has been increased in mass to match the 1618 standard. This might be thought to go against the abolition of trone weight in the 1618 Assize, but it is clear that the attempt at abolition was neither total nor successful – for example, Sir James Hope recorded his own use at the Mint of an unnamed stone of 20 pounds which was specifically for use in lead weighings.[74]

Difficulties are encountered when we examine the incomplete set of 1618 standard weights still preserved in Edinburgh, which now comprises weights of 8, 4 and 2 pounds, and 8, 4, 2 and 1 ounce (Item 37): all show signs of radical adjustment and they are to an ounce of a little over 476 grains. The adjustment takes the form of a brass plate brazed to the ground-down lower surface of each weight. The bases therefore no longer carry the three double-headed eagle stamps which are seen on the base of the 8-pound weight at Lanark and which must originally have been on the Edinburgh weights also. We will discuss in Chapter 9 this readjustment work, which was carried out in the 1790s, following an earlier adjustment made by Alexander Bryce in about 1753.

Jardine's measurement on the 'original' Lanark stone weight, generating an ounce which we shall round up to 475·6 grains, was accepted as the definitive mass of the Lanark stone for the county reports of the 1820s. However, doubt about this mass crept in almost immediately. By the time George Buchanan's digest of the county reports was published in 1829, Jardine had already concluded that the mass he had obtained for the 'original' Lanark stone was unreliable.[75] Robison had previously come to depend on Robert Stewart's calculations, and as a result it was the accessible but readjusted Edinburgh set of 1618 weights that influenced subsequent work. The growth of this scientifically-informed antiquarianism will be addressed in Chapter **9**.

However, we should first establish that this new weight standard of about 475·6 grains to the ounce was generally applied and was not restricted to a few sets of official weights issues in 1618. An illustration of this comes from the impressive but incomplete weights of the Stirling Guildry. These now comprise two large hexagonal and three cylindrical weights (hence perhaps for different classes of use), all with loop handles and of a recognisably Dutch design (Item **40**). They correspond with the set of weights, from 50 pounds down, ordered by the Guildry from Holland in 1676.[76] The cylindrical weights exhibit a certain amount of wear and must have been to an ounce standard of over 474·4 grains; but the two hexagonal weights of 25 and 50 pounds are in good condition and are now to ounces of 475·6 and 475·4 grains respectively, matching the 1618 standard.

As we have seen, after 1618 the ounce in Scotland was about 475·6 grains and had been incorporated into the weights of Edinburgh's dean of guild which were measured by Hope in 1647. Only one further significant change was made: meal, an important component of everyday life, which had previously been measured by the boll was now to be reckoned by weight. In 1624, Hunter had stated that: 'If Meale were solde also by weighte, it might prove proffitable to the Leiges'; as indeed it might, considering the variability in the act of measuring by volume.[77] This suggestion was only acted upon in 1679 when the Privy Council included meal in a directive that bere (or barley) was to be weighed, not measured. They had found that 'inequality of the values of the boll of bear in the severall shires of the kingdome renders it deficult and unequall to appoint one price of drink effeirand [corresponding] to the measure of a boll of bear'. A committee was appointed 'for adjusting the Linlithgow measure to ane equall standard of weight', and after weighing different samples of barley they recommended a compromise weight for the boll's contents. The Privy Council's verdict was that:

> … in time coming the bear shall be sold … by weight and not by measure, and ordaines the magistrats of Linlithgow to prepare and make the standard weights of brasse, containing fifteen stone troas weight as the weigh for the new bear, and another standard of fourteen stone troas weight for the weigh of the old bear [of the previous season] …

And, farder, wee, with advice foresaid, have ordained that in time coming the constant rule of selling meall shall be by weight and not by measure, and that in place of a boll measure the same be sold and received at the weigh of eight stone troas weight. And the saids magistrats of Linlithgow are to make a standard conforme …[78]

Seventeen years later this was confirmed by parliamentary legislation.[79] If the intent of the 1679 act was that Linlithgow should manufacture weights of this magnitude, at present they are conspicuous by their absence; however, there is little doubt the act was put into effect.

There are still several unresolved problems with the seventeenth century weights, as there are with the dry capacity measures of the same period. First, Sir James Hope, writing thirty years after the 1618 Assize, made it clear that *fleur-de-lys* weights were still the common weights of the merchants, and therefore by implication the weights of the Edinburgh merchants.[80] So, even in the principal burgh of the kingdom the new weights may not have been used for more than official functions such as implementing the important Assizes of Bread and Ale (see Appendix E). Second, it is not clear why the *fleur-de-lys* weights examined by Hope should have been half a grain in the ounce lighter than the accepted (albeit nineteenth-century) value of 474·7 grains to the ounce for the troois weights.[81] But we have now been able to make an independent check of some of Hope's 1647 comparisons, and so we must give some authority to the accuracy of his value of about 474·2 grains. An important confirmation of this is also provided by a pristine seventeenth-century trone stone weight preserved at Lanark (Item **42**) which carries a stylised version of the Lanark double-headed eagle. The troy mass basis of this weight is an ounce of 474·2 grains (or 474·3 old grains).

Perhaps the reason for Hope's half-grain difference lies in the type of fine adjustment to localised standards that has been described by Houben.[82] In this connection it should be noted that the Scottish staple port in the Netherlands moved from Bruges in Flanders to the deep-water port of Veere in the neighbouring county of Zeeland in the mid-fifteenth century: although in the third quarter of the century there was no fixed staple, it was re-established at Veere by James IV in 1508.[83] Weights encountered by Scots merchants would therefore be adjusted to Zeeland standards at Middelburg, which did not necessarily correspond precisely with Flemish standards.[84]

Third, by the beginning of the eighteenth century the most significant Continental weight standard for Scots merchants would probably have been the Amsterdam commercial weight, to an ounce of 476·6 grains.[85] It is usually this weight that is referred to as 'Amsterdam' or 'Dutch' weight in Scottish records. It is very close to the somewhat artificial ounce of 476 grains proposed by Stewart, or that of 476·25 grains proposed by Robison, which were obtained (see Chapter **9**) by interpreting the 1618 Act literally for a definition of the ounce as a fifty-fifth part of the pint's contents. But there is no indication in the 1618 Assize that any change at all was being made in the weight basis. Should we then interpret the 1618 ounce of 475·6 grains as a compromise step intended to establish a standard (where there had not previously been one) which was sufficiently close to the existing merchant weights for the change to pass unnoticed, yet one that helped bridge the gap between the Scots weights and a valuable external trading equivalent?

The answer to questions such as these may be found in a better understanding of external standards, or perhaps in a move to bring the Scottish system progressively into line with that of England. There is reason to suppose that the issues addressed at the assize were considered carefully: the 1618 Assize was set up during the first visit to Scotland since the Union in 1603 of the joint monarch of the two kingdoms, and King James and his advisers were actively concerned with fiscal relationships between the kingdoms. As we have seen in Chapter 7, one of his close and early economic advisers was Sir Archibald Napier, later first Baron Napier, who was the son of John Napier, the inventor of logarithms, and grandson of Sir Archibald Napier (the metrologist of the 1587 Assize), and who became Depute Treasurer of Scotland in the 1620s.[86]

One thing is fairly certain: only the most sensitive balances could possibly have discriminated successfully among these closely similar ounce sizes, differing by only about two grains per ounce, and these differences must have been small compared with the degree of latitude in trading weights permitted by the burgh officials. For almost all purposes of trade, and presumably all general trade in produce conducted in the burgh markets, these ounces were indistinguishable. It is of interest to try to understand their relationship, and clearly it was a recurrent concern for the administration in formulating official definitions, but these small adjustments are likely to have affected only a tiny minority of people, probably all of them closely involved in bullion movements.

THE MINT STANDARD AND THE COINAGE OUNCE

Amongst the most demanding issues in weighing are those affecting the weight and fineness of money coined at the Mint; and the control of money supply (as a last resort, for example, through depreciation) was an important element in the monarch's fiscal policy. Merchants trading abroad were quick to exploit any discrepancies in coinage value that could be turned to their advantage. For the same reason that excessively-worn coin was rejected by the observant, over-weight coin was culled out for re-coining if this could be done at a profit. If foreign mints offered a more attractive bullion rate than the domestic rate, as happened at the Dutch mints in the fifteenth century, then Scots merchants would contrive to export large quantities of coin for melting even though this was illegal. The administration's response to a shortage of bullion was to require returning merchants to bring a small set proportion of their cargo's value in the form of bullion, which could then be purchased by the Scottish Mint, but this probably did not significantly reduce the nett outflow of coin while there was still sufficient profit motive and such pronouncements from the administration occur regularly from the fifteenth century.[87]

The Scottish Mint at Edinburgh operated a weight series for bullion purchase which was very similar to but separate from the merchant weight, and had an ounce of about 471 grains. This was slightly lighter than the Paris ounce, which is presumed to be the French ounce to which the coinage was struck. It is plausible that the difference between them may be related to the coinage 'remedy', or permitted weight range for new coin, to prevent the Mint from losing financially in the purchase of light coin for re-coinage. These various Scottish weight systems had ounces which are very close to each other and which are difficult to distinguish, particularly since they are often described only as 'trois' weight. However, the fact that the bullion and coinage ounces relate specifically to the internal operations of the Mint allows the appropriate context to be provided for a few instances where detailed equivalences with other weight systems are known, and from these we can at least demonstrate continuity in the Mint's bullion ounce.

The first accurate comparison of any Scottish weight series in terms of an external system is one given in the context of Scottish Mint operations in 1604, the year after the Union of the Crowns when James VI of Scotland also assumed the English throne as James I. A set of standard English troy weights, up to one troy pound, was sent from the London Mint to Edinburgh in 1604 for official comparison against the Scottish Mint weights. Calculations were performed by Sir Archibald Napier, the General of the Mint, and these reveal a mint ounce of only 471·15 grains.[88] It is quite possible that his son, the mathematician John Napier, was involved in this work (his brother Francis had been appointed assayer at the Mint in 1581, so it was rather a family concern), and his reputation was already sufficient for John Skene to have consulted him on metrology in 1597 and to refer to him as 'sear of *Merchistoun*, ane gentleman of singular judgement and learning, specially in Mathematicque sciences'.[89] The calculation is known from its inclusion in papers connected with a subsequent general of the Mint, Sir James Hope, and was first published by R. W. Cochran-Patrick in 1876.[90] There is, however, also a version in the Scottish exchequer papers, and there can be little doubt about the authority of the calculated value.[91] A marginally different value (clearly based, however, on the same figures) was published by Alexander Hunter in 1624, yielding an ounce of just over 471·2 grains.[92] Hope was still using Napier's 1604 figures in 1647, without modification or qualification.[93]

From 1604 onwards there was an accurate enough knowledge of the

English troy equivalent of the Scottish Mint's troy weights to enable them to be maintained at the appropriate level, but there was also a cogent administrative purpose, which had not been present before the Union of the Crowns, to ensure that no disparities emerged between the coinage quality of the two kingdoms. On one notable occasion in the 1680s the General of the Scottish Mint was found to have been operating outside the agreed conventions, and there is a surviving 1682 report of a commission appointed by James VII and II to investigate the Edinburgh Mint.[94] Four sets of nested troy weights (or 'pyles') were prepared in London in 1662 to the accurate bullion standards of England and Scotland, as part of new controls to regularise the weight and fineness of the separate coinages of the two kingdoms. Two sets (presumably one English and one Scottish) were retained in the Tower of London and the other two sets (again presumably one English and one Scottish) were sent to the Scottish Mint. However, the Scottish bullion weights at Edinburgh had been put to one side by Charles Maitland of Haltoun (d.1691), who had been appointed General of the Mint at the Restoration, and bullion transactions were conducted using a slightly heavier standard, namely that of the Edinburgh dean of guild's weights, which were the legal weights of the 1618 Assize.

The master coiner discovered this situation in 1680 and re-adjusted his weights to the correct standard, but Maitland required him to surrender these weights and have a new set adjusted to the dean of guild's standard for 'his greater gaine'. The particular gain for Maitland was that merchants delivering bullion to the Mint found themselves obliged to sell it according to the heavier dean of guild weights rather than by the 'new pyle of weights sent hither by his Majestie for the common rule of weight', and were therefore underpaid. The difference between the weights was recorded in 1682 as 2½ ounces per stone, which is 4·6 grains per ounce.[95] We have concluded that the ounce size legalised at the 1618 Assize was 475·6 grains, and therefore the Mint's bullion ounce was a little in excess of 471 grains. This is certainly compatible with Napier's value of 471·15 grain per ounce, and the simplest inference is that the bullion ounce had remained unchanged.

Charles Maitland was dismissed in 1682, but his eldest son, Richard, Lord Maitland (1653-95), had been in post as joint General of the Mint since 1668, remaining there until 1689.[96] A new issue of large silver coins was also produced in 1687 and 1688, and a surviving nest of standard troy weights dated 1687 (Item **41**) seems to have been the new Mint standard by which they were regulated.[97] These weights are now preserved in Edinburgh City Museums. In 1789 John Robison (1739-1805), appointed professor of natural philosophy at the University of Edinburgh in 1773, recorded weighings for the Edinburgh dean of guild, and correcting for the ounce basis of Robison's weights (see entry for Item **41**) allows the ounce for this set to be given as 471·0 (modern) grains.[98] The weights can be identified from the inscription Robison recorded, but he did not appreciate the significance of these weights and could not account for the ounce size: this perhaps explains why the weights were not carefully preserved in their original state but were subsequently adjusted to English troy (with an ounce of 480 grains), and this was probably done in the 1820s or 1830s, perhaps to enable them to be used in the preliminary adjustment of weights. The adjustment has been made by the inexpert addition of lead to the base recess of each cup weight. With the agreement of the Director of Edinburgh City Museums, the lead adjustment was removed from one of the larger weights in the set, and this allowed the ounce basis for this single weight to be obtained directly as 470·9 grains. In the Inventory entry for this set we use an earlier set of weighings by Robert Stewart to demonstrate that the mean ounce for the set was 471·0 modern grains, and similarly we can reduce John Napier's equivalence to 471·05 modern grains, representing effectively precise coincidence of these very different sources for the Mint's bullion ounce.

The differences we are encountering between these various values are very small indeed, and almost certainly consequences of the measuring and comparison process. For example, we have had to assume the absolute accuracy of the single small set of English troy weights used in the 1604 comparisons.

Fig. 8.8
Markings on the set of troye nesting weights, pile of 512 bullion ounces (or 32 pounds), 1687, for the Scottish Mint, Edinburgh. (Item **41** in the Inventory.)

Equally, we have been unable to take account of the sensitivity of the balances and beams used in such comparison work, although this clearly imposes a serious limitation on the precision of individual measurements. It seems curious that our understanding of what constituted the bullion weight for Scotland should be more sketchy than our knowledge of the merchant weight: in this period our *direct* knowledge of troy weight is limited, but we have fairly secure details of a merchant weight of 475·6 grains to the ounce from the 1618 Assize and from James Hope's calculations of 1647. This figure is also compatible with the findings of the 1682 report on the Mint.

We can obtain a view of the practical operation of the Scots troy system by examining detailed figures given in the Privy Council minutes relating to carefully weighed quantities of silver table-ware sent from London for the use of James VI and I during his 1617 progress through Scotland. This very valuable consignment, totalling some 35 stone (280 kg), was weighed in England before dispatch, and then re-weighed in Scotland after receipt by the Depute Treasurer. The Edinburgh weighing was apparently conducted at the Mint, and the availability of the masses of the several batches of silver in both weight systems enables us to derive a practical equivalence between them. Noting the presence of rounding errors, particularly in the English weighings, which are only given to the nearest ounce, the mass of the Scots Mint ounce comes out a little higher than might have been expected, at 471·6 grains.[99]

At the very least, this suggests that we should be cautious in interpreting precise levels of equivalence for the troy standard. These slightly different figures may suggest that some small internal inconsistencies existed in the adjustment of operating (as opposed to standard) weights at the Mint, and qualifies to some extent the reliance that can be placed on comparisons between individual and small denomination weights such as those of 1604. In practice we may only be justified in saying that the troy system appeared to operate to a standard of about 471·5 grains to the ounce, although the administration's understanding of the Scottish Mint weight was based on an ounce of about 471·15 grains measured in 1604. Although the Mint's weights regulated its own activities, the 1682 enquiry tells us that they were also intended to form an official troy standard for bullion transactions. This is an important consideration since, as we have said earlier, merchants were often under government pressure to import bullion in the course of their foreign trading. However, the report also implies that this was not a universally-applied troy weight, and that the merchant weights of the burghs were also considered relevant for bullion transactions. These, as we shall see, were based on the troy weights of the Low Countries; and they probably had a greater claim to be described as the troy weight of Scotland than the 'official' weights of the Mint.

A clear distinction must be drawn between the bullion ounce, used to control the bullion purchased by the Mint, and the coinage ounce, which formed the theoretical weight basis of the minted coin. Modern numismatic scholarship accepts that Scottish coins (and certainly those of the sixteenth and seventeenth centuries) were minted in relation to French weight. It seems inescapable that this was initially based on the Paris ounce of 472·5 grains – as we have seen in Chapter 4, this played a central role in early commercial and precision weighing, and we have argued that this ounce formed the basis for Scottish metrology from the early sixteenth century. The implication of this is that the silver coinage remedy in 1604, when John Napier gave the equivalence for the bullion ounce, was about 1·35 grains in the ounce, or about 0·3 per cent. (For comparison, an earlier English silver coinage remedy of one penny in the pound of 240 pence is 0·4 per cent.) In theory, coin weights, such as those specified in 'trois' ounces in parliamentary acts, should be interpreted in Paris ounces; the existence of a permitted remedy inevitably leads to a tendency on the part of Mint officials to produce coins as near as possible to the lower end of the coinage remedy range in order to maximise profits. In practice, therefore, it would merely be necessary to demonstrate that minted coin (in the appropriate multiples) was slightly heavier than the required trois weight in bullion ounces.

In some instances it is possible to come to an independent view about the identity of the ounce used as the weight basis for the coinage, since coins were normally struck with a set number to the ounce: providing sufficiently good examples can be found, it may be possible to deduce the ounce type. Alternatively, if the ounce type is known, the minted weight of the coin can be deduced. Thus, for example, Ian Stewart has considered the 1450 silver coinage of James II, coined at 8 groats to the ounce, and obtained a weight of 58·9 grains, based on use of a French ounce (taking the value given by John Napier).[100] The assumption in recent years has been that both English and Scottish mints were controlled by English troy weight, with an ounce of 480 grains, and therefore necessarily utilising the Tower ounce as a poise weight (at 20 silver pence to the ounce of 450 grains): it followed that a change at the Scottish Mint to the use of French weight should be detectable. The authority usually cited for this is an entry for the Mint's accounts for the period 1436-8, described by Burns as relating to an ounce 'conforme to French Weight'.[101] In fact Burns was incorrect in associating this quotation with the 1436-8 accounts, which are those of Robert Gray, master moneyer, and which record only the use of 'libra continente sexdecim vncias de troya'.[102] There is nothing to indicate the ounce type of this 16-ounce pound, which might be the 472½-grain Paris ounce, but equally might be the 480-grain 'trois' ounce which figures to the exclusion of others in the 1426 Assize. The increase in the weight of the 1450 groat implied by considering this ounce is only about 1½ per cent which is significantly less than the amount of wear that would be expected in surviving examples of the coin.

SUMMARY

The diversity and obscurity of the old Scots weights has presented a serious obstacle to their interpretation. Although the analysis given here has involved considering a great variety of ounce sizes and weight types, the evidence of the records and of the surviving objects points to a comparatively simple evolutionary structure.

In the earliest period we have a general trading pound, of 15 Cologne ounces, each of 450 grains (hence 6,750 grains). This is identical to the contemporary English pound. It was combined in stones of 15 pounds. From at least the second half of the fourteenth century there was also a distinct merchant pound for fine goods (the 'Scots' pound), containing 16 ounces and combined in stones of 16 pounds. This weight also served a troy function for weighing precious metals. The trone pound, appropriate for heavy goods, was probably introduced shortly afterwards, and ultimately replaced the original 15-ounce pound. It is suggested that it was introduced at 1¼ times the mass of the Scottish fine goods weight to mirror English practice.

In 1426 a separate official troy series called trois weight was introduced with a pound of 16 troy ounces of 480 grains, and hence a pound of 7,680 grains. The merchant and troy weight continued unchanged until about 1500, when both were reduced by about one per cent. Although the official level for the ounce was probably the Paris ounce of 472½ grains, in practice the trading burghs adopted an ounce of 474·2 grains, following the troy standard of the Low Countries.

In 1563 the merchant weight was finally raised to the level of the troy weight, in an acknowledgement that merchant trade with the Low Countries had been conducted in Flemish troy weight for at least 50 years and probably for a great deal longer. In effect, the old merchant ('Scots') weight simply faded away. Over the next ten years the heavy-goods or commercial trone weight also rose by the same amount (about 6 per cent) because its size was tied to the merchant series. All the other changes have represented small adjustments or gradual accommodations; but this rise in the size of the trone weights in the 1560s is the only potentially disruptive change encountered.

For the whole of the sixteenth century there had been official standards in the burghs only for the commercial trone weight, but not for the troy weight. At the 1618 Assize an official standard for the combined troy and merchant weight was introduced, at about a quarter of a per cent higher than the Low Countries' standard, but for most mercantile purposes the earlier weights continued to be used. This troy weight, however, was distinct from a separate weight series used for coinage and bullion transactions at the Scottish Mint, which continued to operate using the Paris ounce, with the bullion ounce adjusted to a level representing the coinage remedy below the Paris ounce.

1   T. Thomson and C. Innes (eds), *The Acts of the Parliaments of Scotland [APS]*, 13 volumes (Edinburgh, 1814-75), IV, 538: Act 8, 28 June 1617; ibid., IV, 585-9: quote p. 587.

2   *APS*, IV, 587: 19 February 1618.

3   J. Hill Burton, *et al.* (eds), *The Register of the Privy Council of Scotland [RPCS]*, three series (Edinburgh, 1877-1970), second series, VIII, 333: 6 January 1613.

4   'And that Weght called of old the Trone weght to bee allvtterlie abolisched and discharged': *APS*, IV, 587: 1618.

5   See Items **31** and **32** in the Inventory.

6   *RPCS*, second series, VIII, 333: 6 January 1613.

7   Ibid., first series, I, 373-5: royal contract of 26 August 1565 with John, Earl of Athole, licencing him to extract lead from the mines at Glengonar and Wyndock (in Lanarkshire) for export to Flanders, requiring payment to the Exchequer 'for everie thowsand stane wecht – trone wecht as said is – of the said leid ure, fiftie unces of fyne silvir' (quote p. 374).

8   The issue is discussed in A. D. C. Simpson, 'Scots "Trone" Weight: Preliminary Observations on the Origins of Scotland's Early Market Weights', in *Northern Studies*, 29 (1994), 62-81.

9   Ibid.

10 For example, the following from the minutes of Lanark Council on 22 December 1658: 'The baillies and counsell ordaines the wechts in the wiehouse, the tron stone to be just XX lib. trois and the rest conforme': Robert Renwick (ed), *Extracts from the Records of the Royal Burgh of Lanark [Lanark Extracts]* (Glasgow, 1893), 170.

11 Edward Burns, *The Coinage of Scotland*, 3 volumes (Edinburgh, 1887), II, 56. The reference is from J. D. Marwick, *et al.* (eds), *Records of the Convention of the Royal Burghs of Scotland [RCRB]*, 7 volumes (Edinburgh, 1866-1918), I, 76; also printed in R. W. Cochran-Patrick, *Records of the Coinage of Scotland*, 2 volumes (Edinburgh, 1876), I, lxxviii. We are grateful to Joan Murray and Ian Stewart (Lord Stewartby) for their comments on this issue: personal communications, February 1993.

12 For example, Alexander Stevenson assumes these changes took place at the 1426 Assize in his calculation of the size of the woolsack as 187·58 kg: A. Stevenson, 'Trade with the South, 1070-1513', in M. Lynch, M. Spearman and G. Stell (eds), *The Medieval Scottish Town* (Edinburgh, 1988), 180-206, p. 194. The calculation depends on the sack still containing 24 stones, now of 16 pounds each of 16 ounces, but the ounce has been taken as the later Mint bullion ounce.

13 Ranald Nicholson, *Scotland: The Later Middle Ages* (Edinburgh, 1989), 596-600.

14 J. D. Marwick, *et al.* (eds), *Extracts from the Records of the Burgh of Edinburgh (1403-1718) [Edinburgh Extracts]*, 14 volumes (Edinburgh, 1869-1967), I, 133: 6 May 1511. The earlier decision was not located in the manuscript minutes: personal communication from the City Archivist, February 1993.

15 *Dictionary of the Older Scottish Tongue [DOST]*, 12 volumes (Chicago, Aberdeen and Oxford, 1931-2002), s.v. *minis*.

16 W. Hay (ed), *Charters, Writs and Public Documents of the Royal Burgh of Dundee, 1292-1880* (Dundee, 1880), 26: document 47, 20 March 1511/12.

17 *APS*, II, 246: Act 47, 15 March 1503/4.

18 Ibid., II, 226, Act 15, 18 May 1491.

19 R. D. Connor, *The Weights and Measures of England [WME]* (London, 1987), 127.

20 See, for instance, Alexander Stevenson, 'Medieval Scottish Associations with Bruges', in Terry Brotherstone and David Ditchburn (eds), *Freedom and Authority: Historical and Historiographical Essays presented to Grant G. Simpson* (East Linton, 2000), 93-107.

21 *APS*, II, 541: Act 14, 4 June 1563.

22 The change in the troy weights amounts to an increase of about 3½ gr in the mass of the troy pound, which remains defined as 5,760 gr. In terms of these larger grains, the mass of a stone defined as 141,750 of the old grains weighs about 141,660 modern grains.

23 The size of the Mint weight, and the nature of the corrections are discussed in the final section of this chapter, and also at Item **41** in the Inventory.

24 Joan E. L. Murray, 'The Organisation and Work of the Scottish Mint 1358-1603', in D. M. Metcalf (ed), *Coinage in Medieval Scotland (1100-1600): Second Oxford Symposium on Coinage and Monetary History* (Oxford, 1977), 155-69.

25 *RPCS*, third series, VII, 357-8.

26 Alexander Huntar [*sic*], *A Treatise of Weights, Mets and Measures of Scotland* (Edinburgh, [1624]), 2.

27 *RCRB*, I, 2: 4 April 1552.

28 Ibid.

29 Ibid. There is a void in the Statute Book between 1551 and 1555.

30 We are grateful to John Higgitt, Department of Fine Art, University of Edinburgh, for his advice

in interpreting the name in this inscription: personal communication, January 1994.

31 Sir James Balfour Paul (ed), *Accounts of the Lord High Treasurer of Scotland,* 13 volumes (Edinburgh, 1900-78), VII, 199; X, 234, 301, 355, 402. For details, see the entry for Item **31** in the Inventory.

32 W. B. Cook (ed), *The Stirling Antiquary,* 4 volumes (Stirling, 1893-1908), II, 151.

33 W. B. Cook, 'Notes for a New History of Stirling, Part II: A Municipal Relic of Old Stirling', in *Stirling Natural History and Archaeological Society Transactions,* 20 (1897-8), 65-9.

34 *RCRB*, I, 76: 27 February 1578.

35 Ibid., I, 482: 30 July 1596.

36 Ibid., II, 242: 4 July 1607.

37 *APS*, IV, 587: 1618.

38 Ibid., II, 541: Act 14, 4 June 1536.

39 Ibid., III, 521-2: Act 136, 1587.

40 Ibid., IV, 587: 19 February 1618.

41 For example: *RCRB*, I, 76, last day of February, 1578 at Cupar, Fife: ' … all burrowis within this realme haif thair wechtis, sic as stane, half stane, pund and half pund, the trone wecht, according to the stane of Laneark, to be maid of bras, and markit with the tovnis stamp of the burgh … Ilk trois wecht keip the iust wecht of xvj vnces for the pund, conforme to the Frenche wecht …'

42 *Lanark Extracts*, 37: 22 October 1567.

43 Ibid., 52.

44 *RCRB*, II, 165: 7 July 1603. See also ibid., II, 5: 5 July 1597; II, 26: 1 July 1598.

45 John Skene, *De Verborum Significatione* (Edinburgh, 1597), s.v. *serplaith.*

46 Hunter, op. cit. (26), 2.

47 Cochran-Patrick, op. cit. (11), I, lxxix.

48 *RCRB*, I, 482: 30 July 1596.

49 Cochran-Patrick, op. cit. (11), Introduction, lxxii.

50 Reference to 'libra continente sexdecium uncias ponderis de Troya': ibid., lxxvii.

51 *WME*, 153 (for 1474 and 1497).

52 We acknowledge the help and advice of Marace Dareau, of the Scottish Language Dictionaries, in this discussion.

53 We are grateful to Marace Dareau for this suggestion.

54 K. M. C. Zevenboom and D. A. Wittop Konig, *Nederlandse Gewichten* (Amsterdam, 1970), 19-29, esp. pp. 24-5.

55 Gerrit Moll, 'On the Comparison of British, French and Dutch Weights', in *Journal of the Royal Institution of Great Britain,* second series, 2 (1831), 64-75. The principal exceptions are a group of three standard weights preserved in Bruges, which are discussed in Chapter 4.

56 Cochran-Patrick, op. cit. (11), I, lxxix.

57 Ibid., I, lxxx; ibid., I, lxxix: 20 July 1647. Hope gave 16 ounces English troy as 16 ounces 3 drops 6 grains trose *fleur-de-lys* weight, or 12 ounces English troy as 12 ounces 2 drops 13½ grains trose weight; both yield 1 ounce trose as 474.14 grains English troy. His measurement of trose *fleur-de-lys* in terms of Mint weight gave 16 ounces trose as 16 ounces 2 denier 11 grains Mint weight, where there are 24 (Scots) grains to the denier, and 24 denier to the ounce; yielding the trose ounce as 474.17 grains English troy. The size of the trose ounce as measured by Hope will be given as 474.2 grains English troy. But, recognising that by this date the English (Mint) weight to which Hope would have had access would have been absorbed about two-thirds of the small increase in the size of the troy standard, this should be considered as 474.1 modern grains or 474.4 old (pre-1580s) grains.

58 *RCRB*, I, 76: February 1578 and ibid., II, 242: 4 July 1607.

59 *RPCS*, second series, VIII, 333: 6 January 1613.

60 *APS*, III, 521: Act 136, 29 July 1587. It was suggested in Chapter 7 (in the section on the growth of the firlot in the sixteenth century) that 41 ounces (rather than 55 ounces) had been described to bridge the otherwise difficult gap in the statutory definitions.

61 *APS*, IV, 587: 1618.

62 *Edinburgh Extracts*, V, 6: 3 October 1589.

63 Ibid., V, 332-3: extracts from the Gild Accounts; C. G. Drummond, 'An Historical Account of the Weights, Mets and Measures of Scotland', in *The Chemist & Druggist,* 175 (1961), 675.

64 Cochran-Patrick, op. cit. (11), I, lxxx.

65 *Edinburgh Extracts*, V, 333.

66 Cochran-Patrick, *op. cit.* (11), I, lxxix. The difference of 12 French grains yields an ounce of 473.53 troy grains. Hope's second equivalence, notes a difference from the Mint pound of 2 deniers less 1 French grain and yields 473.56 troy grains; this is not independent but is based on Hope's relationship between the Mint and *fleur-de-lys* weights, described in ref. 57 above. For Hope's calculations for the stone and 4 pound weights, see ibid., lxxix-lxxxii.

67 *APS*, III, 521: 29 July 1587.

68 *RCRB*, III, 71: 15 July 1618.

69 G. Buchanan, *Tables for Converting the Weights and Measures hitherto in Use in Great Britain …* (Edinburgh, 1829), 246. Buchanan's weighing gives a slightly higher ounce basis of 476.97 grains.

70 Ibid., 230-1.

71 National Archives of Scotland (NAS), Court of Session process CS 231/F3/1, bundle 10.

72 Robison's weight for the stone was given by him in

1788 as 16 × 7,615 = 121,840 grains, with an ounce basis of over 475·9 grains, showing that this too had been adjusted upwards.

73 Ibid., 199; Edinburgh Council Archives (ECA), MS town council minutes, 23 October 1805.

74 Hunter, op. cit. (26), 3; Cochran-Patrick, op. cit. (11), I, lxxx.

75 'Minutes of Evidence taken before the Select Committee [of the House of Commons] on the Bill to amend and render more effectual two Acts of the Fifth and Sixth Years of the Reign of his late Majesty King George the Fourth, relating to Weights and Measures', *Parliamentary Papers* 1834, XVIII, 256-64 (evidence of James Jardine, civil engineer): see p. 263; Buchanan, op. cit. (69), 21-2 note. Jardine found the internal adjustment recess empty and deduced that there had originally been adjustment metal present: he therefore assumed the weight would originally have been heavier. In contrast, we have assumed that provision for adjustment was to take account of wear after issue, and that the cavity was empty at the time of issue in 1618. The basis for this assumption is that Jardine's original figure for the ounce matches the 1618 ounce derived from other independent sources.

76 'The treasurer … to send to Holland for a 50 lb, 25 lb, ½ stone, ¼ stone, 2 lb, 1lb, ½ lb and ¼lb brazen weights with a steel balk [beam] for weighing 50 lb weight and scales belonging thereto, and to be kept *in retensis* for adjusting gild brethren's weights': W. B. Cook and D. B. Morris (eds), in *Extracts from the Records of the Merchant Guild of Stirling* (Stirling, 1916), 75: 19 January 1676.

77 Hunter, op. cit. (26), 3.

78 *RPCS*, third series, VI, 366-7: 23 December 1679.

79 *APS*, X, 18, 25, 34 and 75: 15, 19, 25 September and 12 October 1696.

80 Cochran-Patrick, op. cit. (11), I, lxxix.

81 Patrick Kelly, *Universal Cambist and Commercial Instructor,* second edition (London, 1835), 20-1; Zevenboom and Wittop Koning, op. cit. (54), 24-5; D. A. Wittop Koning and G. M. M. Houben, *2000 Jaar Gewichten in der Nederlanden* (Lochem-Poperinge, 1980), 18.

82 Ibid., 67.

83 Stevenson, op. cit. (12), 198-201.

84 For a general comment on such regional variation see Wittop Koning and Houben, op. cit. (81), 67.

85 Zevenboom and Wittop Koning, op. cit. (54), 35.

86 Archibald Napier (1576-1645), eldest son of John Napier (1550-1617), the inventor of logarithms, was created Baron of Merchiston on 18 June 1597 on achieving his majority. He became Depute Treasurer of Scotland in 1622. His grandfather of the same name (1534-1608), had been General of

the Mint in 1576. For details, see the articles by T. F. Henderson in Sidney Lee (ed), *Dictionary of National Biography*, 63 volumes (London, 1894), XL, 34-7, and by W. Rae Macdonald, in ibid., XL, 59-65.

87 See Nicholson, op. cit. (13), 433-8, for numerous references to *APS* on this topic.

88 'The difference of the once English from the once Scottish is 10 grs 19 pr ½ as it wes deliuerit on tryell be the Lord [here 'Laird'] of Merkestoun at his lodging by ordour of the Committes the 22 of October 1604.': Cochran-Patrick, op. cit. (11), I, lxxxiv. There being 24 deniers in the ounce, 24 (Scots or 'French') grains in the denier, 24 primes in the grain, 24 seconds in the prime, etc., 1 ounce English troy of 480 grains is equal to 1·01877 ounces, Scots Mint weight. So the Mint ounce is 480 ÷ 1·01877 or 471·16 English troy grains. The inherent accuracy in this calculation is ± 0·01 English troy grains, and so the ounce is therefore given as 471·15 grains.

89 John Skene, op. cit. (45), s.v. *particata*.

90 Cochran-Patrick, op. cit. (11).

91 We are grateful to Dr Julian Goodare for this information.

92 Hunter, op. cit. (26), 3. Hunter has taken the 1604 equivalence of 12 ounces English troy with 12 ounces 5 deniers 9 grains and 18 primes Scottish Mint weight, and simplified this to 12 ounces 3 drops and 21 grains by converting the deniers into drops and ignoring the 18 primes: 5 deniers (of 24 grains) plus 9 grains is 129 (French) grains, which is the same as 3 drops (of 36 grains) and 21 grains, with the grains in each instance being Scots grains. This allows the Scottish Mint ounce to be calculated as 471·21 English troy grains. Hunter does not indicate that this is the Mint ounce – it is presented merely as 'Scottish weight', but he is presumably using comparison figures provided by the Mint, the significance of which will have been very clear to him from his time as King's Exchanger to the Mint: see Murray, op. cit. (24), 161.

93 Hope's memorandum of 30 July 1647 equates an English pound of 12 ounces English troy to 12 ounces 5 deniers 9 grains 18 primes Scottish Mint weight, referring to a contract between James VI and 'Thomas Achesone, M[aste]r Coinyier' in 1604: Cochran-Patrick, op. cit. (11), I, lxxix. Hope served as General of the Mint from 1641 and throughout the Commonwealth period: ibid., xxiii.

94 Ibid., II, 186-97.

95 Ibid., II, 192.

96 Ibid., I, xxiii. Richard Maitland, the eldest son of Charles, 3rd Earl of Lauderdale (died 1691), succeeded his father as 4th Earl of Lauderdale, but,

being a Roman Catholic, followed James II into exile in 1689: James Balfour Paul (ed), *The Scots Peerage*, 9 volumes (Edinburgh, 1904-1914), V, 307-10.

97 The 1686 Act (*APS*, VIII, 603-8, see p. 604: Act 38), specified the coinage fineness and weights: 'That is to say for each pound Scots of sextein ounces Conforme to the standart pile of Scots weight now in His Majesty's mint.' Coins for 40 and 10 shillings were issued, and the stated weights, assuming an ounce of 472½ grains, were 285·7 and 71·4 grains respectively. Compare Cochran-Patrick, op. cit. (11), II, 208-15, pl. XV.

98 A letter of 17 July 1789 from Robison to the council noted that their standards included 'a nest of Brass box weights marked "16IR87" from cclvi oz down to ⅛ oz. The pound is 7542 English troy grains and corresponds to no weight that I know anything about': ECA, MS council minutes, 19 August 1789. The ounce of this pound is therefore 471·38 English troy grains, about 471·4 gr.

99 *RPCS*, third series, II, 132: 13 May 1617. The weight of the silver was recorded in English and Scottish units in seven portions, amounting in all to 9,006 English troy ounces and 9,170·75 Scottish Mint ounces. In aggregate this gives the Scottish Mint ounce as 471·38 gr. But the English weighings were recorded only to the nearest ounce; the Scottish weight to one drop (¹⁄₁₆ ounce). The given weights ranged from 5,528 to 200 ounces and at least two of the weighings, 5,528 and 1,004 ounces, were far above what could have been weighed on a precision balance. It is likely that the weights given for these heavier batches were the sums of several smaller weighings hence the errors are likely to be larger than expected. Examining the effect of ± ½ oz English and ± 1 drop in Scots weight enables us to estimate the overall likely error. Applying this correction yields a Scottish Mint ounce as 471·6 gr.

100 Ian Stewart, *The Scottish Coinage with Supplement*, revised edition (London, 1967), 46-9 and 196-7.

101 Burns, op. cit. (11), II, 56.

102 Cochran-Patrick, op. cit. (11), I, 22.

9

THE SCIENCE OF
MEASUREMENT

Opposite:
Detail of Fig **9.8**.

THE BACKGROUND TO THE UNION OF THE PARLIAMENTS IN 1707 IS OFTEN
viewed by historians of the period as overtly political; it had, however, an
immediate and practical effect on the weights and measures of Scotland,
which were to be replaced henceforth by those of England. The Act of Union
and its surrounding legislation, it has often been remarked, left Scotland
with her own legal system, her distinctive form of education and her Kirk;
but it also ensured that 'from and after the Union the same Weights and
Measures shall be used throughout the United Kingdom as are now Established in England'.[1]

To allow this to be enforced, sets of measures were sent to Edinburgh
in 1707 for distribution to the principal Scottish burghs. 'Union' sets of standard length, weight and capacity measures were sent initially to Edinburgh
from London, and then on for wider distribution by those burghs associated
with each standard.[2] It appears that the contract for their manufacture was
let to John Snart, an eminent member of the London Blacksmiths' Guild.[3]
Although English weights and measures were to some extent known and used
in southern Scotland, they had hardly penetrated Scotland's internal markets.
As if to emphasise the artificial nature of this administrative imposition, the
new sets of standards included the English wine gallon of 231 in[3] (3.79 litres)
which had been in customary use south of the border for centuries, although
the first English legal definition came only in 1706. This was a consequence
of a test case in which it was ruled that this size of gallon could not be used
for assessing the duty on wine imported into England because there was no
authorised standard in the Exchequer.[4] Standards were produced in the
following year for distribution to English municipalities, and because this
also happened to coincide with the passing of the Act of Union, wine gallon
standards were amongst the sets sent to Scotland. Unfortunately for the
hopes of the legislators, this attempt to impose English measure on Scotland
met with only limited success.

Over the past thirty years, historians of eighteenth-century Scotland have examined closely the reasons for and effects of the parliamentary union of England and Scotland, particularly since a pioneering work entitled *Scotland in the Age of Improvement* was first published in 1970.[5] Broadly speaking, the key to understanding the administration, government and running of every aspect of Scotland during the long eighteenth century, appears to be in the all-enveloping web of patronage which covered positions in all walks of life from the law, through the universities and various agencies, to the encouragement of artists and artisans.[6] Scotland was ostensibly governed from Westminster, where a handful of Scottish nobles were realising their interests and becoming irreversibly Anglicised in the process. However, through the patronage of their political masters, a new class of managers emerged in Scotland, often related to each other through extended family networks, and based mainly in Edinburgh. There, through the Kirk, the law, and various government boards, new structures of administration developed during the eighteenth century.

By mid-century, Edinburgh could be seen as a centre of the establishment in Scotland. Her university, founded in 1583, became a focus of what was subsequently termed the Scottish Enlightenment. Recent work has demonstrated that this was essentially an urban phenomenon, and was centred mainly on the three universities of Aberdeen, Glasgow and Edinburgh.[7] (The fourth, at St Andrews, was based in a town by now too small to provide the urban infrastructure necessary for Enlightenment.) This was no mere local phenomenon, although it has been argued that in its Scottish form, there were marked characteristics. Throughout the eighteenth century a general spirit of enquiry into the natural and human worlds permeated much of intellectual European thought, and came to be known as the Enlightenment. In Scotland it was marked, among other important philosophical pursuits, by a particular interest on the one hand in antiquarianism, and on the other, amidst a whole range of practical scientific activities, by the quest for increasing precision in science and technology.[8]

In metrology this was shown in two ways. First, there was a desire to 'recover' the original measurement standards by locating the ancient primary artefacts and comparing them with the surviving legislation: as we have seen, there was difficulty in making these 'fit' the conventional interpretation of the parliamentary record. Unsuccessful attempts were made to legitimise the standards with reference to ancient Greek and Roman metrology, stretching their pre-history further back than the written evidence. This led to a second step: re-measuring the standards themselves, using modern and therefore ostensibly more precise procedures than could be attained in medieval times. In fact, these attempts to understand Scotland's apparently aberrant metrology further muddied the waters; but the desire to clarify the problem stemmed from a wider European movement to collate and ultimately standardise the basic units of measurement internationally.[9] This impetus came about through scientists wishing to make comparisons and check the results of experiments, and can be seen in specific instances, such as the acquisition by the Swedish scientist Anders Celsius (1701-44) of a yard scale made by the pre-eminent London instrument maker John Bird (1709-76).[10]

Serious scientific attempts to bring the Scottish standards into line with those of England date from as early as 1707 and continued throughout the eighteenth century. These efforts were not mere antiquarianism: under the terms of the Union, Scots law had been protected, but English measure had to be used from 1707, whereas existing contracts and legal documents – such as leases, land purchases, feu-duties and rents – were all couched in the old Scots terms. The imperative of legal cases was perhaps the main practical motive behind these bids to relate Scottish and English metrologies, and the expert consultants chosen for the task were invariably university science and mathematics professors, who presumably found the fee a welcome addition to their basic salary. During this period, there was pressure for greater uniformity in and between both systems; in particular, the work of the Scottish judge, John Swinton (d.1799), and that of the English MP John Proby (1720-72), Lord Carysfort, ensured that proposals for legislation were kept on the legal agenda.

There was also during this period a growing Europe-wide appreciation of the problems involved in increasing precision and the definitions of standards. The attempts to determine the marginally-flattened shape of the Earth had resulted in the production of the metre – a 'natural' unit, making up one ten-millionth of the quadrant of the Earth's meridian through Paris; and as scientific methods demanded, the 'natural' basis of metrology (where new physical standards could be replicated by reference only to units in nature) underwent critical review through the eighteenth century and well into the nineteenth century.[11] By the mid-nineteenth century – really beyond the scope of this study – the state was more than ever interested in standardisation, which moved, as Simon Schaffer has observed, from artefactual standards towards intrinsic standards.[12] During the eighteenth and early nineteenth centuries the problem of precision measurement became more complex and intricate as it was examined in increasing detail.[13]

However, despite arguments at the philosophical level, there remained the practicalities of the market-place, which is what concerns us here. There was a necessity for local units which could be used parochially, and which could produce equivalents in neighbouring and foreign metrologies. For the historian, the 'snapshot' of Scottish metrology provided in the early seventeenth century by Alexander Hunter finds a counterpart in 1779 with John Swinton's *A Proposal for Uniformity of Weights and Measures in Scotland*. In turn, this was followed by that intrepid collector of facts, working from London, but collecting data about the state of Scotland – Sir John Sinclair of Ulbster (1754-1835) – who masterminded the production of county reports on agriculture and the first *Statistical Account*, both of which drew attention to local weights and measures.[14]

All of these publications produced a picture of the disarray of Scottish metrology: but south of the border, the situation was perceived as being little better.[15] The legal standards in use in England throughout the eighteenth century were largely those which had been issued in 1588 and 1601, under Elizabeth I. The most recent addition had been, as we have seen in Chapter 6, the wine gallon, provided as late as 1707. But by the early years of the nineteenth century there had been a sea-change in the philosophy behind the definition of practical physical standards, coming from working scientists, from legislators and from interested amateurs, and much thought went into the framing of such definitions.

With the implementation of the Imperial system in 1824 for the whole of the United Kingdom, a more concerted attempt was made to bring all local weights and measures systems within the jurisdiction into line. However, as we shall see, even this rigorous effort had to be reinforced with subsequent legislation in 1835, and even beyond. Somehow old Scots measures, especially those in everyday use, just refused to lie down and die.

THE UNION OF THE PARLIAMENTS IN 1707

The Union of 1707 – still one of the most controversial topics in Scottish history – is now regarded by many historians as more of a punctuation mark in the long eighteenth century rather than either a full stop or a fresh beginning. Most of the effects of the Union – especially on the economy – had roots in the events of the late seventeenth century.[16] The Union, it has been stated, had as 'a particular English aim … to ensure Scottish acceptance of the Hanoverian succession as laid down in the English Act of Settlement of 1701'.[17] The ultimate failure of the House of Stuart – Scottish monarchs removed to London after the extinction of the (Welsh) Tudor line – to produce male heirs other than those disbarred from the succession by the events of 1688 and the Act of Settlement of 1701, was thus the political catalyst which produced a political union. But this is the English short-term point of view to an economic background, which demonstrates that the late seventeenth century was a particularly difficult time for the Scots. Famine had been frequent in the late sixteenth century, but low prices and a slight decrease in population had meant that it was not until the crop failures in 1674 and in the late 1690s that there were severe effects on the population.[18] These came as an unwelcome surprise after many years of reasonable supply.

In general, economic recovery had been slow after the Restoration of the Stuart Monarchy in 1660: the English Navigation Act passed that same year to protect English trade directly contravened Scottish naturalisation rights granted in 1608. 'The increasing difficulties faced by the Scottish economy in the later seventeenth century were partly due to the fact that while foreign trade was vital to Scotland its contribution in European terms was marginal.'[19] By the end of the century, linen had become Scotland's major export, followed by cattle and wool; whereas coal, salt, fish, woollens and grain were now less important than they had been, although still of significance.[20] Despite the English Navigation Act, by the 1680s a transatlantic trade based in Glasgow had developed with the Caribbean and North American mainland, sugar and tobacco being the main imports, linen and woollen yarn among the chief exports.

In a concerted attempt to emulate English and Dutch mercantilism, which appeared to the economically-depressed Scots as the models of successful European nationhood, Scotland's only bid to become a colonial power was launched in 1693 as what subsequently became known as the Darien scheme. Unable to raise the capital for this in London (thanks to hostility from the East India Company), in Amsterdam (resistance from the Dutch East India Company), or even in Hamburg (adverse pressure from the English consul), the Scots themselves raised the necessary funds.[21] As Rosalind Mitchison has pithily recounted:

> The disastrous history of the 'Company of Scotland trading to Africa and the Indies' has often been told: how the Scots decided to raise funds on their own, and did so, but failed to show much grasp of reality, economic, political, climatic or strategic in their location of a trading base in Panama, or in the goods sent out with which to initiate business. Afflicted by disease, the active hostility of the Spanish government which claimed the territory and the passive hostility of England and the king, the main settlement, founded in 1698, had failed by 1700.[22]

All of this – political necessity on the English side together with economic pressure on the Scottish side – led to a situation where an act of parliamentary union became first the subject of negotiation, then a practicality. The Scots were keen, after the vicissitudes of the 1690s, to receive compensation for the Darien scheme, and exemption from the high levels of tax incurred by England's European wars. A sum of about £400,000 (called the 'Equivalent') was set aside for these national debts, in due course further supplemented, in particular to promote Scottish industry. Rosalind Mitchison characterises an 'eighteenth-century Britain, in which Scotland was now an integrated part, [as] a country where government kept a low profile … [and] home affairs were left to local landowners … and to burgh councils'.[23]

Intellectual circles tended to be wider than this, although intersecting with those of the land and the burghs. It was possible to obtain lucrative appointments through patronage and kinship connections, as well as through meeting the right people at university, whether in Scotland or on the Continent – there was a remarkable network of trained medics, lawyers and scientists who had been to Dutch universities at this time. For instance, during the 1680s, a group of physicians centred around Robert Sibbald (1641-1722), met in Edinburgh every fortnight to discuss scientific ideas. Sibbald had been educated in Leiden, Paris and London, became physician to Charles II by 1682, president of the Royal College of Physicians in 1684 and first professor of medicine at the University of Edinburgh in 1685 (although he did not teach). Unfortunately, in order to remain in favour with his patrons, he converted to Catholicism just as the political wind changed, and was forced to flee the country in 1689.[24] Although he returned, his authority was perhaps never quite so assured again. Among his associates were two intellectual teachers, who subsequently formed their own spheres of influence, and who were deeply impressed by the work of Sir Isaac Newton (1642-1727), particularly by his recently-published *Principia Mathematica*. These were Archibald Pitcairne (1652-1713) and David Gregory (1659-1708), who met in Edinburgh and held much in common. Both received assistance from Newton; both were enthusiastic about his ideas. As Anita Guerrini has noted, 'within a few months of each other in 1692 [they] assumed two of the most prestigious academic chairs in Europe, Gregory of the Savilian chair of astronomy at Oxford and Pitcairne the chair of the practice of medicine at the University of Leiden'.[25] Gregory, in particular, was a member of a dynasty of intellectuals who came from Aberdeenshire. He had been professor of mathematics at Edinburgh between 1683 and 1691, and, as we saw in Chapter 2, was the author of a manuscript, *Treatise of Practical Geometry*, which examined Scots measure. Through Newton's direct influence, Gregory was enabled to escape from probable religious persecution in Edinburgh to a secure post in Oxford. In due course he was able to assist Newton, who was Master of the Mint at the Tower of London at the time of the Act of Union.

Not only was there to be a single (English) system of weights and measures in the newly-united Kingdom, but the coinage and the practices of the Exchequer were also to be integrated along English lines. This led to the immediate post-Union Scottish recoinage, under the supervision of the English mint. In a revealing paper, Athol Murray demonstrates that this successful exercise in unification owed much to Newton's interest in the 'practical and technical problems arising from the recoinage but also the friendly relations that developed between him and the officers of the Edinburgh mint'.[26] The Tower mint had supplied weights and other material to the Edinburgh mint since the Restoration, and a set of weights dating from 1687 is discussed at Item 41 in the Inventory. From after the Union, the Scottish mint was henceforth to use English troy weights instead of Scots weight. Newton also called on David Gregory to go to Edinburgh to assist with the recoinage: 'a tactful choice', remarks Athol Murray, as Gregory had earlier drawn up tables for regulating bullion at the Edinburgh mint, and he had been 'one of those involved in calculating the Equivalent'.[27] Much of the successful practical detail of the recoinage was reported by Gregory to Newton in London.

The contract for producing new sets of weights and measures was let to a London master scale maker, John Snart.[28] It is not known when the contract was placed, although it was presumably not before the crucial initial articles of the bill were approved by the Scottish parliament in November 1706.[29] Once passed, the Act of Union was to come into effect on 1 May 1707, and this date is marked on the 'Union' weights and measures. They had apparently been received in Edinburgh by 18 October 1707, when there is a reference to the gauging of the liquid measures in the council chamber and then in the natural philosophy classroom at the university.[30]

As before the Union, the new standards were to be distributed among the burghs, as custom had dictated. Twenty-one sets were received at Edinburgh, which was to decide the burghs which were to acquire them. Two complete sets were retained by Edinburgh, together with the stock of linear measures; and the remainder were to be divided into three groups comprising the dry measures, the liquid measures, and the weights, and passed on to the former holders of individual Scots standards – Linlithgow, Stirling and Lanark respectively – who wished to preserve their former privilege of making and distributing duplicates to the other burghs. Stirling was so anxious to press its case that its council decided the Stirling Jug should be carried to Edinburgh 'for vindicating the tounes right to the keeping of the liquid measure'; however, it was not until early 1708 that the stock held in Edinburgh was released to the three distributing burghs.[31] The Convention of Royal Burghs, meeting in July 1708,

> found the saids burghs could only retain each of them one or tuo setts, and that the rest ought to be distribute amongst the remnant burghs who want as farr as they will goe, and this being done other new ones must be made by the forsaids burghs who haue right to keep them according to the standart of those they haue already, to be marked with the dean of gilds seall of the respective burrows, keepers thereof, and thereafter to be distribute amongst those who want.[32]

The Convention further decided that:

> the number of seventen of each of the saids severall standards be furthwith delyvered up by the said priviledged burghs to such of the burghs royall as the lord provest of Edinburgh shall nominat for receiving of the same … the said three priviledged burrows to prepare and furnish exact setts of the saids weights, measures, &c., precisely conforme to the standart, with the respective burghs marks stamped upon them, and that each burgh who shall receive of the saids standarts from the said priviledged burghs shall pay thirty pounds Scots for each of the saids setts … And to the effect ther may be ane uniformity of weights and measures, &c., through the wholl burrows, the conventione appoint and ordaine

the wholl royal burrows to provyde themselves with the setts above mentioned betuixt and the first day of November next to come, after which the convetione discharge the wholl royall burrows to make use of any other weights, measures, &c., then those abovementioned, under the penalty of ane hundred pounds Scots … And inregard of the generous offer and concessione of the good toun of Edinburgh to provyde the wholl royal burrows with setts of the eln and yeard, and that gratis without any pryce or compositione, therefor the conventione returned their hearty thanks to the representatives of the said good toun.[33]

In January 1708, Stirling had authorised its officers to receive 'the standard of the liquid measures now come from Engleand in stead of the former jog … grant receipt thereof and secure the same in a convenient place untill the toune call for the same'.[34] The eventual cost of this, including carriage, came to £84 8s, which strongly suggests that no more than a nominal charge was made for the supply of the measures; and it perhaps explains why Stirling was so anxious to revive its privilege, and be able to charge £30 for each set of liquid measures supplied.[35] Since the Convention had required every royal burgh to obtain new weights and measures, and there were then over fifty royal burghs, Stirling and the other two distributing burghs would have the monopoly of supplying duplicates to these additional thirty burghs – work that might otherwise go to Edinburgh. Although the Convention must have anticipated that Edinburgh might make a similar charge for the single linear measures, the waiving of this was probably influenced by the likely difficulty of securing payments for such comparatively slight items (at least in terms of the weight of brass or bronze involved).

In Stirling, intimation was 'made by tuck of drum' in July 1708 that no weights and measures were to be used within the burgh after 1 November except those that conformed to the new standards recently sent from England.[36] Glasgow ordained something very similar in May 1712.[37] Also in July 1708, Stirling's dean of guild was authorised to distribute the capacity standards to the various royal burghs as indicated by the provost of Edinburgh, 'after the saids standarts shall be stamped with the tounes arms, and to receive thirtie pounds Scots for each of the setts'.[38] In September the dean of guild was instructed to write to his opposite numbers in Edinburgh, Linlithgow and Lanark, so that Stirling itself might obtain 'the standarts of the elne, weights, and drye measures'.[39] The total expenditure for a full set of weights, liquid measures and the dry bushel, was presumably £90, plus the cost of carriage. Thus Glasgow's council, which had paid £30 to Stirling in July 1708 for a set of liquid measures, authorised their treasurer to repay their dean of guild £114 4s 6d Scots which was owed to him 'for getting new weights and measures to this burgh, of the English standart'.[40]

The sum was sufficiently large to cause resistance. In July 1709 there was a complaint from the other three burghs 'keepers of the weights and measures' – Linlithgow, Lanark and Stirling – that few, if any, of the royal burghs had acquired sets of their standards, despite the threat of the penalty, which had now been reduced from a hundred to thirty pounds Scots.[41] Two years later, the Convention decreed that if burghs had still failed to take out the sets of standards within the year, they would have to pay the price anyway, whether they took the sets or not.[42]

To reinforce their rights, the distributing burghs each marked the items which they were privileged to distribute. Lanark stamped the weights with the mark 'D·LK·', for Dutch weight, Lanark. Linlithgow stamped the rim of the Winchester bushels with an 'L' five times, and affixed a bronze plaque to the outside of the measures, which carried the seal of Linlithgow and the word 'LINLITHGOW' in relief (these were presumably manufactured and fitted in Edinburgh). The four English capacity measures – wine gallon, corn gallon, and the quart and pint of the larger ale gallon – were stamped inside the base with an impressed mark representing Stirling's seal, incorporating a wolf crouching on rocks and with the letters 'STR.', for Stirling. Edinburgh's dean of guild stamped the combined yard and ell. None of these marks should be counted as verification marks since none of the issuing burghs were in a position to test the accuracy of the weights or measures, but probably this aspect was considered secure enough because they carried the stamps of the English Exchequer.

After the twenty-one 'Union' sets had been distributed, other English standards had to be produced through local manufacture, and a few of these survive. For instance, the Inverkeithing linear measure made in wood (Item **14**) is clearly intended to be a copy of the 1707 bronze measure, but would have been much less expensive to produce, and easier to afford for a small royal burgh such as Inverkeithing. No Lanark-made weights have been noted, but a group of four weights by an Edinburgh hammerman (Item **87**) were verified in Edinburgh in 1708 for another small royal burgh, Culross, but Lanark may have been less protective of its traditional rights. Linlithgow would certainly have produced duplicates at this time, because it was also manufacturing wooden measures to accompany the bronze bushels, but none have been found.[43]

Stirling is also known to have distributed three sizes of locally-made pewter measures in English sizes (see Items **179-197**) – the wine gallon, the ale quart, and the dry gallon – and the presumption is that these were produced to meet demand after the main 'Union' series was distributed. In October 1710 it was recommended that 'dean of gild and conveiner, with the baillies, to meet with Stephen Crawfoord, coppersmith in Glasgow, quhenever he shall come to this burgh, and agrie with him anent the standards of liquid measures made by him for this burgh, to be disposed by them to other burghs'.[44]

Despite all this emphasis on English measure, in 1722 the Convention of
Royal Burghs 'recommended to such of the royal burrows as have not as yet
taken the measure called the jugg from the burgh of Stirling to call for the
same betwixt [this] and the next annual convention'.[45] Indeed, it is very clear
that there continued to be a demand for standards of the old Scottish
measures after the Union.

Although in theory only English measure could be used within the
burghs, enforcement cannot have been strict and market practice clearly
continued much as before. Legislation alone could not change overnight
the mindset of many centuries' custom and usage, particularly in the local
market-place. As Ian Whyte has outlined, the Convention of Royal Burghs
'was unique in Europe, created by the burghs themselves rather than the
crown … [it] apportioned the burghs' contributions to royal taxation,
regulated their affairs, promoted their interests and defended their rights'.[46]
By the late seventeenth century the newer burghs of barony had won most of
the rights to foreign trade formerly apportioned to the royal burghs; and
although there remained export and import trade, in a climate of economic
depression much of Scotland's marketing at this period was internal and
fiercely competitive.[47] This is why local metrology mattered so much.

There is certainly some evidence that there was increasing use of English
weights and measures in Scotland during the eighteenth century. The
advocate John Swinton published *A Proposal for Uniformity of Weights and
Measures in Scotland* in 1779, which provided a valuable picture of the various
measurement standards used in the thirty-two counties of Scotland. He
proposed a number of ways of enforcing the current legislation on weights
and measures to improve the conditions for trade (thereby admitting that
they were indeed not working); and he also provided tables which compared
Scottish and English measure, described the practice in each county, and
finally he gave his assessment of Scots metrological legislation since the
time of David I (and some of his conclusions were discussed in Chapter 5).
His 'County Tables' provide some indication of creeping anglicisation in
Scotland's market-places, particularly in the Border counties and what became
the Lothians (Haddingtonshire, Edinburghshire and Linlithgowshire).
Avoirdupois weight was found in trade here, and elsewhere, although often
described as appropriate for 'English goods' and groceries, and still used along-
side troye and trone weight. In a few instances (for example in Perthshire and
Dumfriesshire) the avoirdupois stone was also specified, but found in the old
Scottish practice as 16 pounds, rather than in the English usage of 14 pounds.

Trone weight remained the basic weight of the markets, and here avoirdupois weight was absorbed as the method of defining the local trone
weights, although it never actually replaced them. Swinton's county reports
show that, by 1779, over two-thirds of the Scottish counties defined their
trone pounds in terms of avoirdupois ounces (or stones in terms of pounds,

with the assumption again that these were reckoned at 16 to the stone), whereas less than a third still defined them in terms of Scots troye weight. The situation was even worse for dry capacity measure, and Swinton made almost no mention of the Winchester bushel, except in Dumfries, where 'the Winchester bushel is now most in use', although, even so, five other customary bolls or pecks were used in country districts.[48] But the strongest demonstration of the failure of Article 17 of the Act of Union, must be the continuing interest in the measurement and construction of standards based on the old Scots system, by scientists, antiquarians and men of law, until the eve of the introduction of the Imperial system in 1824.

THE PHILOSOPHICAL SOCIETY
AND THE CONVENTION OF ROYAL BURGHS

One of the most important arenas for patronage and networking for men of a scientific turn of mind was the Philosophical Society of Edinburgh. This had its immediate origins in Alexander Monro's Medical Society of 1731. As Roger Emerson explains in his detailed study of the Society, its antecedents extended further into the past even than that, and may have had roots in societies formed in the 1680s by Sir Robert Sibbald, in the 1690s by Sir George Mackenzie, and others created in the 1710s and 1720s.[49] Colin Maclaurin (1698-1746) appears to have been the main intellectual moving force behind the formation of the Society in its 1737 form. After ten years (and the death of Maclaurin), it became all but moribund, but flourished sporadically between 1748 and 1768, had a resurgence of activity between 1768 and the mid 1770s, and then again became dormant. As Emerson explains, the Society 'had a long career of booms and near busts caused by the mobility of members, a small recruitment base, a failure of non-resident members to resign and the inability of amateurs to maintain interest in the work of its most productive members, the medical men'.[50] From its ashes arose the Society of Antiquaries of Scotland in 1780, and the Royal Society of Edinburgh in 1783.[51]

There were several members of the early Philosophical Society who were concerned with metrological issues. Colin Maclaurin, the professor of mathematics at Edinburgh from 1725, was the most prominent of these, and his views are seen most clearly in his 1751 edition of the previously unpublished manuscript of David Gregory's *Treatise of Practical Geometry*, written for his university courses when Gregory had taught in Edinburgh in the 1680s. This is largely concerned with linear measure and the measurement of land, but Gregory does briefly refer to Scots pints as having the somewhat unexpected volume of 109 in^3.[52] Maclaurin, in his additions to the 1745 edition of Gregory's work, explains that these are the common 'pewterers jugs', which we have termed brewers' pints, and shows how this size arises from the measurements of the cask conducted in 1707 for the Excise Commissioners.[53]

In 1708 Robert Stewart (1675-1758) became the first professor of natural philosophy at the University of Edinburgh. He had become regent there in 1703, teaching all subjects in the curriculum, but with a radical change in the teaching system introduced in 1708 he was able to specialise in a single area.[54] He was the twelfth and youngest child of Sir Thomas Stewart, first baronet of Coltness by his first wife. Sir Thomas, implicated in the Battle of Bothwell Brig, fled to Utrecht after the Rye House Plot in 1683. Robert received his education at the universities of Utrecht and Edinburgh. His eldest brother, Sir David Stewart, sold the family estate of Coltness in Lanarkshire in 1712 to his uncle, Sir James Steuart of Goodtrees (1635-1713), Lord Advocate. Sir David was succeeded in the Coltness baronetcy by his nephew Thomas in 1723; and Thomas died unmarried in 1737, aged 29. Robert Stewart thus succeeded his nephew as fourth baronet, but having no estate never assumed the title; nor did his only son, John (1715-59), who also succeeded his father as professor of natural philosophy at the University of Edinburgh in 1742.[55] By 1709-10 Robert Stewart, who had obtained his university post at Edinburgh through patronage, was able to purchase instruments to help illustrate his lecture demonstrations, and although he 'was said to have been at first a Cartesian, … he was finally converted to the school of Newton'.[56]

Robert Stewart was a slightly older contemporary of Colin Maclaurin, and like him, a founder member of the Philosophical Society of Edinburgh. Measurements by Stewart of early Edinburgh standards – notably the sets of 1618 standard weights and of the avoirdupois weights supplied at the Union – are quoted by John Robison in his Commonplace Book and dated to 1720.[57] They match additions made by Maclaurin to Gregory's text, where, for example, he notes that 'they who have measured the weights which were sent from London after the union of the Kingdoms, … have found the English Avoirdupois pound (from a medium of the several weights) to weight 7000 grains'.[58] Not only does this help confirm that Stewart's own troy weights were adjusted to the same level as the 1707 issue, it also suggests that Stewart did not act alone. His weighings of the 1618 Scots weights held by the Edinburgh dean of guild showed that the stone was to a noticeably heavier standard than the others in the set. The mean pound for the set was calculated at 7,599½ grains, or 7,600 grains, the figure quoted by Maclaurin, whereas the stone was given as to a pound of 7,619¼ grains, a value later to have significance for the work by Robison.[59]

However, Stewart was clearly also aware of the 1618 standard pint (Item **113**) and the 1555 half-pint (Item **109**) in the care of the Edinburgh dean of guild, because the notes copied by Robison give the capacity of the pint as 3 pounds and 7 ounces (the 55 'French' ounces of the 1618 Assize definition), and he noted that if this was a mean of the Edinburgh jugs then the relevant pound was 7,621·818 grains: the unstated implication is that the pint would contain 5⁵⁄₁₆ such pounds, or 26,200 grains.[60] No explanation for juxtaposing these comments was provided by Stewart, and he may only have intended to show the disparity between these results. At a later date, however, this was to provide a persuasive indication to Robison of how to recover the 'original' standard of the 1618 weight series.

As with many other formal bodies in Scotland at this time, the Philosophical Society of Edinburgh became a vehicle for patronage, in particular for James Douglas, thirteenth Earl of Morton (1702-68), who from 1737 until his death was its president. Morton had become close friends with Maclaurin after his return to Scotland from a Cambridge education and the Grand Tour. Elected a Fellow of the Royal Society of London in 1733, he observed a solar eclipse in 1737 with, among others, Maclaurin and the Edinburgh telescope maker, James Short. All of those involved in this astronomical *soirée* went on to become founder members of the Philosophical Society of Edinburgh, and Maclaurin and Short, in particular, were beneficiaries of Morton's patronage. Morton was a Whig representative peer in the Westminster Parliament, Grand Master of England's Grand Masonic Lodge and, after 1764, President of the Royal Society of London.

Maclaurin (Fig. **9.1**) had been appointed to the Edinburgh chair of mathematics in 1725, alongside the ailing James Gregory (d.1742), who by this date was in such poor health he was unable to teach. So keen was Sir Isaac Newton that Maclaurin should get this post that Newton offered to contribute £20 a year to his protégé until Gregory's death.[61] This appointment has been seen as the start of an enlightened policy by the town council as patrons of the University of Edinburgh to enhance the University's reputation by attracting outstanding and vigorous teachers. This policy, in the capable hands of George Drummond (1687-1766), the highly-influential and long-serving provost of the town council, played a notable part in the extraordinary success of the university in the second half of the eighteenth century. Like Morton (and others), Drummond also used Masonic influence and contacts to further his own and the city's objectives.[62] As Jack Morrell has explained, teaching at Edinburgh occupied a heavy academic session which lasted for the six months between early November and early May, followed by six months' vacation: 'unless a professor chose to deliver a summer course lasting three months, he enjoyed the enviable opportunity of having half the year in which to continue his private practice, to perform consultancy work, to prepare his lectures, to do research, and to publish. Indeed, the inadequacy of their low fixed salaries positively encouraged professors to publish, not only for personal profit but also to increase their reputation and the prestige of the University.'[63]

The proliferation of societies in mid-eighteenth century Edinburgh, together with what T. C. Smout has called 'the golden age of Scottish culture'

Fig. 9.1
Colin Maclaurin (1698-1746),
professor of mathematics at
Edinburgh Univerity, drawing
in chalk and pencil by the
Earl of Buchan, *c*.1780.
(Scottish National Portrait
Gallery)

– which included this scientific academic growth – meant that Edinburgh's
publishing base grew along with the flowering of Scottish imaginative literature; the giants of the age here being Robert Burns and Sir Walter Scott.[64] The
end of the eighteenth century saw the beginning of the age of encyclopaedias,
in which all knowledge was to be found. Although the original *Encyclopédie*
was of course a product of the Enlightenment in France, Edinburgh was
home to both the *Encyclopaedia Britannica* and *Chambers Encyclopaedia*, and
subsequently to the *Edinburgh Encyclopaedia*.[65] Well into the nineteenth
century the publishing houses of Blackwood and Constable provided a huge
range of diverse periodical literature to an eager and literate public.

Maclaurin was an influential teacher, not least in his popular promotion
of Newtonian ideas, but his consultancy work also proved to be of great
importance. It is possible that Maclaurin was already relying on Stewart's
unformed views of the pint's size in an important report on the gauging of
casks written for the Excise Commissioners in 1735, which has recently been
published from the surviving manuscript by Judith Grabiner.[66] In recording
the relationship that had been calculated between the English wine gallon
and the Scots 'Pewtherer's pint Jug', Maclaurin stressed that he was not
speaking of 'the Content of the Standard measures such as are kept by the
Dean of Guild', and he noted in an aside that in 'some Experiments made
with a good deal of Care upon the Standard Juggs kept by the Dean of Guild
of Edinburgh (by some ingenious Gentlemen) the Scots pint was found not

to exceed 102·3 Cubick Inches'.[67] Such a result is explicable in terms of Maclaurin's 1745 statement that the 'Scots Troy-pound (which by the Statute [of 1618] was to be the same as the French) was commonly supposed equal to' the current French pound, whose size was confirmed in 1735 to be 7,560 English troy grains.[68] It must perhaps be assumed that Maclaurin himself did not form part of the investigative group on that occasion, otherwise the experimental result (rather than a theoretical one) would have been quoted.

Such a group may already have included Alexander Bryce (1713-86). This former pupil, and now protégé, of Maclaurin was involved with trigonometric survey work in the early 1740s for the Earl of Morton, then President of the Edinburgh Philosophical Society, who later provided Bryce with a church living at Kirknewton, just west of Edinburgh, in 1744.[69] Bryce soon found himself deputising for Maclaurin in the maths classroom when the latter fell seriously ill in helping prepare the city's defences ahead of the Jacobite army's advance on the capital in 1745. However, after Maclaurin's death, Bryce failed to secure the mathematics chair, which went instead to Matthew Stewart.

Another member of this metrological group was James Gray (d.1761), the Master of the Iron Mill at Dalkeith, and also an active member of the Philosophical Society.[70] It is very likely that his involvement can be taken back to at least 1744, the date of a new linear measurement standard for Dalkeith, whose construction and division was almost certainly supervised by the group and which may well have been manufactured at Gray's foundry. This ell and yard standard (Item **15**) was constructed in steel along recognisably 'scientific' principles of precision and was authorised by Edinburgh's dean of guild. (It was discussed with reference to other Scottish linear measures in Chapter **3**). The commission for this measure may stem from a prohibition of 1737 against trading in yards in Dalkeith, presumably reflecting the absence of an acceptable standard held by the bailie of this burgh of regality, or burgh of second rank, immediately to the south of Edinburgh.[71]

The publication in London of the results of the Royal Society's programme of comparisons of weights and linear measures in 1742 and 1743 had highlighted the necessity of relating new reference weights and measures to historical standards of sufficiently secure legal authority.[72] It was the Stirling Jug, and not the copy at Edinburgh, that had figured in the last major Scottish assize legislation in 1618, and Bryce clearly appreciated the need to check the capacity of this vessel. His initial attempt to locate it in the Stirling Guildhall in 1750 failed, but he was eventually successful in the spring of 1752.[73] A heroic tale has survived of how Bryce discovered this 'precious object of his long research' in the detritus discarded after the public sale of the effects of a Stirling pewterer named Urquhart: he was believed to have had the use of the standard to manufacture copies, but had joined the Jacobite rising in 1745 and had not returned from the campaign. We must assume that this account is substantially correct, although there are now no adequate sources for the story, which first emerged in the literature in 1810, and in its fully developed form was recounted with characteristic style by Robert Chambers in 1835.[74] The satisfaction of Chambers – and, by implication, Bryce – is palpable: 'Thus was recovered the only legal standard of weights and measures in Scotland; after it had been offered, in ignorance, for public sale, and thrown aside unsold as trash, and long after it had been considered by its *constitutional guardians* [Stirling council] as irretrievably lost.'[75]

Another founder member of the Edinburgh Philosophical Society who was deeply interested in metrology was James Stirling of Garden (1692-1770).[76] Coming from a committed Jacobite family, he was educated at the University of Glasgow while his father was in prison on a charge of treason. Subsequently Stirling matriculated at Oxford in 1711, but had to leave as he could not take the oath of allegiance. His promising early mathematical work had been drawn to the attention of Sir Isaac Newton, and he managed through various influential friends to get a position in Venice. By mid-1724 he was back in Scotland, and afterwards settled in London, where in 1726 Newton helped in his election to the Royal Society. Although occupied in mathematical research at this time, Stirling was also undertaking practical work: he was involved in a successful London academy, and produced a popular work on scientific demonstration. Despite publishing the serious mathematical work which made his reputation, he moved to Leadhills in Lanarkshire in 1735, in order to reorganise the Scottish Mining Company in the lead mines there. In 1746, with the death of Colin Maclaurin, Stirling was suggested as a strong candidate to succeed him on grounds of merit, but he was disqualified because of his politics and the chair was initially left vacant.[77] In 1752 he made a survey of the river Clyde, for the town council of Glasgow, but only one of his publications reflects his interest in metrology, and that appeared after his death, in 1792.[78] However, his papers, still in private hands, contain two copies of the two volumes of his unpublished 'Treatise on Weights and Measures', unfortunately undated.[79]

These demonstrate Stirling's evident interest in the classical background, informed by his consultation of Robert Stewart's manuscript treatise. For instance, in this manuscript Stewart recounts that shortly after the Union, standards of troy and avoirdupois weight were sent to Edinburgh; comparing these in 'the experiment room of the college [i.e. the university]' through six trials with six different weights, the avoirdupois pound was found to have a mean of 7,000 troy grains with 'about ⅓ part of a grain over'.[80] Much of the manuscript is taken up with Roman and Arabic measures, in an attempt to vindicate the veracity and venerability of the contemporary measures of Scotland.

Bryce's recovery of a key early standard of apparently unquestioned authenticity opened the prospect of re-establishing the basis of the old Scots system with investigative rigour. Not only was the unit of liquid measure now fixed (and it was soon clear that it matched the Edinburgh measures already examined), but through this the dry capacity and weight units could also be determined.

The evidence appeared to come at just the right time. In mid-1752 official dissatisfaction with the variable sizes of trading weights and measures across the country became a formal issue at the annual Edinburgh meeting of the Convention of Royal Burghs, which acted in effect as Scotland's commercial parliament.[81] In spite of legislation for ensuring 'a uniformity and standart of weights and measure, the same is not only not obtain'd, but that the weights and measures are various and different almost in every burgh and shire in Scotland, which occasions great confusion in trade and renders it impracticable for merchants or others to sell or buy with any degree of certainty'. Every burgh was charged to provide 'an exact accompt' of their weights and measures to the next annual convention; the head burghs of each county had to co-ordinate returns from the lesser burghs of regality and barony; and the sheriffs had to act for those counties without royal burghs. A crucial requirement, and one that underlines the fact that Scots measure continued to dominate trade and contracts despite its theoretical abolition in 1707, was the Convention's insistence that burghs specify 'what proportion [these weights and measures] respectively bear to the legall standarts established by the laws of Scotland before the Union'.[82] The matter was to be kept under review by the Convention's Annual (or standing) Committee.

Little seems to have been forthcoming, however, although at the 1754 Convention it was 'reported that the several Sheriff deputes of Scotland have for some time been deliberating on the most effectual means to bring about ane uniformity of weights and measures', and the Annual Committee was authorised to correspond with them.[83] One of the Convention's continual concerns was to keep lines open to Westminster in order to lobby for legislation and for some influence over bills that were drafted predominately to address English issues. At this period the Convention was frequently presided over by George Drummond, a highly astute political manager who was effectively perpetual Lord Provost of Edinburgh throughout the 1750s and early 1760s, and who became clearly identified with promoting improvements in Edinburgh and with control of the patronage of the university.[84] Drummond was successful in securing a parliamentary act for promoting Scottish linen manufacture in 1754, and his agents in London were also instructed to 'concurr in any bill that may be devised for procuring an act of parliament establishing the uniformity of weights and measures'.[85]

George Drummond is very likely to have been behind the proposal when it was first made at the 1752 Convention, and it must be supposed he knew of Bryce's recovery of the Stirling Jug and would have appreciated that the Convention was now in a position to resolve the issue. He was again in the chair and trying to persuade the burghs to act in 1754, 1755 and 1756.[86] He would certainly have been aware of the House of Commons committee that met in 1758-9 under the chairmanship of Lord Carysfort (an Irish peer, who sat for the Commons constituency of Huntingdon) to consider the equally pressing problem of reforming English weights and measures.[87] It was not, however, a matter of merely waiting for Carysfort's recommendations to be passed into law and then pressing for the Act to extend to Scotland, because Carysfort's Bill fell foul of the Commons timetable and failed. However, when the parliamentary process was believed to have revived again in 1764, Drummond was again president and was able to instruct the Convention's London agent to represent the case of the Scottish royal burghs.[88] Unfortunately, no act was forthcoming, and despite increasing pressure from further parliamentary committees, Carysfort's reforms were not implemented until 1824.

Although collectively the royal burghs may not have felt impelled to undertake the type of research for which Drummond was pressing, some rapid progress was indeed made. The capacity of Stirling's venerable jug was measured, and the result used in the adjustment of Edinburgh's weights. A new precision linear standard was constructed for Stirling (Item **16**). Perhaps of greater significance, key decisions were made about interpreting the 1618 legislation governing the construction of the dry capacity measures, leading to the justices of the peace for the county of Stirling commissioning impressively-engineered standards of the wheat and barley firlots (Items **216** and **217**). All this work was controlled by Bryce and his colleagues in the Philosophical Society, and the leading conclusions were published under James Gray's name in the Society's *Essays and Observations* in 1754.[89]

A number of burgh commissioners to the Convention must have appreciated that this essential preliminary work had been carried out and that it would inevitably form the basis for Drummond's proposed reforms. Certainly this will have been well known to those who had houses in and around Edinburgh and were already members of the Philosophical Society. These included the Excise Commissioner George Clerk-Maxwell (1715-84), commissioner for Sanquhar in Dumfriesshire, and the Edinburgh lawyer John McGowan, commissioner for Whitehorn in Galloway, who was actively involved in adjusting the Stirlingshire firlots.[90] Indeed, the burgh commissioners were not merely aware of the Society, they supported it financially during this period through grants towards the premiums the Society awarded for scientific and technical improvement in manufacture and agriculture. The first such grant, of 100 guineas, was voted by the 1755 Convention on a motion almost certainly drafted by Drummond.[91] In 1757 Drummond involved Alexander Monro, his star professor of anatomy and surgery at Edinburgh University and the joint secretary of the revived Philosophical Society, in making a similar flattering appeal to the liberality of the Convention, in which he pointed out how the success of the Edinburgh venture had led to the more recent copy-cat establishment of a similar society in London.[92] Grants from the Convention continued until at least 1764.[93]

James Gray's published account 'Of the Measures of Scotland, compared with those of England' appears in the first volume of the *Essays and Observations, Physical and Literary, read before a Society in Edinburgh and published by them*. The date when the paper was read is not known, although it might have been as early as 1753. The text is, however, virtually identical to an unsigned manuscript report addressed 'For the Justices of the Peace of The County of Stirling 1754' which accompanied the two firlot measures, and the two accounts are clearly taken from the same source.[94] There are some differences: the printed version provides some information on the testing of the Edinburgh pint and gives the capacity of some of the English liquid measures (including the 224 cubic inch size for the early wine gallon), whereas the manuscript report specifically rejects the linear dimensions for the firlot given in the 1618 Assize as 'erroneous'. Both reproduce the two shields cast in relief on the Stirling Jug, and the careful drawing in the manuscript reconstructs the somewhat worn figure on the seal of Stirling as a young child (see Fig. **9.2**) – very different from the rather crude woodblock engraving of the published account.

The Jug was filled with clear river water, which was found to weigh 26,180 English troy grains, averaging the result of five trials, with the assurance that these differed by only a few grains.[95] A special vessel had been constructed to have a capacity of exactly 100 cubic inches and filled with the same water: repeated trials gave the weight of water, to within a grain, as 25,318 grains, so that the weight of a cubic inch of water was 253·18 grains. This is a good match for Thomas Everard's result of 253·176 grains per cubic inch, undertaken in 1696 and subsequently published in his *Stereometry*.[96] From this, the Jug's capacity can be calculated as the stated value of 103·404 cubic inches, rounded down in the summary table to 103·4 cubic inches. Since the 1618 Assize stated that the pint contained 3 pounds and 7 ounces (or 55 ounces) 'of the clear running Water of Leith', the pound of 16 ounces can be calculated as 7,616 grains.[97]

There is, however, a third discussion of these measurements, given in Robert Chambers's nineteenth-century account of Bryce, which does appear to be based on more accurate and complete details.[98] Thus, for example, Chambers gives the mean aperture of the Jug as 4·17 inches rather than the 4³⁄₂₀ inches of Gray's version; the balance used is stated to turn on a grain when loaded with 96 ounces on each side, whereas Gray's account requires 100 ounces. Chambers's source also adds details of the experimental technique:

> the mouth of the Jug was made exactly horizontal by applying to it a
> spirit level; a minute silver wire of the thickness of a hair, with a plummet

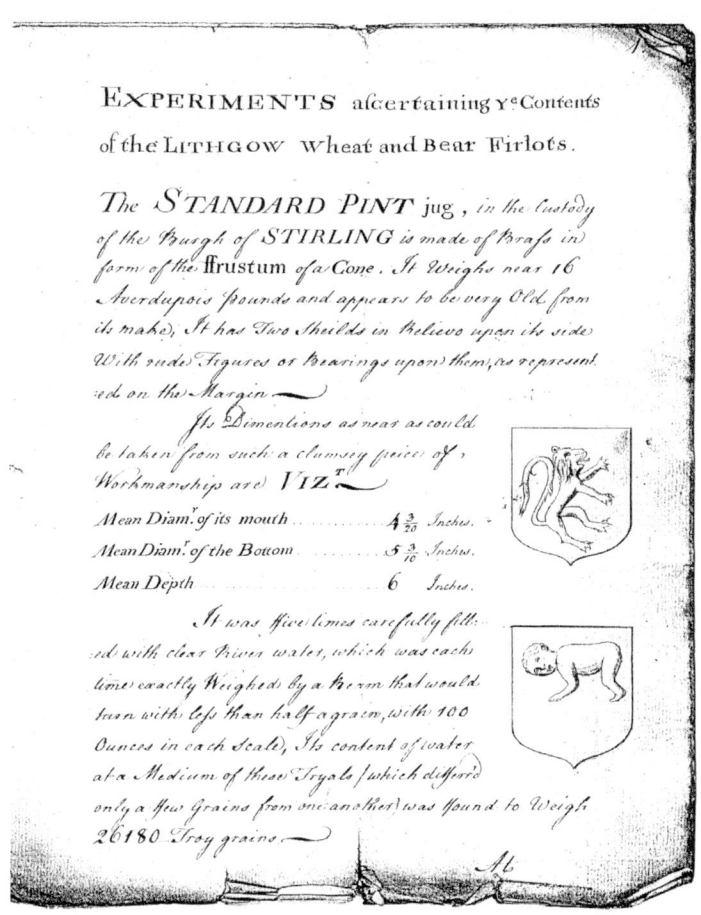

Fig. 9.2
The opening page of the
1754 Report to the Justices
of the Peace of Stirlingshire,
describing the Stirling Jug.
(Stirling Council Archives)

attached to each end, was laid across the mouth, and water poured gently in, till, with a magnifying glass, it was seen just to touch the wire …[99]

Where Gray referred to five trials, Chambers has 'seventeen trials with clear spring and river water, several of which were in presence of the magistrates of Edinburgh'.

Perhaps the most interesting additional information concerns the standard vessel of 100 cubic inches. Chambers's source has this as a cylindrical brass vessel 'made with great accuracy, by a scale of Bird, the celebrated mathematical instrument-maker of London'.[100] Little is known about this period of John Bird's life other than that he had worked for George Graham, the principal mechanician at the Royal Society; he has also been associated with Jonathan Sisson, who made the Royal Society's linear standards for Graham that were compared with the French standard toise, probably in the late 1730s.[101] (This scale by Bird is also discussed at Item **10** in the Inventory.) Gray cited the 1742 published results of these experiments in his account of the Stirling Jug. John Bird's independent activities are usually reckoned from 1758 when, as 'one of the best Artists in London', he was engaged by

Lord Carysfort to measure the standard vessels found in the Exchequer at Westminster using a set of specially constructed square-section brass capacity measures of specified volumes.[102] The use of test vessels of this sort was not new: for example, one by Isaac Carver of London was employed in the 1688 measurement (on the floor of the House of Commons, no less) of the early wine gallon standard found at the Guildhall in London.[103] But the association of Bird with Bryce's work is interesting and indicates that Bryce or Maclaurin had purchased the scale, perhaps through Graham or Sisson, since Bird was certainly not active at the claimed date of Robert Stewart's experiments in 1720. Although no maker of the cylindrical measure is given, it seems reasonable to assume that it was the Edinburgh instrument maker John Miller *senior,* who was responsible for the two Stirlingshire firlots of 1754.[104]

The massive mahogany and brass construction of the two firlots (Items **216** and **217**) made for the Stirlingshire county authorities, the high quality of the engineering and the copious detailed inscriptions, all underline that no trouble has been spared in creating authoritative and durable standards. Engraved and silvered plaques record their construction according to the terms of the 1618 Assize, and their adjustment by John Stewart, Robert Stewart's son and successor in the Edinburgh natural philosophy chair, James Gray, and the Edinburgh lawyer and Royal Burgh commissioner for Whithorn, John McGowan. Great care has been taken to record in the various inscriptions every feature of their dimensions and volume, the volume of the internal support structure, and the weight of water content of its relationship to the pint. All this is presented as conforming exactly to the requirements of the 1618 Assize.

The crucial exception that is not admitted is that the depth and diameter dimensions are different from those in the act. In Chapter **6** we established that the measurement of 21¼ pints for the wheat firlot, and 31 pints for the barley firlot, had to be conducted using grain as the measuring medium in order to match the stated capacity to the dimensional volume. Gray and his colleagues, guided by the experimental standards of their day, used water as the measuring medium and therefore obtained larger volumes because water, unlike grain, does not pack to different densities depending on how it is dispensed. Using grain, as had been intended in 1618, would have given volumes of about 2,110 and 3,020 in³ for the firlots, but using water they generated volumes of 2,197·34 and 3,214·65 in³. Their technique would have been perfectly acceptable if they had been adjusting a new standard against an old one, but it was not appropriate for creating new standards using the dispensing method described in the act. They had presumably been misled, as was suggested in Chapter **6**, by the act's appeal to water as a measuring medium because the pint's capacity was defined by the weight of its water content. They should, of course, have been concerned that their standards

had to be constructed to larger dimensions than specified in 1618, and should
have taken advice. It was well-known that differential grain packing affected
the size of large grain measures: for example, as late as 1811, the Rev. George
Skene Keith described a new wheat firlot made for Aberdeen, adding 'it may
be noted incidentally, that the difference between filling slowly with a small
shovel, and quickly with a large one was found to be 3 per cent: where fraud
is intended, it is more than 6 per cent'.[105]

There remained the additional problem of interpreting the original
dimensions that had been given in the 1618 Assize. Because Gray and his
colleagues had adopted Maclaurin's 37·2-inch ell, they would have supposed
that the use of a Scots inch would increase the capacities of the firlots to 2,145
and 3,070 cubic (English) inches. These were still smaller than the capacities
they generated, but perhaps this was an additional area of uncertainty that
cast doubt on the original dimensions.

The problem was resolved simply by side-stepping the issue and stating
that the dimensions were erroneous. But the result was that this new and
authoritative published description of the dry capacity measures represented
an unwitting break with established tradition. There can be no doubt that
this first clear statement of the volumes of the firlots was influential, and was
accepted by local authorities who found they had to renew their measures in
the period that followed. Linlithgow's cooper was certainly dispensing water
by the pint at a legal examination of the burgh's firlots in 1779, and in the
same year John Swinton published these values as the standards for the whole
country.[106]

The figures published by Gray in 1754, and those engraved on the firlots,
form a self-consistent group.[107] The stated volumes and water contents of the
firlots are precisely 21¼ and 31 times the pint's stated volume of 103·404 in^3
and 26,180 grains contents, and since the pint was supposed to contain 55
troye ounces, the troye pound was 7,616 grains. Finally, a new standard ell of
37·2 inches was adjusted by Alexander Bryce for the town of Stirling in 1755
(Item **16**). With this addition, the Philosophical Society's metrological
group appeared to have recovered secure values for all the old Scottish units,
and constructed standards which could form models for the county and
burgh authorities.

The immediate cause for the re-awakened interest of the Annual Committee of the Convention of Royal Burghs in 1764 may have been the
impending (but abortive) parliamentary activity in Westminster. However,
George Drummond was also able to report that the Board of Police in
Scotland was now involved in moves towards a unified system of weights and
measures and had written to sheriffs and magistrates asking about local units
and their relationship to the legal standards.[108] There is at least some evidence
of responses: in 1759 the Aberdeen magistrates submitted a report to the
Board of Police that was subsequently criticised for its failure to recognise

that the Aberdeen pint was significantly larger than the Stirling Jug.[109] The situation was rectified a few years later when a copy of the Stirling pint, made to James Gray's published capacity, was presented to the town by Alexander Fraser of Moremont, in the parish of Strichen, which is just north of Aberdeen. The pint (Item 200) is inscribed: 'THIS JUGG / Was made by MR. D. Hutchison A.D. 1765 / Which when carefully filled with Clear / Running water weighs 26180 Troy Grains / English, which is exactly the same weight / that the Stirling JUGG Contains.' Hutchison can tentatively be identified as David Hutchison, the nephew of James Gray, who inherited the Ironmill at Dalkeith after the death of his uncle in 1761; plausibly, he may have been involved with Gray's earlier work in the 1740s and 1750s.[110]

The whole issue of uniformity of weights and measures was again on the agenda at the Convention's meeting in Edinburgh in 1774: a committee was established to take the matter further, which included the Lord Advocate, George Clerk-Maxwell and John McGowan.[111] In the following year the new Lord Advocate, Henry Dundas, joined his predecessor Robert Orde, now Lord Chief Baron of the Scottish Exchequer. But perhaps the most significant addition was that of the advocate John Swinton (d.1799), who shouldered the burden of co-ordinating the county-by-county survey of local weights and measures. His 1779 report, entitled *A Proposal for Uniformity of Weights and Measures in Scotland, by execution of the Laws now in force*, was compiled from reports obtained from the magistrates of the royal burghs and the sheriffs of the counties, as well as 'other learned and judicious persons', and must be considered as the Convention's case for reform.

As we have already seen, Swinton's proposals amounted to six points. After a rapid résumé of the Scottish legislation from 1491 to beyond the Union of 1707, and lamenting that Westminster efforts towards reform in the 1750s and 1760s had been abandoned, Swinton made the following suggestions. First, in line with the 1618 legislation, all burghs should 'possess themselves of compleat and accurate sets of the legal standards, both English and Scotch … and settle a method for giving out authentic duplicates in terms of law'. Second, each burgh should appoint an official adjuster. Third, local customary weights should be calculated against the standards, and adjustment tables made publicly accessible. Fourth, an amnesty period should be advertised, in which non-standard weights and measures could be marked correctly against the standard, and afterwards 'all persons who shall use, in buying, selling, or delivering, weights and measures of denominations different from the standards, or disconform thereto, or who shall use false weights and measures in any manner shall be prosecuted and punished according to the law'. Fifth, other metrological 'malpractices' and their punishments should be proclaimed as a deterrent; and, sixth and finally, an officer should be appointed 'for carrying on prosecutions, and [to] advertise a reward for informers, to be paid on conviction'.[112]

Swinton then listed the difficulties his six proposals raise, concluding that, especially with corn measure, 'it is recommended, not to trust to common report, or even to stamped measures, but to try such measures by the weight of water which they contain'.[113] He then produced general tables of Scottish and English weights and measures, with conversions of one to the other, followed by further tables of the customary weights and measures for the thirty-two Scottish counties. 'Considerable pains have been taken to make the calculations exact,' commented Swinton. 'This was the work of an able and ingenious accountant, who gave his assistance to that part.' It is unclear who this unnamed accountant might be. Where the general tables discuss 'Amsterdam weight', Swinton mentions 'one sixteenth part of a standard Dutch stone from Campvere, in the possession of a gentleman in Edinburgh'.[114] It might have been thought that this 'gentleman' was the recently-appointed professor of natural philosophy at the University of Edinburgh, John Robison, who, as we shall see, was already by 1779 involved in important metrological consultancy work. However, Robison's own notebook refers to the 'dutch Stone in Mr Swinton's Custody, being examin'd by me July 15 1778 ...', thus demonstrating that each was aware of the other's work, but neither the owner of this particular weight.[115] The close link between Swinton's work for the Convention of Royal Burghs and the Carysfort Committee at Westminster is alluded to in Swinton's Proposal: 'The materials from which [the county tables] are made have been collected occasionally by a gentleman who was called upon by the chairman of the late committee of the House of Commons, to give his assistance in forming some clauses which were to have been added to the bills above mentioned, had they been resumed, in order to adapt them to this part of the united kingdom.'[116]

Swinton's county tables take up half of his book, demonstrating the diversity of local and customary variation from one Scottish county to another, helping to support the Convention's case that new legislation was required to bring weights and measures into line, not only within Scotland, but by implication (and reinforced by the inclusion of 'standard' English tables in the book) with those of England. Swinton's picture of Scotland's metrological state in 1779 is convincingly chaotic. Indeed, internal trade was not straightforward, and grain dealers were forced to use ready-reckoners each harvest time, as the English (Winchester) bushel was by no means the common unit of exchange between regions.[117] Swinton's book was reprinted in 1789, without further amendment.

Meanwhile, south of the border, Sir John Riggs Miller, member of parliament for Newport, had been gathering information about discrepancies in English weights and measures. He claimed to have received more than a thousand letters on this subject, and he spoke vigorously about the problem to the House of Commons on three occasions during 1790, but lost his seat in the election later that year.[118] However, his activities had brought the question to the attention of Sir John Sinclair of Ulbster (1754-1835) (Fig. **9.3**), who at this point in his political career was well placed to make further use of Miller's information, to gather more through his own initiatives, and, as a person of power and influence moving on the fringes of government, who could lobby effectively for reform. Sinclair, who had estates in Caithness and was an agricultural improver, had also trained in law in both England and Scotland, and sat as a member of parliament between 1780 and 1811. A man of huge energy, and a fascination with fact-gathering, he supervised the compilation of the first *Statistical Account of Scotland* between 1791 and 1799, consisting of a description of every parish in Scotland, chiefly with the help of the minister of each parish. There are only a few incidental mentions of local weights and measures in this twenty-volume work, for by the time his questionnaires were sent out this had not yet become a burning issue for Sinclair, although one minister had commented that 'the variety of weights and measures, which universally prevails both in England and Scotland, demands the attention of the Legislature'.[119] As an agriculturalist, Sinclair introduced improved methods of tillage and new breeds of livestock to northern Scotland, and obtained through political patronage the establishment of the Board of Agriculture in London in 1793, of which he was

Fig. 9.3
Political cartoon of Sir John Sinclair, '"Improvements in Weights & Measures", or Sir John Seeclear [*sic*] discovering the Ballance of the British Flag', by James Gilray, 1798. (Science and Society Picture Library, London)

president between 1793 and 1798, and again from 1806 to 1813. There he supervised the compilation and publication of another multi-volumed work, the county reports on the state of agriculture throughout Great Britain, and these reports discussed local weights and measures in some detail.[120] Julian Hoppit has pointed out that Sinclair was appointed to the Commons Select Committee to consider the returns from the 1790 investigation initiated by Miller. Miller's motives are obscure; like Carysfort (but unlike Sinclair) he appears to have had no political axe to grind.[121] Hoppit suggests that he may have been influenced by Swinton's 1779 work, or possibly by Sir James Steuart's scheme for a decimal system, published in 1790.[122]

At market level it was essential to know the equivalent of local measures in order to trade with neighbouring districts. As Rosalind Mitchison has observed, 'every year, nominally at Candlemas, the sheriff courts of the country "struck" and recorded the "fiars prices", the prevailing prices of the most common sorts of grain, as a basis for the settlement of various fixed payments such as feu duties and rents, and also for legal processes'.[123] This was done once a year, retrospectively, for a country-wide market, and although the measurement units were often local (until 1827), and there was often disagreement between counties about which grain was 'common', it was a necessary part of commercial activity. 'The first movement from rent in grain to rent in money would begin with conversion by the fiars,' comments Mitchison. She goes on to observe that there was no Scottish unity of market before 1685, but by the 1690s there was close agreement between prices in the Lothians and Lanarkshire, joined by the southern uplands in the 1720s and much of Ayrshire, Angus, Aberdeenshire and the Moray Firth by the 1730s: there was a unified market in Scotland for grain before the transport improvements of the late eighteenth century, and indeed, before the measurement was undertaken to a universally-agreed standard.

Fig. 9.4
Measure constructed by Alexander Bryce, Minister of Kirknewton, for 'tallying his Stipend, then partly payable in kind, i.e. oats, bear or barley and wheat'.
(NMS W.MP.32)

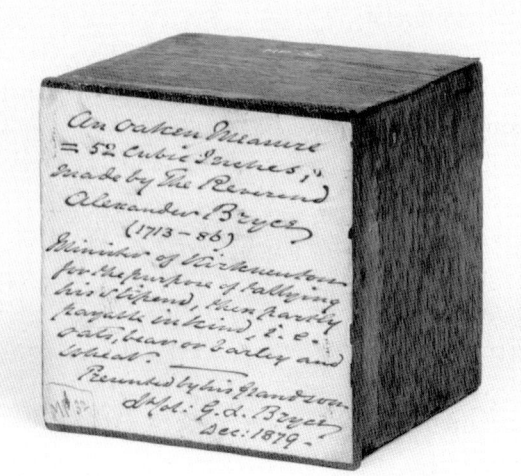

In 1808 the equivalent of grain was accepted formally as the basis for the stipends of the clergy, although in fact this must have been long-established practice. Alexander Bryce, whose metrological work we have already encountered, constructed a half-pint measure for this very purpose (Fig. **9.4**). It has an inscription on the base which reads: 'An Oaken Measure = 52 cubic Inches, made by The Reverend Alexander Bryce (1713-86) Minister of Kirknewton, for the purpose of tallying his Stipend, then partly payable in kind, i.e. oats, bear or barley and wheat.'

JOHN ROBISON AND THE COURT OF SESSION

Between about 1780 and the introduction of the Imperial system in 1824 there were two principal figures involved with the re-measurement of the technically outlawed Scots standards. Both were university professors, but neither published their results: this work appears to have been done on a consultancy basis for different burghs, and perhaps as such, remained a sensitive issue. One was John Robison (1739-1805) (Fig. **9.5**), professor of natural philosophy at the University of Edinburgh; the other was Patrick Copland (1748-1822), who held a similar position at the University of Aberdeen. Copland's work will be discussed later.

John Robison has been seen recently as 'a transitional figure … [who] … illustrates differences between late eighteenth century experimental philosophy and early nineteenth century physics'.[124] His appointment to the Edinburgh chair in 1774 came after the university had already become established as a flourishing centre for science and learning; and his role has been seen by most recent historians, perhaps unfairly, as a supporting bit-player behind the towering figures of Joseph Black and John Playfair: only some of his papers have survived and he lacks a modern biography.[125]

The son of a Glasgow merchant, Robison graduated from the University of Glasgow in 1756, but deciding against going into the church he went to London, where the promise of royal patronage failed to materialise. Instead he became tutor to the son of Admiral Charles Knowles, and became involved in naval affairs. He accompanied the younger Knowles to Canada, was present at the siege of Quebec in 1759, gained a thorough grounding in practical surveying and navigation, and in 1761 was appointed by the Board of Longitude as its representative in the testing of John Harrison's chronometer, devised for determining longitude at sea. Robison went with this device on a voyage to the West Indies, but, disappointed by the Board's subsequent lack of patronage or even remuneration, he returned to Glasgow in 1762 in order to qualify for the church. He had already, before he left Glasgow in 1758, encountered both Joseph Black and James Watt. His friendships with both men now flourished and he became Black's student. In 1766, when Black transferred to the

Fig. 9.5
John Robison (1739-
1805), professor of natural
philosophy at Edinburgh
University, engraving by
Charles Turner from a
portrait by Henry Raeburn,
1805. (Scottish National
Portrait Gallery)

University of Edinburgh, Robison was appointed through Black's influence to lecture in chemistry at Glasgow, a position which was annually renewed. However, in 1770 he once again left Scotland, this time to go as his secretary with Admiral Knowles to St Petersburg, where Knowles was appointed by Catherine the Great as president to the Russian board of the admiralty. Robison worked towards upgrading the construction and navigation capability of Russian warships, and subsequently was selected for the mathematical chair attached to the imperial sea-cadet corps of nobles.

In 1773, again on the recommendation of Black, Robison was offered the chair of natural philosophy at Edinburgh, a position he took up the following year and held until his death.[126] By this time Robison was well-versed in practical, hands-on science; he had been exposed to Continental methodologies and his lectures 'were given with great fluency and precision of language, and with the introduction of a good deal of mathematical demonstration'.[127] However his salary – as Jack Morrell has demonstrated – was all he could rely on, as there were no pensions or superannuation: if a professor, such as Robison, was 'badly placed financially, he held his chair, irrespective of his intellectual competence'.[128] So he was forced to turn to other sources of income: Robison became, in the opinion of John Playfair (1748-1819), his obituarist and successor in the Edinburgh chair of natural philosophy, 'the first contributor who was professedly and really a man of science, and from that time [1793] the *Encyclopaedia Britannica* ceased to be a mere compilation'.[129] It is more than likely that his extensive, but unpublished, metrological work, like his publications for *Britannica*, was done as paid consultancy work during the long vacations. He was also drawn through it towards networks of useful patronage. In 1778 an attempt was made to revive the Philosophical Society of Edinburgh, and younger men, including John Robison, became members and injected some vitality into it. Robison became secretary, but for various reasons – mostly political, as the society was still a forum for patronage, as Emerson has explained – the Society was dissolved in 1783, just at the time when the Royal Society of Edinburgh was formed. Robison became secretary of the new society and the papers of the defunct society were handed over to the new. As Emerson dryly remarks, 'perhaps Robison passed them from his left hand to his right'.[130]

In 1777 there were accusations in Edinburgh that the Linlithgow firlot standards used by the Edinburgh bakers were too large, with the implication that the farmers were being underpaid for the grain they supplied. This developed into a serious legal challenge brought in the following year by a number of farmers in and around Edinburgh against the magistrates and council of the burgh of Linlithgow, and the Incorporation of Bakers in Edinburgh and Leith.[131] The complaint was that Linlithgow had supplied the bakers with firlots which were illegal, and that these had to be regulated according to the 1618 Act.

The case dragged on at the Court of Session in Edinburgh for over thirteen years (including a five-year gap followed by a continuation action brought by further parties) before a final judgement in 1791. Robison was heavily involved; and, from the outset, it must have been clear that this legal test could resolve once and for all the uncertainty about the 'correct' sizes of the standards, with the potential for imposing the type of uniformity for which the Convention of Royal Burghs and others had been pressing. As an indication of the case's implications for the judicial authorities, and possibly also for its bearing on Swinton's work, the Lord Advocate was one of the pursuers in the initial action, which lapsed in 1782. The progress of the trial can be followed through the original bundled process papers, or at least those that have been retained as relevant to the 1791 judgement, in the National Archives of Scotland; and these are mirrored in the court's printed papers in the run preserved at the Library of the Writers to the Signet.[132] A third source is the unpublished 'Commonplace Book' of John Robison in which he recorded his reading round the issue and the results of at least some of the measurements made in the course of this work. It is clearly an incomplete record, but to a limited extent it does allow his thought processes to be followed, although with some difficulty. It allows us to see how he was guided by, and often depended on, Robert Stewart's earlier manuscripts; and also to see his use of other published sources, including George Graham's accounts of comparison of the Royal Society's weights, Thomas Everard's *Stereometry*, the 1758 report of Lord Carysfort's parliamentary committee on weights and measures, and Mathieu Tillet's 1767 weight inter-comparisons for the Académie Royale des Sciences.[133]

The first complaint against the Bakers' Incorporation had the effect of sending a delegation of bakers with their contentious firlots to Linlithgow where, in the presence of the provost, the dean of guild and the cooper of Linlithgow, the measures were found to be 'exact'.[134] The next step, however, was for the Edinburgh dean of guild to send the measures to Robison, who recorded in his Commonplace Book in February 1778 that he had found that the wheat firlot 'complained of' was too large by ⅔ of a Scots pint in 21¼, and the barley firlot too large by 8/11 pint in 31 – representing 1·8 per cent and 2·2 per cent respectively.[135] John Swinton also reported that the 'Linlithgow Measure' firlots available in Edinburgh were larger than those stamped 'Edinburgh Measure': the size he gave for the 'Linlithgow' wheat firlot in Edinburgh precisely matched Robison's determination, indicating perhaps that Robison was involved in supplying data to Swinton.[136] This preliminary legal skirmish at Edinburgh's Dean of Guild Court was perhaps also the source for Swinton's careful comment that there was a popular misconception that Linlithgow and standard measure were one and the same, whereas 'in Edinburgh, Linlithgow firlots are avowedly made above 1½ per cent larger than the legal standard'.[137] Even more telling were Swinton's comments about other recently-acquired over-sized Linlithgow firlots, at Perth and Kinross, tested at the time when Robison was actively interested in the Bakers' case.[138] Robison was certainly acquainted with Swinton because amongst his early notes under 'w', for weight, in his Commonplace Book is an entry describing the source of Swinton's own weights.[139]

So what size of pint was Robison using at this time to denote the 'standard' capacity? The entry in the Commonplace Book shows the pint as only about 103·5 in³ (in fact 103·482 in³), deriving not from a direct measurement but from an early value of 102·3 in³ given in Robert Stewart's notes.[140] Robison noted that Stewart 'supposes a cubic inch of Edr fountain w[ate]r to weigh 256 gr, as appeared by a trial made with a brass cubic inch', but recorded that this had been superseded by Thomas Everard's 'more accurate Expt' of 1696 which gave the density as 253·176 grains/in³. Robison has merely substituted this later density figure to arrive at a revised value of 103·482 in³ from Stewart's mean weight of 26,200 grains for the pint's content.[141] Stewart acknowledged that the figure of 102·3 in³ had previously been published by the Edinburgh scientific instrument maker and almanack compiler, James Paterson, in his 1685 *Scots Arithmetrician*.[142] We referred to this in Chapter 3 as providing an indication of David Gregory's work in the early 1680s (in which he may have had Paterson's assistance), a view reinforced by the use of an early water density.

The larger court action was registered at the Court of Session in November 1778, when the farmers demanded that the Linlithgow standards be rectified and sought damages for their trading loss. Robison appears to have been retained swiftly as an expert, and he began a thorough investigation, including the identification and measuring of surviving artefacts with a view to nailing down the key values of the standards established at the 1618 Assize.

On three days in February 1779, Robison and Alexander Jamieson, the official cooper of the burgh of Linlithgow, solemnly filled the disputed wheat and barley firlots of the Bakers with water, in the presence of the magistrates of Linlithgow, at the Dean of Guild Court in Edinburgh, using the Linlithgow Jug (Item 114). The wooden firlots had been steeped in water to make them water-tight, but even so water was lost from the barley firlot on the first occasion. By comparing Robison's own notes and the preliminary report of 1 March he submitted to the Court of Session, it is quickly apparent where the problem lay.[143] Jamieson, Linlithgow's cooper, 'measured the firlot in the way in which he had been accustomed to measure the firlots delivered by the Burgh to the Lieges', but in Robison's opinion, with 'the Jug heaped and running over.'[144] By weighing the pint 'filled to the Satisfaction of the Cooper of Linlithgow', Robison found in trials over the three days that the Jug exceeded his standard (which he set at 26,197 grains) by 158, 75 and 65 grains. An argument used against him was that these figures were clearly not consistent: 'If the Professor is not exact, it is absurd to expect the Dean of Guild or the Cooper to be exact.'[145]

In addition, the wooden measures were not well finished: the inner edge of the barley firlot was lower than the outer edge, and the rim of the wheat firlot was low where the cross-bar attached and the bar arched upwards. It was difficult therefore to tell when the measure was full, and Jamieson claimed the wheat firlot was full when he had added 21¼ pints, whereas Robison demonstrated it would take more.[146] When Robison repeated the process, he filled the pint 'rather more than I thought exact', and this later enabled the claim to be made that there was no reason to allege that Robison had not filled the jug adequately.[147] Even so, he still reckoned to get an additional 7 gills in the wheat firlot and 9½ gills in the barley firlot, demonstrating that they were both about 2 per cent too large.

Although Robison would have had to agree with Linlithgow's protestations that the dimensions of the firlots had been wrong in the 1618 Act, and therefore measuring by the pint was the only option available, he would recognise, however, that no two individuals following such a procedure would produce the same results and that the method was basically flawed.

Robison's March 1779 report also reveals that he was already committed to a line of attack that would ultimately fail. His initial response to measuring a number of standard pints was to say that they were all closely compatible. However, because they were most conveniently measured by their weight of content, and because the 1618 Assize had stipulated that this should be 55 ounces of French weight, it followed that the volume would most easily be described in terms of the weight series. Unfortunately, the weight did not seem to be a conventional French weight and so he had to rely on any surviving accurate Scots standards issued in 1618. He found no unworn weight in Lanark, but there was a 1618 set in Edinburgh, which he located with Thomas Simpson, the Edinburgh pewterer who adjusted the burgh weights on behalf of the dean of guild. The set had been measured by Robert Stewart in 1720, but when Robison saw them he realised that all the weights except the stone were worn and had been readjusted (probably by Alexander Bryce in the 1750s).[148] These weights survive (Item 37), but have now been subject to more radical adjustment, probably in the 1790s.

However, the stone weight was in remarkably good condition, and the Lanark double-headed eagle stamps on its base were as crisp as the stamps on the body, indicating that it was scarcely worn.[149] Robison came to believe that it had come from a different set and had been the original standard weight held by Lanark, having for some reason been borrowed by Edinburgh's dean of guild and not returned. In his report, Robison recorded that Stewart had found the stone to be to a pound of 7,618 grains, whereas he now found that it was 7,615 grains. Assuming that this represented a progressive loss through wear in use, Robison reasonably proposed that it might have lost about the same amount of weight in its first hundred years, and it would therefore originally have been to a pound of 7,621 grains.[150]

Fixing the pound in this manner allowed the pint to be calculated as 55 ounces, or 26,197 grains. This is the standard pint size quoted in his report, and although he later revised it up to 26,200 grains (matching Stewart's earlier value and generating more convenient numbers of grains for the subdivisions of the pint), this size remained frozen in all Robison's metrological dealings. Similarly his recovered pound of the 1618 Assize remained at 7,621 grains, although from about 1790 (again apparently for computational convenience) he used 7,620 grains as its mass. We will return to this issue when we consider the final phase of Robison's consultancy work, when he produced new standards for Edinburgh's dean of guild in about 1800; and we will argue that Robison had realised that he had boxed himself in a corner and had to manipulate his figures in order to preserve his position.

By the time the initial phase of the Bakers' trial drew to a close in 1780-1, an interdict had been placed on the burgh of Linlithgow preventing the issue of further measures; and Robison and his colleague, the professor of mathematics at Edinburgh, were instructed to advise the court on a more secure way of measuring firlots. Dougald Stewart (1753-1828), who later distinguished himself as professor of moral philosophy, deputised for his father Matthew Stewart in the mathematics chair. He may have been appointed with Robison at the outset, but at this point he excused himself, pleading pressure of time, and George Clerk-Maxwell, now a Commissioner of Customs, was appointed in his place. They reported in November 1781, recommending two possible methods, both involving the construction of special narrow-necked vessels and the filling of a firlot under test through a funnel. In the first method, a large spheroidal measure, closed at the bottom by a stopcock, was carefully adjusted by filing down its neck so that it contained precisely the required number of pints, which had been measured into it with the greatest care. Opening the stopcock would then allow the correct volume of water to flow into the firlot whose capacity was to be checked or adjusted. The second method represented a more rigorous but laborious version of the original 1618 procedure. A pint measure of a similar shape (but with handles and no stopcock) was adjusted with its contents balanced by a standard poise weight, and then the requisite number of fills would be dispensed through a funnel under carefully controlled conditions.

Robison had been a pupil of Joseph Black, who had formulated the concept of specific heat in the 1750s, and so Robison was clearly aware how crucial temperature variation could be in measurement experiments of this type.[151] Not only does the water expand with rising temperature, but the metal vessel that contains it will also increase in size, to some extent compensating for the increase in water volume. Robison certainly possessed a thermometer, because he mentions it in his notes.[152] He recorded the temperatures of his early experiments on the Linlithgow measures and attempted to reduce these to a standard temperature, although his method is opaque. He considered the appropriate temperature at which to measure the Scottish vessels was 50° Fahrenheit, which brings with it an understanding that running river water was used, probably during the winter months. Adam Anderson subsequently demonstrated that the temperature of river water was comparatively constant at the level that Robison had suggested.[153] In the more accurate experiments on pint measures, Robison was not always definite about the water temperature that he used. In fact, the differences to the volumes measured are comparatively small, and this is discussed further in the Introduction to the Inventory. However, for our present purposes, we should note that the type of temperature range that encompasses Robison's experiments is comparable with the accuracy with which individual determinations of the capacities of earlier pints could be made.

The court approved the first method and instructed that firlots should be constructed, stipulating that the letter 'R' was to be impressed on these new firlots. The burgh of Linlithgow had no objection to this, as long as it was not to be put to any expense. However, by February 1783 no new standards had been produced, and Linlithgow was authorised to issue the firlots in the old form.[154]

After another five years, the court action was renewed by two of the farmers. By this time George Clerk-Maxwell had died. Robison had two firlots of their 1782 pattern constructed, and requested that the court approve them. In a summary report of June 1788 Robison re-stated his findings. He had been unable to locate the reference duplicates of the standards, which had been deposited at Dumbarton and Edinburgh castles in 1618. However, he had measured several early standard pints, including the principal standard at Stirling, and found that their capacities did not differ by an eighth of a cubic inch. This volume matched the standard size of 26,200 grains, which he had derived independently from the original stone weight. Finally, he had now constructed and adjusted the two standard firlots to the design he and George Clerk-Maxwell had proposed, and they were 'in my class room in the University awaiting the sanction of the court'. In 1792 Robison reported that they were 'Lodged in the Custody of the Magistrates of Edinburgh', but after this we lose sight of them. One example of this type of measure now survives – a wheat firlot at Aberdeen (Item **222**, and Fig. **9.6**), later adapted to act as an Imperial bushel. Circumstantial evidence suggests that it was one of the original pair adjusted by Robison, presumably made at the financial risk of his coppersmith, who no doubt anticipated substantial sales when the court ultimately sanctioned its use.[155]

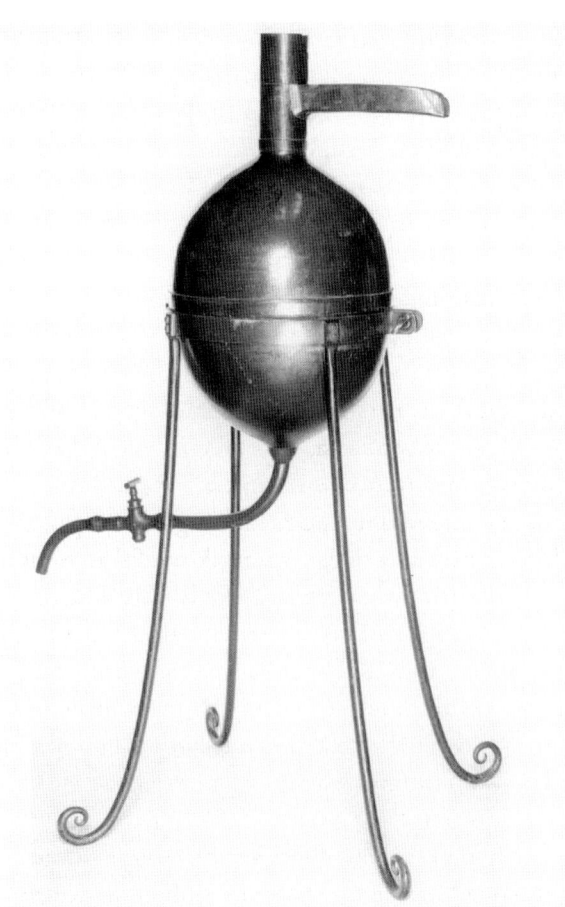

Fig. 9.6
Wheat firlot designed by John Robison and George Clerk-Maxwell, and supplied to the town of Aberdeen in 1811. (Item **222** in the Inventory.)

Lord Alva's final judgement in May 1791 must have come as a bitter disappointment to Robison. The contentious firlots of the Bakers' Incorporation were indeed to be destroyed, but the magistrates and town council of Linlithgow were merely ordered to issue only legal firlots which fully complied with the requirements of the 1618 Act. It was found that there was no legal necessity to comment on technique: 'The Lords find that it is not necessary in this case to enquire into the usage adopted in issuing the firlot measure, in regard that the rule followed by the Magistrates and Town Council of the burgh of Linlithgow, and the Cooper of said Burgh, is erroneous and disconform to the rule pointed out by the Act.'[151] The only aspect in which any change had been introduced was the specification of a temperature at which the measurement was to be undertaken, a feature which in 1618 was not recognised as of importance. But this was hardly the specific temperature as Robison would have wished: the judgement imprecisely required the measuring water to be 'in a moderate state of temperature', or 'of a medium temperature'.

In the circumstances it is unlikely that Linlithgow would acquire a set of Robison's apparatus, but they do appear to have been stung into some action. A new lightweight copper pint was made in November 1791, again with the traditional wide mouth; and, to reinforce its legality it was engraved to show that it had been adjusted against the Stirling Jug in the presence of the provosts of Stirling and Linlithgow (Item 203). The comparatively large capacity of this pint – about 200 grains in excess of the Stirling Jug – suggests, however, that the same practice of over-filling the jug, of which Robison had been so critical, was used again in dispensing from the Stirling Jug. There is no satisfactory way of testing whether Linlithgow firlots changed at all as a result of the judgement. Only one set of measures now survives with the full range of Linlithgow markings – an undated set in Cupar (Items 214 and 215), which have provisionally been dated to the late eighteenth century because the calculated capacities closely match those of Gray and also Robison. They certainly do not display, for example, the 3 per cent and 4 per cent excess in volume found in the Perth measures acquired from Linlithgow in 1774, and so perhaps pre-date Jamieson's tenure as Linlithgow's cooper.[152]

The Perthshire county commissioners took careful account of the judgement, noting that 'those standards constructed by the Professor altho' repeatedly approved of do not appear to have received the ultimate sanction of the Court, or to have been directed to be used as standards'.[153] Nonetheless they commissioned a pair which Robison adjusted. He had previously provided them with the pint and mutchkin standards and counterweights needed to create firlot standards according to Robison and Clerk-Maxwell's second and more laborious method. Although none of this apparatus now survives, two wooden firlots adjusted by this process in 1793 are in Perth Museum (Items 223 and 224). Robison's own pint standard (Item 202), made in the same manner and carrying a 1789 inscription relating its contents to the 1618 Act, together with Robison's signature in facsimile, was purchased by Edinburgh's dean of guild at the sale of Robison's effects after his death in 1805.[154]

As Robison became more involved in the practical overseeing of such precision measurement, he needed on hand an engineer capable of the exactitude he demanded in re-creating new standards to what he regarded as historic accuracy. He found such a person in John Milne, an Edinburgh founder, who was in business with his father, also named John Milne. The elder John Milne had begun as a founder in 1742, and was in partnership with his son by 1767: the partnership continued until just before the father's death in 1810 and the younger Milne continued on his own account until 1824. Perhaps as a consequence of the Linlithgow Court of Session case, Robison was again involved in consultancy work in 1800, this time for the Edinburgh dean of guild. In 1798 the dean of guild was told that his verification rights were being eroded and should be reinforced. In 1800 he asked the council for authority to examine the physical standards to make the necessary replacements, and to make an annual inventory. Amongst surviving examples of these additions is a new set of troy weights made by John Milne, based on Robison's recovered value for the 1618 standard. The four largest are also inscribed 'Adjusted in terms of Act of Par^l. 1618 / by Jo. Robison LLD Prof. Nat. Ph. Edinburgh / March 1800', with the equivalent English troy weight (Item **95**). Similarly, the Scots pint and its fractions (Items **204** and **205**) and the English gallon and gill, both dated 1800, and the half-gallon and its fractions, dated 1801 (Items **230**, **231** and **232**), all with Robison's name, and made at the same time, survive. Milne went on to make standard weights for Cupar dated 1812 (Items **98** and **99**).

One of the problems of interpreting Robison's figures is that we have inadequate information about the precise basis for the weights used in his various comparisons. Sometimes these were the Edinburgh dean of guild's weights (in particular, a set of 1707 nested troy standards, which does not survive), sometimes they were his own weights, and sometimes a mixture of the two. In 1800 he made the specific claim that he had always used the same standard, and that it was the one used previously by James Gray. Since Gray cites the results of the comparisons made by the Royal Society in the 1740s it is almost inescapable that the Edinburgh group were using weights based on the Royal Society's standards. Robison also believed that his weights were of the same family: when describing the Royal Society's 16-ounce nest of troy weight, he commented, 'I suppose mine to be a copy of this'.[160] Indeed the Royal Society's weights remained the standard for precision scientific work until the end of the century – they were, for example, the basis of the weights supplied in the 1790s by the premier instrument engineer of the day, Edward Troughton, for an important series of mass and buoyancy experiments which were central to the evolution of the Imperial system.[161] It is not surprising, therefore, to find that John Swinton's weights were also of the Royal Society standard, and were in fact made by John Reid of London, who had constructed the original weights for the Royal Society in the 1730s.[162]

We discussed in Chapter 4 some of the implications of the establishment of the Imperial troy pound at about 3½ grains heavier than the early troy pound embodied in Reid's weights for the Royal Society. The difference in terms of the Scots pint's capacity is relatively small – only about 15 grains in the pint of 26,200 grains – even less than the inherent uncertainty in the value arising from temperature variations. However, the effect of the change is readily apparent in the set of 1800 weights by John Milne for Edinburgh (Item **95**), which are engraved with Robison's derived weight of the 'Lanark Troy' described in terms of English troy grains. Weighing them in modern grains reveals a nearly uniform difference between the two sizes of about 6 grains in the pound.

Although we might have expected a difference of 4½ to 5 grains in the 16-ounce pound, Robison was already making internal allowances in his calculations to compensate for wear in his weights, and so the agreement is acceptable. From this we can tell that Robison's derived pound of 7,620 grains is equivalent to 7,614 modern grains, and this is borne out in Robison's own standard of the pound, of about 1790 (Item **92**).

Further evidence of Robison's thinking comes from a set of seven lead poise weights representing the water contents of Scots and English liquid measure, adjusted by Robison for the Edinburgh dean of guild in 1800 (Items **93** and **105**). These are stamped with Robison's conventional weights of 26,200 grains for the Scots pint and 14,609 grains for the English wine quart, and with an official-looking Scots shield devised by Robison (also on the 1800 weights), to convey authenticity and to demonstrate that the weights have not been adjusted. Because the metal was soft, suspicions of unevenness were raised in 1817, but in the event only the English weights were re-made in more durable metal (Item **106**). The wine quart can be shown to have been slightly under standard, but the remainder of these show a reasonably consistent amount of wear (typically about 4 grains) on each weight when Robison's figures are corrected to modern grains and compared with the present mass.

However, there is one telling exception to this. The weight for the pint's contents, claimed to be 26,200 grains, is in fact about 40 grains larger. This is certainly too large to have been overlooked by Robison, and it is equally clear that the weight has not been altered. The strong suspicion has to be that Robison recognised that his computed weight for the pint was too small, and it was likely that a poise weight supplied to Edinburgh would be tested against an authentic standard such as the Edinburgh 1618 pint (Item **113**).

It has already been explained how Robison had adapted at an early stage in the court proceedings his standard capacity of the pint to Robert Stewart's mean weight of 26,200 grains, with a derived volume of 103·5 in³. Although Robison's own pint vessel, inscribed for him in 1789 (Item **202**), also claimed to contain 26,200 grains, in practice it contains about 103·6 in³; yet even this is noticeably smaller than the 1618 Edinburgh pint (Item **113**) at about 103·7 in³ or the Stirling Jug (Item **108**) at about 103·8 in³.

It now seems obvious that the error had been to interpret the 1618 Assize literally, so that the pint of '55 French ounces' was taken to mean exactly 55 ounces of the weight defined in 1618.[163] Our own analysis of the 1618 weight, coupled with seventeenth-century references which relate this to the bullion weight at the Mint, leads us to an ounce of about 475·6 (modern) grains, and a pound of 7,610 grains, rather smaller than the figure of 7,614 (modern) grains that arises from Robison's work. If Robison's computed size for the pint is too small, it follows either that the prescribed 55 ounces must have been of a different 'French' type, or that the multiple is not intended to be exact and is greater than 55. Taking the pint at about 26,250 grains and the ounce at 475·6, the pint contains about 55·2 ounces.

But Robison's dilemma was that he could not depart from his method of approaching the stone's weight through consideration of cumulative wear, or of calculating the pint as 55 ounces of the stone, without risk of losing his credibility as the court's expert on metrology. Even after the verdict in 1791 he was involved with trying to sell his system to local authorities, so its physical basis had to remain unquestioned. Robison appears to have resolved this in his dealings with Edinburgh, simply by silently increasing the size of the poise weight for the pint (Item **93**). Even so, in the following year, when he was adjusting the set of liquid measures made for Edinburgh (Item **205** and **232**), he found the size of the Scots pint showed the greatest discrepancy, as it was 54 grains too small.[164]

In January 1805 the mysterious and painful illness which had afflicted him intermittently for nineteen years suddenly caused Robison's death. Shortly afterwards, the Edinburgh dean of guild reported that he had paid £11 12s at the auction of Robison's effects for three standard measures (his Scots pint is Item **202**) and a Lanark troy pound weight adjusted by Robison himself (Item **92**).[165] His post at the university was filled by John Playfair, who was himself inevitably called upon subsequently to comment on metrological issues. Playfair had been professor of mathematics at Edinburgh since 1785, and one of his more promising pupils was James Jardine (1776-1858), who attended his classes but did not graduate. Playfair encouraged Jardine to become a civil engineer, a career in which he was extremely active and successful. He, too, became involved in precision metrology, initially in Aberdeen, and subsequently in Edinburgh.

PATRICK COPLAND, JOHN PLAYFAIR
AND THE INTRODUCTION OF THE IMPERIAL SYSTEM, 1824

Edinburgh and Stirling were not the only centres of this metrological insecurity: in Aberdeen various individuals had been expressing qualms about the state of affairs. An anonymous detailed summary of Aberdeen metrological events appeared in William Kennedy's *Annals of Aberdeen* in 1818, and this has been identified by John Reid as a printing of George Skene Keith's unpublished 1795 report to the dean of guild, which also appears in the Guild Court Book.[166] George Skene Keith (1752-1823) graduated from Marischal College at the University of Aberdeen in 1770, and was licensed to preach in 1774; after a problem over his appointment to the living at Keith-Hall and Kinkell, Aberdeenshire, was resolved in the courts, he took up his position as minister there in 1778. Towards the end of his life he claimed that he had spent 'a period of above thirty years' investigating methods of equalising the weights and measures systems of Great Britain, and strongly supported the adoption of the seconds pendulum as the fundamental standard of length.[167] As he succinctly wrote, demonstrating the need for reform:

> In addition to the confusion that has arisen among our legal and national standards, and the weights and measures derived from them, a number of provincial ones, equally imperfect, or more exceptionable, has been established by practice, in opposition to law, in the different Counties, or inferior Districts of Great Britain. England contains about *two hundred and thirty*, and Scotland above *seventy*, of these provincial weights and measures; so that there are above three hundred of them used in different parts of the island. Frequently, four or five different ones are found in the same county; and (what completes the disorder among them) there are different weights and measures established by law or practice, for different purposes; one gallon for ale, another for wine; and a third for corn; one weight for the precious metals, gold and silver, another for iron and copper, and a third for lead; one pound for the Apothecary, another for the Baker, and a third for the Grocer – all used in the same county or city.[168]

Keith had laid his plan for a 'natural' metrology system based on the pendulum before a committee of the House of Commons in January 1790; this was headed by Sir John Riggs Miller, who intended to bring it in as a bill, but lost his seat at the following general election. Keith published these proposals in 1791.[169]

In 1794, as Keith himself recounts, he was employed by the dean of guild, Thomas Bannerman, to examine the standards kept by Aberdeen, and found that the 1707 Union bushel 'contained only 77 lbs. 2 oz. 7 grains of spring or fountain water, in a heat of 50° ... This, and the other experiments about the weight of the measures of Aberdeen, were made with all possible accuracy; and as both the national and provincial standards were weighed with the same water, and in the same temperature, the proportions between the two here given may be depended upon'.[170] In this report, Keith mentions that 'the Scotch Troy pound has been fixed by Professor Robison, in his report to the court of session, at 7621⁹⁄₁₁ grains [this is just Stewart's earlier value of 7,621·818 grains]… The Aberdeen pint is 400 grains weightier than the Stirling jug, as fixed by Professor Robison, and contains 26600 English Troy grains …', and he comments on the figures 'fixed by Sir Robert Stuart [*sic*] in his manuscript on weights and measures'.[171] From these remarks, it is evident that Keith was in contact with Robison and his metrological work. Subsequently, Keith was to assist Patrick Copland in his efforts to provide Aberdeen with a comparison between old and new measures, and this is discussed below.

In Edinburgh Robison's investigation of the standards had been prompted by a Court of Session case, and possibly also influenced by pressure from the Convention of the Royal Burghs which had resulted in Swinton's *Report*. In Aberdeen another university professor became involved in overseeing a similar exercise, also prompted by a Court of Session case, in which the court had turned initially to John Playfair (and this is discussed in Chapter 3). Patrick Copland (1748-1822), as John Reid explains, was professor at Marischal College at the University of Aberdeen between 1775 and 1822. Although he taught natural philosophy for all that time except for one year, in title he was professor of mathematics from 1779 to 1817. Born locally, educated locally, yet 'in spite of this narrow experience he was just the right man to widen the horizons of Marischal College students and the citizens of Aberdeen to the growing involvement of science in society'.[172] One of the areas in which he became practically involved was in the acquisition of new measures in 1801, and the re-measurement, together with Keith, of existing standards held by the Aberdeen dean of guild in 1811.[173]

Copland's measurement of the old Edinburgh ell standard (Item 2) was made against one of the most accurately-graduated scales then in existence, made for him in 1801 by Edward Troughton of London.[174] The Edinburgh measures were carried north not by Playfair himself, but by his young mathematical protégé, James Jardine, who assisted Copland with the comparisons. Jardine, who had studied under Playfair, initially became a mathematics teacher, and then at Playfair's urging set up as a civil engineer. He made a considerable name for himself, working on canals, railways, bridge and river projects, acting in addition from 1815 as the observer to the Astronomical Institution of Edinburgh, of which Playfair was president. Jardine's work with Copland identified him as the person best equipped to co-ordinate the re-measurement of the old Scots units when the new Imperial system standards were introduced in the 1820s.

The reforming weights and measures Act of 1824 was the result of a period of increased parliamentary awareness of the problems that had already been highlighted by, amongst others, Swinton, Riggs Miller and Sinclair.[175] Against a background of much more radical (indeed, revolutionary) changes in weights and measures in France, the British proposals were for a simplified and unified system which was essentially more conservative although still based on modern scientific principles.[176] Agitation in Scotland had continued, and in 1811 a committee of the Highland Society, with the involvement of Sir John Sinclair, investigated the possibility of reforming weights and measures across the United Kingdom to incorporate the type of change they deemed necessary for Scotland. Playfair had a significant input to this, and the committee got as far as preparing an outline specification for legislation. Their 1813 report was sent to magistrates and members of parliament across the country, and it has been speculated that it was this that led to the formation of a parliamentary select committee in 1814, and possibly also to the Royal Commission established in 1819.[177] Indeed, one of the prime movers for the legislation that was eventually successful was the member of parliament Davies Gilbert, who was a member of the Board of Agriculture, and had also become closely involved in the affairs of the Royal Society (of which he subsequently became president).[178]

However, in his analysis of the political background to this episode, Julian Hoppit has stressed the key role played by Sir George Clerk (1787-1867), a Lord of the Admiralty and M.P. for Edinburghshire, who, he believes, was given the government task of piloting the measure through the House of Commons.[179] Clerk was a member of the select committee, and also of the Royal Commission, and as sixth baronet of Penicuik, he was clearly familiar with the work of his predecessor and grandfather, Sir George Clerk-Maxwell (the fourth baronet), and of the equally-pressing case north of the border.

The act was originally to come into effect in 1825, but this period was extended to 1826, apparently because of the difficulty in supplying all the local authorities across Britain with the entire set of new brass standards engineered to the level of precision now expected (see Fig. **9.7**).[180] Consequently, the most important authorities have standards dated 1824, the date of the act, but those of smaller authorities, which had to wait longer, are dated 1825 or 1826, although some authorities clearly dragged their feet and ordered late, and a number did not receive sets until after further legislation in the 1830s. The instrument supplier who obtained the initial Treasury contract was Robert Brettell Bate, a London scientific instrument maker and entrepreneur.[181]

Unfortunately the best intentions of the reformers were thwarted by the 1824 Act's admission that customary measures and customary practices such as heaping could be used providing their relationship with the new standards was sufficiently well-understood.[182] The effect of this admission was that the use of customary measure, particularly in the local market economies in most

Scottish counties, continued largely unchanged, and it was not until two subsequent acts were introduced in the mid-1830s that local customary units were progressively eliminated.[183] In spite of this, some local units continued in use and examples are prominently found in the returns of the Scottish Poor Law Commission in the 1840s and in other sources, which typically use Scots bolls and 'Dutch' stone, sometimes even referring to 'old bolls'.[184]

The 1824 Act specifically directed that the relationship of all local units to Imperial units throughout the United Kingdom should be recorded; in Scotland this was ostensibly for dealing with payments such as the stipends of clergy, feu-duties, rents, tolls and customs, which were payable in kind, so that in future these could be converted into the new Imperial units. However, in practice it had the effect of allowing the continued use of the old units. The information was to be gathered by 'inquisitions' before the county sheriff-deputes or equivalents. The results were to be sent to the Court of Exchequer in Edinburgh, and tables for conversion were to be published.[185]

A number of Scottish counties commissioned special reports from experts to guide the inquisition juries. The most influential of these was undoubtedly the report done by James Jardine, working with the instrument maker Alexander Adie, and the excise accountant David Murray. Jardine and his colleagues reported in December 1825 the results of examination of the surviving 1618 Lanark weights and the Stirling Jug, in order to provide a historical basis for the customary measures then in widespread use, but not surprisingly they took on trust Jardine's earlier measurement (with Copland) of the Edinburgh ell at Aberdeen. There were, however, a number of other such reports, where juries took advice from local specialists, and in particular Adam Anderson, the Rector of Perth Academy, conducted further tests on the Stirling Jug, reporting in 1827 to the Stirlingshire jury. The findings of the various juries were collated and general conclusions drawn about the original sizes of the Scots standards. These, together with abstracts of the jury reports discussing the local variant measures of those counties, and extended tables of conversion between the various units, were published in an accessible form by the civil engineer George Buchanan in 1829, with a further edition (very largely unchanged) appearing in 1838, presumably in response to the later legislation enforcing the Imperial system.

Although the wealth of detail provided by Buchanan has encouraged some commentators, such as Ronald Zupko, to interpret this as representing an apparently chaotic administrative situation, exhibiting little evidence of central control, in fact the variety of weights and measures is merely an exhaustive response to the requirement in the 1824 Act that local measures must be related to their Imperial equivalents. The jury reports give no indication about the extent to which more controlled units were used for trade between areas; it must be appreciated that a great deal of Buchanan's variety of measures relates only to internal Scottish markets, which were already beginning to break down.

Buchanan's volume contains Jardine's report to the Edinburghshire jury, with the details of his re-measurement of the 1618 Lanark weights, and further details are contained in the Lanark jury report. Jardine's attitude was to base his deductions about the original Lanark weight directly on the surviving weights and to make no allowance for wear, as Robison had done earlier. His measurements are given to an extreme level of accuracy, well beyond what was reasonable, bearing in mind the age and state of the surviving weights, and indeed his results are undercut by his admission that doubts subsequently emerged about whether small adjustment weights had been present in the surviving Lanark stone.[186] By the time Buchanan's tables were approaching press, Jardine had admitted that he was conducting new enquiries into the original weight of the Lanark stone. In this he was clearly unsuccessful, because he made a pressing request in his evidence to the Parliamentary Select Committee in 1834 that the proposed act should specify a size for the French weight on which the Lanark stone was based, because of the administrative need for a value for a weight that 'affects most of the clergy in Scotland, and many of the landlords, and their feuars and tenants'.[187]

Perhaps significantly, Adam Anderson's report to the Stirlingshire jury was not printed in Buchanan's compilation, because Anderson came to quite different conclusions about the Lanark stone (discussed in the Inventory, Item 85). Also, his more considered and careful measuring of the Stirling jug (discussed in the Inventory, at Item 108) produced results which were incompatible with those of Jardine. The result of this work was to freeze ill-considered and over-engineered values for the early Scots standards into the official papers. As we have demonstrated, early Scottish standards were not produced to mid-nineteenth century standards of precision, and it is clearly preposterous to represent them in this way.

In Glasgow these matters were in the hands of James Cleland (1770-1840), a statistician, who from 1814 had been superintendent of public works in Glasgow. He was already advising the Glasgow dean of guild on weights and measures by 1816 and in 1821 he undertook a comprehensive examination of all the standard weights and measures in Glasgow, followed by an inspection of over 20,000 weights and measures in the city.[188] In 1826, in conjunction with Andrew Ure (1778-1857), professor of chemistry at Anderson's College, and others including the Glasgow instrument maker James Crichton, he advised the Lanarkshire jury and prepared some precise liquid capacity measures along the lines first proposed by John Robison for use in testing trading measures; two of these standards constructed for Glasgow by Stephen Millar survive (Item **241**).[189] Cleland was subsequently appointed as inspector-general of weights and measures for Lanarkshire, and also for Dunbartonshire and Renfrewshire, and was able to oversee the implementation of the 1834 and 1835 Acts. In 1835, he produced his *Conversion of the Weights and Measures of the County of Lanark*, which contained instructions for the new inspectors of weights and measures. In this he noted that the Exchequer standards should remain in the custody of the clerk of the justices of the peace, and that the inspectors therefore had to have duplicate measures, and he specified a conical form which was clearly derived from Andrew Ure's earlier standards for Glasgow. A number of these sets survive (Items **246-253**), in the specified range of two gallons to half-a-gill, and all engraved as he had stipulated: 'Duplicate of imperial standard (the denomination) for the County of Lanark, adjusted by James Cleland, LL.D., Inspector-General of Weights and Measures for the County, 1835.'[190]

This requirement for duplicates was clearly appreciated at an earlier date when the 1824 Act was passed. Locally-engineered narrow-necked designs appeared as early as 1824. Those of Russell for Kirkcaldy (Item **240**) may

Fig. 9.8
Commercial cast-iron weights of 1826, on which Lanark continued to claim its traditional role through the use of its seal. (South Lanarkshire Museum Services, Hamilton)

perhaps have been influenced by Bate's prototype of the gallon measures which were also conical, and may have been in turn influenced by the published design of Robison's firlot. Other examples for Stirlingshire almost certainly by Andrew Boyle of Perth (Item **245**) had cylindrical bodies. The maker of Cleland's 1835 measures (Items **246-253**) is still to be uncovered.

CONCLUSION

The old and distinct Scottish system of weights and measures survived in practical market-place use until the end of the nineteenth century. As late as 1894, a Scottish court ruled in favour of allowing 'Sutherland stones' to be used in a contract, providing only that the stones were described as amounting to 25 pounds avoirdupois.[191] It might be thought that the passage of one hundred and fifty years since the advent of effective controls under the 1834 and 1835 Weights and Measures Acts – combined with over a hundred years of diligent efficiency by specialist weights and measures inspectors – would have finally eliminated the last traces of the old Scots metrology. Yet surprisingly this is not quite true.

Some isolated words still survive, even in central Scotland. The Scots traditional singer Sheena Wellington, brought up in Dundee, clearly remembers her granny's concern that her 'metty of coal' should not turn out to contain half a metty of dross.[192] The mett, or capacity measure, for coal was by then a weight – the so-called 'imputed bushel' of 84 pounds – and a similar transformation had overtaken the two specific volume terms that remain in occasional use. It was recently still possible to buy a 'forpet of potatoes' in Edinburgh, originally a capacity unit of a quarter-peck, a term transferred to a quarter-stone. The equivalent expression in Fife and Perthshire would be a 'lippie' (or 'lippy'), which was frequently found as a named measure in Scots dry capacity sets in the mid- to late nineteenth century.

The original Scots weights and measures had been largely imposed from outside by the influence of Scotland's trading partners and the practical necessity of the market-place. If there is a lesson from history, the replacement of Scotland's old system in 1707 was quite as problematic as the introduction of the metric system proved in France, and so the Scots at least should not be surprised that modern moves towards metrication have dragged on for decades longer than originally anticipated: it takes more than an administrative act to change centuries of customary use. But over and above that, there remain necessary and practical requirements to use and understand the old terms, because these continue to be embodied in early documents which still have legal currency, notably in land transfer, or sasines. It was, of course, for this very reason that the eighteenth-century proceedings drew Robison and others into considering the Scots weights and measures.

Unravelling the story of Scotland's weights and measures has helped fill gaps in our understanding of English metrology, and has taught us to respect the customary procedures for trading in the market-place. The system of allowances, which must reflect endemic practice elsewhere in Europe, brings with it a realisation that Scotland's gross domestic product has been consistently undervalued in the past, but above all, it provides a warning about the complexities of interpreting such regulations as survive. The analysis here has provided a reasonably self-consistent framework for the evolution of Scotland's trading units, demonstrating a high degree of continuity where previous commentators have often found the type of change and disruption which implied administrative weakness. In contrast, we have found a strong and flexible system which operated happily in conjunction with those of other emerging European economies.

1 T. Thomson and C. Innes (eds), *The Acts of the Parliaments of Scotland [APS]*, 13 volumes (London, 1814-75), XI, 201-5: Act of Union, Article XVII, p. 203: 'That from and after the Union the same Weights and Measures shall be used throughout the United Kingdom, as are now Established in England, and Standards of Weights and Measures shall be Kept by those Burroughs in Scotland to whom the Keeping the Standards of Weights and Measures now in use there do's of Special Right belong, All which Standards shall be sent down to such respective Burroughs from the Standards kept in the Exchequer at Westminster Subject nevertheless to such Regulations as the Parliament of Great Britain shall think fit.'

2 Ibid., op. cit. (1).

3 H. W. Turnbull, *et al.* (eds), *The Correspondence of Isaac Newton,* 7 volumes (Cambridge, 1959-77), VII, 456: Letter X.735, Newton to Godolphin, 21 January 1707/8, listing 'severall Tools Utensills & other Materialls … provided & Conveyed to ye said Mint at Edinburgh … bought of Mr John Smart [*sic*].'

4 R. D. Connor, *The Weights and Measures of England [WME]* (London, 1987), 162-4.

5 N. T. Phillipson and Rosalind Mitchison (eds), *Scotland in the Age of Improvement* (Edinburgh, 1970).

6 See Eric Cregeen, 'The Changing Role of the House of Argyll in the Scottish Highlands', in Phillipson and Mitchison, op. cit. (5), 5-23; Rosalind Mitchison, 'The Government and the Highlands, 1707-1745', in ibid., 24-46; John M. Simpson, 'Who Steered the Gravy Train, 1707-1766?', in ibid., 47-72; John Stuart Shaw, *The Management of Scottish Society 1707-1764* (Edinburgh, 1983); John Stuart Shaw, *The Political History of Eighteenth-Century Scotland* (Basingstoke, 1999). For artists, see James Holloway, *Patrons and Painters: Art in Scotland 1650-1760* (Edinburgh, 1989). For artisans, the cases of the telescope maker James Short and the clock-maker Alexander Cumming are discussed by A. D. Morrison-Low, '"Feasting my Eyes with the View of Fine Instruments": Scientific Instruments in Enlightenment Scotland 1680-1820', in C. W. J. Withers and Paul Wood (eds), *Science and Medicine in the Scottish Enlightenment: a Reassessment* (East Linton, 2002), 17-53.

7 Roger L. Emerson, *Professors, Patronage and Politics: the Aberdeen Universities in the Eighteenth Century* (Aberdeen, 1992); Joan Pittock and Jennifer Carter (eds), *Aberdeen and the Enlightenment* (Aberdeen, 1987); Andrew Hook and Richard B. Sher (eds), *The Glasgow Enlightenment* (East Linton, 1995); D. Daiches, P. Jones and J. Jones (eds), *A Hotbed of Genius: The Scottish Enlightenment, 1730-90* (Edinburgh, 1986).

8 For Scottish antiquarianism, see Stuart Piggott, 'The Ancestors of Jonathan Oldbuck', in S. Piggott, *Ruins in a Landscape: Essays in Antiquarianism* (Edinburgh, 1976), 133-59; I. G. Brown, *The Hobby-Horsical Antiquary: A Scottish Character, 1640-1830* (Edinburgh, 1980); and A. S. Bell (ed), *The Scottish Antiquarian Tradition* (Edinburgh, 1981). There is a vast and growing literature concerning Scottish science at this time, but see Anand Chitnis, *The Scottish Enlightenment: a Social History* (London, 1976), especially Chapter 6, 'Universities: Medicine and Science', pp. 124-94; Roger L. Emerson, 'Science and the Origins and Concerns of the Scottish Enlightenment', in *History of Science*, 26 (1988), 333-66; and Withers and Wood, op. cit. (6).

9 John Heilbron discusses the improvements in precision which took place in scientific instrumentation during the eighteenth century, together with a concomitant interest by the state in numbers, culminating in the reform of weights and measures by the French Revolutionary government: J. L. Heilbron, 'Introductory Essay', in T. Frängsmyr, J. L. Heilbron and R. E. Rider (eds), *The Quantifying Spirit in the Eighteenth Century* (Berkeley, 1990), 1-23, and J. L. Heilbron, 'The Measure of Enlightenment', in ibid., 207-42. Details about how the French instrument trade responded to the difficult circumstances of supplying the state with precision measures are discussed by A. J. Turner, *From Pleasure and Profit to Science and Security: Etienne Lenoir and the Transformation of Precision Instrument-Making in Paris 1760-1830* (Cambridge, 1989). The initial failure of the metric system in France is discussed by Ken Alder, 'A Revolution to Measure: the Political Economy of the Metric System in France', in M. Norton Wise (ed), *The Values of Precision* (Princeton, 1995), 39-71, and Ken Alder, *The Measure of All Things: the Seven-Year Odyssey that Transformed the World* (London, 2002).

10 Museum Gustavianum, University of Uppsala, inventory no. 26. For Bird, see ref. 101.

11 See A. D. C. Simpson, 'The Pendulum as the British Length Standard: a Nineteenth Century Legal Aberration', in R. G. W. Anderson, J. A.

Bennett and W. F. Ryan (eds), *Making Instruments Count: Essays in Historic Scientific Instruments presented to Gerard L'Estrange Turner* (Aldershot, 1993), 174-90.

12   Simon Schaffer, 'Modernity and Metrology', in Luca Guzzetti (ed) *Science and Power: the Historical Foundations of Research Policies in Europe* (Luxembourg, 2000), 71-91, especially p. 90. Schaffer has also discussed the later problems of nineteenth-century metrology in 'Accurate Measure is an English Science', in Wise, op. cit. (9), 135-72; and 'Metrology, Metrication, and Victorian Values', in Bernard Lightman (ed), *Victorian Science in Context* (Chicago and London, 1997), 438-74.

13   See, for instance, the papers in English by Gerrit Moll, 'On the Comparison of the British, French and Dutch Weights', in *Journal of the Royal Institution of Great Britain,* 2 (1831), 64-75; and 'Comparison of the Kilogramme with English Weights', in *Astronomische Nachrichten,* 9 (1831), col. 74-6. In both of these papers Moll looks at the work of his predecessors and contemporaries and finds that there is no such thing as absolute equivalence in metrology. He looks as far back as similar work by the Frenchman Mathieu Tillet in 1767, and includes the more recent efforts of his countryman Jan van Swinden. Our thanks to Dr Anita McConnell for these references.

14   Alexander Huntar [*sic*], *A Treatise of Weights, Mets and Measures of Scotland* (Edinburgh, 1624); [John Swinton], *A Proposal for Uniformity of Weights and Measures in Scotland* (Edinburgh, 1779); Sir John Sinclair (ed), *The Statistical Account of Scotland,* 21 volumes (Edinburgh, 1791-99); [various authors], *General View of the Agriculture [of counties for the] Board of Agriculture,* 57 volumes (Edinburgh and London, 1794-1814).

15   See, for example, Richard Sheldon, Adrian Randall, Andrew Charlesworth and David Walsh, 'Popular Protest and the Persistence of Customary Corn Measures: Resistance to the Winchester Bushel in the English West', in Adrian Randall and Andrew Charlesworth (eds), *Markets, Market Culture and Popular Protest in Eighteenth-Century Britain and Ireland* (Liverpool, 1996), 25-45. Our thanks to Dr Simon Schaffer for this reference.

16   David Allen, *Scotland in the Eighteenth Century: Union and Enlightenment* (Harlow, 2002); for the economy, see Bruce Lenman, *An Economic History of Modern Scotland* (London, 1977), 44-55; Ian D. Whyte, *Scotland's Society and Economy in Transition c.1500-c.1760* (Basingstoke, 1997), 140-65; for the long term nature of Scottish scientific ideas, see Emerson, op. cit. (8), and Paul Wood, 'The Scientific Revolution in Scotland', in Roy Porter

and Mikuláš Teich (eds), *The Scientific Revolution in National Context* (Cambridge, 1992), 263-87.

17   Shaw, *Political History,* op. cit. (6), 1.

18   R. Mitchison, *Lordship to Patronage: Scotland 1603-1745* (London, 1983), 108-9; Whyte, op. cit. (16), 151.

19   Whyte, op. cit. (16), 152.

20   Ibid., 153.

21   Ibid.; Allen, op. cit. (16), 6; A. J. C. Cummings, 'Scotland's Links with Europe, 1600-1800', in Jenny Wormald (ed), *Scotland Revisited* (London, 1991), 142-50, esp. p. 146.

22   Mitchison, op. cit. (18), 109.

23   Ibid., 136.

24   See A. D. C. Simpson, 'Sir Robert Sibbald – the Founder of the College', in R. Passmore (ed), *Proceedings of the Royal College of Physicians of Edinburgh Tercentenary Congress 1981* (Edinburgh, 1982), 59-91; for the background see Hugh Ouston, 'York in Edinburgh: James VII and the Patronage of Learning in Scotland, 1679-1688', in John Dwyer, Roger A. Mason and Alexander Murdoch (eds), *New Perspectives in the Politics and Culture of Early Modern Scotland* (Edinburgh, 1981), 133-55.

25   Anita Guerrini, 'The Tory Newtonians: Gregory, Pitcairne and Their Circle', in *Journal of British Studies,* 25 (1986), 288-311, quote, p. 290.

26   Athol L. Murray, 'Sir Isaac Newton and the Scottish Recoinage, 1707-10', in *Proceedings of the Society of Antiquaries of Scotland,* 127 (1997), 921-44, quote p. 921.

27   Ibid., 924.

28   Much of the information about Snart has come from personal communications from Diana Crawforth-Hitchins and Dr P. Buchanan; see also D. F. Crawforth, 'London Scale Makers (Part I) The Transmission of Scalemaking Knowledge from 1632-1800', in *Equilibrium* (Spring 1980), 207-15, esp. 210-12; Gloria Clifton, *Directory of British Scientific Instrument Makers 1550-1851* (London, 1995), 257-8, 205; G. Clifton, 'Scientific Instrument Makers at the Royal Mint', unpublished paper presented to the XIX Scientific Instrument Commission Symposium, Oxford, September 2000.

29   David Daiches, *Scotland and the Union* (London, 1977), 150-1.

30   'Professor [John] Robison's Commonplace Book', 2 volumes (subsequently Robison, Commonplace Book): St Andrews University Library MS Q171.R8, I, p. 442. The arrival of the consignment at this date is confirmed in Stirling Council Archive, MS Stirling council minutes, 18 October 1707: '… a letter from the provost of Edr signifying that the weights and mesures were now come from England.'

31 R. Renwick (ed), *Extracts from the Records of the Royal Burgh of Stirling [Stirling Extracts]*, 2 volumes (Glasgow, 1887-9), II, 113: 1 November 1707.

32 J. D. Marwick, *et al.* (eds), *Records of the Convention of the Royal Burghs of Scotland (1295-1779) [RCRB]*, 7 volumes (Edinburgh, 1866-1918), IV, 459-60: 13 July 1708.

33 Ibid., 462-3: 15 July 1708. The fifteen royal burghs who received sets from the four distributing burghs can be identified because all have retained some parts of their holdings (listed in the Inventory): Aberdeen, Banff, Crail, Cupar, Dumfries, Dunfermline, Dundee, Glasgow, Haddington, Jedburgh, Kirkcaldy, Perth, St Andrews, Selkirk, Wigtown. Additionally, two sets were retained by Edinburgh, and one each was held by Lanark, Linlithgow and Stirling, with additional sets divided between them (thus providing Lanark with two sets of weights, etc.). This is a further six sets, making a total of twenty-one.

34 *Stirling Extracts*, II, 113: 24 January 1708.

35 Ibid., 114: 26 April 1708.

36 Ibid., 116: 19 July 1708.

37 J. D. Marwick and R. Renwick (eds), *Extracts from the Records of the Burgh of Glasgow [Glasgow Extracts]*, 6 volumes (Glasgow, 1908), IV, 476-7: 27 May 1712.

38 *Stirling Extracts*, II, 116: 24 July 1708.

39 Ibid., 116: 24 July and 4 September 1708.

40 Ibid., 116: 24 July 1708; *Glasgow Extracts*, IV, 430: 2 October 1708.

41 *RCRB*, IV, 483: 7 July 1709.

42 Ibid., V, 19: 19 July 1711.

43 In July 1709 Banff received its measures by sea from Leith, including 'ane bushell from Linlithgow, English measure, of metal, ane peck of oak wood and fourth part thereof, houped with iron': W. Cramond (ed), *Annals of Banff*, New Spalding Club nos. 8 and 10, 2 volumes (Aberdeen, 1891-3), II, 352.

44 *Stirling Extracts*, II, 123: 7 October 1710.

45 *RCRB*, V, 315: 6 July 1722.

46 Whyte, op. cit. (16), 118.

47 Ibid., 115-29.

48 Swinton, op. cit. (14), 72-3.

49 R. L. Emerson, 'The Philosophical Society of Edinburgh, 1738-1747', in *British Journal for the History of Science*, 12 (1979), 154-91, pp. 155-8; also, 'The Philosophical Society of Edinburgh, 1748-1768', in ibid., 14 (1981), 133-76; 'The Philosophical Society of Edinburgh, 1768-1783', in ibid., 18 (1985), 255-303; also 'Sir Robert Sibbald, Kt., the Royal Society of Scotland and the Origins of the Scottish Enlightenment', in *Annals of Science*, 45 (1988), 41-72.

50 R. L. Emerson, 'The Scottish Enlightenment and the End of the Philosophical Society of Edinburgh', in *British Journal for the History of Science*, 21 (1988), 33-66, quote pp. 34-5.

51 Ibid.; Steven Shapin, 'Property, Patronage and the Politics of Science: the Founding of the Royal Society of Edinburgh', in ibid., 7 (1974), 1-41.

52 David Gregory, *A Treatise of Practical Geometry … by the Late Dr David Gregory, sometime Professor of Mathematicks in the University of Edinburgh … Translated from the Latin; with Additions [by Colin Maclaurin]* (Edinburgh, 1745), 110.

53 Ibid., 112.

54 A. Grant, *The Story of the University of Edinburgh during its First Three Hundred Years*, 2 volumes (London, 1884), II, 348-9; D. B. Horn, *A Short History of the University of Edinburgh* (Edinburgh, 1967), 40.

55 James Dennistoun (ed), *The Coltness Collections 1608-1840* (Edinburgh, 1842), 66-7; G. E. C[ockayne], *Complete Baronetage 1611-1800*, 5 volumes (Exeter, 1900-9), IV, 375-6.

56 Emerson, op. cit. (8), 342 and 360 note 49; quotation from Grant, op. cit. (54), 349. For more about Stewart, see Emerson, 'Philosophical Society, 1738-1747', op. cit. (49), 190.

57 Robison, Commonplace Book, II, p. 393.

58 Gregory, op. cit. (52), 116.

59 Ibid., 116; Robison, Commonplace Book, I, p. 446; II, p.393. Stewart's measurement was quoted by Robison in his 19 June 1788 report to the Court of Session as the crucial starting point in his determination of the size of the pound and therefore the volume of the pint: National Archives of Scotland (NAS), MS Court of Session papers 231/F3/1, bundle 32. The draft of this is in his Commonplace Book, II, p. 391.

60 Robison, Commonplace Book, II, p. 393.

61 Ronald M. Birse, *Science at the University of Edinburgh 1583-1993* (Edinburgh, 1994), 21. James was the younger brother of David Gregory, and had previously held the chair of mathematics at St Andrews University.

62 J. B. Morrell, 'The Edinburgh Town Council and its University, 1717-1766', in R. G. W. Anderson and A. D. C. Simpson (eds), *The Early Years of the Edinburgh Medical School* (Edinburgh, 1976), 1-26; Walter H. Makey, 'George Drummond's New Edinburgh', in [A. H. B. Masson and A. D. C. Simpson (eds)], *Edinburgh's Infirmary: A Symposium … * (Edinburgh, 1979), 19-22.

63 J. B. Morrell, 'The University of Edinburgh in the Late Eighteenth Century: Its Scientific Eminence and Academic Structure', in *Isis*, 62 (1970), 158-71, quotation p. 164.

64 T. C. Smout, *A History of the Scottish People 1560-1830* (Glasgow, 1969), 451-69.

65 Richard Yeo, *Encyclopaedic Visions: Scientific Dictionaries and Enlightenment Culture* (Cambridge, 2001), esp. pp. 170-92, 'The *Encyclopadia Britannica* and the Scottish Enlightenment'.

66 Judith V. Grabiner, 'A Mathematician among the Molasses Barrels: Maclaurin's Unpublished Memoir on Volumes', in *Proceedings of the Edinburgh Mathematical Society*, 39 (1996), 193-240. The manuscript is 'Memorial offered to the Honourable Commissioners of Excise concerning the Mensuration of Tuns or Backs ... 1735': National Library of Scotland (NLS) Adv. MS 23.1.13.

67 Ibid., 236.

68 Gregory, op. cit. (52), 116.

69 For Bryce, see Robert Chambers, *A Biographical Dictionary of Eminent Scotsmen*, 4 volumes (Glasgow, 1835), IV, 488-93; Emerson, 'Philosophical Society, 1737-1747', op. cit. (49), and 'Philosophical Society, 1748-1768', op. cit. (49).

70 For Gray, see Emerson, op. cit. (49); H. W. Thompson (ed), *The Anecdotes and Egotism of Henry Mackenzie 1745-1831* (London, 1927), 221; David R. Smith, 'The Ironmill', *Old Dalkeith*, no. 2 (1985), 7-9. Gray was appointed the king's Master Smith and Iron Monger in Scotland in 1757: NLS MS 17504, f.87.

71 Edinburgh Council Archives (ECA) MS council minutes, 26 October 1737.

72 [George Graham], 'An Account of the Proportions of the English and French Measures and Weights, from the Standards of the same, kept at the Royal Society', and 'An Account of a Comparison lately made by some Gentlemen of the Royal Society, of the Standard of a Yard, and the several Weights lately made for their Use; with the Original Standards of Measures and Weights in the Exchequer, and some others kept for public Use at Guild-hall, Founders-hall, the Tower, &c.', in *Philosophical Transactions of the Royal Society*, 42 (1742-43), 185-8 and 541-56.

73 Chambers, op. cit. (69), 490.

74 First referred to in the 1810 second volume of George Chalmers, *Caledonia; or an Account ... of North Britain*, 3 volumes (London, 1807-24), II, 40. Robert Chambers's first account is in his *Picture of Stirling* (Edinburgh, 1830), 44-6, and the fullest version is in Chambers, op. cit. (69), 490-2.

75 Ibid., 490.

76 Emerson, 'Philosophical Society, 1748-1768', op. cit. (49), 174.

77 For biographical details, see the biography in Sydney Lee (ed), *Dictionary of National Biography [DNB]*, 63 volumes (London, 1885-1904), LIV, 379-

80; the biography by P. J. Wallis, in C. C. Gillispie (ed), *Dictionary of Scientific Biography [DSB]*, 16 volumes (New York, 1970-80), XIII, 67-70; the introduction to Ian Tweddle, *James Stirling: 'This about series and such things'* (Edinburgh, 1988); and William Fraser, *The Stirlings of Keir and their Family Papers* (Edinburgh, 1858), 91-102.

78 James Stirling of Leadhill, 'An Account of the Money, Coins, and Weights, used in England, during the Reigns of the Saxon Princes', in *Transactions of the Society of Antiquaries of Scotland*, 1 (1792), 216-33.

79 National Register of Archives (Scotland) (NRA(S)), 2362; National Register of Archives, 14810/ 137, 138, 139 and 140. When consulted at the NAS in 1989, these were given the temporary reference number TD/89/29. We would like to thank Dr Ian Grant, Secretary of the NRA(S), and the Garden Trustees.

80 NRA(S), 2362, 138, p. 3 and 139, p. 4.

81 *RCRB*, VI, 401: 8 July 1752.

82 Ibid.

83 Ibid., VI, 447: 3 July 1754.

84 On Drummond, see Makey, op. cit. (62).

85 *RCRB*, VI, 465: 26 November 1754.

86 Ibid., VI, 447, 486, 511.

87 'Report from the [Carysfort] Committee appointed to Inquire into the Original Standards of Weights and Measures in this Kingdom; and to consider the Laws relating thereto, 26 May 1758' [Carysfort 1st Report], and 'Report from the [Carysfort] Committee, Appointed (upon the 1st Day of December 1758) to Inquire into the Original Standards of Weights and Measures in this Kingdom; and to consider the Laws relating thereto, 11 April 1759' [Carysfort 2nd Report], *Reports from Committees of the House of Commons*, 3 volumes (1715-1801), II, 411-51 and 453-63.

88 *RCRB*, VII, 180: 8 March 1764.

89 James Gray, 'Of the Measures of Scotland, compared with those of England', in *Essays and Observations, Physical and Literary, read before a Society in Edinburgh*, 1 (1754), 200-4.

90 McGowan was involved with Professor John Stewart, son and successor of Professor Robert Stewart, in the 1754 reconstructions of the 1618 firlots (see the entries for Items **216** and **217** in the Inventory). Clerk-Maxwell's subsequent involvement with John Swinton and John Robison is discussed later in this chapter.

91 *RCRB*, VI, 493, 496: 10 July 1755.

92 Ibid., VI, 553, 555: 6 and 7 July 1757.

93 Ibid., VI, 579; VII, 42, 147, 165; meetings of 13 July 1758, 9 July 1760, 13 July 1763 (on a motion by George Clerk-Maxwell), and 11 July 1764.

94 'Experiments ascertaining The Contents of the

Lithgow [*sic*] Wheat and Bear Firlots: for the Justices of the Peace of the County of Stirling. 1754': Stirling Council Archive MS SB/T/1. We are grateful to George Dixon for drawing this manuscript to our attention.

95 Gray, op. cit. (89), 201.

96 Thomas Everard, *Stereometry: or the Art of Gauging made easy, by the help of a Sliding-Rule*, tenth edition (London, 1738), 192-3. See Chapter **6**, ref. 58, for a discussion of Everard's work.

97 Gray, op. cit. (89), 203.

98 Chambers, op. cit. (69), 491-2.

99 Ibid., 491.

100 Ibid.

101 On Bird, see Clifton, op. cit. (28), 30, and Anita McConnell, 'John Bird's Work', unpublished paper presented to the XX Scientific Instrument Commission Symposium, Stockholm, October 2001. For Graham's metrological work, see A. D. C. Simpson and R. D. Connor, 'The Mass of the English Troy Pound in the Eighteenth Century', in *Annals of Science*, 61 (2004), 323-51.

102 Carysfort 1st and 2nd Reports, op. cit. (87), 434, 456.

103 William Hunt, *A Mathematical Companion, or the Description and Use of a New Sliding-Rule* (London, 1697), 164.

104 See entry for Items **216** and **217** in the Inventory. On Miller, see T. N. Clarke, A. D. Morrison-Low and A. D. C. Simpson, *Brass & Glass: Scientific Instrument Making Workshops in Scotland* (Edinburgh, 1989), 26.

105 George Skene Keith, *A General View of the Agriculture of Aberdeenshire [for the] Board of Agriculture* (Aberdeen, 1811), 553.

106 Robison, Commonplace Book, I, p. 450; Swinton, op. cit. (14).

107 There are two minor errors on the barley firlot: see the entry for Item **216**.

108 *RCRB*, VII, 180: 8 March 1764.

109 William Kennedy, *Annals of Aberdeen*, 2 volumes (London, 1818), II, 295.

110 'Dalkeith', *New Statistical Account*, 15 volumes (Edinburgh, 1845), I, 503.

111 *RCRB*, VII, 99: 13 July 1774.

112 Swinton, op. cit. (14), 15-16.

113 Ibid., 21.

114 Ibid., 39.

115 Robison, Commonplace Book, II, p.400.

116 Ibid., 20.

117 For instance, Alexander Bald, *The Farmer and Corn-dealer's Assistant, or the Knowledge of Weights and Measures made easy by a Variety of Tables* (Edinburgh, 1780).

118 J. R. Miller, *Speeches in the House of Commons upon the Equalization of Weights and Measures of Great Britain* (London, 1790).

119 Rev. John Scott, 'United Parishes of Twyneholm and Kirk-Christ (County and Presbytery of Kirkcudbright, Synod of Galloway', in Sinclair, op. cit. (14), XV, 91.

120 For a biography, see Rosalind Mitchison, *Agricultural Sir John: the Life of Sir John Sinclair of Ulbster 1754-1835* (London, 1962).

121 Julian Hoppit, 'Reforming Britain's Weights and Measures, 1660-1824', in *English Historical Review*, 108 (1993), 82-104.

122 Steuart was a distant cousin of Robert Stewart. His work was posthumously published by his son as *A Plan for Introducing an Uniformity of Weights and Measures over the World*, in Sir James Steuart, *The Works, Political, Metaphysical, and Chronological: Collected by his Son Sir James Steuart*, 6 volumes (London, 1805), V, 379-415.

123 Rosalind Mitchison, 'The Movements of Scottish Corn Prices in the Seventeenth and Eighteenth Centuries', in *Economic History Review*, 18 (1965), 278-91, quote p. 278.

124 Robinson M. Yost, 'Pondering the Imponderable: John Robison and Magnetic Theory in Britain (*c.*1775-1805)', in *Annals of Science*, 56 (1999), 143-74, p. 144.

125 The exceptions are: Yost, op. cit. (124); J. B. Morrell, 'Professors Robison and Playfair, and the *Theophobia Gallica:* Natural Philosophy, Religion and Politics in Edinburgh, 1789-1815', in *Notes and Records of the Royal Society of London*, 26 (1971), 43-63; J. R. R. Christie, 'Joseph Black and John Robison', in A. D. C. Simpson (ed), *Joseph Black 1728-1799: A Commemorative Symposium* (Edinburgh, 1982), 47-52. See also the entry by Harold Dorn, *DSB*, XI, 495-8.

126 Ibid.

127 John Playfair, 'Biographical Account of the late John Robison', in *Transactions of the Royal Society of Edinburgh*, 7 (1815), 495-539, p. 514.

128 Morrell, op. cit. (63), 164.

129 Playfair, op. cit. (127), 521.

130 Emerson, op. cit. (50), 47.

131 NAS CS 231/F3/1, process McKie, Sharp and Others v. The Magistrates and Town Council of Linlithgow and the Incorporation of Bakers of Edinburgh and Leith, bundle 32: Report of J. Robison, 19 June 1788.

132 NAS CS 231/F3/1; Signet Library, Court of Session papers, volumes 191 (1779-82) and 192 (1786-90).

133 Graham, op. cit. (72); Everard, op. cit. (96), probably consulted in the fifth or subsequent edition; Carysfort, op. cit. (87); M. Tillet, 'Essai sur le rapports des poids étrangers avec le marc de

France', in *Memoires de l'Académie Royale des Sciences* (1767), 350-408.

134 Signet Library, Court of Session papers, volume 191, 2 October 1779.

135 Robison, Commonplace Book, I, p. 445.

136 Swinton, op. cit. (14), 75.

137 Ibid., 100.

138 Ibid., 91, 100, 110-11. Measures supplied to Perth in 1774 were tested in March 1778, and others supplied to Kinross-shire in 1775, were also tested in 1778.

139 Robison, Commonplace Book, II, p. 400.

140 Ibid., I, p. 445.

141 Ibid., I, p. 441. Since 102·3 × 256 = 26,189 in³, a volume of 102·344 in³ has been used to bring the pint's volume up to 26,200 grains. As a check for his gauging of the firlots, Robison measured the Union bushel (Item 171) with the Edinburgh pint (Item 113) as 20 pints 3 mutchkins and ¼ gill, generating 2,148·9 in³, against a theoretical Winchester bushel capacity of 2,150·3 in³.

142 Ibid.; James Paterson, *The Scots Arithmetrician, or Arithmetick in all its Parts* (Edinburgh, 1685), 7.

143 Robison, Commonplace Book, I, p. 450; NAS CS 231/F3/1, bundle 10.

144 Ibid.

145 Ibid., bundle 14.

146 Jamieson's measurements of the barley firlot at 31¼ pints were disregarded because Linlithgow's agent claimed the firlot had leaked: ibid., bundle 10.

147 Ibid., bundle 16.

148 Ibid., bundle 10.

149 Stewart's measurements of the pounds represented by each of the weights showed the stone as being to a pound of 7,619¼ grains, whereas the majority of the smaller weights were to pounds of between 7,586 and 7,596 grains: Robison, Commonplace Book, II, p. 393.

150 NAS CS 231/F3/1, bundle 10. Robison found this suggestively close to types of French weights listed by Picard and others.

151 Henry Guerlac, 'Joseph Black's Work on Heat', in Simpson, op. cit. (125), 13-22.

152 Robison, Commonplace Book, I, p. 397: '... but this part of my thermr was not very exact ...'. A thermometer made by Alexander Wilson of Glasgow and dated 1778 (Robison began work on this court case in 1778), and now in the National Museums of Scotland (NMS T.1975.56), has been associated with the natural philosophy class of the University of Edinburgh, and may well have belonged to Robison.

153 Adam Anderson and Thomas Bruce, *Report on the Stirlingshire Weights and Measures* ([Stirling], 1827), 13.

154 NAS CS 231/F3/1, bundles 30 and 31.

155 See discussion at Item 222 in the Inventory.

156 Perth and Kinross Council Archive, MS CC1/2/1/2, Perthshire county commissions minutes, 1 May 1792.

157 Swinton, op. cit. (14), 100.

158 See ref. 156.

159 See Item 202.

160 Robison, Commonplace Book, II, p. 395.

161 George Shuckburgh, 'An Account of Some Endeavours to ascertain a Standard of Weights and Measures', in *Philosophical Transactions*, 88 (1798), 133-82.

162 Robison, Commonplace Book, II, p. 400; see also Simpson and Connor, op. cit. (101).

163 *APS*, IV, 585-9: 19 February 1618; see Appendix A.9.

164 Robison, Commonplace Book, I, p. 456: 'Scots pint 54 gr. too small.'

165 For the auction in 1805, see Item 92.

166 Kennedy, op. cit. (109), II, 293-300; discussed by John Reid, 'Patrick Copland 1748-1822: Connections outside the College Courtyard', in *Aberdeen University Review*, 51 (1985-6), 226-50, esp. pp. 231-4. A copy of the original printed report, 'Account of the Weights and Measures used in Aberdeen', is in NLS, shelf mark X.225.a.1(63).

167 George Skene Keith, *Different Methods of Establishing an Uniformity of Weights and Measures Stated and Compared* (London, 1817), 32.

168 Ibid., 1-2.

169 G. S. Keith, *Tracts on Weights, Measures and Coins …* (London, 1791).

170 Keith, op. cit. (105), 554.

171 Kennedy, op. cit. (109), II, 294, 295 and 298, quoting Keith.

172 John S. Reid, 'Patrick Copland 1748-1822: Aspects of his Life and Times at Marischal College', in *Aberdeen University Review*, 50 (1984-5), 359-79, p. 360.

173 Reid, op. cit. (166), 231-4.

174 Aberdeen University Library MS M167, reproduced in John S. Reid, 'Patrick Copland (1748-1822)', unpublished University of Aberdeen MLitt thesis, 1983, appendix 2, 4-5.

175 Act 5 George IV Cap. 74, 1824.

176 For technical details, see *WME*, 279-88; and for the French changes, see Alder, *Measure of All Things*, op. cit. (9).

177 *Report of a Committee of the Highland Society of Scotland on Weights and Measures* (Edinburgh, 1811); *Report on Weights and Measures by Committee of the Highland Society of Scotland, and of Gentlemen from the Different Counties of Scotland, and from the Convention of the Royal Burghs* (Edinburgh, 1813); discussed by Meredyth Somerville, *The*

Standardization of Weights and Measures in Scotland, Department of Geography, University of Edinburgh, Occasional Paper no. 11 (Edinburgh, 1989), 42-4.

178 For Davies Gilbert, see Hoppit, op. cit. (121), 101.

179 Ibid., 101.

180 *WME*, 255-60.

181 Anita McConnell, *R. B. Bate of the Poultry 1782-1847: The Life and Times of a Scientific Instrument Maker* (London, 1993), 19-28.

182 Act 5 George IV Cap. 74, 1824, section 16 made it lawful to buy and sell 'by any weights or measures established either by local custom, or founded on special agreement'.

183 Acts 4 and 5 William IV Cap. 49, 1834 and Acts 5 and 6 William IV Cap. 63, 1835. Specifically, the Winchester bushel and the Scottish ell were abolished.

184 Ian Levitt and Christopher Smout, 'Some Weights in Scotland, 1843', in *Scottish Historical Review*, 56 (1977), 146-52.

185 Act 5 George IV Cap. 74, sections 18 and 19. The originals of the verdicts are preserved in the Scottish Exchequer papers at NAS E 349/8.

186 This item no longer survives, but for a discussion, see Chapter **8**, ref. **75**.

187 'Minutes of Evidence taken before the Select Committee [of the House of Commons] on the Bill to amend and render more effectual two Acts of the Fifth and Sixth Years of the Reign of his late Majesty King George the Fourth, relating to Weights and Measures', in *Parliamentary Papers* 1834, XVIII, 280.

188 James Cleland, *Exemplification of the Weights and Measures of the City of Glasgow* (Glasgow, 1822).

189 Patrick Auld, 'An Account of the Imperial Standard Measures made by and for Messrs. J. & H. Wardrop … under the direction of Andrew Ure, MD, FRS, Professor in the Andersonian Institution, Glasgow', in *Glasgow Mechanics' Magazine,* 4 (1826), 457-8, 453; Stephen Millar is identified as the maker of the latter measures in James Cleland, *An Historical Account of the Local Weights and Measures of Lanarkshire, and an Inventory of those belonging to the Corporation of Glasgow* (Glasgow, 1832), 34.

190 James Cleland, *Conversion of Weights and Measures for the County of Lanark* (Glasgow, 1835), 25.

191 G. A. Owen (revised by A. W. L. Poole), *Law Relating to Weights and Measures* (London, 1947), 65.

192 Personal communication, May 2002.

PART II

INVENTORY OF SURVIVING STANDARDS

Introduction

THIS INVENTORY REPRESENTS THE FIRST ATTEMPT TO GATHER TOGETHER information about the surviving early standards of Scottish weights and measures. In the main they have been restricted to the official standards of administrative districts, used as the legal basis for controlling the practices of trade.

Most of these items are the standards of the former royal burghs, which not only had the right to hold fairs and markets but also had the monopoly of conducting foreign trade. Disputes were normally decided in the burgh courts, and it was burgh officials, notably the Dean of Guild, who ensured that proper weights and measures were used and who could prosecute miscreants and 'break' any weights and measures which were found to be deficient against the burgh's retained standards. From the mid-eighteenth century the increased responsibilities of county commissioners led to a separate need for county standards, although they were often in the control of the same officials who had charge of the standards of the chief burgh in the county. In addition a few items in this inventory were the standards of recognised trade incorporations which had traditional rights within particular burghs and their liberties, or over larger areas.

Inspectors of weights and measures were widely appointed in the 1830s; and, with the creation of county and burgh police forces, inspection control passed to police officers, and then progressively to qualified full-time local authority inspectors as a result of legislation in the 1880s. Some of the material we have surveyed has remained in the cupboards of trading standards inspectors, but much has long since ceased to have any legal relevance and has been located in the displays and stores of the burgh museums that often inherited this material when it had become redundant.

In theory at least, these standards had been compared and approved against the standards of burghs with higher authority or against national standards held by royal officials – for example, during the Chamberlain's 'ayre', or progress round the principal burghs. Unfortunately, we know nothing definite about the accuracy or consistency of such comparisons; and as no attested national standards now survive, the evidence of local standards is all that we can use. In some instances we know that standards were poorly controlled: for example, there is significant variation in the capacities of some of the copies of the pint issued by Stirling as a result of legislation passed in 1618. However, we can see that this is simply because some of the surviving cast vessels have never been subjected to individual adjustment. In other instances the adjustment has clearly been careful and conscientious.

The Inventory forms an integral part of this study, because the evidence contained in the objects themselves provides essential information for the development and testing of our arguments. The silent testimony of the standards allows us to understand more of how the practitioners of the day interpreted the legal responsibilities placed on them, and this in turn provides a revealing corrective for too-literal an interpretation of assize legislation. It also enables us to assess the more dogmatic statements of earlier commentators and to provide a more secure evidence context in which to place the items.

Sometimes individual pieces have presented crucial evidence, which has allowed us to fill gaps in the legislative record with greater confidence. For example, the recognition that the Inverkeithing ell of 1500 (Item 1 in the Inventory) was also a capacity gauge for the firlot dimensions, enabled us to confirm a step-like growth sequence for the dry capacity measures – arguably the most testing battle-ground in Scots metrology. Although this is a particularly striking example, the evidence from a large number of specimens has contributed significantly towards the detail of arguments in the book. This evidence has demonstrated a level of consistency and accuracy that lends support for our contention: first, that early Scots metrology had as sophisticated and flexible a structure as was necessary; second, that it shared important features of principle and practice with those of its external trading partners; and, finally, that it was far from being the hopeless muddle that its detractors and reformers have often claimed.

STRUCTURE OF THE INVENTORY

The Inventory is divided into three principal sections by three categories of measurement unit: Section A covers surviving linear measurement standards; Section B is for weight standards; and Section C includes both liquid and dry capacity standards. Within these sections the material is treated chronologically and grouped in a number of sub-sections. The first begins with the earliest surviving material, from the 1500 firlot gauge (which is now the only extant example of what must have been a basic and widespread type of measurement standard), and it includes a handful of other sixteenth-century standards. The second covers the century from the reforming weights and measures assize of 1618 up to the Union of the Scottish and English parliaments in 1707. The standards distributed to the Scottish burghs as a consequence of the requirements of the Act of Union are treated in the third group. The fourth includes standards of the old (and theoretically abolished) Scots system, up to its effective demise in legal metrology in the mid-nineteenth century; and the fifth is of standards based on the former English system which was imposed in 1707 and was at least partially absorbed in the century or so that followed.

The new Imperial system of weights and measures was designed to unify, simplify and regularise what was eventually recognised as an imprecise and irrational method of controlling measurement standards across England, but the legislation applied also to Scotland. Introduced in 1824, the Imperial system had limited impact north of the border until reinforced in new legislation in the mid-1830s. The Scottish burghs and counties were obliged to acquire new Imperial standards, and often did so from London instrument makers, such as Robert Bate, who obtained the principal government contracts for their manufacture. These patterns, which are well represented in collections throughout the United Kingdom, are not noted in this Inventory. However, in some instances local initiative led to authorities commissioning from Scottish instrument makers standards which conformed to the new Imperial system (or indeed to newer metric units) but which were sometimes of innovative construction. Standards made by Scottish firms before 1900 have been described in the sixth and final group.

To qualify for inclusion in the Inventory, an item must have been held by a responsible organisation, normally a local authority, with the intention of checking the competence and legal acceptability of measurement units used in trade or in other contracts. This does not, for example, include items issued by a local authority or verified by it as authorised for trade, unless they are known to have been used by that authority (or another) as standards. Thus many of the characteristic Scottish cast iron commercial weights, through which the Dutch and trone systems continued in retail use in local markets in the nineteenth century, are excluded here, although examples associated with Lanark's traditional authority are illustrated in Chapter **9**.

NAMES OF UNITS

Some Scottish measurement units have the same names as English units – such as weight units of ounces, pounds and stones, and capacity units of gallons and pints – yet their sizes may differ greatly and there may also be several types of each. Other units have distinctive Scots names for which there is no English variant. Thus the half and quarter of the Scots pint are a

Fig. I.1
The traditional name 'lippy'
being used for a quarter-
peck measure in an Imperial
capacity set for the County
of Perth, mid nineteenth
century. (Perth Museum and
Art Gallery)

'chopin' and a 'mutchkin' (which respectively have French and Dutch origins). Similarly, although the peck is the quarter of the Scots 'firlot', as it is of the equivalent English 'bushel', the quarter of the Scots peck is a 'lippie' (or 'lippy') or 'forpet' (literally a fourth part). The simplified classification scheme in the Inventory distinguishes between the principal types of each unit name. For example, conventional yards and ells are separated from the slightly longer 'plaiding' yards and ells (usually a sixteenth longer than the basic unit), where the enhanced measure is usually to allow for cloth shrinkage, but may sometimes also be for other purposes such as for calculating building work.

Occasionally there may be no accepted term for a distinct measure type, and in two instances we have proposed a working name which reflects the measure's origin and is used for convenience only. One of these is the 37·2-inch 'linen ell' (discussed in Chapter 3), which appears to arise in the context of yarn measuring in the late eighteenth century, but came to be applied in much wider use as an alternative to the conventional 37-inch ell. The name could perhaps equally have been linked to the mathematicians David Gregory or Colin Maclaurin, who were associated with its early promotion, but we have preferred a descriptive term.

The other is the type of enhanced pint, again incorporating an allowance of a sixteenth, which regulated the sale of beer and wine in burghs. These were working standards used for checking trade measures and were often self-adjusting. They are quite distinct from standards of the legally specified volume, which may have had a narrower role (and perhaps only a reserve role at that) in the definition of the dry capacity measures. These larger pints are not usually recognisable in the literature until the eighteenth century when they are sometimes called 'pewterer's pints', possibly reflecting the status and adjustment of conventional 'tappit hen', or crested, pewter measures. However, since the extra quantity appears to relate originally to an uncharged allowance for sediment, we have opted to call them 'brewer's pints'.

Where there is a broadly accepted classification this has been used. Thus English 'bushels' are distinguished from Scots 'firlots' and the latter are divided simply into 'wheat' and 'barley' firlots. The 1618 Assize recorded that the smaller firlot was for wheat, but also for rye, beans, peas, meal and salt; so all these are described as wheat firlots, even though they may have been given or have acquired other names, such as the 1766 Stirlingshire 'bean firlot' (Item **218**). The larger measures for malt (that is, malted barley), beare or bere barley and oats, are similarly named barley firlots. Later measures described as for 'corn' are in the same category. The same distinction is made for measures with specific local names, such as the Selkirkshire 'peck', which is a half (and not a quarter) wheat firlot, and the 'full', which is a barley firlot.

Weights based on English practice are divided into the two accepted categories of 'troy' and 'avoirdupois', but the Scottish weight series present more problems. In the Inventory we have followed the convention used by John Swinton in his 1779 *Proposal for Uniformity of Weights and Measures in Scotland*, where the principal weight types are given as 'troye' and 'trone'.[1] This is despite the fact that we have shown in the text that there are several evolutionary stages in both weight series, and indeed that different varieties were sometimes in use at the same time. It has sometimes been convenient to separate these in the discussion – for example, the 'trois' weight of the 1426 Assize, the 'troys' weight of the 1618 Assize (also marked on the surviving weights), and the contemporary Flemish 'troois' or 'troas' merchant weights – in order to draw out particularly characteristics. However, it is equally clear that, particularly in the earlier period, there was no accepted correct spelling and that there were several variant spellings, sometimes used in plural forms (including the terminal contraction 'e' for 'is').

Here, the single descriptive name 'troye' has been used in the Inventory, principally as a means of gathering several weight series together and distinguishing them from English troy, and with no further linguistic inference. It is intended to encompass all the Scottish troy-measuring and merchant functions, including weight series from the Low Countries and those specifically described as 'Dutch'. Similarly, in using 'trone' in preference to the more modern 'tron' or earlier variants such as 'troyn', we are following Swinton's lead and also the spelling used in the only official references found to these weights, around the time of the 1618 Assize. It should be stressed that this approach is different from that of the post-1824 Scottish county juries, whose

reports and the general conclusions drawn from them were published by
George Buchanan: there 'troyes' and 'tron' were adopted – the former influenced by a misreading, presumably by James Jardine, of the markings on the
surviving 1618 weights.[2]

LOCATIONS

Many of the standards listed here are known, from their inscriptions or otherwise, to have been made for a specific burgh or county. These are described
here as 'for' or 'from' that local authority, depending on whether or not
they are still located in a museum or elsewhere in the same locality. Other
standards, such as the 1707 Union sets, were supplied in effectively identical
forms to a number of burghs, and in these instances standards are merely
described as 'at' that authority (or 'from' it if the standard has been transferred from a known original burgh).

The present location is given for all items in public collections. This is not
intended to indicate that they are on display, although items in storage are
likely to be accessible to enquirers giving adequate notice and quoting the
museum inventory numbers given in the entries to this Inventory. One of the
purposes of publishing information of this sort is to encourage increased use
of museum collections.

The names of the governing bodies responsible for these collections are
given in brackets after each institution's names. Many of these have changed
in recent years as a result of two rounds of local government re-organisation
in Scotland. In 1975 the royal burghs and a number of smaller counties lost
their independent status. Responsibility for museums was passed to 53 new
district councils, whereas weights and measures administration became a
function of 12 larger regional and islands councils. In 1996 this two-tier
structure was dismantled and 32 new unitary councils were established.

MARKINGS

The retained standards of a local authority had to carry appropriate marks to
show their status, but for early Scottish standards there was no simple scheme
of using the monarch's mark and that of the principal legal repository, as was
the case with English standards. Instead we are left with an array of burgh marks
of differing status and the knowledge that the true national standards, equivalent to those held in the English Exchequer, do not survive. These English
markings do appear in Scotland, on the London-made 1707 Act of Union
standards, which carried the crowned royal cipher and the square chequer-
board of the Exchequer, but with additional Scottish distribution marks.

The same type of marking, incorporating the royal cipher, is found on at least some of the post-1824 Imperial standards which were produced in Scotland, but in general burgh marking prevailed. Regular dated marking of all standards by Board of Trade officials began only in the 1880s, with the Westminster 'portcullis' mark replacing that of the Exchequer. However, by this time earlier weights which were of a form that remained legally acceptable (such as the 1707 avoirdupois bell-weights) were no longer kept in full verification as local standards and were merely used as working standards. As such they required marking with the local inspectors' own stamps, which from about 1880 normally were the uniform verification number marks, incorporating the crowned royal cipher and a unique number identifying an individual inspector's location. Where standards have been kept in verification (or been pressed into subsequent use) as local Imperial standards, they have been incorporated in the numbered indenture documents, issued initially by the Exchequer in 1824, which record the composition and state of each of the authorised local standards. Where relevant this information is provided. For details of burgh and county marking and of inspection procedure, the recent digest by Carl Ricketts and John Douglas should be consulted.[3]

REFERENCES IN THE LITERATURE

Published and manuscript references to individual pieces are given wherever possible, particularly where items are illustrated or if information needs to be corrected or qualified. Some items have been included in special exhibitions, and where catalogue information is available this can throw light on contemporary understanding of the pieces exhibited.

A series of three major exhibitions held in Kelvingrove Park in Glasgow in 1888, 1901 and 1911, featured increasingly ambitious historical displays for which catalogues were published. At the International Exhibition of 1888 a collection of around 1,500 archaeological and historical relics were housed in a full-sized reconstruction of the Bishop of Glasgow's palace or castle, the original of which had been demolished in 1792. This is catalogued as *The Book of the Bishop's Castle and Handbook of the Archaeological Collection* (Glasgow, 1888). More than twice as many items were shown in part of the new, and permanent, Kelvingrove Galleries, which formed a section of the International Exhibition of 1901, catalogued in the *Official Catalogue of the Scottish History and Archaeology Section* (Glasgow, 1901). This was dwarfed, however, by the two-volume catalogue of the displays of the specially erected Palace of History at the Scottish Exhibition of National History, Art and Industry of 1911, which featured material lent by 1,400 owners. References cited here are restricted to entries in the 'Burghal Relics' section of the *Palace of History: Catalogue of Exhibits*, in two volumes (Glasgow, 1911), II, pp. 942-52.[4]

Other exhibitions catalogues cited are *Catalogue of the Special Loan Collection of Scientific Apparatus at the South Kensington Museum, 1876,* third edition (London, 1877); *'Old Dundee': a Pictorial and Historical Exhibition … of the Ancient Burgh of Dundee, 1892-93,* second edition (Dundee, 1893); *The Memorial Catalogue of the Old Glasgow Exhibition 1894* (Glasgow, 1894); *Catalogue of [the] Exhibition of Local Relics and Antiquities shown in the Sandeman Gallery, January and February 1903* (Perth, 1903); *Treasures of South Lanarkshire: Low Parks Museum, Hamilton, 2001-2002* (Hamilton, 2001). Details of exhibits in the St Andrews Preservation Trust's Exhibition of Weights and Measures, Scottish Pottery, and Paisley Shawls, St Andrews, Summer 1968, and of the Royal Scottish Museum's Exhibition of Historical Weights and Measures, Edinburgh, February-April 1972, are held in the Library of the National Museums of Scotland.

MEASUREMENT PRACTICE

We know almost nothing definite about how the early standards were used in practice. Such procedures seldom get recorded, although they were certainly familiar to those involved in measurement transactions at the time. Thus we have no clear idea how standards were adjusted, how comparisons were made, or what degrees of tolerance were permitted and under what circumstances. Such details as we have turn out to derive from the assumptions made by eighteenth- and nineteenth-century commentators. At least some of these can safely be discounted, but fortunately we can extract useful information from examination and measurement of the standards that still survive. Measurements have been made in metric quantities, and normally expressed in millimetres (mm), in grammes (g) and in litres or millilitres (ml). However, we have also followed historical precedence by additionally giving them in Imperial units of inches, troy grains (gr) and cubic inches (in³), since this enables more ready comparisons to be made with earlier commentaries.

Our general approach has been to aim for a level of precision that is commensurate with the engineering quality of the items. For example, the sixteenth-century Edinburgh ell bed (Item 2) is of great significance in terms of its authority for the size of the ell. But it is now in a comparatively distressed and corroded condition and the opposing faces of the measure are certainly not flat or parallel to each other. It may be possible, by making assumptions about its original condition, to determine its length visually to within about half a millimetre (0·02 inches) using conventional steel tapes; but it clearly is not appropriate to measure it, as was done in 1811, against one of the most accurate contemporary scales in existence to a precision of one-hundredth of this quantity (or to four decimal places). This was particularly inappropriate since the result was not accurately reproducible even at the time, and the precise relationship of this contemporary scale to others then in use had not been established. The apparently authoritative value for the length of the ell derived from these same 1811 measurements was eventually accepted in 1826 as 37·0598 inches.[5] But the measure was certainly not constructed to this level of accuracy against its predecessor, nor will its uncertain contents admit such precision. Sadly, such a minutely defined result has encouraged others to believe that it can be relied on with confidence. This is all the more unfortunate because we established in Chapter **2** that the wrong measurement was made and that the ell unit is exactly 37 inches long.

Weighing was performed to 0·1 g using a 16 kg capacity Mettler electronic scale with excellent linearity characteristics and calibrated with an appropriate commercial reference weight. A standard weight, or indeed any other weight, which has been in active use will show signs of damage and wear, and its original intended mass will necessarily be slightly larger that the measured mass. It may have been adjusted subsequently, usually by the addition of lead to the base, in which case it may not be possible to suggest its original mass. For weights which form a set, there will inevitably be some variation in the adjustment at the time of manufacture, since this depends on the finite sensitivity of the balances used and on the type of internal comparisons required; but in addition, different weights in the set will also have been subjected to different levels of wear. Eighteenth-century commentators such as John Robison tended to take a weighted mean of the weights in a set to represent the size of its underlying pound or ounce unit, but this must produce a low result. In most cases we have preferred to give less emphasis to original adjustment variation and have proposed a minimum figure for the underlying unit since this takes better account of changes due to wear.

In the past, the masses of weights have also been quoted to unrealistic levels of apparent precision. It was shown in Chapter **9** how John Robison took the 1618 stone that he believed to have been Lanark's reference copy, and explained the difference of several grains in the pound between his own and Robert Stewart's earlier measurements as the result of wear: assuming that it had already been worn at the time of Stewart's work in 1720, he proposed that it had originally been based on a pound of 7,620 grains (or about 494 g). This contrasts sharply with James Jardine's approach in 1825: he weighed the same stone weight and assumed this mass was its true original size, giving the pound size to a preposterous 11 significant figures as 7,608·9496875 grains.[6] This is the size that was eventually adopted in the analysis of the county jury reports under the terms of the 1824 Weights and Measures Act. However, within a few years Jardine was quietly expressing doubts about whether the weight had been 'entire' (probably a result of finding nothing in the adjustment recess under the handle) and was proposing to revise the troye pound upwards to as much as 7,656·25 grains.[7]

However, even Robison's more cautious figures require careful interpretation. His own set of weights were ultimately adjusted against the Royal Society's standards in London, which are now known to have been about 0·1 per cent lighter than the 1824 Imperial weights, and therefore to have had a lighter grain.[8] This is equivalent to a difference of about 5 grains in the troye pound. But if all Robison's figures require to be adjusted, we are left wondering about the bases of the work of Robert Stewart and James Gray. Bearing this sort of uncertainty in mind, we have tried not to imply greater accuracy than is justified in individual cases; and, where appropriate, conversions have been made to allow comparisons.

The volumes of liquid capacity measures have been measured by first weighing them empty, and then full of water. It is clear that this method has been used from the time of the earliest Scottish definitions, which are always in terms of the weight of the water contents of the gallon or pint. Our procedure for measuring a typical cast pint was first to place the balance on a very stable surface on a solid floor, so that an observer could view it from all sides without disturbing the reading on the balance, and it was well lit from above. The interior of the pint was then carefully cleaned, and it was wetted inside and allowed to drain but not become dry. It was next set on the balance and levelled by metal shims so that an averaged surface for the rim was horizontal. The weight reading was set to zero and the vessel was carefully filled with

Fig. I.2
Measuring the volume of the Linlithgow Pint (Item **114**), from its water capacity, with the Dunfermline Pint behind (Item **115**): R. D. Connor (left) with A. D. C. Simpson (right).

water up to the rim. A syringe was used to add or remove water while the surface was observed at a shallow angle above the rim and against a light card. The rim was often not accurately flat, so the water was brought to what was judged to be the best visual compromise position which represented a mean surface. It was then carefully emptied and allowed to drain as before, and the balance checked to read the residual weight. Normally three measurements were made, and averaged if they were within acceptable limits. Measurements were taken in cool surroundings using (soft) tap water. The measure was then carefully washed again and dried out with warm air to ensure that any casting imperfections or cracks in the metal were properly dry.

A cast bronze pint measure typically has an aperture of 100-105 mm, weighs in the region of 7,000 g, and contains a further 1,710 g of water. If the rim has been well adjusted, particularly at the sides of the handle, the contents can be judged to within ± 0·5 g. There are, however, two further limiting factors. First, the inner surface of these vessels is not perfectly smooth, so different draining times will leave different amounts of residual water adhering to the sides – the variation often being several tenths of a gramme. Second, we have taken no account of temperature, partly because thermal expansion was unknown when these vessels were made and used.[9] It was only in the early 1730s that the differing expansion rates of materials was demonstrated; and accurate thermometers were still rare in the 1750s when William Cullen and Joseph Black did their pioneering work on heat and its measurement, using thermometers made by their colleague at the University of Glasgow, Alexander Wilson. There is no reference to temperature measurements in the accounts of Alexander Bryce or James Gray in the 1750s, but there are in John Robison's work in the 1770s. Taking into account the thermal expansion of the measuring water in the pint, but also the expansion of the vessels itself, a temperature drop of 1°C would require the addition of about ¼ g of water to bring it back to full capacity; and a temperature range of about 10°F, which figures in Robison's measurements, is equivalent to about 1·5 g per pint. In the circumstances it is clearly not appropriate to consider an error margin of less than ± 0·5 g per pint, which is equivalent to ± 0·05 in³, and so Scots pint volumes in cubic inches will be given to one decimal place only.

Again it is instructive to contrast this approach with that of the reporters for the post-1824 county juries. James Jardine measured the Stirling Jug at 62°F (the temperature at which the Imperial standards had been fixed) as 26,306·982 grains.[10] But since this measure has an uneven rim and Jardine filled to the underside of a glass plate placed over the rim, the measured volume was necessarily too high. Adam Anderson reported to the Stirlingshire jury that the value to the lowest diameter across the aperture was 26,274·77 grains.[11] He then improbably calculated the value of the non-flat surface, comprising two slightly inclined semi-circular planes that closely matched the rim, obtaining 26,286·41 grains; and finally repeated Jardine's measurement with a glass cover plate, getting 26,302·26 grains, which was 4½ grains less than Jardine. This last difference is as likely to reflect the quality of adjustment of their respective comparison weights as anything else. In no way is the uncertainty in interpreting the measure compatible with the apparent accuracy of these various results.

Robison described the method used to fill the pint by the cooper at Linlithgow in the late 1770s, which involved the pint being filled to over-flowing and transferred with water surface which was 'convex' over the rim, and thus clearly over-full.[12] Robison, like Gray before him, stretched a fine silver wire over the aperture and filled the pint until the wire was just covered. Our approach was to fill the pint to a visual average of the rim, and we can be confident that this is close to the method originally used because of the behaviour of David Rowan's beautifully-engineered pint for St Andrews (Item III): this sixteenth-century brewer's pint has a self-adjusting spout which reduces the water to be as accurately co-incident with the plane of the rim as we could detect.

Measuring the larger dry capacity measures presents different problems. John Robison described how the Linlithgow standard firlots had to be soaked for extended periods to let the wood expand enough to prevent water leaking in a wet determination of their values.[13] For surviving early measures in museum collections, this is no longer an option and volumes can only be calculated from measured dimensions. However, progressive shrinkage, warping and general deterioration often make it hard to establish dimensions with sufficient confidence and accuracy. For example, an error of only one millimetre in assessing the original diameter of a wheat firlot alters the computed volume by nearly one per cent (or 20 in³), and often the cumulative uncertainty in the volume is greater than this.

The National Museums of Scotland would be very glad to hear of any other measures which come within the scope of this inventory.

1   [John Swinton], *A Proposal for Uniformity of Weights and Measures in Scotland* (Edinburgh, 1779), 38.

2   George Buchanan, *Table for converting the Weights and Measures ... also ... The Local Measures throughout the Different Counties of Scotland ...* (Edinburgh, 1829), 21, 23, 199.

3   Carl Ricketts and John Douglas, *Marks and Marking of Weights and Measures of the British Isles* (Taunton, 1996).

4   For a discussion of these exhibitions, see P. Kinchin and J. Kinchin, *Glasgow's Great Exhibitions: 1888, 1901, 1911, 1938, 1988* (Glasgow, [1988]).

5   Buchanan, op. cit. (2), 17.

6   Ibid., 200.

7   Ibid., 21-2; 'Minutes of Evidence taken before the Select Committee on ... Weights and Measures' (HC 464), *Parliamentary Papers* 1834, XVIII, 263 (p. 19).

8   This issue is discussed in Chapter **4**.

9   Robison's work is discussed in Chapter **9**.

10 Buchanan, op. cit. (2), 201.

11 Adam Anderson and Thomas Bruce, *Report on the Stirlingshire Weights and Measures* (Stirling, 1827), 8-10.

12 See Chapter **9**.

13 See Chapter **9**.

SECTION A

LINEAR STANDARDS

A1: PRE-1618 LINEAR STANDARDS

1 ELL BED AND CAPACITY GAUGE, 1500, AT INVERKEITHING

Capacity measure gauge and linear measure, in brass and iron. Comprising a broad flat strip of wrought brass, with various lengths represented on it, which are defined by step cuts of about 10 mm deep in the edges of the strip, and with a simple engraved line following the edge on both sides, except in the squared cuts. It is strengthened by being clamped along its centre-line between two wrought iron bars. Six square-headed iron bolts passing through the iron bars and the brass strip are spaced approximately equally along their length and arranged symmetrically, with the second and fifth bolts reversed in relation to the others. Three of the round decorated brass nuts are now missing and the

bolts have been riveted over. At one end of the brass strip there are two holes, perhaps for suspension.

One edge of the brass strip has been cut back to form a bed measure of one ell in length. To one side of this is an inscription cut in Gothic black-letter minuscule: 'Willm̅s Carmichel : anno d̅n̅i millesimo qu̅ingentesm.' Possibly a superscript raised terminal letter 'o' (completing 'quingen-tesimo') is now missing where the edge of the brass appears to have been rounded down at the end face of the ell bed (and this has resulted in the loss of part of the engraved border on both sides of the step at this point). On the reverse side there is an engraved decoration of three leaves indicating the end of the ell. A punch or notch in the edge marks the centre point of the ell. We are grateful to John Higgitt of the Department

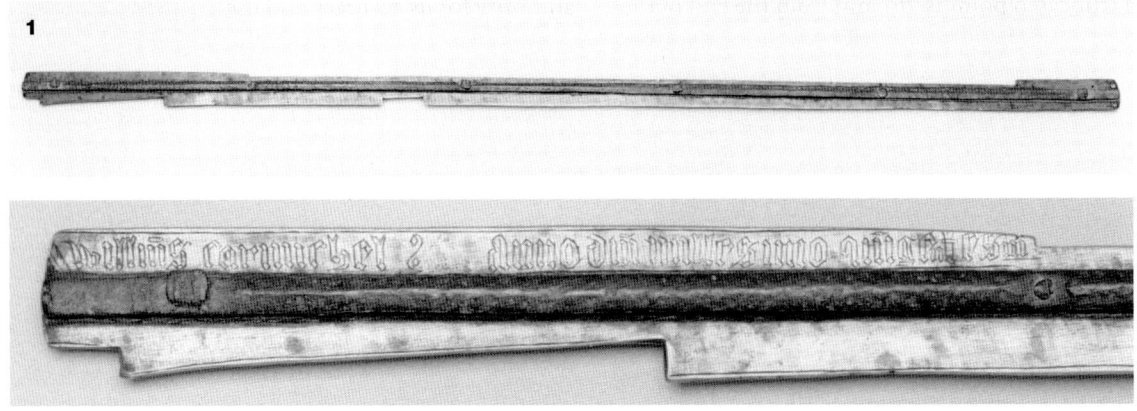

of Fine Art, University of Edinburgh, for his comments and advice on the inscription.

The other edge of the brass strip carries two stepped cuts near one end of the bar. One marks a length of 179 mm from the extremity of the bar for measuring the height of the rim of vessels 179 mm deep: a scratched line 6 mm beyond this may indicate a revised depth of 185 mm. The other stepped cut, closer to the end of the bar, is 429 mm to the near side of another cut (about 50 mm wide) further along the bar: this enables the two to be used as an inside gauge for the internal diameter of a vessel 429 mm across.

SIZE: overall length 1,350 mm, maximum width 45 mm; thickness of brass 2 to 3 mm; width of the iron bars 11 to 14 mm, and thickness 8 mm.

MEASURED DIMENSIONS: ell length 941 mm (37·05"), sufficient to accommodate a 37" ell standard with a clearance play of about 0·05"; with the half-ell 471 mm. Measured firlot diameter 429 mm and depth 179 mm.

There are signs of minor alterations to the measure. The rounding-down of one of the end faces of the ell measure may be associated with the addition of a small notch on this shoulder about 5 mm beyond the end face: this may be to accommodate the later 'linen' ell of 37·2". A notch on the ell bed (again displaced by 5 mm) may indicate the quarter of this ell. A further mark on the ell bed in the form of an arrow with a truncated point is 719 mm from the end of the bar: conceivably, this is a diameter measurement, used in conjunction with a depth of 285 mm, which is the length from the end of the inscribed portion of the bar, to give the dimensions of a boll of about 115·7 litres, equivalent to 96 pints (the water boll, defined in Edinburgh in 1504). But if the length of the inscribed portion of the bar may have some significance, this in turn suggests that the cutting of the ell bed was a later modification and the bar may originally have carried firlot markings on one edge and boll markings on the other. Thus there would probably have been a step out at the end of the inscribed portion (where the ell is now cut back), and there would have been a square-cut notch located a boll's diameter away. Since the gauge could not have been altered successfully when the boll was next enlarged, an ell bed was cut in its place.

Depth and diameter figures of a cylindrical vessel of 179 and 429 mm respectively give a volume of 25·875 litres or 1,580 in³. Using the 2,110 in³ volume derived in Chapter 5 for the 1618 firlot of 21¼ pints, we can obtain the volume of the firlot of one-sixteenth larger than 15 pints, projected in Chapter 6 for the assize of the late 1490s, as 1,582·5 in³. This is only 0·2 per cent more than the volume of 1,580 in³ generated from the dimensions on the Inverkeithing gauge, and is well within the intrinsic uncertainty in the size of the 1618 volume.

William Carmichael was admitted a burgess of Edinburgh in 1494, becoming a Bailie in 1498 and Treasurer in 1500 (and again in 1501): Margaret D. Young, *The Parliaments of Scotland: Burgh and Shire Commissioners*, 2 volumes (Edinburgh, 1992-3), I, 107-8. He was a Lord of the Articles (a member of the parliamentary committee which drew up the parliamentary agenda) and in 1504 was Commissioner to Parliament for Edinburgh. He subsequently represented Dundee in Parliament and later became Provost of Dundee, dying in about 1540. His name is noted on the measure as providing Edinburgh's authority for its accuracy and use.

Discussed in Chapters 2 and 7. Described and illustrated by J. Balfour Paul, 'Notes on Old Scottish Measures, with a Notice of the Inverkeithing Ellwand', in *Proceedings of the Society of Antiquaries of Scotland*, 31 (1896-7), 210-15, p. 215; but Paul's conclusion was that 'it does not look as if there was any standard here except the Scotch ell, which was a fraction over 37 English inches'. Paul's description was referred to in James L. Anderson, 'An Old Chapmen's Standard

Yard-Measure from Ceres, Fife', ibid., 57 (1922-3), 167-70, p. 168. His illustration block was re-used in J. Paton (ed), *Scottish History and Life* (Glasgow, 1902), 201, fig. 242. It is noted in W. Stephen, *History of Inverkeithing and Rosyth* (Aberdeen, 1921), 27, and W. Stephen, *The Story of Inverkeithing and Rosyth* (Edinburgh and London, 1938), 56, in both of which it is incorrectly described as being of steel. Illustrated in A. D. C. Simpson, 'Grain Packing in early Standard Capacity Measures: Evidence from the Scottish Dry Capacity Standards', in *Annals of Science*, 49 (1992), 337-50, p. 348.

Exhibited at the International Exhibition, Glasgow, 1901, item 2478, the description in the *Official Catalogue* being taken from that of Paul.

Inverkeithing Museum, Inverkeithing (Fife Council, Fife Museums West, Dunfermline: Inv. DUFDM 1988.281. Transferred by Inverkeithing Town Council in 1974, in advance of local government re-organisation in 1975.

2   ELL BED, MID-SIXTEENTH CENTURY(?), AT EDINBURGH

Linear measure of one ell, in wrought iron. Oblong section bar with chamfered edges, and with raised projections or lugs at both ends forming a bed measure; the lug at one end tapering down, and at the other end thickening and pierced for a suspension ring with two links of a chain and an attachment ring. The upper face of both lugs with deep authorisation stamps showing, raised in the upper centre, a castle with a central tower (Edinburgh) and perhaps with characters (indecipherable, but possibly a date) underneath. The bar is heavily pitted and there are traces of cream paint on the surface. There are no divisions marked on the bed itself, but there is a defining line cut on the sides and top of the upper lug about 2 mm beyond the upper face of the bed measure, and an angled notch at

the sides of the lower lug with its edge at the defining face of the bed measure: the distance between these lines forms a standard of the linen ell. Sub-division of the linen ell are cut in one side of the bar at a half, quarter, eighth, sixteenth and thirty-second part of the ell, all measured from the suspension end. A modern printed label pasted to bar reads 'SCOTS ELL 16th Century / (The property of the Corporation of Edinburgh.)' and another (also modern) in MS on a label with a printed border is damaged and indecipherable.

SIZE: overall length 1,038 mm, including chain 1,370 mm; cross section of main part of bar 25 × 15 mm; authorisation stamp recess (top to bottom) 12 mm.

MEASURED DIMENSIONS: The angles between the two defining faces of the bed measure and the bar are well formed and are cut at right angles to the length of the bar. The lower face is not flat but has a slight concavity near the angle, so the two faces incline slightly away from each other. Over most of their area the separation of the faces is 942 mm (37·09"), but the separation at the two angles between the faces and the bar is 941·5 mm (37·07"). This latter size – 37·07" – is taken as the measured dimension of the ell bed. This is considered to be the 37" standard plus a conventional clearance of $\frac{1}{16}$" for play. The linen ell is 945 mm (37·20"). The fractional parts (measured from the upper line) are: half 472·5, quarter 236, eighth 117, sixteenth 58, and thirty-second 29 mm: these dimensions are all accurately compatible with an ell of 945 mm. It is presumed that these markings were made at a much later date, and they are taken to be exact.

The measure is discussed in Chapter 2. Comparisons of this measure made by Patrick Copland, professor of natural philosophy at Aberdeen University, against his own 1801 scale by Edward Troughton of London are known from a copy by William Knight of Copland's 23 September 1811 report with subsequent notes: Aberdeen

University Library MS M167, reproduced in John S. Reid, 'Patrick Copland (1748-1822)', unpublished University of Aberdeen MLitt thesis, 1983, appendix 2, 4-5. The arrangements were set up by the mathematician John Playfair, professor of mathematics at Edinburgh, who was advising the Court of Session in a case (discussed in Chapter 3) which appeared to require an authoritative determination of the length of the Scots ell. Copland recorded that the faces of this early ell were not perpendicular to the bar, but set the amount of difference between the separation at the top and bottom of the faces at 0·08", much larger than measured by us. His principal measurements were taken at a height of 0·05" above the corner angles of the faces. The average of measurements made at 55°F on 3 and 4 September 1811 was 37·0603", and the average of further measurements made at 60°F on 26 and 27 September 1811 (after the submission of the report) was 37·0707": ibid., appendix 2, 4-5. The result presumably reflects the fact that all Troughton's scales of this type were severely constrained on wooden supports to limit scale expansion: A. D. C. Simpson, 'The Pendulum as the British Length Standard: a Nineteenth-Century Legal Aberration', in R. G. W. Anderson, J. A. Bennett and W. F. Ryan (eds), *Making Instruments Count: Essays on Historical Scientific Instruments presented to Gerard L'Estrange Turner* (Aldershot, 1993), 174-90, p. 188. The Edinburgh civil engineer James Jardine had taken the ell to Patrick Copland in

1811 and had assisted with its measurement. He did not repeat this when he conducted the experiments for the 1825 report to the Edinburgh county jury, convened by the Sheriff-Depute of Edinburgh to determine the relationship of the old Scottish units to the new Imperial weights and measures: *An Abstract of the Act 5. George IV. Cap. 74 … for ascertaining and establishing Uniformity of Weights and Measures; also of the Act 1618, Establishing the late Scottish Standards* … (Edinburgh, 1827), 23. He accepted the result he had obtained with Copland in 1811, correcting it, nevertheless, to 37·0598", and this was the value that was subsequently adopted as the definitive measurement of the ell: George Buchanan, *Tables for Converting the Weights and Measures hitherto in Use in Great Britain* … (Edinburgh, 1829), 28, 199. It is argued in Chapter 2, however, that the error in all these measurements has been to assume that the standard is accurately represented in the bed measure, which appears to incorporate an additional sixteenth of an inch (0·06") of clearance play. In the circumstances the match with the recent measurement of 37·07" is good, but we must consider that the form of the bed measure does not allow its length to be determined to better than 0·01".

In July 1552, and presumably also in July 1553, the ell standard was taken by the Provost of Edinburgh to Stirling for meetings with representatives of Lanark, Linlithgow and Stirling, about the distribution of burgh standards: J. D. Marwick, *et al* (eds), *Records of the Convention*

2

of the Royal Burghs of Scotland, 7 volumes (Edinburgh, 1866-1918), I, 2. It is not known whether this was the Chamberlain's standard bar, this Edinburgh ell-bed, or its predecessor. In December 1566 the bed was physically attached to the Tolbooth in Edinburgh, and a payment is recorded 'to Nicoll Andersoun, maissoun, to hing the yrne elwand in the nether Tolobuith', together with an allowance of lead for securing the chain in the stonework: R. Adam (ed), *Edinburgh Records: the Burgh Accounts*, 2 volumes (Edinburgh, 1899), II, 236. John Robison, professor of natural philosophy at Edinburgh University found it fixed 'on the N^o^ Side of the N^o^ Window of the Council chamber' in the Tolbooth in 1778. He dated it to 1555 and said that it was 1·0825" longer than the 1707 brass standard yard (Item **10**): St Andrews University Library, MS Q171.R8, 'Professor [John] Robison's Commonplace Book', I, p. 443. However, he noted at the same time that the Union yard had been measured as 36·015", so his measurement of the ell should be given as 37·09", which matches the separation we found for the upper part of the faces. Robison gave no source for his statement about the dating of the measure, but it should not be dismissed, because at the same time he unexpectedly dated the companion 1663 Edinburgh ell (Item **4**) with the operative date of the Act (March 1664) rather than the date of its passing. Although he may have been prompted by a knowledge that the Edinburgh chopin (Item **109**) was made in 1555, it is quite plausible that information about the date of the early ell was still available at the time, and possibly displayed with the measure. If so, it was almost certainly lost after the Council's move to new premises in 1811 and the subsequent demolition of the Tolbooth: Royal Commission on the Ancient and Historical Monuments of Scotland (RCAHMS), *Tolbooths and Town-houses: Civic Architecture in Scotland to 1833* (Edinburgh, 1996), 82.

Illustrated (together with Items **4** and **10**) in R. W. Cochran-Patrick, *Mediaeval Scotland* (Glasgow, 1892), figure opposite p. 158, where it is described as the 'Ancient Scottish Ell Standard' of 37 inches. He incorrectly gives its size as 37·001". The block is re-used in J. Paton (ed), *Scottish History and Life* (Glasgow, 1902), 200, fig. 240.

Exhibited at the Scottish Exhibition of National History, Glasgow, 1911, item 76.

The Museum of Edinburgh, Edinburgh (City of Edinburgh Council, Museums and Art Galleries): Inv. EDNMG HH 5976/2000. Transferred from Edinburgh Town Council by 1899. Since August 1992 on loan to the Linlithgow Heritage Trust, for 'The Linlithgow Story', Linlithgow.

3   ELL BED, EARLY SEVENTEENTH CENTURY(?), FOR FALKLAND

Linear measure of one ell, in wrought iron, subsequently converted into a yard measure. Square section bar with raised projections or lugs at both ends forming a bed measure; the lower end formed by a right-angle turn in the bar, the upper projection rounded at the end, pierced and with a suspension ring. One face at the upper end is cut with the letter 'F' (presumably for Falkland). The bar has been deeply corroded but has been re-forged to hammer out the ell markings and form a yard measure, and has been re-surfaced to provide well defined edges. The original lower lug has been removed, the bar bent up to provide a new end stop, and it has been adjusted by cutting back at the upper measuring face. Both sides of the bar are marked with the origin of the ell measure at the upper end (at 3 mm and 5 mm from the yard face). The one thirty-second ell mark, which is discernible on the upper face, and the sixteenth, eighth, quarter and half-ell marks on the side of the bar, are hammered out: this re-working has resulted in the bar becoming slightly elongated. A half-yard mark has been made on the upper surface.

3

SIZE: overall length 988 mm, including suspension ring 1,050 mm, cross-section of main part of bar 17 × 17 mm.

MEASURED DIMENSIONS: yard bed measured between the lugs is 915 mm (36·00"), and half-yard is cut at 457 mm. Distances of the defaced ell divisions from the centre of the origin line at the suspension end are: half-ell 474, quarter-ell 239, eighth-ell 120, sixteenth-ell 59, thirty-second part of the ell 29 mm. Re-working of the bar to remove these divisions is most pronounced at the half, quarter, eighth and sixteenth-ell marks. Assuming an equal elongation is introduced at these four divisions, the dimensions quoted are compatible with an elongation of about 2 mm at each division. This implies that the original divisions were at 467, 234, 117, 58, 29 mm, with

an ell of 935 mm (36·80"). This provides adequate confirmation that the measure started life as an ell bed (presumably incorporating a tolerance allowance) with an ell division scale marked along the bed.

Town House Museum, Falkland (National Trust for Scotland, Edinburgh): Inv. NTS 52.1003. Presented to the Trust by Falkland Burgh Council at Falkland Palace in August 1973. The measure had previously been on loan from 1969 to the Fife Folk Museum, Ceres, Fife (and still carries a painted inventory number '69·126'), having been lent through the Town Clerk of Falkland, who was also a founder of the Fife Folk Museum: it was withdrawn by the Burgh in 1973 for presentation to the National Trust for Scotland.

4   Ell bed, 1663, at Edinburgh

Linear measure of one ell, in copper. Oblong section bar with raised projections or lugs at both ends forming a bed measure, the lower end finished square and the upper end with a stepped tapering profile and pierced for an iron suspension ring to which three iron links are attached. The under surface of the measure is engraved 'CHARLES · II · PARL · 3 · ACT · XVII · ANNO · 1663' with red infill of the lettering. The bed is divided by cuts into 12 inches (at the suspension end) and at the 2- and 3-foot marks, with an additional cut across half the width at 1' 6". The foot markings are engraved '1', '2' and '3', with red infill of the numerals; the cuts have subsequently been filled with white colouring but some have remains of red filling. It can be seen that the final 8 mm of the bed in the final interval beyond the 3-foot mark has been filed down when the defining face of the lower lug has been adjusted.

SIZE: overall length 1,014 mm, including chain 1,245 mm; cross section of main part of bar 18 × 18 mm.

MEASURED DIMENSIONS: The defining faces of the bed measure are not accurately parallel, and in particular the upper face is inclined so that the separation of the faces is greater on the right hand side of the measure. The minimum clearance between the faces (at the left hand side) is just over 940·5 mm (37·04"), whereas the separation at the centre of the faces readily allows the insertion of a bar of 941·0 mm (37·05"). Neither face is flat – both are slightly convex, apparently as a result of progressive wear at the opening edges, and the clear separation of these faces at their opening is 942 mm (37·09"). The effective size of the bed is therefore at least 941 mm and perhaps as much as 941·5 mm, and hence between 37.05" and 37·07" (even though this would not quite pass between the full depth of the faces at the left hand side of the measure). This is again considered to be the 37" standard plus a conventional allowance of about 0·05" (or perhaps 1/16") for play. The distances of the lower section of the defining face at the suspension end to the cuts for inches 1 to 11 (measured only to 0·5 mm) are: 26·0, 51·5, 76·5, 101·5, 127·0, 152·0, 178·5, 203·0, 229·0, 255·0, 279·5 mm. Although these show some variation (largely within the level of accuracy of the measurements), the more significant markings for 1', 1'6", 2' and 3' are more consistent, at 305·5, 458·0, 610·5, 915·0 mm. The theoretical lengths for these latter divisions are 304·9, 457·2, 609·6, 914·3 mm – all smaller than the measured dimensions by approximately

the same amount of 0·7 mm, indicating that the origin of the scale is about 0·7 mm away from the upper face of the bed. The same is true for the origin of the inch divisions, although some do not appear to have been placed with such care.

Although there is some minor inaccuracy in the marking of individual line positions, these divisions are compatible with the use of an inch which is identical to the English inch. Had we assumed that the full bed length was the ell, and it was divided into 37 conjectural Scottish inches, then the distance of the 3-foot mark would have been 916·0 mm. This is sufficiently different from the measured distance of 915·0 mm to discount this alternative explanation. Instead, the measured dimensions clearly point towards the origin of the scale being a small distance away from the upper face of the bed, and the divisions appear to have been copied directly from those on a standard bar measure of 37 (English) inches placed in the bed measure. A small gap of around 0·7 mm between the end faces of the bar and bed at the top is matched by a similar gap at the bottom, so that the total tolerance or play in the bed is about 1·5 mm or about 0·05" (or perhaps ¹⁄₁₆").

John Robison recorded this measure as 'the brass Ellwand on the Nᵒ Side of the Sᵒ Window' of the Council chamber in the Tolbooth in 1778, and said that its length was greater than the yard on Edinburgh's brass Union standard (Item 10) by 1·0525": St Andrews University Library, MS Q171.R8, 'Professor [John] Robison's Commonplace Book', I, p. 443. He dated it to March 1664, which is the date stipulated in the 1663 Act by which burghs had to obtain and display copies: T. Thomson and C. Innes (eds), *The Acts of the Parliaments of Scotland*, 13 volumes (Edinburgh, 1814-75), VII, 488: Act 57, 29 September 1663. Robison also noted that the Union yard (bed) had been measured as 36·015", so his measurement of the ell should perhaps be given as 37·06".

Illustrated in R. W. Cochran-Patrick, *Mediaeval Scotland* (Glasgow, 1892), opposite p. 158, fig. 3. The illustration is inaccurate in that the 3-foot marking is shown at the end of the bed, apparently indicating that the measure is only 36" long. The block was re-used in J. Paton (ed), *Scottish History and Life* (Glasgow, 1902), fig. 240.

Exhibited in the Historical Weights and Measures exhibition, Royal Scottish Museum, Edinburgh, 1972.

The Museum of Edinburgh, Edinburgh (City of Edinburgh Council, Museums and Art Galleries): Inv. EDNMG HH 5976/2000. Transferred from Edinburgh Town Council by 1899.

5   FOOT BED, 1663, FOR DUMFRIES

Only one of the standard measures of one foot established under the 1663 Act has survived: the measure held by the Royal Burgh of Dumfries. The fact that the foot was never subsequently one of the required standards almost certainly led to its being considered redundant and therefore discarded by those burghs that had received copies. Even the Dumfries standard was not retained and its provenance is now unknown beyond the fact that it was acquired by a collector who presented it to the National Museum of Antiquities of Scotland (now part of the National Museums of Scotland) in 1957.

It is a conventional bed measure, in a wrought copper alloy, and the raised projections or lugs which form the ends of the bed have a stepped tapering profile similar to the suspension end of the longer and more substantial Edinburgh ell of the same series (Item 4). One end is pierced for a suspension ring, and the other is stamped (and mis-struck) on the side with a single mark of authority comprising a crown over the initials 's·H' and a further indistinguishable symbol (perhaps a hammer), presumed to be the mark of Samuel Hoyll (*fl.*1653-68?), founder, brasier, coppersmith and also Hammerman of Edinburgh: C. E. Whitelaw, *Scottish Arms Makers* (London, 1977), 131. (This work was edited by Sarah Barter from Whitelaw's 1939 manuscript.) The underside of the measure is engraved, in a very similar fashion to the 1663 ell '* CHARLES * II * PARL * 3 * ACT * XVII * ANNO * 1663 *', and the side of the measure is engraved in the same hand 'FOR * DUMFRIES'. The bed is undersized at 302 mm (11·9"); however, it has sub-divisions of inches cut (with minor unevenness), and apparently counting from the lower end (i.e. away from the suspension point), where a half-inch is also indicated. However, the origin of this scale is not at the lower defining face of the measure, but at about 1 mm into the measure. This is the approximate amount that would be anticipated to provide some acceptable play when inserting a candidate bar measure in the bed. This in turn suggests that the measure has not been adjusted, and that it should have been cut back at the suspension end by approximately 5 mm to provide a full foot plus about 1·5 mm (0·05" or ¹⁄₁₆") as an allowance for tolerance play, as provided in the longer ell measures, making a total of about 306·5 mm.

SIZE: measured bed length 302 mm (11·9"); overall length 337 mm; width 11 mm at suspension end, 8 mm at other end; length of bar 7 to 9 mm; length of lugs 18 mm.

National Museums of Scotland, Edinburgh (Trustees of NMS): Inv. NMS H.VH.141. Acquired by the National Museum of Antiquities of Scotland (now part of NMS) from a private collection in Port Seton, East Lothian, in 1957. On loan since 1967 to Dumfries Museum (now under Dumfries and Galloway Council Museums Service): their loan 337 (inv. 14,175).

6   ELL, 1668, FOR KIRKWALL

Cylindrical measure in hardwood with rings of twelve inset brass nails indicating the half-, quarter-, eighth- and sixteenth-ell measured from one end, and with brass endcaps with the seals of Kirkwall (a three-masted ship) and Edinburgh (a castle with three towers). The latter is engraved around the collar: 'This Ell-Wand being the just measure conform to the Ell or Standard of Edinburgh, made for the Burgh of Kirkwall in Orknay by Arthur Baikie, Bailie there Anno, 1668'. The endcaps have been damaged, particularly at the end that bears the inscription, and have been repaired and reset, apparently without sufficient regard to the original overall length.

5

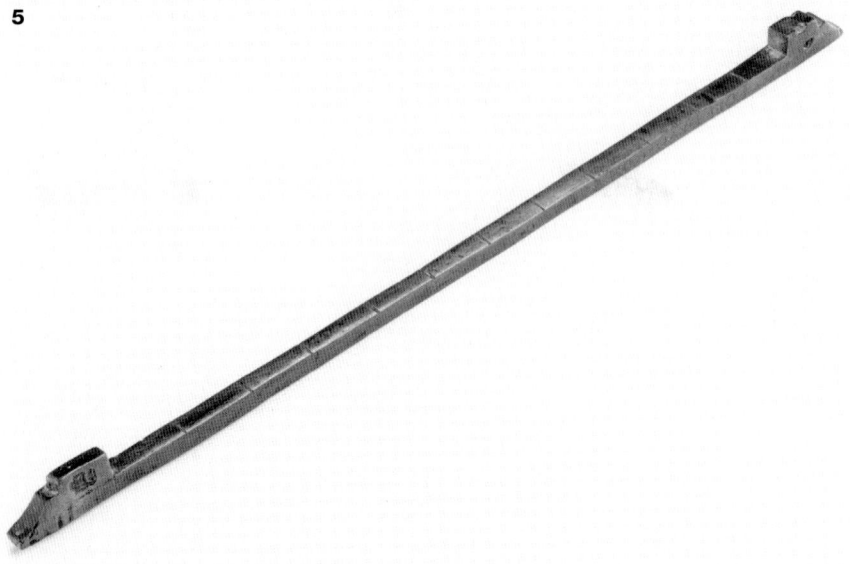

SIZES: overall length 948·5 mm (37·4"), rod diameter 32 mm. Distance to the sixteenth, eight, quarter and half-ell 57·0, 117·5, 235·0, 473·0 mm. The distortion of the inscribed end-caps has certainly increased the apparent length. Unaccountably, the centre point seems to be marked about 3 mm too close to the damaged end, but otherwise the internal divisions are compatible with an original overall length of 940 mm or 37·0".

Arthur Baikie is recorded in James Mackenzie's account of the so-called 'Pundlar Process' (for which see Appendix D): James Mackenzie, *The General Grievances and Oppression of the Isles of Orkney and Shetland* (Edinburgh, 1750, rep. 1836), 55. The Commissioners who undertook this enquiry for the Earl of Morton dispatched Baikie to Edinburgh in 1664 to acquire a standard weight to represent the lispund or setten. Presumably he ordered this item at the same time.

The Science Museum, London (Trustees of the National Museum of Science and Industry):

Inv. NMSI 1932-327. Purchased in 1932 from the dealer T. J. Rochelle of the Georgian Galleries, 10-12 King Street, St James's, London.

7   ELL BED, *c.*1670(?), FOR FETTERCAIRN

Linear measure for testing market traders' ellwands, cut into the stone shaft of the mercat or burgh cross in Fettercairn village. No other example of such an inscised ell-bed is known. The shaft itself has a square section at either end, but for the major portion of its length the corners are bevelled to give an octagonal section. There is now marked weathering and frost damage to the surface, including the portion into which the ell-bed has been cut. The shaft is mounted on a series of five octagonal steps on a built pedestal, and is surmounted by a moulded capital with cubical head and ogee cap finial. The head carries a vertical dial (south face), the coroneted initials 'E / IM' for John, Earl of Middleton, and his arms (east and west faces respectively), and the date 1670 (north face).

6

SIZE: The measure comprises a simple channel, about 37½ inches long, and approximately 35 mm across, the width narrowing down to 30 mm at the base of the channel. Although the stonework at the top of the measure is in comparatively good and square condition, the edge at the bottom is now very worn, but assuming that it was originally square the length of the bed is estimated at 952 mm (37·50 inches).

A. C. Cameron recounted the popular belief that the cross was the old market cross of the decayed burgh of Kincardine, former county town of Kincardineshire, removed to Fettercairn in 1730, but he noted that there was no surviving record of this: A. C. Cameron, *The History of Fettercairn: a Parish in the County of Kincardine* (Paisley, 1899), 144-5. A royal license to erect the former 'Kirkton of Fetlircarn' into a free burgh, with a market

cross, a weekly market and an annual fair, was granted to Adam Hepburn of Craggis in 1504. The license was renewed to the Earl of Middleton in 1670, and it is this that is presumably commemorated on the cross head: ibid., 45. Cameron speculated that the shaft may have been that of the original cross erected in 1504. The charter making Kincardine the capital of the county was granted to the Earl Marischal in 1532, but it continued as such only until 1607. In the absence of any indication of the age of the ell or even of the origin of the shaft, it is assumed that it was cut in or after 1670. Nor is there an adequate explanation for the ell's length, which is a little longer than expected. John Swinton recorded the late eighteenth-century use in Kincardineshire only of the English yard for all English cloth and a plaiding ell of 38½" for home manufacture: [John Swinton], *A Proposal for Uniformity of Weights and Measures in Scotland* (Edinburgh, 1779), 89.

Fettercairn market cross is a Scheduled Ancient Monument (scheduled 7 November 1936) and a Category A protected structure in Historic Scotland's Statutory List of Buildings of Architectural and Historic Importance (listed 18 August 1972; Fettercairn Parish, no. 18). On the Statutory List, see Item **20**.

The cross is recorded in Rev. Alexander Whyte, 'Parish of Fettercairn', in *The New Statistical Account of Scotland*, 15 volumes (Edinburgh and London, 1845), XI, 116-17; and some fifteen years later also by Queen Victoria: A. Helps (ed), *Leaves from the Journal of our Life in the Highlands* (London, 1868), 148. The cross and measure are described and illustrated in D. MacGibbon and T. Ross, *Castellated and Domestic Architecture of Scotland,* 5 volumes (Edinburgh, 1892) V, 399-400; J.W. Small, 'The Market Crosses of Scotland [II]', in *Transactions of the Stirling Natural History and Archaeology Society*, 18 (1895-6), 96-9, and plate; J. W. Small, *Scottish Market Crosses* (Stirling, 1900), plate 38. The dial is separately recorded in Andrew R. Somerville, 'The Ancient Sundials

7

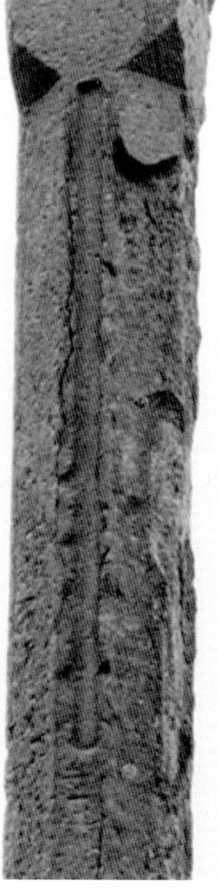

of Scotland', in *Proceedings of the Society of Antiquaries of Scotland*, 117 (1987), fiche D1.

Cut into Fettercairn Cross, Fettercairn (Angus Council, Forfar).

8 ELL, EARLY EIGHTEENTH CENTURY(?), FOR THE INCORPORATION OF WEAVERS, STIRLING

Linear end-measure in wood (box?), of rectangular section and with brass endcaps, one engraved 'STIRLING / WEVERS / ELL'. The endcap length indicates the thirty-second division of the ell, and inset brass strips are placed for the half-, quarter-, eighth- and sixteenth-ell, each measured from the inscribed end. The wood has now warped. Subsequently accommodated with Item 13 in the fitted case for Item 16.

SIZE: cross-section 25 × 12 mm; length along the outer circumference 941 mm, so original length estimated as 940 mm (37.0").

Smith Art Gallery and Museum, Stirling (Trustees of the Stirling Smith Art Gallery): Inv. STIGM 10171. Probably acquired from Stirling Town Council at least twenty years before Items 13 and 16, but the acquisition date is not recorded.

9 ELL BED AND PLAIDING ELL BED, 1706, FOR DUNKELD

Combined linear measure of one ell and one plaiding ell, in wrought iron, attached vertically to the corner of the former St George's Hospital building in the central square of Dunkeld. The ell bed is on the front surface of the measure between two defining projections or lugs, and the longer plaiding ell is similarly defined on the left hand side. On the metal block above the beds is inscribed an ornate 'A' (for 'Anno' or perhaps for the Duke of Atholl, the Superior of the Burgh) and the date '1706'. The lower block is inscribed 'D', which may signify 'Domini' or 'Dunkeld'. Pairs of angled arms at the top and bottom of the measure have returned points which are secured between the masonry quoins. Cuts on the front face of the ell bed indicate the half, quarter, eighth, sixteenth and thirty-second of the ell, all measured from the upper end. The bed has been enlarged by subsequent adjustment to the size of the 'linen ell'. The side face has a single division of the plaiding ell at its half point. The wrought iron shows some signs of de-lamination and one upper arm is fractured.

SIZE: overall height 1,350 mm; width across top fork 140 and across bottom fork 160 mm; cross-section of central bar approximately 18 × 18 mm.

MEASURED DIMENSIONS: ell bed length: 944.5 mm (37.19"); divisions of thirty-second to half ell, measured from the upper face: 35.0, 64.5,

121·5, 240·0 and 475·0 mm. Plaiding ell bed length: 992 mm (39·06"); division of half-ell at 495 mm from the upper face. These measurements were made in 1991 immediately after the National Trust for Scotland had removed thick paint accumulations which had reduced the apparent lengths of the two bed measures by about 2 mm.

Within the limits of accuracy of the measurement, the ell length is undoubtedly to the larger 37·2" length associated with standards of the mid-eighteenth century, such as Items **15** and **16**; and like these, it has been given no tolerance allowance for play. The plaiding ell is similarly 39¹⁄₁₆", a length probably raised as an ell of 37", with the 2-inch allowance of the 1693 Act and with what appears to be the conventional tolerance play of ¹⁄₁₆": T. Thomson and C. Innes (eds), *The Acts of the Parliaments of Scotland*, 13 volumes (Edinburgh, 1814-75), IX, 311-13: Act 48. The scale of fractional parts of the ell, marked in the principal

ell bed, can be shown to have an origin about 5·0 mm below the top face of the bed, from which the measured ell and its fractions are 939·5, 470·0, 235·5, 116·5, 59·5, 30·0 mm. These correspond accurately to the divisions that would have been transferred from an inserted ell bar standard of precisely 37", with positional errors for only the three smallest divisions, of −1·0, +0·5 and +0·5 mm. The overall length of the bed has therefore been enlarged by about 3·5 mm (0·15") to increase the original size of an ell plus a tolerance allowance to the larger linen ell of 37·2". This is likely to have been done when the ell was moved to its present position, which may have been at the time of the rebuilding of *c.*1750. There is indeed some indication that the two faces have been finished in a different manner: the lower face is accurately square to the bar and its outer edge is rounded by wear to a radius of about 2 mm, whereas the upper face is similarly square at its junction with the bar, but becomes

inclined near the outer edge and ends in a sharp corner which exhibits little sign of wear. The top face of the plaiding bed on the side of the measure has also been cut back by about 11 mm: this was either the result of the initial adjustment, or it may indicate that the bed has been enlarged (perhaps by adjustment at both faces) from a plaiding yard of 36" plus a sixteenth, or 2¼".

The original buildings of St George's Hospital, an early charitable foundation of the Bishops of Dunkeld, were largely destroyed when Dunkeld was burnt by the Cameronian covenanters in 1689, and the present structure was described as 'newly-built' in 1757 (Statutory List entry, see below). The town had recovered sufficiently for an attempt to be made in 1704 to raise its status from a burgh of barony (under the Marquis, and subsequently Duke of Atholl) to a royal burgh, but this lapsed: G. S. Pryde, *The Burghs of Scotland: A Critical List* (London, 1965), 33. The ell measure, which dates from this period, was presumably fixed near to its present location as a standard for the regular markets and cloth fairs attended by the 'chapmen' or itinerant merchants of the region. The Society of Chapmen in Angus and Perthshire, established in the early sixteenth century, held its annual court meeting in Dunkeld until 1776, subsequently alternating between Dunkeld and Coupar-Angus. Its members were obliged to keep their trading weights and measures in adjustment 'with the standards kept for the purpose', of which, at Dunkeld, 'the standard for the measures is a fixed iron bar, placed on the wall of a house near the market-place, from which a part of the street takes the name of The Gauge': 'City of Dunkeld and Parish of Dowally', in Sir John Sinclair (ed), *The Statistical Account of Scotland*, 20 volumes (Edinburgh, 1791-9), XX, 432-3. See also Items **18**, **22** and **23** for other measures associated with chapmen societies. For accounts of the chapmen of East Lothian, see 'Parish of Prestonpans', in ibid., XVII, 78-80; and more generally, Roger Leitch,

'"Here chapman billies tak their stand": a pilot study of Scottish chapmen, packmen and pedlars', in *Proceedings of the Society of Antiquaries of Scotland*, 120 (1990), 173-88, and A. Durie, *The Scottish Linen Industry in the Eighteenth Century* (Edinburgh, 1979), 41.

The Ell House is a Category B protected building in Historic Scotland's Statutory List of Buildings of Architectural and Historic Importance (listed 5 October 1971; Dunkeld and Dowally Parish, no. 16). On the Statutory List, see Item **20**.

Illustrated and described by H. Coates in James L. Anderson, 'An Old Standard Yard-Measure from Ceres, Fife', in *Proceedings of the Society of Antiquaries of Scotland*, 57 (1922-3) 167-70, p. 169. The measure is described as being of bronze, dated 'A.D. 1706' and incorporating standards of 37⅛" (identified as the 37" ell with some wear at each end) and 39". He notes the rebuilding of the St George's Hospital in 1757 and presumes it was transferred from the original building. The photograph was republished in Elizabeth Stewart, *Dunkeld: an Ancient City* (Dunkeld, 1926), opposite p. 48. It is described and poorly illustrated by line drawing in R. M. Wells, 'Old Scottish Linear Measure', in *Libra*, 6 (1967), 21: Wells records the length as 37·2" (which he equates to 37 Scots inches), gives the 1706 date and notes a 'Queen Anne cypher', but this is a misreading of 'AD'. Also illustrated by a line drawing in Craig Mair, *Mercat Cross and Tolbooth* (Edinburgh, 1988), 122. The cleaned measure is illustrated in the *Dundee Courier and Advertiser*, 27 September 1991, and in Alan B. Lawson, 'Pecks and Ploughgates', in *Scots Magazine*, new series, 137 (1992), 61-5, p. 63.

Mounted outside The Ell Shop, The Square, Dunkeld (National Trust for Scotland, Edinburgh). The property was presented to the Trust by Atholl Estates in 1954 and subsequently restored.

10   Union yard bed and English ell bed, by John Snart, 1707, at Edinburgh

Following the practice established in the 1618 Act when the Edinburgh ell was designated as the principal reference standard, the English linear measures sent from London in 1707, to comply with the clause in the Act of Union that required English weights and measures to be used in Scotland, were distributed from Edinburgh. As with the other 1707 standards, 21 examples were to have been sent. Perhaps recognising the difficulty experienced by the other distributing burghs in extracting payments, Edinburgh decided to send out the comparatively inexpensive linear measures gratis: J. D. Marwick, *et al* (eds), *Records of the Convention of the Royal Burghs of Scotland,* 7 volumes (Edinburgh, 1866-1918), IV, 462-3: 13 July 1708. In addition to the expected yard measure, these standards also include the English cloth ell of 45" – this latter measure had no relevance to Scotland, except in the restricted sense that it was the English equivalent to the short reel (see Chapter 3), and this is the only issue in which it is found. The Edinburgh standard is one of only two complete examples known to survive, although two others have been located which have been cut down to remove the superfluous English ell, leaving only the shorter yard.

The yard and English ell linear measure is cast in brass, as an oblong section bar with raised projections or lugs on two opposite faces forming the end stops of the two bed measures, with recessed channels cut in the faces of the bar between the end stops. At the upper end of the bar is the lug for the longer English ell measure, and this is pierced for a suspension shackle and a chain of six links. On the opposite face of the

10

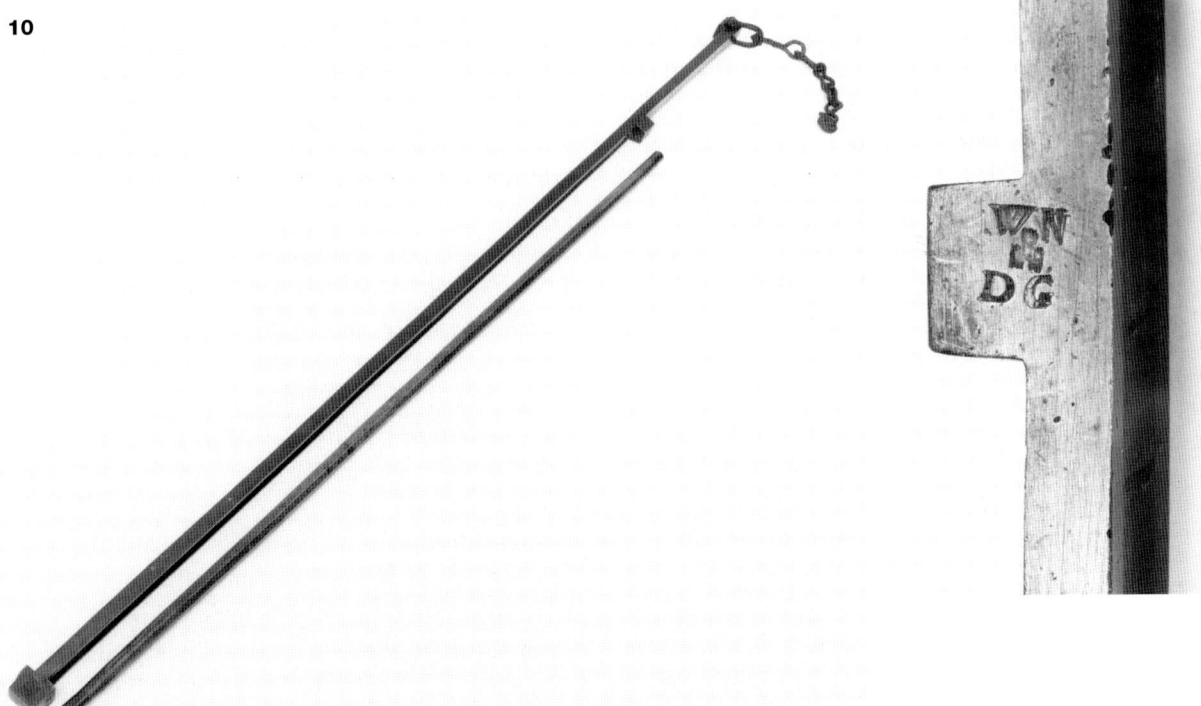

bar, on the blank section beyond the upper lug of the smaller yard measure, is stamped '[crown] / IOHN SNART'. Each of the four lugs is stamped with the Exchequer mark and crowned 'AR' marks, and the defining face of each lug is also stamped with the crowned 'AR'. The sides of the lugs are stamped in two places 'WN / [castle with tower] / DG', for William Neilson, Edinburgh Dean of Guild, 1706-8. The bar is slightly bent at the end nearest the suspension chain. MS labels with printed borders pasted on the bar read 'Imperial Yard / 36·028^{in}' and 'English Ell = / 45·0[paper tear]^{in}'. In the illustration, the bed is shown with the later yard from Item 12.

SIZE: overall length 1,189 mm; length including chain 1,430 mm; cross section of main part of bar 24 × 13 mm.

MEASURED DIMENSIONS: yard 915 mm (36·02"); ell 1,144 mm (45·04"). These dimensions presumably represent the standard lengths plus a small amount of tolerance play.

The metal alloy composition of one example of this issue (the former Linlithgow standard, Item 12) was examined by Dr Paul Wilthew, Head of the Analytical Research Section of the National Museums of Scotland, using X-ray fluorescence spectroscopy. It was found to be a typical brass of the period, with small amounts of lead and tin, perhaps introduced through the use of scrap material, and the composition was significantly different from the leaded bronzes of the other weights and measures of the Union series, which were of nominally uniform composition: Paul Wilthew and Phyllis Herd, 'Analysis of Five English Weights and Measures issued from the Exchequer in 1707', NMS Analytical Research Section Report 92/26 (C&AR 6245), November 1992.

John Snart was a prominent member and later Master of the City of London Blacksmiths' Company, and he supplied scales and weights for use at the London Mint, and also the Mint

in Edinburgh, during the period of reorganisation that followed Isaac Newton's appointment as Master of the Mint in 1699. It was probably this that entitled him to use a crown over his name in marking weights and measures (see also Item 43). He was the son of Richard Snart, a draper in Coventry, and in June 1685 he was bound apprentice in the Blacksmiths' Company to Joseph Hart, becoming a freemen of the company in October 1692 and starting to trade on his own account shortly afterwards. By 1698, Snart was an established master scale maker, and during his working life, which lasted until the mid-1720s, he trained nine apprentices, of whom four became master scale makers themselves. His premises were at the 'Heart & Scales in Maiden Lane over against Goldsmiths Hall London', and he was succeeded at this address by another scale maker, Henry Oven, who was scale maker to the English Mint between 1723 and 1727: see D. F. Crawforth, 'London Scale Makers (Part I). The Transmission of Scalemaking Knowledge from 1632-1800', in *Equilibrium* (Spring 1980), 207-15, especially pp. 210-12; Gloria Clifton, *Directory of British Scientific Instrument Makers 1550-1851* (London, 1995), 257-8, 205; and G. Clifton, 'Scientific Instrument Makers at the Royal Mint', unpublished paper presented to the XIX Scientific Instrument Commission Symposium, Oxford, September 2000.

John Robison examined the measure in 1778, at which time it was fixed in the council chamber in the Tolbooth, adjacent to the two earlier Edinburgh ells (Items 2 and 4), and he recorded the lengths of these two ells in terms of this yard bed. He noted that the yard 'is agreeable to the Exchequer Standard [since it carries Exchequer verifications], but when compared with a measure corrected by Mr Geo. Graham, in the possession of Col. Roy, it contains 36·015 Inches': St Andrews University Library, MS Q171.R8, 'Professor [John] Robison's Commonplace Book', I, p. 443. Roy's scale, signed by Jonathan Sisson, but known to have been divided by John

Bird of London, had been Graham's own, before being passed to a protégé of Colin Maclaurin, the telescope maker James Short, latterly based in London. The military engineer William Roy had acquired it at the sale of Short's effects in 1768. The comparison may have been made in the early 1770s when Roy was in Scotland undertaking barometric survey work with Dr James Lind, an Edinburgh physician and astronomer, who was a close associate of Alexander Bryce: Yolande O'Donoghue, *William Roy 1726-1790: Pioneer of the Ordnance Survey* (London, 1977), 40; T. N. Clarke, A. D. Morrison-Low and A. D. C. Simpson, *Brass & Glass: Scientific Instrument Making Workshops in Scotland* (Edinburgh, 1989), 29. It is described by J. E. Insley in N. Cossons, A. Nahum and P. Turvey (eds), *Making of the Modern World* (London, 1992), 32-3.

Illustrated in R. W. Cochran-Patrick, *Mediaeval Scotland* (Glasgow, 1892), fig. 2 opposite p. 158. The block was re-used in J. Paton (ed), *Scottish History and Life* (Glasgow, 1902), fig. 240.

The Museum of Edinburgh, Edinburgh (City of Edinburgh Council, Museums and Art Galleries): Inv. EDNMG HH 5977/2000. Transferred from Edinburgh Town Council by 1899.

11 UNION YARD BED AND ENGLISH ELL BED (MODIFIED), BY JOHN SNART, 1707, AT JEDBURGH

Another example, as Item 10, but reduced in length to act as a horizontal yard bed by cutting off the suspension end of the measure above the upper projection or lug of the yard bed and by removing the lower lug of the English ell. One end of the remaining yard bed has been broken off and repaired, and the measure has then been reinforced by adding a brass strip underneath, which covers the groove running along the former ell bed. Cuts have been made across the yard bed at a sixteenth, eighth, quarter and half of the full length measured from one end. Now with a later tapering cylindrical pearwood rod, divided into eighths, with a Victorian London Guildhall verification mark and stamped 'ROX 73' (Roxburghshire, 1873).

SIZE: overall length 955 mm; cross-section of bed 23 × 13 mm; bed length 916 mm.

Jedburgh Castle Jail and Museum, Jedburgh (Scottish Borders Council Museum Service, Selkirk): Inv. JEDMG 370a and b. Transferred by Jedburgh Dean of Guild to the museum after 1898.

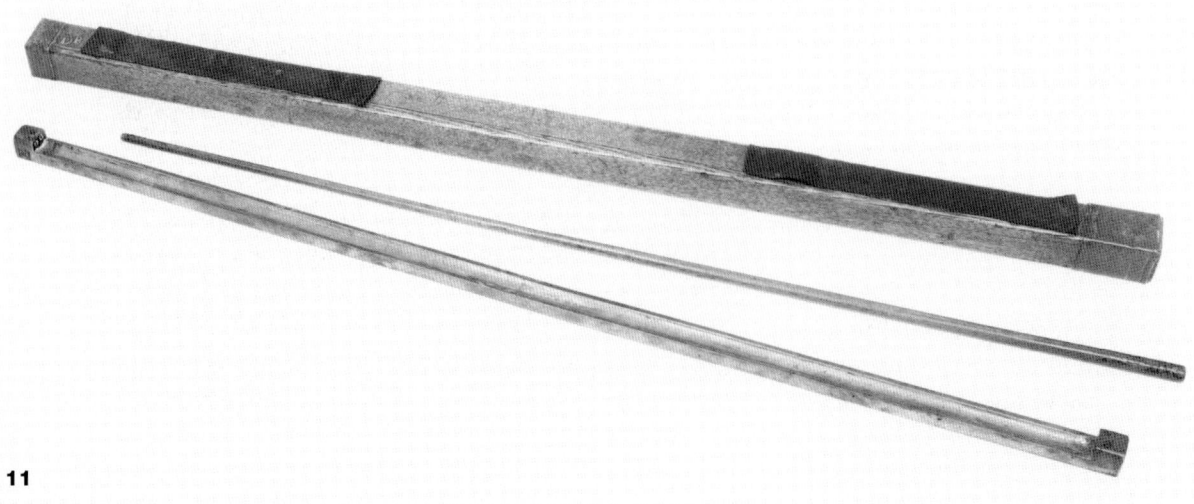

11

12 UNION YARD BED AND ENGLISH ELL BED (MODIFIED), BY JOHN SNART, 1707, FROM LINLITHGOW

Another example, as Item **10**, but converted into a horizontal yard bed in a similar manner to Item **11**. It also has a break near one end, but the end has been replaced. Now with a later brass-tipped wooden hexagonal yard rod, stamped 'C. / LI W. / B.' (Linlithgow County and Burgh).

For the alloy composition of this measure, see Item **10**.

National Museums of Scotland, Edinburgh (Trustees of the NMS): Inv. NMS T.1973.142. Presented in May 1973 by the Council of the Royal Burgh of Linlithgow to the Royal Scottish Museum (now part of NMS) in advance of local government reorganisation in 1975.

13 UNION YARD BED AND ENGLISH ELL BED (MODIFIED WITH THE ADDITION OF A SCOTS ELL), BY JOHN SNART, 1707, AT STIRLING

Another example, as Item **10**, is as issued except that the suspension swivel has been cut from the upper end. The bar is stamped '[crown] / JOHN SNART', and the lug common to both beds and the lower lug of the English ell each stamped on both sides 'WN / [castle] / DG'. An additional lug has been attached to the English ell bed to provide an inner bed for a Scots ell, but is now missing; however, the scribed end of the ell is visible next to the screw attachment hole for the end stop. Cuts on the bed define a sixteenth, eight, quarter and half of an ell, measured from the lower end. Subsequently accommodated with Item **8** in the fitted case for Item **16**.

Exhibited at the Scottish Exhibition of National History, Glasgow, 1911, item 38.

Smith Art Gallery and Museum, Stirling (Trustees of the Stirling Smith Art Gallery): Inv. STIGM 19817/15/1.3. Transferred from Stirling Guildhall, 1990.

13

The fixing hole for an end-piece added to form a bed for the Scots ell.

14 YARD BED AND ENGLISH ELL BED, 1707-8, AT INVERKEITHING

Yard and English ell measure in stained pine, constructed as two back-to-back bed measures, recessed into opposite faces of a single oblong-section wooden bar; the rounded ends of the bar reduced from the full width by a stepped ogee profile. Four small angled brass plates (one now missing) protect the definition faces of the measures from damage in use and from unauthorised adjustment; each is nailed in place and impressed on both exposed faces with the stamp 'WN / DG' (for William Neilson, Edinburgh Dean of Guild, 1706-8) incorporating the seal of Edinburgh (a castle with a central tower and gateway). The origins of the two measures coincide near one end of the bar, and the base of each bed is marked with a series of lines indicating the sixteenth, eighth, quarter and half part of the measure from that end. The measure has been broken at approximately its mid-point and has been splinted between two iron plates approximately 125 mm long, riveted together.

The form of the combined measure clearly follows that of the bronze 1707 linear standards by John Snart. It is likely that low-cost replicas of this sort were constructed and authorised at Edinburgh for those burghs that were not supplied with one of the bronze measures. As such, it is in the same category as Edinburgh-made weight at Culross (Item **87**) and the Stirling pewter gallons and quarts (Items **179** to **197**).

Some burghs also acquired additional capacity measures in the form of conventional iron-bound oak-coopered vessels to complement English bronze measures – Banff obtained a wooden peck and quarter-peck from Linlithgow in 1709: William Cramond, *The Annals of Banff*, Spalding Club, new series nos. 8 and 10, 2 volumes (Aberdeen, 1891-3), I, 352. Presumably, therefore, minor royal burghs would also have lower-cost examples of the bushel.

SIZE: measured length of the yard 915 mm (36·0"), and of the ell 1,145 mm (45·05"); overall length 1,232 mm; cross-section at ends 40 × 25 mm, and at centre 18 × 25 mm.

Included amongst material listed as being in the town house in William Stephen, *History of Inverkeithing and Rosyth* (Aberdeen, 1921), 27, and W. Stephen, *The Story of Inverkeithing & Rosyth* (Edinburgh and London, 1938), 36.

Exhibited at the Glasgow International Exhibition, 1901, item 2479, where it was described merely as 'wooden standard ell measure'.

Inverkeithing Museum, Inverkeithing (Fife Council, Fife Museums West, Dunfermline): Inv. DUFDM 1988.275. Transferred by Inverkeithing Town Council in 1974, in advance of local government reorganisation in 1975.

14

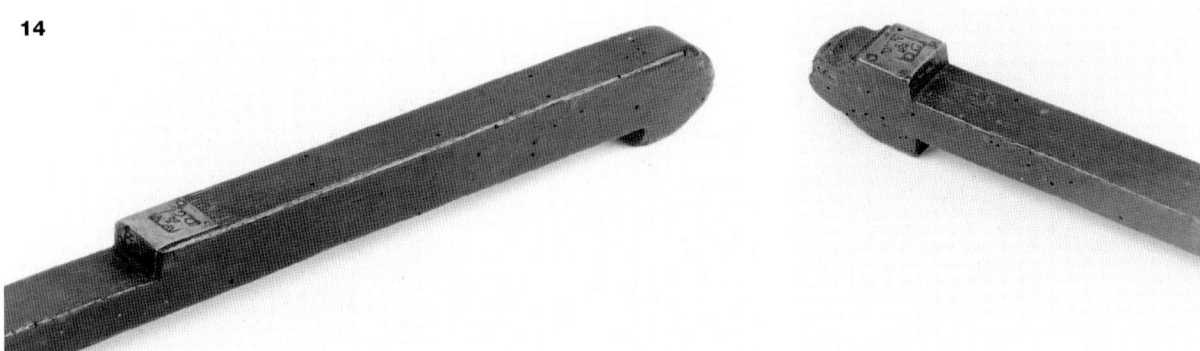

15 Ell bed and yard bed, 1744, for Dalkeith

Linear measure of one ell and one yard in steel, in the form of two opposed bed measures, with a suspension shackle at one end. At the base of the measure is a triangular bracket with curved mouldings which defines the lower ends of the beds and is engraved 'DALKEITH'. Two projections defining the other end of the bed measures are engraved 'YARD' and 'ELL' and between them run two scales, identified at the centre as 'English Inches' and 'Scots Inches', such that the ell is 37 (Scots) inches or 37·2 (English) inches, and the yard is 36 (English) inches, each scale being divided into individual unnumbered inches, with only inches 12, 24, 36 and 37 numbered, and the portion beyond 36 inches being divided into respective tenths. Within the bed for the ell is engraved 'Scots Ell', and numbered divisions for the sixteenth-, eighth-, quarter- and half-ell from the suspension end are shown; the other side, for the 'English Yard', is treated similarly. The same divisions of the yard and ell, but unnumbered, are engraved alongside each other on the reverse of the measure, which also carries at the upper end a large stamp showing a crenelated castle and tower (for Edinburgh); and the base of the measure is stamped with 'HH / 1744' and the interlocking letters 'DG', for Hugh Hawthorn, Edinburgh Dean of Guild, 1742-4. The measure is contained in a contemporary felt-lined oak case, with a sliding lid.

SIZE: overall length, including suspension shackle, 1,098 mm; the ell bed 944·5 mm (37·2"); the yard bed 914·5 mm (36·0"); width across base 70 mm; cross-section of bar 19 × 26 mm. Case size: 1,140 × 115 × 68 mm high.

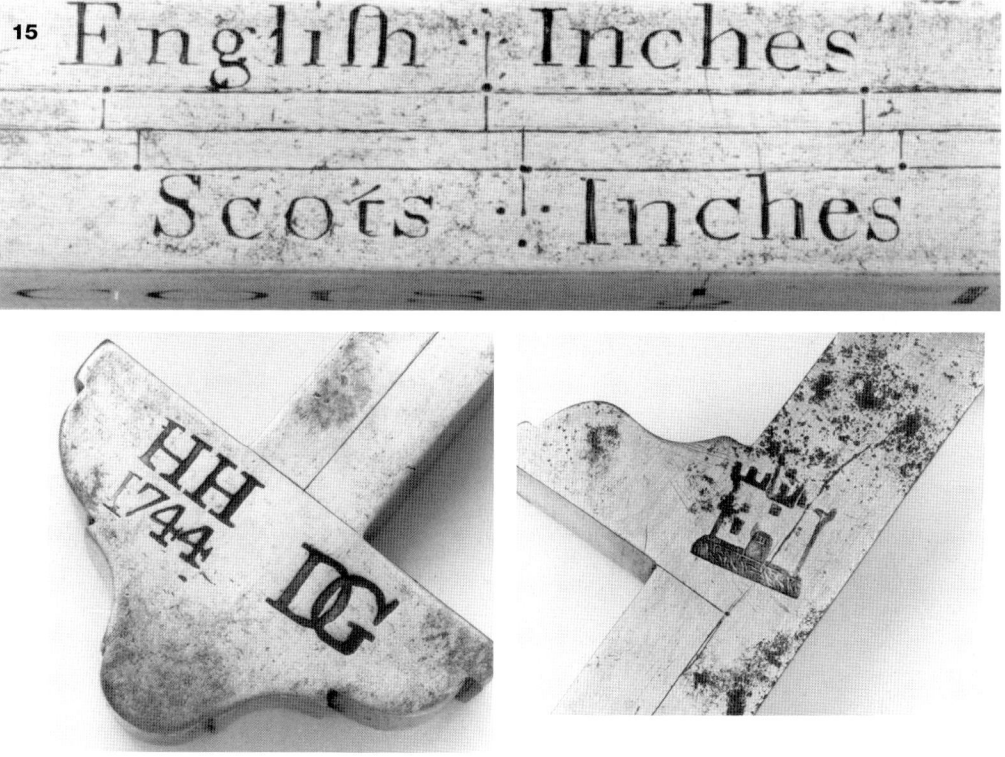

The measure dates from the time when Alexander Bryce was associated with Colin Maclaurin in teaching mathematics at the University of Edinburgh, and they formed part of a group of members of the Edinburgh Philosophical Society who were taking an active interest in the recovery and measurement of the Scottish standards. One of these was James Gray, the Master of the Dalkeith Iron Mill, and the presence of the name 'DALKEITH' on this measure suggests that it may have been made by Gray and its construction supervised by Bryce. The specific requirement for such a measure may have stemmed from a prohibition of 1737 issued by Edinburgh town council against trading in yards in Dalkeith, presumably indicating that Dalkeith, as a subsidiary burgh of regality, did not possess acceptable linear standards: Edinburgh City Archives, MS council record, 26 October 1737.

Described in the *Edinburgh Evening News*, 1 December 1958, the *Scotsman*, 2 December 1958, and described and illustrated in R. W. Plenderleith, 'An Old Scottish Yard and Ell Measure', in *Scottish Studies*, 3 (1959), 105-6.

Exhibited at the Historical Weights and Measures exhibition, Royal Scottish Museum, Edinburgh, 1972.

National Museums of Scotland, Edinburgh (Trustees of the NMS): Inv. NMS T.1958.43. Presented in 1958 by Matthew Manuel, Dalkeith.

16   ELL BED AND YARD BED, BY GEORGE WATT, 1755, FOR STIRLING

Linear measure of one ell and one yard in brass, in the form of two opposed bed measures. One face engraved 'This STANDARD for the Royal Burgh of Stirling, Was adjusted at Edinburgh, the 26th of Feby 1755. With great Care by the Revd Mr Alexr Bryce'. The other face engraved 'George Watt, fecit' and 'William McKillop, [Stirling] Dean of Guild'. The ell bed engraved 'The Scots Ell Consisting of 37 Inches & 2 Tenth parts.' and divided in a binary fashion 'HALF ELL' 'QUARTER' 'HALF QUAR' 'NAIL', the final 'Inch' sub-divided into tenths alternately marked 2, 4, 6, 8, 10. The yard bed engraved 'The English Yard Consisting of 36 Inches' and divided in a binary fashion 'HALF YARD' 'QUARTER' 'HALF QUAR' 'NAIL', the final 'Inch' subdivided into tenths alternately marked 4, 6, 8, 10. Accommodated in a fitted case, together with Items **8** and **13**.

16

SIZE: cross-section of the bed 31 × 18 mm; the width across the defining lugs 70 mm; measured length of the ell 946 mm (37·25"), and of the yard 914 mm (36·00"); overall length, including suspension shackle, 1,065 mm. Case: 1,255 × 230 × 55 mm.

The guildry members were ordered to adjust their measures according to the new standard in March 1755, with the stipulation that these measures should be shod with brass or iron at the ends to prevent fraud: W. B. Cook and D. B. Morris (eds), *Extracts from the Records of the Merchant Guild of Stirling 1592-1846* (Stirling, 1916), 106: 2 March 1755.

Exhibited at the Loan Exhibition of Scientific Apparatus, London, 1876, item 315a, at the Glasgow International Exhibition, 1888, item 1045, at the Glasgow International Exhibition, 1901, item 2439, and at the Scottish Exhibition of National History, Glasgow, 1911, item 39.

Smith Art Gallery and Museum, Stirling (Trustees of the Stirling Smith Art Gallery): Inv. STIGM 19817/15/1.2 (the fitted case is 19817/15/1.1). Transferred from Stirling Guildhall, 1990.

17 CHAIN OF TWO ENGLISH ELLS, 1704, FOR THE INCORPORATION OF WEAVERS, STIRLING

Linear measure of two English ells, comprising four iron bars of approximately 6 mm square section hinged together at their extremities, the two outer bars being slightly shorter, each terminating in a circular chain link.

SIZE: overall length, across all four long links and the two rings is 2,310 mm (90·9"), which suggests an intention to match 2 English ells (90") plus a small allowance. However, the construction of the chain suggests that a slightly different measure was intended. The use of four long hinged links implies that the lengths of the links themselves are significant. Although one would expect that the links would be of the same length, this is not the case: whereas the second and third links are both 565 mm (22·25") between hinge centres, the equivalent lengths of the first and fourth links are only 548 and 550 mm. Measuring between the extreme ends of the four links gives 2,240 mm (88·2"). Almost certainly the measure was used by stretching it between two people, each holding one of the end rings by a forefinger. The measure would be held over a length of cloth, but the extreme position of the rings would be obscured by the holders' fingers. The two likely defining lengths would be the overall lengths of the four unequal links (i.e. 2,240 mm), or the distance between the two knuckles. Depending on the size and orientation of the knuckles, the distance between them would vary from about 2,265 to 2,275 mm (89·2" to 89·6"). This implies an effective length for the end links of about 570 mm, which is very close to the 565 mm of the other two links. Some support for this interpretation is provided by the size of the end rings, which are such that the knuckle almost coincides with the centre of the ring: the distance between the ring centres is 2,270 mm (89·4"). In the absence of further information about the use of the chain it will assumed that its length is defined in this way.

The specification for this measure is given in the MS minutes of the Trade Incorporation of Weavers, 6 May 1704: '… instead of former chain … the new one to be made consisting of four links and two rings … one of the rings of the old chain shorter than the old chain was. … John Finlayson being Deacon': Stirling Council Archives, Incorporation of Weavers Minutes 1700-92, PD7/11/4. The new measure is probably of the same form as the previous chain, but is being defined by abolishing an earlier permitted allowance of perhaps 2 inches.

17

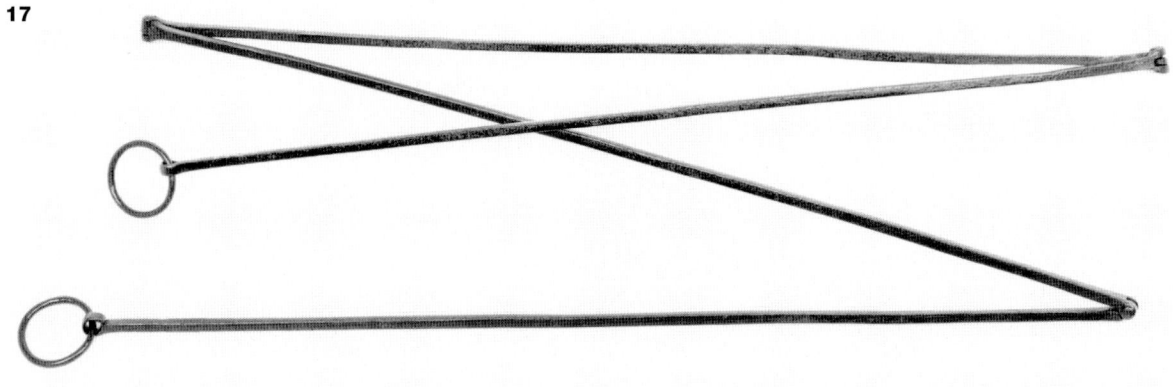

Smith Art Gallery and Museum, Stirling (Trustees of the Stirling Smith Art Gallery): Inv. STIGM 8944. Probably transferred from Stirling Town Council before 1970, but the date of acquisition is not recorded.

18   YARD BED, 1705, FOR THE FIFE CHAPMAN COURT

Two-part folding yard measure in wrought iron. Two projecting orthogonal lugs at each end on the inner face and one side of the bar, the pair on the inner face interlocking when the measure is folded; with a simple three-leaf riveted hinge and an extension on one arm to prevent over-opening. The only divisions visible are on the side face, which carries marks for the quarter and eighth of the yard, and on the lower face which is similarly marked for the eighth-, quarter- and also half-yard. The inside face has a residual inscription on the two halves of the measure, which appears to be '… ER …)) EP …' and 'RO … ND … ON CLERK 1705'. This was tentatively interpreted as '… CERES …' and 'ROBERT ANDERSON CLERK 1705' by James Anderson, writing in 1923: James L. Anderson, 'An old Chapman's Standard Yard-Measure from Ceres,

Fife', in *Proceedings of the Society of Antiquaries of Scotland*, 57 (1922-3), 167-70.

SIZE: The measured length of both beds is 915 mm (36·0"); overall length open 927 mm, and folded 495 mm; width 88 mm; bar dimensions 15 mm across and 10 to 12 mm thick.

The measure has a north Fife provenance, and at the beginning of the nineteenth century was held to have been the standard of the Chapmen of Fife for the district of Ceres. Chapmen, or semi-itinerant small traders, played a significant part in the economy of rural districts and in supplying the hinterland of the larger burghs. Roger Leitch's recent study of chapmen has noted that in the eighteenth century a great deal of their trade was in all kinds of cloth and haberdashery and he has emphasised their role as middle-men in the distribution of home-produced linen and other cloth at rural fairs and to merchants further afield: Roger Leitch, '"Here chapman billies tak their stand": a pilot study of Scottish chapmen, packmen and pedlars', in *Proceedings of the Society of Antiquaries of Scotland*, 120 (1990), 173-188; and see A. Durie, *The Scottish Linen Industry in the Eighteenth Century* (Edinburgh, 1979), 41. Chapmen Societies were set up in a number of

18

lowland and east-coast counties and flourished in the early eighteenth century. Often styled as 'Incorporations of Chapmen' in emulation of burgh guild structures, these county-wide bodies were divided into district associations, they elected officers, they issued regulations which purported to control members, and they held secret court meetings. The accuracy of trading standards was one area in which Leitch suggests the societies sought to improve their respectability: an accessible reference standard sanctioned by the society would enable the length measures of individual members to be regulated, but presumably in the context of burgh markets and fairs only with the toleration of the guildry officials who held the existing burgh standards.

Three chapmen measures survive: this one, associated with the Ceres branch of the Fife Chapmen, is the earliest at 1705, and two later measures survive for the Fife Chapmen Court (Item 22) and the Angus Chapmen Court (Item 23) of 1739 and 1758. It is initially surprising why all three should be yards and not the ells which remained firmly entrenched in market practice for much of the eighteenth century (although, of course, it may be that no more than the greater relevance of the yard in the nineteenth century that ensured that these examples were retained and not discarded). The existence of the 1704 two-yard chain measure (Item 17) of the Stirling Weavers Incorporation, for use alongside their ell standard (Item 8), demonstrates a need to measure cloth bolts in English measure and presumably reflects the importance of the export trade in linen. These chapmen measures may similarly represent the growing eighteenth-century role that Durie and Leitch describe for travelling merchants in moving cloth from rural districts towards the fairs that supplied English markets. However, the virtually identical form and decoration of the Fife and Angus measures (Items 22 and 23), separated in date by nearly 20 years, may mean that they cannot be interpreted in terms of their practical use in controlling trade in the eighteenth century, and instead suggests

that the measures played a more symbolic role in the ritual of these societies, which Leitch describes as 'functioning along the lines of a pseudo guild with freemasonry undertones': Leitch, p. 183.

James Anderson, who presented this measure to the Society of Antiquaries of Scotland in 1923, had been given it in 1914 by George Millar-Bowman of Logie (or Logie-Murdoch) in Fife, whose mother came from Ceres. She had purchased it in about 1860 from David Henry, a merchant in Ceres who had been the last boxmaster (or treasurer) and clerk of the Ceres chapmen. It was this initial link that prompted Anderson to associate the measure with a society of chapmen.

National Museums of Scotland, Edinburgh (Trustees of the NMS): Inv. NMS H.VH.42. Presented to the Society of Antiquaries of Scotland by James L. Anderson, February 1923, and forming part of the collections of the National Museum of Antiquities of Scotland (now part of NMS).

19   YARD BED, WITH ADDED ELL BED, 1719, FOR CUPAR

Constructed as a bed measure for one yard, in wrought iron, and modified to act also as an ell bed measure. One end is pierced for suspension and has a single brass suspension ring, and the other end carries an indistinct stamp showing a castle with a central tower and crenellations (presumably for Edinburgh). Inset in fine brass strips on the side of the bar are marks for the half, quarter, eighth and sixteenth of the yard measured from upper defining face of the measure, and also 'CUPAR 1719'. On the under-side of the bar two additional iron pieces, forming the defining faces for a bed measure of one ell, have been dovetailed into the bar and brazed in place, and cuts on the other side face of the bar mark the half, quarter, eighth and sixteenth of the ell measured from the upper face.

19

SIZE: the measured ell is 944 mm (37·17"), and the yard is 916 mm (37·06"). Overall length (excluding ring) 970 mm; bar thickness 15 to 18 mm, width 16 mm.

Illustrated in L. Burrell, 'The Standards of Scotland', in *The Monthly Review: the Journal of the Institute of Weights and Measures Administration*, 69 (1961), 49-62, p. 57.

Exhibited at the St Andrews Preservation Trust Exhibition of Weights and Measures, St Andrews, 1968, item 6.

East Fife Museums Service, Cupar (Fife Council, Fife Museums East): Inv. CUPMS 1984.105. Transferred from Cupar Town Council to North East Fife District Council at local government reorganisation in 1975 (Museums Service established in 1983).

20   PLAIDING YARD BED, *c.*1720, AT DUMFRIES

Bed measure in wrought iron attached to the wall above the external staircase on the south side of the Mid Steeple building in the High Street, Dumfries. The measure is a single flat bar stepping to two shallow wedge-shaped ends where the steps form the defining faces of the measure, separated by 973 mm (38·31"). This probably represents a plaiding yard of one yard, with a cloth shrinkage allowance of a sixteenth (2¼") and a small additional quantity for play in the fitting of measures to be tested. The only division is a cut marking the half measure: this is approximately 1 mm too low, indicating that the quantity allowed for play is about 2 mm at the top of the measure, which is compatible with the suggested size for the shrinkage allowance at a sixteenth. Both edges of the bed measure are marked with an apparently regular series of notches in groups of three: these are, however, not regular enough to represent sub-divisions of the measure and must therefore be assumed to be decorative. Three flat-headed bolts hammered into lead plugs secure the bar to the wall.

SIZE: measured length of yard bed 973 mm (38·3"); overall length 1,098 mm; bar width 24 to 28 mm, and thickness 8 to 9 mm.

The Mid Steeple was initially designed in 1704 by John Moffat of Liverpool to provide a council chamber, offices for the burgh records, a prison and a steeple, but executed by Tobias Bachup of Alloway. The building was completed in 1708 with the addition of a fine wrought iron balustrade, by Patrick Sibbald of Edinburgh, on the external forestair to the principal floor, which was initially used as the circuit court house, and from 1830 as the council chamber. The burgh weigh-house was housed in part of the ground floor: The Royal Commission on Ancient and Historical Monuments and Constructions of

Scotland, *Seventh Report with Inventory of Monuments and Constructions in the County of Dumfries* (Edinburgh, 1920), 48. A new town hall was acquired in 1866, and much of the Mid Steeple was subsequently used as shops and warehouses. It is tentatively concluded that the measure was moved from a position in or near the weigh house to the south wall of the building at about the time of the council's move in 1866. The measure is illustrated in detail, and a position to the right of the principal doorway (approximating to its present position) is indicated, in W. R. McDairmid, 'Notes on the Old Town hall of Dumfries, commonly called the Mid Steeple', in *Proceedings of the Society of Antiquaries of Scotland*, 20 (1885-6), 186-9. Although the drawing reproduced by McDairmid was based on a view of about 1780 and shows the principal doorway before a classical pediment was added in 1830, the measure may have been inserted by him to match its 1885 position. Certainly, a drawing by G. H. Johnson of 1830 in Dumfries Museum (Inv. DUMFM 85·3) shows a nameboard in this position. However, an undated photo-

graph of *c*.1860 in the same collection shows it in the position indicated by McDairmid, as does a photograph by Valentine of Dundee (no. 1823) datable to late 1879: we are grateful to Ian Fisher, Royal Commission on the Ancient and Historical Monuments of Scotland, Elaine Kennedy, Dumfries Museum, and Robert Smart, St Andrews University Library, for their help and advice on this question. The measure was lowered slightly to its present position when the stone of the building was refaced by James Barbour in 1909. With the assistance of David Lockwood, then Cultural Services Manager for Nithsdale District Council, accumulated paint was removed and the surface protected before the yard was remeasured in 1990.

The Mid Steeple is a Category A protected building in Historic Scotland's Statutory List of Buildings of Architectural and Historic Importance (listed 11 July 1961; Dumfries Parish, no. 141). For a recent account, which also records the measure, see Royal Commission on the Ancient and Historical Monuments of Scotland (RCAHMS), *Tolbooths and Town-houses: Civic Architecture in Scotland to 1833* (Edinburgh, 1996), 72. On the background to the Statutory List, see David M. Walker, 'Listing in Scotland: Origins, Survey and Reserving', in *Transactions of the Ancient Monuments Society*, 38 (1994), 31-96.

Mounted on the south wall of the Mid Steeple, High Street, Dumfries (Dumfries and Galloway Council).

21 PLAIDING YARD BED, EARLY EIGHTEENTH CENTURY(?), AT DORNOCH

Bed measure formed between two small wrought iron staples or hoops set into a large slab of sandstone located in the burial ground at Dornoch Cathedral. The slightly tapering but approximately rectangular stone block, which gives the appearance of having been the cover for a stone coffin, is now horizontal in the church yard about 10 metres east of the cathedral. A modern (1950s?) notice alongside describes it as a 'plaiden ell (tailor's measure) for measuring cloth at the fairs and markets held on this site since medieval times'. The iron hoops form end stops, each of which has been shaped into a triangular hoop standing about 20 mm proud of the stone surface, with the two legs embedded together in a single hole and leaded in place. The upper surface of each hoop has been hammered thin and finished by file to be approximately flat and about 30 mm long and 12 mm wide but only about 2 to 3 mm thick. The inner edges of the thin centre sections of the hoops act as the defining edges of the bed measure, and the measure has

been adjusted by cutting back these edges. The measured length of 975 mm (38·4″) probably represents a plaiding yard of one yard and a cloth shrinkage allowance of 2¼″ (a sixteenth of a yard) plus a small additional quantity for play. The iron upper surfaces have been painted white.

SIZE: the distance between the staples is 970 mm, but between the defined length between the adjusted centre sections is 975 mm (38·4″). The stone slab is 1,770 mm long, 770 mm wide, tapering down to 570 mm at the east end, and 140 mm thick.

The churchyard had historically been used as the burgh's marketplace, surrounding the cathedral. The tolbooth, together with the council chamber above it, were constructed on the east end of the cathedral, being renewed in the 1730s and finally demolished in 1813 and the materials sold. The stone slab carrying the ell bed, which is immediately adjacent to the site, may earlier have been attached to the tolbooth's outer wall. The ell is noted, but without information about its size or particular location, in a number of local histories including Donald F. Sage (ed), *Memorabilia Domestica; or parish life in the north of Scotland, by the late Rev. Donald Sage* (Wick,

Photograph: Daniel Dennett

21

1889); H. M. Mackay, *Old Dornoch; its trad-itions and legends* (Dingwall, 1920); Charles D. Bentinck, *Dornoch Cathedral and Parish* (Inverness, 1926).

The church is a Category A protected building in Historic Scotland's Statutory List of Buildings of Architectural and Historic Importance (listed 18 March 1971; Dornoch Burgh, no. 10), and the measure is similarly protected as an item within the curtilage of the listed building. The grave-yard has recently (24 March 2003) been desig-nated as a Scheduled Ancient Monument.

Mounted on a horizontal stone in the precincts of Dornoch Cathedral, Dornoch (The Church of Scotland, Edinburgh).

22   YARD BED, 1739, FOR THE FIFE CHAPMEN COURT

Two-part folding yard measure in brass, which is clearly very closely related both in structure and decoration to the later yard of 1758 for the Angus Incorporation of Chapmen (Item **23**). Cross pieces screwed into the ends of the measure form projecting lugs which define identical bed measures on either side of the bar; with three-leaf hinge and pin to locate the two halves when folded, and with a plate screwed onto one half at the hinge to prevent over-opening. Engraved on the side face: 'THE GAUGE BELONGING TO THE CHAPMAN COURT IN FYFE – 1739 –' and 'IOHN GREIG LORD IAMES CUTHBURT DEPUT IAMES PRYD CLARK'. The opposite face engraved with a

foliated design with animal heads, and engraved near the centre with a decorative '4' merchant symbol with crossed ends. The upper face with a divided inch scale, divided to eighths, the inches numbered 1 to 3[6]; the under surface divided to indicate the half, quarter, eighth, sixteenth and thirty-second of the yard from each end.

SIZE: length of bed measures 914 mm; overall length open 925 and closed 481 mm; bar thick-ness 11 mm square; width across lugs 45 mm.

East Fife Council Museum Service, Cupar (Fife Council, Fife Museums East): Inv. CUPMS 1984.106. Transferred by Cupar Town Council to North East Fife District Council at local govern-ment reorganisation in 1975 (Museums Service established in 1983).

23   YARD BED, 1758, FOR THE ANGUS INCORPORATION OF CHAPMEN

Two-part folding yard measure in brass, which is clearly very closely related both in structure and decoration to Item **22**. Two projecting lugs at each end of the measure form identical bed measures on either side of the bar; with three-leaf hinge and pin to locate the two halves when folded, but now lacking the extension at the hinge to prevent over-opening. Engraved on the side face: 'THE GAUGE BELONGING TO THE INCORPORA-TION OF CHAPMEN IN ANGUS 1758' and 'IOHN CONSTABLE. LORD: DAVID DAKERS. DEPUT: ANDREW DALL. TREASURER: WILLM ORR CLERK'.

22

23

The opposite face engraved with the same foliated design with animal heads used on Item **22**, and similarly engraved near the centre with a decorative '4' merchant symbol and the letter 'M'.

SIZE: length of bed measures 914 and 915 mm; overall length open 937 and closed 477 mm; bar thickness 11 to 13 mm square; width across lugs 31 mm.

Meffan Institute, Museum and Art Gallery, Forfar (Angus District Council, Cultural Services, Forfar): Inv. ADMUS F.1978.61. Known to have been displayed in Forfar Library in the 1960s, and subsequently transferred by the Town Council to the museum.

24   YARD AND PLAIDING YARD BEDS, *c*.1810(?), FOR ABERDEEN

Bed measure in iron or steel, specifically designed to mark the divisions of the yard on the faces of wooden traders' yards and plaiding yards. The bar is set into a length of mahogany and screwed in place from below; a second length is recessed along its length to clear the protruding parts of the measure and hinged to first piece to form a protective case. A brass plate on the case describes it as 'ABERDEEN DEAN OF GUILD'S / STANDARD / for YARD and ELL'. The measure itself is a square-section bar, with a stepped end-stop to locate and support a wooden yard

measure above a series of steel knife edges 5 mm high and set across the bar at quarter-yard intervals, together with additional knife edges at the opposite end for an eighth, sixteenth and thirty-second of the yard. These divisions would be cut into the face of a square wooden yardstick by striking the stick down onto the knife-edges. The correct yard length could have been determined initially by striking the wooden stick on a higher knife-edge (16 mm high): once the wooden measure had been cut at the marked line, the vertical face of this larger knife-edge acts as the further end stop of the bed measure. By hinging up the stepped end-stop, a longer rod can be inserted against a fixed end-stop to mark the length of a plaiding yard.

SIZE: yard measured as 915 mm (36·0"); plaiding yard measured as 975 mm (38·41"); overall length of bar 1,041 mm, width 15 mm; case 1,132 × 69 × 50 mm.

Although the plaiding yard is larger than the yard by very slightly more than the sixteenth of a yard (2¼") that might be expected, it accurately matches the figure of 38⁵/₁₂", or 38·42", given in George Skene Keith's 1795 report to the Aberdeen Dean of Guild: W. Kennedy, *Annals of Aberdeen*, 2 volumes (London, 1818), II, 294. Although this is properly a plaiding yard, Keith associates it with the 'Scots ell' of 37⅓" as the 'Aberdeen plaiding [ell]', which presumably explains why the measure is described on the case as the

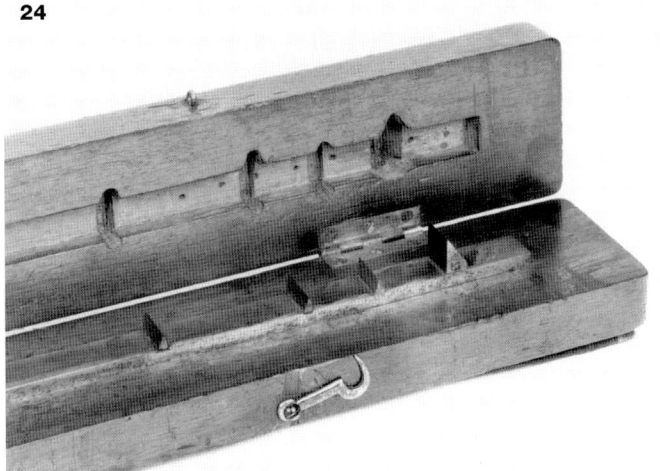

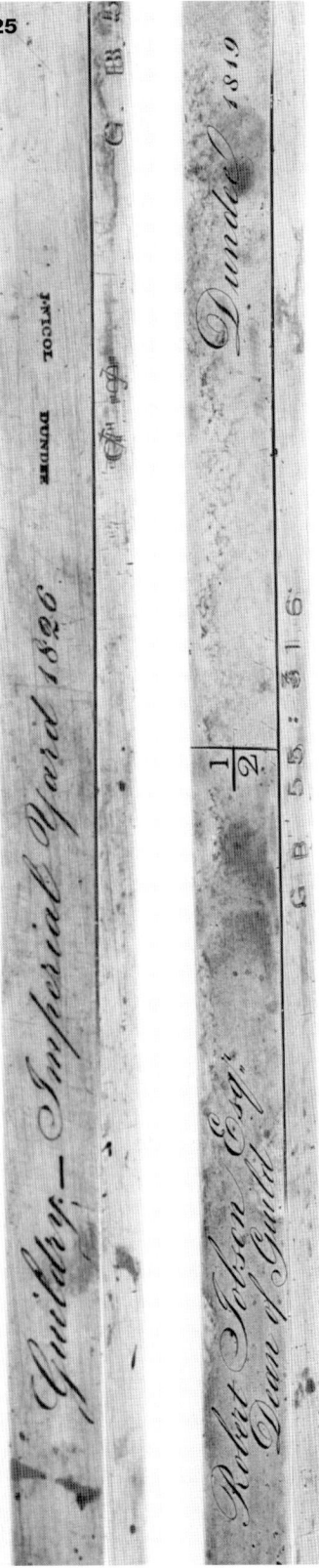

standard for the 'yard and ell'. John Swinton gave the same length in 1779, describing it as a 'plaiding yard', but the 1825 county report for Aberdeenshire follows the inscription on the measure and refers to it as a 'plaiding ell': [John Swinton], *A Proposal for Uniformity of Weights and Measures in Scotland* (Edinburgh, 1779), 53; G. Buchanan, *Tables for converting the Weights and Measures hitherto in use in Great Britain …* (Edinburgh, 1829), 174. Patrick Copland measured the hinged iron bed measure at Aberdeen in 1811 as 38·65" and noted that this did not amount to '38 and 5/12 which it ought to do', which was possibly a reference to the statements of Swinton and Keith: John S. Reid, 'Patrick Copland (1748-1822)', unpublished University of Aberdeen MLitt thesis, 1983, appendix 2, p. 4. This early standard no longer survives. Possibly the present measure was made at this time to provide a standard that Copland believed was of the correct length.

Exhibited at the International Exhibition, Glasgow, 1901, item 2415, 'Yard and Ell Measure, Aberdeen Dean of Guild Standard: used until 1825 for measuring certain coarse cloths'.

Town House, Aberdeen (Aberdeen City Council): Inv. THA 3039.

25 YARD BED, BY JAMES NICOL, 1819, FOR DUNDEE

Flat brass bar with two right-angled bends at the ends, and steel faces screwed to these upstands to form a yard bed measure. Divided with binary divisions of half-, quarter-, eighth-, sixteenth- and thirty-second of a yard from one end, and with a scale of 12 inches, subdivided to ⅛" from the other. Engraved at the centre 'Robert Jobson Esqr. / Dean of Guild' and 'Dundee 1819'. The underside stamped 'J.NICOL DUNDEE' and engraved in a different hand 'Guildry. – Imperial Yard 1826'. The bed upstands stamped with the Exchequer mark and '[crown] / WR / IIII'. In fitted oak case.

SIZE: length measured as 915 mm (36·0"); overall length 934 mm, bar cross section 17 × 8 mm; height over upstands 30 mm.

McManus Galleries, Dundee (Dundee City Council, Arts and Heritage): Inv. DUNMG 1955.316 (the code 'G. B. 55:316' is stamped into the sides of the measure). Transferred in 1955 from the Weights and Measures Department, Dundee Corporation.

26 WOODEN YARD, BY DONALDSON, EARLY NINETEENTH CENTURY, FROM KINGHORN

Yard measure of oblong section in boxwood, with inset T-section up-stands, forming the defining faces of the yard measure. Stamped at both ends with the County of Fife mark and the initials 'J.T.', and stamped at one end 'DONALDSON EDIN'. In fitted oak box.

SIZE: yard measured as 914 mm; overall length 953 mm; bar cross section 33 × 25 mm; height of end stops above bar 21 mm.

The initials 'J.T.' as verification marks have been identified with J. Tod, the Chief Magistrate who received Exchequer standards in 1826 on behalf of Pittenweem, a Fife coastal burgh 20 miles to the east of Kinghorn: Carl Ricketts and John Douglas, *Marks and Marking of Weights and Measures of the British Isles* (Taunton, 1996), 124.

Exhibited at the Glasgow International Exhibition 1901, item 2488 (with Item **28**), and at the Scottish Exhibition of National History, Glasgow, 1911, item 41, in both cases as from Kinghorn Town Council.

Kirkcaldy Museum and Art Gallery, Kirkcaldy (Fife Council): Inv. KIRMG 1972.6. Transferred from Kinghorn Town Council, 1972.

26

27 YARD BED, INCORPORATING AN ELL AND PLAIDING YARD, BY ALEXANDER ADIE, *c.*1822(?), SUBSEQUENTLY FOR THE COUNTY OF EDINBURGH

Yard measure in square-section brass, with projecting rectangular steel end pieces forming the faces of a yard bed, and with further cylindrical steel extensions. The measure is undivided, although it carries two longitudinal lines scribed on the upper face, as if provision had been made for a engraved scale. It is engraved at the centre of the bar 'Imperial Standard yard / COUNTY OF EDINBURGH, 1860 / A. Adie Edinburgh', with the second line falling within the two scribed lines on the face. However, these lines of text are not correctly centred, and it is apparent that it was originally engraved 'Standard yard // A.Adie Edinburgh', with 'Imperial' and 'COUNTY OF EDINBURGH' added later, and finally the date '1860'. Adie ceased to trade as Miller & Adie in 1822 (although his uncle John Miller had in fact died in 1815), and Imperial measure was introduced in 1824, which suggests a very narrow range for the likely construction date. The upper surface is stamped at the ends with the Exchequer mark and '[crown] / VR', implying that the first verification date may have been 1860. (At any rate, a Victorian verification would necessarily come after the business had begun trading as Adie & Son in 1835, demonstrating that at least part of the inscription has been added.) The side of the measure carries what may be a serial number '10' at either end, together with later Board of Trade verification marks.

SIZE: yard measured as 915 mm (36·0"); bar cross-section 24 × 24 mm; steel endstops 48 mm high; length over these 944 mm (37·2"); cylindrical end-stops diameter 19 mm; length over these 981 mm (38·6").

It is possible that Adie had intended the steel extension pieces to preserve values for earlier traditional linear measures. The distance across the outer faces of the rectangular plates is 944 mm, which is the longer eighteenth-century ell of 37·2". (Noticeably this is not the ell of 37·06" found in Jardine's 1811 work with Copland which was later accepted by the 1825 Edinburgh jury, which was advised by Jardine and Adie: G. Buchanan, *Tables for converting the Weights and Measures ... hitherto in use in Great Britain* (Edinburgh 1829), 198-9.) The length over the cylindrical extensions is 981 mm, or 38·6", which possibly represents a plaiding yard of a sixteenth (2¼") more than the yard.

National Museums of Scotland, Edinburgh (Trustees of the NMS): Inv. NMS T.1977.L163.50. Lent by Lothian Regional Council to the Royal Scottish Museums (now part of NMS) in February 1977 (authorised in December 1976).

27

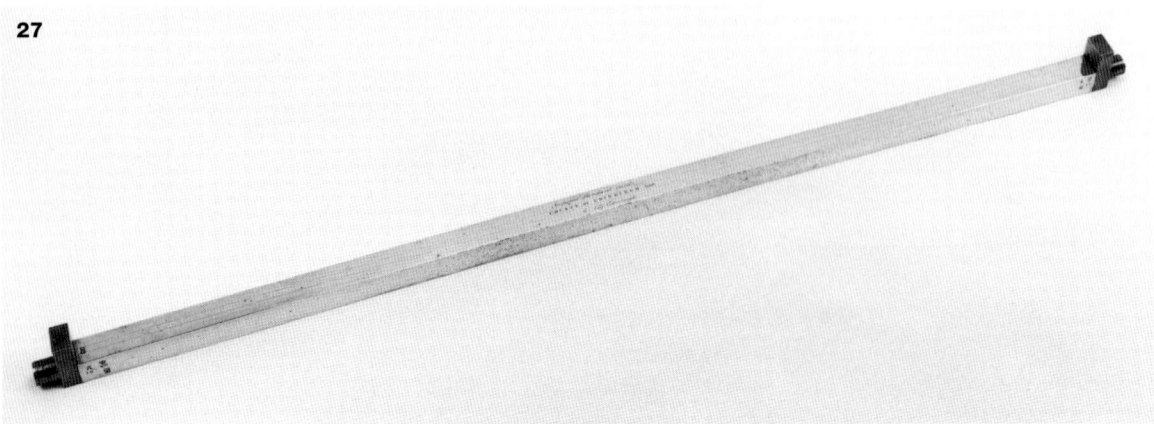

A6: Scots manufactured Post-Imperial Linear Standards

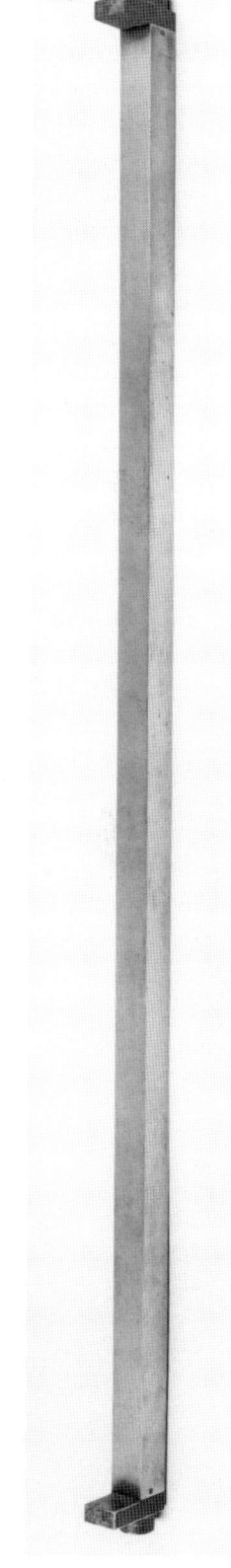

28 Yard bed, incorporating an ell, attributed to Alexander Adie, *c.*1830, from Kinghorn

Another example, of the same cross-section and terminal construction as Item **27**, and from its style certainly attributable to the Edinburgh Adie workshop. It is, however, entirely unmarked except for 1830-7 Exchequer and William IV marks of initial verification. It differs from Item **27** in that the length over the rectangular steel extension pieces is 37·0", which may indicate that an ell of 37" was specified in the order for the piece. However, the length over the cylindrical extensions is 37·8", which does not appear to have any Kinghorn or Fifeshire significance. In a fitted mahogany case.

SIZE: yard bed measured as 914 mm; overall length 960 mm; cross section 24 × 24 mm; steel end-stops 49 mm high; length over these 938 mm.

Exhibited at the Glasgow International Exhibition, 1901, item 2488 (with Item **26**); and at the Scottish Exhibition of National History, Glasgow, 1911, item 40, in both cases as from Kinghorn Town Council.

Kirkcaldy Museum and Art Gallery, Kirkcaldy (Fife Council): Inv. KIRMG 1972.7. Transferred from Kinghorn Town Council, 1972.

29 Yard scale, by Adie & Son, *c.*1860, for Leith Dock Commission

Linear measure in square-section tubular brass with solid brass ends; carrying yard scales in inches, sub-divided to tenths of an inch, on two faces, and engraved 'STANDARD YARD. / ADIE & SON, EDINBURGH.' and 'LEITH DOCK COMMISSION'. In fitted case.

SIZE: overall length 923 mm; cross-section 20 × 20 mm; case 953 × 50 × 45 mm.

Commissioners for the Harbour and Docks of Leith were first appointed under a parliamentary Act of 1826, and commonly called the Leith Dock Commission from that date.

Part of the loan collection of the Edinburgh and East of Scotland Association of the Institution of Civil Engineers formed from 1971 at the Royal Scottish Museum (part of the National Museums of Scotland from 1986), and described in C. Roe, *Museum Collection 1971-1981, Catalogue [of the] Exhibition at the Royal Scottish Museum* (Edinburgh, 1981), item 33. The loan was terminated in 1989, and the collection subsequently lent to the Mechanical Engineering Department of Heriot-Watt University.

Department of Mechanical Engineering, Heriot-Watt University, Riccarton (loan collection of the Edinburgh and East of Scotland Association of the Institution of Civil Engineers). Presented to the Association in July 1973 by Leith Dock Commission (subsequently incorporated in the Forth Ports Authority).

30   METRE SCALE, BY ADIE & SON, *c.*1840, FOR THE NORTHERN LIGHTHOUSE BOARD

Linear measure in flat-section oxidised brass with a bright chamfered edge carrying a divided scale, and with two lifting pieces. Divided into 100 numbered centimetres, and sub-divided into millimetres. Engraved at the centre 'Adie & Son / EDINBURGH'.

SIZE: overall length 1,008 mm; width 32 mm.

The provenance of this item suggests that it was commissioned as a standard scale for use in significant optical engineering work undertaken for the Northern Lighthouse Board in Edinburgh in about 1840. The Commissioners for the Northern Lights were established in 1788 with responsibility for constructing and maintaining lighthouses around the storm-swept coasts of Scotland and the Isle of Man. Unlike their English counterparts, the Brethren of Trinity House, the Northern Lighthouse Board retained professional engineers to carry forward their programme of work, and successive generations of the Edinburgh civil engineering firm of Stevensons held the appointment of Engineer to the Board until the late 1930s. The innovative approach of the Stevensons, and particularly of Robert Stevenson (1772-1850) and his sons Alan, David and Thomas, quickly established the Northern Lighthouse Board as the leading authority in this new engineering field and the Stevensons at its principal exponents. Although the Northern Lighthouse Board was the largest client of Robert Stevenson & Sons, occupying at least one of the partners full-time, the firm was also engaged in other major projects such as harbour and bridge construction work and specialist engineering consultancy.

454

In 1958 the engineer and lighthouse historian D. Alan Stevenson, nephew of David A. Stevenson, who had been the family's last resident Engineer to the Board, presented this scale and four other instruments by the Adie business to the Royal Scottish Museum. Two, by Alexander Adie, were for adjusting arrays of argand lamp reflectors in lighthouse optics; the other two, by the post-1835 partnership of Adie & Son, were for aligning Fresnel lens panels for the optic designed by Alan Stevenson for the Board's new lighthouse built by him on the hazardous Skerryvore reef to the west of Mull. The optic had been commissioned from François Soleil in Paris, and was received in Edinburgh in 1843.

It is assumed that this metre scale was constructed by Alexander Adie or his second son and partner John, for use in preparing the specification or installation drawings for the Soleil optic. Although the instruments in this group would have been commissioned from Adie by the Stevensons in the Board's name, they had remained in the hands of the Board's Engineers.

On the Northern Lighthouse Board and the Stevensons' work for them, see R. W. Munro, *Scottish Lighthouses* (Stornaway, 1979); and on the Adies see T. N. Clarke, A. D. Morrison-Low and A. D. C. Simpson, *Brass & Glass: Scientific Instrument Making Workshops in Scotland* (Edinburgh, 1989), 25-74.

National Museums of Scotland, Edinburgh (Trustees of the NMS): Inv. NMS T.1958.51. Presented by D. Alan Stevenson FRSE, Civil Engineer, 90A George Street, Edinburgh.

SECTION B

WEIGHTS

B1: PRE-1618 WEIGHT STANDARDS

31 TRONE RING WEIGHT OF ONE STONE, THE 'CRAIGENGELT WEIGHT', BY HANS COCHRAN, 1553, AT STIRLING

Trone weight in cast bronze, of tapering cylindrical form with a domed top, with iron eyelet and ring handle. The body of the weight is divided by projecting beads into three bands: these carry a continuous Scots relief inscription in Gothic black-letter minuscule, formed by cutting away the background on the casting master. The inscription begins at a banded ornament in the top line and continues from *fleurs-de-lys* in the second and third lines, and with a final *fleur-de-lys* separating the names of the maker. It reads, 'iohn cragingelt of yat ilk me cōding / maid quhē he ves prouest of striviling / anno dñi m d l iii : m : hanes coqhren'. The first two lines form a rhyming couplet, with ten syllables in each line. Contractions are recorded conventionally with a macron-like abbreviation mark over the word, that at the end of the first line being for 'conding', or condign, in the sense that the weight has been made as a suitable or corresponding copy which conforms to the standard. However, no abbreviation is marked at the initial name of the maker, whose surname is clearly intended to be understood as 'Cochran'. Possibly it should read 'in : hanes', for Jonhanes or John, but equally it might be 'm : hanes', or

Master Hans, using the Germanic equivalent to John, but reflecting Cochran's actual status as a master gunner in the Royal Artillery factory at Edinburgh Castle. Taking the latter possibility, the inscription can be read as: 'John Craigengelt, Laird of Craigengelt, had me made conformable [to the Standard] when he was Provost of Stirling, A.D. 1553, Master Hans Cochran.' We are grateful to John Higgit of the Department of Fine Art, University of Edinburgh, for his comments and advice on this item.

SIZE: base diameter 130 mm; height of main body of weight 130 mm; height to top of shackle 160 mm.

MEASURED WEIGHT: 9,173·0 g, or 141,558 gr; representing a trone pound of 573·3 g or 8,847 gr. Providing the traditional relationship is preserved, whereby the trone stone is 1¼ troye stones or 20 troye pounds, the weight represents a troye pound of 458·7 g or 7,078 gr, with an ounce of 442·4 gr. The merchant pound of 15 Paris ounces, each of 472½ gr, would be 7,087½ gr. But since this weight clearly pre-dates the small increase in the English troy grain discussed in Chapter 4, the weight of the pound would be 7,083 modern troy grains. If this is the correct basis of the weight, it has lost only about 110 gr, but it presumably remained in use only until

replaced by the much larger stone introduced in 1563 (see Chapter **8**). The alternative possibility of using the Scottish Mint weight, with an ounce of about 471·0 modern troy grains (see Item **41**), would generate a stone of only about 141,300 gr. As this is smaller than the Craigengelt Weight, the latter cannot be based on the Mint system.

The Craigengelt estate is a few miles south-west of Stirling. John Craigengelt became a Bailie of Stirling town council in 1533, and Provost on many occasions between 1541 and 1567; he was also a Commissioner to Parliament for Stirling in 1543 and 1567: Margaret Young (ed), *The Parliaments of Scotland: Burgh and Shire Commissioners,* 2 volumes (Edinburgh, 1992-3), II, 149. He was Convenor of the Convention of Royal Burghs meeting at Stirling in 1553 when new standards of weights and measures were due to be ordered for the burghs: J. D. Marwick, *et al.* (eds), *Records of the Convention of the Royal Burghs of Scotland,* 7 volumes (Edinburgh, 1866-1918), I, 2. It is in this capacity, rather than narrowly as Provost of Stirling, that his name is

recorded on the weight, which is presumably only one of several that may have been made and issued to any one of the burghs represented at the Convention.

Hans Cochran first appears in the Lord Treasurer's accounts as 'Maister Johne Cocherane maister gunnar to the Kingis grace', drawing £100 as his pension for the year past in 1539: J. Dickson, *et al.* (eds), *Accounts of the Lord High Treasurer of Scotland,* 12 volumes (Edinburgh, 1877-1970), VII, 199. In the same year he was co-ordinating the extensive work involved in casting at least ten large guns, including two double culverins, in a period of great activity in the Artillery Works (the 'grete munitioun hous') in Edinburgh Castle and defensive preparations at Edinburgh and Leith: ibid., VII, 341-61, especially pp. 343, 345, 349. After James V's death in 1542, Cochran was heavily involved in works for the Earl of Arran, the Lord Governor or regent during the early minority of the infant Mary, Queen of Scots. In 1543 he was moving heavy artillery at Dundee, Forfar and Brechin in the tense period which led up to Henry VIII's punitive 'rough wooing' campaign in Scotland, and in 1544 he was active in the defence of Edinburgh Castle against the English army under the Earl of Hertford: ibid., VIII, 253, 292, 293. After peace was concluded in 1551, Cochran makes appearances in the Treasurer's records only to draw his fees between 1554 and 1558: ibid., X, 234, 301, 355, 402.

The weight was purchased at auction in Glasgow on 31 July 1889 for the Kelvingrove Museum, Glasgow, as one of a number of items from the collection of A. Russell Pollock of Renfrewshire. The pieces acquired by the museum were largely foreign ethnographic items, but they also included the unprovenanced Scots pint of c.1622, which is still at Glasgow (Item **126**). James Paton, the curator at Glasgow, identified the weight as the Stirling trone stone and corresponded about the inscription with W. B. Cook, a council member of the Stirling Natural History and Archaeological Society, who also contributed a column to the *Stirling Sentinel* newspaper under the name of 'The Stirling Antiquary'. Cook borrowed the weight and described it to the Society, publishing an account in the *Sentinel* a week later, on 25 January 1898, where it was first described as 'the Craigengelt Weight'. He concluded that it could not represent the size of the Lanark trone stone, because he associated the latter with the smaller merchant stone of the 1618 Assize. He was unable to explain its weight of 'exactly' 20 pounds and 3½ ounces avoirdupois (which would be equivalent to 9,171 g).

Cook's paper, with a woodblock engraving of the weight and its inscription, and a photograph, was subsequently published in the Society's *Transactions*, 20 (1897-8), 65-9. An abbreviated account, with the photograph, was published as 'A Municipal Relic of Old Stirling' in the April 1898 issue of the *Scottish Antiquary*, 12 (1897-8), 164-6. Alexander Hutcheson, an architect in Broughty Ferry who was a Fellow of the Society of Antiquaries of Scotland, responded in the following (July) issue of the *Scottish Antiquary*,

13 (1898-9), 5-6, with a corrected reading of the inscription. This last contribution, together with the three earlier pieces from the *Sentinel*, were gathered in W. B. Cook (ed), *The Stirling Antiquary*, 4 volumes (Stirling, 1893-1908), II, 148-53, 259-62.

In September 1898 the Provost of Stirling led a deputation to Glasgow Town Council's Galleries Committee, where they successfully argued that the weight should be transferred to Stirling's Guildhall or to the Smith Institute (now the Smith Art Gallery and Museum, Stirling). This was reported in a piece illustrated by a second woodblock engraving, in the *Supplement to the Sentinel*, 4 October 1898, with the successful result announced on 22 November 1898. The weight was sent to Stirling in late December 1898, after a plaster cast was taken. This cast remains at Kelvingrove Museum, Glasgow, with the original inventory number of the weight (inv. GLAMG 1898·56 g); the suspension ring is now detached. More recently, the weight has been illustrated and discussed in A. D. C. Simpson, 'Scots "Trone" Weight: Preliminary Observations on the Origins of Scotland's Early Market Weights', in *Northern Studies*, 29 (1992), 62-81.

Smith Art Gallery and Museum, Stirling (Trustees of the Stirling Smith Art Gallery): Inv. STIGM 2980 (previously UD.31). Presented by Glasgow Town Council to Stirling Town Council in 1898, and transferred by them to the Smith Institute.

32   TRONE RING WEIGHTS, FROM ONE STONE, LATE SIXTEENTH CENTURY, FROM LANARK

Group of four trone weights, in cast bronze, of a tapering square section, with iron ring-handles. The three larger weights, for 1 stone, 4 pounds and 2 pounds, have a stylised double-headed eagle, representing the crest of Lanark cast to the same size in relief on one face, and are clearly part

of a set which presumably at one time included an 8-pound and 1-pound weight. The 1-stone weight has weight adjustment rings attached to the ring handle, in iron and lead. The smallest weight of 4 ounces has cut into one face '4 v / TROYN' and a brass adjustment screw attached to the handle support; again, presumably, it formed part of a larger range of fractional parts of the pound, and is provisionally associated with the larger weights.

SIZES: widths 115, 65, 50, 27 mm; heights (without ring) 148, 115, 83, 39 mm

MEASURED WEIGHTS: 9,799·3, 2,453·5, 1,220·8, 153·0 g, or 151,223, 37,862, 18,839, 2,361 gr; these represent troye ounces (at 16 to the pound and 20 troye pounds to the trone stone) of 472·6, 473·3, 471·0 and 472·2 gr respectively. Thus the weights were adjusted to an ounce of at least 473·3 modern grains (or 473·6 old grains) – hence possibly a Flemish ounce of about 474½ gr.

The three larger weights are illustrated in A. D. Robertson (ed), *Lanark: the Burgh and Its Councils 1496-1880* (Lanark, 1974), plate opposite

p. 177, and the stone in J. Paton (ed), *Scottish History and Life* (Glasgow, 1902), 199, fig. 236. More recently, the weights have been discussed and illustrated in A. D. C. Simpson, 'Scots "Trone" Weight: Preliminary Observations on the Origins of Scotland's Early Market Weights', in *Northern Studies*, 29 (1994), 62-81, fig. 2.

The stone weight was probably the one exhibited by the Town Council at the Glasgow International Exhibition, 1901, item 2473, as 'Lanark Stone Weight (Ancient)', and at the Scottish Exhibition of National History, Glasgow, 1911, item 148, 'Lanark Stone Weight'. Also exhibited at the Treasures of South Lanarkshire exhibition, Hamilton, 2001, item 63d.

Low Parks Museum, Hamilton (South Lanark-shire Council, Museums Service): Inv. SLCMD 2001.75, 74, 73 and 138. Apparently packed and transferred from the Burgh of Lanark's Weights and Measures Department to the Town Council's offices in 1932, and passing to Clydesdale District Council (Museum Service established 1993) at local government reorganisation in 1975.

33 TROYE NESTING WEIGHTS, PILE OF 16 POUNDS, FROM EIGHT POUNDS DOWNWARDS, BY GEORG WEINMANN, 1572, AT PERTH

Incomplete set of nesting cup weights in turned brass, of the type associated with manufacture in Nuremberg, comprising a decorative outer weight with lid and handle, and two original surviving weights, and now with six unrelated inner weights from different sets (one in copper). The outer cup, with banded decoration and stamped stars and roses, also has a crudely incised letter 'P' (presumably indicating Perth). The zoomorphic hinge of the lid is in the shape of two sea-monsters, with the plain (replacement?) clasp supported by another with two opposing heads. Two male heads support an iron handle, which is almost certainly a replacement. Stamped on the lid are a *fleur-de-lys*, and the hand-mark of Antwerp, with a group of three negro heads.

These last have been identified as the master sign of Georg Weinmann, d.1604, weight maker of Nuremberg: D. A. Wittop Koning and G. M. M. Houben, *2000 Jaar Gewichten in de Nederlanden: Stelsels, Ijkwezen, Vormen, Makers, Merken, Gebruik* (Lochem-Poperinge, 1980), 108, 111, mark 24a. Cut round the lid is the inscription 'HENRICVS ADEMSONE DECANVS GILDE + 1572 +'. The two original inner turned weights hang by their lips, and have simple banded decoration on the inside, outside and rims, and have adjustment recesses in their bases and some added lead. The larger is stamped with a hand, and the smaller with the hand and *fleur-de-lys*. The next two weights are from another set and are stamped '1' and '16' inside. The final four are unmarked, assorted weights; however, all six of these have lead adjustments.

SIZES: The outer cup diameter 122 mm; overall width 165 mm; overall height excluding handle 135 mm; original inner cups diameter 105, 85 mm; heights 75, 60 mm.

MEASURED WEIGHTS: 3,952·6, 1,961·7, 980·1 g, or 60,997, 30,273, 15,125 gr. All these weights have lead adjustment, but the ounce basis of the outer case (476·5 gr) is considerably greater than that of the others, indicating perhaps the subsequent replacement of the handle and clasp. The ounce bases of the cups are 473·0 and 472·7 gr, so in their unaltered and unworn state they may have been adjusted to a Flemish troois ounce of about 474½ gr.

33

Illustrated in L. Burrell, 'The Standards of Scotland', in *The Monthly Review: the Journal of the Institute of Weights and Measures Administration*, 69 (1961), 58. Described in the 1826 Perthshire jury verdict as 'the Stone Scotch Troyes' in the possession of the Guildry of Perth, and quoted as having engraved on it 'Henricus Ademsone, Decanus Gilde, 1572', and representing a pound of 7,652·424 grains: G. Buchanan, *Tables for Converting the Weights and Measures …* (Edinburgh, 1829), 246.

Perth Museum and Art Gallery (Perth and Kinross Council): Inv. PERGM 104. Presented by J. Murray Graham of Murrayshall in 1862.

34

34 TRONE RING WEIGHT OF EIGHT POUNDS, EARLY SEVENTEENTH CENTURY(?), FROM LANARK

Trone weight in cast bronze of tapering square section with ring handle. The four faces of the weight are unmarked, but a design cut into the upper surface, and perhaps representing a double-headed eagle, is now largely obscured by a lead adjustment support for the ring handle.

SIZE: base 85 x 80 mm; height without ring: 145 mm

MEASURED WEIGHT: 4,923·2 g or 75,975 gr; representing a trone pound of 9,497 gr and a troye ounce (at 20 to the trone pound) of 474·8 gr. It is suggested that this was originally adjusted

(like Item 32) to match a Flemish ounce standard of about 474½ gr, but was increased in weight with the addition of lead at the shoulder to match the heavier 1618 standard.

South Lanarkshire Museums Service, Hamilton (South Lanarkshire Council): Inv. SLCMD 2001.III. Apparently packed and transferred from the Burgh of Lanark's Weights and Measures Department to the Town Council's offices in 1932, and passing to Clydesdale District Council (Museum Service established 1993) at local government reorganisation in 1975.

B2: Post-1618 Weight Standards

35 TROYE WEIGHT OF ONE STONE, 1618, AT CULROSS

Bronze cylindrical troye weight of 1 stone, cast and with a domed top and bar handle. Stamped on the side '·STONE·TROYS· / ·1618·' and on the opposite side with a double-headed eagle (for Lanark) and '·LANERK·'. The base is not finished-off square, giving the weight a tilt of about 5°. A line running 30 mm above the edge of the base and discernible for three-quarters of the circumference indicates a casting discontinuity or joint. At some time the handle has been detached and is repaired with solder.

SIZE: diameter 100 mm; maximum height 166 mm.

MEASURED WEIGHT: 7,908·5 g or 122,044 gr; representing a pound of 7,628 gr and an ounce basis of 476·7 gr – and therefore probably the result of eighteenth-century adjustment.

35

For the two additional sets of weights supplied by Lanark to the Privy Council for safe keeping at Edinburgh and Dumbarton castles in connection with the 1618 Assize, see Chapter 2, ref. 90 and also Item 113.

This item was at Culross in 1826 and is recorded in the verdict of the Perthshire jury (drawing on the report by Adam Anderson, Rector of Perth Academy) as weighing 122,105 gr: G. Buchanan, *Tables for Converting the Weights and Measures hitherto in use in Great Britain …* (Edinburgh, 1829), 246. On the 1826 report by Anderson and his mathematics assistant at the Academy, Thomas Bruce, see Kenneth J. Cameron, *The Schoolmaster Engineer: Adam Anderson of Perth and St. Andrews c.1780-1846* (Dundee, 1988), 20-1.

Illustrated in L. Burrell, 'The Standards of Scotland', in *The Monthly Review: the Journal of the Institute of Weights and Measures Administration*, 69 (1961), 57, and in the Royal Commission on the Ancient and Historical Monuments of Scotland (RCAHMS), *Tolbooths and Town-houses: Civic Architecture in Scotland to 1833* (Edinburgh, 1996), 6.

Culross Town House Museum, Culross (The National Trust for Scotland, Edinburgh): Inv. NTS 33.222. In the Town House at its transfer to the Trust from the Town Council of Culross in advance of local government reorganisation in 1975.

36 TROYE WEIGHT OF EIGHT POUNDS, 1618, FROM LANARK

Bronze cylindrical troye weight of 8 pounds, cast and with a domed top and bar handle. Stamped on the side '·8·P·TROYS· / ·1618·' and on the opposite side with a double-headed eagle (for

Lanark) and '·LANERK·' The same eagle stamp is impressed three times on the base. The handle assembly apparently screws into the collar on the top of the weight, and presumably gives access to an adjustment recess within. The base of the weight seems to have been formed by soldering on a plate about 12 mm thick, before turning.

SIZE: diameter 80 mm; height 173 mm.

MEASURED WEIGHT: 3,933·2 g or 60,697 gr; representing a pound of 7,587 gr and an ounce basis of 474·2 gr.

This is all that remains of a full set of weights from 1 stone down, which were originally held at Lanark, and of which Lanark was empowered to issue commercial copies. The stone appears to have been borrowed by the authorities at Edinburgh during the eighteenth century, and was apparently the weight measured by Robert Stewart. It was identified as the Lanark copy by John Robison in the 1780s when he discovered it together with the Edinburgh set (Item 37) in the care of Thomas Simpson, the Edinburgh

36

pewterer, to whom it appears to have been issued by Edinburgh Dean of Guild. The stone, which similarly had stamps on the base, was subsequently marked as having been inspected by Robison on 24 February 1800; it was to have been carefully preserved by the Edinburgh Dean of Guild, but it no longer survives, and has at some stage been replaced by a more recent copy (Item 94). Robison's work is discussed in Chapter 9.

The stone was weighed again by James Jardine in 1825, and the rest of the set (including this 8-pound weight) was sent from Lanark to Edinburgh to be weighed at this time: G. Buchanan, *Tables for Converting the Weights and Measures hitherto in use in Great Britain …* (Edinburgh, 1829), 199-200. Although the Lanarkshire jury was aware that the stone at Edinburgh had been identified as Lanark's standard, it was still retained in Edinburgh when the jury issued their verdict in 1827: ibid., 230. The weighings recorded by the Lanarkshire jury were presumably made by Jardine and his colleagues, and are as follows: 60,703·7 (8 pound), 30,407·4 (4 pound), 7,608·7 (1 pound), 3,791·0 (8 ounce), 1,890·4 (4 ounce), 943·4 (2 ounce), 471·4 gr (1 ounce): ibid., 231. Interestingly, these appear to divide into two distinct groups: the ounce basis of the 4-pound and 1-pound weights are 475·1 and 475·5 gr respectively; whereas the ounce basis for the 8-pound weight is only 474·2 gr, and those for the weights under a pound are all less than 474 gr.

Remembering that Lanark would originally have held at least one full set of standards, and probably also a residue of undistributed weights (including presumably more stones, since these were issued singly), then it is suggested that these residual weights were of two different types. Some (including here the 4 pound and 1 pound) were of the full size authorised by the 1618 Assize, which we have deduced were to an ounce of 475·5 (modern) grains. The others (including the 8-pound weight) were adjusted down to match the original (pre-1618) standard of 474·2 (modern) grains, which were in the correct proportion to

36

the early Lanark trone stone submitted to the 1618 commissioners. For the larger weights, like the 8-pound weight, the extra mass could have been removed from the recess under the handle, which would leave the base stamping intact.

South Lanarkshire Museum Services, Hamilton (South Lanarkshire Council): Inv. SLCMD 2002.227. Apparently packed and transferred from the Burgh of Lanark's Weights and Measures Department to the Town Council's offices in 1932, and passing to Clydesdale District Council (Museum Service established 1993) at local government reorganisation in 1975.

37 TROYE WEIGHTS, FROM EIGHT POUNDS, 1618, FOR EDINBURGH

Part set of brass cylindrical troye weights (8 pounds to 1 ounce), which has been completed by the addition of later 1-stone and 1-pound weights (Items **94** and **92** below). The weights have a turned finish, domed tops and bar handles, or knob handles below 1 pound. The 8-pound weight is stamped on one side with a double-headed eagle (for Lanark) and 'LANERK', and stamped on the other side '·8·P·TROYS· / ·1618·';

the 4-pound, 2-pound, 8-ounce and 4-ounce weights are similarly stamped, with the denominations 4P, 2P, 8V, 4V; the weights for 2-ounce and 1-ounce are unmarked. The base of each weight is finished with a sheet of soldered brass, presumably covering an adjustment recess: this is approximately 2 mm thick for the three weights with bar handles, and approximately 1 mm thick for the four smaller weights with knobs. The weights therefore carry no stamped markings on their bases.

SIZES: diameters 79, 65, 52, 36, 27, 21, 17 mm; heights 137, 109, 64, 49, 39, 36, 26 mm.

MEASURED WEIGHTS: 3,948·9, 1,973·4, 987·0, 246·8, 123·5, 61·8, 30·9 g, or 60,939, 30,454, 15,231, 3,809, 1,906, 954, 477 gr. Taking the four largest weights (8 pounds to 8 ounces), these have been adjusted to a pound of at least 7,618 gr, with an ounce basis of at least 476·1 gr. This is certainly well in excess of the expected seventeenth-century level for the ounce and clearly reflects a subsequent adjustment.

These are presumably the same weights that John Robison recorded in 1778 being found in the care of Thomas Simpson, pewterer, who was using them on behalf of the Dean of Guild for adjusting weights in the burgh. Robison described them as 'marked with the spread eagle 1618' and noted that all of them 'have been much used, except the Stone [but] the others have been frequently adjusted by it': St Andrews University Library, MS Q171.R8, 'Professor [John] Robison's Commonplace Book', I, p.446. The ounce basis here is larger than that recorded by Robison for the stone (which he took to be Lanark's original 1618 standard), but it does match approximately his projected weight for the standard in 1618, extrapolating from the amount of wear he detected in the stone between his own measurements and the earlier ones of Robert Stewart (see Chapter **9**). On this basis the adjustment has probably been carried out in the 1790s, and has presumably eliminated any trace of the adjustment (to a lighter ounce basis) carried out by Alexander Bryce in about 1755, and subsequent alterations.

Robison copied from Robert Stewart's notes his measurements of the Edinburgh Dean of Guild's 1618 weights, supposedly made by Stewart in 1720: ibid., II, p.393. These show the stone as being to a pound of 7,619¼ gr, which was clearly larger than the pounds of the other weights,

apparently supporting the idea that it may have been from a different set or preserved under better circumstances. The smaller weights (except for the 2- and 8-ounce) were all to pounds of between 7,586 and 7,596 gr. Without knowing the precise basis of Stewart's weights, it should be stressed that his measurements for the smaller weights were all less than the recent measurements by a factor of about 1·003; whereas his

37

weight for the stone is *greater* than Jardine's (which was also in modern grains) by a factor of 1·001, acknowledging that some of this will be accounted for by wear. It is apparent that all the weights have been substantially altered since Stewart's time. He also included a measurement for the original 1-pound weight, which was presumably discarded at the much later adjustment when it was replaced by Robison's pound (Item 92).

The Museum of Edinburgh, Edinburgh (City of Edinburgh Council, Museums and Art Galleries): Inv. EDNMG HH 5978/2-4, 6-9/2000. Transferred from Edinburgh Town Council by 1899.

38

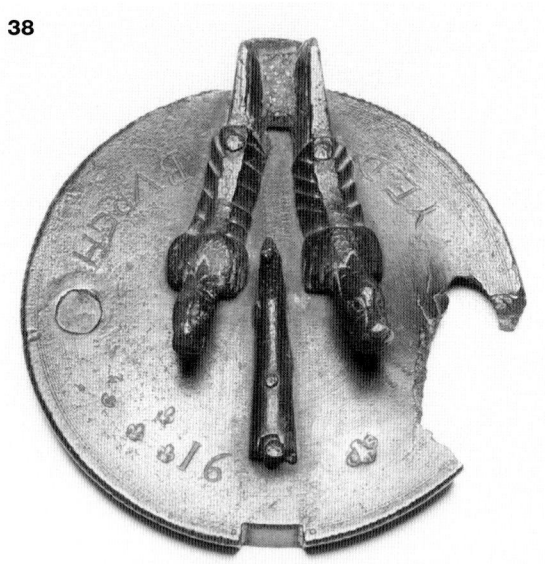

38 TROYE NESTING WEIGHTS, PILE OF 16 POUNDS, FROM EIGHT POUNDS DOWNWARDS, BY CHRISTOPHE SCHIRMER, LATE SEVENTEENTH CENTURY, AT JEDBURGH

Incomplete set of nesting cup weights in turned brass, of the type associated with manufacture in Nuremberg, comprising a decorative outer weight, damaged and with its lid detached and now missing its handle, and two original surviving inner weights, and a third, unrelated, in copper. The outer cup, with raised bands and stamped decoration. The zoomorphic hinge of the lid is in the shape of two sea-monsters, and the support for the clasp is another, broken at the centre. Part of the lid has been torn away with one of the handle supports; the other support trimmed off flush with the lid. Stamped on the lid is the weight figure '16', a group of three *fleurs-de-lys*, an unidentified mark apparently showing a hunting horn, and another mark showing a bell flanked by the letters 'C' and 'S'. This last has been identified as the master sign of Christophe Schirmer (1656 to after 1690), weight maker of Nuremberg: D. A. Wittop Koning and G. M. M. Houben, *2000 Jaar Gewichten in de Nederlanden: Stelsels, Ijkwezen, Vormen, Makers, Merken, Gebruik* (Lochem-Poperinge, 1980), 107, 111, mark 16a. Cut round the lid is the inscription 'YED BVRGH'. The inside of the outer cup is stamped '8' and with a *fleur-de-lys*. The two

466

original inner turned weights hang by their lips, and have simple banded decoration on the inside and outside, and are similarly stamped with a *fleur-de-lys* and the figures '4' and '2' respectively.

SIZES: outer cup diameter 128 mm; height to lid excluding handle 95 mm; original inner cup diameters 106, 85 mm; heights 72, 56 mm.

MEASURED WEIGHTS (of first two inner cups): 1,965·4 g (four pound weight), 981·9 g (two pound weight), or 30,330 and 15,153 gr. This indicates that these weights have been adjusted to a pound of over 7,583 grains, or a troois ounce of at least 473·9 grains.

Jedburgh Castle Jail and Museum, Jedburgh (Scottish Borders Council Museum Service, Selkirk): Inv. JEDMG 373. Transferred by Jedburgh Dean of Guild to the museum after 1898.

39   TROYE NESTING WEIGHTS, PILE OF EIGHT POUNDS, FROM FOUR POUNDS DOWNWARDS, BY G. B. WEINMANN, 1672, AT DUNDEE

Incomplete set of nesting cup weights in turned brass, of the type associated with manufacture in Nuremberg, comprising a decorative outer weight, with lid and handle, and three original surviving inner weights. The outer cup is decorated with raised bands and stamped decoration, and is cut with the inscription: ' * THE GILDRIE OF DVNDIE / PRVDENTIA · ET · CANDORE * ' (motto of the Burgh of Dundee). Stylised hinge, clasp and handle with formalised zoomorphic elements, the hinge itself broken and repaired. Stamped on the lid is the weight figure '8', a group of three *fleurs-de-lys*, and a bow-and-arrow mark. This last has been identified as the master sign of Georg Bernhard Weinmann (1656 to after 1685), weight maker of Nuremberg: D. A. Wittop Koning and G. M. M. Houben, *2000 Jaar*

39

467

Gewichten in de Nederlanden: Stelsels, Ijkwezen, Vormen, Makers, Merken, Gebruik (Lochem-Poperinge, 1980), 109, 111, mark 36. The lid additionally inscribed with the date '1672'. The inside of the outer cup is stamped with the weight figure '4' and with three *fleurs-de-lys*. The three inner turned weights hang by their lips, and have simple banded decoration on the inside, outside, and rims. All are stamped inside with a *fleur-de-lys* and the figures '2' and '1' and '16' respectively.

SIZES: outer cup diameter 100 mm; height to lid excluding handle 120 mm; original inner cups diameter 83, 67, 54 mm; heights 59, 45, 35 mm.

MEASURED WEIGHTS: 1,963·3, 982·8, 491·3, 245·6 g, or 30,298, 15,167, 7,582, 3,790 gr. The outer cup, which shows signs of alteration, now has an ounce basis of only 473·4 gr, but the inner cups indicate adjustment to a pound of at least 7,583 gr, and a troois ounce of at least 474·0 gr.

Exhibited at the 'Old Dundee' exhibition, Dundee, 1892, item 857.14, 15, 17 (presumably considered as three weights and a case).

McManus Art Gallery and Museum, Dundee (Dundee City Council, Arts and Heritage): Inv. DUNMG 1978.1083 (1-4), but previously 1874.322-325 (the first part of a group of '16 standard weights used by the Burgh Magistrates for testing accuracy of weights used by traders in Dundee, date 1672'). Presented by the Guildry Incorporation and Magistrates of Dundee, November 1873.

40   TROYE BELL-SHAPED WEIGHTS, FROM 50 POUNDS, 1676, FOR STIRLING

Part set of cast bronze weights, originally extending to eight weights, of which five survive. They carry no denominations, but by calculation, these represent 50, 25, 8, 4 and 1 pounds. The largest two weights are in the form of tapering hex-

40

agonal prisms, with a rounded moulding at the top and an octagonal looped handle. These are the only examples of hexagonal-shaped standards known in Scottish collections, and the only ones to represent weight multiples of 25, more normally found in Continental practice. It is likely that they were purchased in the Low Countries. The other three weights are cylindrical in section and have plain oval-shaped handles. The weights are all entirely unmarked, except for two individual letter stamps 'G' and 'S' in shield-shaped fields on the shoulders (not present on the smallest). This presumably denotes owner-ship by the Guildry of Stirling, although the marks have not been found on any other stan-dards at Stirling. A slip of bronze is attached to the top of the handle of the largest cylindrical weight as adjustment. A fitted oak chest accommodates the weights in hexagonal and circular wells, with three spaces unfilled: these are compatible with cylindrical weights of the same form for 16, 2 and ½ pound.

SIZES: maximum base width or diameter 167, 135, 78, 61, 35 mm; heights 285, 224, 153, 123, 81 mm. Case: 725 x 250 x 300 mm high.

MEASURED WEIGHINGS: 24,643·4, 12,326·7, 3,930·6, 1,967·5, 488·5 g; or 380,297, 190,226, 60,657, 30,362, 7,539 gr. This represents pounds of 7,606 and 7,609 gr respectively for the two hexagonal weights, indicating they were to a pound standard of not less than 7,609 gr or an ounce of 475·6 gr. The three cylindrical weights represent pounds of 7,582, 7,591, 7,539 gr respectively, indicating a pound standard of not less than 7,582 gr, or a 'trose' merchant ounce of 474·4 gr.

These weights are presumably those ordered by the Merchant Guild in January 1676, when the Guild's Treasurer was instructed 'to send to Holland for a 50lb., 25lb., ½ stone, ¼ stone, 2lb., 1lb., ½lb., and ¼lb. brazen weights with a steel balk [beam] for weighing 50lb. weight and

scales belonging thereto to be kept in retentis for adjusting gild brethren's weights': W. B. Cook and David B. Morris (eds), *Extracts from the Records of the Merchant Guild of Stirling AD 1592-1846* (Stirling, 1916), 75. The omission of the stone may simply reflect the fact that Stirling had obtained a stone standard in 1618, and this or a copy will have been held by the Guildry – in the event, however, the stone was obtained as well. In 1724, the Treasurer was instructed 'to send the standard [*recte* 'stand' or set?] of the Guildry's old weights to Edinburgh to be adjusted conform to the weights used for weighing Scotch and Dutch merchant goods': ibid., 89. This may explain the differences in the ounce bases of the weights – the larger weights have been adjusted to the current Scots troye ounce (represented by the Edinburgh Dean of Guild ounce as 475·6 gr), and the smaller weights to the old merchant 'trose' ounce of 474·4 gr (see Chapter **8**).

The case for the weights probably dates from April 1724, when the Treasurer was instructed to replace the 50-pound capacity beam and to obtain 'two small chests, one for holding the standard [*recte* 'stand'?] of the English and the other for the Dutch [weights]': ibid., 90. The 'English' weights are presumably the 1707 avoirdupois bell-weights (Item **58**). There is another empty fitted case which came from the Guildhall and is of closely comparable style (inv. STIGM 19817/15/3): this will accommodate the 1707 weights but the 56-pound must have been separately cased.

Smith Art Gallery and Museum, Stirling (Trustees of the Stirling Smith Art Gallery): Inv. STIGM 19817/15/2.1-2.5; the case is 2.0. Transferred from Stirling Guildhall, 1990.

41 TROYE NESTING WEIGHTS, PILE OF 512 BULLION OUNCES (OR 32 POUNDS), 1687, FOR THE SCOTTISH MINT, EDINBURGH

Part set of seven nested troy cup weights (256 ounce to 4 ounce) in bronze, with a turned finish, stepped base recesses (and with subsequent leaded adjustment, probably of the 1820s or 1830s). The outermost cup with plain banded decoration cut in the upper surface and engraved 'ANᵒ. [crown] DOM: / 1687 / I² [rose] R' and on the opposite side 'I² [crown] R / [thistle]'; the outer surfaces of all the other cups decorated with double line banding. The cups engraved on their upper rims

'CCLVI oz', and similarly the CXXVIII, LXIIII, XXXII, XVI, VIII, IIII oz, and stamped with a crowned rose and a crowned thistle indicating the English and Scottish mints. The four cups, 64 to 8 ounce, are also stamped 'I·M' on the base – perhaps the mark of John Mushatt, Counterwarden at the Edinburgh Mint, 1686-90: R. W. Cochran-Patrick, *Records of the Coinage of Scotland*, 2 volumes (Edinburgh, 1876), I, xxv. A small and badly struck late-nineteenth century Edinburgh city verification stamp on the lead infill on the bases of the 128, 64 and 16 ounce cups. The bases of the 8- and 4-ounce cups have been filed down before adjustment.

SIZES: diameter of completed stack 157 mm, height 133 mm; diameters of inner cups 127, 106, 83, 67, 53, 43 mm.

MEASURED WEIGHTS: The 128-ounce cup weighs 3,905·5 g, or 60,270 gr, representing an ounce of 470·9 gr. This is the only cup from which the rather clumsy addition of lead adjustment in the turned recesses of the bases has been removed. This particular weight was chosen because there was apparently poor adhesion to the bronze surface, with residual flux visible. The lead was excavated by John Hazle of the National Museums Laboratories in May 1992 in conjunction with Huntly House staff. When originally

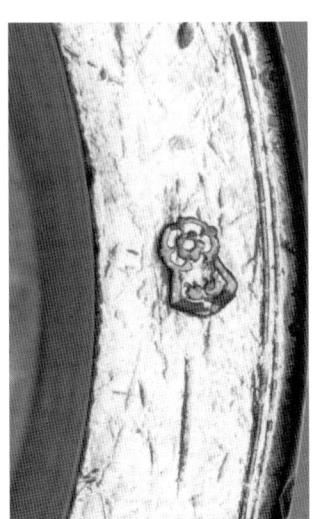

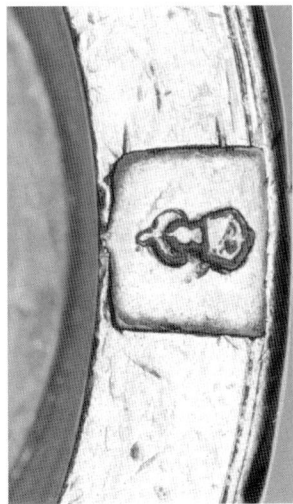

470

encountered, the cups (all with lead adjustment present) were weighed at 7,968·1, 3,983·6, 1,991·7, 995·9, 497·9, 248·8, 124·4 g; or 122,964, 61,475, 30,736, 15,369, 7,684, 3,839, 1,920 gr. This represents a (12-ounce) troy pound of 5,764 gr and an ounce of 480·3 gr, which is very slightly over-standard and may therefore have been used in the preliminary adjustment of weights.

This set of weights is one of the most interesting items to emerge from the recent survey, and it is a rare survivor of an official administration standard. The use of the cypher of James II rather than James VII indicates that the weights were made in England, and therefore presumably in London. The crowned rose was used by the Mint in the Tower of London to mark standards of 1707 and the rose has also been found on standards verified at the Tower in 1588: Allen Simpson, 'English Standard Troy Weight verified at the Mint, 1588', in Silke Ackermann (ed), *Humphrey Cole: Mint, Measurement and Maps in Elizabethan England*, British Museum Occasional Paper no. 126 (London, 1998), 105-6. The thistle has been noted as the distinguishing mark of the Edinburgh Mint in the period after 1604: Joan L. Murray, 'The Organisation and Work of the Scottish Mint 1358-1603', in D. M. Metcalf (ed), *Coinage in Medieval Scotland (1100-1600)*, British Archaeological Reports no. 45 (Oxford, 1977), 155-69, p. 169 note 99.

These weights have therefore been identified as a set provided by the London Mint to replace the previous Scottish Mint standards of 1662, which had not been correctly used: the circumstances are discussed in the final section of Chapter 8. The reference set to the Scottish standard, retained in London in 1662, has not survived, so its precise ounce basis is not known; but the ounce of this replacement set is very similar to the ounce calculated in English troy grains by the mathematician John Napier in 1604, when it became essential to provide an accurate relationship between coinage matters in the newly-united kingdoms. Napier's result

was a Scottish Mint ounce of 471·15 English troy grains. Assuming that the English comparison weights used by Napier derived from the London Mint and had therefore already absorbed the small increase of about 0·2 gr in the size of the troy ounce which occurred in the 1580s (this is discussed in Chapter 4), the equivalence expressed in modern troy grains would be about 471·05 gr. If we use the measurements of this 1687 set made by John Robison in 1792, these showed that the 128-ounce cup was slightly light in relation to the rest of the set, and gave an ounce which was lighter than the mean by 0·1 gr: St Andrews University Library, MS Q171.R8, 'Professor [John] Robison's Commonplace Book', II, p. 408. Hence the mean ounce in modern grains would be 471·0 gr, against 471·05 gr from Napier's calculation.

Robison's results for these weights are difficult to interpret. A weighing of the 16-ounce cup in February 1778 gave an ounce of 471·4 gr: ibid., II, p. 394. However, this seems to have been measured with his own weights, which derived from the Royal Society's standard of the 1730s and can be shown to be about 4 grains in the troy pound light (see Chapter 8 and discussion under Item 95). This would give an equivalent figure in modern grains of about 471·0 gr, which matches our recent value for the Mint ounce. However, a set of measurements of April 1792, made in terms of his own weights and those of a Mr Gardner (probably the pewterer John Gardner of Edinburgh, who is known to have sold coin-weights and scales), cannot readily be related to other standards, but do at least enable the individual weights to be compared: ibid., II, p. 408.

This set of weights clearly puzzled Robison greatly, and he described them as follows in a letter to Bailie William Creech in 1789: 'There is a nest of brass box weights marked *161R87* from CCLVI ounces down to ⅛ ounce. The pound of this set is 7542½ Eng troy grains [hence ounce of 471·4 gr] and corresponds to no weight that I know anything about. That prince [James] perhaps was informed that the Scotch weight was French and has got a standard on that supposition. But it is too light for even the French pound, which is 7560. The English troy is 7680 and the Lanark tron [*recte* troye] is 7618, 19, 20 or 21. This box should be kept apart from all the rest, because it certainly formed on some erroneous calculation and may create mistakes': printed in Meredyth Somerville, *The Standardization of Weights and Measures in Scotland*, Department of Geography, University of Edinburgh, Occasional Publications no. 11 (Edinburgh, 1989), 27; copy of the original in Edinburgh City Archives, MS council minutes, 19 August 1789. From references in the 1680s, the weight is identified as the bullion receipt weight, and the ounce size is described in the final section of Chapter **8** in terms of a Paris coinage ounce (of 472·5 gr), reduced by the coinage remedy.

Exhibited at the Scottish Exhibition of National History, Glasgow, 1911, item 85, at which time ten weights were recorded (although only seven were noted in the Huntly House display location list of 1899).

The Museum of Edinburgh, Edinburgh (City of Edinburgh Council, Museums and Art Galleries): Inv. EDNMG HH 5979/1-7/2000. Transferred from Edinburgh Town Council by 1899.

42

crest of Lanark, which has been somewhat crudely cut from a raised area on the casting.

SIZE: base 128 x 124 mm; height (without ring) 180 mm.

MEASURED WEIGHT: 9,832·7 g, or 151,738 gr; representing a trone pound of 9,484 gr and a troye ounce basis of 474·2 modern grains (474·3 old grains).

Illustrated in A. D. Robertson (ed), *Lanark: the Burgh and Its Councils 1496-1880* (Lanark, 1974), plate opposite p. 177.

South Lanark Museums Service, Hamilton (South Lanark Council): Inv. SLCMD 2001.110. Apparently packed and transferred from the Burgh of Lanark's Weights and Measures Department to the Town Council's offices in 1932, and passing to Clydesdale District Council (Museum Service established 1993) at local government reorganisation in 1975.

42 TRONE RING WEIGHT OF ONE STONE, LATE SEVENTEENTH CENTURY(?), FROM LANARK

Trone weight in cast bronze of tapering square section with iron ring handle, with a square recess in the base. One face carries a stylised double-headed eagle in relief, representing the

43 UNION AVOIRDUPOIS BELL-SHAPED WEIGHTS, STAND OF 112 POUNDS, FROM 56 POUNDS TO ONE POUND, BY JOHN SNART, 1707, FROM LANARK

Set of English weights, originally seven in number (of 56, 28, 14, 7, 4, 2 and 1 pounds), supplied from London in 1707 for the principal Scottish burghs to act as legal standards under the terms of the 1707 Act of Union. These are 'bell' weights, in the shape traditionally associated in England for the 'avoirdupois' series, used from the late middle ages for the weighing of heavy goods, and allow any weight in pounds to be made up by using the weights in different combinations, up to a total of one hundredweight for this series, namely 112 pounds, and using multiples of the 7-pound clove (or one-sixteenth) of the hundredweight. Supplied with these weights were a stack of avoirdupois disc weights, divided in a binary fashion from 8 pounds down to 1 pound and its fractions (see Item **61**), and a pile of troy, or bullion, weights similarly divided from 256 troy ounces to 1 ounce and its fractions (see Item **75**). These weights presumably saw comparatively little use initially, but after the introduction of the Imperial system from 1824 the avoirdupois weights were often retained in use, and some were kept in adjustment as verified working standards until comparatively recently.

As with the other Union weights and measures, 21 sets were received in Edinburgh in 1708. Two sets were retained and the remainder were dispatched to the burghs associated with the various categories of Scottish standards. The weights went to Lanark, which had traditionally been responsible for supplying duplicate standards of the stone. Again, two sets were retained, and the remainder were distributed to other principal burghs who purchased sets. These distributed sets were stamped with the issue mark 'D·LK·', which is presumed to stand for Lanark Dutch weight (for which see Chapter **9**). Of the 21 sets originally supplied, parts of 18 distinct sets have been located. The particular set described in this entry is one of the retained sets at Lanark, so it carries no issue marks.

This set originally comprised the full seven pieces when first identified in 1987, but the 2-pound weight (and the same weight in Item **44**) remained lost after they were stolen in 1990. The weights have been well preserved and are largely in bright original condition. They are cast in bronze, with a turned finish, and have pierced triangular handles rising from a shallow domed shoulder. They have been adjusted down in

43

weight by turning a recess in the base. The waisted exterior engraved between inscribed lines with a crown over the inscription 'ANNA * REGINA * PRIMO * MAII * 1707 * LVI[LI] * AUERDUPOIS *' for the largest weight, and similarly for those of 28, 14 and 7 pounds, 'A * R * PRIMO * MAII * 1707 * AUER[S] *' for the 4- and 2-pound weights, '17 A * R 07' for the 1-pound weight. Exchequer and '[crown] / AR' verification marks are stamped on the shoulders and handles of all the weights, and in addition the four largest weights are stamped '[crown] / IOHN SNART' on the handles.

SIZES: diameters 187, 150, 115, 83, 74, [57], 45 mm; heights 280, 242, 182, 152, 120, [100], 77 mm.

Only the larger bell weights carry the name of the maker, John Snart of London, a prominent member and later Master of the City of London Blacksmiths' Company, whose name also appears on the linear measure of this series. He is discussed in the entry for the 1707 combined yard and English ell measure, Item 10. In all probability, the two other bronze weight types, and indeed the bronze capacity measures, were also made by Snart. Little is known about the conventions for naming the maker on standard weights and measures, although particular London guild companies (such as the Coopers and the Founders) reserved the right of their own members to mark their names on trade weights and measures which were subsequently sealed at Goldsmiths' Hall or the Guildhall. Possibly the large avoirdupois weights (and the linear measures) were the only categories on which Snart, as a member of the Blacksmiths' Company, could normally apply his name. Snart's position was, however, slightly unusual in that he was supplying scales and weights for use at the Mint, in the Tower of London, during the period of reorganisation instituted by Isaac Newton after his appointment as Master of the Mint in 1699. As such he may be considered as holding a Royal appointment, of a type first explicitly acknowledged for Samuel Freeman (II),

as scale and weight maker to the Mint, in the 1730s, and this may have enabled Snart to avoid the restrictions of other guilds. It may therefore be significant that Snart's name appears under an impressed crown. The weights and measures supplied for the Scottish burghs are, however, sealed with Exchequer rather than Mint marks.

Some support for the idea that Snart made all the weights and measures in the Union series can be drawn from an analysis carried out by Dr Paul Wilthew, Head of the Analytical Research Section of the National Museums of Scotland, using X-ray fluorescence spectroscopy, which showed that the alloys of items selected from the former Linlithgow standards at NMS had the same nominal composition. These were all leaded bronzes, containing, however, significant small amounts of nickel, zinc, arsenic and antimony; and this composition was very different from the brass used by Snart for the linear measures: Paul Wilthew and Phyllis Herd, 'Analysis of Five English Weights and Measures issued from the Exchequer in 1707', NMS Analytical Research Section Report 92/26 (C&AR 6245), November 1992. The pieces tested were Item 12, single weights from Items 56, 72 and 82, and the pint from Item 127.

Located with other early weights and measures in December 1986 in a wall safe in Lanark offices used by Clydesdale District Council and examined in detail in February 1987. Moved to another council store, from which material was stolen in May 1990 but recovered after auction at Christie's, Glasgow, Sale of Clocks, Scientific Instruments, etc., 11 July 1990. The material sold (Items 61, 62, 63, 75, 76, 137, 158 and 175) was subsequently restored to Clydesdale District Council. The 2-pound weights from this set and from Item 44 may have been taken as a preliminary to test whether their loss would be detected: they were not recovered. The legal proceedings were reported in the *Lanark and Carluke Advertiser*, 24 August 1990 and the *Lanark Gazette*, 31 August 1990.

Exhibited at the Treasures of South Lanarkshire exhibition, Hamilton, 2001, item 63b.

South Lanarkshire Museum Service, Hamilton (South Lanarkshire Council): Inv. SLCMD 2001.102-107. Apparently packed and transferred from the Burgh of Lanark's Weights and Measures Department to the Town Council's offices in 1932, and passing to Clydesdale District Council (Museum Service established 1993) at local government reorganisation in 1975.

44   UNION AVOIRDUPOIS BELL-SHAPED WEIGHTS, BY JOHN SNART, 1707, FROM LANARK

Another incomplete set of six weights, as Item 43, in similarly fine condition, but also lacking the 2-pound weight, and without Lanark issue marks. On the loss of the 2-pound weight in 1990, see Item 43.

South Lanarkshire Museum Service, Hamilton (South Lanarkshire Council): Inv. SLCMD 2001.96-101. Apparently packed and transferred from the Burgh of Lanark's Weights and Measures Department to the Town Council's offices in 1932, and passing to Clydesdale District Council (Museum Service established 1993) at local government reorganisation in 1975.

45   UNION AVOIRDUPOIS BELL-SHAPED WEIGHTS, BY JOHN SNART, 1707, AT BANFF

Another set of seven weights, as Item 43. Countermarked 'D·LK·' for issue by Lanark.

Warrants were issued by Banff Town Council for receiving the weights and measures from the burghs of Stirling, Linlithgow and Lanark on 22 January 1709; on 16 July various capacity measures arrived from Stirling and Linlithgow,

and, 'Now arrived the weights and measures shipt by the town's agent in Edinburgh, on board Robert Duncan's ship: – two sets of brass weights from Lanrick …': W. Cramond, *Annals of Banff*, 2 volumes, New Spalding Club nos. 8 and 10 (Aberdeen, 1891-3), I, 180, 352.

Banff Museum, Banff (Aberdeenshire Council, Aberdeenshire Heritage, Aberdeen): Inv. PEHMS B2360/1-7. Purchased by the Banffshire Field Club from Banff Town Council at some point between 1880 and 1919, and then presented to Banff Museum in 1919: A. E. Mahood, *Banff and District* (Banff, 1919), 272-3.

46   UNION AVOIRDUPOIS BELL-SHAPED WEIGHTS, BY JOHN SNART, 1707, AT CRAIL

Another incomplete set of two weights, as Item 43, but comprising only the 56- and 28-pound weights. Countermarked 'D·LK·' for issue by Lanark.

Crail Museum and Heritage Centre, Crail (Crail Museum Trust, previously Crail Preservation Society): Inv. CRLPS 1978.9 i and j. Transferred from Crail Town Council before local government reorganisation in 1975; museum opened in 1979.

47   UNION AVOIRDUPOIS BELL-SHAPED WEIGHTS, BY JOHN SNART, 1707, AT CUPAR

Another incomplete set of one weight, as Item 43, but comprising only the 28-pound weight. Inscribed in error 'XXVII' pounds. Countermarked 'D·LK·' for issue by Lanark.

Exhibited at the St Andrews Preservation Trust Exhibition of Weights and Measures, St Andrews, 1968, item 34.

East Fife Council Museums Service, Cupar (Fife Council, Fife Museums East): Inv. CUPMS 1984.173. Transferred by Cupar Town Council to North East Fife District Council at local government reorganisation in 1975 (Museums Service established in 1983).

48 UNION AVOIRDUPOIS BELL-SHAPED WEIGHTS, BY JOHN SNART, 1707, AT DUMFRIES

Another incomplete set of six weights, as Item 43, but lacking the 1-pound weight. Countermarked 'D·LK·' for issue by Lanark. With lead adjustment in the base recesses and carrying later verification marks.

Dumfries and Galloway Council, Trading Standards office, Dumfries (Dumfries and Galloway Council, Environmental and Consumer Services).

49 UNION AVOIRDUPOIS BELL-SHAPED WEIGHTS, BY JOHN SNART, 1707, AT DUNDEE

Another set of seven weights, as Item 43. Countermarked 'D·LK·' for issue by Lanark, and engraved 'GUILDRY'. The original stamped markings on the handles, shoulders and bases of the weights have been filed off (except for the issue stamps on the base of the 56-pound weight), and compensating adjustment plugs and leading added in the base recesses. Single Exchequer and William IV verification marks of c.1835 have been added to the shoulders, and the denominations stamped on the handles, as '56 lb. AVOIR.', etc. The 56-pound weight with dated 1974 and 1985 verification stamps.

Dundee City Council, Trading Standards office, Dundee (Dundee City Council, Environmental and Consumer Protection). It is likely that these weights formed part of the group of 16 early weights initially transferred to Dundee Museum on its foundation in 1873 by the Guildry Corporation and Magistrates of Dundee, Inv. 1874.314-321, the others being the nine weights comprising Items 39, 67 and 79. They were presumably returned for use in the Weights and Measures Department before 1892, as they were not included in the 'Old Dundee' exhibition that year.

50 UNION AVOIRDUPOIS BELL-SHAPED WEIGHTS, BY JOHN SNART, 1707, FROM DUNFERMLINE(?)

Another incomplete set of five weights, as Item 43, but lacking the 7- and 1-pound weights. Countermarked 'D·LK·' for issue by Lanark. Screw-capped adjustment plugs have been drilled in the bases of all three weights (with an additional lead plug in the handle of the 14-pound weight) to bring them up to weight at the introduction of the Imperial system, and the handles are stamped '56.LB^S' etc, and 'Imp^l.' or 'Imperial'.

When initially examined in 1987, these weights were with the former Fife County and Burgh standards at the Consumer Protection Department, Fife Regional Council, Glenrothes. Since this collection also included the early pint (Item 115) associated with Dunfermline, these weights will also tentatively be identified as the Dunfermline set.

Fife Council, Trading Standards office, Glenrothes (Fife Council, Trading Standards Service).

SIZES: (the final 8-dram weight from Item 72) diameters 145, 118, 97, 79, 65, 52, 40, 34, [25] mm; heights 32, 24, 18, 14, 11, 9, 7, 5, [4] mm.

Stolen from the former Clydesdale District Council's premises in Spring 1990, but recovered after its sale at Christie's, Glasgow, 11 July 1990, part of lot 119, where it was referred to as 'reproduction'. See Item 43.

South Lanarkshire Museum Service, Hamilton (South Lanarkshire Council): Inv. SLCMD 2001.154-147. Apparently packed and transferred from the Burgh of Lanark's Weights and Measures Department to the Town Council's offices in 1932, and passing to Clydesdale District Council (Museum Service established 1993) at local government reorganisation in 1975.

62   UNION AVOIRDUPOIS STACKING WEIGHTS, 1707, FROM LANARK

Another incomplete set of eight weights, as Item 61, but lacking the 2-ounce weight and those less than 8 drams, and in similarly fine condition. This is the second set retained by Lanark, and also lacks the 'D·LK·' countermark.

Stolen from the former Clydesdale District Council's premises in Spring 1990, but recovered after its sale at Christie's, Glasgow, 11 July 1990, part of lot 119, where it was referred to as 'reproduction'. See Item 43.

South Lanarkshire Museum Service, Hamilton (South Lanarkshire Council): Inv. SLCMD 2001.145-142, 171, 141-139. Apparently packed and transferred from the Burgh of Lanark's Weights and Measures Department to the Town Council's offices in 1932, and passing to Clydesdale District Council (Museum Service established 1993) at local government reorganisation in 1975.

63   UNION AVOIRDUPOIS STACKING WEIGHTS, 1707, FROM LANARK

Another incomplete set, as Item 61, but with only the 8-pound weight now surviving. This is presumably from an unissued set, but lacks the 'D·LK·' countermark. The rim with four, rather than the usual eight, verification marks.

Stolen from the former Clydesdale District Council's premises in Spring 1990, but recovered after its sale at Christie's, Glasgow, 11 July 1990, part of lot 119, where it was referred to as 'reproduction'. See Item 43.

South Lanarkshire Museum Service, Hamilton (South Lanarkshire Council): Inv. SLCMD 2001.146. Apparently packed and transferred from the Burgh of Lanark's Weights and Measures Department to the Town Council's offices in 1932, and passing to Clydesdale District Council (Museum Service established 1993) at local government reorganisation in 1975.

64   UNION AVOIRDUPOIS STACKING WEIGHTS, 1707, AT BANFF

Another incomplete set of six weights, as Item 61, but lacking the weights of less than 4 ounces. Countermarked 'D·LK·' for issue by Lanark.

Warrants were issued by Banff Town Council for receiving the weights and measures from the burghs of Stirling, Linlithgow and Lanark on 22 January 1709; on 16 July various capacity measures arrived from Stirling and Linlithgow, and, 'Now arrived the weights and measures shipt by the town's agent in Edinburgh, on board Robert Duncan's ship: – two sets of brass weights from Lanrick …': W. Cramond, *Annals of Banff*, 2 volumes, New Spalding Club nos. 8 and 10 (Aberdeen, 1891-3), I, 180, 352.

Banff Museum, Banff (Aberdeenshire Council, Aberdeenshire Heritage, Aberdeen): Inv. PEHMS B2362/1-6. Purchased by the Banffshire Field Club from Banff Town Council at some point between 1880 and 1919, and then presented to Banff Museum in 1919: A. E. Mahood, *Banff and District* (Banff, 1919), 272-3.

65 UNION AVOIRDUPOIS STACKING WEIGHTS, 1707, AT CUPAR

Another incomplete set of four weights, as Item **61**, but lacking the weights of less than 8 ounces. Countermarked 'D·LK·' for issue by Lanark. The 8-pound weight is stamped 'C / D G' for Cupar, Dean of Guild.

The 'C / D G' mark is incorrectly identified by Carl Ricketts and John Douglas, *Marks and Marking of Weights and Measures of the British Isles* (Taunton, 1996), 122, as the mark for the Royal Burgh of Culross. However, a Cupar standard Imperial bushel at East Fife Museums Service (Inv. CUPMS 1984.161) is clearly marked 'C. D. G. / CUPAR'.

65

The 8-ounce weight exhibited at the St Andrews Preservation Trust Exhibition of Weights and Measures, St Andrews, 1968, item 25.

East Fife Museums Service, Cupar (Fife Council, Fife Museums East): Inv. CUPMS 1984.180, 179, 178, 177. Transferred from Cupar Town Council to North East Fife District Council at local government reorganisation in 1975 (Museums Service established in 1983).

66 UNION AVOIRDUPOIS STACKING WEIGHTS, 1707, AT DUMFRIES

Another incomplete set of two weights, as Item **61**, but now comprising only the weights of 8 and 4 pounds. Countermarked 'D·LK·' for issue by Lanark.

Exhibited at the Scottish Exhibition of National History, Glasgow, 1911, item 56, described as 'Four iron weights: Linlithgow stamp, date 1707'. (A list of items in Dumfries Burgh Museum which were on loan from Dumfries Town Council records the continued presence of only four weights in this set in 1936: Dumfries Museum Archives.)

Dumfries Museum, Dumfries (Dumfries and Galloway Council, Museums Service): Inv. DUMFM 1936.31.9 (previously acq. 207.k). Acquired before 1911 from the Dumfries Burgh Weights and Measures Department. The museum, based at the Maxwelltown Observatory, Dumfries, was operated by the Dumfries and Maxwelltown Observatory Society until 1936.

67 UNION AVOIRDUPOIS STACKING WEIGHTS, 1707, AT DUNDEE

Another incomplete set of two weights, as Item **61**, but now comprising only the weights of 4 and 2 pounds. Countermarked 'D·LK·' for issue by Lanark.

McManus Galleries, Dundee (Dundee City Council, Arts and Heritage): Inv. DUNMG 1955.315. Transferred in 1955 from the Weights and Measures Department, Dundee Corporation. However, they were previously part of the group 1874.222-237 (with Items **39**, **49** and **67**), transferred to Dundee Museum on its foundation in 1873 by the Guildry Incorporation and Magistrates of Dundee, and apparently withdrawn (with Item **49**) before 1892 to be used as working standards in the Weights and Measures Department.

68 UNION AVOIRDUPOIS STACKING WEIGHTS, 1707, AT EDINBURGH

Another incomplete set of four weights, as Item **61**, but lacking the weights of less than 1 pound. This is one of two sets retained by Edinburgh when the remainder were sent to Lanark for distribution: it therefore does not carry the Lanark issue mark. The base of the 1-pound weight has been stamped 'RAR'.

The Museum of Edinburgh, Edinburgh (City of Edinburgh Council, Museums and Art Galleries): Inv. EDNMG HH 5981/1-4/2000. Transferred from Edinburgh Town Council by 1899, when the 8-ounce weight was still present.

69 UNION AVOIRDUPOIS STACKING WEIGHTS, 1707, AT EDINBURGH

Another incomplete set, as Item **61**, but with only the 4-pound weight now surviving. This is the residue of the other set retained by Edinburgh when the remainder were sent to Lanark for distribution: it therefore does not carry the Lanark issue mark. Adjustment lead has been added to the base.

The Museum of Edinburgh, Edinburgh (City of Edinburgh Council, Museums and Art Galleries): Inv. EDNMG HH 5981/5/2000.

70 UNION AVOIRDUPOIS STACKING WEIGHTS, 1707, AT GLASGOW

Another incomplete set of three weights, as Item **61**, but lacking the weights of less than 2 pounds. Countermarked 'D·LK·' for issue by Lanark.

People's Palace Museum, Glasgow (City of Glasgow Council, Museums and Art Galleries): Inv. GLAMG 1902.84.k, 84.l, 1897.161.c. The first two presented in 1902 by Alexander Wood, Partick (see Item **101**); the other understood to have been transferred from the city's Weights and Measures Department.

70

71 UNION AVOIRDUPOIS STACKING WEIGHTS, 1707, AT HADDINGTON

Another incomplete set of six weights, as Item **61**, but lacking the weights of less than 4 ounces. Countermarked 'D·LK·' for issue by Lanark. All the weights have subsequent lead adjustment, but in addition the 8- and 4-ounce weights are stamped with the Haddington burgh marks 'MD', and the Haddington seal (a horned goat) over the letters 'GS'.

The 2-pound weight is illustrated in Sharon Barron, *A History of Shops and Shopping in Scotland* (Hamilton, 2002), 29.

East Lothian Museums Service, Haddington (East Lothian Council): Inv. HADDM 1996.96.6-1 (previously a.14). Transferred from Haddington Town Council *c.*1930 and displayed in the former public library. To East Lothian District Council at local government reorganisation in 1975 (Museums Service established in 1990).

72 UNION AVOIRDUPOIS STACKING WEIGHTS, 1707, FROM LINLITHGOW

Another incomplete set of nine weights, as Item **61**, but lacking the weights of less than 8 drams. Countermarked 'D·LK·' for issue by Lanark.

The 1-pound weight from this set was one of the Union pieces whose composition was analysed in 1992: see the discussion under Items **10** and **43**.

National Museums of Scotland, Edinburgh (Trustees of the NMS): Inv. NMS T.1973.140.a-i. Presented in May 1973 by the Council of the Royal Burgh of Linlithgow to the Royal Scottish Museum (now part of NMS) in advance of local government reorganisation in 1975.

73 UNION AVOIRDUPOIS STACKING WEIGHTS, 1707, AT STIRLING

Another incomplete set of eight weights, as Item **61**, but lacking the weights of less than 1 ounce. Countermarked 'D·LK·' for issue by Lanark. Subsequent lead adjustment is present on the bases of all the weights, and the largest four are stamped with the verification mark '[crown] / VR / 362' (Burgh of Stirling, post-1880).

Smith Art Gallery and Museum, Stirling (Trustees of Stirling Smith Art Gallery): Inv. STIGM 2972-2979 (previously UD.23-30). Presumably transferred from Stirling Town Council, and first recorded in the *Catalogue of Collections … of the Smith Institute Stirling*, third edition (Stirling, 1934), 166.

74 UNION AVOIRDUPOIS STACKING WEIGHTS, 1707, FROM WIGTOWN

Another incomplete set, as Item **61**, but now comprising only the 8- and 4-pound weights. Countermarked 'D·LK·' for issue by Lanark.

Stranraer Museum, Stranraer (Dumfries and Galloway Council, Museums Service, Dumfries): Inv. WIWMS 1988.470 and 1986.466. Transferred from Wigtown County Council at local government reorganisation in 1975.

75 UNION TROY NESTING WEIGHTS, PILE OF 512 OUNCES, FROM 256 OUNCES DOWNWARDS, 1707, FROM LANARK

Alongside the avoirdupois trade standards (Items **43** and **61**) distributed after the 1707 Act of Union, there were also standards of the English troy system, used principally for gold and silver. Like the avoirdupois weights, the troy weights were similarly distributed from Lanark and the larger weights carry the 'D·LK·' countermark.

The troy weights are in what had become the standard English pattern of un-lidded cup weights, hanging from their rims and with a uniform upper surface carrying the denomination and authorisation marks, and with inscriptions also on the outer faces of the majority of the cups. The full set comprised 14 weights, in binary divisions down to a sixteenth of an ounce and including an additional sixteenth-ounce plug in the centre. This brought the total weight to 512 ounces, or twice the weight of the outer cup, enabling the internal consistency of the set to be checked since each cup weighed the same as the sum of the cups it contained. Of the twelve sets that survive, only two are virtually complete – at Perth (Item **83**) and at Stirling (Item **85**). The Stirling set is contained in an early (and perhaps original) folding wooden case. Part of a simpler leather-bound case survives at Kirkcaldy (Item **81**). One further set has been recorded recently, but is no longer located – three weights remaining from the St Andrews set were exhibited at the St Andrews Preservation Trust Exhibition of Weights and Measures, St Andrews, 1968, item 10.

This particular set, now of only eight weights, lacks the weights of less than 2 ounces, although described as 'complete' when exhibited at the Glasgow International Exhibition, 1901, item 2475. It is one of the two sets retained by Lanark at the time of distribution, and it therefore does not carry the Lanark distribution mark, which was normally stamped on the top rim of the three outer cups. The weights have been well preserved and are largely in bright original condition. They are cast in bronze and have a turned finish. The outer cup is decorated with incised lines, above and below. The outer surface of the cups down to 2 ounces engraved with a crown over 'A:R', and with the additional flanking inscription 'PRIMO MAII / AN^O. D^ONI. / 17. .07 / A^O REGNI VI.º' on the 256- to 8-ounce weights; the weights for 1 ounce and below are unmarked. The outer cup stamped on the upper rim 'CCLVI OZ', and similarly for the weights down to 4 ounces; with the 'TR' (troy) monogram, together with Exchequer and '[crown] / AR' verifications, normally down to the 2-ounce weight. Where present, the 1-ounce weight is only marked 'I' on the rim, and smaller weights are unmarked.

The Union troy weights are unlikely to have seen much practical use initially, and this is presumably why fewer sets were retained or survived beyond their replacement by Imperial troy standards of a different approved pattern in the early nineteenth century. Even in the case of Edinburgh, only one set is detectable in the council's inventory of weights and measures in 1800, where it is perhaps the 'set of Troy Weights nine in number from one Stone [i.e. in the Scots sense of 16 × 16 = 256 ounces] downwards in the Custody of Mr John Hutton master of Police': Edinburgh City Archives, Dean of Guild Court sederunt book, 23 October 1805. This set no longer survives.

Although the principal impetus for issuing these weights was to meet the requirement of the Act of Union, at least two sets are known in English use. Considering John Snart's close connection

75

with the Mint in London, it is perhaps not surprising that there was a set at the Mint, described in 1758 as 'of the 6th [year] of Queen Anne, 1707', and that a successor of Snart's as scale and weight maker to the Mint had another (perhaps originally Snart's own set) 'by which he makes weights for sale': 'Report from the Committee [chaired by Sir John Proby, first Baron Carysfort], appointed to inquire into the Original Standards of Weights and Measures in this Kingdom', *Reports from Committees of the House of Commons*, 3 volumes (1737-1801), II, 437.

SIZES (based on the nearly complete set, Item **85**, and estimating the size of the central plug): diameters 165, 130, 106, 83, 67, 54, 42, 35, 29, 23, 18, 15, 12, [10] mm; heights 132, 100, 77, 61, 46, 35, 26, 20, 16, 12, 9, 6, 4, [3] mm.

Stolen in 1990, and sold at auction by Christie's, Glasgow, 11 July 1990, lot 117, where they were referred to as 'reproduction'. See Item **43**.

Exhibited at the Glasgow International Exhibition, 1901, item 2475, and probably also the set exhibited at the Scottish Exhibition of National History, Glasgow, 1911, item 87 (described only as 'Set of Weights, date 1707').

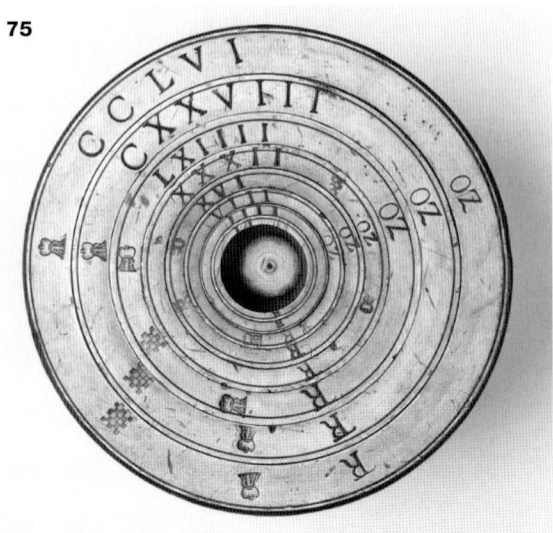

75

South Lanarkshire Museum Service, Hamilton (South Lanarkshire Council): Inv. SLCMD 2001.162-155. Apparently packed and transferred from the Burgh of Lanark's Weights and Measures Department to the Town Council's offices in 1932, and passing to Clydesdale District Council (Museum Service established 1993) at local government reorganisation in 1975.

76   UNION TROY NESTING WEIGHTS, 1707, FROM LANARK

Another incomplete set of eight weights, as Item **75**, but lacking the weights of less than 2 ounces, and in similarly fine condition. This is the second set retained by Lanark, and also lacks the Lanark countermark.

Stolen in 1990, and sold at auction by Christie's, Glasgow, 11 July 1990, lot 118 (illustrated), where they were referred to as 'reproduction'. See Item **43**.

South Lanarkshire Museum Service, Hamilton (South Lanarkshire Council): Inv. SLCMD 2001.163-170. Apparently packed and transferred from the Burgh of Lanark's Weights and Measures Department to the Town Council's offices in 1932, and passing to Clydesdale District Council (Museum Service established 1993) at local government reorganisation in 1975.

77   UNION TROY NESTING WEIGHTS, 1707, AT BANFF

Another incomplete set of eight weights, as Item **75**, but lacking the weights of less than 2 ounces. Countermarked 'D·LK·' for issue by Lanark.

Warrants were issued by Banff Town Council for receiving the weights and measures from the burghs of Stirling, Linlithgow and Lanark on 22 January 1709; on 16 July various capacity measures arrived from Stirling and Linlithgow, and, 'Now arrived the weights and measures shipt by the town's agent in Edinburgh, on board Robert Duncan's ship:– two sets of brass weights from Lanrick …': W. Cramond, *Annals of Banff*, 2 volumes, New Spalding Club nos. 8 and 10 (Aberdeen, 1891-3), I, 180, 352.

Banff Museum, Banff (Aberdeenshire Council, Aberdeenshire Heritage, Aberdeen): Inv. PEHMS B2361/1-8. Purchased by the Banffshire Field Club from Banff Town Council at some point between 1880 and 1919, and then presented to Banff Museum in 1919: A. E. Mahood, *Banff and District* (Banff, 1919), 272-3.

78   UNION TROY NESTING WEIGHTS, 1707, AT DUMFRIES

Another incomplete set of seven weights, as Item 75, but lacking the weights of less than 4 ounces. Countermarked 'D·LK·' for issue by Lanark.

Exhibited at the Glasgow International Exhibition, 1901, part of item 2451, and at the Scottish Exhibition of National History, Glasgow, 1911, item 88.

Dumfries Museum, Dumfries (Dumfries and Galloway Council, Museums Service): Inv. DUMFM 1936.31.10 (prev. acq. 207.l). Acquired before 1911 from the Dumfries Burgh Weights & Measures Department. The museum, based at the Maxwelltown Observatory, Dumfries, was operated by the Dumfries and Maxwelltown Observatory Society until 1936.

79   UNION TROY NESTING WEIGHTS, 1707, AT DUNDEE

Another incomplete set of three weights, as Item 75, but lacking the weights of less than 64 ounces. Countermarked 'D·LK·' for issue by Lanark.

Exhibited at the 'Old Dundee' exhibition, Dundee, 1892, part of item 857, and probably also at the Scottish Exhibition of National History, Glasgow, 1911, item 86.

McManus Galleries, Dundee (Dundee City Council, Arts and Heritage): Inv. DUNMG 1978.1084, but previously part of the group 1874.322-337. Presented by the Guildry Incorporation and Magistrates of Dundee, November 1873 .

80   UNION TROY NESTING WEIGHTS, 1707, AT HADDINGTON

Another nearly complete set of nine weights, as Item 75, but lacking the weights of less than 1 ounce. Countermarked 'D·LK·' for issue by Lanark.

East Lothian Museum Service, Haddington (East Lothian Council): Inv. HADDM 1996.95.9-1 (previously A.15). Transferred from Haddington Town Council c.1930 and displayed in the former public library. To East Lothian District Council at local government reorganisation in 1975 (Museums Service established in 1990).

81   UNION TROY NESTING WEIGHTS, 1707, AT KIRKCALDY

Another incomplete set of five weights, as Item 75, but lacking the weights of less than 16 ounces. Countermarked 'D·LK·' for issue by Lanark. The set has a baize-lined leather-covered tapered cylindrical wooden container, now lacking its lid.

Kirkcaldy Museum, Kirkcaldy (Fife Council): Inv. KIRMG 1925.90.6-10; the container being 1925.90.5. Transferred by Kirkcaldy Town Council, 1925.

81

82 UNION TROY NESTING WEIGHTS, 1707, FROM LINLITHGOW

Another nearly complete set of ten weights, as Item **75**, but lacking the weights of less than ½ ounce. Countermarked 'D·LK·' for issue by Lanark.

The 8-ounce weight from this set was one of the Union pieces whose composition was analysed in 1992: see the discussion under Items **10** and **43**.

National Museums of Scotland, Edinburgh (Trustees of the NMS): Inv. NMS T.1973.139.a-j. Presented in May 1973 by the Council of the Royal Burgh of Linlithgow to the Royal Scottish Museum (now part of NMS) in advance of local government reorganisation in 1975.

83 UNION TROY NESTING WEIGHTS, 1707, AT PERTH

A virtually complete set of 13 weights, as Item **75**, but lacking only the central sixteenth-ounce plug. Countermarked 'D·LK·' for issue by Lanark.

Examined by Adam Anderson, Rector of Perth Academy, and mentioned by him in 'Weight of the Dutch Pound Troy', in *Edinburgh Philo-*

sophical Journal, 4 (1820-1), 449-50. See Item **85**. Illustrated in L. Burrell, 'The Standards of Scotland', in *The Monthly Review: the Journal of the Institute of Weights and Measures Administration*, 69 (1961), 58. The avoirdupois bell weights presumably remained with Perth Guildry in 1862 and they have not been found. It is possible, however, that two of the avoirdupois stacking weights may have been transferred as items 109 and 110 in the Literary and Antiquarian Society's inventory, but they are not located.

Perth Museum, Perth (Perth and Kinross Council): Inv. PERMG 103. Presented by Perth Town Council in 1862 to the Literary and Antiquarian Society of Perth (established 1784), whose collection from 1914 onwards formed part of Perth Museum.

84 UNION TROY NESTING WEIGHTS, 1707, AT SELKIRK

Another nearly complete set of nine weights, as Item **75**, but lacking the weights of less than 1 ounce. Countermarked 'D·LK·' for issue by Lanark.

Scottish Borders Council, Museum and Gallery Service, Selkirk (Scottish Borders Council, Life-

presumably the ounce of the Amsterdam commercial weight, normally accepted as 476·6 gr.

Swinton did not give a troye size for Dunbartonshire, other than infer it was his general troye pound of 7,616 gr; and by the time of the 1827 Dunbartonshire jury report the pound had become a conventional one of 17½ avoirdupois ounces, or 7,656 gr: George Buchanan, *Tables for Converting the Weights and Measures hitherto in use in Great Britain …* (Edinburgh, 1829), 191.

West Dunbartonshire Museum Collection, Library Headquarters, Dumbarton (West Dunbartonshire Council, Education and Cultural Services): Inv. DMBDM 1982.82. Displayed in the former burgh library, perhaps from the 1930s.

90 TRONE WEIGHTS OF EIGHT POUNDS AND TWO POUNDS, 1767, AT PAISLEY

Part of a set of trone weights, probably from 16 pounds to 1 pound, but now comprising only the 8-pound and 2-pound weights, in cast brass. Tapering cylindrical weights with flattened spheroid knob handles. The larger weight

90

stamped round the top of the body of the weight 'VIII POWND TRON 1767', and similarly for the 2-pound weight; and the handle impressed 'IW'. This maker's mark is unidentified: it may conceivably represent Andrew Watson, John Willson, James Whytlaw or James Witherspoon, who became free of the Glasgow Hammermen between 1724 and 1729, but no minutebooks survive for the period 1733-75: Harry Lumsden and P. Henderson Aitken, *History of the Hammermen of Glasgow* (Paisley, 1912), 293.

SIZES: diameters 91, 60 mm; heights: 147, 84 mm.

MEASURED WEIGHTS: 5,077·9, 1,266·9 g; or 78,362, 19,551 gr. This implies a trone pound of not less than 9,795 gr. This is not a pound of 20 troye ounces (which would give an ounce of 490 gr). However, Swinton gave the Renfrewshire trone pound (for butcher-meat, fish, cheese, butter and tallow) as a conventional avoirdupois multiple of 22½ ounces; refined to 22⁴/₁₀ ounces, or 9,800 gr, by the 1827 Renfrewshire jury, probably informed by a report from James Cleland: [John Swinton], *A Proposal for Uniformity of Weights and Measures in Scotland* (Edinburgh, 1779), 114; George Buchanan, *Tables for Converting the Weights and Measures hitherto in use in Great Britain …* (Edinburgh, 1829), 250.

Paisley Museum and Art Galleries, Paisley (Renfrewshire Council): Inv. PSYMG 1908/688a and b. Transferred from Paisley Town Council in December 1908.

91 TRONE BAR-HANDLED WEIGHTS OF EIGHT POUNDS AND FOUR POUNDS, BY COLIN BLYTH, 1786, FOR PAISLEY

Part of a set of trone weights, probably from 16 pounds to 1 pound, but now comprising only the 8-pound and 4-pound weights, in cast bronze. Tapering cylindrical weights with circular-

91

section T-bar handles, and with a slight turned recess in the base. Engraved between two incised lines on the body of the weights 'PAISLEY STANDART / 1786'; engraved on the shoulder 'VIII TRONE' and similarly for the 4-pound weight. The seal of Glasgow stamped twice on the shoulder, three times on the base, and on both ends of the bar handle; and the top of the handle stamped 'C · BLYTH'. For Blyth, see the entry in Appendix G.

SIZES: base diameters 83, 70 mm; heights 163, 120 mm.

MEASURED WEIGHTS: 5,078·4, 2,536·8 g; or 78,370, 39,148 gr. This implies a trone pound of not less than 9,796 gr, which is the Renfrewshire trone pound of 22·4 avoirdupois ounces, or 9,800 gr, for which see Item **90**.

Paisley Museum and Art Galleries, Paisley (Renfrewshire Council): Inv. PSYMG 1908/688c and d. Transferred from Paisley Town Council in December 1908.

**92 TROYE WEIGHT OF ONE POUND,
*c.*1790, FOR EDINBURGH**

Brass cylindrical weight with knob handle and turned base recess, retained in association with the Edinburgh 1618 standard troye set. Engraved on the side (in a late eighteenth-century hand) '1 Pound Lanark Troy / 7620 Gr. English Troy'. Made for and adjusted by John Robison, and representing his computed value for the 1618 troye pound.

SIZE: diameter 44 mm; height 60 mm.

MEASURED WEIGHT: 493·4g, or 7,614 gr; indicating an ounce of 475·9 gr. From the bias of Robison's own weights (see Chapter **9** and the entry for Item **95**), this would have been measured by him at 5 to 6 grains heavier, namely at 7,619 to 7,620 gr, and hence presumably at 7,620 gr when new.

This is almost certainly the Lanark troye pound weight in John Robison's personal possession that was purchased with other items after his death by the Edinburgh Dean of Guild on the town council's behalf. The council recorded in June

494

1805 that, 'having seen in the list of Articles for sale which belonged to the late Professor John Robison three standard Measures and a Lanark troy pound weight adjusted by the Professor, he thought it proper that these should be in the possession of the City … [and decided to] purchase them at the Auction at the price of £11. 12. - ¾. Of which the magistrates and Council Approved … And Appointed these Articles to be lodged in the Dean of Guild Office and to be added to the Inventory of adjusted weights and measures belonging to the Community': Edinburgh City Archives (ECA), MS council minutes, 12 June 1805. The items were perhaps withdrawn before the auction and may be the missing lot 125 in the printed *Catalogue of the Optical and Astronomical Instruments … sold* by Elliott on 28 May 1805. An annotated copy of this is in NAS GD113/5/342/74. The weight is not separately identified in the annual inventory of the Dean of Guild's weights and measures in the council minutes. Robison requested authorisation to have a pound weight made by the Edinburgh brass founder John Milne, to be marked 'Lanark Pound – 7620 grains Troy', which he would adjust, but this was to be a bell weight which was to form the basis for a new set of standards for the City, which are Item **95**: James Cleland, *Exemplification of the Weights and Measures of the City of Glasgow* (Glasgow, 1822), 77; the original transcription of Robison's letter of 25 February 1800 to James Jackson, Edinburgh Dean of Guild, is in ECA, MS council minutes, 16 September 1801.

Illustrated in H. J. Chaney, *Our Weights and Measures* (London, 1879), 29.

The Museum of Edinburgh, Edinburgh (City of Edinburgh Council, Museums and Art Galleries): Inv. EDNMG HH 5978/5/2000. Transferred from Edinburgh Town Council by 1899, and earlier purchased by the Edinburgh Dean of Guild at the sale of Robison's effects in 1805.

93

93   WEIGHTS FOR THE WATER CONTENT OF CAPACITY MEASURES, FROM ONE PINT TO HALF-MUTCHKIN, *c.*1800, FOR EDINBURGH

Set of four check-weights for the weight of water in Scots standard capacity measures (pint to half-mutchkin), accompanying a similar set (Item **105** below) for the English measures. Rounded bell-shape in lead, and cast around iron keys, the loops of which protrude to form the handles of the weights. Stamped with a shield with the Scottish lion *rampant* (the die damaged at the lower point of the shield), twice on the shoulder and twice on the base (only once on the base for the two smallest weights). Stamped on the waist of the weights 'SCOTCH PINT / 26200 G^R', 'CHOPIN / 13100 G^R', 'MUTCHKIN / 6550 G^R', '½ MUTCHKIN X3275 G^R'.

SIZES: diameters 62, 47, 41, 33 mm; heights 115, 92, 77, 58 mm.

MEASURED WEIGHTS: 1,699·1, 848·2, 424·0, 211·7 g; or 26,221, 13,089, 6,543, 3,267 gr.

These items are marked with conventional weights, which are fractions of the Scots pint size deduced by John Robison as 26,200 gr. Since Robison's own weights can be shown to have been light by about 6 grains in the troye pound of 7,614 gr, the equivalent modern troy masses will be about 26,180, 13,090, 6,545 and 3,272 gr (see Chapter 9 and Item 95). The three smaller weights exhibit some slight wear, but the surprise is that the weight for the pint's contents is about 40 gr larger than indicated, and it does not show signs of having been altered. The suspicion is that Robison recognised that his computed value for the pint (which depended on the weight and capacity sizes being fully integrated at the 1618 Assize) was too small, and so this weight had to be made larger to balance the contents of an authentic standard such as the Edinburgh 1618 pint.

Robison's 26,200-gr size was certainly established by 1788, and it is possible that these weights may have been produced under his direction by Thomas Simpson, the Edinburgh pewterer who was then the adjuster of the town's weights, and who held the 1618 weight set (Item 37). They are described as having been adjusted by Robison in 1800 in the Dean of Guild's inventory of weights and measures in 1818: Edinburgh City Archives (ECA), MS council minutes, January 1818. The computed sizes for the Scots liquid measures were rehearsed by Robison in a letter of August 1801 relating to the construction of new Scots and English capacity measures for the Edinburgh Dean of Guild (Items 205 and 232); at which time Robison also supplied four check-weights for the Scots measures (the present set) and three for the English measures (Item 105), requesting that they be 'stamped (not engraved)' with their weights and noting 'I have put a very faint stamp of the Scotch Arms on every part that could be filed; this protects their authenticity': James Cleland, *Exemplification of the Weights and Measures of the City of Glasgow* (Glasgow, 1822), 86; copy of the original is in ECA, MS council minutes, 16 September 1801. The check-weights

are recorded in the Dean of Guild's inventory of weights and measures simply as 'lead check-weights' of these four sizes: ECA, MS council minutes, 20 January 1817. They were not yet present in the inventory for 1800, which is printed in Meredyth Somerville, *The Standardization of Weights and Measures in Scotland*, Department of Geography, University of Edinburgh, Occasional Publication no. 11 (Edinburgh, 1989), 28-9.

The Museum of Edinburgh, Edinburgh (City of Edinburgh Council, Museums and Art Galleries): Inv. EDNMG HH 5982/1-4/2000. Transferred from Edinburgh Town Council by 1899.

94   TROYE WEIGHT OF ONE STONE, c.1800, FOR EDINBURGH

Bronze cylindrical weight with domed top, bar handle and turned base recess, associated with the incomplete Edinburgh 1618 standard troye set. Engraved on the side (in an early 19th-century hand) '16 POUNDS Lanark Troy / 15 Pounds 14 Ounces English Troy'. The weight exhibits casting flaws on the shoulder.

SIZE: diameter 95 mm; height 183 mm.

MEASURED WEIGHT: 7,886·7 g, or 12,1710 gr; representing a pound of 7,607 gr and an ounce of 475·4 gr.

It is likely that this weight was constructed to complete the readjusted 1618 Edinburgh weights (Item 37) after the stone which had been kept with these weights was laid aside in 1800 as the lost Lanark standard, and after Robison's pound (Item 92) had replaced the original pound in 1805. It is under-weight, and this may perhaps be a consequence of being chemically cleaned at the former National Museum of Antiquities of Scotland in 1950-1. No correspondence covering this now exists, but the work is known from an annotated photograph of the weight after cleaning found in NMAS files in 1993. Unfortunately, this does not record the mass of the weight before the work began.

Illustrated in H. J. Chaney, *Our Weights and Measures* (London, 1879), 28. Chaney was misled by the absence of a date on the weight to suggest that this was made before 1618.

The Museum of Edinburgh, Edinburgh (City of Edinburgh Council, Museums and Art Galleries): Inv. EDNMG HH 5978/1/2000. Transferred from Edinburgh Town Council by 1899.

95 TROYE BELL-SHAPED WEIGHTS, FROM 64 POUNDS, BY JOHN MILNE & SONS, 1800, FOR EDINBURGH

Set of seven bronze bell-shaped troye weights (64 pounds to 1 pound) with triangular handles, turned and with flat or recessed bases, but with lead adjustment only in the 8-pound weight. Engraved on the handles '64 POUNDS / Lanark Troy', and similarly for the 32- and 16-pound; '8 Pounds / Lanark Troy', and similarly for the 4-, 2-, 1-pound. Stamped twice on the handle 'I.MILNE / ED^R', or once on the shoulder for the 2- and 1-pound; and with a shield bearing the Scottish lion *rampant* (the die damaged at the lower point of the shield) twice on the shoulder (64-, 32-, 16-pound) or once on the shoulder (8-, 4-, 2-pound), and also on the base six times (64-, 32-pound), five times (16-, 8-pound), four times (4-pound), twice (1-pound) or once (2-pound). Engraved on the waists of the four largest weights 'Adjusted in terms of Act of Par.! 1618 / by Jo. Robison LLD Prof. Nat. Ph. Edinburgh / March 1800', and in addition on the 16-pound alone 'Ordered by Ja.^s Jackson Esqr D.G.'. The English equivalent weights are engraved round the upper edge of the body '63 Pounds 8 Ounces English Troy' and similarly 31 pounds 12 ounces, 15 pounds 14 ounces, 7 pounds 15 ounces; then '30480 Grains English Troy', 15,240 grains, 7,620 grains. The 8- and 2-pound weights do not have an incised line at the lower part of the body. For Milne, see the entry in Appendix G.

94

95

A small manuscript label is pasted on the base of the 4-pound weight, and three other labels were found detached with the weights. It reads: 'Lanark Troy / - / 4 lbs.= / 4 lbs. 5 oz. 10 dr [but dram, not drop] / Avoirdupois', and similarly 69 lbs 10 oz 12[*recte* 6] dm (for 64-pound), 34 lbs 12 oz 15 dm (for 32-pound), [8] lbs 10[*recte* 11] oz 4 dm (8-pound).

SIZES: diameters 220, 170, 118, 95, 74, 53, 45 mm; heights 310, 260, 190, 160, 132, 108, 81 mm.

MEASURED WEIGHTS (from the 32 pound weight): 15,789·4, 7,894·4, 3,946·7, 1,973·5, 986·9, 493·4 g; or 243,662, 121,826, 60,905, 30,455, 15,230, 7,614 gr. This indicates a pound of 7,614 gr (ounce 475·9 gr).

From the denominations engraved on the weights, they have been constructed as exact multiples of the pound of 7,620 gr, doubling it at each step. This was how John Robison specified them when he proposed the production of such a set to Edinburgh's council in February 1800: 'Give me authority to order Mr. Mylne to make a pound Weight, bell-shaped, which I shall adjust to 7620 grains. This should be stamped "Lanark Pound – 7620 grains Troy." Cause Mr. Mylne to make a two pound, four pounds, eight

pounds and sixteen pounds, by doubling continually this unit. I undertake that they shall be more accurately fixed than any set you got from the Exchequer': James Cleland, *Exemplification of the Weights and Measures of the City of Glasgow* (Glasgow, 1822), 77; copy of the original in Edinburgh City Archives (ECA), MS council minutes, 16 September 1801. The base pound was copied from his own pound, Item **92**. He described the form of inscription he wanted on the 16-pound weight in a subsequent letter of 3 March 1800, and this was to be '16 Pounds / Lanark Troy / 15 lbs 14 oz English Troy / Adjusted in terms of the Act of Parliament, 1618, by Jo. Robison, LLD. Prof. Nat. Ph. Edinburgh, March 1800': Cleland, op. cit., 78. Milne was to leave the weights a few grains heavy so that Robison could direct the final adjustment. Robison records adjusting these on 5 April 1800 and noted the additional inscription marked on them as 'ordered by Jas. Jackson Esqr. D.G.': St Andrews University Library, MS Q171.R8, 'Professor [John] Robison's Commonplace Book', II, p.413. The set was extended to 64 pounds, and the total cost was £17 / 7 /-, of which the 64-pound weight alone cost £6 / 8 / -: ibid. Originally, the bell weights were accompanied by a nested set of fractional weights totalling 1 pound. Robison adjusted these a few days later, noting an error he had made in interpreting his own weights, which were light against the correct English troy standard, saying ruefully 'I also examined a Box [or nest] of Lanark troy which I had adjusted yesterday, when unwell, and much disturbed': ibid.

Robison's own weights had been adjusted against weights taken from the Royal Society's standards of the 1730s (see Chapter **9**), which were about 3 to 4 grains in the (12-ounce) English troy pound lighter than the weights produced by Carysfort's commission on which the Imperial standards were based. We can therefore compare the masses recorded by Robison for these weights with the masses in terms of modern grains (the latter appear smaller because the modern grains are larger). The differences become progressively larger as we move from the 1-pound to the 32-pound weights, and are 6, 10, 25, 55, 94 and 178 gr, which averages at 6 grains per troye pound. This confirms the basis of his weights and enables those given by him (other than those made with the Dean of Guild's weights) to be checked – as, for example, his troye pound, Item **92**. Robison noted the 6-grain difference between his own weights and a new troy set acquired by Milne from the London Guildhall in 1792: ibid., II, p. 408.

The weights were recorded in the inventories of the Dean of Guild's weights as a 'Set of Standard Lanark Troy Adjusted and marked by Professor Robison and contained in a Box marked 1800', comprising weights of 64 to 1 pound, together with a 1-pound 'shell weight': ECA, MS council minutes, 23 October 1805.

The use of the lion *rampant* shields is presumably meant to indicate that these were considered to be accurate (indeed, definitive) representatives of the Scots standards. The identical stamp also appears on the liquid capacity check-weights of the same year (Items **93** and **105**), and when Robison sent these to the Dean of Guild in August 1801, he explained that he had 'put a very faint stamp of the Scotch Arms on every part that could be filed; this protects their authenticity': James Cleland,

Exemplification of the Weights and Measures of the City of Glasgow (Glasgow, 1822), 86; copy of the original is in ECA, MS council minutes, 16 September 1801.

The Museum of Edinburgh, Edinburgh (City of Edinburgh Council, Museums and Art Galleries): Inv. EDNMG HH 5983/1-7/2000. Transferred from Edinburgh Town Council by 1899.

96 TRONE BELL-SHAPED WEIGHTS, FROM 16 POUNDS, *c.*1810, FOR EDINBURGH

Part set of bronze bell-shaped trone weights (16 pounds to 1 pound, and 4 ounce) with triangular handles, turned and with base recesses. Engraved on the handles '16 lib: TRON', and similarly for the 8-, 4-, 2-, 1-pound; the 4-ounce engraved '4^{OZ} TRON'; and all stamped on the shoulder 'WT / DG', for William Tennant, Edinburgh Dean of Guild, 1809-10. The shapes of the weights are uniform, with the exception of the 1-pound (which has a sharp lower edge and is extensively surface marked), but the 4- and 2-pound weights are of a brassier alloy and the 1-pound is more copper coloured. The centre of the base of the 2- pound weight is a soldered insert with lead adjustment in the hollow beyond. The 1-pound also has a base insert and has a lead infill.

SIZES: diameters 120, 107, 84, 67, 48, 32 mm; heights 210, 170, 136, 120, 90, 66 mm.

MEASURED WEIGHTS: 9,980·7, 4,991·0, 2,495·5, 1,247·1, 623·4, 156·1 g; or 154,022, 77,021, 38,511, 19,245, 9,620, 2,409 gr. These weights show more variation than might be expected, and certainly more than the set made by Milne for Cupar in 1812 (Item **98**), for which these weights may have been the pattern. However, this may simply be because greater tolerance may have been allowed for trone weights than for troye weights, and the weights may have been left mar-

ginally over-sized to allow for wear. They appear to be to a standard of about 9,627 gr to the pound, or at least to a pound of between 9,623 and 9,628 gr. This does not relate to a known troye ounce (at 20 to the trone stone). John Robison claimed in a letter to the Edinburgh Dean of Guild in 1800 that he could 'give no help as to the trone weight': Edinburgh City Archives, MS council minutes, 16 September 1801, copy of letter to James Jackson of 3 March 1800; also printed in James Cleland, *Exemplification of the Weights and Measures of the City of Glasgow* (Glasgow, 1822), 78.

In 1825, James Jardine found the Dean of Guild's trone weights (presumably this set, Item **96**) to be to a pound of 7,622·7 gr, but provided no equivalence to any other weights. However, Buchanan noted that it was very close to a conventional avoirdupois level of 22 ounces avoirdupois to the trone pound: George Buchanan, *Tables for Converting the Weights and Measures hitherto in use in Great Britain ...* (Edinburgh, 1829), 23, 203. This in turn probably arose because it was 1¼ times the conventional troye equivalent of 17½ avoirdupois ounces: ibid., 21 note. A trone pound of 22 avoirdupois ounces is 9,625 gr. The 1-pound weight, which is badly scratched

on its base, is indeed underweight; and the 2-pound weight, which is the only other weight with a lead adjustment recess, is to a pound of 9,622·6 gr, and is probably the specific source of Jardine's value.

The Museum of Edinburgh, Edinburgh (City of Edinburgh Council, Museums and Art Galleries): Inv. EDNMG HH 5984/1-6/2000. Transferred from Edinburgh Town Council by 1899 (at which time the location inventory recorded weights from 64 pounds to 4 ounces).

97 TRONE BELL-SHAPED WEIGHTS, FROM 32 POUNDS, *c.*1810, FOR EDINBURGH

Another part set (32 pounds to 1 pound), as Item **96**, differing only in having no verification marks.

SIZES: diameters: 160, 123, 100, 80, 63, 50 mm; heights: 280, 223, 175, 130, 98, 80 mm.

MEASURED WEIGHTS: [19,975], 9,983·4, 4,990·6, 2,495·0, 1,247·2, 623·4 g; or [308,254], 154,064,

77,015, 38,503, 19,247, 9,620 gr. See the comments for Item **96**. The mass of the 32-pound weight is taken from the weighing of '44lbs 1oz 8dr Avoirdupois' on a slip of paper attached to its base (similar late nineteenth-century MS labels were found with Item **95**).

Only one of the two sets of trone weights, Items **96** and **97**, is identifiable in the inventories of the Dean of Guild's weights and measures, and this may be the unmarked set because a set used externally would almost certainly require to be marked. It is recorded as eight pieces of 32, 16, 8, 4, 2, 1, ½ and ¼ pound: Edinburgh City Archives, MS council minutes, 20 January 1817.

The Museum of Edinburgh, Edinburgh (City of Edinburgh Council, Museums and Art Galleries): Inv. EDNMG HH 5984/7-12/2000. Transferred from Edinburgh Town Council at an unknown date.

98 TRONE BELL-SHAPED WEIGHTS, FROM 16 POUNDS (WITH NESTING FRACTIONS), BY JOHN MILNE & SONS, 1812, FOR CUPAR

Set of five bell-shaped weights in turned brass, 16 pounds to 1 pound, with cased set of nested fractional parts from 8 ounces to ¼ ounce. The largest weight engraved '16 pound. TRON' on the handle and 'CUPAR 1812' on the side, and stamped 'I.MILNE / EDIN:' on the shoulder of the weight, and similarly for the smaller weights. The set of six nested cup weights, and central plug, for the fractional parts of the pound in turned copper and contained in a turned wooden case with lid. The 8-ounce cup stamped 'CUPAR 1812' between incised rings, and stamped on the rim 'VIII OZ TRON' and 'T'; the other five cups, and the ¼-ounce central plug, stamped only 'T' and with the weight in ounces (IV, II, I, ½ and ¼). For Milne, see the entry in Appendix G.

SIZES: bell weight diameters 120, 101, 75, 60, 50 mm; bell weight heights 220, 181, 141, 100, 79 mm; nested weights diameter 50 mm, height 43 mm.

MEASURED WEIGHTS: 9,979·4, 4,990·0, 2,494·5, 1,246·9, 623·3, 311·7, 155·6, 77·6, 39·0, 19·2, 9·6, 9·6 g; or 154,002, 77,006, 38,495, 19,242, 9,619, 4,810, 2,401, 1,198, 602, 296, 148, 148 gr. This indicates a maximum pound of 9,625-9,626 gr, which corresponds well with a conventional multiple of 22 avoirdupois ounces to the trone pound (9,625 gr), and is a good match for the 1810 set for Edinburgh (Item **96**), which is less closely adjusted, and which presumably forms the model for this set.

No value for a trone standard for Fife is recorded in the very brief verdict of the 1825 Fife county jury: George Buchanan, *Tables for Converting the Weights and Measures hitherto in use in Great Britain …* (Edinburgh, 1829), 211-12.

Illustrated in L. Burrell, 'The Standards of Scotland', in *The Monthly Review: the Journal of the Institute of Weights and Measures Administration*, 69 (1961), 55.

All except the 16-pound weight exhibited at the

St Andrews Preservation Trust Exhibition of Weights and Measures, St Andrews, 1968, items 39 and 23.

East Fife Council Museum Service, Cupar (Fife Council, Fife Museums East): Inv. CUPMS 1984.115-119, 132-139. Transferred from Cupar Town Council to North East Fife District Council at local government reorganisation in 1975 (Museums Service established in 1983).

99   **98**

98

99 TROYE BELL-SHAPED WEIGHTS, FROM 16 POUNDS (WITH NESTING FRACTIONS), BY JOHN MILNE & SONS, 1812, FOR CUPAR

Set of five bell-shaped weights in turned brass, 16 pounds to 1 pound, with cased set of nested fractional parts from 8 ounces to ¼ ounce. The largest weight engraved '16 lb. DUTCH' on the handle and 'CUPAR 1812' on the side, and stamped 'I.MILNE / EDIN:' on the shoulder of the weight, and similarly for the smaller weights. The set of six nested cup weights, and central plug, for the fractional parts of the pound in turned copper and contained in a turned wooden case with lid. The 8-ounce cup stamped 'CUPAR 1812' between incised rings, and stamped on the rim 'VIII OZ DUTCH' and 'D'; the other five cups, and the ¼-ounce central plug, stamped only 'D' and with the weight in ounces (IV, II, I, ½ and ¼). For Milne, see the entry in Appendix G.

SIZES: bell weight diameter 113, 95, 72, 58, 44 mm; bell weight heights 210, 176, 131, 104, 82 mm; nested weights diameter 53 mm, height 36 mm.

MEASURED WEIGHTS: 7,895·6, 3,948·0, 1,974·1, 987·0, 493·3, 246·6, 123·5, 61·8, 30·9, 15·4, 7·7, 7·7 g; or 121,845, 60,926, 30,464, 15,231, 7,613, 3,806, 1,906, 954, 477, 238, 119, 119 gr. This indicates a troye pound of 7,616 gr (ounce 476·0 gr), in reasonably close agreement with Milne's 1800 set for Edinburgh (Item **95**, to a measured pound of 7,614 gr), which may have been the pattern for this set.

The exceedingly brief 1825 jury report for Fife gives little useful information on the weights – the standard Scots pound (presumably the pound given by Buchanan as 7,608·95 gr in the general tables) is supposed to relate to the Dutch pound as 1:1·09214: George Buchanan, *Tables for Converting the Weights and Measures hitherto in use in Great Britain …* (Edinburgh, 1829), 212. If this is a misprint for 1:1·0009214, it would give the Dutch pound as 7,616 gr.

Illustrated in L. Burrell, 'The Standards of Scotland', in *The Monthly Review: the Journal of the Institute of Weights and Measures Administration*, 69 (1961), 55.

99

All except the 16-pound weight exhibited at the St Andrews Preservation Trust Exhibition of Weights and Measures, St Andrews, 1968, items 40 and 24.

East Fife Council Museum Service, Cupar (Fife Council, Fife Museums East): Inv. CUPMS 1984.114, 120-31. Transferred from Cupar Town Council to North East Fife District Council at local government reorganisation in 1975 (Museums Service established in 1983).

100 TROYE WEIGHT OF ONE STONE 'DUTCH', *c.*1820, FROM LINLITHGOW

Weight in cast bronze with turned finish, in the form of a domed cylinder topped by a circular-section T-bar handle. Engraved round the body of the weight '1ST LANARK DUTCH'.

SIZE: diameter 102 mm; height 160 mm.

MEASURED WEIGHT: 7,961·5 g or 122,861 gr,

100

representing a pound of 7,679 gr and an ounce of 479·8 gr.

This weight poses a difficult problem of identification. Although made in the form of the 1618 standards, its simple and clean construction in bronze suggests an early nineteenth-century date. It is clearly not based on the Scots troye or a Dutch ounce, which appears to rule out a late eighteenth-century origin, when the views of John Swinton and John Robison, establishing a troye pound of about 7,620 gr, were well-known. Nor does it follow the findings of James Jardine for the 1825 Edinburgh county jury.

However, it does follow the conclusion promoted by Adam Anderson, Rector of Perth Academy, based on his study of the 1707 English troy weights at Perth (Item **83**), and subsequently also those at Stirling (Item **85**), that these weights were distributed at the Union as examples of the Scottish troye series. He had already adopted this view in a paper read to the Literary and Antiquarian Society of Perth, and reported in April 1821: 'Weight of the Dutch Pound Troy', in *Edinburgh Philosophical Journal*, 4 (1820-1), 449-50. He reinforced it in his 1827 report to the Stirlingshire jury on weights and measures, and probably also in his report to the Perthshire jury, who cite this in their verdict as one of four expressions of Scots troye: Adam Anderson and Thomas Bruce, *Report on the Stirlingshire Weights and Measures* ([Stirling], 1827), 12-16; George Buchanan, *Tables for Converting the Weights and Measures hitherto in use in Great Britain...* (Edinburgh, 1829), 246.

It is presumed that Linlithgow, influenced by Anderson's argument, commissioned through him a weight representative of the supposed troye standard. Anderson is likely to have had contacts with Linlithgow's Dean of Guild at about this time, and in 1827 he published tables for converting Linlithgow measure to Imperial: Kenneth J. Cameron, *The Schoolmaster Engineer: Adam Anderson of Perth and St Andrews c.1780-1846* (Dundee, 1988), 20-1, citing A. Anderson,

Appendix to the Report on the Weights and Measures of Perthshire (Perth, 1827).

Illustrated in J. Calder (ed), *The Wealth of a Nation in the National Museums of Scotland* (Edinburgh and Glasgow, 1989), 187.

National Museums of Scotland, Edinburgh (Trustees of the NMS): Inv. NMS T.1973.138. Presented in May 1973 by the Council of the Royal Burgh of Linlithgow to the Royal Scottish Museum (now part of NMS) in advance of local government reorganisation in 1975.

101 TRONE BELL-SHAPED WEIGHTS (UNFINISHED?), FROM 16 POUNDS, 1821, FOR GLASGOW

Part-set of bronze bell-shaped trone weights, the largest engraved on the handle '16 libs / Standard Tron for / Butcher Meat 1821.' and similarly for the 4, 1, ½, ¼ pound weights. Each weight also stamped twice with the Glasgow seal, on the handle for the 16 pound weight and on the shoulders for the others.

SIZES: diameters 130, 82, 56, 40, 35 mm; heights 225, 115, 100, 67, 45 mm.

MEASURED WEIGHTS: 10,238·5, 2,558·5, 638·9, 319·2, 159·7 g; or 158,000, 39,483, 9,860, 4,926, 2,464 gr. These present a problem of interpretation because the pounds represented in the first two weights are quite large, at 9,875 and 9,871 gr respectively, whereas the pound is 9,860 gr, and the pounds of the smaller weights are 9,852 and 9,858 gr.

The 1827 Lanarkshire jury report made it clear that the trone weights for fish, cheese and butter in Glasgow was to a pound of about 9,819 gr, but 'by an agreement entered into about a hundred years ago between the Magistrates of Glasgow and the Incorporation of Fleshers, Beef, Veal, Mutton, Lamb, and Fresh Pork, have, since that time, been sold by a weight also called Tron, containing 22½ Ounces Avoirdupois', namely about 9,844 gr to the pound: George Buchanan, *Tables for Converting the Weights and Measures hitherto in use in Great Britain …* (Edinburgh, 1829), 232.

In 1821 James Cleland, Superintendent of Public Works for Glasgow, undertook a comprehensive examination of all the standard weights and measures in Glasgow for the Dean of Guild in order to establish their correct values, and this was followed by an inspection of over 20,000 traders' weights and measures in the city: James

101

Cleland, *Exemplification of the Weights and Measures of the City of Glasgow* (Glasgow, 1822). Cleland was also a member of the expert committee that reported to the 1827 Lanarkshire jury, and he noted that the conclusions they reached in 1827 were identical to his earlier findings: James Cleland, *Enumeration of the Inhabitants of the City of Glasgow and County of Lanark for the Government Census of [1831]* (Glasgow, 1832), 188 note. An inventory of the Dean of Guild's weights and measures, prepared by Cleland as early as 1816, noted only one unmarked set of large brass trone weights, with a set in lead for adjusting the flesh-market weights: Glasgow City Archives, MS A2.1.3 (XIII (5)3), 'Reports and Memorials 1814-1824', p.138.

A new set of brass trone weights specifically for butcher-meat was produced by the Glasgow smiths James Liddel & Co. (for which, see Appendix G), inevitably matching Cleland's pound of 22½ avoirdupois ounces. However, they were not bell weights as here, but made to the traditional bar-handled pattern, described as 'with cross handles', to match the earlier weights (and as used for the Paisley weights, Item **91**); they were included in subsequent inventories by Cleland, who noted that they were inscribed 'STANDARD TRON FOR BUTCHER MEAT, I. L. & CO.': James Cleland, *An Historical Account of the Local and Imperial Weights and Measures of Lanarkshire, and an Inventory of those belonging to the Corporation of Glasgow* (Glasgow, 1832), 31.

The Glasgow smith, Alexander Wood, was also the city's Adjuster of Weights and Beams at this time, and he was based at James Liddel & Co. until he took over their premises in 1823 (see Appendix G). Wood certainly had access to the Dean of Guild's standards, and indeed six of the city's 1707 avoirdupois weights (from Item **54** and **70**), which were included in Cleland's 1832 inventory, were subsequently in the possession of Wood's firm. One (part of Item **54**) was exhibited by them in the 1894 'Old Glasgow' exhibition (item 1587: 'Tron Weight, "John Snart, 1707"') and then presented with a number of others;

including the present 1821 butchers' trone weights, to the city's museum in 1902.

The explanation for the unexpected size of these 1821 butcher-meat weights may therefore be that they had been officially stamped but had not yet been fully adjusted by Wood. A distinct possibility may be that Cleland objected to their construction in the shape traditionally associated with avoirdupois weight on the grounds that this would lead to confusion and error, and that he insisted that they be re-made in the 'cross handle' trone form. They may have remained in Liddel's premises or with Wood, to be discovered much later. The collection of over 20 weights presented to the city in 1902 by the elderly Alexander Wood, junior, appears to represent historical mementoes gathered by the business over the years. It includes an interesting group of three unmarked eighteenth-century iron bar-handled trone weights to a pound of 9,800 to 9,805 gr, which may be to the Glasgow or the Renfrew trone standard (Inv. GLAMG 1902.84.q,r,s).

The 16-pound weight was exhibited at the 'Old Glasgow' exhibition, Glasgow, 1894, item 1570, lent by Alexander Wood & Sons. A printed label with the number '1554' on the base presumably indicates the provisional exhibit number.

The People's Palace Museum, Glasgow (City of Glasgow Council, Museums and Art Galleries): Inv. GLAMG 1902.84.e, g, h, i, j. (The 8-pound weight, 1902.84.f, was unlocated.) Presented by Alexander Wood of Partick.

102 TRONE NESTING WEIGHTS, PILE OF EIGHT POUNDS, FROM FOUR POUNDS DOWNWARDS, EARLY NINETEENTH CENTURY, BY A. COATS & CO., AT DUMBARTON

Set of nesting cup weights in turned brass, derived from the type associated with manufacture in Nuremberg, comprising an outer lidded case with handle, and nine inner weights. Inscribed on the outer case '8 Lib. Tron' and 'Dumbarton', and on the base 'A. Coats & Co. / Glasgow', with turned adjustment recess in the base. There are no further markings, but the inner weights form a binary series down to ⅛-trone ounce, suggesting that it is only lacking a cup and central plug, each of 1/16 ounce. For Coats, see the entry in Appendix G.

SIZES: outer cup diameter 105 mm; overall height (excluding handle) 118 mm. Inner cup diameters: 91, 74, 60, 48, 39, 32, 26, 22, 19 mm; heights: 69, 56, 41, 33, 23, 18, 14, 11, 7 mm.

MEASURED WEIGHTS: 2,610·9, 1,305·1, 652·3, 325·8, 162·7, 81·1, 40·8, 20·8, 10·4, 5·3 g; or 40,291, 20,140, 10,066, 5,028, 2,511, 1,252, 630, 321, 160, 82 gr. This indicates that these weights have been adjusted to a pound of about 10,070 gr (ounce of about 629·4 gr).

John Swinton noted that the trone weight in Dunbartonshire, for butter, cheese, butcher-meat, fish and lint, was a conventional pound of 23 avoirdupois ounces, which (if exact) would be 10,062½ gr, with an ounce of 628·9 gr, and this was repeated in the 1827 Dunbartonshire jury verdict: [John Swinton], *A Proposal for Uniformity of Weights and Measures in Scotland* (Edinburgh, 1779), 71; George Buchanan, *Tables for Converting the Weights and Measures hitherto in use in Great Britain …* (Edinburgh, 1829), 191.

West Dunbartonshire Museum Collection, Library Headquarters, Dumbarton (West Dunbartonshire Council, Education and Cultural Services): Inv. DMBDM 1982.81. Displayed in the former burgh library, perhaps from the 1930s.

102

103 AVOIRDUPOIS NESTING WEIGHTS, PILE OF EIGHT POUNDS, FROM FOUR POUNDS DOWNWARDS, MID-EIGHTEENTH CENTURY, AT DUMBARTON

Incomplete set of nesting cup weights in turned brass, of the type associated with manufacture in Nuremberg, comprising a decorated outer weight with lid and handle, and five original surviving inner weights. The outer cup with banded and stamped decoration. Stylised hinge and handle, with a horse-headed clasp. Stamped on the lid are two small Glasgow burgh marks and '8E', where the 'E' denotes English (avoirdupois) weight. The five inner turned weights hang by their lips, and have simple banded decoration on the inside, outside and on the rims. The largest three inner cups stamped 'II', 'I', '8' on the rims; the largest inner cup stamped 'I' inside the base.

SIZES: outer cup diameter 100 mm; overall height (excluding handle) 113 mm. Inner cup diameters: 84, 67, 55, 44, 33 mm; heights: 56, 45, 33, 26, 19 mm.

MEASURED WEIGHTS: 1,815·5, 907·6, 453·8, 226·6, 113·2, 56·5 g; or 28,017, 14,006, 7,003, 3,497, 1,747, 872 gr. This indicates that these weights

have been adjusted to an avoirdupois pound of about 7,003 gr (ideally 7,000 gr), or an ounce of about 437·7 gr.

It was eventually required that all retail weights (except the smallest) should be stamped with a mark to indicate the weight type, to reduce the danger of fraud. In 1764, Edinburgh's Dean of Guild ruled that troye weights be stamped 'T', trone weights 'Tr', and English or avoirdupois 'E', the stamps being held by Thomas Simpson, the pewterer who was authorised by the council to adjust weights: Edinburgh City Archives, MS council minutes, 25 January 1764. The potential for confusion with the English 'A' and 'TR' markings for avoirdupois and troy does not seem to have been considered. John Swinton noted that avoirdupois weight was used in Dunbartonshire for 'English Goods and Groceries': [John

103

Swinton], *A Proposal for Uniformity of Weights and Measures in Scotland* (Edinburgh, 1779), 71.

104 AVOIRDUPOIS BELL-SHAPED WEIGHTS OF 56 AND 28 POUNDS, BY COLIN BLYTH, 1786, FOR PAISLEY

Part set of bell-shaped weights, now comprising only the 56-pound and 28-pound weights, in cast brass or bronze with a turned finish, and with pierced rectangular handles. Inscribed round the larger weight 'PAISLEY STANDART / 1786', and on the shoulder 'LVI' and 'AVOIRDUPOIS', and similarly for the 28-pound weight. Stamped 'C · BLYTH' on the handles, and with the seal for Glasgow on the shoulders and bases (and on the handle of the larger weight). Numerous lead adjustment sites on the bases. With subsequent verification and identification marks, including a working standard set number 'C 72' on the larger weight. For Blyth, see the entry in Appendix G.

SIZES: diameters 185, 140 mm; heights 290, 233 mm.

105 WEIGHTS FOR THE WATER CONTENT OF CAPACITY MEASURES, QUART TO HALF-PINT, *c.*1800, FOR EDINBURGH

Set of three check-weights for the weight of water in English standard capacity measures (quart to half-pint) accompanying a similar set (Item **93** above) for the Scots measures. Rounded bell-shape in pewter, and cast around iron keys, the loops of which protrude to form the handles of the weights. Stamped with the same Scottish shield as the companion weights, Item **93**, and similarly stamped on the waist of the weights 'QUART / 14621', 'PINT 7310½', '[½] PINT / 3655¼'.

SIZES: diameters 51, 42, 33 mm; heights 101, 77, 61 mm.

MEASURED WEIGHTS: 946·7, 473·1, 236·4 g; or 14,609, 7,301, 3,648 gr.

These items are marked with conventional weights which are fractions of the English wine gallon of 231 in³, which at 253·18 gr/in³ contains 58,484 gr, or 14,621 gr per (wine) quart. John

Robison re-established these figures for the Edinburgh Dean of Guild in August 1800 by direct measurement of the city's two 1707 Union wine gallon standards (Items 153 and 154), which he measured as 235·4 in³ and 231 in³: St Andrews University Library, MS Q171.R8, 'Professor [John] Robison's Commonplace Book', I, p. 474. The sizes for the Scots and English liquid measures were rehearsed by Robison in a letter to the Dean of Guild, and he noted that he was providing four check-weights for the Scots measures (Item 93) and three for the English measures (the present set), requesting that they be 'stamped (not engraved)' with their weights and noting, 'I have put a very faint stamp of the Scotch Arms on every part that could be filed; this protects their authenticity': James Cleland, *Exemplification of the Weights and Measures of the City of Glasgow* (Glasgow, 1822), 86; copy of the original is in Edinburgh City Archives (ECA), MS council minutes, 16 September 1801. The check-weights for the English and Scots measures were recorded in the Edinburgh Dean of Guild's inventory of weights and measures in 1818 as having been adjusted by Robison in 1800, but 'now laid aside being defective from long use and the softness of the metal': ECA, MS council minutes, January 1818. In the previous year, it had been noted that 'some small deficiencies' had been discovered in the standard measures and that James Jardine had been asked to adjust them; however, in the event, only the English check-weights were renewed (by Item 106): ibid., 20 January 1817. The check-weights were not yet present in the inventory for 1800, which is printed in Meredyth Somerville, *The Standardization of Weights and Measures in Scotland*, Department of Geography, University of Edinburgh, Occasional Publication no. 11 (Edinburgh, 1989), 28-9.

The Museum of Edinburgh, Edinburgh (City of Edinburgh Council, Museums and Art Galleries): Inv. EDNMG HH 5982/5-7/2000. Transferred from Edinburgh Town Council by 1899.

106 WEIGHTS FOR THE WATER CONTENT OF CAPACITY MEASURES, FROM GALLON TO GILL, 1817, FOR EDINBURGH

Set of five simple cylindrical check-weights in a white brass alloy, made as a replacement for Item 105. The largest engraved 'ROBERT JOHNSTON, D.G. / 1817. / ENGLISH GALLON / lb. 7. oz. 9¾ Gr. 44. or, 58,484 TROY GRAINS / Compared with the Standard / Adjusted By Prof.ʳ JOHN ROBINSON [*sic*] / 1800. / BY JAMES JARDINE / CIVIL ENGINEER. 1817.', and similarly for the quart, pint, half-pint and quartern at 1 pound 14¾ ounces 41 grains or

105

14,621 grains, 15⅛ ounces 50½ grains or 7,310½ grains, 7½ ounces 55 grains or 3,655¼ grains, and 3¾ ounces 28 grains or 1,828 grains.

SIZES: diameters 83, 52, 41, 33, 27 mm; heights 83, 52, 41, 32, 25 mm.

MEASURED WEIGHTS: 3,788·7, 947·1, 473·5, 236·7, 118·4 g; or 58,467, 14,616, 7,307, 3,653, 1,827 gr.

Like the weights in Item **105**, these are marked with the same conventional weights for the capacities of the English wine gallon and its fractions. As in earlier instances, the troy pound has been considered as one of 16 and not 12 troy ounces, in the Scottish tradition. The individual weights are marginally larger than Robison's, and indicate that Jardine has been adjusting them against more accurate standards. The set was described in the Dean of Guild's inventory of weights and measures in 1818 as 'Hard compound metal check wts in a small box' and noted that they were marked 'Compared with the standard adjusted by Prof John Robison 1800 by James Jardine, Civil Engineer 1817': Edinburgh City Archives, MS council minutes, January 1818. An example of the quartern weight, bearing the same inscription was in the possession of James Cleland, the Glasgow metrologist: James Cleland, *Exemplification of the Weights and Measures of the City of Glasgow* (Glasgow, 1822), 85 note.

The Museum of Edinburgh, Edinburgh (City of Edinburgh Council, Museums and Art Galleries): Inv. EDNMG HH 5985/1-5/2000. Transferred from Edinburgh Town Council at an unknown date.

107 AVOIRDUPOIS BELL-SHAPED WEIGHTS OF 14 AND SEVEN POUNDS, 1819, FOR ARBROATH

Part set of bell weights, now comprising only the 14-pound and 7-pound weights, in cast bronze with a turned finish, and with pierced rectangular handles. Engraved round the lower part of the weights 'ARBROATH GUILDRY 1819', marked on the shoulder AVOIR / 14 LB' and similarly for the 7-pound weight. Stamped 'AS / DG' for Alexander Smellie, Edinburgh Dean of Guild 1819-21, with the seal for Arbroath (the abbreviation 'ARB^TH' and a portcullis) on the shoulders, and a post-1890 Board of Trade verification mark for Arbroath.

SIZE: diameters 107, 91 mm; heights 205, 53 mm.

Angus Council, Trading Standards office, Arbroath (Angus Council, Environmental and Consumer Protection).

106

107

SECTION C

CAPACITY MEASURES

C1: PRE-1618 CAPACITY STANDARDS

**108   PINT, THE 'STIRLING JUG',
EARLY SIXTEENTH CENTURY,
FOR STIRLING**

By long tradition, this undated vessel is held to be the national standard, entrusted to the keeping of the Royal Burgh of Stirling. Its status was confirmed by the 1618 Assize, at which time copies were made for the other principal burghs. These copies, and a more extensive series made for lesser burghs, carry representations of the shields on this pint. It is variously known as the 'Stirling Jug' or the 'Stirling Stoup', and its

capacity is considered to be one Scots or Stirling pint. This is the earliest example of the 'large' pint, whose introduction is discussed in Chapter 7, and for which a conventional capacity has been set at 103·7 in³.

The vessel has apparently been sand-cast in bronze, and has a single handle forged from a flat strip and attached by two rivets. The base has been cast as a separate piece and brazed to the body – the residual square-section stump of the casting riser for the base is visible inside the pint. The joint is thin and well made, except at one position where an apparent leak has been filled

108

with solder. The straight tapering conical sides are rounded at the base and have been finished by file. On the front of the vessel are two shields cast in high relief. The area between them shows signs of having been worked back to the level of the main surface and there is a slight indication that there may have been raised letters or perhaps the numerals of a date which has subsequently been removed. The upper shield carries a lion *rampant*, with its tail inverted, or *coward*. The beast is incorrectly shown facing *sinister*, indicating that the shield, on the pattern from which the casting was made, was pressed in a mould with the correct *dexter* orientation and that the maker has not allowed for the inversion of the motif. The lower shield provides some problems of interpretation. It appears to show the figure of a child on all fours and with its head turned to the viewer. It is suggested in Chapter 7 that this may represent the Christ child in a pre-Reformation version of Stirling's seal, subsequently depicted as a Paschal lamb. Only the upper part of the shield is occupied. The upper rim of the vessel is not accurately flat, limiting the accuracy with which its content can be determined. It is unclear whether this has been caused by subsequent adjustment, but this

must be a distinct possibility. The lower rivet of the handle is now slightly loose, so that the vessel has a slow leak: this was sealed on the inside with tape during our measurements. The present rivets have substantial cylindrical heads with crisp edges inside the vessel, and may therefore not be original. The two series of measures, supposedly duplicates of the Stirling pint and produced as a result of the 1618 legislation, carry motifs which are apparently based directly on those of this measure (Items **114-126**). This lends support to the assumption that this vessel is the town's standard which was still in force in 1618. The upper shields of the 1618 measures also carry a lion *passant* (for Scotland), but the lower shields have a long-tailed beast *couchant guardant*, possibly a wolf, over the letter 's' (for Stirling).

SIZE: base diameter 160 mm; top diameter 119 mm; height to rim 168 mm; height to top of handle 185 mm; aperture 105 mm; depth 155 mm.

MEASURED CAPACITY: 1,701·4 g, or 26,256 gr, equivalent to 1·701 litres, or 103·8 in³. (This was taken as the average of three determinations of 1,701·4, 1,701·2 and 1,701·5 g, each to ± 0·5 g, and each made after leaks were satisfactorily stopped and when the amount of residual water left after draining had been clearly established.) At 15·432 gr/g, this is 26,256 gr (± 10 gr); and assuming a water density of 252·9 gr/in³, the volume is 103·82 in³ (± 0·05 in³). Within the temperature uncertainty, we will give this as 103·8 in³ (± 0·1 in³). If the handle was originally held with flush rivets, rather than the present cheese-headed rivets (see above), the volume would be increased to 26,270 gr, or 103·85 in³.

The large pint probably did not come into formal use until after 1508, at which date the Edinburgh water measures were still being defined in term of the small pint: J. D. Marwick, *et al.* (eds), *Extracts from the Records of the Burgh of Edinburgh*, 14 volumes (Edinburgh, 1869-1967), I, 118. Presumably a few were made at the same time as

standards for the principal burghs, of which only Stirling's example survives. By analogy with the pint standards by David Rowan (Items **109**, **111** and **112**) and the stone standard by Hans Cochran (Item **31**), it seems very likely that the Stirling Jug was also constructed in the royal ordnance workshops, which had transferred to a major new foundry at Edinburgh Castle by 1511: David. H. Caldwell, 'Royal Patronage of Arms and Armour in Fifteenth and Sixteenth-Century Scotland', in David H. Caldwell (ed), *Scottish Weapons and Fortifications 1100-1800* (Edinburgh, 1981), 73-93, p. 76. If so, the Stirling Jug may tentatively be attributed to Robert Borthwick, who by 1513 was described in the Treasurer's accounts as 'maister meltare of the Kingis gunnis'; he was first recorded as being issued with his royal livery in 1511, but was dead by 1532 when he was succeeded by David Rowan's father: C. E. Whitelaw, *Scottish Arms Makers* (London, 1977), 136.

There are numerous references in the literature to the Stirling Jug, principally because it figured by name in the 1618 Assize. For example, it was recorded by Sir Robert Sibbald in his *History Ancient and Modern of the Sheriffdoms of Linlithgow and Stirling* (Edinburgh, 1710), the Stirlingshire portion being more accessibly reprinted as *Sibbald's History & Description of Stirlingshire, Ancient and Modern, 1707* (Stirling, 1892), see p. 44. The earliest published description of the vessel was produced by James Gray in 1754, when its water capacity was measured as 26,180 grains, or 103·404 in³ based on a separate determination of the weight of water in a standard 100 in³ capacity measure: James Gray, 'Of the Measures of Scotland compared with those of England', in *Essays and Observations, Physical and Literary*, I (1754), 200-4. Gray did not speculate about the jug's origin, beyond referring to the 1618 Act and noting that the jug 'appears, by its make, to be very old'. The text of his brief article matches that of the first part of an unsigned manuscript report 'Experiments ascertaining the Contents of the Lithgow [*sic*] Wheat and Bear Firlots, for

the Justices of the Peace of the County of Stirling, 1754', which described the gauging of the two surviving 1754 standard firlots (Items **216** and **217** in the Inventory), adjusted in Edinburgh by John Stewart (professor of natural philosophy at the University of Edinburgh), James Gray and John McGowan. Whereas Gray's article is illustrated by rather crude wood-block illustrations of the shields on the pint measure, the manuscript has careful drawings which show an optimistic amount of detail, perhaps in an attempt to recreate the original appearance of the shields. We are grateful to George A. Dixon, formerly Archivist to Central Regional Council, for drawing this manuscript and other sources to our attention.

Before this period, however, the pint does not have a continuous association with Stirling Council and it must remain only reasonable conjecture that this pint is the pre-1618 standard. The story of the discovery and identification of the vessel in 1752 by the mathematician and antiquary, the Rev. Alexander Bryce, is recounted in Chapter **9**. The earliest reference we have found to Bryce's role in recovering the pint is in the 1810 second volume of George Chalmers, *Caledonia; or an Account of North Britain*, 3 volumes (London, 1807-24), II, 40. This was the source for an editorial comment by William MacGregor Stirling in the 1817 second edition of William Nimmo's *History of Stirlingshire* (Stirling, 1817), 339. Chalmers noted merely that the Scots standards 'seem not to have been very carefully kept; the Stirling jug … was actually lost, till it was discovered by the Reverend Alexander Bryce of Kirknewton'.

A greatly extended account by the author and publisher Robert Chambers, in which he associated this pint with 1437 legislation (an error for the 1457/8 Assize) appeared in his *Picture of Stirling* (Edinburgh, 1830), 44-6. This has formed the basis of subsequent accounts, although greater numerical detail is given in his notice on Bryce: Robert Chambers, *Biographical Dictionary of Eminent Scotsmen*, 4 volumes (Glasgow, 1835),

IV, 488-93. Chambers did not give the source of his information on Bryce. Although some of the antiquarian material he published was certainly collected orally, that appears not to be the case here and he has presumably been working from a manuscript source. Specific details are given of how Bryce undertook his measurements of the pint in Edinburgh in 1752-3, presumably in conjunction with Gray (who is not mentioned by Chambers). Many of these details are not in Gray's account, and it appears they were taken from a more extended version which is now unlocated. An analysis of the measurements made by Bryce and Gray is given in Chapter **9**. The story of Bryce's recovery of the Stirling Jug was of sufficient importance to Robert Chambers for his brother William to reprint it from the 1830 *Picture of Stirling* in his *Memoir of Robert Chambers* (Edinburgh and London, 1872), 205-7. The account is also reprinted, with a small wood-cut illustration, in 'Strevlina' (pseudonym), 'The Stirling Jug', in *The Monthly Review: the Journal of the Society of Inspectors of Weights and Measures*, 17 (1909), 90-1.

A few additional details are supplied in the version given by the Stirling local historian James Shirra, who also noted rescuing the pint from obscurity yet again when he located it in 1852 'lying in the sill of a window in a room above the chamber where the Town Council usually met, all covered with dust, and seemingly as a thing utterly worthless in its associations': J. S., 'The Stirling Jug', in the *Stirling Observer*, 29 July 1880. Gray's capacity of 103·404 in³ is recorded for the Stirling Jug by John Swinton in his *Proposal for Uniformity of Weights and Measures in Scotland* (Edinburgh, 1779), 29, 121, and in Patrick Graham's *General View of the Agriculture of Stirlingshire [for the] Board of Agriculture* (Edinburgh, 1812), 338, 341.

The 1824 Weights and Measures Act, which introduced the new and uniform Imperial system, required (paragraph 18) the sheriffs of the Scottish counties to call specialist juries to determine the relationships of the Imperial standards to the old standards, in order to ensure continuity in the settling of contracts of all sorts. At the request of the Sheriff-Depute of Stirlingshire, the Stirling Jug was sent to Edinburgh to be measured in the new units. The results of the experiments conducted by James Jardine, Alexander Adie and David Murray were published in their 1825 report to the Edinburgh jury and reprinted in George Buchanan's collected version of the jury reports and tables of equivalents: 'Report to Adam Duff, Esq., his Majesty's Sheriff-Depute of the County of Edinburgh, regarding the Weights and Measures heretofore in use in said County', in *An Abstract of the Act 5. George IV. Cap. 74 … for ascertaining and establishing Uniformity of Weights and Measures; also of the Act 1618, Establishing the late Scottish Standards …* (Edinburgh, 1827), 25-7; George Buchanan, *Tables for Converting the Weights and Measures hitherto in Use in Great Britain …* (Edinburgh, 1829), 200-2.

The Stirling Jug was subsequently re-measured by Adam Anderson, Rector of Perth Academy, and Thomas Bruce, for their report to the Stirlingshire jury: *Report on the Stirlingshire Weights and Measures* ([Stirling], 1827), 1-5, 8-10. Anderson's report was largely reprinted by Parliament in the 1834 Report of the Select Committee on the Sale of Corn, as appendix 11, with an account of the earlier work of Gray and Bryce in appendix 9: *Parliamentary Papers* 1834, VII, 455-6, 451-4.

Various sizes have been given for the Stirling Jug. James Jardine's 1825 determination of its capacity for the Edinburgh county jury used a glass cover-plate over the Jug and was 26,307 gr at 62°F. This approach was repeated by Adam Anderson in his 1827 report to the Stirlingshire jury, at 26,302 gr, but he said that such a result would necessarily be too large since the glass sat on the highest points of the uneven rim. Anderson's value for the volume up to the low points of the rim was 26,275 gr. The difference between these two extremes is equivalent to a volume change of 0·1 in³. However, Anderson's preferred value, taking an improbably bent surface that followed the rim and had to be determined by calculation, was 26,286 gr. From these figures, the value that would have reflected our assessment of the mean level would probably have been about 26,280 gr. This represents a good match for our 26,270 gr.

Because the density of water was given in the 1824 Act as 252·5 gr/in³ at 62°F (it is now understood to be somewhat greater, at 252·9 gr/in³), and because Jardine's measurement was taken as authentic, the Stirling Jug's capacity emerged as 104·2 in³. This is the nineteenth-century figure that has entered the literature (and contrasts with Gray's eighteenth-century 103·4 in³). However, using the modern density value, Jardine's volume reduces to 104·0 in³. But we have argued that Jardine over-filled the Jug and that a better estimate is close to Anderson's preferred value, which with the modern density becomes 103·9 in³.

Anderson also provided a volume at the reduced temperature of 45°F, which he described as the mean temperature of river water in Scotland, and therefore approximating to the conditions of the original definition. A volume containing 26,280 gr at 62°F, would (by his conversion) contain 26,310 gr at 45°F, and the density would give 253·2 gr/in³. This matches the water density determined by Thomas Everard in 1696 and used by James Gray and Alexander Bryce in their c.1754 measurement of the Stirling Jug. (Everard's experiment was claimed in 1834 to have been at 55°F, but in practice this cannot be established: *Parliamentary Papers* 1834, VII, p. 452.)

John Robison provided volumes of a range of different pints in his Commonplace Book and Court of Session reports in the 1770s to 1790s. In spite of this he never gives a direct measurement of the Stirling Jug itself. Its volume is only given in terms of the pint's 55-ounce capacity (from the 1618 Assize), and follows his early conclusion that the original mass of the 1618 pound could be settled at 7,621 gr. The pint is therefore 55 ounces of this pound, or 26,197 gr, which he normally gives as 26,200 gr: National Archives of Scotland (NAS), Court of Session process CS 231/F3/1, bundle 10. Presumably, this is equivalent to about 103·5 in³, although a value in cubic inches is never stated.

Robison measured the 1618 Linlithgow pint (Item **114**) several times when overfilled by the Linlithgow official cooper in his presence in 1779; and the volumes given are indeed at or above the recently measured volume of this pint: ibid., bundle 10. These volumes are additionally given as specific quantities 'over standard', where the 'standard' is one of 26,197 gr, showing that Robison had already committed himself to this size in his 1779 report to the Court of Session. A measurement of the Edinburgh 1618 pint (Item **113**), using the Dean of Guild's 1707 troy standards in 1780, gave the Edinburgh pint as 26,274 gr at 42°F, and using a density of 253·2 gr/in³ for this temperature gives a capacity of 103·8 in³: St Andrews University Library, MS Q171.R8, 'Professor [John] Robison's Commonplace Book', I, p. 450. Measurements of the 1555 Edinburgh chopin (Item **109**) in 1789 and 1792 similarly gave volumes of 13,163 gr at 43°F, equivalent to pints of between 103·9 and 104·0 in³: ibid., I, p. 441. The 1622 Dumfries pint (Item **118**), measured by him in 1778 against the Dumfries troy standards (Item **77**), was found to be 26,424 gr, equivalent to 104·4 in³: ibid., I, p. 450. These are all compatible with our recent measurements.

However, Robison also stated in his 1788 report to the Court of Session that the jugs at Stirling, Linlithgow, Edinburgh, Glasgow (Item **121**) and Dumfries 'do not differ from each other ⅛ of a cubic inch': ibid., II, p. 391; NAS CS 231/F3/1, bundle 32. This clearly does not square with a Stirling Jug postulated as 103·5 in³. Robison must surely have been aware that the Stirling Jug was larger than this, and probably about 103·9 in³. He no doubt felt constrained by the implication of his 1779 evidence to the Court of Session to continue to argue for a 'true' original value of 26,200 gr. Presumably he felt that changing the basis of his argument away from a dependence of the mass of the Lanark stone would have threatened his professional credibility in court.

The Jug was re-measured in 1877 by A. Stephen Wilson of Aberdeen, and described and illustrated in his *Botany of Three Historical Records* (Edinburgh, 1878). Wilson was the first to note the leak, which is still present, but after the leak had been stopped an average of five trials with fresh water from the River Tay gave the capacity as 26,272 gr, and a further six with water from the Water of Leith gave 26,288 gr (both corrected to 62°F): ibid., 118. In spite of this small difference, he made it clear that 'both these waters are almost absolutely of the same specific gravity as distilled water': letter of Wilson to the Stirling Town Clerk quoted in the *Stirling Observer*, 30 August 1877; the original letter of 21 August 1877 is in Stirling Council Archives, SB1/10/37.

It remains to be noted that no account is taken here of the small changes in the size of the English troy pound which we described in Chapter 4, amounting to about 3½ grains in a troy pound of 5,760 grains. It was shown in Chapter 9 that Bryce and Robison were both using weights related to the Royal Society's standards of the 1740s, which were based on the smaller and earlier (medieval) standard rather than the slightly larger modern standard (embodied in the Imperial system). The difference in a pint containing about 26,300 grains is only 15 grains, representing a volume change of 0.06 in^3 if the capacity is fixed at a set number of ounces. This is smaller than the existing uncertainties inherent in measuring these vessels and assessing their temperature dependence, and so for our purposes here it is ignored.

The Stirling Jug was illustrated in R. W. Cochran-Patrick, *Medieval Scotland* (Glasgow, 1892), frontispiece, the block being re-used in J. Paton (ed), *Scottish History and Life* (Glasgow, 1902), 198. More recently, the measure was described by Lawrence Burrell, 'The Standards of Scotland', in *The Monthly Review: the Journal of the Institute of Weights and Measures Administration*, 69 (1961), 49-62, especially pp. 49, 55 and illustrated on pp. 50, 52. Burrell adopted the conventional view of dating the pint to 1457, justifying such an early date by claiming that casting an integral handle was not possible until later, and he argued for continuity in the size of the pint. Dr Elizabeth Gemmill correctly rejected this view, dating the Jug to the sixteenth century: E. Gemmill and N. Mayhew, *Changing Values in Medieval Scotland: a Study of Prices, Money, and Weights and Measures* (Cambridge, 1995), 94-5. Less critically, a photograph of the Jug was reproduced by J. T. Graham in an article directed to collectors of measures, where it was bizarrely described as a 'leathern tankard' valued at £70: Tom Graham, 'A Quantity of Measures', in *Antique Collecting*, 10 (1975), 10-15, p. 10. It was still described as a 'leather mug' in his *Weights and Measures*, Shire Album no. 44 (Aylesbury, 1979), 22. Also illustrated by Craig Mair, *Mercat Cross and Tolbooth* (Edinburgh, 1988), 121, and Alex Neish, 'The Treasures of the Smith Institute, Stirling', in *Journal of the Pewter Society*, 9 (1993), 63-9, fig. 2. Illustrated in the Scottish Museums Council's *Scotland's National Audit Full Findings Report: A Collective Insight* (Edinburgh, 2002), 11, again with the date 1457.

Exhibited at the Loan Exhibition of Scientific Apparatus, London, 1876, item 324, and described (following Chambers) as entrusted to Stirling in 1437. The Science Museum, London, retains an electrotype copy made by Elkington & Co. at the end of this exhibition and initially forming part of the South Kensington Museum Education Collection (E44-1878): Inv. NMSI 1878.19. The Stirling Jug was shown at the Glasgow International Exhibition, 1888, item 1040, where it is described as the standard deposited in Stirling according to the Assize of 1457, and at the Glasgow International Exhibition, 1901, item 2436, which repeats this description.

Smith Art Gallery and Museum, Stirling (Trustees of the Stirling Smith Art Gallery): Inv. STIGM 1238 (previously UA. 1). Transferred by Stirling Town Council, and listed in the published *Descriptive Catalogue [of the] Smith Institute, Stirling* (Stirling, 1882), 48, in case 17.

109   HALF-PINT, BY DAVID ROWAN, 1555, AT EDINBURGH

This is the earliest dated measure by the bronze-founder David Rowan and is an impressive example of his work. It was presumably made in connection with the activities of the 1555 Parliamentary Commission on Weights and Measures to complement the existing Edinburgh pint, which at this period may have been considered the national standard: this pint does not survive, and in 1618 it was replaced as the Edinburgh standard by Item **113**, described below.

The vessel has a capacity of one chopin, or half a (large) pint, and is in cast bronze with a single handle; the external sides have a slight inward taper, rising to a plain rim with a projection at the handle. It has been sand-cast: flash lines are visible on the inside, and filed-down lines on the exterior and base are discernible. It has apparently been cast in one piece,

with the handle extending from its upper end into the mould as a tapering horizontal bar, which was subsequently bent down and brazed at its lower end to the body of the vessel. Cast in relief on the front of the vessel and heavily chased are a crowned shield with the Scottish lion *rampant*, the date '1555' and a castle with a central tower (for Edinburgh); and to the left of the date the monogram 'DR' (for David Rowan) is chased in an incised shield border.

SIZE: diameter 118 mm; width across handle 163 mm; height 136 mm; height to rim 129 mm; aperture 186 mm; internal depth 117 mm.

MEASURED CAPACITY: 851·5 g average, or 13,140 gr, equivalent to 51·95 in^3 or a pint of 103·9 in^3.

The 1555 measures are not mentioned in the published extracts from the Edinburgh Town Council minutes, although a proclamation was issued against the use of unjust weights and measures in October 1554. Rowan was paid by the Dean of Guild in March 1555 for the casting of a bell: J. D. Marwick, *et al.* (eds), *Extracts from the Records of the Burgh of Edinburgh*, 13 volumes (Edinburgh, 1869-1967), II, 201, 355. Rowan's work is discussed further in Chapter 7.

109

The measurements of this vessel made in the eighteenth-century by Robert Stewart, and subsequent workers, are described in Chapter **9**. It is listed as '1 Scots chopin cast in metal 1555' in the Edinburgh Dean of Guild's inventory of weights and measures in 1800, reproduced in Meredyth Somerville, *The Standardization of Weights and Measures in Scotland,* Department of Geography, University of Edinburgh, Occasional Paper no. 11 (Edinburgh, 1989), 28-9.

Illustrated in R. W. Cochran-Patrick, *Mediaeval Scotland* (Glasgow, 1892), frontispiece, fig. 2 (the block re-used in J. Paton (ed), *Scottish History and Life* [Glasgow, 1902], 199), and in H. J. Chaney, *Our Weights and Measures* (London, 1879), 28.

Exhibited at the Scottish Exhibition of National History, Glasgow, 1911, item 105 and at the Historical Weights and Measures exhibition, Royal Scottish Museum, Edinburgh, 1972.

The Museum of Edinburgh, Edinburgh (City of Edinburgh Council, Museums and Art Galleries): Inv. EDNMG HH 5986/2000. Transferred from Edinburgh Town Council by 1899.

110 BREWER'S PINT, THE 'JEDDART JUG', 1563, FOR JEDBURGH

The Jedburgh pint, or 'Jeddart Jug', is unique amongst the surviving early pints in being shaped more like a conventional jug, with a comparatively narrow aperture. It is one of only two existing standards of the larger pints which were used to check burgh vessels for the sale of ale, and which were made with the enhanced capacity – which for convenience we have described as the 'brewer's pint' – containing a sixteenth more than the pint defined in the legal assizes. The deans of guild in all the royal burghs must at one stage have held such vessels. The construction of the spout allows excess water to run off in a controlled fashion, leaving the same amount of water

110

present in repeated fills, a feature that is of importance in the regular trial of a large number of submitted trade vessels.

The vessel is a single-piece casting in bronze, produced by the lost-wax process. Its cylindrical body, tapering to a neck and expanding above, is reminiscent of the traditional shape of the characteristic Scottish pewter measures known as 'tappit hens'. A simple flat-sided triangular spout with a run-off groove, is formed at one side of the jug, and at the other side there is a plain functional handle with chamfered edges and a projecting thumb-piece. There is some evidence of the attachment of the spout component to make up the wax master. Since such a complex shape would have been difficult to adjust, the relative location of the inner and outer casting moulds was critical: the iron pins used to locate them have been found in five positions embedded in the bronze. Above the handle is a flat shield-shaped area, presumably intended to take the burgh's seal, as in other examples. Instead, it has been cut with the date '1563' over a small saltire. Round the rim of the vessel is cut 'THIS ∘ IS ∘ YE ∘ COMVN ∘ MVSVR ∘ OF ∘ IEDBVRGHT'. Stamped near the base of the handle is '3:146', being a subsequent (eighteenth-century?) reference to its capacity in English ale pints of the type distributed at the 1707 Union (see Item **127**). Traces of black paint are visible on parts of the surface.

SIZES: base diameter 130 mm; maximum external width 177 mm; height 228 mm; aperture 95 mm.

MEASURED CAPACITY: 1,781·5 g average, or 27,492 gr, equivalent to 108·7 in³ (raised from a legal pint of 102·3 in³ or 25,870 gr, with an incorporated allowance of one-sixteenth).

Swinton incorrectly described this pint as dated 1556, but noted its capacity as 'above a gill more than the legal standard', where the gill is the sixteenth part of the pint: [John Swinton], *A Proposal for Uniformity of Weights and Measures in Scotland* (Edinburgh, 1779), 117. The report on the weights and measures of Roxburghshire, made in the terms of the 1824 Weights and Measures Act, was prepared by the astronomer, optician and mechanic James Veitch of Jedburgh: for Veitch, see T. N. Clarke, A. D. Morrison-Low and A. D. C. Simpson, *Brass & Glass: Scientific Instrument Making Workshops in Scotland* (Edinburgh, 1989), 16-24. In his report Veitch noted that 'the only Standard measure of capacity used in the County of Roxburgh, was the Pint Jug in custody of the Magistrates of Jedburgh, bearing the following inscription:- "This is the Common Measure of Jedburgh – 1563"': James Veitch, *Tables for Converting the Weights and Measures hitherto used in Roxburghshire, into the Imperial Standards …* (Jedburgh, 1826), 5. The county jury's verdict, printed by Buchanan, paraphrases Veitch's report, but provides a more accurate transcription of the lettering as 'This . is . ye comvn . mvsvr . of Jedburght . 1563': George Buchanan, *Tables for Converting the Weights and Measures hitherto in use in Great Britain …* (Edinburgh, 1829), 256-7. Veitch's measured capacity of 27,346·1522 gr, equivalent to about 108·1 in³, seems low in comparison with the recently measured volume which matches that of one of the Rowan pints (Item **112**), and is compatible with the sixteenth-century sizes discussed in Chapter 7.

Discussed by Rosalind Mitchison, 'Measuring by Jedburgh Jug: Rampant Scots Individualism', in the *Scotsman*, 28 December 1963. Illustrated in L. Burrell, 'The Standards of Scotland', in *The Monthly Review: the Journal of the Institute of Weights and Measures Administration*, 69 (1961), 49-62, p. 50.

Exhibited at the Scottish Exhibition of National History, Glasgow, 1911, item 107, illustrated in plate opposite p. 948.

Jedburgh Castle Jail and Museum, Jedburgh (Scottish Borders Council Museum Service, Selkirk): Inv. JEDMG 368. Transferred by Jedburgh Dean of Guild to the museum after 1898.

III   BREWER'S PINT, BY DAVID ROWAN, 1574, FOR ST ANDREWS

The quality of this, the finest of the surviving pint measures by David Rowan, reflects the status of the medieval town of St Andrews as one of the principal centres of learning and ecclesiastical patronage in the kingdom, and as a focus for the landed and commercial interests of the east coast. The significant feature of this measure is that it does not operate as a brim-measure, but is self adjusting, simplifying the testing of quantities of submitted trade vessels. Like the Jedburgh pint (Item 110), this vessel is a brewer's pint, containing a sixteenth allowance beyond the legal pint.

The vessel is a single-piece lost-wax casting in bronze, finished by file, and with a raised portion of the metal at the front cut back in lateral extent to form a shield, on which has been engraved the seal of the burgh of St Andrews. A shaped bronze handle has been riveted in place at its extremities, and a carefully-formed notch, chamfered at the rear, has been cut in the upper rim to allow for the overflow of excess water, leaving the remaining 'brewer's pint' quantity. A hole of approximately 7 mm diameter 30 mm below the rim, perhaps the position of a casting flaw, has been plugged. The shield which carries the burgh's arms (a boar against a tree) is flanked by the initials 'S' and 'A', and above it is cut 'PINTA · SANCTI · ANDREAE'. Below the shield is cut within a border the following inscription: 'RECEPTAE · IST · HEC / PINTA · SCOTICE · MEN / SV RA · DE · STIRVILINGO · / PER · PATRICIVM · LERMO / NTH · DE · DERSIE · MELITEM / PREPOSITVM · CIVITAT / IS · SANCTI ANDRIE · 1574'. Beneath this is repeated the date '1574', and the stamped monogrammed initials 'DR' of David Rowan.

SIZE: base diameter 160 mm; width across handle 210 mm; aperture 106 mm; external height 164 mm; depth 144 mm.

MEASURED CAPACITY: 1,787·9 g average, 27,591 gr, equivalent to 109·1 in^3 (raised from a legal pint of 102·7 in^3 or 25,970 gr, with an incorporated allowance of one-sixteenth).

The self-levelling nature of this vessel allows us to check whether or not our assumption is correct that the level of fill should be coincident with the rim of the vessel. Filling this vessel demonstrates clearly that the correct level is coincident with the rim (even though surface tension will allow the vessel to be filled above the rim), since repeated tests showed that the capacity obtained was highly consistent, and was identical (within the limits of our technique) to the capacity measured with the notch sealed off and the vessel used as a brim measure. Rowan emerges from this examination as an engineer of considerable skill. Rowan's work, and the relationship of the sixteenth-century capacity measures carrying his mark, are discussed further in Chapter 7, and see Item 109.

David Learmont of Clatto, member of Parliament for St Andrews in 1524, apparently acquired the estate of Dairsie in 1520. His son, Sir James Learmont, succeeded in 1526 and was provost of St Andrews, frequently representing the burgh in Parliament; he was also master of the Household to James V, and Treasurer to the infant Queen Mary after her father's death. 'His son, Patrick Learmont, was made baillie of the legality of St Andrews in 1568, and he and his direct descendants were representatives of St Andrews in Parliament almost continuously from 1567 to 1625': A. H. Millar, *Fife: Pictorial and Historical; its People, Burghs, Castles, and Mansions*, 2 volumes (Cupar, Fife, 1895), I, 166; a parliamentary biography is given in Margaret D. Young (ed), *The Parliaments of Scotland: Burgh and Shire Commissioners*, 2 volumes (Edinburgh, 1992-3), II, 418.

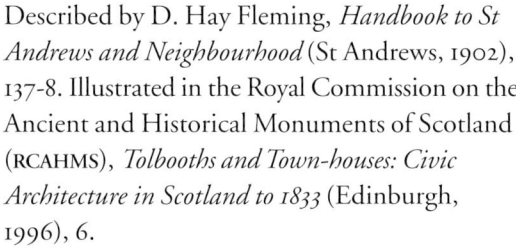

111

Described by D. Hay Fleming, *Handbook to St Andrews and Neighbourhood* (St Andrews, 1902), 137-8. Illustrated in the Royal Commission on the Ancient and Historical Monuments of Scotland (RCAHMS), *Tolbooths and Town-houses: Civic Architecture in Scotland to 1833* (Edinburgh, 1996), 6.

Exhibited at the Glasgow International Exhibition, 1888, item 962, and at the St Andrews Preservation Trust Exhibition of Weights and Measures, St Andrews, 1968, item 13.

St Andrews Museum, St Andrews (Fife Council, Fife Museums East, Cupar): Inv. CUPMS 1990.365. Transferred from St Andrews Town Council to North East Fife District Council in 1975 (Museums Service established 1983) and then to St Andrews Museum (opened 1990). Previously on loan to the St Andrews Preservation Trust, North Street, St Andrews.

112 PINT, BY DAVID ROWAN, LATE SIXTEENTH CENTURY, AT DUNDEE

Another example of a pint by Rowan, of virtually identical design to the St Andrews pint (Item **111**) but to the smaller legal standard, rather than the enhanced brewer's standard. The raised area on the front of the vessel has similarly been cut back to form a shield-shape, but this is blank, and the vessel carries no inscription. It is, however, marked with David Rowan's 'DR' monogram stamp beneath the shield. As with the St Andrews pint, there is a single plugged hole, 8 mm diameter, 17 mm down from the rim, again perhaps the original position of a casting flaw. The vessel does not sit quite flat, because its lower surface is slightly rounded.

SIZE: base diameter 160 mm; width across handle 210 mm; height to rim 168 mm; aperture 105 mm.

MEASURED CAPACITY: 1,677·0 g average, or 25,879 g, equivalent to 102·3 in³.

Rowan's work and the relationship of the capacity measures by Rowan are discussed in Chapter 7; see also Item **109**.

114

MEASURED CAPACITY: 1,702·5 g average, or 26,273 gr, equivalent to 103·9 in³.

The report of the 1826 Linlithgow county jury gave the capacity of the Linlithgow pint at 62°F as 26,297·159 gr, 10 gr (or 0·004 in³) less than James Jardine's reported capacity for the Stirling Jug, noting that the difference 'arises obviously from errors of adjustment': George Buchanan, *Tables for Converting the Weights and Measures hitherto in use in Great Britain …* (Edinburgh, 1829), 236-7. The Linlithgow pint was one of five pints – the other being those for Stirling, Edinburgh, Glasgow and Dumfries – that John Robison said differed by less than one-eighth of a cubic inch (or 0·1 in³): St Andrews University Library, MS Q171.R8, 'Professor [John] Robison's Commonplace Book', II, p. 391. It has already been noted in relation to the Stirling Pint (Item 108) that this comment was optimistic.

The alloy composition of this pint was analysed in 1992 and compared with one of the second (1622) series: see discussion under Items 113 and 116.

Illustrated in A. D. C. Simpson, 'Grain Packing in Early Standard Capacity Measures: Evidence from the Scottish Dry Capacity Standards', in *Annals of Science*, 49 (1992), 339-50, p. 341.

National Museums of Scotland, Edinburgh (Trustees of the NMS): Inv. NMS T.1973.135. Presented in May 1973 by the Council of the Royal Burgh of Linlithgow to the Royal Scottish Museum (now part of NMS) in advance of local government reorganisation in 1975.

115   PINT, 1618, FROM DUNFERMLINE

Another example of this issue, as Item 113 (noting, however, some differences in the form and positioning of the central bands). The measure has apparently not been adjusted – its capacity is too large and there is no indication of the rim having been filed down – and it is possible there-fore that it was part of an intermediate issue or a surplus vessel that was distributed with the second issue. This measure appears to be of the full height of the master, with the rim about 20 mm higher than that of the Edinburgh example.

SIZE: diameter 158 mm; width across handle 203 mm; height to rim 175 mm; aperture 100 mm; internal depth 165 mm; base thickness 10 mm.

MEASURED CAPACITY: 1,712·0 g average, or 26,420 gr, equivalent to 104·5 in³.

The Royal Burgh of Dunfermline was not one of the burghs listed in the 1618 Act as holding one of the standards, and it is not clear why it should have received one of the first issue of pint measures. One possible reason may be the enhanced status of Dunfermline at that time. The Abbey had been a significant royal foundation and an important pilgrimage centre before the Reformation. James VI annexed the remaining lands in 1587 as a royal estate and presented the Lordship of Dunfermline to his queen, Anne of Denmark, who remodelled the Abbey's former royal guest house as a palace: Richard Fawcett, *The Abbey and Palace of Dunfermline* (Edinburgh, 1990). The future Charles I was born at Dunfermline Palace in 1600, and James stayed there during his 1617 Scottish progress.

This may be the measure referred to in the 1825 Fife county jury verdict on weights and measures, which noted that the firlot was 'adjusted by the Jug, commonly called the Stirling Jug, which was kept and used for that purpose: and which Stirling Jug contains 105 cubic inches': George Buchanan, *Tables for Converting the Weights and Measures hitherto in use in Great Britain …* (Edinburgh, 1829), 211. This is equivalent to 104·8 in³ using the modern density value.

Exhibited at the St Andrews Preservation Trust Exhibition of Weights and Measures, St Andrews, 1968, item 14, where it was described in the exhibition catalogue as the Dunfermline pint and dated to *c.*1580. The pint came from the collection of early standards held by Fife County Council's Weights and Measures Department, and photographs taken by the Trust at the exhibition allowed it to be positively identified as the standard retained by its successor, Fife Regional Council, when examined in 1987. Dunfermline ceased independent inspection of weights and measures *c.*1910 when authority passed to the combined County and Burghs inspection service: Carl Ricketts and John Douglas, *Marks and Markings of Weights and Measures of the British Isles* (Taunton, 1996), 123. Presumably the pint's association with Dunfermline was familiar to

115

subsequent Chief Inspectors for the county, and in particular to David Reid who assisted the Trust in preparing the 1968 exhibition.

Dunfermline Museum, Dunfermline (Fife Council, Fife Museums West): Inv. DUFDM 2000.0052. Transferred from the Consumer Protection Department, Fife Regional Council, Glenrothes, 1996.

116   PINT, 1622, AT DUMFRIES

A second series of pint measures was distributed by the Stirling town council to royal burghs in about 1622, in its capacity as holder of the standard authorised by the legislation of 1618. The Stirling minutes record in September 1622 that £806 Scots was paid for 34 jugs of brass to be sold to other Royal Burghs: *Extracts from the Records of the Royal Burgh of Stirling 1519-1666* (Glasgow, 1887), 157. These seem to be in addition to the initial copies issued to the four principal burghs designated in the assize, and perhaps also to a number of others, such as Dunfermline. This second series is of markedly poorer quality of decoration and finish, and appears crude in comparison with the first issue. Of the 34 understood to have been issued, 11 have been traced. The pint for Dumfries is described here, and the following entries (Items 117-125) are for the standards of Culross, Dunbar, Forfar, Forres, Glasgow, Haddington, Kinghorn, Kirkcaldy and Perth. A further example (Item 126) has now lost its provenance, but it may be one of five others which are identifiable from the nineteenth-century literature, but which are no longer located.

A preliminary analysis of the alloy composition of a pint from this issue (the Dunbar standard, Item 118) has been carried out by Dr Paul Wilthew, Head of the Analytical Research Section of the National Museums of Scotland, using X-ray fluorescence spectroscopy, and he has found a leaded bronze which differed signific-

antly from the composition of a pint from the first (1618) series (Item 114): Paul Wilthew and Phyllis Herd, 'Analysis of Two Scots Pint Measures', NMS Analytical Research Section Report 92/25 (C&AR 6242-3), November 1992. Measurements at three sites showed that lead, the principal copper alloying component, was very inhomogeneously distributed but the estimated total lead content (21%) was twice that found in the first issue pint, whereas the tin content was about half that found in the other pint.

The vessel has a single handle and is a lost-wax casting in bronze from a master built up from separate standardised components. A raised band runs round the vessel below the rim, and another at the middle separates two shields cast in relief with the royal arms (a lion *passant*) and the symbol of Stirling (a beast *couchant guardant*) with the letter 's' – both apparently identical to those on the first issue of the pint standards. The outer surface tapers in from a slight flare at the base, becomes cylindrical at the upper bands and then turns outwards. As with the first issue, some castings, such as the Dunbar pint (Item 118), and those for Culross (Item 117) and Perth

116

(Item **125**), are truncated at the base and do not extend to the lower flare: it is suggested that they are cast with the mould inverted and the metal added only until the inner core is adequately covered. The slightly irregular form of the two raised bands appears to be the same in each example, implying that they are moulded on the outer surface of the master and not applied separately. In building up the wax master, there is some variety in the placing of the shields and of the handle component. The form and position of the inner core in relation to the outer mould determines the capacity of the vessel: some, such as the Dunbar, Culross and Perth pints, have had their rims substantially lowered to reduce the volume. However, it is clear that the capacities have not been carefully controlled and there is a significant range of volumes.

SIZE: base diameter 165 mm; width across handle 214 mm; height to rim 169 mm; top diameter 116 mm; aperture 100 mm; depth 152 mm; base thickness 17 mm; height to top of handle 190 mm. These sizes are fairly typical, but there is some variation. For example, the Haddington pint (Item **122**) is 160 mm to the rim, but has an aperture of 106 mm and a base thickness of only 5 mm. The Perth pint (Item **125**) is only 145 mm to the rim.

MEASURED CAPACITY: 1,710·8 g, or 26,401 gr, equivalent to 104·4 in³. Of the eleven surviving examples, seven have been measured. Their capacities, in ascending order, are as follows: Haddington 103·7 in³; Dumfries 104·4 in³; Dunbar 104·7 in³; Kinghorn 104·8 in³; Forfar 105·3 in³; Forres 106·3 in³; Kirkcaldy 106·5 in³.

Two of these results can be checked against figures given in the county jury reports for the 1820s, where the Dumfries jug is given as 104·2 in³ and the jugs for Elgin and Forres as 106·2 in³: George Buchanan, *Tables for Converting the Weights and Measures hitherto in use in Great Britain …* (Edinburgh, 1829), 197, 204. John Robison measured the Dumfries pint in 1778

and gave a result equivalent to 26,424 gr at an unspecified temperature, but probably equating to 104·4 in³: St Andrews University Library, MS Q171.R8, 'Professor [John] Robison's Commonplace Book', I, p. 452. As an additional check when the recent measurements were made, the Kinghorn and Kirkcaldy pints were measured together, as were the Haddington and Dunbar pints, so we can be certain that there were substantial real differences in their values. It is reassuring to note that the Haddington pint has the same capacity as the Edinburgh 1618 pint (Item **113**), since a discrepancy between these would not have passed unnoticed.

A variation in capacity of 2 to 3% is much greater than would be expected in a single issue of standards, and does seem to indicate that at least some may not have been adjusted at all. Certainly, the rim of several are remarkably uneven and still have rounded edges that would have been lost if they had been filed down in the course of careful adjustment. The level of variation was commented on in the 1826 jury report for Forfarshire, a county that contained several burghs with early standards: 'the copies of the Stirling Jug in the custody of the Magistrates of Forfar, Dundee and Brechin … differ from the Stirling Jug … in so much that some of them contain somewhat less, and others somewhat more than the Standard; and … these discrepancies must have arisen from errors in the adjustment of the copies': Buchanan, op. cit., 125. (The low value will have been the early Dundee pint, Item **112**.) Of the examples not measured above, the county jury reports provides the capacities of the Culross, Glasgow and Perth pints as 105·3 in³, 105·1 in³ and 105·5 in³ respectively: ibid., 244, 232, 243.

The prevalence of the larger size of about 105 in³ has caused comment in the past but has never been satisfactorily explained. Colin Maclaurin, in a report of 1735 to the Excise Commissioners, noted the difference between the official standards and the ordinary measures for the sale of drink (which he termed the 'Pewtherer's pint Jug') where the latter 'are generally found to hold about 105 Cubick inches', against the current value of 102·3 in³ for the Edinburgh Dean of Guild's pint: Judith V. Grabiner, 'A Mathematician among the Molasses Barrels: Maclaurin's Unpublished Memoir on Volumes', in *Proceedings of the Edinburgh Mathematical Society*, 39 (1996), 193-240, p. 236. By 1745, in his editorial additions to David Gregory's *Practical Geometry*, Maclaurin was using the revised description of the Edinburgh pint as 103·4 in³ and noting that the Pewterers' jugs 'are said to contain sometimes betwixt 105 and 106 cubic inches', whilst at the same time casting some doubt on Gregory's account of a 1707 measurement which had set it at about 109 in³: [Colin Maclaurin] (ed), *A Treatise of Practical Geometry … by the late Dr David Gregory … Translated from the Latin; with Additions* (Edinburgh, 1745), 113, 110. The version of Gregory's text that survived in the *Encyclopaedia Britannica* recorded that the Stirling Jug 'was supposed to contain 105 cubic inches; and though, after several careful trials, it has been found to contain only about 103½ inches; yet, in compliance with established custom, founded on that opinion, the pint stoups are still regulated to contain 105 inches, and the customary ale measures are about ¹⁄₁₆ above that standard': 'Measure' in *Encyclopaedia Britannica*, fifth edition, 20 volumes (Edinburgh, 1817), XIII, 37.

So, whereas Maclaurin had implied that the 105 in³ size included some enhancement (although clearly this was smaller than the brewers' sixteenth), it was now being accepted that this represented the basic legal size – presumably because the 1622 pints were supposed to be duplicates of the Stirling Jug – and a further level of enhancement was taken above this. An additional sixteenth over a capacity of 105 in³ was still recorded at Glasgow in 1827: George Buchanan, *Tables for Converting the Weights and Measures hitherto in use in Great Britain …* (Edinburgh, 1829), 232, 233.

There are two distinct interpretations. Either the 1622 issue was very inadequately controlled, with many pints being issued over-sized without proper adjustment, leading to a belief that the Stirling measure was larger; or, by the 1620s there may have been greater latitude in the allowances permitted, and this may be reflected in the sizes the burghs demanded. Although no evidence for the latter has been found, if the Excise Commissioners (on the advice of Maclaurin and others) were accepting a Pewterer's Jug of 105 in³, then there would be no effective pressure to revise the size.

The county jury reports of the 1820s identify at least four other examples of this issue (or the 1618 issue) that are now unlocated. These were at Brechin ('the copy of the Stirling Jug'), at Dumbarton ('a massive brass Jug … having two armorial shields in front, the upper one bearing in relief a lion rampant, and the lower the letter S underneath the figure of an animal'), at Renfrew ('[a jug] of massive thickness and ancient appearance … which from its shape and devices evidently appears to have been intended as a duplicate of the Standard Scots Pint or Stirling Jug'), and at Selkirk ('Pint Jug, marked with a Lion Rampant and a Deer at rest, below which is a large capital s'): ibid., 215, 191, 250, 261. Early pints are also indicated, but not described, at Banff, Elgin, Nairn and Tain: ibid., 180, 209, 240, 187. Additional examples (other than early brewers' pints) are indicated by John Swinton at Ayr and Peebles: [John Swinton], *A Proposal for Uniformity of Weights and Measures in Scotland* (Edinburgh, 1779), 57, 108.

More recently, James Shirra of Stirling recounted the tale of the discovery in 1852 of the Dysart standard, which he described as 'a rather massive, antique-looking jug, adorned with the rampant Scottish Lion', showing nonetheless 'a more modern style of workmanship' than the Stirling Jug: J. S., 'The Stirling Jug', in the *Stirling Observer*, 29 July 1880; and see 'Stirling Relics in the Glasgow International Exhibition', the *Scotsman*, 21 April 1888. In 1880, the Dysart pint (discussed at Item **126**) was in the possession of the nephew and successor of John Dick of Craigengelt House, Stirlingshire.

Described and illustrated in James Williams, 'An Example of the Stirling Pint of 1622', in *Transactions of the Dumfriesshire and Galloway Natural History and Antiquarian Society*, third series, 47 (1970), 194-5, and plate 11. He over-estimates the capacity as 105·9 in³.

Dumfries Museum, Dumfries (Dumfries and Galloway Council, Museums Service): Inv. DUMFM 1936.31.5 (previously acq. 207e). Acquired before 1911 from the Dumfries Burgh Weights and Measures Department. The museum, based at the Maxwelltown Observatory, Dumfries, was operated by the Dumfries and Maxwelltown Observatory Society until 1936.

117   PINT, 1622, AT CULROSS

Another example of the same series, as Item **116**. The rim has been cut down to the level of the first band, and the casting at the base stops before the outer flare. A large letter 'C' (presumably for Culross) scratched on base.

That the tiny Royal Burgh of Culross held a stone from the 1618 distributed weight standard (see Item **35**), and a pint from this series, demonstrates that although it may have subsequently gone into economic decline, at this point the burgh still regarded itself as a sufficiently prosperous fishing and trading community to acquire a set of standards.

The pint is recorded in the 1826 Perthshire jury report and measured as 26,592 gr, or 105·3 in³ (equivalent to 105·1 in³ using the modern density value): George Buchanan, *Tables for Converting the Weights and Measures hitherto in use in Great Britain …* (Edinburgh, 1829), 244. Swinton noted that Perthshire standards were held at Culross and Dunblane as well as at Perth itself: [John Swinton], *A Proposal for Uniformity of Weights and Measures in Scotland* (Edinburgh, 1779), 111.

Culross Town House, Culross (National Trust for Scotland, Edinburgh): an unnumbered item on the inventory of property 33. In the Town House at its transfer to the Trust from the Town Council of Culross in advance of local government reorganisation in 1975.

117

118 PINT, 1622, FROM DUNBAR

Another example of the same series, as Item **116**. The rim has been adjusted down to the level of the top band.

The alloy composition of this pint was analysed in 1992 and compared with one of the second (1622) series: see discussion under Items **113** and **116**.

The Dunbar pint was listed as 'the jug stoup' amongst the burgh standards held in the Charter House of Dunbar in an inventory of January 1700 published in the *Haddingtonshire Courier*, 29 November 1912.

Exhibited at the Historical Weights and Measures exhibition, Royal Scottish Museum, Edinburgh, 1972.

National Museums of Scotland, Edinburgh (Trustees of the NMS): Inv. NMS H.MC.25. It was presented by the Magistrates of Dunbar in 1828 to the Society of Antiquaries of Scotland, and

was in the Society's collection in the National Museum of Antiquities of Scotland, from 1985 in the holdings of the National Museums of Scotland.

119 PINT, 1622, AT FORFAR

Another example of the same series, as Item **116**.

MEASURED CAPACITY: 105·3 in³ (see Item **116**).

Illustrated in L. Burrell, 'The Standards of Scotland', in *The Monthly Review: the Journal of the Institute of Weights and Measures Administration*, 69 (1961), 49-62, p. 52. The pint is mentioned in the 1826 Forfarshire jury report, but its capacity is not given: George Buchanan, *Tables for Converting the Weights and Measures hitherto in use in Great Britain ...* (Edinburgh, 1829), 215-18.

Meffan Institute, Museum and Art Gallery, Forfar (Angus District Council, Cultural Services, Forfar): Inv. ADMUS F.1978.63. Known to have

118

119

been displayed in Forfar Library in the 1960s, and subsequently transferred by the Town Council to the museum.

120   PINT, 1622, AT FORRES

Another example of the same series, as Item **116**. Stamped 'FORRES' on the handle, near the rim, and on the base; stamped 'DG' and 'WB' at either end of the handle; and 'WB' on the base.

MEASURED CAPACITY: 106·3 in³ (see Item **116**).

Illustrated in the *Aberdeen Press and Journal*, 30 March and 1 April 1996. The 1826 jury report for Morayshire (then the County of Elgin, but termed there 'Elgin and Moray') indicates that identical pint standards were held by both the ancient royal burghs of Elgin and Forres, although it was the 'Elgin Jug, commonly called the Pint Stoup' that was considered the county standard: George Buchanan, *Tables for Converting the Weights and Measures hitherto in use in Great Britain ...* (Edinburgh, 1829), 209. Both were 106·2 in³. Swinton's information on the county ('Elgin and Forres shire') mentions only the Elgin pint, at 105·4 in³: [John Swinton], *A Proposal for Uniformity of Weights and Measures in Scotland* (Edinburgh, 1779), 77.

Falconer Museum, Forres (Moray Council, Elgin): Inv. FOREM L.1996.1. Transferred at local government reorganisation in 1996; formerly in the care of Grampian Regional Council, Department of Consumer Protection, Aberdeen, which in 1975 took over from, amongst other authorities, the Moray, Nairn and Elgin Joint Inspection Committee (established in 1930).

121   PINT, 1622, AT GLASGOW

Another example of the same series, as Item **116**.

Included in James Cleland's 'Inventory of weights and measures belonging to the City of Glasgow', 22 November 1816, as 'Scotch standard Stirling Ale pint, composite metal in form of a tapering jug with a handle ornamented in front with a lion and another animal with the letter S underneath them': Glasgow City Archives, MS A2.1.3 (Inv. XIII (5)3), 'Reports & Memorials 1814-1824'. The 1827 Lanarkshire jury report records the pint at 105·1 in³: George Buchanan, *Tables for Converting the Weights and Measures hitherto in use in Great Britain ...* (Edinburgh, 1829), 232.

Exhibited at the 'Old Glasgow' exhibition, Glasgow, 1894, item 1583, and at the Glasgow International Exhibition, 1901, item 2212.

Kelvingrove Art Gallery and Museum, Glasgow (Glasgow City Council): Inv: GLAMG 1882.82c. On loan since August 1992 to 'The Linlithgow Story', Linlithgow (Linlithgow Heritage Trust).

122

122   PINT, 1622, AT HADDINGTON

Another example of the same series, as Item **116**. The base of the measure is thin: it is now somewhat dished, and the metal is cracked (and had to be sealed during measurement), probably resulting from an attempt to increase its capacity by striking it on a vertical post.

MEASURED CAPACITY: $103 \cdot 7$ in^3 (see Item **116**).

East Lothian Museum Services, Haddington (East Lothian Council): Inv. HADDM 1996.102 (previously A.38). Transferred from Haddington Town Council c.1930 and displayed in the former public library. To East Lothian District Council at local government reorganisation in 1975 (Museums Service established in 1990).

123   PINT, 1622, FROM KINGHORN

Another example of the same series, as Item **116**. Stamped on the base 'THE / STIRLING JUG / OF KINGHORNE 4 NOV. / 1618'.

MEASURED CAPACITY: $104 \cdot 8$ in^3 (see Item **116**).

Exhibited at the Scottish Exhibition of National History, Glasgow, 1911, item 99.

Kirkcaldy Museum and Art Gallery, Kirkcaldy (Fife Council, Fife Museums Central): Inv. KIRMG 1972.13. Transferred from the Royal Burgh of Kinghorn in January 1972.

124   PINT, 1622, AT KIRKCALDY

Another example of the same series, as Item **116**, but a rather more crisp casting than others. The base is dished, perhaps indicating an attempt upward adjustment in the capacity.

MEASURED CAPACITY: $106 \cdot 5$ in^3 (see Item **116**).

Kirkcaldy Museum and Art Gallery, Kirkcaldy (Fife Council, Fife Museums Central): Inv. KIRMG 1986.368.

124   **123**

125   PINT, 1622, AT PERTH

Another example of the same series, as Item **116**. The rim has been adjusted down to the level of the top band.

The 1826 Perthshire jury report records the Perth pint at 104·2 in³, whereas Swinton has 104·3 in³: George Buchanan, *Tables for Converting the Weights and Measures hitherto in use in Great Britain …* (Edinburgh, 1829), 243; [John Swinton], *A Proposal for Uniformity of Weights and Measures in Scotland* (Edinburgh, 1779), 110.

Perth Museum and Art Gallery (Perth and Kinross Council): Inv. PERGM 106. Presented by Perth Town Council in 1862 to the Literary and Antiquarian Society of Perth (established 1784), whose collections from 1914 onwards formed part of Perth Museum.

126   PINT, 1622

Another example of the same series, as Item **116**, but with no provenance.

Acquired by Glasgow Museums at auction in 1889 from the same source as the Craigengelt Weight. It is not known how or when the Craigengelt Weight (see Item **31**) came into the possession of A. Russell Pollok of Renfrewshire, but it is at least possible that the name on the weight had led to it being drawn to the notice of John Dick of Craigengelt. Recalling that Dick had acquired the Dysart Jug through his Stirling connections and that this had been inherited by his nephew before 1880, it is plausible that both items were associated at this stage and had passed to Pollok: see [James Shirra], 'The Stirling Jug', in the *Stirling Observer*, 29 July 1880. Alternatively, Pollok may have been able to acquire the Renfrew jug, recorded in the 1827 Renfrew county jury report, but no longer located: George Buchanan, *Tables for Converting the Weights and Measures hitherto in use in Great Britain …* (Edinburgh, 1829), 250.

Exhibited at the Glasgow International Exhibition, 1901, item 2441 (where the Glasgow pint, Item **121**, was also shown as item 2212).

Kelvingrove Art Gallery and Museum, Glasgow (Glasgow City Council): Inv. GLAMG 1889·56 g. Purchased at auction in 1889 from the collection of A. Russell Pollok of Renfrewshire.

125

127 UNION GALLON, QUART AND PINT SET, 1707, FOR LINLITHGOW

Set of three English liquid capacity measures, supplied from London in 1707 for the principal Scottish burghs to act as legal standards under the terms of the 1707 Act of Union. They represent a Winchester or corn gallon (about 270 in³, or 4·42 litres), and the quart and pint of the larger ale gallon (of about 280 in³, 4·58 litres), and they carry the operative date of the Act, namely 1 May 1707. Twenty-one sets were received in Edinburgh and dispatched to Stirling in 1708 for distribution to the other burghs, in recognition of Stirling's earlier role as holder of the standard of Scots liquid measure and distributor of authorised copies. However, two sets were retained by Edinburgh (Items **133** and **134**) and these are the only surviving standards of this series that do not carry Stirling's identifying countermark. All the others are stamped inside the base with an impressed mark incorporating a wolf crouching on rocks and the letters 'STR.' This was based on the council's small seal, which was provided in July 1708 'in order to the causeing of workmen

make ane litle stamp conforme theirto for marking of the setts of the severall liquid measurs': R. Renwick (ed), *Extracts from the Records of the Royal Burgh of Stirling*, 2 volumes (Glasgow, 1887-9), II, 116. Of the material distributed by Stirling, ten distinct provenanced sets have been located, together with a number of stray items with no known history, and all carry the Stirling mark.

A number of the distributed sets will not have been retained by burgh councils after the introduction of Imperial standards with different gallon and pint sizes between 1824 and 1835, and some of these will inevitably have moved into the collector's market. Four pints and a quart which have now lost their original provenance have been recorded (Items **141-145**), and there must be a number of other such pieces in private hands. Sadly, it will be apparent that some items which have passed into museum collections have also disappeared over the years. For example, the *Catalogue* of the Smith Institute, Stirling, recorded additional quart (UA.28) and wine gallon measures of 1707 in the 1934 edition, but of these only the wine gallon (Item **148**) now survives.

The three measures are cast in bronze on the

127

127

lost-wax process, the wax masters taken from tapering cylindrical patterns with moulding top and bottom, and with applied decoration and single handles. The surviving examples show some variation in the precise placing of decoration and lettering panels. The relief decoration is cast and chased round the outer circumference. For the gallon, reading from left to right from the handle, this is the royal cypher 'AR', rose, thistle, *fleur-de-lys* and harp, each of these crowned, and under them normally two scrolled labels with the legend 'ANNA : MAG : BRIT · FRA' and ': & HIBERN : REGINA : 1707'. The wax lettering is apparently moulded as a single strip but usually divided at the same place before the scroll ends are added. In one instance (Item **137**) it is spread over three labels and it can be seen that part of the inscription is normally lost, since it reads: 'ANNA : MAG : BRIT' AND 'FRANC : & HIBERN : REGINA', with the date '1707' at a lower position. The quart and pint similarly carry the crowned 'AR', rose and thistle on one side, and on the other a scrolled label with 'PRIMO : MAII : 1707'. There are, however, two variant forms of the quart and pint depending on whether the handle of the wax master was assembled on the left or the right of the measure. The interior, the base and the rim are lathe-turned after casting. The rims have been stamped with the English Exchequer 'chequer' and crowned 'AR' marks, four of each on the gallon and two of each on the quart and pint.

In English usage, the gallon standard would have been considered a dry rather than a liquid gallon and indeed its decoration and inscriptions match those of the 1707 bushel standard, described at Item **163** below. However, the perception of the Scottish administrators was presumably that it was a liquid measure, hence its assignment to Stirling rather than Linlithgow, which handled the bushel measures, and so the gallon, quart and pint are treated here as a group.

SIZES: outside diameters 185, 138, 107 mm; width across handle 225, 170, 140 mm; heights 255, 135, 105 mm; apertures 162, 118, 94 mm; depths 250, 128, 99 mm.

MEASURED CAPACITIES: The capacities of this particular set were measured to provide confirmation of the gallon type, and these measurements showed a clear distinction between the size of the gallon measure (269·5 in³), and the gallons of the quart and pint (279·6 and 278·0 in³ respectively) which were considerably larger. The capacities of the quart and pint were 69·9 and 34·8 in³. These are a close match for the Elizabethan physical standards of 1601 and 1602 at the Exchequer, identified for the Carysfort Commission in 1758, and remeasured in 1931-2, with capacities of 269.0 in³ (and another at 270.6 in³), 70.2 in³ and 34.5 in³: R. D. Connor, *The Weights and Measures of England* (London, 1987), 160; 'Report from the [Carysfort] Committee appointed to Inquire into the Original Standards of Weights and Measures in this Kingdom; and to consider the Laws relating thereto, 26 May 1758', *Reports from Committees of the House of Commons*, 3 volumes (1715-1801), II, 411-51, p. 433.

John Robison in 1800 measured the two Edinburgh quarts as 69·83 and 69·95 in³ and the two pints as 34·56 and 34·68 in³: St Andrews University

Library, MS Q171.R8, 'Professor [John] Robison's Commonplace Book', I, p. 456. He had previously commented on a set of earlier measurements – apparently those conducted for the Commissioners of Excise in the Edinburgh Council Chamber in October 1707 and repeated in the 'College Expr. Room' (the University laboratory) – and presumably extracted from the notes of Robert Stewart, one of the professors of philosophy: ibid., I, p. 442. These gave the gallon, quart and pint as 268·217, 70·11 and 33·45 in³, with the first two re-measured at the College as 267·539 and 69·92 in³, and allowed Robison to demonstrate the gallon was a corn gallon, whereas the quart and pint were fractions of the English ale gallon.

The pint from this set was one of the Union pieces whose composition was analysed in 1992: see discussion under Items **10** and **43**.

National Museums of Scotland, Edinburgh (Trustees of the NMS): Inv. NMS T.1973.134.a-c. Presented in May 1973 by the Council of the Royal Burgh of Linlithgow to the Royal Scottish Museum (now part of NMS) in advance of local government reorganisation in 1975.

128 UNION GALLON AND QUART, 1707, AT STIRLING

Another (part) set, as Item **127**, but lacking the pint measure. The quart with the Stirling countermark stamped twice.

The gallon is illustrated in A. Neish, 'The Stirling Corn Measures', in *Journal of the Pewter Society*, 6 (1987), 25-7.

Exhibited at the Glasgow International Exhibition, 1888, item 1043, and at the Glasgow International Exhibition, 1901, item 2433.

Smith Art Gallery and Museum, Stirling (Trustees of Stirling Smith Art Gallery): Inv. STIGM 1243 and 1245 (previously part of UA.3). Transferred by Stirling Town Council and exhibited by 1882; itemised in the *Catalogue of the Collections … of the Smith Institute Stirling*, second edition (Stirling, 1898), 101.

129 UNION GALLON AND QUART, 1707, AT STIRLING

Another (part) set, as Item **127**, but lacking the pint measure. The quart has been cleaned.

Smith Art Gallery and Museum, Stirling (Trustees of Stirling Smith Art Gallery): Inv. STIGM 1244 and 1246 (previously part of UA.3). Transferred by Stirling Town Council and exhibited by 1882; itemised in the *Catalogue of the Collections … of the Smith Institute Stirling*, second edition (Stirling, 1898), 101.

130 UNION GALLON, 1707, AT CUPAR

Another (part) set, as Item **127**, but lacking the quart and pint measures.

East Fife Museums Service, Cupar (Fife Council, Fife Museums East): Inv. CUPMS 1984.176. Transferred from Cupar Town Council to North East Fife District Council at local government reorganisation in 1975 (Museums Service established in 1983).

131 UNION GALLON AND QUART, 1707, AT DUMFRIES

Another (part) set, as Item **127**, but lacking the pint measure.

Exhibited at the Glasgow International Exhibition, 1901, item 2451 (incorrectly described as a gallon and half-gallon).

Dumfries Museum, Dumfries (Dumfries and Galloway Council, Museums Service): Inv. DUMFM 1936.31.7 and 6 (previously acq. 207.g and f). Acquired before 1911 from the Dumfries Burgh Weights and Measures Department. The museum, based at the Maxwelltown Observatory, Dumfries, was operated by the Dumfries and Maxwelltown Observatory Society until 1936.

132   UNION GALLON, QUART AND PINT, 1707, AT DUNDEE

Another set, as Item 127.

Exhibited at the 'Old Dundee' Exhibition, Dundee, 1892-3, under item 857 ('Eight Standard Bronze Measures … 1707'), and at the Scottish Exhibition of National History, Glasgow, 1911, items 104, 93 and 95.

McManus Galleries, Dundee (Dundee City Council, Arts and Heritage): Inv. DUNMG 1978.1355(2-4), but previously (with Item 152) 1874.318-321. Presented by the Guildry Incorporation and Magistrates of Dundee, November 1873.

133   UNION QUART AND PINT, 1707, AT EDINBURGH

Another (part) set, as Item 127, but lacking the gallon. Differing from Item 127 also in that the Stirling countermark is not present, presumably because the measures were extracted from the delivered standards before the remainder were sent on to Stirling. The quart stamped on the base 'WC / DG' (see Item 226).

The two gallons were still present with the other 1707 standards in 1789 when Robison pointed out that they were English dry gallons: Edinburgh City Archives, MS council minutes, 19 August 1789. Presumably, since they were not relevant, they were not retained and are not included in the inventories of the Deans of Guild reported to the council from 1800. For early measurement of these vessels, see Item 127. Additionally, Robison re-measured the two quarts and two pints of Items 133 and 134 in August 1800, with the quarts at 69·83 and 69·95 in^3 and the pints at 34·59 and 34·68 in^3, and the vessels identified respectively by 3, 4, 5 and 6 notches marked on their bases: St Andrews University Library, MS Q171.R8, 'Professor [John] Robison's Commonplace Book', I, p. 456.

The Museum of Edinburgh, Edinburgh (City of Edinburgh Council, Museums and Art Galleries): Inv. EDNMG HH 5988/1,2/2000. Transferred from Edinburgh Town Council by 1899 (marked with the 1899 display location 'HH 25').

134   UNION QUART AND PINT, 1707, AT EDINBURGH

Another (part) set, as Item 127, but lacking the gallon. Lacking the Stirling countermark, as Item 133, but in the variant form of the measures, with the handles reversed.

The Museum of Edinburgh, Edinburgh (City of Edinburgh Council, Museums and Art Galleries): Inv. EDNMG HH 5988/3,4/2000. Transferred from Edinburgh Town Council by 1899 (marked with the 1899 display location 'HH 25').

135   UNION GALLON, QUART AND PINT, 1707, AT HADDINGTON

Another set, as Item 127. The Stirling countermark is stamped three times in the gallon and pint, and twice in the quart.

East Lothian Museums Service, Haddington (East Lothian Council): Inv. HADDM 1996.98, 100, 97 (previously A.19, 17, 16). Transferred from

Haddington Town Council *c.*1930 and displayed in the former public library. To East Lothian District Council at local government reorganisation in 1975 (Museums Service established in 1990).

136   UNION GALLON, QUART AND PINT, 1707, AT KIRKCALDY

Another set, as Item **127**.

Kirkcaldy Museum and Art Gallery, Kirkcaldy (Fife Council, Fife Museums Central): Inv. KIRMG 1925.90.3, 2 and 1. Transferred from Kirkcaldy Town Council, 1925.

137   UNION GALLON AND QUART, 1707, FROM LANARK

Another (part) set, as Item **127**, but lacking the pint. For the variant inscription on the gallon, see Item **127**.

Stolen from the former Clydesdale District Council's premises in Spring 1990, but recovered after their sale at Christie's, Glasgow, 11 July 1990,

part of lot 120 (illustrated), where they were described as 'reproductions'. See Item **43**.

Exhibited at the Glasgow International Exhibition, 1901, item 2475, and at the Scottish Exhibition of National History, Glasgow, 1911, items 94 and 103.

South Lanarkshire Museum Service, Hamilton (South Lanarkshire Council): Inv. SLCMD 2001.94 and 95. Apparently packed and transferred from the Burgh of Lanark's Weights and Measures Department to the Town Council's offices in 1932, and passing to Clydesdale District Council (Museum Service established 1993) at local government reorganisation in 1975.

138   UNION QUART AND PINT, 1707, AT PERTH

Another (part) set, as Item **127**, but lacking the gallon.

Exhibited at the Exhibition of Local Relics and Antiquities, Perth, 1903, item 221 ('Four Measures, 1707'), when the gallon of the set was apparently also present. It is also possible that the bushel

136

may also have been acquired at the same time as this group in 1862, and may be identified as item 107 in the collection.

Museum and Art Gallery, Perth (Perth and Kinross Council): Inv. PERGM 105. Presented by Perth Town Council in 1862 to the Literary and Antiquarian Society of Perth (established 1784), whose collection from 1914 onwards formed part of Perth Museum.

139 UNION GALLON, QUART AND PINT, 1707, AT ST ANDREWS

Another set, as Item **127**. The quart and pint with St Andrews 1848 verification marks and stamped 'QUART' and 'PINT' respectively.

Described by D. Hay Fleming, *Handbook to St Andrews and Neighbourhood* (St Andrews, 1902), 137-8.

Exhibited at the St Andrews Preservation Trust Exhibition of Weights and Measures, St Andrews, 1968, item 8.

St Andrews Museum, St Andrews (Fife Council, Fife Museums East, Cupar), Inv. CUPMS 1990.325, 327 and 328. Transferred from St Andrews Town Council to North East Fife District Council in 1975 (Museums Service established 1983) and then to St Andrews Museum (opened 1990).

140 UNION GALLON, QUART AND PINT, 1707, FROM WIGTOWN

Another set, as Item **127**.

Stranraer Museum, Stranraer (Dumfries and Galloway Council, Museums Service, Dumfries): Inv. WIWMS 1986.474, 471, 472. Transferred from Wigtown County Council at local government reorganisation 1975.

141 UNION QUART, 1707

Another, as the quart of Item **127**, with no provenance.

Private collection, outside the United Kingdom. Sold at auction by Boardman, Clare, Suffolk, 22 February 1995, lot 98, having previously been in

159   WINE GALLON, 1707, FROM LINLITHGOW

Another example, as Item **146**.

Illustrated in *Things to See in the Royal Scottish Museum,* second edition (Edinburgh, 1980), 14, and in J. Calder (ed), *The Wealth of a Nation in the National Museums of Scotland* (Edinburgh and Glasgow, 1989), 187.

National Museums of Scotland, Edinburgh (Trustees of NMS): Inv. NMS T.1973.133. Presented in May 1973 by the Council of the Royal Burgh of Linlithgow to the Royal Scottish Museum (now part of NMS) in advance of local government reorganisation in 1975.

160   WINE GALLON, 1707, AT PERTH

Another example, as Item **146**.

Exhibited at the Exhibition of Local Relics and Antiquities, Perth, 1903, item 221 ('Four Measures, 1707'), along with the gallon, quart and pint (Item **138**).

159

Perth Museum and Art Gallery, Perth (Perth and Kinross Council): Inv. PERGM 108. Presented by Perth Town Council in 1862 to the Perth Literary and Antiquarian Society (established 1784), whose collections from 1914 onwards formed part of Perth Museum.

161   WINE GALLON, 1707, AT ST ANDREWS

Another example, as Item **146**.

Described by D. Hay Fleming, *Handbook to St Andrews and Neighbourhood* (St Andrews, 1902), 137-8.

Exhibited at the St Andrews Preservation Trust Exhibition of Weights and Measures, St Andrews, 1968, item 9.

St Andrews Museum, St Andrews (Fife Council, Fife Museums East, Cupar): Inv. CUPMS 1990.326. Transferred from St Andrews Town Council to North East Fife District Council in 1975 (Museums Service established 1983) and then to St Andrews Museum (opened 1990).

162   WINE GALLON, 1707, FROM WIGTOWN

Another example, as Item **146**.

Illustrated in the Royal Commission on the Ancient and Historical Monuments of Scotland (RCAHMS), *Tolbooths and Town-houses: Civic Architecture in Scotland to 1833* (Edinburgh, 1996), 6.

Stranraer Museum, Stranraer (Dumfries and Galloway Council, Museums Service, Dumfries): Inv. WIWMS 1986.473. Transferred from Wigtown County Council at local government reorganisation in 1975.

163 UNION BUSHEL, 1707, FROM LINLITHGOW

English dry capacity measure of one 'Winchester' or corn bushel, supplied from London in 1707 for the principal Scottish burghs to act as legal standards under the terms of the 1707 Act of Union. As with the other measures in this series, the bushels were received initially in Edinburgh and then sent on for wider distribution to Linlithgow, the burgh traditionally associated with issuing authorised replicas. The bronze plaques riveted to the side of the measures, and carrying the seal of Linlithgow and the word 'LINLITHGOW' in relief, were presumably manufactured and fitted in Edinburgh, and they are not present on the two examples (Items **171** and **172**) which were retained by Edinburgh council when the consignment arrived from London.

This particular measure is one of the two reference standards held by Linlithgow, and unlike all the bushels distributed to other burghs, it does not carry Linlithgow's authorisation marks of the letter 'L' impressed five times on the rim of the measure, exactly as specified for firlot measures in the 1618 Assize.

Sixteen bushel measures of this issue have been recorded by us, although only fourteen of these are currently located. At least one other is known from the literature. In 1822 the Glasgow metrologist James Cleland provided an exact description, down to the details of the 'Linlithgow' plaque and the stamped 'L' on the rim, of an example that 'deserves to be preserved' which he had found 'several years ago, in the storehouse of a friend' and which was probably the Glasgow bushel: James Cleland, *Exemplification of the Weights and Measures of the City of Glasgow* (Glasgow, 1822), 64 note.

The measures are cast in bronze using the lost-wax process as substantial shallow rounded cylindrical vessels on three feet and with two carrying handles. Cast in relief and chased round the outside are the royal cypher 'AR', rose, thistle, *fleur-de-lys* and harp, each repeated and beneath

a crown, and the legend 'ANNA · D · G · MAG · BRIT · FRANC · ET · HIBERN · REGINA · I · MAII · 1707 · ETREGNI · VI ·'. The rims have been stamped with the English Exchequer 'chequer' and crowned 'AR' marks, each four times.

SIZE: outside diameter 520 mm; height 250 mm; aperture 495 mm; depth at centre 205 mm.

MEASURED CAPACITY: The considerable weight of these measures when full makes them difficult to gauge. A single example (Item **166**) was measured for us by the Department of Consumer Protection of Grampian Regional Council in 1987 at 35·1 (± 0·1) litres, without use of a glass strike. This compares with the accepted size of the pre-Imperial bushel as 35·21 litres (2,148·3 in³), or 35·24 litres (2,150·4 in³) for the simplified dimensioned form specified in the 1696/7 Act. One of the two Edinburgh bushels (Items **171** and **172**) was measured by John Robison in 1778 and found to contain 2,148·9 in³: St Andrews University Library, MS Q171.R8, 'Professor [John] Robison's Commonplace Book', I, p. 445.

Exhibited at the Glasgow International Exhibition, 1901, item 2468.

National Museums of Scotland, Edinburgh (Trustees of the NMS): Inv. NMS T.1973.132. Presented in May 1973 by the Council of the Royal Burgh of Linlithgow to the Royal Scottish Museum (now part of NMS) in advance of local government reorganisation in 1975.

164 UNION BUSHEL, 1707, AT LINLITHGOW

Another example, as Item **163**, being the other retained bushel standard at Linlithgow, and similarly lacking the 'L' authorisation marks.

The Linlithgow Story, Linlithgow (Linlithgow Heritage Trust): not on inventory. Previously presented in May 1973 by the Council of the Royal Burgh of Linlithgow to the Royal Scottish Museum (now part of the National Museums of Scotland), Inv. NMS T.1973.X127, and held against the establishment of a local museum in Linlithgow. Returned to West Lothian District Council in March 1991 and lent by them in the same year to the Linlithgow Heritage Trust.

165   UNION BUSHEL, 1707, FROM LINLITHGOW

Another example, acquired with Items **163** and **164**, but stamped on the rim with Linlithgow's authorisation mark, and therefore assumed to be the residue of unissued stock. For full description, see Item **163**.

National Museums of Scotland, Edinburgh (Trustees of the NMS): Inv. NMS T.2001.225 (previously T.1973.X126). Presented in May 1973 by the Council of the Royal Burgh of Linlithgow to the Royal Scottish Museum (now part of NMS) in advance of local government reorganisation in 1975.

166   UNION BUSHEL, 1707, AT ABERDEEN

Another example, as Item **165**.

Described in George Skene Keith, *A General View of the Agriculture of Aberdeenshire* (Aberdeen, 1811), 553-4 and mentioned in W. Kennedy, *Annals of Aberdeen*, 2 volumes (London, 1818), II, 295.

Exhibited at the Glasgow International Exhibition, 1901, item 2406.

Town House, Aberdeen (Aberdeen City Council): Inv. THA 128b.

167   UNION BUSHEL, 1707, AT BANFF

Another example, as Item **165**.

Warrants were issued by Banff Town Council for receiving the weights and measures from the burghs of Stirling, Linlithgow and Lanark on 22 January 1709; on 16 July 'Received … ane bushell from Linlithgow English measure of metal …': W. Cramond, *Annals of Banff*, 2 volumes, New Spalding Club nos. 8 and 10 (Aberdeen, 1891-3), I, 180.

Banff Museum, Banff (Aberdeenshire Council, Aberdeenshire Heritage, Aberdeen): Inv. PEHMS B2364. Purchased by the Banffshire Field Club from Banff Town Council at some point after 1880, and on exhibit in Banff Museum by 1891: W. Cramond, ibid., I, 352 note; A. E. Mahood, *Banff and District* (Banff, 1919), 272-3.

168   UNION BUSHEL, 1707, AT CUPAR

Another example, as Item **165**.

Exhibited at the St Andrews Preservation Trust Exhibition of Weights and Measures, St Andrews, 1968, item 19.

East Fife Museums Service, Cupar (Fife Council, Fife Museums East): Inv. CUPMS 1984.174. Transferred from Cupar Town Council to North East Fife District Council at local government reorganisation in 1975 (Museums Service established 1983).

169   UNION BUSHEL, 1707, AT DUMFRIES

Another example, as Item **165**.

Exhibited at the Glasgow International Exhibition, 1901, item 2451, and at the Scottish Exhibition of National History, Glasgow, 1911, item 161 (incorrectly described as dated 1704).

180   GALLON, *c*.1710, AT STIRLING

Another example, as Item **179**.

Described and illustrated in L. Ingleby Wood, *Scottish Pewter-ware and Pewterers* (Edinburgh, 1905), 127 and plate 21 (lower right).

Smith Art Gallery and Museum, Stirling (Trustees of the Stirling Smith Art Gallery): Inv. STIGM 2963 (previously R.174 and UA.30). Transferred by Stirling Town Council and first recorded in the *Catalogue of the Collections … of the Smith Institute Stirling*, third edition (Stirling, 1934), 165.

181   GALLON, *c*.1710

Another example, as Item **179**, with no provenance.

Kelvingrove Art Gallery and Museum, Glasgow (Glasgow City Council, Museums and Art Galleries): Inv. GLAMG A.1939-25u. Ex Carvick-Webster Collection, presented by Mrs Agnes Carvick-Webster, London, in 1938.

182   GALLON, *c*.1710

Another example, as Item **179**, with no provenance.

Kelvingrove Art Gallery and Museum, Glasgow (Glasgow City Council, Museums and Art Galleries): Inv. GLAMG A.1949.92. Presented by G. L. Farqhuar, Helensburgh, in 1949.

183   GALLON, *c*.1710

Another example, as Item **179**, with no provenance.

183

Exhibited in *The Daily Telegraph Exhibition of Antiques and Works of Art, Olympia … 1928, Catalogue of Exhibits* (London, 1928), 149, item P203. (The measure is incorrectly marked P98, which was another item lend to this exhibition by the same owner.)

National Museums of Scotland, Edinburgh (Trustees of the NMS): Inv. NMS A.1945.4603. Ex Young Collection: donated by Dr A. J. Young, Christchurch, Hampshire, in 1945.

184   GALLON, *c*.1710

Another example, as Item **179**, with no provenance.

Described and illustrated in J. H. Myrtle, 'The Agnes Carvick Webster Gift of British Pewter', in *Art Gallery of New South Wales Quarterly*, 7 (1965-6), 290-311, pp.300, 304, 307, 310, and H. H. Cotterell, *National Types of Old Pewter* (New York, 1972), 36 and fig. 198.

New South Wales Art Gallery, Sydney, Australia: Inv. M62.1938. Ex Carvick-Webster Collection, presented by Mrs Agnes Carvick-Webster, London, in 1938.

185 GALLON, c.1710

Another example, as Item **179**, with no provenance.

Described and illustrated in A. Neish, 'The Stirling Corn Measures', in *Journal of the Pewter Society*, 6 (1987), 25-7.

Harvard House Pewter Museum, Stratford upon Avon (Shakespeare Birthplace Trust): pewter collection, inv. SBT 1996-44/518. Ex Neish Collection (reference 518 E), acquired from Alex Neish of Edinburgh and Barcelona, in 1997. Previously in the collection of the late Rupert Smith of Edinburgh (and subsequently of Denholm, Roxburghshire); sold by Phillips Scotland, Edinburgh, 27 June 1986, lot 13 (illustrated), and acquired by David Ingram on behalf of A. Neish.

186 GALLON, c.1710

Another example, as Item **179**, with no provenance.

Private collection, St Albans. Acquired by the Glasgow dealer P. J. O'Loughlin at a house auction in Glasgow in early 1992, but the provenance otherwise unknown. Sold to the present owner in February 2002.

187 ALE QUART, c.1710, AT STIRLING

Ale quart, constructed in a similar fashion to match the gallon measure, but with only one handle.

SIZE: diameter 120 mm; width over handle 172 mm; height 128 mm; aperture 107 mm; internal depth 125 mm.

CALCULATED CAPACITY: approximately 70 in³ (calculated from measurements), corresponding

187

to the quart of the English ale gallon and making this the direct analogue of the quart in the Union set (see Item **127**).

Described (incorrectly as a half-gallon) and illustrated in L. Ingleby Wood, *Scottish Pewterware and Pewterers* (Edinburgh, 1905), 127 and plate 21 (top right).

Smith Art Gallery and Museum, Stirling (Trustees of the Stirling Smith Art Gallery): Inv. STIGM 2966 (previously R.177 and UA.33). Transferred by Stirling Town Council and first recorded in the *Catalogue of the Collections ... of the Smith Institute Stirling*, third edition (Stirling, 1934), 165.

188 ALE QUART, c.1710, AT STIRLING

Another example, as Item **187**.

Described (incorrectly as a half-gallon) and illustrated in L. Ingleby Wood, *Scottish Pewterware and Pewterers* (Edinburgh, 1905), 127 and plate 21 (top left).

Smith Art Gallery and Museum, Stirling (Trustees of the Stirling Smith Art Gallery): Inv. STIGM 2965 (previously R.176 and UA.32). Transferred by Stirling Town Council and first recorded in the *Catalogue of the Collections … of the Smith Institute Stirling*, third edition (Stirling, 1934), 165.

189   ALE QUART, *c.*1710

Another example, as Item **187**, with no provenance.

190

190

Kelvingrove Art Gallery and Museum, Glasgow (Glasgow City Council, Museums and Art Galleries): Inv. GLAMG A.1939-25a. Ex Carvick-Webster Collection, no. 265, presented by Mrs Agnes Carvick-Webster, London, in 1939.

190   ALE QUART, *c.*1710

Another example, as Item **187**, with no provenance.

National Museums of Scotland, Edinburgh (Trustees of the NMS): Inv. NMS A.1945.4591. Ex Young Collection, donated by Dr A. J. Young, Christchurch, Hampshire, in 1945.

191   WINE GALLON, *c.*1710, AT STIRLING

Wine gallon, constructed in a similar fashion to match the gallon measure, as Item **179**, and with two handles.

191

SIZE: diameter 195 mm; width over handles 320 mm; width over handles 320 mm; height 164 mm; aperture 171 mm; internal depth 160 mm.

CALCULATED CAPACITY: approximately 225 in³ (calculated from measurements). Within the limits of accuracy of the measured dimensions, this is considered an adequate match for the wine gallon of 231 in³.

Described and illustrated in L. Ingleby Wood, *Scottish Pewter-ware and Pewterers* (Edinburgh, 1905), 127 and plate 21 (lower left).

Smith Art Gallery and Museum, Stirling (Trustees of the Stirling Smith Art Gallery): Inv. STIGM 2964 (previously R.175 and UA.31). Transferred by Stirling Town Council and first recorded in the *Catalogue of the Collections … of the Smith Institute Stirling*, third edition (Stirling, 1934), 165.

192   WINE GALLON, *c.*1710, AT STIRLING

Another example, as Item **191**.

Smith Art Gallery and Museum, Stirling (Trustees of the Stirling Smith Art Gallery): Inv. STIGM 19817/15/7-1. Transferred from Stirling Guildhall, 1990.

193   WINE GALLON, *c.*1710, AT STIRLING

Another example, as Item **191**.

Smith Art Gallery and Museum, Stirling (Trustees of the Stirling Smith Art Gallery): Inv. STIGM 19817/15/7-2. Transferred from Stirling Guildhall, 1990.

194   WINE GALLON, *c.*1710, AT STIRLING

Another example, as Item **191**.

Smith Art Gallery and Museum, Stirling (Trustees of the Stirling Smith Art Gallery): Inv. STIGM 19817/15/7-3. Transferred from Stirling Guildhall, 1990.

195   WINE GALLON, *c.*1710, AT STIRLING

Another example, as Item **191**.

Smith Art Gallery and Museum, Stirling (Trustees of the Stirling Smith Art Gallery): Inv. STIGM 19817/15/7-4. Transferred from Stirling Guildhall, 1990.

196   WINE GALLON, *c.*1710, AT STIRLING

Another example, as Item **191**, but now lacking the circular seal.

Smith Art Gallery and Museum, Stirling (Trustees of the Stirling Smith Art Gallery): Inv. STIGM 19817/15/7-5. Transferred from Stirling Guildhall, 1990.

197   WINE GALLON, *c.*1710

Another example, as Item **191**, with no provenance.

Described and illustrated in H. H. Cotterell, *Old Pewter, its Makers and Marks* (London, 1929), 116.

Kelvingrove Art Gallery and Museum, Glasgow (Glasgow City Council, Museums and Art Galleries): Inv. GLAMG A.1948-14cp. Ex Clapperton Collection, no. D/115 (purchased by him from W. R. Macgregor, 1917), and presented by Lewis Clapperton, Glasgow, in 1948.

198 CHOPIN AND FRACTIONS, EARLY EIGHTEENTH CENTURY, FOR LINLITHGOW

Set of six Scots liquid capacity measures (chopin to quarter-gill) of cast bronze. Two-part vessels with a tapering body and brazed base, and each with a brazed handle. Attached circular relief medallions in two sizes, carrying the crest of Stirling within the words 'STIRLINI OPIDVM', are applied to all but the quarter-gill measure, although the medallion is now lacking from the gill measure. The medallions differ in some respects from those used on Items **179** to **197**, notably in the spelling of Stirling.

SIZES: diameters 120, 101, 76, 65, 51, 42 mm; heights 127, 94, 70, 48, 42, 32 mm; apertures 75, 62, 53, 43, 33, 25 mm.

MEASURED CAPACITIES: 850·8, 428·4, 215·1, 107·2, 54·5, 27·2 g, or 13,130, 6,611, 3,319, 1,654, 841, 420 gr; equivalent to 51·9, 26·1, 13·1, 6·5, 3·3, 1·6 in³. These indicate a spread in pint sizes (103·9, 104·4, 104·8, 104, 105, 105 in³) suggesting that the smaller measures may not have been adjusted adequately, but implying construction against a standard of 103·9 in³.

198

The largest measure is illustrated in J. Calder (ed), *The Wealth of a Nation in the National Museums of Scotland* (Edinburgh and Glasgow, 1989), 187.

Exhibited at the Glasgow International Exhibition, 1901, item 2465.

National Museums of Scotland, Edinburgh (Trustees of the NMS): Inv. NMS T.1973.136.a-f. Presented in May 1973 by the Council of the Royal Burgh of Linlithgow to the Royal Scottish Museum (now part of NMS) in advance of local government reorganisation in 1975.

199   CHOPIN AND FRACTIONS, 1737, FOR EDINBURGH

Set of three Scots liquid capacity measures (chopin, mutchkin and gill) of copper, deeply patinated. Conical vessels beaten from single sheets of copper, with plain sheet bases attached by a turned lip from the base of the cone; the chopin with two handles, the others with one each. Engraved around the upper edge of all three measures 'M^R THO: HERIOT DEAN OF GUILD: J737'. The mutchkin and gill have their original tapered sheet copper handles, riveted at top and bottom for the mutchkin and at the top only for the gill.

The chopin has distorted replacement handles soldered in place, attached at the top to extension lugs projecting above the rim of the vessel.

SIZES: diameters 128, 101, 68 mm; width across handles 188, 130, 88 mm; heights 151, 102, 64 mm; apertures 78, 64, 39 mm; internal depths 118, 97, 61 mm.

Apparently identical measures, clearly by the same manufacturer, are illustrated in *The Connoisseur*, 13 (1905), 52, where they are described as having engraved on each: 'This measure is the same as used at Edinr., 1777'. They are shown together with a tall two-handled copper pint measure with 'INVERNESS' stamped on its base. The group, whose present location is unknown, is said to have been purchased in Inverness many years beforehand. See Allen Simpson, 'Where are they now?', in *Journal of the Antique Metalware Society*, 9 (2001), 27-8.

The Museum of Edinburgh, Edinburgh (City of Edinburgh Council, Museums and Art Galleries): Inv. EDNMG HH 5991/1-3/2000. Transferred from Edinburgh Town Council by 1899 (marked with the 1899 display location 'HH 25'). Unidentified small paper labels marked in ballpoint 'L59' on all three vessels.

199

200 PINT, BY D. HUTCHISON, 1765, FOR ABERDEEN

Scots pint liquid capacity measure of brass; conical vessel, fabricated with turret joint behind the handle and turned, with a solid base and curved solid handle riveted at top and bottom. Engraved on the face 'THIS JUGG / Was made by MR D. Hutchison A.D. 1765 / Which when carefully filled with Clear / Running water weighs 26180 Troy Grains / English, which is exactly the same weight / that the Stirling JUGG Contains. / Which JUGG is appointed by Act of / Parliament of the 19th Feby 1618 to be / the STANDARD for Settling the / MEASURES of SCOTLAND'. Added in a script hand: 'This Jugg was Gifted by Alex^r Fraser, Esqr / of Moremont to the Town of Aberdeen / Anno 1765'.

SIZE: diameter 145 mm; width across handle 190 mm; height 169 mm.

MEASURED CAPACITY: 1,700·6 g, or 26,244 gr; equivalent to 103·8 in³.

200

The 'D. Hutchison' in the inscription has been tentatively identified as David Hutchison, the nephew of James Gray, who inherited the Iron-mill at Dalkeith after the death of his uncle in 1761: 'Dalkeith', in *New Statistical Account*, 15 volumes (Edinburgh, 1845), I, 503. The stated capacity of 26,180 gr is the figure given in Gray's 1754 published account, where the controlling act is similarly given as 'act. parl. of 19 February 1618, anent settling the measures and weights of Scotland': James Gray, 'Of the Measures of Scotland, compared with those of England', in *Essays and Observations, Physical and Literary*, I (1754), 200-4, pp. 201, 202. Since Gray's association with the practical aspects of this work was well appreciated, it is entirely plausible that a commission should have been prompted by his publication. The commission may perhaps have resulted from renewed pressure from the Convention of the Royal Burghs at this time (see Chapter **9**).

The 'Alex^r Fraser, Esqr … of Moremont' also presents difficulties. Moremont, or Moremond, is in the parish of Strichen, to the north of Aberdeen. Alexander Fraser, seventh of Strichen, took his seat in the Court of Session as Lord Strichen in June 1730, and died at Strichen in February 1775, aged 76; his eldest son, also Alexander Fraser, eighth of Strichen, 'received the family estates from his father during the latter's lifetime by disposition dated 5 February 1759 … He died at Strichen 17 December 1794': Sir James Balfour Paul, *The Scots Peerage*, 9 volumes (Edinburgh, 1904-14), V, 545.

The Rev. George Skene Keith, in his 1795 report to the Aberdeen Dean of Guild on the town's weights and measures, identified this measure of 26,180 grains as 'the Stirling jug', presented by Fraser in 1765; and he distiguished it from the 'Aberdeen pint', about 400 grams heavier, which had been used to set the dry capacity measures and which was the pint recorded in a report of 1759 from the Aberdeen magistrates to the board of police: W. Kennedy, *Annals of Aberdeen*, 2 volumes (London, 1818),

II, reprinting Keith's report on pp. 293-300, especially p. 295. This larger pint (of about 105 in^3) was subsequently recorded by Keith as the standard for both dry and liquid measure: George Skene Keith, *A General View of the Agriculture of Aberdeenshire [for the] Board of Agriculture* (Aberdeen, 1811), 555. This may have been one of the poorly-adjusted 1622 pints (see Item **116**), but cannot have been an Edinburgh brewers' pint as Keith speculated: Kennedy, op. cit., II, 298. By contrast, Swinton noted that Aberdeenshire liquid measure was by the general Stirling pint, presumably by reference to the Hutchison vessel, with a 'tin Pint-stoup' of about 108·9 in^3 for liquors: [John Swinton], *A Proposal for Uniformity of Weights and Measures in Scotland* (Edinburgh, 1779), 53. Keith said in 1811 that the use of this last pint had ceased 20 years beforehand and it no longer existed: Keith, op. cit., 555.

Exhibited at the Glasgow International Exhibition, 1901, item 2406.

Town House, Aberdeen (Aberdeen City Council): Inv. THA 112.

201 CHOPIN AND MUTCHKIN, EIGHTEENTH CENTURY, AT DUNDEE

Pair of Scots liquid capacity measures in bronze; single-piece castings of uniform design with residual marks of hand-finishing on the exterior, and a turned finish on the inside and on the recessed base. The tapering cylindrical body with a single raised band at the centre, and the round-sectioned handles with scrolled ends and the upper extension forming a hollowed thumb-piece. Neither vessel carries any marking. The crispness of their finish suggests that they may have been made as companion pieces for Dundee's pint by David Rowan (Item **112**), which similarly carries no markings.

SIZES: diameters 129, 102 mm; width across handles 165, 130 mm; heights to rim 122, 100 mm; apertures 88, 68 mm; depths 113, 92 mm.

MEASURED CAPACITIES: 866·6, 442·5 g; or 13,373, 6,829 gr. This indicates pints of 105·8 and 108·0 in^3.

Exhibited at the 'Old Dundee' exhibition, Dundee, 1892, as items 857.3 and 4, and at the

201

Scottish Exhibition of National History, Glasgow, 1911, item 106 ('3 Copper Measures': comprising these two and the Rowan pint, Item 112).

McManus Galleries, Dundee (Dundee City Council, Arts and Heritage): Inv. DUNMG 1874.316 and 317 (but subsequently 1961-666-2 and 3). Presented by the Guildry Incorporation and Magistrates of Dundee, November 1873.

202   PINT, 1789, FOR EDINBURGH

Scots pint capacity measure of copper, deeply patinated, with single handle; ovoidal cylindrical vessel with conical taper to a cylindrical neck; of beaten copper sheet with turret-brazed joints; the handle brazed at the top and riveted at the bottom. Engraved on the body 'Edinʳ July 15ᵗʰ 1789 / This measure contains 26200 English Troy Grains of the clear water of the water of Leith, when of the Temp= / rature 50° of Fahrenheits Thermometer which is – / equivalent to 3 pounds seven ounces of the Lanark Trois / weight declared by Stat 1618 to be the contents of the / Scots pint / [signature in facsimile] John Robison'.

SIZE: diameter 122 mm, width across handle 163 mm, height 211 mm, aperture 44 mm.

MEASURED CAPACITY: 1,698·5 g, or 26,211 gr, equivalent to 103·6 in³. For the relationship between this (small) size and the capacity of the Stirling Jug and other early pints, see Chapter 9.

John Robison was appointed professor of natural philosophy at the University of Edinburgh in 1773 and held the chair until his death in 1805. He was one of two Edinburgh academics retained to provide technical evidence in a Court of Session case beginning in 1778 which turned on determining the true size of the firlot measures defined at the 1618 Assize. These still had important relevance in market practice and contracts,

and once legal doubts about the interpretation of the legislation had emerged, a secure resolution was crucial. The case is discussed in Chapter 9. Robison soon decided that the size of the Lanark stone could be determined by measuring an existing 1618 stone in Edinburgh (which he believed was the principal standard that had been held by Lanark) and relating his measurements to earlier ones carried out by Robert Stewart in the 1720s. By interpreting the 1618 Act literally, the pint's water contents would be 55 ounces of this stone, and his preferred result at 50°F was 26,200 grains of water, which he stated matched the capacity of the Edinburgh and Stirling jugs. This result was already implicit in a preliminary report of March 1779 to the Court of Session: National Archives of Scotland (NAS), MS Court of Session process 231/F3/1, bundle 10.

Robison had been very critical of the measuring procedure used by the official cooper of Linlithgow in the adjustment of firlots. In November 1781, he recommended to the Court a novel method of testing firlots with great accuracy by dispensing water into firlot measures

202

under test from a narrow-necked self-adjusting spherical vessel containing precisely the 21¼ or 31 pints stated in the 1618 Assize, but based on his newly-determined pint size: NAS, MS CS 231/F3/1, bundle 31. Robison confirmed his confidence in his deduced size of the pint and the advantage of adopting his proposed test method in a further report to the Court in June 1788: NAS, MS CS 231/F3/1, bundle 32. Final judgement was made in May 1791, but the Judge avoided endorsing Robison's firlot measuring technique.

This pint vessel is Robison's personal standard, constructed to embody his 'recovered' volume of the Scots pint, dated after this final report to the Court of Session and carrying his signature in facsimile. Robison claimed in 1792 that he had his standard 'ready to present to and deposit with the Chamber of Commerce at Edinburgh as a monument of his anxiety and attention to the interest of the Country, and to be an Evidence of a correct and accurate standart in a matter of great and essential public concern': Perth and Kinross Council Archives, MS CC 1/2/1/2, Perthshire County Commissioners minutes, 1 May 1792. However, it appears that he did not do so, and instead Edinburgh commissioned a set of standards based on this pint (Item 205). After Robison's death, it was one of the lots offered in the sale of his effects, where it was purchased by the Edinburgh Dean of Guild to be retained as an inventoried item in the list of the burgh's standards: Edinburgh City Archives (ECA), MS council minutes, 12 June 1805. Robison's Lanark pound weight (Item 92) was purchased on the

same occasion. The items were perhaps withdrawn before the auction, and may be the missing lot 125 in the printed *Catalogue of Optical and Astronomical Instruments ...* sold by Elliott on 28 May 1805. An annotated copy of this is in NAS GD113/5/342/74.

The pint's particular and novel characteristic is a narrow aperture, which in principle allows much greater accuracy and consistency in filling the vessel to its correct capacity. This is a feature of all the measures designed and adjusted by Robison, first seen in the 1781 design for his enclosed spherical firlot, and followed in his 1801 measures of the English and Scottish series for Edinburgh (Items 205 and 232). Duplicates of this type were ordered by a number of Scottish authorities (see Items 209, 213, 233, 235). Such a marked design improvement – not previously used in Britain, although later employed for some of the principal copies of the Imperial standard – must have exercised considerable influence.

It is not known when or by whom this pint was constructed. An isolated reference in Robison's surviving notes for this period describes a pint of this same weight made in January 1785 for use in comparing a Linlithgow standard against a 1707 bushel, but this may not be the same vessel: St Andrews University Library, MS Q171.R8, 'Professor [John] Robison's Commonplace Book', I, p. 454. The pint was not measured by James Jardine and his colleagues for their 1825 report to the Edinburgh county jury.

Probably the 'Scotch Pint Measure' exhibited by Edinburgh Corporation at the Scottish Exhibition of National History, Glasgow, 1911, item 98 (and certainly distinct from the Edinburgh 1618 pint which was exhibited as item 102).

The Museum of Edinburgh, Edinburgh (City of Edinburgh Council, Museums and Art Galleries): Inv. EDNMG HH 5992/2000. Transferred from Edinburgh Town Council by 1899 (marked with the 1899 display location 'HH 25').

203 PINT, 1791, FOR LINLITHGOW

Scots pint liquid capacity measure of copper, tinned inside, with a brass collar. Cylindrical vessel with slight inward taper, brazed turret joints with two tubular handles attached by rivets. The brass band engraved 'This Jug Contains the Exact Quantity of the Stirling Jug and was adjusted at Stirling on the 2ᵈ day of Novʳ 1791 / years In Presence of Henry Jaffray Esqʳ Provost of Stirling and James Andrew Esqʳ Provost of Linlithgow / Conform to Act of Parliament James VIᵗʰ 19 of Febʸ 1618.'

SIZE: diameter 140 mm; width across handles 275 mm; height 143 mm; aperture 110 mm.

MEASURED CAPACITY: 1,714·5 g, or 26,458 gr, equivalent to 104·6 in³.

The procedures used by the official cooper at Linlithgow in constructing authorised firlot measures raised from the 1618 Linlithgow pint (Item 114) were scrutinised by John Robison on behalf of the Court of Session in the late 1770s and 1780s (see Chapter 9). Robison was critical of the measurement practice used and made a number of recommendations for improving accuracy and consistency in the measurements. It appears to be as a consequence of these findings that the Linlithgow Dean of Guild was obliged

to order this replacement pint. It was adjusted against the Stirling pint (Item 108), which would be considered to have legal precedence, although Robison himself favoured the Edinburgh 1618 pint. The unexpectedly large capacity found here suggests that some of the practices of which Robison disapproved, such as filling from an over-filled dripping vessel with a convex water surface, may have been employed again.

Exhibited at the Glasgow International Exhibition, 1901, item 2466.

National Museums of Scotland, Edinburgh (Trustees of the NMS): Inv. NMS T.1973.137. Presented in May 1973 by the Council of the Royal Burgh of Linlithgow to the Royal Scottish Museum (now part of NMS) in advance of local government reorganisation in 1975.

204 GILL, 1800, FOR EDINBURGH

Scots gill capacity measure of brazed copper, deeply patinated; cylindrical vessel with conical taper to cylindrical neck. Engraved on the taper

204

203

'J JACKSON D.G. / 1800' and on the body 'SCOTS GILL / OZ. 3⅜ Gr. 18.. or 1637½ GRAINS of WATER / Adjusted by Prof Joⁿ. Robinson [*sic*]'.

SIZE: diameter 51 mm, height 90 mm, aperture 15 mm.

The Museum of Edinburgh, Edinburgh (City of Edinburgh Council, Museums and Art Galleries): Inv. EDNMG HH 5994/1/2000. Transferred from Edinburgh Town Council by 1899 (marked with the 1899 display location 'HH 25').

205 PINT AND FRACTIONS, 1801, FOR EDINBURGH

Set of five Scots liquid capacity measures (pint to half-gill) of copper, deeply patinated; the pint with two handles, the chopin with one handle, the remainder without; ovoidal cylindrical vessels with conical tapers to cylindrical necks; of beaten copper sheet with turret-brazed joints; the handles brazed at the top and riveted at the bottom. All engraved on the taper 'J.. JACKSON D..G.. / 1801'; and engraved on the body 'SCOTCH PINT / lb. 3.. OZ.. 6½.. Gr. 40.. or 26200 TROY GRAINS / Adjusted by Prof.ʳ. John Robinson [*sic*]', and similarly for the chopin at 1 pound 11¼ ounces 20 grains or 13,100 grains, mutchkin

at 13⅜ ounces 10 grains or 6,550 grains, half-mutchkin at 6¾ ounces 35 grains or 3,275 grains, half-gill at 1⅝ ounces 39 grains or 819 grains. (The set is completed with the 1800 gill measure, Item **204**.)

SIZES: diameters 130, 100, 85, 67, 40 mm; width across handles 220 mm (pint), 147 mm (chopin); heights 187, 154, 129, 110, 67 mm; apertures 42, 36, 20, 15, 15 mm.

MEASURED CAPACITY (of pint): 1,702·0 g, or 26,265 gr, equivalent to 103·8 in³.

The Museum of Edinburgh, Edinburgh (City of Edinburgh Council, Museums and Art Galleries): Inv. EDNMG HH 5995/1-5/2000. Transferred from Edinburgh Town Council by 1899 (marked with the 1899 display location 'HH 25').

206 PINT, *c.*1805, FOR KINCARDINESHIRE

Scots pint capacity measure, in cast bronze, with turned finish, constructed in clear imitation of the Edinburgh 1618 pint (Item **113**), and stamped twice on the rim 'WC / DG' for William Coulter or William Calder, Edinburgh Deans of Guild for 1805-7 and 1807-9 respectively. A substantial

205

vessel with single handle with scrolled ends, soldered and attached by rivets. Raised bands define two exterior fields, each with an engraved shield, the upper representing a Scottish lion *rampant* and the lower a beast from the seal of Stirling and the letter 's' below; and engraved on either side of these: 'For The Use of / Kincardine Shire'.

SIZE: diameter 172 mm; width across handle 215 mm; height 167 mm; aperture 107 mm; depth 147 mm.

MEASURED CAPACITY: 1·71 litres, or 104·3 in³ (undertaken by Grampian Regional Council, Department of Consumer Protection, 1986).

It is likely that this vessel was constructed at the same time as the similarly-inscribed half-forpet measures for Kincardineshire (Items **226** and **227**), which also have the same Edinburgh Dean of Guild stamp. The commission may perhaps have been placed as a result of the work carried out for the Aberdeen Dean of Guild by George Skene Keith in 1794. The naïve style of engraving is somewhat similar to that found on the 1811 liquid capacity measures by Charles Lunan of Aberdeen (Item **207**).

Banchory Museum, Banchory (Aberdeenshire Council, Aberdeenshire Heritage, Aberdeen): Inv. PEHMS S265. Transferred from Kincardineshire County Council to the Tolbooth Museum, Stonehaven in 1963; moved to Banchory Museum in 1994.

207   PINT AND FRACTIONS, BY CHARLES LUNAN, 1811, FOR ABERDEEN

Incomplete set of four liquid capacity measures in copper, originally from pint to half-gill, now lacking the gill and half-gill. Each made in two parts, with a hemispherical base and turret-jointed tapering ogee top, joined at a brass circumferential band, with an attached circular brass foot and brass handle. Each is engraved with the arms and supporters of Aberdeen, with the motto 'BON ACCORD' above and the largest measure with 'PINT 60^{OZ} 12^{DR}' in an engraved label and with the date '1811' below. Below this, on the central band, is engraved 'C. LUNAN'. The chopin, mutchkin and half-mutchkin are similarly engraved for 30 ounces 6 drops, 15 ounces 3 drops, and 7 ounces 9½ drops, except that the chopin is named 'CHOPINE', Lunan signs only 'C.. L..' on the mutchkin, and the date is

206

207

placed after Lunan's name on the half-mutchkin. The brass lip of the vessel has a rounded cross-section externally, but a lower edge at the aperture, making it unclear whether the vessel should be filled to this inner level or whether the meniscus should be allowed to rise to the level of the outer lip. Although these vessels remain in the Council Chamber, the fitted case is in the collection of Aberdeen Museums (inv. ABDMS 10609). A brass label on this box is in the same engraver's hand, carries an identical coat of arms, and is marked 'Distilled / WATER', 'Avoir-dupois / WEIGHT' and 'Standard / Liquid Measures / OF / ABERDEEN'.

SIZES: diameters 165, 130, 102, 85 mm; heights 225, 177, 138, 112 mm; apertures 30, 25, 18, 17 mm.

George Hogarth, Aberdeen Dean of Guild, ordered a county standard firlot for oats and bear and a set of small liquid capacity measures from the Aberdeen watch and instrument maker Charles Lunan, who also worked for Patrick Copland; they were tested by Hogarth, Copland, and George Skene in September 1811 at Marischal College: John Reid, 'Patrick Copland (1748-1822)', unpublished University of Aberdeen MLitt. thesis, 1983, appendix 2, pp. 8-10. Lunan's account for the firlot and six liquid measures was

£62 2s: personal communication from Judith Cripps, Aberdeen City Archivist, 25 January 2002.

Exhibited at the Glasgow International Exhibition, 1901, as part of item 2406, at which time both the gill and half-gill were still present.

Town House, Aberdeen (Aberdeen City Council): Inv. THA III a,b,c,d.

208   PINT, 1814, FOR CUPAR

Scots pint capacity measure of brazed copper. Tapering cylindrical vessel, with a single handle from sheet metal attached by rivets. Engraved on the body 'Stirling 1ˢᵗ Novʳ. 1814 / Adjusted by the Stirling Jug in presence of John Jaffray / Dean of Guild'. Two companion measures with English capacities are described as Item 234.

SIZE: diameter 139 mm; width across handles 207 mm; height 142 mm; aperture 111 mm.

Exhibited at the St Andrews Preservation Trust Exhibition of Weights and Measures, St Andrews, 1968, item 31, where two companion English measures are referred to as a mutchkin and Scots gill (quarter-mutchkin).

East Fife Council Museums Service, Cupar (Fife Council, Fife Museums East): Inv. CUPMS 1984.151. Transferred from Cupar Town Council to North East Fife District Council at local government reorganisation in 1975 (Museums Service established 1983).

209   CHOPIN AND FRACTIONS, 1817, FOR FORFAR

Set of four Scots liquid capacity measures (chopin, mutchkin, half-mutchkin and gill) in turned brass or copper. Ovoidal form with ogee neck, turned ring foot and attached base. To a very similar design, and probably by the same manufacturer, as the four Edinburgh, 1817, English measures (Item 235). The chopin engraved 'FORFAR / SCOTCH CHOPIN / lb 1. OZ 11¼. GR 20. OR 13100 TROY GRAINS / COPIED FROM THAT / ADJUSTED BY PROFᴿ JOHN ROBISON / BELONGING TO THE CITY OF / EDINBURGH / 1817.' The mutchkin similarly engraved 13⅝ ounces 10 grains or 6,550 grains; the half-mutchkin, 6¾ ounces 18 grains or 3,275 grains; and the gill, 3⅜ ounces 18 grains or 1,637 grains. Each stamped twice on the neck 'RJ / DG', for Robert Johnston, Edinburgh Dean of Guild, 1815-17.

208

209

SIZES: diameters 100, 87, 70, 57 mm; heights 171, 132, 103, 78 mm; apertures 29, 26, 21, 17 mm.

Meffan Institute Museum and Art Gallery, Forfar (Angus District Council, Cultural Services, Forfar): Inv. ADMUS F.1978.53, 51, 49, 47. Known to have been displayed in Forfar Library in the 1960s, and subsequently transferred by the Town Council to the museum.

210   PINT AND FRACTIONS, *c.*1820, FOR ABERDEEN

Incomplete set of Scots liquid capacity measures (pint, chopin, half-mutchkin and gill) of sheet copper, tinned inside. Originally five measures from one pint downwards, but now lacking the mutchkin. Each tapered cylindrical jug with a curved spout soldered in place and enclosing a rectangular cut below the rim to establish the water level, and opposite this a tubular handle riveted in place. Engraved on the side with label carrying the 'BON ACCORD' motto of Aberdeen, and the capacity ('PINT', 'CHOPIN', '½ MUTCHKIN' and 'GILL').

SIZES: diameters 128, 111, 77, 61 mm; width across handle and spout 235, 192, 115, 93 mm; heights 212, 152, 93, 73 mm; apertures 89, 74, 52, 39 mm.

Aberdeen Art Gallery and Museums, Aberdeen (Aberdeen City Council): Inv. ABDMS 1248, 12369, 12370, 12371.

211   PINT, BY J. & W. GRIEVE, *c.*1820, FOR SELKIRK

Scots pint measure in sheet copper with single handle; tapered conical vessel with brazed turret joints and inset base, and tinned inside; tapered hollow handle riveted in place. Stamped on the base 'J & W GRIEVE / EDIN.R' For J. & W. Grieve, see the entry in Appendix G.

SIZE: diameter 163 mm; height 146 mm; aperture 107 mm.

Scottish Borders Council, Museum and Gallery Service, Selkirk (Scottish Borders Council, Life-long Learning, Newtown St Boswells): Inv. ETLMS 1534 (previously S77). Displayed (from the late nineteenth century?) at the former Selkirk Burgh Museum (closed 1983); transferred to Ettrick and Lauderdale Council at local government reorganisation in 1975 (and in the care of the Council's Museums Service from 1978), and in 1996 to the new Scottish Borders Council.

211

212

211

213   PINT, *c.*1821, FOR PEEBLES

Scots pint measure of copper, with two handles; ovoidal cylindrical vessel with conical taper to a cylindrical neck; of beaten copper sheet with turret-brazed joints and tinned inside; the handles brazed at the top and riveted at the bottom. Engraved on the body 'PEEBLES STANDARD / SCOTCH PINT. / 3 lb 6½ oz. 40gr. or 26,200 / TROY GRAINS. / Stamped by the DEAN OF GUILD, / from the Edin.ᵣ Standard / Which was all adjusted by / Prof.ᵣ John Robinson [*sic*]. –', and stamped on the neck 'IT / DG', for John Turnbull, Edinburgh Dean of Guild, 1821-2. Acquired with a cap which is a later addition.

SIZE: diameter 127 mm; height 200 mm; width across handles 205 mm; aperture 37 mm.

Described in *Proceedings of the Society of Antiquaries of Scotland*, 64 (1929-30), 11.

Exhibited by the then owner, William Young,

212   MUTCHKIN, BY J. & W. GRIEVE, *c.*1820, FOR SELKIRK

Mutchkin ovoidal measure in turret-jointed copper, tinned inside, and with a fitting cap; engraved 'Selkirk Standard / Scots Mutchkin' and stamped on the base 'J & W GRIEVE / EDIN.ᴿ'.

SIZE: diameter 104 mm; height 121 mm; aperture 26 mm.

Private collection, Selkirk.

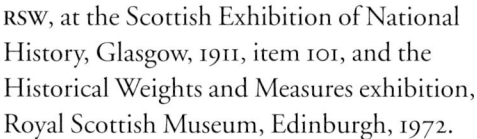

RSW, at the Scottish Exhibition of National History, Glasgow, 1911, item 101, and the Historical Weights and Measures exhibition, Royal Scottish Museum, Edinburgh, 1972.

National Museums of Scotland, Edinburgh (Trustees of the NMS): Inv. NMS H.MC.39 (1929.538). Presented by Miss Cumming, Glasgow, in 1929, to the National Museum of Antiquities of Scotland, now part of NMS.

214 WHEAT FIRLOT AND COMBINED PECK AND HALF-PECK, LATE EIGHTEENTH CENTURY, AT CUPAR

Pair of dry capacity coopered measures, the firlot measure being of conventional shape with a bar across the aperture, and two hinged iron handles, and the peck and half-peck measure being a smaller double-ended vessel. The firlot is a cylindrical vessel with slight inward taper, comprising 15 oak staves and a five-part oak base, iron-bound with two external bands and one narrow inner band at the rim; with a narrow triangular-section iron bar across the aperture

attached to a vertical rod rising from the centre of the base and riveted to crossed-iron reinforcing bars under the base. Branded on the outside 'LINLITHGOW DG H' and similarly with the crown and seal of Linlithgow (a hound bound to a tree) twice, each on both sides of the base. The rim of the measure is stamped 'L' (for Linlithgow) six times. In addition, there is an unidentified mark 'A I' impressed on the body at one of the handles. The second vessel is similarly slightly-tapering and comprises 12 oak staves and a three-part oak divider, iron-bound with three external bands; and with the same markings as the firlot, but the crown and Linlithgow seal marked once only on each side of the divider. Each rim is stamped 'L' four times.

214

SIZES: firlot: external diameter 560 mm; height 233 mm; aperture 484 mm; depth 185 mm. Double-sided measure: external diameter 357 mm; height 195 mm; apertures 330 mm (peck) and 300 mm (half-peck); depths 118 mm (peck) and 68 mm (half-peck).

CALCULATED CAPACITY (of the firlot): 2,185 in^3. This represents a close match to the value of 2,197 in^3 obtained by Gray (see Chapter **9**).

Items **214** and **215** are the only surviving measures carrying the full marking for official capacity measures issued by Linlithgow.

The branding irons for these five marks (together with a sixth, used for barley measures) survived at Linlithgow, where these two measures were made and verified, and were gifted to the Royal Scottish Museum (now part of the National Museums of Scotland) in 1973 in advance of local government reorganisation (inv. NMS T.1973.143). Although strictly outside the scope of this Inventory, they are described and illustrated here under Item **214** for completeness. They were exhibited (together with a separate brand used for the Linlithgow barley measures, see Item **215**) at the Glasgow International

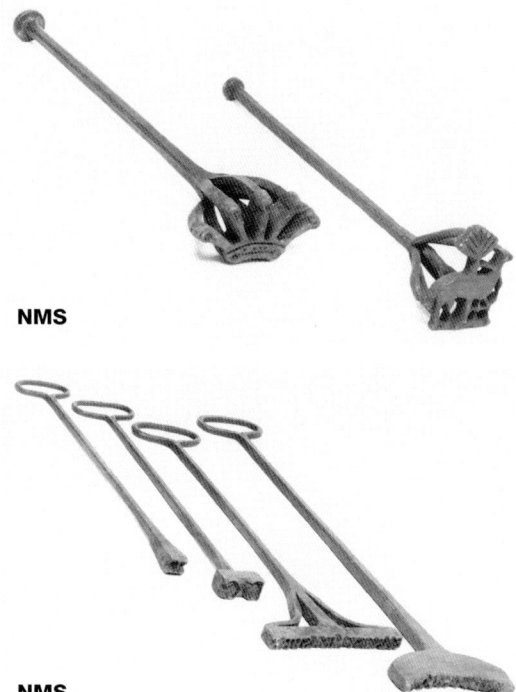

NMS

NMS

Exhibition, 1888, item 1032, but described as 'seven branding irons'; and at the Glasgow International Exhibition, 1901, item 2467; also, the *Official Catalogue of the Dawson Collection and Burgh Museum in the Council Chamber, Linlithgow* ([Linlithgow, 1907]) records items 9-15 as 'Branding Irons for the Linlithgow Firlot, viz. – 2 with Burgh Arms, 1 with Crown, 1 with letters D.G., 1 with "Barley Measure", 1 with letter "H" for Firlot of Malt, Bear, and Oats, and 1 with "Linlithgow"': another, presumably smaller, example of the burgh crest, is no longer with the group.

The measures were exhibited at the St Andrews Preservation Trust Exhibition of Weights and Measures, St Andrews, 1968, item 17.

East Fife Museums Service, Cupar (Fife Council, Fife Museums East): Inv. CUPMS 1984.162 and 165. Transferred from Cupar Town Council to North East Fife District Council at local government reorganisation in 1975 (Museums Service established 1983).

215 BARLEY FIRLOT AND COMBINED PECK AND HALF-PECK, LATE EIGHTEENTH CENTURY, AT CUPAR

Pair of dry capacity coopered measures, being the companions for the wheat measures (Item **214**), but of the separate and larger barley series introduced at the 1618 Assize. The construction of the barley firlot differs from the wheat firlot in having three rather than two external bands, and having 17 vertical staves, and the branded inscription includes the additional marking 'BARLEY ♦ MEASURE'. The barley peck and half-peck measure similarly matches the wheat measure, but now lacks one of its external bands. It is also additionally stamped 'BARLEY ♦ MEASURE'. Branding marks are as for Item **214**.

SIZES: (firlot) external diameter 570 mm; height 310 mm; aperture 490 mm; depth 267 mm. Double-sided measure: external diameter 380 mm; height 237 mm; apertures 355 mm (peck) and 315 mm (half-peck); depths 142 mm (peck) and 78 mm (half-peck).

CALCULATED CAPACITY (of the firlot): 3,210 in³. This represents a close match to the value of 3,205 in³ obtained by Gray (see Chapter **9**).

Exhibited at the St Andrews Preservation Trust Exhibition of Weights and Measures, St Andrews, 1968, item 7.

East Fife Museums Service, Cupar (Fife Council, Fife Museums East): Inv. CUPMS 1984.163 and 164. Transferred from Cupar Town Council to North East Fife District Council at local government reorganisation in 1975 (Museums Service established 1983).

215

575

216 WHEAT FIRLOT, BY JOHN MILLER, DAVID ROBERTSON AND JAMES STARK, 1754, FOR STIRLINGSHIRE

Dry capacity coopered measure of high precision quality with two hinged brass handles; cylindrical vessel in mahogany and brass formed from 28 narrow tapering staves and a single-piece base, brass-bound with three external bands and a recessed inner band at the rim secured by countersunk flat-headed brass nails; with a triangular-section brass bar across the aperture, attached at the centre to a vertical rod rising through the base and bolted to a three-legged reinforcing brace under the base. A silvered brass plate attached to the front of the vessel is engraved with the following, all within a decorative border depicting cereal crops: 'This Standard Wheat Firlot, / Linlithgow Measure, for the use of the County of Stirling, / Was Adjusted at Edinburgh the 14th of August 1754 by / Dr John Stewart Professor of Natural & Experimental Philosophy in the University / of Edinr Messrs James Gray of the Iron Mill near Dalkeith & John McGowan Writer / in Edinr When the same was ffound to Contain accurately 21 Pints & 1 Muchkin of the / Stirling Jugg, Or 73 pounds & ¾ of an Ounce French Troyes Weight of Edinburgh / Fountain Water, which after repeated Experiments was found to be precisely of the same Spe- / cifick gravity, with the clear running Water of Lieth [*sic*]. Therefore this Firlot / contains the Exact Quantity and Weight of Water Prescribed by Act of / Parliament IAs. VI 19 February 1618'. The cross bar is engraved at the centre 'DON'T LIFT / BY this BAR.' To the left of this prohibition is engraved: 'This Firlot, Wheat Measure is a Cylinder whose Diamr is

216

576

19 Inches, Its depth 7⁷⁸⁄₁₀₀ Inches, Contains of Cubical Inches 2206·18 / Deduct the Content of the Cross Bar & its Supporter, the sides of the One and yᵉ circumference of the other being of one Inʰ contains of Dᵒ Inʰ 8·84 /Remains the Content of the Wheat Firlot in Cubicall Inches 2197³⁴⁄₁₀₀'; and on the right of the cross bar: 'This Firlot, contains exactly 21 & ¼ of the Stirling Jug, Or of clear ffountain Water of Edinʳ 73 pounds & ¾ of an Ounces / French Troyes Weight, Ordain'd to be the Weight of Scotland, by Act of parliament James VI 19 Febʳʸ 1618 or 79 pounds / & 7⁶⁄₁₀ Ounces Avoirdupoise, Or 1159 Ounces English Troy.' The three arms of the support structure under the base are engraved with the names of the three craftsmen who produced the measure: 'John Miller Turner Edinʳ', 'David Robertson Smith Edinʳ', and 'Jaˢ Stark Joiner at Bristow near Edinʳ'.

Accompanying the firlot is a substantial cylindrical mahogany or rosewood strike, with fabric-covered ends, and held in a pine baize-lined trough. Circular brass end plates are engraved at one end 'STRAIK FOR THE STANDARD FIRLOT' and at the other 'STIRLING SHIRE 1754'.

SIZE: base diameter 580 mm; top diameter 546 mm; height 248 mm; aperture 482 mm (19·0"); depth 198 (7·8"); wall thickness at top about 27 mm. Cross bar width 35 mm, depth 16 mm; spindle diameter 8 mm. Strike diameter 50 mm (2·0"), length 762 mm; container length 815 mm.

A brief unsigned manuscript account of 'Experiments ascertaining The Contents of the Lithgow Wheat and Bear Firlots for the Justices of the Peace of the County of Stirling, 1754', now at Stirling Council Archives (ref. SB/T/1), was retained with the 1754 wheat and barley firlots (Items **216** and **217**) and the 'ancient Stirling pint jug' (Item **108**) at the Town House, Stirling, and as such the manuscript is recorded in an inventory on its final page as a 'paper describing the jug and firlots'. Notes added in 1801 and 1805 show that these standards were remitted to the care of the keeper of the prison. The text matches that of James Gray's published description 'Of the Measures of Scotland, compared with those of England', in *Essays and Observations, Physical and Literary*, 1 (1754), 200-4, except where Gray described the Edinburgh pint (Item **113**) and the English capacity measures. This paper was read to

the Edinburgh Philosophical Society, precursor of the Royal Society of Edinburgh, which had been revived in 1752. The capacities recorded on the measures themselves, namely 2,197^{34}/$_{100}$ and 3,205^{57}/$_{100}$ cubic inches, are those given in the manuscript and printed accounts. The volume of the wheat firlot is 21¼ pints, using Gray's value of 103·404 cubic inches for the pint, and the weight of the water contents is based on the pint holding exactly 55 ounces 'French troy', where the pound is Gray's 'Scotch pound' of 7,616 English grains. However, the 1754 manuscript also contains an important 'N[ota] B[ene]', not present in the published article, stating that 'As the dimensions prescribed by the Act of Parliament for the Diameter and Depth of these Firlots are Erroneus, they are not to be observed, the only certain method to follow in adjusting Firlots with accuracy, is by the Wet measure conformable to the said Act'. The consequences of this error of interpretation are discussed in Chapter **9**.

John Miller, the scientific instrument engineer who probably co-ordinated the work on the firlot standards for Stewart, subsequently worked as assistant and demonstrator to John Robison, who was appointed as professor of natural philosophy at Edinburgh in 1774. He was the father of the John Miller who established the scientific instrument business in Edinburgh, continued by his nephew Alexander Adie, which became the most notable and inventive scientific workshop in Scotland, and which continued to provide specialist services to the Edinburgh professoriate. The business is discussed in T. N. Clarke, A. D. Morrison-Low and A. D. C. Simpson, *Brass & Glass: Scientific Instrument Making Workshops in Scotland* (Edinburgh, 1989), 25-65.

Described and illustrated in A. D. C. Simpson, '"Handle with Care": Handling Warnings on Early Scientific Instruments and Two Early Scottish Grain Measures', in *Bulletin of the Scientific Instrument Society*, no. 30 (1991), 3-4.

The firlots and their measurement context are also discussed in A. D. C. Simpson, 'Grain Packing in Early Standard Capacity Measures: Evidence from the Scottish Dry Capacity Standards', in *Annals of Science*, 49 (1992), 337-50. Also mentioned by [John Swinton], *A Proposal for Uniformity of Weights and Measures in Scotland* (Edinburgh, 1779), 122; and Patrick Graham, *General View of the Agriculture of Stirlingshire* (Edinburgh, 1812), 339.

One of the two firlots (probably the wheat firlot, because it is in much better condition) was exhibited at the Glasgow International Exhibition, 1888, item 1042, and at the Glasgow International Exhibition, 1901, item 2438.

Smith Art Gallery and Museum, Stirling (Trustees of the Stirling Smith Art Gallery): Inv. STIGM 1239 (previously UA.2), with the strike as 1239a. Transferred from Stirling Town Council, and first recorded in *Descriptive Catalogue [of] The Smith Institute, Stirling* (Stirling, 1882), 49.

217   BARLEY FIRLOT, BY JOHN MILLER, DAVID ROBERTSON AND JAMES STARK, 1754, FOR STIRLINGSHIRE

Dry capacity coopered measure of high precision quality, in mahogany and brass, of identical construction to the wheat firlot above (Item **216**) but made taller to accommodate the larger barley firlot generated by 31 (rather than 21¼) water fills of the Stirling pint. The inscriptions are modified accordingly, and differ otherwise only in minor details.

The engraved plate on the front of the vessel describes it as the 'Standard Bear Firlot, Linlithgow Measure', for the county, and records that Stewart, Gray and McGowan found it to contain 31 pints, or 106 pounds 9 ounces 'French Troyes Weight of Edinburgh Fountain Water', which was the same density as Water of Leith, so that the firlot complied with the 1618 Act. The cross-bar inscriptions note that it is 19 inches in diameter and 11·37 inches deep, containing 3,214·65 in³, with the cross-bar and support accounting for 9·11 in³, leaving the measure's capacity as 3,205·54 in³. This is 31 pints of clear water by the Stirling Jug, weighing 106 pounds and 9 ounces 'French Troyes', 115 pounds 15 $^{87}/_{177}$ ounces avoirdupois, or 1,690 ounces 380 grains English troy.

SIZE: base diameter 572 mm; top diameter 543 mm; height 340 mm; aperture 482 mm (19·0"); depth 287 mm (11·3"); wall thickness at top about 27 mm.

There appear to be two small slips in the calculation or the engraving of the figures above. First, the stated depth (11·37 inches) is larger than observed and would result in a calculated volume of 3,224·14 in³, where the quoted 3,214·65 in³ is the correct volume of 31 pints and the support structure. (This uses Gray's result for the pint's capacity, and his approximate value of π = 3·142, found to apply in the calculations for Item **216**.) The correct volume would be obtained by using a depth of 11·337 inches, and it is suggested that 11·37 has been engraved in error rather than 11·34. Second, the somewhat eccen-

tric fraction $^{87}/_{177}$ in the avoirdupois weight is clearly an error: the correct answer is obtained with $^{7}/_{177}$, but we have been unable to reconstruct the calculation that would lead to a denominator of 177.

Smith Art Gallery and Museum, Stirling (Trustees of the Stirling Smith Art Gallery): Inv. STIGM 9136 (previously UA.24). Transferred by Stirling Town Council, and first recorded in the *Catalogue of Collections … of the Smith Institute, Stirling,* third edition (Stirling, 1934), 164.

218 WHEAT FIRLOT, 1766, FOR STIRLING

Dry capacity coopered measure, with two hinged iron handles; cylindrical vessel with slight inward taper comprising 16 oak staves and a four-part oak base, iron-bound with three external bands and one recessed inner band at the rim; with a narrow triangular-section iron bar across the aperture, attached to a vertical rod rising from the centre of the base and bolted to crossed iron reinforcing plates under the base. Stamped on the outside 'D:G' and 'P' (perhaps for 'pease' or meal); the under side of the base inscribed '1766

217

218

'/ RH / AC', and with an impression in red wax of the seal of Stirling. Adam Anderson and Thomas Bruce, in their *Report on the Stirlingshire Weights and Measures* ([Stirling], 1827), 11, name this as the 'Stirlingshire Bean Firlot' and record the base being marked 'R. H. / A. C. / with the date 1766'. The presence of the Stirling seal suggests, however, that it was made principally for the burgh and only subsequently considered a county standard. Beans and pease are both appropriate for measurement by the wheat firlot. The companion barley firlot, which Anderson and Bruce described as being unmarked except for an inscription on the cross-bar reading 'STIRLING-SHIRE BARLEY FIRLOT', has not been located.

SIZE: diameter 564 mm; height 245 mm; aperture 462 mm; inside base diameter 505 mm; depth 210 mm.

CALCULATED CAPACITY: 38·48 litres or 2,355 in³. From Anderson's quoted capacity in terms of Imperial bushels (measured as the weight of its contents in water), the firlot's volume equates to 39·20 litres or 2,390 in³. Considering that Anderson's experiment had been conducted with a 60-year old vessel that had been used exclusively as a dry measure, it must be expected

that it had absorbed some water during his determination. The dimensions are also to some extent uncertain because the wood has now shrunk and the vessel is distorted. The calculated volume is therefore considered an acceptable match to Anderson's volume, and the difference can be used to inform the identification of Items **219** and **238**.

John Swinton recorded that the customary firlot for wheat, pease, beans and rye was one of 2,378·292 in³, based on an (enhanced) firlot of 23 pints: [John Swinton], *A Proposal for Uniformity of Weights and Measures in Scotland* (Edinburgh, 1779), 122. Swinton's tabulation of this was repeated by Patrick Graham, *General View of the Agriculture of Stirlingshire [for the] Board of Agriculture* (Edinburgh, 1812), 340. The Clackmannanshire jury report of 1826 noted that pease and beans were customarily sold by the 1766 'Stirlingshire Bean Firlot', and that its volume had been measured in April 1826 by David Murray and Alexander Adie, two of the authors of the 1825 report to the Edinburgh county jury: George Buchanan, *Tables for Converting the Weights and Measures hitherto in Use in Great Britain …* (Edinburgh, 1829), 189. Their volume (given in terms of Imperial bushels) was 2,400 in³.

Smith Art Gallery and Museum, Stirling (Trustees of the Stirling Smith Art Gallery): Inv. STIGM 19817/15/8.1. Transferred from Stirling Guildhall, 1990.

219 WHEAT QUARTER-BOLL(?), EIGHTEENTH CENTURY, AT STIRLING

Dry capacity coopered measure, very similar to the Stirling Guildry wheat firlot of 1766 (Item **218**), but in poorer condition, now lacking the cross-bar and pillar, and with formed sheet metal handles. It is similarly from the holdings of the Stirling Guildry, but lacks any identification or markings.

SIZE: diameter 556 mm; height 232 mm; aperture 446 mm; inside base diameter 490 mm; depth 209 mm.

CALCULATED CAPACITY: 40·5 litres or 2,470 in³.

It is considered plausible that this is a Guildry working standard for the wheat firlot, augmented by a sixteenth, for the traditional extra charity of a peck to the boll. Had Adam Anderson provided a volume for a quarter-boll, including the charity, in his *Report on the Stirlingshire Weights and Measures* ([Stirling], 1827), it would presumably have equated to about 41·6 litres or about 2,540 in³. The difference approximately matches that between the two measurements recorded for Item **218**.

Smith Art Gallery and Museum, Stirling (Trustees of the Stirling Smith Art Gallery): Inv. STIGM 19817/15/8.2. Transferred from Stirling Guildhall, 1990.

220 BARLEY HALF-FORPET, 1783, FOR EDINBURGH

Half-forpet dry capacity measure of copper, deeply patinated, with two handles; conical measure of copper turret-jointed from a single sheet, with a flat base with turned down edge brazed into the lower part of the cone, the handles of sheet copper riveted top and bottom. Engraved 'CORN / HALF FORPET', and round the upper edge 'THOS CLEGHORN ESQr DEAN OF GUILD 1783'.

SIZE: diameter 152 mm; width over handles 216 mm; height 180 mm; aperture 73 mm; depth 163 mm.

The companion wheat half-firlot is recorded in the Dean of Guild's inventory of standards: Edinburgh City Archives, MS council minute, 23 October 1805. John Robison commented on their capacity in a letter of July 1789, copied in ibid., 19 August 1789. For similar vessels, see Items **226** and **227**.

The Museum of Edinburgh, Edinburgh (City of Edinburgh Council, Museums and Art Galleries):

220

Inv. EDNMG HH 5997/2000. Transferred from Edinburgh Town Council by 1899.

221 WHEAT FIRLOT, LATE EIGHTEENTH CENTURY, FOR DUNDEE

Dry capacity coopered measure, with two hinged iron handles. The vessel is cylindrical with a slight inward taper, constructed with 15 oak staves and a four-part oak base, with two external bands and one narrower inner band at the rim. The top bands riveted together in four places and elsewhere nailed. The exterior painted with wood grain effect, and lettered 'GUILDRY DUNDEE'. Impressed brand mark of Dundee's seal (a pot of lilies) four times on the rim and a slightly larger size four times on the inside base. Cut under the base is the letter 'A'.

SIZE: diameter 528 mm; height 265 mm; aperture 425 mm; depth 230 mm.

CALCULATED CAPACITY: 36·2 litres or 2,210 in³.

This measure has been associated with the so-called 'coal mett' (Item **239**) which is about twice the capacity of this measure, and it has therefore been familiarly described as the 'half-

221

mett'. Although the measure has no capacity marked on it, it is clearly a wheat firlot because it is of the standard firlot form, and its capacity matches the Dundee wheat firlot capacity of 2,240 in³ derived from the Forfar county jury report of 1826: George Buchanan, *Tables for Converting the Weights and Measures hitherto in use in Great Britain …* (Edinburgh, 1829), 216. It is therefore unrelated to the 'coal mett'. The most obvious difference is that this vessel has been constructed with a wall thickness which is about twice that of the 'coal mett'.

McManus Galleries, Dundee (Dundee City Council, Arts and Heritage): Inv. DUNMG 1978.1357. However, the associated coal mett (1978.1358) is a re-registration of 1937.263.1, suggesting that this measure was originally 1937.236.2. The likely source is therefore the Weights and Measures Department of Dundee Corporation .

222 WHEAT FIRLOT OF THE ROBISON FLASK TYPE, *c*.1788, AT ABERDEEN

Copper vessel in the form of a prolate spheroid, for checking the capacity of conventional wheat firlots using water. The vessel is filled through a cylindrical neck at the top, up to a lip cut in the rim, and excess is carried off by a spout. A pipe, with a stop-cock, extends from the bottom of the measure and allows the contents to be let into an adjacent firlot. The vessel is beaten from copper sheet in two hemispheres, with a horizontal overlapped joint. It is supported on four iron legs, joined to a brass band which is clamped just below this joint. The cylindrical neck has subsequently been extended upwards with an additional notched aperture, to convert the vessel to the marginally-larger size of the Imperial bushel. The upper hemisphere is engraved on one side with the arms and sup-porters of the City of Edinburgh and the motto 'NISI DOMINUS FRUSTRA'. The other side is

222

is 0·325 litres or 20·0 in³ (78 mm diameter × 68 mm high). Assuming that the bushel is accurate, this provides an independent check that the firlot is about 2,199 in³ and that Robison's pint is about 103·5 in³; this is certainly less than the Stirling Jug with which it is ostensibly compatible. The stated size of Item **223**, taken from Robison's standard is 2,199 in³.

This pattern of measure was first proposed by John Robison, professor of natural philosophy at the University of Edinburgh, and Sir George Clerk-Maxwell, then a Commissioner of Excise and of Customs, in response to a commission from the Court of Session in February 1781 to resolve the correct method of raising a firlot from the standard Scots pint, to accord with the 1618 legislation. The background to this issue and Robison's interpretation of the legislature's intentions is discussed in Chapter **9**. The report by Robison and Clerk-Maxwell was received in November 1781 and printed in 1782: copies in the National Archives of Scotland (NAS), MS Court of Session process CS 231/F3/1, bundle 31, and Signet Library, Court of Session papers, volume 192. The report was subsequently reprinted in the Highland Society of Scotland's *Report on*

engraved, in a different hand, with the arms and supporters of the City of Aberdeen and the motto 'BON ACCORO [*sic*]', and beneath this is engraved: 'STANDARD WHEAT FIRLOT / made from the Standard Scots Pint in the possession of / THE CITY OF EDINBURGH / Containing 21¼ Scots pints and weighing 79 libs 7.oz. 3.dr. / (Avoirdupois) pure River Water / ANNO 1811.'

SIZE: overall height 1,390 mm; central flask diameter 370 mm; height 500 mm; outlet extending 450 mm beyond flask axis; aperture 78 mm; distance between firlot and bushel overflow levels 68 mm.

CAPACITY: a check of the capacity by Grampian Regional Council, Department of Consumer Protection in 1987 gave 36·4 litres or 2,220 ± 2 in³, against a theoretical size of 2,219 in³ for the Imperial bushel. The volume of the extension piece which increased the capacity to one bushel

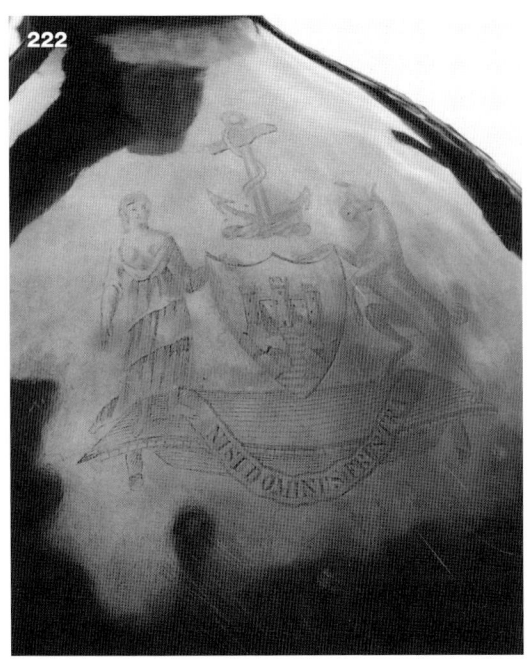

Weights and Measures (Edinburgh, 1813), 74-6, and the 'Report from the Select Committee on Weights and Measures' (HC 290), *Parliamentary Papers* 1813-14, III, appendix 4. Their proposed method of filling firlots under test from an adjusted spherical container with a stop-cock was approved by the Court in July 1782, and standards were ordered to be made. However, it was not until June 1788 that Robison reported that he had adjusted two standard firlots, for wheat and barley, and they were 'in my class room in the University awaiting the sanction of the court': NAS, CS 231/F3/1, bundle 32. Nevertheless, in the final judgement in May 1791, Lord Alva failed to sanction Robison's method. In spite of this, the county commissioners for Perthshire decided to purchase two duplicates of these firlots, to be adjusted by Robison (see discussion at Item **223**), and their initial report of April 1792 on this proposal noted that Robison's two standards were then 'lodged in the Custody of the Magistrates of Edinburgh': Perth and Kinross County Archives MS CC1/2/1/2, Perthshire county commissioners minutes, 1 May 1792.

It is not known who financed these initial firlots – it may have been Robison or his copper-smith, banking on future sales stimulated by the endorsement of the Court. Nor is it known if they were ever used by Edinburgh's Dean of Guild, who certainly did not include them on his annual inventory of standards supplied to the Council from 1800, although these inventories included no other large dry measures except for the two 1707 Winchester bushels. It is conceivable that they may have been the other two unnamed measures acquired by the Dean of Guild in 1805 at the sale of Robison's effects, together with Robison's 1789 pint (Item **202**), but these are more likely to have been two small unfinished measures now in Edinburgh City Museums (unregistered): Edinburgh City Archives, MS council minutes, 12 June 1805.

The Aberdeen wheat and barley (or rather pease and beans) firlots were reckoned in George Skene Keith's 1795 report for the Dean of Guild to be 26 and 34 pints respectively, where the pint was criticised for being larger than the Stirling Jug (and see Item **200**): William Kennedy, *Annals of Aberdeen*, 2 volumes (London, 1818), II, 295. Writing fifteen years later, he noted that with the growing importance of wheat cultivation, the magistrates had concluded that a new firlot should be made, but one based specifically on Linlithgow measure: George Skene Keith, *General View of the Agriculture of Aberdeenshire [for the] Board of Agriculture* (Aberdeen, 1811), 553. In 1811 Patrick Copland, professor of natural philosophy at Marischal College, Aberdeen, measured the Edinburgh length standards (Items **2** and **4**) against his own standard scale (made by Edward Troughton, the eminent London instrument maker), apparently at the request of the mathematician John Playfair, who had succeeded Robison in the Edinburgh chair of natural philosophy. The measures were taken to Aberdeen by James Jardine, who assisted Copland in their measurement; but discussion had clearly extended to the renewal of Aberdeen town council's weights and measures because at the same time new standard weights were received from Edinburgh, as were a pair of Robison's pattern of copper firlot measures. The Dean of Guild's accounts for Michaelmas 1810-11 record a payment of £42 5s 2d to the Edinburgh coppersmith Henry Armstrong for barley and wheat firlots: personal communication from Judith Cripps, Aberdeen City Archivist, 25 January 2002.

The wheat firlot remains at Aberdeen, but probably only because it could so easily be converted to measure the new Imperial bushel size, first introduced in 1824, by making a small extension to its neck. The larger barley firlot does not survive. They were both tested by Keith, Copland and George Hogarth, Aberdeen's Dean of Guild, in September 1811, and the weight of water content of the wheat firlot is what is now engraved under the Aberdeen coat of arms: John Reid, 'Patrick Copland (1748-1822)', unpublished University of Aberdeen MLitt thesis, 1983, appendix 2, pp. 8-10. The cost of this engraving seems to be the sum of £1 0s 1d paid to John Leith, coppersmith in Aberdeen, in the Dean of Guild's accounts (see above).

The surprising aspect of the wheat firlot is that it also carries the engraved arms and supporters of Edinburgh, but no explanatory inscription. We conclude that this is one of the two original firlots produced for the Court of Session, for which a ready sale to the town council was anticipated following the formal approval and sanction that the Court never gave for the method. The fact that it was bought from Armstrong suggests that the measure had been made at his risk and had remained on his hands.

Aberdeen Art Gallery and Museums, Aberdeen (Aberdeen City Council): Inv. ABDMS 16014.

223 WHEAT FIRLOT, 1793, FOR PERTHSHIRE

Dry capacity coopered measure, with two riveted iron handles; heavily constructed oak vessel with 16 oak staves tapering in thickness to form a cylindrical interior, with five part oak base, iron-bound with two external bands and an iron ring inset into the thickness of the wooden rim; shallow triangular-section iron cross-bar across the aperture, attached at the centre to an iron spike rising from the centre of the base. Deeply carved on the outer surface 'PERTH / WHEAT FIRLOT', with the apparent contemporary addition of the words 'For the County of' before 'PERTH'; on the inside of the base 'F.F. / AD NORMAM / JOA: ROBISON / IN ACAD. EDIN. / PHYS. PROF. / MDCCXCIII' [made to the standard of John Robison, University of Edinburgh, professor of physics, 1793]; and engraved on the cross-bar 'Solid Contents = 2199 Inches'. There are no verification marks.

SIZE: diameter 585 mm; height 250 mm; aperture 484 mm; depth 197 mm.

CALCULATED CAPACITY: 36·17 litres or 2,205 in³ (taking account of the cross-bar and support). This is an acceptable match to the stated volume of 2,199 in³.

Following the conclusion in 1791 of the Court of Session case against the burgh of Linlithgow, which had emphasised the illegal variation in the size of the grain firlots, the Perthshire county commissioners set up a committee to investigate their position in relation to the judgement. The committee's report, considered in May 1792, indicated that they had consulted John Robison, and that he had already adjusted a duplicate of his own standard pint (Item **202**) for the county, which had now been stamped at Stirling; and that he had also provided a mutchkin (enabling a firlot of 21¼ pints to be measured), check-weights for the pint and its contents and a set of weights representing his recovered 1618 standard: Perth and Kinross Council Archives, MS CC1/2/1/2, Perthshire county commissioners minutes, 1 May 1792. Robison had also offered to order and adjust two of his copper firlot gauges, and this was agreed in April 1793: ibid., 30 April 1793. In October a committee was instructed 'to prepare Duplicates of the standards to be Transmitted to the Districts': ibid., 1 October 1793. None of these standards have survived, but the two wooden firlots, Items **223** and **224**, are presumably two of the 'Duplicate' measures, and they are certainly marked to indicate that they follow Robison's standard.

The purchase of the standards (for £24 7s 9d) was arranged through an Edinburgh solicitor, so the manufacturer's name is not known.

Perth Museum, Perth (Perth and Kinross Council): Inv. PERGM SHNN 532. Provenance unknown.

224   BARLEY FIRLOT, 1793, FOR PERTHSHIRE

Barley firlot to match the wheat firlot (Item **223**), of near identical construction, but with 18 staves and a six-part base, and with three external bands; inscriptions as above, but with 'PERTH / BEAR FIRLOT' and with the contents given as 3,208 cubic inches.

SIZE: diameter 586 mm; height 310 mm; aperture 502 mm; depth 265 mm.

Perth Museum, Perth (Perth and Kinross Council): Inv. PERGM SHNN 530. Provenance unknown

223

224

225 POTATO FORPET AND HALF-FORPET, 1801, FOR EDINBURGH

Double-ended dry capacity coopered measure for potatoes (forpet and half-forpet); slightly tapering, with oak staves and one-piece divider, iron-bound with three bands and with two iron handles. Stamped in three positions on each rim 'J·J'; branded on the divider 'DG' on the forpet side and with a large ornate 'E' (Edinburgh?) on the other; branded on the external surface 'POTATOE MEASURE / FORPET 7^{LB} AVOIRDUPOIS [this line added subsequently] / J JACKSON D.G / 1801 / JB'.

SIZE: diameter 232 mm; width over handles 295 mm; height 256 mm; apertures 206 mm (forpet)

and 185 mm (half-forpet); depths 159 mm (forpet) and 90 mm (half-forpet); base diameter 195 mm.

The Museum of Edinburgh, Edinburgh (City of Edinburgh Council, Museums and Art Galleries): Inv. EDNMG HH 5998/2000. Transferred from Edinburgh Town Council at an unknown date.

226 WHEAT HALF-FORPET, *c.*1805, FOR KINCARDINESHIRE

Half-forpet dry capacity measure of copper, with two handles. Conical vessel of copper formed from a single turret-jointed sheet; a flat base

225

with turned down edge brazed into the lower part of the cone, the handles of sheet copper riveted top and bottom. Engraved 'WHEAT / & / PEASE / HALF / FORPET' and round the upper edge 'FOR THE USE OF KINCARDINESHIRE'. Stamped twice at the top 'WC / DG' for William Coulter or William Calder, Edinburgh Deans of Guild for 1805-7 and 1807-9 respectively. Subsequently stamped 'IMP! QT' and with an 1882 verification mark for the County of Kincardineshire.

SIZE: diameter 133 mm; width across handle 195 mm; height 168 mm; aperture 72 mm; depth 152 mm.

CAPACITY: 1·14 litres or 69·6 in³, from a firlot of 2,220 in³ (measured by Grampian Regional Council, Department of Consumer Protection, 1986).

It is likely that this vessel was constructed at the same time as the similarly-inscribed barley half-forpet measure, and the pint, for Kincardineshire (Items **227** and **206**), which also have the same Edinburgh Dean of Guild stamp. The commission may perhaps have been placed as a result of the work carried out for the Aberdeen Dean of

Guild by George Skene Keith in 1794. The vessel has been stamped with the mark of the closely-approximating Imperial size, the quart being 69·3 in³.

Banchory Museum, Banchory (Aberdeenshire Council, Aberdeenshire Heritage, Aberdeen): Inv. PEHMS S267. Transferred from Kincardineshire County Council to the Tolbooth Museum, Stonehaven, in 1963; moved to Banchory Museum in 1994.

227 BARLEY HALF-FORPET, *c.*1805, FOR KINCARDINESHIRE

Another example as above, but engraved 'CORN / HALF FORPET' and round the upper edge 'FOR THE USE OF KINCARDINESHIRE', stamped 'WC/DG' but without any subsequent marks or verification.

SIZE: diameter 153 mm; width across handle 208 mm; height 178 mm; aperture 85 mm; depth 161 mm.

CAPACITY: 1·64 litres or 100·1 in³, from a firlot of 3,200 in³ (measured by Grampian Regional

226

227

Council, Department of Consumer Protection, 1986).

Banchory Museum, Banchory (Aberdeenshire Council, Aberdeenshire Heritage, Aberdeen): Inv. PEHMS S268. Transferred from Kincardine-shire County Council to the Tolbooth Museum, Stonehaven, in 1963; moved to Banchory Museum in 1994.

228 WHEAT HALF-FIRLOT, BY JOHN BLACKIE AND WILLIAM EVANS, 1823, FOR SELKIRKSHIRE

Dry capacity coopered measure with two riveted iron handles; cylindrical vessel with 10 oak stakes and a three-part base, iron-bound with two external bands (the top one now missing) and a recessed inner band at the rim; with a narrow triangular-section iron bar across the aperture, attached to a vertical rod rising from the centre of the base and bolted to crossed iron reinforcing plates under the base. Painted externally, and lettered in gilt 'SELKIRK-SHIRE OLD STANDARD / WHEAT PECK, of 1707, / Restored, & Adjusted in / 1823, by Alex.ʳ Kinghorne Engin[ee]ʳ.' and marked 'W*H / D.G', for Walter Hogg, Dean of Guild for Selkirk, 1821-3.

228

SIZE: diameter 430 mm; height 200 mm; aperture 375 mm; internal base diameter 368 mm; depth 174 mm.

CALCULATED CAPACITY: The volume of the wheat peck from the measured dimensions, less the internal structure, is 18·8 litres or 1,145 in³ (namely half a firlot of 37·6 litres or 2,290 in³), compared with Kinghorne's figure of 1,140 in³.

Alexander Kinghorne, civil engineer, was commissioned by Selkirk's Dean of Guild to replace the old county measures, then kept in the burgh of Selkirk, reporting to the Magistrates and Town Council of Selkirk in April 1827: Scottish Borders Council Museum and Gallery Service, Selkirk, report of 29 April 1827, copied in MS Selkirk Town Council minute book 10, pp. 76-80. He quoted Swinton's account of 1779 to confirm the correct relationship between the pint and the 'old standards' of wheat and barley measure, namely the wheat 'peck' and the 'half fow' or 'half full' (each given as half of the respective firlots). It is not clear how he proceeded, merely that he followed 'proper and accurate principles' in supervising the work of John Blackie, carpenter, and Williams Evans, cooper, who also witnessed his report. The Dean of Guild had reported that proper standards 'were greatly wanted, the old Half full Measure belonging to the Burgh having been entirely decayed through age': ibid., MS council minutes, 29 May 1823.

Although a pint following the pattern of the 1618 issue was recorded only four years later in the 1827 county report on the relationship of the old standards to the new Imperial standards, it was not mentioned in Kinghorne's report, and it appears that his approach was to construct vessels that would have the volumes given by Swinton, namely about 1,140 in³ for the wheat measure and 1,615 in³ for the barley measure.

Scottish Borders Council, Museum and Gallery Service, Selkirk (Scottish Borders Council, Lifelong Learning, Newtown St Boswells):

Inv. ETLMS 01.0071 (previously S75). Displayed (from the late nineteenth century?) at the former Selkirk Burgh Museum (closed 1983); transferred to Ettrick and Lauderdale Council at local government reorganisation in 1975 (and in the care of the Council's Museums Service from 1978), and in 1996 to the new Scottish Borders Council.

229 BARLEY HALF-FIRLOT, BY JOHN BLACKIE AND WILLIAM EVANS, 1823, FOR SELKIRKSHIRE

Dry capacity coopered measure with two riveted iron handles; cylindrical vessel with 13 oak stakes and a three-part base, iron-bound with three external bands (the top one now missing) and a recessed inner band at the rim; with a narrow triangular-section iron bar across the aperture, attached to a vertical rod rising from the centre of the base and bolted to crossed iron reinforcing plates under the base. Painted externally, and lettered in gilt 'SELKIRK-SHIRE OLD STANDARD / BARLEY, & OAT HALF-FULL, of 1707, / Restored, & Adjusted in 1823, / by Alex.ʳ Kinghorne Engin[ee]ʳ.' and marked on the other side 'W∗H / D.G', for Walter Hogg, Dean of Guild for Selkirk, 1821-3.

SIZE: diameter 500 mm; height 210 mm; aperture 436 mm; internal base diameter 440 mm; depth 180 mm.

CALCULATED CAPACITY: The volume from the measured dimensions, less the internal structure, is 27·0 litres or 1,640 in³ (namely half a firlot of 54·0 litres or 3,280 in³), compared with Kinghorne's figure of 1,615 in³.

Scottish Borders Council, Museum and Gallery Service, Selkirk (Scottish Borders Council, Lifelong Learning, Newtown St Boswells): Inv. ETLMS 01.0070 (previously S74). Displayed (from the late nineteenth century?) at the former Selkirk Burgh Museum (closed 1983); transferred to Ettrick and Lauderdale Council at local government reorganisation in 1975 (and in the care of the Council's Museums Service from 1978), and in 1996 to the new Scottish Borders Council.

C5: English Pre-Imperial Capacity Standards

230 GALLON, 1800, FOR EDINBURGH

English gallon capacity measure of sheet copper, deeply-patinated; squat ovoidal conical vessel with no neck and with a separate base soldered in. Two riveted handles. Engraved 'ENGLISH GALLON / lb.. 7.. OZ.. 9¾.. GR.. 44.. or 58484 GRAINS TROY / Adjusted by Prof.ʳ John Robinson [sic] / 1800.' Subsequently engraved above this 'ROBERT JOHNSTON. D.G. / 1817.' And engraved below the main inscription 'Readjusted, / By JAMES JARDINE / CIVIL ENGINEER. / 1817.'

SIZE: diameter 205 mm; width across handles 212 mm; height 225 mm; aperture 45 mm.

Robison has taken the 1706 statutory size of the English wine gallon, namely 231 in³, and has used Everard's 1696 water density result of 253·18 troy grains per cubic inch to obtain the gallon size of 58,484 grains. Robison presumably chose the wine gallon as the basis for his English measures because this reflected the likely area of application. The same figures are used for measures which survive at Dumfries and Forfar, and the whole series of sizes from the gallon to half-quartern (or half-gill) are quoted in Robison's letter of 11 August 1801: Edinburgh City Archives, MS council minutes, 16 September 1801. (James Cleland's 1822 published version of this correspondence does not detail the sizes of the measures.)

The Museum of Edinburgh, Edinburgh (City of Edinburgh Council, Museums and Art Galleries): Inv. EDNMG HH 5993/2000. Transferred from Edinburgh Town Council by 1899 (marked with the 1899 display location 'HH 25').

230

231 GILL, 1800, FOR EDINBURGH

English gill capacity measure of bronzed copper, deeply patinated; in three parts with cylindrical centre section attached by overlapping joints to a curved base with flat centre section and to a conical upper part, which tapers to a cylindrical neck. Engraved 'ENGLISH QUARTERN / 1828. TROY GRAINS [subsequently added at this point: 'or oz 3¾ Gr 28'] / Adjusted by Prof.ʳ John Robinson [sic]' and above this engraved 'J. JACKSON D.G. / 1800.'

SIZE: diameter 57 mm; height 90 mm; aperture 17 mm.

The volume of this gill has been obtained as a fraction from the English wine gallon of 231 in³. Further measures (Item **232**) were made in 1801 to provide a complete range of sizes from the 1800 English (wine) gallon to this gill.

The Museum of Edinburgh, Edinburgh (City of Edinburgh Council, Museums and Art Galleries): Inv. EDNMG HH 5994/2/2000. Transferred from Edinburgh Town Council by 1899 (marked with the 1899 display location 'HH 25').

232 HALF-GALLON AND FRACTIONS, 1801, FOR EDINBURGH

Set of four English liquid capacity measures in sheet copper, deeply patinated; cylindrical turret-jointed vessels comprising half-gallon, quart, pint, half-quartern (half-gill), with conical shoulders and narrow cylindrical necks, and with riveted handles. The largest engraved: 'ENGLISH HALF GALLON / lb 3.. oz 12⅞ Gr 22 or 29242 TROY GRAINS / Adjusted by Prof.ʳ. Joⁿ Robinson [sic]' and 'J. JACKSON, D.G. / 1801.'; and similarly for the quart, pint and half-quartern as 1 pound 14¾ ounces 41 grains or 14,621 grains, 15⅞ ounces 50½ grains or 7,310½ grains, 1⅞ ounces 14 grains or 914 grains.

SIZES: diameters 128, 98, 84, 43 mm; width across handles 212, 150 mm; heights 208, 159, 141, 67 mm; apertures 38, 35, 22, 14 mm.

Made in 1801 to complement the gallon and gill (Items 230 and 231) to form a series derived from the English wine gallon.

The Museum of Edinburgh, Edinburgh (City of Edinburgh Council, Museums and Art Galleries): Inv. EDNMG HH 5995/6-9/2000. Transferred from Edinburgh Town Council by 1899 (marked with the 1899 display location 'HH 25').

233 GALLON AND FRACTIONS, c.1801, FOR DUMFRIES

Incomplete set of four liquid capacity measures (gallon, half-gallon, pint and half-pint) in sheet copper, deeply patinated. Squat ovoidal gallon with no neck and on a ring foot, and cylindrical vessels with conical shoulders and narrow cylindrical necks for the half-gallon, pint and half-pint; with turret joint on the body sections and two riveted handles each of the gallon and half-gallon. Lacking the quart and quartern (or gill) and probably a half-quartern. The gallon engraved 'ENGLISH GALLON / lb 7 oz 9¾ Gr 44 OR 58484 TROY GRAINS // DUMFRIES STANDARD' and stamped twice near the aperture 'TH / DG', for Thomas Henderson, Edinburgh Dean of Guild, 1801-3; and similarly for the half-gallon at 3 pounds 12⅞ ounces 22 grains or 29,242 grains, the pint at 15⅛ ounces 50½ grains or 7,310½ grains, and the half-pint at 7½ ounces 53 grains or 3,655¼ grains.

233

SIZES: diameters 209, 130, 87, 67 mm; heights 212, 205, 125, 112 mm; apertures 44, 41, 26, 23 mm.

The half-gallon measure exhibited at the Scottish Exhibition of National History, Glasgow, 1911, item 100.

Dumfries Museum, Dumfries (Dumfries and Galloway Council, Museums Service): Inv. DUMFM 1990.1, 1936.31.1, 31.3, 31.2 (the smaller measures previously acq. 207.a-c). Acquisition date of the gallon measure is unknown (although tentatively thought to be 1953). The other measures were acquired by the Maxwelltown Museum before 1911 from Dumfries Burgh Weights and Measures Department. The museum, based at the Maxwelltown Observatory, Dumfries, was operated by the Dumfries and Maxwelltown Observatory Society until 1936.

234 PINT AND GILL, 1814, FOR CUPAR

Pair of English liquid capacity measures (pint and gill) of brazed copper. Tapering cylindrical vessels, with single handles from sheet metal attached by rivets. Both engraved on the body

'Stirling 1ˢ.ᵗ Nov.ʳ 1814 / Adjusted by the Stirling Jug in presence of John Jaffray / Dean of Guild'. Although this implies that these companions to the Scots pint (Item 208), are to a Scottish standard, in fact they represent English measure.

SIZES: diameters 91, 56 mm; width across handles 136, 86 mm; heights 93, 63 mm; apertures 76, 47 mm.

Exhibited at the St Andrews Preservation Trust Exhibition of Weights and Measures, St Andrews, 1968, item 31, where they are described as a mutchkin and Scots gill (quarter-mutchkin).

East Fife Council Museums Service, Cupar (Fife Council, Fife Museums East): Inv. CUPMS 1984.150, 149. Transferred from Cupar Town Council to North East Fife District Council at local government reorganisation in 1975 (Museums Service established 1983).

235 QUART AND FRACTIONS, c.1815, FOR FORFAR

Set of four liquid capacity measures (quart, pint, half-pint and gill) in sheet copper, deeply

234

patinated. Rounded cylindrical vessels with conical shoulders and narrow cylindrical necks with turret joints on the body. To the same design, and with the same style of inscription, as the smaller Edinburgh, 1801, Scots measures (Item 205). The quart engraved 'FORFAR / – / ENGLISH QUART / lb 1. OZ. 14⅜. Gr 41 or 14621 TROY GRAINS, / AJUSTED [*sic*] BY PROF.^R / JOHN ROBINSON [*sic*]'. The pint similarly engraved 15⅛ ounces 50½ grains or 7,310½ grains; the half-pint, 7½ ounces 55 grains or 3,655¼ grains; the quartern (gill), 3¾ ounces 28 grains or 1,828 grains. Each stamped at the neck 'RJ / DG', for Robert Johnston, Edinburgh Dean of Guild, 1815-17.

SIZES: diameters 108, 88, 69, 57 mm; heights 171, 150, 110, 97 mm; apertures 31, 23, 17, 14 mm.

Meffan Institute, Museum and Art Gallery, Forfar (Angus District Council, Cultural Services, Forfar): Inv. ADMUS F.1978.54, 52, 50, 48. Known to have been displayed in Forfar Library in the 1960s, and subsequently transferred by the Town Council to the museum.

236 QUART AND FRACTIONS, 1817, FOR EDINBURGH

Set of four liquid capacity measures (quart, pint, half-pint and gill) in cast bronze, with turned finish, each with a base flange, narrow neck and reinforcing ring near the rim. The largest engraved 'ROBERT JOHNSTON, D.G. / 1817. / ENGLISH QUART / lb. 1. OZ 14⅝. Gr. 41 or 14621 TROY GRAINS, / Compared with the Standard / Adjusted by Prof.^r JOHN ROBINSON [*sic*] / 1800, / By JAMES JARDINE / CIVIL ENGINEER / 1817.'; and similarly for the quart, pint, half-pint and quartern as 15⅛ ounces 50½ grains or 7,310½ grains, 7½ ounces 55 grains or 3,655¼ grains, 3¾ ounces 28 grains or 1,828 grains.

SIZES: diameters 106, 85, 68, 57 mm; heights 173, 141, 117, 90 mm; apertures 27, 25, 22, 17 mm.

These were made at a time when Professor John Playfair was advising the Edinburgh Dean of Guild after the death of John Robison. Playfair's protégé James Jardine had detected errors in the measures adjusted by Robison in 1800 and 1801 (Items 230, 231 and 232), and Jardine was commissioned to act on Playfair's behalf in constructing and adjusting this set of measures

235

to replace them. They are recorded in the Dean of Guild's Inventory for 1818, where they are described as being made of 'hard compound metal': Edinburgh City Archives, MS council minutes, January 1818.

The Museum of Edinburgh, Edinburgh (City of Edinburgh Council, Museums and Art Galleries): Inv. EDNMG HH 5996/1-4/2000. Transferred from Edinburgh Town Council by 1899 (marked with the 1899 display location 'HH 25').

237 BUSHEL, BY JOHN SAVIDGE, 1770, FOR HADDINGTONSHIRE

Dry capacity bent-wood vessel of oak. Single band of thin oak overlapped for about a fifth of the circumference, with top and bottom copper bands connected by six vertical copper strips, two of which form extensions of a reinforcing strip passing under the base, and all held by copper rivets, and additionally iron strips forming the support to the two handles. Inner inset copper band at rim; four-part wooden base supported originally by twenty metal angled brackets riveted through the wall. Stamped on the cross-piece under the measure 'I · SAVIDGE TOWER STREET LONDON' and engraved on the outer bands '[crown] / HADDINGTON SHIRE / 1770' and on the opposite side '[crown] / G / THE STANDARD WINCHESTER BUSHEL'. Branded four times inside the measure near the upper rim '[crown] / G' with the London Guildhall mark (St Paul's sword).

SIZE: diameter 495 mm; height 222 mm; aperture 475 mm; depth 190 mm.

John Savidge appears to have had a working life between 1757 and 1773. He was made free from Joseph Johns, in the Coopers' Company, in June 1757: Corporation of London Record Office, CF28/6. He registered his mark, an intertwined 'I' with three short cross-bars and an 'S', also in June

1757: Guildhall Library, MS 5636 vol. 1 (Coopers' Company). Entered in *Baldwin's Complete Guide* in the 1768 and 1770 editions as 'Jn Savidge, cooper, 26 Tower St'. He had property in Mercers' Court, which lay on the north side of Tower Street, between Mincing Lane and Mark Lane. His Will was proved June 1773: The National Archives PROB 11/989 sig.264. Our thanks to Dr Gloria Clifton and Dr Anita McConnell for this information. There is another example in NMS of a 1770 Winchester bushel by Savidge, signed as above on the cross-piece under the base, and engraved on the outer bands '[crown] / COLCHESTER IN ESSEX / 1770': Inv. NMS T.1994.31.

National Museums of Scotland, Edinburgh (Trustees of the NMS): Inv. NMS W.MP.233, alternatively H.VH.24. Purchased in 1896, provenance unknown.

237

238 BUSHEL(?), EIGHTEENTH CENTURY, AT STIRLING

Dry capacity coopered measure, very similar to the Stirling Guildry wheat firlot of 1766 (Item **218**), similarly from the holdings of the Stirling Guildry, but lacking any identification or markings.

SIZE: diameter 556 mm; height 237 mm; aperture 446 mm; inside base diameter 490 mm; depth 209 mm.

CALCULATED CAPACITY: 36·0 litres or 2,190 in³.

It is assumed that this is a Guildry working standard. Although it is unidentified, it is plausible that it is intended to represent the Winchester bushel, introduced in 1707, for which the conventional struck volume was 35·25 litres or 2,150 in³.

Smith Art Gallery and Museum, Stirling (Trustees of the Stirling Smith Art Gallery): Inv. STIGM 19817/15/8.3. Transferred from Stirling Guildhall, 1990.

239 COAL BUSHEL, *c.*1820, FOR DUNDEE

Dry capacity coopered measure with two wrought-iron handles; cylindrical vessel with a slight outward taper, constructed with 18 oak stakes and a two-part oak base, iron-bound with three riveted external bands and one recessed inner band at the rim. The exterior of the measure has been varnished and lettered 'GUILDRY DUNDEE' and above this subsequently painted 'IMP.ᴸ COAL or LIME', and beneath this 'METT'.

SIZE: diameter 560 mm; height 530 mm; aperture 525 mm; depth 495 mm.

CALCULATED CAPACITY: 93·3 litres or 5,700 in³.

The term 'mett of coal' is still encountered in Dundee, although it is inevitably associated with the later imputed bushel of 84 pounds avoirdupois. Although this measure has a tapered form, its average diameter of 19½ inches and depth of 19½ inches matches the sizes specified for the English coal bushel defined in 1719: R. D. Connor, *The Weights and Measures of England* (London, 1987), 182. This was redefined in 1819, but abolished in 1824. It is thought that this bushel was constructed as a result of the 1819 legislation and the subsequent naming of it as an 'Imperial' bushel was to give it authority, even though this was not a permitted Imperial size. The County of Forfar jury verdict in 1826 noted that English coal was sold in Dundee by measure as well as by weight; 'and that of a Met of such Coals contains fifty-four Standard Scots Pints, stricken measure': George Buchanan, *Tables for Converting the Weights and Measures hitherto in use in Great Britain …* (Edinburgh, 1829), 218. They had previously found such variety in the various Forfar county capacity measures that they decided to quote capacities in terms of the pint, although in this instance the measure clearly relates to the English coal bushel. The 1826 jury also found that lime was sold by a 63-pint boll in Dundee, so perhaps this lime boll was phased out in favour of a unified measure at a later date.

McManus Galleries, Dundee (Dundee City Council, Arts and Heritage): Inv. DUNMG 1978.1358, but previously 1937.263.1. Transferred in 1937 from the Weights and Measures Department, Dundee Corporation.

C6: Scots manufactured
Post-Imperial Capacity Standards

**240 GALLON AND HALF-GALLON,
BY ALEXANDER RUSSELL, 1824,
FOR KIRKCALDY**

Pair of liquid capacity measures (gallon and half-gallon) in brass; conical form turned from the solid and with heavy bases soldered in place and single curved solid handles riveted in place. The larger is engraved with the seal of Kirkcaldy

(a three-spired building) and beneath this 'IMPERIAL STANDARD GALLON, / BURGH OF KIRKALDY [*sic*] / 1824' and 'Alex^r.. Russell Kirkaldy [*sic*]' at the base. The half-gallon measure has not yet been engraved but has defining lines for the engraving to be added.

SIZES: diameters 303, 237 mm; heights 246, 190 mm; apertures: 30, 27 mm.

These measures presumably match the unsigned cast-iron Imperial bushel standard (Item **256**) for Kirkcaldy. Alexander Russell is known to have owned an iron foundry in Kirkcaldy at this time: see entry in Appendix G.

Fife Council, Trading Standards Service office, Glenrothes (Fife Council, Trading Standards Service).

240

241 TWO-GALLON AND GALLON, BY STEPHEN MILLAR & CO., 1826, FOR GLASGOW

Pair of liquid capacity measures (two gallon and one gallon) in thick sheet copper; tall conical bodies rising from a short cylindrical lower rim enclosing an inset flat base, tapering to rounded brass ring collar and projecting narrow cylindrical neck with a pierced plug; curved tubular handle, with flat thumb piece, riveted over the vertical turret joint on the body. Lacking the plug for the one-gallon measure. Engraved on the larger measure is a shield with the seal of Glasgow (a fish and bell against a tree) and the motto 'LET GLASGOW FLOURISH' set in a label, and 'DUPLI-CATE OF / IMPERIAL TWO GALLONS / FOR THE CITY OF GLASGOW / Adjusted by / DR MEIKLEHAM / DR THOMSON / DR URE / JAMES CLELAND / JAMES CRICHTON / 1826', and similarly on the smaller gallon measure. No other markings present except that the gallon is stamped 'G' on the rim.

SIZES: diameters 290, 235 mm; width across handle 370, 300 mm; heights to rim 460, 365 mm; height of two-gallon measure to top of plug 486 mm; apertures at rim 28, 26 mm.

These measures formed part of a set of eight from 2 gallons to half-gill, which are described by James Cleland, *An Historical Account of the Local and Imperial Weights and Measures of Lanarkshire, and an Inventory of those belonging to the Corporation of Glasgow* (Glasgow, 1832), 34, where he noted that they were made by Stephen Millar & Co., Glasgow. For Millar, see entry in Appendix G.

William Meikleham (1770/1-1846) was assistant to John Anderson, professor of natural philosophy at the University of Glasgow, and was his executor at Anderson's death in 1796:

241

Anderson's bequest established a rival college in his name – Anderson's Institution – which ultimately became the University of Strathclyde. Meikleham gained the regius chair of practical astronomy at the University of Glasgow in 1799, transferring to Anderson's former chair of natural philosophy in 1803. Dr Andrew Ure (1778-1857), FRS, from 1804 the first professor of natural philosophy in Anderson's Institution, was an ambitious chemist with a particular interest in the application of science to industry, eventually moving to London in 1830: see John Butt, *John Anderson's Legacy: the University of Strathclyde and its Antecedents 1796-1996* (East Linton, 1996). On Thomas Thomson (1773-1852), appointed first regius professor of chemistry at the University of Glasgow, see Jack Morrell, *Science, Culture and Politics in Britain, 1750-1870* (Aldershot, 1997). For James Cleland and James Crichton, see Chapter 9 and Appendix G respectively.

Museum of Transport, Glasgow (City of Glasgow Council, Museums and Art Galleries): Inv GLAMG T.1952.147.V.

Photograph: Ken Daly

peck) by De Grave of London, 1835. For Jamieson, see entry in Appendix G.

Dundee City Council, Trading Standards office, Dundee (Dundee City Council, Environmental and Consumer Protection).

243   GALLON AND FRACTIONS BY THOMAS CALLAM, 1826, FOR ANSTRUTHER WESTER

Set of seven liquid capacity measures (one gallon to half-gill) in brass; all with simple conical shapes, with a collar at the top, on bases with a recess underneath, and the gallon with two handles. The gallon is engraved '[crown] / ONE . IMPERIAL . GALLON / [seal of three interlocking fish] / ANSTRUTHER / WESTER / JAMES YOUNG ESQ^R / JAMES RAIKER ESQ^R / BAILIES / 1826'; the others all similarly engraved. Inscribed round the necks with the maker's name 'T. CALLAM, LEITH.' and 'FECIT.', and the bases stamped 'T. CALLAM / LEITH.' Verification marks for George IV.

SIZES: diameters 273, 210, 170, 140, 111, 92, 76 mm; heights 250, 227, 195, 145, 113, 87, 75 mm; apertures 53, 40, 30, 29, 24, 21, 18 mm.

For Callam, see entry in Appendix G.

242   GALLON AND FRACTIONS, BY J. JAMIESON, 1826, FOR DUNDEE

Set of six capacity measures (gallon to gill) in turned brass, possibly missing a half-gill measure. Plain cylindrical vessels with an upper flange at the rim and the base integral with a lower flange; the gallon with two brass handles riveted to a central band. Engraved on the gallon '1826. / IMPERIAL / GUILDRY OF DUNDEE. / GALLON. / J. Jamieson, Dundee.'; and similarly for the smaller measures. Stamped on the rim with Exchequer and William IV marks of *c*.1835.

SIZES: diameters 195, 155, 127, 98, 77, 60 mm; heights 181, 153, 120, 103, 84, 65 mm.

Possibly verified to complement a similarly marked set of dry capacity standards (bushel to

Scottish Fisheries Museum, Anstruther (Trustees of the Scottish Fisheries Museum): Inv. ANFFM.1975.124. Transferred by Anstruther Town Council at local government reorganisation in 1975.

244 GALLON AND FRACTIONS, BY THOMAS CALLAM, *c*.1826, FOR LEITH

Set of six liquid capacity measures (one gallon to gill) in sheet copper with brass collars, turret-jointed vertical seams and soldered inset bases. Gallon with tubular handle and two side handles, the half-gallon and quart with only two side handles. The largest three measures of bottle shape with cylindrical necks, the smaller three of conical flask shape with cylindrical necks. The largest engraved 'GALLON / BURGH / OF / LEITH' on the front, the maker's name 'T. CALLAM / LEITH', and Board of Trade indenture number '1791'; the brass collar is stamped 'GALLON' and with other verification marks from 1884. The smaller measures, for half-gallon, quart, pint, half-pint, and gill are similarly marked.

SIZES: diameters 200, 165, 135, 123, 100, 76 mm; heights 305, 230, 180, 175, 130, 100 mm; apertures 52, 43, 36, 29, 24, 24 mm.

The quart measure was exhibited at the Historical Weights and Measures exhibition, Royal Scottish Museum, Edinburgh, 1972.

The Museum of Edinburgh, Edinburgh (City of Edinburgh Council, Museums and Art Galleries): Inv. EDNMG HH 5981/1-4/2000.

245 GALLON AND FRACTIONS, 1827, FOR STIRLINGSHIRE

Set of five liquid capacity measures (gallon to half-gill) in brass; tall enclosed cylindrical vessels fabricated in sheet brass with projecting narrow cylindrical necks with small filling apertures. Engraved on the body of the measure is the seal of Stirling (a seated wolf within a border inscribed 'STIRLINI OPPIDUM') and 'Stirlingshire / IMPERIAL GALLON / 1827', and similarly for the quart, 'PINT OR MUTCHKIN' [*sic*], gill and half-gill. Presumably constructed to complement the larger coopered and brass dry capacity measures (Items **259** and

vertical turret joint on the body. Engraved (and black wax filled) on the front of the largest measure is the seal of Dumbarton (an elephant with a tower on its back) and 'Duplicate of Imperial Standard / TWO GALLONS / - for - / THE COUNTY OF DUMBARTON / ADJUSTED BY JAMES CLELAND LL.D. / Inspector General of Weights & Measures / 1835.'; and the gallon, half-gallon, quart, pint, half-pint, gill and half-gill are similarly inscribed. Stamped below this inscription is the Board of Trade indenture number 1804 (of c.1884); and the vessels carry verification stamps for 1884.

260), all of 1827. The engraving matches that on the latter brass measures.

SIZES: diameters 167, 107, 86, 60, 47 mm; heights 340, 204, 185, 100, 89 mm; apertures 36, 24, 23, 17, 16 mm.

SIZES: diameters 305, 245, 196, 159, 124, 103, 85, 68 mm; width across handles 350, 285, 225, 190, 143, 123, 103, 82 mm; heights 430, 350, 282, 232, 178, 148, 118, 98 mm; apertures 38, 35, 30, 29, 23, 22, 19, 17 mm.

Smith Art Gallery and Museum, Stirling (Trustees of the Stirling Smith Art Gallery): Inv. STIGM 9093-9097 (previously UA.34-38). Transferred by Stirling Town Council and first recorded in the *Catalogue of the Collections ... of the Smith Institute Stirling*, third edition (Stirling, 1934), 165.

246 TWO-GALLON AND FRACTIONS, 1835, FOR DUNBARTONSHIRE

Set of eight liquid capacity measures (2 gallons to half-gill) in thick sheet copper, presumably based on the design of the standards created by Meikleham, Thomson, Ure, Cleland and Crichton for Glasgow in 1826 (Item **241**), but water filled to the level of the rim, rather than to the lower surface of a pierced plug. Tall conical bodies rising from a short cylindrical lower rim enclosing an inset flat base, tapering to rounded brass ring collar and projecting narrow cylindrical neck; a curved tubular handle, with flat thumb piece on the four largest vessels, riveted over the

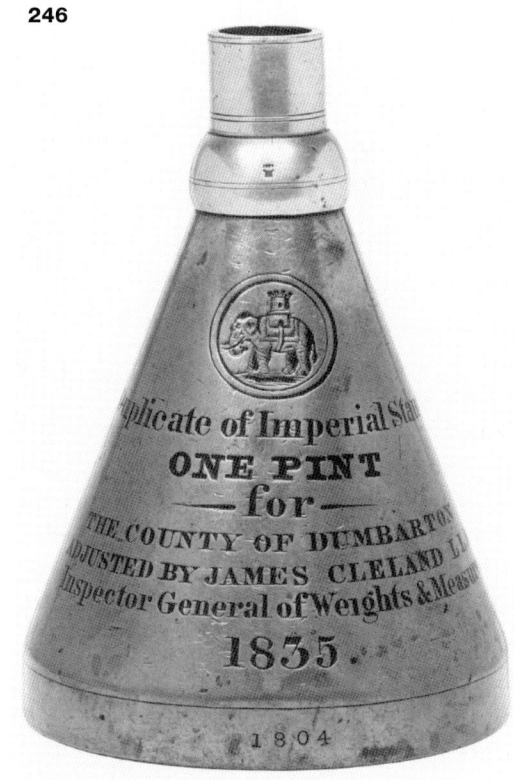

National Museums of Scotland, Edinburgh (Trustees of NMS): Inv. NMS T.1988.32. Purchased from Strathclyde Regional Council, Consumer Protection Department, July 1988.

247 TWO-GALLON AND FRACTIONS, 1835, FOR DUNBARTONSHIRE

Another set of eight measures (2 gallons to half-gill) of the same form as Item **246**, and similarly inscribed as Imperial standard duplicates adjusted by Cleland for the County of Dumbarton in his capacity as Inspector General for the county, 1835. With Board of Trade indenture number 1805, and 1884 verification marks.

West Dunbartonshire Council, Trading Standards office, Clydebank (West Dunbartonshire Council, Protective Services, Dumbarton).

248 GALLON AND FRACTIONS, 1835, FOR LANARKSHIRE

Another, incomplete, set of seven measures (gallon to half-gill), now lacking the two-gallon measure, of the same form as Item **246** (except that the vessels terminate in a conical tapered collar with a short cylindrical projection). The largest measure is engraved with the seal of Lanark (a double headed eagle) and 'Duplicate of Imperial Standard / ONE GALLON. / FOR THE COUNTY OF LANARK, / ADJUSTED BY JAMES CLELAND LL.D. / Inspector General of Weights & Measures / FOR THE COUNTY. / 1835.'; and the smaller vessels are similarly inscribed.

South Lanarkshire Museums Service, Hamilton (South Lanarkshire Council): Inv: SLCMD 2001.115-121. Apparently packed and transferred from the Burgh of Lanark's Weights and

Measures Department to the Town Council's offices in 1932, and passing to Clydesdale District Council (Museum Service established 1993) at local government reorganisation in 1975.

249 TWO-GALLON AND FRACTIONS, 1835, FOR LANARKSHIRE

Another set of eight measures (2 gallons to half-gill), and with a later quarter-gill, of the same form as Item **246**, and similarly inscribed as Imperial standard duplicates adjusted by Cleland for the County of Lanark in his capacity as Inspector General for the county, 1835. The

248

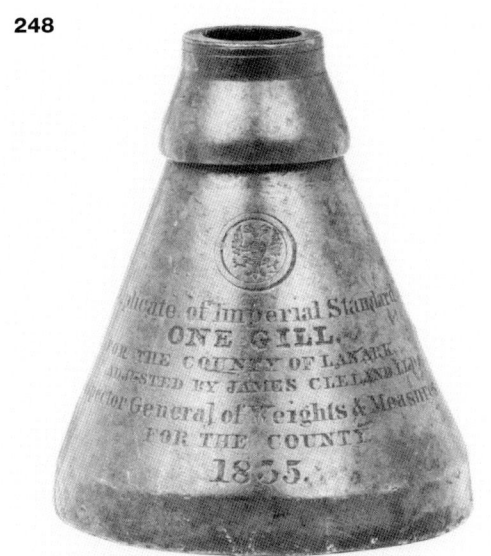

quarter-gill, without handle, is stamped 'COUNTY OF LANARK / QUARTER / GILL', and underneath 'ALEX^R RAMSAY / MAKER / GLASGOW'. Stamped with Board of Trade indenture number 1908, and verification stamp for 1887.

Lanarkshire was issued with a full set of standards in 1860 (Board of Trade indenture 1301); four further sets of weights only were issued in the same year (1302-3, 1307-8) and another nine sets in 1887 (1906-14): Carl Ricketts and John Douglas, *Marks and Marking of Weights and Measures of the British Isles* (Taunton, 1996), 128. A number of the 1835 Cleland sets were incorporated in these indentures, to complete sets of working standards.

National Museums of Scotland, Edinburgh (Trustees of the NMS): Inv. NMS T.1988.81. Purchased from Strathclyde Regional Council, Consumer Protection Department, August 1988.

250 TWO-GALLON AND FRACTIONS, 1835, FOR LANARKSHIRE

Another part set of seven measures (2 gallons to half-gill, but lacking the pint), of the same form as Item **246**, and similarly inscribed as Imperial standard duplicates adjusted by Cleland for the County of Lanark in his capacity as Inspector General for the county, 1835. In fitted case.

248

Sold by Strathclyde Regional Council, Consumer Protection Department, at Christie's Glasgow, 15 February 1989, lot 69. Present location unknown.

251 TWO-GALLON AND FRACTIONS, 1835, FOR CALTON, LANARKSHIRE

Another set of eight measures (2 gallons to half-gill), with a later quarter-gill, of the same form as Item **246**. The largest measure is engraved 'CITY OF GLASGOW' over the seal of Lanark (a double headed eagle) and 'Duplicate of Imperial / TWO GALLONS / - FOR - / THE BURGH OF CALTON / ADJUSTED BY JAMES CLELAND LL.D. / Inspector General of Weights & Measures / FOR THE COUNTY OF LANARK / 1835.'; and the smaller vessels similarly inscribed. The quarter-gill is stamped 'CITY OF GLASGOW / Imperial Standard / QUARTER GILL / 1696' and underneath 'ALEX^R RAMSAY / MAKER / GLASGOW'. Stamped with Board of Trade indenture number 1696 (of *c.*1880) and verification stamp.

Museum of Transport, Glasgow (Glasgow City Council, Museums and Art Galleries): Inv. GLAMG T.1966.28.f. Transferred from Glasgow Town Council Weights and Measures Department, 1966.

252

252 TWO-GALLON AND FRACTIONS, 1835, FOR PAISLEY

Another set of eight measures (two gallons to half-gill), of the same form as Item **246**, and with a later quarter-gill measure. The largest engraved with the seal of Paisley (an abbot with crozier) flanked (except on the smallest two measures) by 'W.' and 'IV' (for William IV) and 'Duplicate of Imperial Standard / TWO GALLONS / - for - / THE BURGH OF PAISLEY / ADJUSTED BY JAMES CLELAND LL.D. / Inspector General of Weights & Measures. / 1835.'; and the smaller vessels similarly inscribed. The later quarter-gill marked '¼ GILL / BURGH OF PAISLEY'.

Paisley Museum and Art Galleries, Paisley (Renfrewshire Council): Inv. PSYMG 1996.195-188. Transferred from Strathclyde Regional Council, Department of Consumer Protection, at local government reorganisation in 1996.

253 TWO-GALLON AND FRACTIONS, c.1835, FOR AYR

Another, incomplete, set of seven measures (2 gallons to half-gill), now lacking the pint measure, of a form which is clearly strongly influenced by the design of the Cleland-adjusted vessels (Item **246**), but differs slightly in shape and constructional details. The largest measure is engraved 'IMPERIAL STANDARD / 2 GALLON / BURGH OF AYR'; and the smaller vessels are similarly inscribed, but with the quart marked as '¼ GALLON' and the half-pint marked as '½ MUTCHKIN'. Stamped beneath the inscription with Board of Trade indenture number 268 and 'SET 1'. The half-pint above is stamped '[crown] / VR' on the rim.

Set 268 was issued to Ayr Royal Burgh in 1826 and presumably comprised Imperial weights only, to which these measures have subsequently been added: Carl Ricketts and John Douglas, *Marks and Marking of Weights and Measures of the British Isles* (Taunton, 1996), 127. For another recorded use of the description 'mutchkin' on an Imperial pint, see Item **245**.

253

SIZES: diameters 300, 240, 195, 152, 102, 80, 66 mm; width across handle 375, 300, 235, 193, 130, 108, 87 mm; heights 458, 368, 300, 230, 157, 127, 103 mm; apertures 30, 28, 24, 22, 19, 18, 17 mm.

National Museums of Scotland, Edinburgh (Trustees of the NMS): Inv. NMS T.1988.33. Purchased from Strathclyde Regional Council, Department of Consumer Protection, July 1988.

254 GALLON AND FRACTIONS, 1835, FOR POLLOCKSHAWS

Cased set of eight liquid capacity measures (gallon to quarter-gill) in sheet copper; curved conical bodies rising from a cylindrical lower rim enclosing an inset flat base and tapering to a brass neck, to contain a solid turned brass plug with pierced hole for overflowing water. All have a single curved tubular handle riveted over the vertical turret joint on the body. Engraved on the largest measure 'POLLOKSHAWS / Duplicate, / Gallon / 1835'. Fitted case to take the eight measures with separate storage for the pierced plugs.

254

SIZES: diameters 255, 210, 165, 135, 107, 90, 67, 52 mm; width across handles 287, 238, 180, 145, 120, 97, 77, 62 mm; heights (without plug) 280, 248, 205, 172, 140, 110, 86, 60 mm; apertures 39, 33, 30, 25, 21, 17, 15, 10 mm. Case: 680 × 390 × 320 mm high.

These measures are somewhat similar to the Cleland-adjusted measures (Item **246**) but are clearly by a different maker. They differ also in using pierced plugs to Andrew Ure's design as applied to the 1826 Glasgow set (Item **241**) and described in Patrick Auld, 'An Account of the Imperial Standard Measures made by and for Messrs. J. & H. Wardrop … under the direction of Andrew Ure, MD, FRS, Professor in the Andersonian Institution, Glasgow', in *Glasgow Mechanics' Magazine*, 4 (1826), 457-8, 453. Pollokshaws was created a parliamentary burgh in 1832 and standards were first issued in 1835: Carl Ricketts and John Douglas, *Marks and Marking of Weights and Measures of the British*

Isles (Taunton, 1996), 127. This set was presumably acquired as working measures to complement the local standards.

Museum of Transport, Glasgow (Glasgow City Council, Museums and Art Galleries): Inv. T. 1966.28b. Transferred from Glasgow Town Council Weights and Measures Department, 1966.

255 FOUR-GALLON AND FRACTIONS, BY BAIRD & TATLOCK, 1894, FOR GOVAN

Set of eleven liquid capacity measures, closely following the design of the Cleland-adjusted measures (Item **246**), with additional measures for three and four gallons and for the quarter-gill. Differing from the earlier design in having a second handle at the front of the three largest measures, and in dispensing with the flattened

section on the main handles. The largest engraved 'IMPERIAL FOUR GALLONS / – / Burgh of Govan / 1894'. The smaller measures similarly inscribed. Engraved at the foot 'BAIRD & TATLOCK / LONDON & GLASGOW'. Board of Trade stamped indenture number 2389, and a small applied label stamped to indicate working standard set 'B4'. Board of Trade verification marks from 1894 (except on gallon and half-gallon).

SIZES: diameters 395, 366, 310, 240, 185, 155, 118, 98, 77, 62, 48 mm; width across handles 440, 400, 350, 280, 225, 185, 150, 130, 105, 85, 67 mm; heights 510, 445, 400, 335, 277, 210, 177, 127, 100, 78, 65 mm; apertures 50, 38, 37, 31, 30, 27, 27, 24, 23, 19, 18 mm.

For Baird & Tatlock, see Appendix G.

Museum of Transport, Glasgow (City of Glasgow Council, Museums and Art Galleries): Inv. GLAMG T.1966.28.a. Transferred from

Glasgow Town Council Weights and Measures Department, 1966.

256   BUSHEL, BY ALEXANDER RUSSELL, 1824, FOR KIRKCALDY

Dry capacity bushel measure in cast iron; broad cylindrical vessel with turned finish on the interior, with two handles on ogee brackets riveted in place; cast in relief on side 'IMPERIAL STANDARD BUSHEL / BURGH OF KIRKALDY. [*sic*] / 1824', with the date flanked by two relief representations of the Kirkcaldy seal (a three-spired building).

SIZE: diameter 512 mm; width across handles 698 mm; height 222 mm; aperture 471 mm; depth 208 mm.

This measure presumably matches the signed brass gallon and half-gallon capacity measures of 1824 (Item 240), for Kirkcaldy. Alexander Russell is known to have owned an iron foundry in Kirkcaldy at this time. There is a horizontal sundial at Kirkcaldy Museum (inv. KIRMG 1964.29), signed 'ALEX^R RUSSELL / KIRKALDY [*sic*] FOUNDERY [*sic*] Lat. [56°] 16' 8"'.

Kirkcaldy Museum, Kirkcaldy (Fife Council, Fife Museums Central): Inv. KIRMG 1980.1636. Transferred from Fife Regional Council, Consumer Protection Department in 1980.

257   BUSHEL, BY ANDREW BOYLE, 1826, AT PERTH

Dry capacity coopered measure, with two brass handles; cylindrical vessel with slight outward taper comprising 19 oak staves and a three-part oak base, brass-bound with three external and two internal bands. The upper external band engraved 'IMPERIAL STREAKED BUSHEL 1826', and at the rear 'A^W Boyle Maker Perth'. The measure has been restored.

SIZE: diameter 460 mm; width over handles 562 mm; height 315 mm; aperture 420 mm; depth 281 mm.

Museum and Art Gallery, Perth (Perth and Kinross Council): Inv. PERGM SHNN 531. Provenance unknown.

256

257

258

258 HEAPED BUSHEL, BY ANDREW BOYLE, 1826, AT PERTH

Dry capacity coopered measure, with two mahogany handles on curved brass supports; shallow cylindrical vessel with 22 oak staves and a three-part oak base, brass-bound with two external and two internal recessed bands. The upper external band engraved 'IMPERIAL HEAPED BUSHEL 1826', and at the rear 'Aᵂ Boyle Maker Perth'. No verification stamps on the rim or elsewhere. The measure has been restored.

SIZE: diameter 503 mm; width over handles 703 mm; height 253 mm; aperture 456 mm; depth 220 mm.

Museum and Art Gallery, Perth (Perth and Kinross Council): Inv. PERGM SHNN 527. Provenance unknown.

259 BUSHEL AND HALF-BUSHEL, BY ANDREW BOYLE, 1827, FOR STIRLINGSHIRE

Two dry capacity coopered measures, each with two mahogany handles on curved brass supports; shallow cylindrical vessels with 21 and 17 oak staves respectively, and three-piece oak bases, brass-bound with two external and two internal

bands. The upper external band on the bushel being engraved 'STIRLINGSHIRE IMPERIAL BUSHEL 1827', and at the rear 'Aᵂ Boyle Maker Perth', and similarly for the half-bushel. No verification stamps on the rim or elsewhere. Presumably constructed to complement the smaller brass dry measures (Item **260**) and liquid measures (Item **245**).

SIZES: diameters 495, 412 mm; width over handles 710, 560 mm; heights 253, 190 mm; depths 221, 162 mm.

Smith Art Gallery and Museum, Stirling (Trustees of the Stirling Smith Art Gallery): Inv. STIGM 9137-9138 (previously UA.19, 20). Transferred by Stirling Town Council and first recorded in the *Catalogue of the Collections … of the Smith Institute Stirling*, third edition (Stirling, 1934), 164.

260 PECK AND FRACTIONS, 1827, FOR STIRLINGSHIRE

Set of three dry capacity measures (peck, half and quarter-peck) in brass, each with two mahogany handles on curved brass supports; shallow cylindrical vessels fabricated in thick sheet brass with turned finish. The peck engraved with the emblem of Stirling (a seated wolf), 'Stirlingshire

259

/ IMPERIAL PECK / 1827 / Adjusted under the DIRECTION of Adam Anderson A.M. F.R.S.E. Rector of PERTH Academy', and similarly for the half and quarter-peck. No verification stamps on the rim or elsewhere. Presumably constructed to complement the larger coopered vessels by Boyle (Item **259**) and the smaller brass liquid measures (Item **245**), all of 1827. The engraving matches that on the liquid measures, although Anderson's name is not recorded on these.

SIZES: diameters 317, 250, 202 mm; width over handles 450, 370, 300 mm; heights 140, 117, 88 mm; apertures 302, 235, 188 mm; depths 130, 108, 81 mm.

Smith Art Gallery and Museum, Stirling (Trustees of the Stirling Smith Art Gallery): Inv. STIGM 8799-8801 (previously UA.21-23). Transferred by Stirling Town Council and first recorded in the *Catalogue of the Collections ... of the Smith Institute Stirling*, third edition (Stirling, 1934), 164.

261 HEAPED BUSHEL AND FRACTIONS, BY ROBERT HOOD, FROM LANARK

Set of five dry capacity coopered measures (bushel to quarter-peck) in oak with riveted iron bands, for measuring heaped materials;

261

the bushel and half-bushel with two iron handles. The largest stamped on the base 'R · HOOD' and branded 'IMPERIAL / BUSHEL / HEAPED' and similarly for the half-bushel, peck, half-peck and quarter-peck, some additionally stamped 'R · HOOD' inside. The rims all stamped with 'D.G.' and 'RH'.

SIZES: diameters 510, 410, 320, 280, 110 mm; width across handles 610, 510 mm; heights 260, 215, 185, 155, 120 mm; apertures 455, 355, 275, 215, 170 mm; depths 220, 175, 150, 120, 95 mm.

South Lanarkshire Museum Service, Hamilton (South Lanarkshire Council): Inv. SLCMD 2001.126-122. Apparently packed and transferred from the Burgh of Lanark's Weights and Measures Department to the Town Council's offices in 1932, and passing to Clydesdale District Council (Museum Service established 1993) at local government reorganisation in 1975.

PART III

APPENDICES

APPENDIX A

THE CENTRAL SCOTTISH METROLOGICAL ACTS

1   The Assize of David I, attributed to the twelfth century (Thomas Thomson and Cosmo Innes (eds), *The Acts of the Parliaments of Scotland [APS]*, 13 volumes (Edinburgh, 1814-75), I, 673-4).

1a The Assize of David I in the Latin version (*APS*, I, 673-4).

2   Three Acts preliminary to the Assize of 1426 printed by William Robertson in *The Parliamentary Records of Scotland, 1240-1571* [Edinburgh, 1804], from National Archives of Scotland (NAS) MS PA.5/6/2.

3   The Assize of James I, 11 March 1426, as contained in the contemporary manuscripts in Register House, Edinburgh and printed verbatim by William Robertson, in *The Parliamentary Records of Scotland 1240-1571* [Edinburgh, 1804]. The text given is that of NAS MS PA.5/6/2, with differences contained in the slightly later version, NAS MS PA.5/6/1, given in parentheses.

3a The Assize of James I, 11 March 1426 as given in the first printing of the acts of the parliaments of Scotland, James Balfour's 'Black Acts' of 1566. The changes of the Thomson-Innes edition of 1814 (*APS*, II, 12) are given in parentheses.

4   The Act of James II, 6 March 1458 (*APS*, II, 50: Act 18).

5   The Act of James IV, 15 March 1504 (*APS*, II, 246: Act 47).

6   The Act of Mary, Queen of Scots, 20 June 1555 (*APS*, II, 496: Act 20).

7   The Act of Mary, Queen of Scots, 4 June 1563 (*APS*, II, 540: Act 14).

8   The Act of James VI, 29 July 1587 (*APS*, III, 521-2).

9   The Act of James VI, 19 February 1618 (*APS*, IV, 585-9).

NB: Marked contractions in the printed transcriptions have been expanded.

APPENDIX A.1

The Assize of David I, attributed to the twelfth century. (Thomas Thomson and Cosmo Innes (eds), *The Acts of the Parliaments of Scotland [APS]*, 13 volumes (Edinburgh, 1814-75), I, 673-4)

THE ASSIZE OF KYNG DAUID OF MESURIS AND WECHTIS.

Of the eln ·

The eln aw to conteyn in lenth · xxxvii · Inch met with the thowmys of iii men that is to say a mekill man and of a man of messurabill statur and of a lytill man bot be the thoume of a medilkinman it aw to stand or ellis efter the lenth of iii bear cornys gud and chosyn but tayllis · the thoum aw to be messurit at the rut of the nayll ·

Of the wecht of the stane ·

Item the stane for weying of woll and uther geir aw to wey · xv · pund · Item the stane of wax aw to conteyne · viij · pund · Item the vaw aw to conteyn · xii · stan ·

Of the wecht of the pund ·

Item the pund in King Dauidis dayis weyit xxv · schillingis · Now the pund aw to wey in siluer xxvi schillingis and iij sterling penijs and that for the mynoratioun of the peny that is in the tym now · Item the pund sould wey · xv · vncis ·

Of the wecht of the unce ·

The unce contenit in King Dauidis time xx gude and sufficient sterling penijs and now it sall wey · xxi · penijs for the demynicioune of the mone ·

Of the wecht of the sterling ·

King David ordanyt at the sterlyng suld wey xxxij cornys of gude and round quhete ·

Of the boll ·

Item the boll sall contene a sexterne viz · xii gallonis of aile and it sould haue in deipnes ix inches and in wydenes abone xxiiij inches by the thicknes of the trie and in the roundnes and circumference abone lxxij and in the roundnes at the boddom lxxi inches ·

Of the gallon ·

Item the galloun suld be sax inches and ane half in deipnes and in breid of the boddom viij inches and a half with the thicknes of the trie on baith the sides and in rowndnes abone xxvij inches and a half and in rowndnes below xxiij inches ·

Item the gallon aw to conteyn · xij · pundis of watir that is for to say iiij pundis of salt watir of the see, iiij pundis of standande watir and iiij pundis of rynnand watir ·

APPENDIX A.1a

The Assize of David I, attributed to the twelfth century in the Latin version (*APS*, I, 673-4). Additional wording from the late fourteenth-century Bute manuscript (National Library of Scotland MS 21246, ff. 119r-119v) is given in round brackets. The portion in square brackets has been added to the *APS* printed text but is not present in the Bute manuscript.

ASSISA REGIS DAVID DE MENSURIS ET PONDERIBUS.

De ulna ·

Vlna Regis Dauid debet continere in se · xxxvij · pollices mensuratas cum pollice trium hominum · scilicet ex magno · ex medio · et ex paruo · Et ex medio pollice hominis debet stare · aut ex longitudine trium granorum boni ordei sine caudis · Pollex autem debet mensurari ad radicem vnguis pollicis ·

De pondere petre ·

Item lapis ad lanam et ad alias res ponderandas debet ponderare · xv · libras · Item vaga debet continere · xij · petras cuius pondus continet · viij libras · (Petra cere viii libras, xii librae facut petra Londonie.)

De pondere libre ·

Item libra debet ponderare · xxv · solidos · hoc erat in illo tempore assise pretacte · [Nunc autem libra ponderabit] · xxvj · solidos · et · iij · denarii · quoniam in tantum nunc miniuitur noua moneta ab antiqua tunc vsitata · Item · libra debet ponderare · xv · vncias ·

De pondere uncie ·

Item vncae continebat in se tempore Regis Dauid · xx · denarii sterlingorum · Nunc autem vncea debet continere in se · xxj · denarii · sterlingorum · quia noua moneta nunc in tantum minuitur ·

De pondere sterlingi ·

Item sterlingus debet ponderare · xxxij · grana boni et rotundi frumenti ·

De bolla ·

Item bolla debet continere in se sextarium viz · xij · lagenas seruicie · Et bolla ex profunditate debet esse · ix · pollicium · In latitudine superiori debet esse · xxiiij · pollices cum spissitudine ligni vtriusque partis · In rotunditate superiori debet esse · lxxij · pollicium in medio ligni vtriusque partis superioris · In rotundine inferiori debet esse · lxxj · pollicium ·

De quantitate lagene ·

Item lagena debet esse sex pollicium et dimidii in profunditate · In latitudine inferiori debet esse · viij · pollicium et dimidii cum spissitudine lingi vtriusque partis · In rotunditate superiori debet esse · xxvij · pollicium et dimidii · In rotunditate inferiori debet esse · xxiij · pollicium ·

APPENDIX A.2

The three Acts of 1426 preliminary to the Assize of that year, as transcribed by William Robertson in *The Records of the Parliament of Scotland* **[Edinburgh, 1804], from the near-contemporary manuscript copy of these acts preserved as MS PA.5/6/2 in the National Archives of Scotland, Edinburgh.**

ITEM the king with the consent of the thre estatis has ordanit in this parliament that there salbe maid certane mesouris of bol furlate & half furlate peke ande galone the quhilk salbe generally vsit throu all the realme in al partis of it the quhilk salbe gevin furth at Edinburgh at the ische of this parliament thidder continewit ande thai mesouris salbe gevin to be vsit at the frist day of the moneth of September nixt to cum ande fra thin furth the king forbiddis ony vther mesouris to be vsit in this realme vnder the payn that may folow

ITEM the king ande the perliament has ordanit that thare be maid a stane for gudis saulde & boucht be wecht the quhilk wey xv lele Troyis pundis ande at stane be devisit in xvi Scottis pundis ande of it thare salbe ordainit half stane quarter half quarter pund ande halfe pund ande vthir less weyis accordande tharto with the quhilk al byaris ande sellaris of guids within the realme & with nane vther wechtis fra Witsonday nixt to cum fra thin furth thir forsaide wechtis sal hafe courss

ITEM it is ordanit at the watter mettis at now ar sal remayn & be vsit throcht the realme in tym to cum and in ilk place & toun quhare the gudis ar saulde ande mete be the watter thare be ordanit be the Alderman & balzeis a lele man sworne to mete all gudis sellabil made be watter als wele colis as vthir gudis ande at the sellaris na nane of thare behalf entermete thaim in the metting of sic gudis

APPENDIX A.3

The Assize of James I, 11 March 1426, as contained in the contemporary copy manuscripts in the National Archives of Scotland MSS PA.5/6/2 and PA.5/6/1. The former, near-contemporary copy of the lost original, is reproduced in Fig. 5.4 and the text below is as transcribed by William Robertson in his *Records of the Parliament of Scotland 1240-1571* [Edinburgh, 1804], 63. Differences in the slightly later MS PA.5/6/1 are shown here in parentheses.

ASSISA regis Jacobi de ponderibus et mensuris per totum regnum Scotie generaliter constituendis facta apud Perth in perliamento tento ibidem xi^{mo}, die mensis Martii anno regni sui xxi° per consensum et assensum trium statuum ibidem existentium

IN the frist thai ordanit ande deliuerit the Elne to contene xxxvii [xxxviii] inche as is contenit in the statute of king Dauid the frist playnly maide

ITEM thai ordanit and statute the stane to wey Irne woll ande vthir marchandice to conten xv [xvi] pundis trois ilk trois punde to conten xvi vnce ande that stane to be deuidit in half stane quarter half quarter ande punde &c

ITEM thai ordanit the bol to met with all vitall to be deuidit in foure partis viz four furlatis to contene a bol ande that furlate nowther to be maid eftir the frist mesoure than vsit bott in a medful mesoure betuix the twa

ITEM the mesoure of the furlote is this It sal contene in breid evin ourethort xvi inche vnder & abone within the burdis & in depnes vi inche the thicknes of bath the burdis sal contene ane inche and a halfe the half furlote & the pek thare eftir folowande &c

ITEM the bol sal contene in breid xxix inche within the burdis & abufe xxviii [xxvii] inche & a half evin ourethort ande in depnes ix inche the furlote sal contene twa galonis ande a pynte ande ilk pynt sal contene be wecht of cleir watter of Tay xli vnce that is for to say ii pundis & ix vnce troyis swa weyis the galone xx punde & viii vnche swa weyis the furlote xli pundis. Ande the bol contenande four furlotis weyis viii^{xx} & iiii pundis [This last sentence is omitted in PA.5/6/1] the old bol frist maide be king Dauid contenit a sextarn the saxtarn contenit xii galonis of the auld mete ande ilk galone weyit ten pundis trois & foure vnce of diuerse vatteris swa weyit ye bol vi^{xx} iii pundis sua this new bol new maid weyis mar than the auld boll be xli *lib* quhilkis makis twa galounis & a half & a chopyn of the auld mete ande of the new mete new ordanit ix pyntis & thre muchekynis

APPENDIX A.3a

**The Assize of James I, 11 March 1426, as given in the first printed edition of
the Acts of the Parliaments of Scotland, James Balfour's 'Black Acts',
published as *The Actis and Constitutionis of the Realm of Scotland …*
(Edinburgh, 1566). This is reproduced in Fig. 5.2. Differences found in the
Thomson-Innes 'Record' edition of 1814 (*APS*, II, 12) are given in parentheses.**

ASSISA REGIS IACOBI DE PONDERIBVS ET MENSVRIS
PER TOTVM REGNVM SCOTÆ GENERALITER
Constituendis, facta apud Perth in Parliamento tento ibidem
vndecimo die Mensis Martij. Anno Regni sui vigesimo primo
per consensum & assensum trium Regni statuum ibidem existentium.

Anent the Mesoure of the Elne.

IN THE first thay ordanit and
deliuerit, that the Elne sall contene
xxxvii inche, as is contenit in the
statute of King Dauid the first maid
thairupone.

Quhat the stane sall contene.

ITEM Thay ordanit and statute the
stane to wey Irin, woll, and other
Merchandice with, to contene xv.
[xvi] pund trois, ilk trois pund to
contene xvi. unce, and that stane to
be deuidit in half stane, quarter, half
quarter, pund, half pund, and uther
smailar.

Of the Deuisioun of the boll and the
mesoure of the Fyrlot and the boll.

ITEM Thay ordanit the boll to met
victuall with, to be deuidit in foure
partis, videlicet, foure fyrlottis to
contene a boll, and that fyrlot not to
be maid efter the first Mesoure, na
efter the Measure now usit, bot in
middill Measure betuix the twa.

ITEM The boll sall contene in breid
xxix. inchis within the buirdis, &
abone xxvij. inchis and a half euin
ouerthort, and in deipnes xix inchis.

ITEM The fyrlot sall contene in
breid euin ouerthort xvi · inchis
under and abone within the buirdis,
the thicknes of baith the burdis sall
contene ane inche and ane half, and in
deipnes it sall contene ix · [vj] inche,
the half fyrlot, and the peck thairefter
followand, as effeiris. The fyrlot sall
contene twa gallownis & a pynt, and
ilk pynt sall contene be weicht of cleir
watter of Tay xlj · unce, that is to say,
twa pundis and ix · unce trois. Swa
weyis the gallowne xx · pund and viij ·
unce. Swa weyis the fyrlot xlj · pundis,
and the boll contenand foure fyrlotis,
weyis viij · scoir iiij · pund. The auld
boll first maid be King Dauid, contenit a Sexterne, a Sexterne contenit
xij · gallownis of the auld met, and ilk
gallowne weyit x · pund trois, and
foure unce of diuers watters. Swa weyit
the boll vi · scoir thre pundis, swa weyis
this boll new maidmair, than the auld
boll xli pund, qyhilk makis twa gallownis and a half, and a chopin of the
auld met, and of the new met ordanit,
ix pyntis and thre Muchkinnis.

APPENDIX A.4

The Act of James II, 6 March 1458 (*APS*, II, 50: Act 18)

ITEM anent mettis and mesuris it is sene speidfull that sen we haif bot a king and a lawe vniuersale throuout the Realme we sulde haif bot a met and mesuris generale to serue all the Realme That is to say ane pynt quhilk was giffin be the ordinance of the thre estatis by Johnne forester that tyme beande chavmerlane to the burgh of striuelling as for standart sall remane vniuersale throughout the Realme And the ferlot salbe maide perfest That is to say ilk ferlot sall contene xviij pyntis of the sammyn mesuris rovnde and elik wyde vnder and abvne the twa burdis contenande evyne our in thikness ane inche and a half ande the breide our within the burdis xvj inchis ande a half Ande the half ferlot and pek to folowe theftir in the sammyn kinde And of thir saide mesuris that is to say pynt and ferlot ther salbe new maide iij standartis ane to be sende till abyrdene ane vther to perthe and the thride to Edinburghe to remane and now to be proclamyt

that fra the fest of sanct michaell nixt to cum ther mesuris of pynt and ferlot haif courss and nane vther Sua that in the menetyme all maner of personnis that thinkis to vse the said mesuris may cum to the saide placis and furnyss thame with the saide mesuris beande brynt and selyt with the selys of thai stedis as thaj will vse and be peruit thaireftir And gif ony personnis efter the saids termis vsys vther mesouris than thir and ther may be taynt gottin therof be dittay or vthur wayis thai sall pay the vnlawe of the chavmerlane ayre doublyt And gif ony man be foundyt of aulde or of newe of fyrmess of uthir mesuris than thir abone writtyne ther foundacione sall stande in effect not ganestanding this statute And the mesouris of that foundacione salbe proporcionyt to this mesure that now is sua that the sammyne quantite sall remane with the giffare and the ressauere but præiudice of ony of thame

APPENDIX A.5

The Act of James IV, 15 March 1504 (*APS*, II, 246: Act 47)

ITEM It is statute and ordanit that all mesouris and wechtis baith pynt quart ferlot pec elwand stane and pund be of ane quantite and mesour quhilk salbe ordanit in Edinburgh be our souerane and his chavmerlane and consale and that euerilk burgh cum and fech ther mesouris furtht of Edinburgh selit and maid and keip the samyne And quhar ther Is ony fermez aucht in heritage of the auld mete that the said fermez be perportionat to the quantite of the auld mete and pait with the new mete to the avale of the auld mete perportionaly And gif ony personnis vse ony othir messouris or wechtis in tyme tocum bot the messouris and wechtis now to be maid as said Is It salbe ane point of dittay and thai to be Inditit thairfoir fra thin furtht

APPENDIX A.6

The Act of Mary, Queen of Scots, 20 June 1555 (*APS*, II, 496: Act 20)

ITEM Forsamekill as be vmquhile our Souerane Ladyis maist Nobill predecessouris Kingis James the First and Feird It was statute and ordanit that all mesouris baith pynt quart fyrlot peck elnwand stane and pund to be of ane quantitie to by with and that na Burgh haue ane wecht to by with and ane vther to sell different in wecht thairfra bot that all wechtis mesuris and mettis for bying and selling to be vniuersall baith to Burgh and to land in all tymes thairefter quhilkis actis as ȝit hes not bene put to dew executioun Thairfoir it is statute and ordanit in this present Parliament that thir persounis vnder writtin or ony thre of thame that is to say Williame Bischop of Dumblane Robert Bischop of Orknay Maister Abraham Creichtoun Prouest of Dunglas Schir Williame Hammiltoun of Sanchar Knycht Schir Richard Maitland of Ledingtoun Knycht Maister Thomas Marioribankis of Ratho and Thomas Meinȝeis Prouest of Abirdene conuene in the Burgh of Edinburgh and cause the elnwand the quart pynt fyrlot peck stane and pund be brocht to thame fra the townis of Striuiling Linlithquo Lanerk and be the samin as thay find to mak an vniuersall wecht of the stane and pund ane vniuersall mesoure of the quart pynt fyrlot peck elnwand conforme to the act maid be King James the Feird thairupone except the watter met to remane according to the vse of the cuntrie to be direct furth to the haill liegis of this Realme with the quhilkis thay salbe haldin to by sell met mesoure wey ressaif and deliure and be na vther met mesoure nor wecht and quha dois in the contrare heirof salbe punist for falset conforme to the Law And this ordour to be maid be the persounis foirsaidis or ony thre of thame betuix this and the Feist of Allhallowmes nixt tocum but ony further dilay and thairefter publicatioun to be maid in all partis of this Realm as accordis

APPENDIX A.7

The Act of Mary, Queen of Scots, 4 June 1563 (*APS*, II, 540-1: Act 14)

ITEM Forsamekill as in our Souerane Ladyis maist Nobill predecessouris King James the Fyft and Feird tymes respectiue It was statute and ordanit that all mesouris pynt quart fyrlot peck elnwand stane and pund to be of ane quantitie and mesoure and that na Burgh haue ane wecht to by with and ane vther to sell with different in wecht thairfra bot all wechtis mesouris and mettis for bying and selling to be vniuersall baith to burgh and to land in all tymes thairefter quhilkis actis as ȝit hes not bene put to dew executioun Thairfoir it was statute and ordanit in the Parliament halden in Edinburgh the twentie sax day of Junij the ȝeir of God ane thousand fyue hundreth fyftie fyue ȝeiris that thir persounis vnder writtin or ony thre of thame That is to say Reuerend Fathers in God Williame Bischop of Dumblane vmquhile Robert Bischop of Orknay Maister Abraham Creichtoun Prouest of Dunglas vmquhile Schir Williame Hammiltoun of Sanchquhare Knycht Maister Thomas Marioribankis of Ratho Schir Richard Maitland of Lethingtoun Knycht and Thomas Meinȝeis Prouest of Abirdene Sould haue conuenit in the Burgh of Edinburgh and thair to haue causit the elnwand the quart pynt fyrlot stane and pund be brocht to thame fra the townis of Striuiling Linlithquho Lanark and be the samin as thay stand to haue maid ane universall wecht of the stane and pund ane vniuersall mesoure of the quart pint fyrlot and elnwand conforme to the act maid be the said King James the feird thair-upone except the watter mett to

remane according to the vse of the cuntrie To haue bene directit furth to the haill liegis of this Realme with the quhilkis thay sould haue bene haldin to by sell met mesoure wey ressaif and deliure and be na vther met mesoure and wecht And quha did in the contrare heirof sould haue bene punist for falset conforme to the Law And this ordour to haue bene maid be the persounis foirsaid or ony thre of thame betuix the dait of the said act and Alhallowmes nixt thairefter but ony further dilay lyke as the samin act proportis And because it is vnderstand be the Quenis grace and thre Estatis in this present Parliament that the said act tuke nane effect nor the foirsaidis persounis contenit thairin performit not the contentis of the said act swa that the samin mycht haue takin full effect And thairfoir hir hienes with the auise of the Estatis foirsaidis statutis and ordanis ȝit as of befoir that all mesouris pynt quart fyrlot peck elnwand stane and pund be of ane quantitie and mesoure and that na Burgh haue ane wecht to by with and ane vther to sell with different in wecht thairfra Bot all wechtis mesouris and mettis for bying and selling be vniuersall baith to Burgh and land in all tymes heirefter equallie that ane with the vther and to the effect that this present act and actis foirsaidis maid heirupone of befoir may be put to further execution That thir persounis vnder writtin That is to say ane Reuerend Father in God Williame Bischop of Dumblane Maister Abraham Chreichtoun prouest of Dunglas Schir Richard Maitland of

Lethingtoun Knycht Maister James
Makgill of Rankelour nether Clerk of
Registre Schir Johne Bellenden of
Auchinnoull Knycht Thomas Mein3eis
of Petfoddellis Prouest of Abirdene
Conuene in the Burgh of Edinburgh
and cause the elnwand the quart pint
fyrlot stane and pund to be brocht to
thame fra the townis foirsaidis and be
the samin as thay find to mak ane vniuersall wecht of the stane of the wecht
· xvj · pund trois wecht allanerlie the
pund and vther smaller wechtis equivalent thairto And siclyke ane vniuersall mesoure and met of the quart pynt
fyrlot and elnwand conform to the
wechtis and mesouris vsit in the saidis
townis respectiue according to the
standis thairof Except the watter met
to remane as of befoir And as the foirsaidis persounis or ony thre of thame
directis furth the foirsaidis mettis
mesouris and wechtis that the haill
liegis of this Realme salbe haldin to
by sell met mesoure wey ressaif and
deliuer and be na vther wecht mesoure
nor met and as thay find that thay mak
ane met mesoure and wecht of the
foirsaidis stane fyrlot quart pynt and
pund and vthers smallar mesouris
effeirand thairto of brasse to remane
in the Register that eurie heid Burgh
of this Realme haue the iust mesoure
and quantitie of the samin in sembla-
bill mettell to be vsit vniuersallie in
this Realme quha salbe halden to gif
and deliuer the iust mesoure and
quantitie thairof in maner abone
writtin to euerie Schire nixt adiacent
to thame to be vsit in maner befoir
reheirsit vniuersallie and na vther-
wayis And this ordour to be maid to
the persounis foirsaidis or ony thre of
thame betuix this and the first day of
October nixt to cum And efter the
said day to mak publicatioun and
intimitioun to all and sindrie our
Souerane Ladyis liegis of quhat-
sumeuer estate degre or conditioun
thay be of that cummis in the contrare
heirof thay salbe punist as committaris
of thift to the deid

APPENDIX A.8

The Act of James VI, 29 July 1587 (*APS*, III, 521-2)

MESURIS and wechtis and the Just quantitie thairof

FORSAMEKLE as in our souerane lordis lait parliament halden at edinburgh the xxix day of Julij last bypast his hienes and his thrie estaitis convenit thairin moved be the generall complaint of all his loving and guid subiectis bot speciallie the guid firmoraris and laubouraris of the grund Gaif grantit and committit full pouer and commissioun of parliament To his trustie and weilbelouit counsalouris and wtheris eftir spelit Thay ar to say Mr Dauid makgill of neisbit his hienes aduocat Mr Dauid carnegie of colluthie Robert fairly of braid Sir archibald naper of Edinbillie knyght generall of his hienes cun3ehous Johne arnot commissionar of Edinburgh williame flemyng commissioner of perth Robert forrester prouest and commissioner of striuiling hew campbell prouest and commissioner of Irving Or ony sewin sex or fyve of thame coniuctlie the said lord aduocat being ane To convene in his lugeing within the burght of edinburgh als oft as thay sall think expedient betuix and the last day of august bipast and thair eftir sicht and consideratioun of the lawis and actis of parliament maid anent mettis messores and wechtis in tyme bygane and grundis quhairin thay haif procedit haveand regaird to equitie and indifferance To set mak and establische ane mett measor and wecht to be commoun and vniuersall amangis all our souerane lordis liegis

To by sell ressave and gif out And to present the same to his hienes and his privie counsall betwix & the xxvj day of Maij nixtocum That the same micht be notefiet to all our souerane lordis liegis with all conueneint speid and diligence thaireftir as the saidis commissionaris wald answer to his hienes vpoun thair obedience Ordaning lettres to be direct to chairge and compell hierto gif neid be in forme as effeiris lyk as at mair lenth is contenit in the said commissioun gevin be actis of parliament in maner foirsaid According to the qlk ane sufficient number of the saidis commissionaris convening and haifinge sene and considderit the saidis lawis and actis of parliament maid anent mettis measoris and wechtis in tyme bygane and grundis quhairon thay haif procedit haifand regaird to equitie and indifferance Hes set maid and establischit ane met measor and wecht To be commoun and vniuersall amangis all our souerane lordis liegis To by sell ressaue and gif vp and thairvpoun hes delyuerit the repoirt and conclusioun subscryvit with thair handis Beirand in effect That thay haif sichtit and red and considerit the saidis lawis and actis of parliament maid anent mettis measoris and wechtis in tyme bigane The perfyt grundis quhairon thay haif proceidit That is to say the Elvand the pund trois and the stane proportionat and effeiring thairto The boll mett firlot and peck The pynt quart and galloun Euerie ane in thair awin degrie and proportionis And hes fund that maist wyslie the proportionis and grundis of all thir wechtis mettis and measoris hes bene sa establischit of auld that euerie ane of thame comptrollis wtheris and be iust conferance makis and establischis ane certane measor and wecht And thairfoir hes thocht maist expedient that the same proportioun and comptrolment be obseruit in all tyme cuming and first hes fund the eln and stand thairof committit to edinburgh contening threttie sevin Insches The stane contening sextene pund trois Ilk trois pund contening sextene wnce The pynt of stirling contening tua pund and nyne vnce trois wecht of cleir watter all the premiss to obserue ane iust proportioun according to the lawis and actis of parliament And as to the firlot quherof thir hes bene mentione maid in the actis of king James the secund to haif bene of auchtene pyntis having tryit and comptrollit the same in deipnes and breid be the Elnvand and in quantitie and wecht be the said stoipe of stirling hes fund the same les in proportione nor it aucht to be beand comptrollit be the rest of the wechtis and measuris abonewritten and this as appeiris earest be errour of the prentair Be ressoun that be iust calculatioun tryall and comptrolment the same extendis to nynetene pyntis and a Jowcat And for eschewing of fraud hes thocht expedient that all wictuall in tyme cuming salbe mesorit be straik And be ressoun that malt beir and aittis hes bene wsit to be measorit be heip hes fund be examinatione and tryell that the heip in proportione is the just thrid of the firlot or peck · Thairfoir remittit to the kingis Majestie and the lordis of secreit counsall quhither thay thocht it maist expedient to caus mak a particulair measour for metting malt beir

and aittis be straik Or that ane measour suld stand vniuersallie to ressaue and delyuer thrie for tua or sex for four of malt beir and aittis according to the proportione of wictuall and stuff wsit in tymes bipast to be met be straik sic as quheit ry beanis peiss meill and quheit salt sauld in the mercattis or in the cuntry Except the watter met to remaine according to the vse of the cuntry Siclyke the saidis commissionaris hes thocht expedient that thair be double standertis of all the foresaidis wechtis mettis and measoris maid of braiss The ane of euery ane of thame to remaine in the register and the wther with the tounis to quhome thay haif bene committit of auld To be direct furth to the haill liegis to be wsit vniuersallie And this without preiudice to ony personis quha ar foundit infest or addettit be tak or contractis of auld or new formes of wtheris measoris Bot that thair fundatoun infeftment tak or contract sall stand in effect and the measor of ther fundatoun infeftment tak or contract salbe proportionat to the measoris now to be establisched be his Majestie his parliament and counsall Sua that the samen quantite sall remane with the gevar and the ressauer but preiudice of ony of thame as the said repoirt exhibit to the Lordis of secreit counsall at mair lenth proportis · Qlk being sene considerit and allouit be thame Thay thairfoir according to the pouer and aucthoritie of the said parliament maid report of the saidis commissionaris Haif decernit and ordanit and be thir presentis decernis and ordanis The firlot to be augmentit and the standert thairof to be of the forme eftirspelit and to contene nynetene pyntis and tua Jowcattis and this to be the measour of all wictuall and stuff wsit in tymes bipast to be sauld be straik sic as quheit ry peiss bennis meill and quheit salt sauld in mercattis or in the cuntry The wydnes and braidnes of the qlk firlot wnder and abone ewin ovir within the burdis sall contene auchtene insches and sext part of ane insche The deipnes sevin insches and half insche and the peck to be maid effeirand thereto And that the steppis of the said firlot be of the auld proportione in thiknes of bayth the burdis ane insche and ane half That the bottom thairof be corssit with Irne naillit to the same and to the ryng of the firlot and the edge of the bottom entering within the laggyne be pairit outwith toward the nethir syde and to be maid Inwith plane and iust rewll richt That the mouth be Reyngit about with a circle of girth of irne Inwith and outwith Haveing a croce Irne bar passing ovir fra the ane syd to the wther thrie squarit ane edge doun and a plane syde up Qlk sall gang rewll richt with the edge of the firlot and euery squair salbe a iust insche of breid And that thir be a prik of Irne ane insche in roundnes with a schulder under and abone Ryssing vpricht out of the centrie or middis of the bottom of the firlot and passing throw the middis of the said ovir corss bar Ruissit bayth onder and abone and that the couper caus the richt straik of the said firlot pas fra the ane end of the said ovir Irne bar to the tothir And be ressoun that malt beir and aittis hes bene wsit to be mesorit be heip And for eschewing of fraude Thinkis expedient and ordanis that all wictuallis salbe measorit be the straik in tyme cuming Seing be iust tryell and examinatioun the heip in proportioun is fund the thrid of the firlot or peck And that the foirsaid measoris of the firlot and peck stand vniuersallie

Ressaveing and delyuering thrie for tua or sex for four of malt beir and eittis according to the proportioun abone wrettin Except the watter met to remane according to the vse of the cuntrey And that ther be double standertis of the firlot maid of brass be the burgh of linlithgow and of the foirsaidis wtheris mettis and wechtis be the utheris burrowis to quhome thay haif bene committit of auld And ane of euery ane of thame to remane in the register And the wther with the saidis burrowis and the iust measuris and quantitie of the same To be directit furth be the saidis burrowis to the haill liegis of this realme to be wsit vniuersallie in maner and forme befoir rehersit And that the foirsaidis wechtis mettis and measouris With the qlkis all and quhatsumeuir personis salbe haldin To by sell mett measoure we ressaue and delyuer haif cours allanerlie within this realm Eftir the tuentie sext day of maij nixtocum and na wtheris wechtis mettis messores vnder the panis conteinit in the actis of parliament maid heiranent In his wmqll darrest moderis dayis And that all firlottis to be wsit in mercattis baith to burgh and land be brint and seillit with the birning Irne of the heid burgh of the schire quhare the saidis mercattis ar halden And that all prouestis and baillies of burrowis and cities baith of regalitie and Ryaltie and als the baillies of burowis in barronie and wtheris quhatsumeuir places quhair mercattis of wictuall ar halden Salbe bund that all messoris to be wsit salbe of ane forme and quantitie According to this present act And gif ony different measoris be fund in ony of the places abone wrettin The saidis provestis and ballies of Ryalteis and regalities and baronies to be

accusable theirfoir conforme to the
saidis actis and lawis maid be his hienes
darrest mother of befoir Provyding
alwayis gif ony personis be funden or
infeft Or addettit to tak or contract of
auld or new fermes of wther measoris
thane ar abone wrettin Thair founda-
tioun infeftmen tak or contract quhid-
der it be mair or less salbe proportionat
to this measore that now is Sua that
the same quantitie sall remane with
the gevar and ressaver but preiudice
of ony of thae And ordanis the clerk
of register to insert this present act in
the buikis of parliament ad futuram
rei memoriam And that lettris be
direct for publicatoun of the premisess
be oppin proclamatioun at the mercat
croces of the heid burrowis of this
realme and wther places neidfull That

nane pretend ignorance of the same
Commanding and charging all and
sindrie the saidis pruestis and ballies
of burrowis and cities baith of regalitie and ryaltie and als the ballies of
burrowis in baronie and wtheris
quhatsumeuir places quhar mercattis
of wictuall ar haldin To put in executioun this present act and euery pairt
therof sa far as concernis thame Sua
that the same may tak full effect eftir
the said xxvj day of maij nixtocum
Eftir the forme and tennour of the
saidis actis of parliament and wnder
the panis abone wrettin contenit
thairin With certificatoun to thame
and thay fail3ie Thay salbe callit accusit
and the panis of the saidis actis salbe
execut wpoun thame with all rigor In
example of wtheris

APPENDIX A.9

The Act of James VI, 19 February 1618 (*APS*, IV, 585-9)

THE ACT ANENT THE SETLING OF MEASURES AND WEIGHTS
CONCLUDED AT EDINBURGH THE 19 DAY OF FEBRUAR 1618 YEERES ·
BY THE COMMISSIONERS HAVING POWER TO DOE THE SAME
BY ACT OF PARLIAMENT MADE THE TUENTIE EIGHT DAY OF JUNE 1617·

FORSOMUCH as in Our Soverane Lords laite Parliament holden at Edinburgh the xxviij day of June last bypast his Highnes and Estates conveined therein Mooved by the generall complaint of all his louing Subjects: and in respect of their sensible prejudice seene and felt through many parts of this Kingdome by reason of the diversitie of Measures and Weghts used within the same ·

THEREFORE Our said Soveraine Lord with advise of his saids Estates; For removing of all abuses which may ensue in any tyme to come thereby hath found expedient and by Decreet and Statute of the said Parliament; Descerned Statut and Ordeined That there shall bee but one just Measure and Weght through all the partes of this Kingdome; which shall Universallie serve all his Highnes Leiges by the which they shall buy sell receiue and giue out in all tyme to come · Which Measure his Majestie with advise foresaid Fand; should be That Measure of Linlithgo which is now commonlie used and which hath beine used most customablie through the greatest part of this Kingdome these fiftie or thriescore yeeres bypast · And for the setling of a perfyte order whereby all the Measures that are now used may be reduced to the conformitie of the said measure now authorized and for

making of proportioun answerable betwix the Lesser Measurs and Weghts and the Greatest · HIS MAJESTIE with advise foresaid granted full power and Commission to Sr James Weemes of Bogie knight; Sr George Auchinlek of Balmanno knight · Sr James Fowles of Colingtoun knight · Sr Robert Stewart of Shillinglaw knight · Sr Jhone Waws of Barnbarro knight · Sr William Grierson of Lag knight · And to James Nisbit Bailyie and burges of Edinburgh · Mr Alex^r Wedderburne Clerk of Dondie · Sr Thomas Menzeis Provest of Aberdeine · James Hamilton Provest of Glasgow · John Oisburne burges of Aire and Sr George Bruce of Carnok knight · burges of Culros · Whome or any eight of them His Majestie with advyse foresaid Ordeined to meet and conveine together at such tyme and place as they should think expedient · And to consult and advyse together and to appoint and determine vpon the most convenient meanes how the saidis Measures and Weghts might bee reduced to the conformitie foresaid · As in the said laite Act of Parliament at more length is conteined ·

WHICH whole Commissioners foresaids hauing mett and conveined within the Burgh of Edinburgh upon the twentie ane day of Januar last by past and the most part of them upon

diuers and sundry otheris dayes thereafter in the said moneth of Januar and
Februar instant · And hauing read and
considered the foresaid Act of Parliament anent the saidis measures
and Weghts and finding that It is
Ordeined that their shall be onlie Ane
just Measure and Weght through all
the parts of this Realme which shall
universallie serve all his Maiesties
Leigis (by the which and no other)
they shall buy and sell in all tyme
comming · And that it is declared by
the said Act that the foresaid Measure
and Firlot of Linlithgo which is now
commonlie used and which hath bene
most customablie used through the
greatest part of this Kingdome these
fiftie or thriescore yeeres bygone shalbe the foresaid just Measur and Firlot
which shall be receiued and used by all
his Majesties Lieges in all tyme comming · And that Commission is giuen
bee virtue of the said Act to the saids
Commissioners for settling of a perfect
order whereby all the saids Measures
that are now used may be reduced to
the conformitie of the said measure of
Linlithgo · And for making of a proportion betwixt the lesser measures
and weghts and the greatest · Haue
first thoght it meit and expedient that
the Provest and Baillyes of Linlithgo
who are keepers of the said measure
should produce before them the said
measure which bath bene given owt
be them to the Burrows and all others
his Majesties Leiges these fiftie or
thriescore yeeres bygone with their
Jedges and warrands which they haue
for the same · Who being cited for
that effect Produced before the saids
Comissioners their said Measure and
Firlot with the Jedge which is their
warrand thereof · And the same
measure and firlot being fund agreable

with the said Jedge the saids Commissioners caused præsentlie fill the same
with water which being full they fand
that the same conteined Twentie ane
pincts and ane mutchkin of just
Sterline Jug and measure and that the
foresaid jug containes within the same
Thrie punds and seauen unces of
frensh Troys weght of clear running
water of the water of Leith · And because the saids Commisioners could
find out no other meane whereby they
might trye the warrand of the Quantitie of the said measure and firlot
of Linlithgo which hath bene in use
these fiftie or thrie-score yeeres bygone But be taking of the oaths of the
saids Provest and Baillyes thereanent ·
They tooke the oathes of Andro Milne
Provest of the said Burgh of Linlithgo
· Andro Bell and James Glen Baillyes
thereof who being with all requisite
Solemnitie Sworne : Deponed vpon
their consciences ; That the foresaid
Firlot and Measure produced by
them was the verie true and Just
measure which hath bene given owt
to his Majesties Leigis by them and
their prædicessors these fiftie or thrie-
score yeeres bygone and that the same
by their knowledge hath never been
altered in any sort dureing the tyme
foresaid and siklike declared upon
their consciences that so far as they
could trye by the most ancient and
aged persons of their Burgh that the
foresaids Jedges are of great antiquitie
and haue neuer bene altered or
changed in any tyme bygone : And
that they never had nor hes any other
Measure or Iedge to their knowledge ·

WHICH Firlot the saids Commissioners Haue Fund and Declared Statute
and Ordeined to be the Just and onlie
Firlot which shall be received and used

by all his Majesties Lieges in all tyme
comming : For metting of Wheat Rye
Beines Peas Meal Whyt Salt and such
other stuff and Victuall as before this
tyme hath beine in use to bee measured
by straik Mett within this Kingdome
The wydnes and Breadnes of the which
Firlot vnder and above euen over within the buirds shall contein nyneteen
Inches and sext parte inche ; and the
deipnes seaven Inches and ane thrid
part of ane inche : and the Peck halfe
Peck and fourt part Peck to bee made
effeirand thereto ; And the steppes of
the said Firlot to be in thiknes one
Inche at the least : That the Bottome
thereof bee crossed with Iron nayled
to the same and to the Ring of the
Firlot ; and the edge of the bottome
entring within the lagene bee pared
outwith towards the nether-syde and
to be made inwith plaine and just reul-
right ; That the mouth bee ringed
about with ane croce or girth of Iron
inwith and outwith haueing a croce
Iron barre passing ouer from the one
syde to the other thrie squared and
edge doun and a plaine syde up which
shall goe ruel-right with the edge of
the Firlot and euerie square shall be
ane just Inche of breadth · And that
their be ane prick of Iron one Inche
in roundness with ane shoulder vnder
and aboue and rysing vpright out of
the Centre or midst of the bottome of
the Firlot and passing through the
midst of the said over croce-barre
rooved both vnder and aboue · And
that the said Cowpar cause the ring-
straik of the said Firlot passe from the
one end of the said over Iron barre to
the other : And the same to bee brunt
and sealled with the mark of foure
Crownes vpon both the sydes of the
bottome with fyve impressions of the
letter L. vpon the lippes thereof · And

for eschuing of fraud in all tyme comming ; The saids Commissioners all in
one voice but discrepance or variance
Haue thought expedient Statute and
Ordeined by vertue of the foresaid
Commission granted to them by the
said laite Act of Parliament That all
Victuall and stuff shall bee measured
by straik through all the parts of this
Kingdome in all tyme comming · And
by reason that Mault Beare and Aites
haue euer beene used to bee measured
by heape ; and that by the meaning of
severall preceiding Acts of Parliament
IT hath beine thought that Heapes in
proportion was the just thrid of the
Firlot and Peck So that thrie straiked
Firlots for two heaped Firlots Sex
straiked Firlots for foure heaped
Firlots was thought to bee a just proportion the one agreable to the other
· And the saids Commissioners by
tryall and examination hauing found
that the Heape in proportion IS not
the just thrid part of the Firlot and
Peck but that there is a great difference therein and no small præjudice
both to the giver and receiver of thrie
straiked Firlots or Pecks for two heaped
firlots or pecks and consequentlie of
sex for foure the Heape being always
the lesse measure as said is ·

THEREFORE they haue found it
expedient to cause make ane particular Measure or Firlot for metting of
Mault Beare and Aites by straike in all
tyme comming which being made and
produced in their presence and after
tryall and examination thereof Haue
found the same in proportioun neirest
to the said Heape so that foure streaked
Measures or Firlots thereof conteines
in just proportion (and to the lesse
præjudice of all his Majesties Lieges)
foure heaped Firlots · Which the said

Commissioners hauing caused fill with cleare running water of the water of Leith They find the same to conteine Thrittie one Pincts of the just Sterline Jugge and measure ilk Pinct conteining the Weght foresaid · And the same to be in wydnes and breadnes equall and conforme to the former Firlot and in deipnes Ten Inches and ane halfe inche ·

WHICH they Find Statute and Ordeins To remaine as ane just Measure and Firlot to bee Used for metting and measuring of Mault Beare and Aittes by straike in all tyme comming · And that the Pecks halfe Pecks and fourt part Pecks thereof be made conforme in proportione to the same last Firlot : Which new Firlot in all other respects shall be agreable in forme with the old straike Firlot aboue written having one Iron girth more in the midst thereof outwith and marked with the impression of the letter H. on the outmost fyds thereof ·

AND the same with the foresaid other Firlot conteining twentie ane pincts & ane mutchkin To be giuen out by the saids Provest and Baillyes of Linlithgow to whose custodie the same was committed of old To the Burrowes and all others his Majesties Lieges for that effect betwix the date hereof and the twentie day of Apryle nixt-tocome · And that four fulles of either of the foresaid Firlots conteine and bee repute to bee ane just BOLL in all tyme comming allenerlie · Siclyke the saids Commissioners hauing considered the great Præjudice susteined by all OUR Soveraine LORDS Lieges through the diversitie of Weghts used within all the parts of this Realme ·

THEREFORE and conforme to their said Commission and Act of Parliament foresaid and for eschuing of all fraude Haue thought expedient Statute and Ordeined That there shall bee onely one Just Weght through all the parts of this Kingdome which shall Universallie serue all his Majesties Lieges (by the which and no other) they shall buy and sell all and whatsomever Wares accustomed to be boght and sauld by Weght als weell Foraine as Cuntrie-Wares; in all tyme hereafter : to wit The frensh Troys Stone Weght containing Sexteine Troys Pounds in the Stone and Sexteine Troys Unces in the Pound and the lesser Weghts and Measures to be made in proportioun conforme thereto : (And that Weght called of old the Trone weght to bee allvtterlie abolisched and discharged and neuer hereafter to be received nor vsed ·) And in respect that the keiping and out-giuing of the Weghts of old to the Burrowes and others his Majesties Lieges within this Kingdome was committed to the Burgh of Lanerk : Therefore the saids Commissioners haue committed the keeping and out-giuing of the said frensh Troys Stone Weght now established to the foresaid Burgh of Lanerk and their Successors to be given out by them and their saids successors to the Burrowes and others his Majesties Lieges betwix the date hereof and the First day of Maij next to come and in all tyme comming ·

AND lykewayes Statuts and Ordeins that there bee double Standards of the foresaids Firlots and Measures and Jedges thereof and of the foresaids Weghts Two of euerie one of them to remaine in the Register within the Castell of Edinburgh and other two within the Castell of Dunbritane therein to remaine as a warrand for the measures allenerlie · And the other in the Townes to whom they haue beine committed of old as said is to be direct foorth to the whole Lieges to be vsed universalie · And this without præjudice to any persons who are founded infeft or addetted by Tak or contract of old or new fermes of others Measures and Weghts ; but that their Foundation Infeftment Tack or Contract shall bee proportioned to the Measure and Weght now established so that the same quantitie shall remaine with the giuer and receiuer · but præiudice to any of them · Sicklyke haue found and declared That the Elne and Stand thereof committed to the keiping of the Burgh of Edinburgh conteineth Thrittie seven Inches · And the Pinct Stowp committed to the keiping of the Burgh of Sterline conteineth the weght of Thrie Pounds seaven unces of frensh Troys weght cleare runnng water of the watter of Leith · Which Elne and Stowp They Statut and Ordaine to remaine and abyde in the same integritie as they are now and that no other Elne nor Stowp or greater or lesse proportion conteining the said weght shall be receiued by none of his Majesties Leiges in any tyme comming to buy or sell with in any part of this Kingdome · And the halfe and quarter Elnes and halfe quarters and Nails : Quart Chopin Mutchkin and halfe mutchkine Stowps bee made in proportion conforme thereto · And the Burghes of Edinburgh and Sterline to whome the keiping thereof hath beine committed of old haue the out-giuing of the same to the rest of the Burrous and all others his Majesties Lieges to that effect betwixt and the First day of

Apryle next to come : And that they haue dowble Standards of the saids Elne and Stowp Two of euerie one of them To remaine in the Register within the Castell of Edinburgh and within the Castell of Dunbritane for a warrand as said is : and the other with themselfes and their successors to whome they haue beine committed of old : And that the foresaids Measures Mets and Weghts with the which all and whatsomeuer persons shall bee holden to buy sell mett measure wegh and deliuer haue course allenerlie within this Realme ; after the dayes respute after following viz. The Weghts after the First day of May nexttocom and the Measures of Firlots and Pecks and the rest of that degrie after the First day of June next thereafter and no other Weghts Metts nor Measures to bee received nor vsed in any tyme hereafter in any part of this Kingdome vnder the paines contained in the Acts of Parliament made thereanent : and that all Firlots to be vsed in Markets both to burgh and Land be brunt and sealed either with the marks and Sealls of Linlithgo in maner foresaid or with the burning Iron of the head burgh of the shyre wherein the saids markets are holden And that the Provest and Baillyes of Royall Burrowes and Cities both Regalitie and Royaltie & als the Baillyes of Burrowes in Baronnie and Justices of Peace in whatsomeuer places where markets of Victuall are holden or others Foraine or Cuntrie Wares shall be boght sauld and weghed mett and measured shall be bound That all measures and weghts to be vsed shall bee of one forme and quantitie according to this present Act : And if any different measures and weghts be found in any of the places aboue mentioned · The saids Provest and Baillyes of Royalties Regalities Baronies and Justices of Peace ; shall take order therewith and if need be shalbe holden to informe the Kings Majesties Counsell thereof that they may take order thereanent as appertaineth ·

PROVYDING alwayes as is before provyded if any persons bee founded or infeft and adetted by Tak or contract of old or new Ferme of measures and weghts then are before written in maner foresaid · Their foundation Tack or Contract whether it be more or lesse shall be proportioned to this measure and weght which now is established so that the same quantitie shall remaine with the giver and receiver but prejudice to any of them in maner particularlie before expressed · And because by the Provisions immediatlie before written divers Pleyes and questions may aryse betwix parties receivers and deliverers Maisters and tenents Fewers and their Superiors anent Fermes and Victuall and siclyk anent other stuff and dueties adetted and bound to be payed and delivered by weght either by infeftments Tacks foundations bandes or contracts whatsomeuer made before the date hereof anent the conforming and proportionating of the measures and weghts contained in the saids infeftments Tacks foundations and others Securities foresaids with the measures and Weghts now established if sure notice and tryall be not taken of the just measure and quantitie of the measures and weghts which haue beine most customablie vsed and received these fiftie or thrie-score yeeres bygone within the Shirefdomes vnderwritten viz. Lanerk Wigtoun Drumfreis Roxburgh and Bervik ·

In the which the saids Commissioners considering evidentlie the greatest diversitie of measures and weghts from the measures and weghts now established to bee for the most part : So that these fyve shyres being broght to the conformitie aforesaid the rest of the Shyres within this Kingdome may be easilie reduced to the same comform to this present Act.

THEREFORE the saids Commissioners Finds it meet and expedient and by these presents Statuts Decernes and Ordeins That the Shiref of everie one of the foresaids fyve shyres of Lanerk Wigtoun Drumfreis Roxburgh and Bervik ; or their Deputs shall warne the Baillyes of Regalities within the same Shyres and Steuarts of Steuartries thereof if any bee Justices of Peace and Magistrates of Burrowes To convein ilk one of them within the heade Bourgh of the same Shyre wherein they are Magistrates within twentie dayes after the Counsels pleasure shall be signified to them thereanent : and there not onlie to receiue and embrace the saids measurs and weghts from the Provest & baillyes of Linlithgo & Lanerk to whome the keiping thereof is concredit in manner foresaid & which are established by this present Act : But also to take tryall and cognition of the difference betwix the saids old measures and weghts and the measures and weghts now established · And to appoint conclude & determine ilkone of them within their owne bounds what proportion lesse or more shall bee giuen and received in tyme comming for conforming of their Fermes and deuties adetted by former infeftments fondations Tacks contracts bands and Securities : to the foresaids Measures and Weghts now established and to insert the same in their Registers and Court books ; To remain with them for decisioun of such controversies as may aryse in those bounds hereafter anent the disconformitie foresaid : and to report their diligence thereanent and conclusion in writte authenticklie subscryved by the saids Shirefs of Shyres Magistrats of Royall Burrowes Baillyes of Regalities Burghs of Baronnies and Justices of Peace ; conveining within ilkane of the foresaids Shirrefdomes : And to present the same to the Lords of his Majesties Counsell and Sessioun before the first day of Julie next-to-come ; To the effect the same may bee delyvered to the Clerke of Register to bee insert in the bookes of Counsell (ad futuram rei memoriam ·) And that none of the foresaids fyve shyres nor no others his Majesties Lieges within this Kingdome presume nor tak vpon hand in tyme to come ; To buy Sell blok bargane contract or sett in Tack to or with others for receipt or delyverie with any other Weght Mett or Measure nor the same which now by this present Act is approoved and established · And this for report of the Commission aboue written Requyring the Lords of Our Soveraine

LORDS Counsell and Sessioun That Letters may bee direct for publication of the præmisses by oppen Proclamation at the market Croces of the head Burrowes of this Realme and others places neidfull that none prætend ignorance of the same Commanding and charging all and sundrie the saids Provests and Baillyes of Burrowes and Cities both of Royaltie and Regalitie and also the Baillyes of Burrowes in Baronnies Justices of Peace and others whatsomeuer in places where markets are holden To put in execution this present Act and euerie part thereof in so farre as concerneth them so that the same may take full effect after the dayes respute foresaids : With certification to them & they faillye they shall be called and accused and the paines contained in the Acts of Parliament shall bee execute vpon them in all rigour in example of others · In witnessing of the which the saids Commissioners haue subscryved these presents with their hands Day Yeere and Place fore-saids · And Ordeins these presents to bee delyvered to the Clerke of Register to the effect that hee may cause insert the same in the Register of Parliament ·

APPENDIX B

Land Measure
and Assessment
in Regional Scotland

TODAY, THE LAND WE NOW CALL SCOTLAND CONSISTS OF A GOOD DEAL OF mountainous terrain and moorland. The rest is forest and cultivated land in approximately equal parts, making up about half of Scotland. It was not always so. In early times there had been much more forest and on either side of the major rivers lay wide tracts of reed and marsh. This was particularly true of the River Forth, the so-called 'Scots Water'. The estuary was much wider than that at present with so much marsh that only a narrow neck of dry land existed between the Campsie Hills and the marsh in the neighbourhood of Stirling. The holders of the strong point atop the rock there commanded the only north-south passage by land.

The most fertile land, then as now, was that of Southern Scotland up to the line of the Clyde and Forth and thereafter a zone to the east of a line from Glasgow to Stirling, and on to Fraserburgh. This included the rich lands of Perthshire and Angus and all of the 'Kingdom of Fife' (the last clearly defined remnant of the seven kingdoms of Scotland or, more correctly, of Pictland, north of the Forth-Clyde line that existed in the seventh and eighth centuries).[1] From Fraserburgh, the fertile belt ran westwards to the Black Isle, then northwards to Caithness.

The mountainous highlands to the west had but little cultivation and was given over to grazing, sometimes very rough grazing. With so much of her best arable land immediately adjacent to England, Scotland suffered greatly when invaded by armies from the South, and to a greater extent than England when roles were reversed, because so much of northern England was uncultivated.[2] The ruins of the border abbeys bear silent testimony to the destruction visited upon the Scots by repeated England attacks over many centuries.

Once settlement had been firmly established, land and what the land could carry, became the principal consideration especially in the Highlands where so much was only fit for grazing. Animal husbandry was dominant, as is revealed by the substantial export of hides, wool and woolfels as well as meat and some live animals. In 1498 the Spanish ambassador to the court of James IV reported that:

> … the Scotch are not industrious, and the people are poor. They spend all their time in wars, and when there is no war they fight with one another … They have more meat, in great and small animals, than they want, and plenty of wool and hides.[3]

Throughout the sixteenth century the export of hides steadily decreased while that of live cattle steadily increased. By 1600 the lowlander had in general turned to cereals, especially oats for sustenance, with meat a luxury for the upper classes. Meat was an important element of diet for working people only in the hilly areas where arable land was in short supply.[4] Elsewhere the poor in particular lived for the most part on oats, consuming prodigious quantities daily. We are told that for breakfast some 15½ ounces of oatmeal would be consumed by one man probably as brose.[5] Meat was only reinstated as a significant part of the diet towards the end of the eighteenth century. It is possible that this emphasis on oats led Dr Samuel Johnson in his *Dictionary* of 1755 to define the word as 'a grain which in England is generally given to horses but in Scotland supports the people'.[6]

A factor in the move from a diet of meat to one of cereals was that a similar nutritional value could be obtained from cereals for one-fifth or so of the cost.[7] Potatoes came in somewhat later to supplement the diet and if near the sea there was likely to be fish, principally herring. The extent of the contribution from potatoes to the daily diet has been estimated to have been 75 per cent in the Highlands and one-third of that in the Lowlands as late as 1846.[8] Oatmeal was so central an element of diet that it became traditional at Edinburgh University to grant a Monday holiday halfway through the academic year to enable the students to go home over a weekend and return with the bag of meal that would see them through to the end of classes. With no one today on an oatmeal diet this academic holiday, 'Meal Monday', was abolished a few years ago.

Land was the index of wealth, but it did not actually have to be measured in early days, even if it could have been. It was enough to know that this hill would support a hundred sheep or that that field could grow enough cereal crops to maintain the people. If the local woods yielded enough fuel year round with adequate pannage for the pigs of the settlement, then there was adequate woodland. Nearby there had to be enough meadow for pasturing the oxen who pulled the heavy wooden plough together with the attendant cows and their calves. All that was needed was the assurance that there was enough; the physical area did not matter.

But this changed when common ownership was replaced by private ownership, when one man's field had to be set off from another's in order to levy rent or to receive a share of the fruits of the land. One method for defining the extent of arable land was that sowable by so much seed. Another, for grazing, was to state the number of animals it could support throughout the year. These concepts were prevalent in sixteenth-century Scotland and later, although they have the flavour of belonging to a much earlier age; and presumably such primitive methods of estimation originally had a source in Anglian tradition. For example, a grant of 1574 mentioned two riggs of land extending to the sowing of a firlot, and another area extending to the sowing of three pecks; and in 1681 the Scottish Parliament ratified a grant given by Barbara, dowager Countess of Seaforth to 'the poore and indigent persons presently residing or that shall happen to reside in time comeing within the Towne and burgh of the Chanonrie [Canonry] of Ross', six pecks of land and another peck lying elsewhere.[9] This act goes on to describe another piece of land extending to the sowing of half a boll of barley.

Pasture land could be estimated by the number of 'soums' of cattle, sheep, and other livestock which it could support. Thus, in Inverness-shire, a soum was the grass land of a cow or ox while a horse was taken as two soums. From this concept the Scottish legal action of 'souming and rouming' arose, whereby each tenant's soums and extent of arable lands were determined.[10] Although originating in early centuries, they continued into modern times. A land grant dated 1653 under the Great Seal gave to the Earl of Callander and Lord Hay of Yester, among many other lands and privileges,

> the Templar tenements of Pittencrieff, with the pasturage of three soumes of cattle yearly, and one brood sow and one brood goose with their followers …[11]

As late as 1884, the process of souming and rouming was still being raised.[12]

In many Highland districts where sheep rather than large animals were run, four or sometimes five sheep were counted as one soum. In Harris in 1792, the souming of a farthing-land (one-fourth of a pennyland) was four milk cows, three or four horses, and as many sheep as the tenant could rear.[13] In Kintyre in 1794 the *Statistical Account* mentions that rents are fixed by the number of bolls sown or the soums of cattle the land will support, with a cow or ten sheep (sometimes fewer) making a soum while a horse counted as two soums.[14] Calves, lambs and foals were counted fractionally according to their size relative to the full grown state as late as 1808.[15] These measures were of necessity rough and ready, and bore more relation to the agreement of landlord and tenant than to the superficial area of the land.

No area was mentioned when king or noble made a donation of land to the church or to an individual, but the local people knew the extent of the donated lands and in the eleventh century it was common to find the charter was quite unspecific, stating 'my lands of so and so' or 'the lands of such and such a place'.[16] No bounds were given – far less an estimation of the area – but a little later, in the twelfth century or at the beginning of the thirteenth, detailed descriptions are given of the markers which defined the perimeters of the estates in question.[17] Several of the mid to late twelfth-century charters use natural features, such as rivers, streams, and roads to define the land.[18] In about 1150 the grant of David I of the land of Mobbiscroft, Stirling, describes it as 'lying between the Forth and the road descending from Stirling to the ships as far as the stream descending from the king's mill into the Forth'.[19] Sometimes a portion of land was defined by having the sheriff perambulate the bounds.[20]

ACRES, PLOUGHGATES AND CARUCATES
IN SOUTHERN SCOTLAND

It was in the more fertile southern zones of Scotland, particularly in Lothian, that we first hear of acres of arable land. This is not surprising for the old kingdom of Northumbria extended to the Forth (and for a time well beyond) from the late sixth to the early eleventh centuries and the acre was a Saxon unit introduced into what was to become England at the time of the invasions from Northern Europe in the late fifth and sixth centuries. The written record shows the word 'acre' appearing in English documents at the beginning of the eighth century while the larger unit, the 'hide' (also known in Scotland as the 'carucate' or 'ploughgate') was being mentioned in the seventh.[21]

Notionally, the acre was the land ploughable in a day by a team of eight oxen.[22] In England, in Aelfric's *Dialogue* of about AD 1000, we read of the ploughman saying that '… having yoked my oxen and fastened my share and coulter I am bound to plough, every day, a full acre or more'.[23] As a work unit, the extent of the acre would be quite variable depending on the climate, the nature of the soil, and the animals and equipment to do the job. It was estimated by looking and if buyer and seller, donor and receiver agreed it was an acre, then the deed of transfer would show the land in question to be an acre regardless of the actual superficial extent. In determining the land ploughable in a day, value judgements were called for, not mensuration. When the concept evolved into that of a two-dimensional area, the figure was shaped as a rectangle, its area being the length multiplied by its breadth.

Acres and their supposed variation according to the accepted length of the ell have been discussed earlier, in Chapter 3. The terminology (if not the actual lengths and areas) was virtually the same in England and in Scotland. Walter of Henley's *Husbandry*, an English manuscript of about 1286, tells us that a quarentine (acre) ought to have 4 rods in breadth and 40 rods in length with the rod 16½ feet (5½ yards) in length.[24] Corresponding to the English rod, the Scots had the 'fall', that is the length which literally fell under the measure when it was laid on the ground. Sir John Skene in 1597 wrote that 'sa meikle [so much] lande, as in measuring falles vnder the rod, or raip [rope], in length is called ane fall of measure, or ane lineall fall'.[25]

Six ells made a fall and the acre was (as before) thought of as a rectangle 40 falls by 4 falls (5,760 square ells). Of course few acres would be a neat 40 by 4 rectangle, so tables had to be prepared showing various lengths and widths which together would produce an acre. The English table is to be found in the *Statutum de Admensuratione Terre*, attributed to 1305.[26] The Scottish table is given most conveniently perhaps in Balfour's *Practicks*.[27] Because the fall (of 6 ells, or 222 inches) was two feet longer than the English rod (198 inches), the Scottish acre was larger than that in England by about a quarter.

Although knotted rope was used by land surveyors for measuring the dimensions of plots of land, chains with uniform wire links were in use in both England and Scotland in the 1590s. The calculation of acreage using chains was described in Chapter **3**. The English chain of 66 feet divided into 100 links was introduced by the mathematician Edmund Gunter in the early seventeenth century; and this had the effect of decimalising area calculation since the furlong was 10 chains long and the acre (of 4,840 square yards) was therefore 10 square chains. The equivalent Scottish chain was 6 falls, or 74 feet long, and the acre of **10** square chains was 6,084·4 square yards. The use of a chain of 74·4 feet, based on the ell of 37·2 inches, was also widespread in the late eighteenth century, giving an acre of 6,150·4 square yards. (However, map scales marked with the chain based on the variant ell adopted in 1826 are seldom encountered.)

The figures of early English and Scottish land measures given in Table **B.1** below are taken from the English manuscript of Walter of Henley's *Husbandry* dating from about 1286, and from the Scottish legal *Fragmenta* from the later thirteenth or fourteenth centuries.[28]

The hide is given in the twelfth-century *Dialogus de Scaccario* as being 'a hundred acres' but the long hundred of 120 is intended.[29] In Balfour's *Practicks* of c.1580, the oxgang is given, apparently erroneously, as 12 acres; but it is correctly given in Thomas Hope's *Major Practicks* of 1608-33 as 13 acres (although he errs in giving the ell only 36 inches).[30] We will see shortly

Table B.1

Land measures: England and Scotland.

ENGLAND	SCOTLAND
5½ yards = 1 rod, pole or perch	6 ells = 1 fall
40 square rods = 1 rood	40 [square] falls = 1 rood
4 roods = 1 acre	4 roods = 1 acre
1 oxgait (oxgang) = ⅛ hide (15 acres)	13 acres = 1 oxgang (later 'oxgate'), or 'bovate'
1 virgate = ¼ hide (30 acres)	8 oxgangs = 1 ploughgate or carucate
120 acres = 1 hide, or carucate	104 acres = 1 ploughgate

that there is some early indication that the oxgang may have been considered as 12 acres, but for much of the historical period it was certainly accepted as 13 acres. The ploughgate of 8 oxgangs is therefore 8 × 13 = 104 acres. In the past, some authors have declared the Scottish ploughgate of 104 Scottish acres to be equal to the hide of 120 English acres.[31] In fact, the ploughgate was larger than the hide by over 10 English acres, but perhaps it was considered equivalent in that both represented the land notionally ploughed by a team of 8 oxen in a season (and hence the Scots oxgang could be considered the contribution of a single ox).

What is true is that the word 'carucate' or 'carrucate' (from the Latin *carruca,* a plough), was used for both the Scottish ploughgate and the English hide in charters, in Scotland going back to the early twelfth century and in England, going back at least to the days of the Danelaw in the late ninth century, and in the form 'cassatorum', to the seventh.[32] In the charter of Alexander I to Scone Priory, of about 1120, there are grants of land totalling 36 'carrucatis' and the word is used again in about 1128 when David I granted a carucate of land to the Church in the castle of Roxburgh.[33] That these were grants of arable land is made plain when William the Lion granted half a carucate of land with common pasturage to the monastery at Cambus-kenneth in about 1174.[34] Similarly, an act of William the Lion in 1200 granted '2 bovates [of land] and 2 acres of meadow and pasture for 400 sheep', and here the meadows and pastures were separated from sowable lands.[35]

In the twelfth century this pattern of chartered land grant was not uncommon, with crop-bearing land stated in current units of area based on ploughing, together with associated pasture. Frequently these were essential complements since the ploughing of the arable land was only possible because there was adequate grazing land for the animals which performed the work. So, for example, in about 1160, Malcolm IV granted to the Hospital of St Margaret of Huntingdon from his English lands at Great Stukeley, one virgate of arable land and pasture for four cows.[36] Similarly, in about 1180, William the Lion confirmed to the lepers of St Nicholas's Hospital, St Andrews, Fife, two oxgangs of land with pasture for twenty beasts and six horses in Perthshire.[37] William also granted to Cambuskenneth Abbey four oxgangs of arable land in Fife with pasture for 500 sheep, twenty cows, one plough team of oxen, and for horses.[38] Finally, between 1173 and 1177 a charter to Paisley Priory (later Paisley Abbey) confirmed a ploughgate (carucate) of land in Roxburghshire, measured by the marches and 'perambulated', with pasture for 500 sheep and for as many animals as were allowed with one ploughgate.[39]

The important point was always what the land could produce, because that represented its value. Unproductive land such as marsh was worth nothing; but if poor land such as moorland could yield peat for fuel, then it might be mentioned in the granting charter, as was the case with the 30 acres

of moorland in Northumberland given to the monks of St Cuthbert by Earl
Henry (son of David I) in about 1141.[40]

In England, acres are of considerable antiquity, but in southern Scotland
they appear with carucates in twelfth-century charters and documents. It is
most likely that they were in use much earlier (but the relevant documents
have not survived), and were presumably a remnant of the common heritage
of the early period when Scotland, south of the Forth, was part of Anglian
Northumberland.

Among the earliest surviving Scottish documents which mention acres is
David I's so-called 'Great Charter' to Holyrood Abbey, dating between 1128
and 1136, where he speaks of the gift of land at Crorstorfin (Corstorphine
near Edinburgh) which comprised two oxgangs and six acres, together with
a salt pan and 26 acres of land at Airth, near Stirling, and a carucate of land
and 40 shillings from 'my burgh' of Edinburgh.[41] The same charter also
provides an early reference to the 'rood' or quarter-acre – a term used in both
England and Scotland – when it recorded the gift of a toft (a homestead with
attached arable land) of five roods in Renfrew.[42] Another of David's charters,
this time to Dryburgh Abbey between 1152 and 1156, included a grant of 'a
manor in my burgh of Crail … with three roods of land'.[43] The dimensions
of smaller plots of land within burghs were sometimes given in feet at this
early period: this is discussed in the Land Survey section of Chapter **3**.

The 1152-3 charter of Geoffrey de Percy, just at the end of the reign of
David I, gives the size of the corucate or ploughgate as 104 acres when it
mentions a ploughgate of land in Heton, Roxburghshire, of 5 times 20 plus 4
acres.[44] The acreage is given in this form to emphasise that the hundred in
question is five score and not the long hundred of 120, which would be the
expected meaning at this time.[45] The 104-acre ploughgate is shown again
clearly in a late thirteenth-century charter: in 1296 a tenant's holding near
Aberlady in East Lothian was found to have an estate of 10 carucates and 54
acres of arable land. The annual value of each acre with the grazing was given
as 21 pence, where the total annual revenue was £95 14s 6d, which is 22,974
pence.[46] Dividing by 21, the acreage comes to 1,094 acres, which makes each
carucate 104 acres.

David's Holyrood charter had referred to 26 acres at Airth, which can be
recognised as a quarter of a ploughgate of 104 acres; and this size of 26 acres
or 2 oxgangs was called a 'husbandland'. It is not uncommon to encounter
four husbandmen banding themselves together to maintain a common
plough, each man providing his two oxen to make up the team of eight; just
as eight oxgangs, or in the latinised form bovates, made up the ploughgate.[47]
Of course, the team did not invariably consist exclusively of oxen: on
occasion horses would have been substituted for some of the oxen.[48]

A note on the last leaf of the fourteenth-century Monynet manuscript
collection of the Auld Lawes describes the husbandland:

… ilk husband land xxvj akkeris quhair pluk and syth may gang [where plough and scythe may go, namely, arable land]. Item xiij akker of land is callit ane ox gang Tua ox gang is ane husband land Ane akker of land is of lenth xv[XX] and xii [300 + 12] ellis and ane ell of breid [breadth].[49]

The last sentence of this is corrupt, for an acre that is one ell broad must have a length of 5,760 ells – a quite unworkable ribbon of land. Perhaps the text that was copied should have read '… lenth xv[XX] and xii [300 + 12] and i[XX] [20] in breid', the superscript [XX] having been accidentally omitted by the scribe. This would give the acre an area of 6,240 square ells rather than the usually encountered 5,760 square ells. However, 12 of these larger 'acres' is the same area as 13 acres of the familiar 5,760 square ells. If this interpretation is correct, a 12-acre oxgang may have existed, of limited applicability, in the early centuries. It would have been of the same size, but divided into 12 rather than 13 units. Possibly, in a similar vein, a 1291 inventory of properties held by the deceased Countess of Mar listed 13 *husbandi* each holding 24 (perhaps not yet updated to 26) acres.[50]

Since these charters of the mid-twelfth century are couched in terms of carucates (or ploughgates) of 104 acres, and other sub-units of 52 and 26, as well as the oxgang of 13 acres, it must surely mean that the acre, as well as its multiples up to the ploughgate, was firmly established long before the twelfth century in southern Scotland. The system remained entrenched, and as late as 1667 and 1668 lands in Aberdeenshire and Berwickshire were still being measured in husbandlands and oxgates (the spelling used at a later period).[51]

But what were the Scottish ploughgate and the English hide? It is clear from Table **B.1** that they were intended to serve similar purposes. Like the acre, the hide was a Saxon land unit, and it dated from no later than the seventh century, for there is a charter of AD 673 granting 15 hides of land in Egham, Surrey.[52] In Scotland the ploughgate, like the acre, first appears in the records of the twelfth century, but we have assumed it to be more ancient than that. Both the hide and the ploughgate were intended to be the land needed for the maintenance of a family, probably an extended family, for a whole year, and from this we suppose these terms could be of even more ancient origin than the concept of the acre. Like the acre, the hide and ploughgate began without any definite area associated with them. In Saxon and Norman times the hide was used in England as the unit by which service or taxation of the lord to the king could be assessed. If a noble was granted a hide of land by the king, whatever that territory was, it was called a hide and recorded as such without much regard for its area. The noble would be called on either to pay or to serve the monarch the full levy then currently associated with a hide.[53] Only in later years would the hide become standardised as so many acres. There are Anglian records of hides of 40, 60 or 160 acres, but the most frequently encountered acreage is one of 120, the Mercian hide, also known as the Domesday hide.[54] Presumably the Scottish ploughgate had a very similar evolution.

We have here been considering the occurrence of essentially English units and terminology, initially in southern Scotland but later spreading northwards. But these Anglian influences were not the only ones to leave their marks on Scottish lands and their assessments. Gaelic and Norse influence played their part too, introducing terms and concepts, especially in assessments and levies on land; for while the king might grant land to the church (to provide an income to pay for masses for the repose of his soul in time to come), if he made grants to an individual he wanted something in return, be it military service, money or produce. The services renderable for lands granted are considered in Appendix C. Meanwhile, we will look at the Gaelic and Norse influences, but it should be stressed that there were many local variations of land assessment, beyond the scope of this book. We have endeavoured to follow some of the main developments. Perhaps inevitably, historians have not always been as one when it comes to interpreting these early records.

CELTIC AND NORSE UNITS

The interpretation of Celtic and Norse units of land measure has been under debate for an extended period, but there have been significant advances in recent years in understanding their meaning.[55] Recalling that the Scots came from Ireland to the South-West Highlands in the early sixth century we should not be surprised to find Irish practice translated from Dalriada in Ireland to Scottish Dalriada (the Gaelic kingdom of Dál Riata), located between the Picts to the north and the Strathclyde Britons to the south. John Bannerman's examination of the seventh-century documents *Senchus Fer nAlban* written in Irish and compiled in Dalriada has revealed what must be regarded as the most ancient assessment of lands in Scotland.[56] There were three tribal divisions in Dalriada: the *Cenél nOengusa* occupied the island of Islay; the *Cenél nGabráin* occupied the Kintyre and Cowal peninsulas and the islands of Jura, Arran and Bute; while the *Cenél Loairn* held Lorn and the island of Colonsay together with Mull and mainland areas to the north. These lands of Dalriada were described in the *Senchus* in terms of the number of 'houses' which they contained and it is clear that the 'house' was the rental unit and the basis of levying tribute and producing men for military or naval service. The owning of a house was the property qualification of freemen, as we learn from the eighth-century Irish document *Críth Gablach*.[57] The houses were gathered for administrative purposes, most often into groups of 20 but occasionally into lots of 30, and the various leaders named in the *Senchus* had 60, 30, 20, 15 or 5 houses. Five houses seem to have been the smallest unit recorded, and this is one-quarter of the most frequently encountered number of houses, namely groups of 20. Similar units are found for other Celtic areas: the corresponding Welsh unit was frequently divided into four, as was the Irish unit, the *baile biataigh*.[58] Imposts were levied on the 'houses' individually and also on the group of 20 collectively.

The *Senchus* described the naval levy for the three principal 'tribes' as:

> The expeditionary strength of the hostings of the Cenél nOengusa .i. five hundred men. The expeditionary strength of the Cenél nGabráin .i. three hundred men … The expeditionary strength of Cenél Loairnd, seven hundred men, but the seventh hundred is from the Airgialla [a number of subservient states in central and southern Ulster].[59]

The number of houses is also given, as 430, 560, and 420, respectively, and from each tribe, for a naval expedition, two 7-bench boats from every 20 houses, where there were two men to a bench.[60] This yields 602, 784 and from *Cenél Loairn* itself, 588 men, probably intended to be in even hundreds, 600, 800 and 600. Bannerman has pointed out that the army levy was inconsistent with the naval levy since the largest tribe, the *Cenél nGabráin*, provided the smallest number of men. He suggests that the naval levy is correct, since the figures bracket 700, which was the average number of men an Irish king could raise from his people. Additionally, in Ireland, units of 100 men were frequently used, as witnessed by the term 'the constable of the hundred'.[61]

When the Norse arrived as settlers in the ninth century, they found the Dalriadic 20-house system in place, with the Gaelic word *davoch* (*davach, dauch*) being used to describe that unit.[62] The davoch has been recorded in the *Notitiae* of the *Book of Deer*, written probably in the interval 1130-50.[63] It has been found to be associated with Pictish territory generally, for it has been pointed out that the geographical distribution in the written record and in place-names closely parallels that of Pictish symbol stones.[64] Davochs appear in the place-names of the west and the Western Isles, and subsequently in Moray, Mearns, Angus, Fife, Aberdeenshire, Banffshire and generally throughout the land between the Tay and the Dornoch Firth. The davoch's absence from parts of the far north mainland and totally from Orkney and Shetland can be attributed to its displacement by Norse units beginning in the ninth century.[65]

The davoch began as a measure of agricultural capacity, but it was not long before the produce of the land was subject to a fiscal levy of one sort or another.[67] When the Norse arrived as settlers in the north in the ninth century they found the 20-house system in place and the term davoch and its quarter being the fiscal units employed.[67] The names 'ounceland' and 'penny-land' were given to lands of the house group and to those of the individual house respectively, for the group was assessed at one ounce of silver and the house at a penny, or the equivalent in kind. The Gaelic word for ounceland was *tirunga* (*tir*, or land; *unga* from *uncia*, or ounce), the *urisland* in Norse, the Latin *unciata terre*.[68] At one time it was thought that the origin of the 20:1 system came from there being 20 English pennies to the ounce, but it seems that the basis was the Dalriadic house system, with the Norse adopting this existing system to their own fiscal purposes. Indeed, there is a considerable concentration of pennylands in old Dalriada. Bannerman has identified the davoch, urisland and tirunga with the 20-house grouping and has noted that the characteristic division is into quarters, with the smallest number of houses being five.[69] Earlier writers had concluded that the pennyland and davoch were Norse, originating in the ninth century, but present understanding is that they were in use by the Scots of Dalriada of the seventh century, and were agricultural and fiscal units.[70]

The original meaning of the word 'davoch' changed with time. It has been the subject of much research, with little consensus until recently and then only imprecisely if it is supposed to support an extended family and adherents. In 1886 F. L. W. Thomas gave no fewer than 14 interpretations.[71] No less an authority than Cosmo Innes gave up on the etymology after advancing some of the more popular ideas.[72] The word stems from the Gaelic *dabhach* meaning 'seed-vat', and it has been identified

> … as the amount of land necessary to produce a fixed amount of grain, sufficient to fill a vat of some specified size; and secondly as the area which could be sown with seed from such a vessel. If the former interpretation is favoured, then the amount in question must represent only a proportion of the yield contributed as a fixed levy.[73]

The early davoch could hardly be described as a well-defined unit of land, and by the twelfth century it had lost any connection with grain and was understood solely as a measure of arable land, albeit ill-defined.[74] Its rental and imputed area varied with locality and with time.

In the west there is no doubt, as Bannerman indicated, that the terms 'davoch' and 'ounceland' were synonymous in virtually all respects.[75] Here in the west, the davoch was effectively superceded by the ounceland. The davoch appears universally except in southern Scotland (where acres and bovates prevailed), and in some parts of the northern counties. There was, however, at least one mention of the davoch in southern Scotland, although connected with a grant to a northern abbey: a thirteenth-century confirmation by Pope Honorius III to Arbroath Abbey in Angus recorded the gift of a davoch of land in Tweeddale, latinised in the document to *davata*, a term with which the clergy at Arbroath would have been completely familiar.[76] In Tweeddale in the Borders, carucates and acres had been in use long before the time of Honorius.

Ouncelands and pennylands appear in land charters as named localities, which implies that the lands were consolidated in blocks and not dispersed piecemeal in riggs or strips. An example of the scattered nature of some early grants of land is in the charter, dated between 1153 and 1162, of Malcolm IV confirming the grant of half a carucate of arable land to Kelso Abbey by David I, which in David's day was dispersed throughout the field and thereby inconvenient; it was now to be exchanged for half a carucate all in one piece.[77] The davoch (and its equivalent, the ounceland), and pennylands were particularly units of arable land. Frequently, grants of davochs and pennylands specified that additional meadow and woodlands were to go with the territory, as when Robert II granted 60 davochs 'with pertinances' and forest land.[78] Arable land was not worked in isolation without meadow, pasture and woodland: the animals which worked the land needed pasture as did the cows and calves as food producers, and as fuel was needed there also had to be woodland near the arable section.[79]

The terminology for land measurement and assessment was different in specific geographical areas of the country, and had different regional connotations.[80] It is now apparent that we must consider separately the west (that is the west coast and the Western Isles), the northern mainland, the northeast and the south-west; the situation in southern Scotland has already been dealt with briefly. As well as the meanings of land measurement and assessment changing in a geographical sense, we must expect to find some change in usage over the centuries.

THE WEST

The ounceland appears for the first time in the Latin form as *uniciate terre* in a charter of 1343 when David II granted to Reginald, son of Roderick of the Isles, land from the islands of Uist, Barra, Eigg, Rhum, and the land in Garmoran (west Inverness-shire) described as *'octo vnnctiatas terre de Garw Morwarne'*, or eight ounces of Garmoran.[81]

The conventional Gaelic and the Latin terms for the ounceland (*tirunga* and *unciata terre*) could be used in parallel, as in 1498 in two land grants on the island of Skye, which were later described as 'two unciate of the lands of Trouternes … Skye' granted to Alexander Macleod; and 'four marks of the terunga of Duntullyn' granted to Torquil Macleod of Lewis.[82] Indeed, when the Lord of the Isles granted 'the *unciata* of Baegastallis' in the fifteenth century, we find it in 1655 described as the '*teiroung* of Beagistill'.[83]

Similarly, the words 'davoch' and 'tirunga' are synonymous. So, for example, in 1505 in a listing of properties for North Uist, is found:

> … the davach called in Scotch *le terung* of Yllera, the davach called in Scotch the terung of Paible, the davoch called in Scotch the terung of Pablisgerry, the davach called the terung of Bailranald …[84]

As might be expected, many charters for lands in the west and the Western Isles show the davoch was equal to 20 pennylands, with 'ouncelands' or 'davochs' being progressively replaced by their equivalent in 'pennylands'. One such example, for Knoydart, describes 'the grant of King Robert the Bruce to Roderick, son of Alan in 1309 of lands resigned by Christian of Marr including 3 davochs of Knodworath'; these three davochs were granted again in 1536 by James V to Donald Camroun as 'the nonentry and other dues of the 60 pennylands of Knodart'.[85] In another instance, records for 1373 reveal that 'three unciata of Swynwort and Lettirlochetle', had become by 1723 'the tenements of Swenard … in all 60 pennylands'.[86] In 1583 a grant was made to Janet Campbell by her husband of a series of lands valued at 10 or 5 pennies each. In part, the document reads:

> 10 pennylands of Arnistill called a half davach, 10 pennylands of Aichaglyn called half a davach, 5 pennylands of Lekewuir and 5 pennylands of Meillarie called half a davach …[87]

Frequently, in such documents, lands are designated as 5, 10, or 15 pennylands, which stresses the basic unit of the davoch or ounceland as 20 pennylands.

The records reveal many instances of the groupings of 'houses' into lots of 20 in old Dalriada and military and naval service was assessed either on the 'house' with its pennyland or on the group of 20 or so houses with its associated davoch. The later levy was frequently one man per 'house'. An example of this is the two men with their food which were to be sent when needed to the army of Argyll in the thirteenth century from the two pennylands of Kames and Achadochoun.[88] Another, from the reign of Robert I, is the service of a ship of 26 oars, manned, with food for 15 days, from an ounceland.[89] This charter shows the davoch to be 20 pennylands in describing:

> … half a davach of Ayrssayk, namely, five pennylands of Gedeuall, and five pennylands of Glenbressell and Bethey.[90]

Here 26 oarsmen are required from the davoch which with 2 steersmen bring the total to 28. Further, in 1304, the Lord of Garmoran demanded a ship of 20 oars from each davoch, but sometimes many fewer men were required.[91] In 1309 this land in Arisaig was assessed at 2 davochs, providing one ship of 26 oars.[92] In 1343 the 4 davochs of Assynt were to provide one ship of 20 oars and in the same year 5¼ davochs of Gleneld were to provide one ship of 26 oars.[93] These reduced levels from the presumed norm of 28 or 20 men from each davoch is reminiscent of the English practice of 'beneficial hidage' when for particular reasons, a lighter burden was placed on the land. This frequently occurred during the time taken to bring more arable land into service from what had been previously woodland or waste, or following some natural disaster and was a temporary benefit.[94] Perhaps similar thinking was prevalent in the west and the Western Isles in these early centuries.

In the very early references the davoch, or its equivalents, is never described as so many acres. Rather it is subdivided into simple fractions, such as halves, quarters, or eighths. Only latterly, and in regions exposed to Anglian measures, do we find attempts to associate a definite area with either a davoch or a pennyland.

If it is true that the pennyland is to be associated with the Dalriadic 'house' of an extended family in early times, then it must have been 'a unit of reasonable size'.[95] We may also quote Richard Oram in saying:

> As with the davach itself, there appears to have been no fixed acreage for the pennyland in its purely fiscal form. If it is to be equated with the basic element in the Dalriadic ship levy system, theoretically it should have contained sufficient arable land and pasture to support a 'household' i.e. an enlarged family group plus dependents and followers. This argues against it being a small unit. In the Isles in a later period it may have undergone a transformation from an annual render levied on land of unspecified extent into a form of rent in return for a fixed amount of land.[96]

Indeed it is very likely that the pennyland was of significant size, since grants of farthinglands appear in the records.

The concept of pennyland and davoch eventually changed from being a purely fiscal and military unit to something corresponding to a rental for specified and named lands – it became a measure of arable capacity. The first attempts to determine their approximate areas were made in the eighteenth century. In 1747 a davoch in the west, at Lochbroom, was given as 192 acres.[97] This size was sufficiently near to 200 acres to presume a 10-acre pennyland at this late date. In eighteenth-century Lewis, it has been shown that about 10 bolls were needed to sow a pennyland.[98] With one boll per acre, this also suggests a pennyland of about 10 acres.[99]

We get an early view of rental assessment in the particular case of the island of Islay. Here, the Dalriadic house was the direct equivalent of the house of the lowest grade of freeman in old Ireland, the *Ocaire* (*Og Aire*).[100] The Irish document *Críth Gablach* identifies the house of an *ocaire* as one of 7 *cumals,* where one *cumal* was the name of the unit of land sufficient to maintain three cows with their calves, and was therefore three cow soums: thus the house of 7 cumals held pasture land of 21 soums. A land of 7 soums paid one cow as annual rent; so, the *ocaire* having driven 21 cows to pasture, drove back to the barn only 18, having left his rent for his superior. Such a land paying one cow per year as rent, was, not unnaturally, termed a 'cowland'.[101] The Irish house was therefore a holding of three cowlands. The same pattern appears in Islay, except that some lands are shown as 'horsegangs', two of which equal three cowlands. The boundaries of a number of early holdings on Islay are identifiable on modern maps, and these reveal that about 110 Imperial acres can be taken as a fair representation of the

3-cowland holding of the 'house'.[102] This unit of 110 acres is reminiscent of the Anglian hide of 120 acres, which was considered to be the land of an extended family.

From the ninth to the twelfth century the Western Isles were under Norse control, or ruled by the king of Man under Norse overlordship. In the mid-twelfth century, Somerled, Earl of Argyll, himself of Norse-Celtic origin, seized the islands.[103] By the early thirteenth century, the western regions were divided between Scottish and Norse suzerainty, with the Manx king ruling Skye and the Outer Hebrides and the sons of Somerled ruling mainland Argyll and the Inner Hebrides from Islay to Rhum or Uist.[104] Somerled, who died in 1164, was destined to be the founder of the dynasty which came to be known as the Lords of the Isles. The family names MacDougal, MacDonald and MacRuari arose from branches of the family, with the MacDougals in Lorn and surrounding parts, the MacDonalds on Islay, and the MacRuaris in Garmoran and Knoydart with the islands of Eigg, Rhum, Barra and Uist. John of Islay (a MacDonald) was the first to call himself Lord of the Isles (*Dominus Insularum*) in 1336.[105] The lords acted as kings in their own right, going so far as to enter into alliances with England against the Scottish crown. The notorious example of this was treaty concluded with Edward IV of England in 1462 by John, earl of Ross and last of the four lords of the Isles, whereby the earls of Ross and Douglas proposed to divide Scotland and then transfer allegiance to England, although the result was the forfeiture of the earldom of Ross.[106] Following frequent insurrections in the north by members of his family, decisive action was eventually taken: a number of partial forfeitures were followed in 1493 by MacDonald being deprived of all his lands and titles.[107] He retired in seclusion in the Lowlands and died in 1503, but his extended family continued to create unrest. Finally in 1540 the lordship was annexed inalienably to the crown. However, effective crown control over these remote parts was not achieved until the seventeenth century.

Davochs, pennylands and ouncelands are recorded in profusion throughout Dalriada and nearly all the Western Isles (including those of the lordship), but it is significant that they are totally absent from Islay itself and from its neighbouring isles, Colonsay, Jura and Gigha during the period of the lordship.[108] It has been thought that this may be because these lands were the seat of the Lords of the Isles, and were therefore held in demesne and not divided.[109] Only in the sixteenth century – after the forfeiture – were the lands of Islay divided into money units.

The thirteenth-century lordship, although virtually independent, was nonetheless an appendage of the Scottish crown. Before that century, the Dalriadic 'house' held 3 cowlands or equivalently 2 horsegangs. Two such 'houses' (a tenth of the 20-house unit) or 6 cowlands was the usual 'small ploughgate' of Islay, otherwise called the 'quarter' land of the old system, the rental of which was £1; that is, 1 cowland rented for 3s 4d. Four such quarters rented for £4 or 6 marks: we have the evidence from details of the islands of neighbouring Eigg and Rhum in 1309 which are given as 6 davochs, with every 20 pennylands (or davoch) having an 'extent' (or rental) of 6 marks.[110] The land of one house – an 'eighth-land' – rented for 10 shillings; and from the fifteenth century the rental terms 'quarter-land', 'eighth-land' and 'cowland' were in regular use.[111]

In the reign of Alexander III (1249-86) there was a re-evaluation of land, later to be known as the 'Old Extent', and this is discussed more fully below. The valuation has been ascribed to Alexander's reign largely because when the 'tenth penny' (a tenth of all rentals) was granted to Robert I by the Scottish Parliament in 1326, it was to be 'according to the Old Extent of lands and revenues of the time of the illustrious King Alexander, last deceased'.[112] Land valuation on Islay was also influenced by this, but because of the semi-independent nature of the lordship, it is not clear how vigorously it was enforced. Aspects of the 'ancient' valuation continued in use, and it was not until a further valuation was carried out in the 1490s that the situation was regularised.

Under the Old Extent, 20 houses totalling 60 cowlands were rated at 200 shillings, which was £10 or 15 marks. For example, in the local rental of 1722 for the 20 houses at Losset, near Port Askaig in Islay, lands were shown as having an old extent of £10.[113] This gave a cowland the same rental as before of ¼ mark or 3s 4d, and since the rent was paid by the surrender of a cow, the value of the animal then was 3s 4d.[114] If the rent from 20 houses was £10 under the Old Extent, then one house again paid 10s rent. Money values rapidly displaced what had been before, with the result that the term 'house' fell into disuse by the end of the thirteenth century being replaced by its associated 'pennyland'.[115] However, the cowland survived in charters and records as 'vaccatas terrarum'.[116]

Units, multiples and sub-multiples of the full rental of £10 appear frequently in the records of later times for central Scotland, which gave the value in terms of the Old Extent. Thus the quarter-rental was 50 shillings, and in the Register of the Great Seal for 1547 we are told of the 50-shilling land of Glencroce and Drumfadzeane of Old Extent, and in the same year of the 50-shilling land near Lanark of 'ant. ext.'.[117] Equally, the half of a full rental, namely £5, appears frequently for holdings in Islay in 1506, but William Lamont has cautioned that

> … in all these uses the £5 extent of the lands in question is probably much earlier than the breakdown into 3 Quarters at 33s 4d each.[118]

Although the rentals are those of the Old Extent, the further division into new quarters of 33s 4d (2½ marks) was a new introduction.

Following immediately on the forfeiture to the crown of the Lordship of

which is almost exactly the rental derived from the davoch in the north-east region of Scotland. It has been pointed out that these pennylands yield nearly ten times as much as the greatest rental found for the west.[148] This indicates that by the thirteenth century these pennylands had lost their fiscal connotation and represented definite parcels of land of varying worth; they may have included large tracts of hill grazing which produced acreages many times that of the pennyland in its early form. This is an example of the original assessment being retained when the arable component was expanded, or areas of grazing were extended, making the whole more valuable.[149]

In summary, then, we have seen the regional variations of the davoch and pennyland throughout the centuries. The western system was adapted to meet the needs of the regions into which it spread, with Galloway adopting the 20-pennyland concept as a vehicle for taxation, later becoming a unit of productive capacity, and later still, as a recognisable land area which could vary enormously in extent from sector to sector, depending on land quality and the amount of grazing accompanying the arable. While the davoch and ploughgate are related, earlier declarations that the davoch was four ploughgates predominantly in the east and one ploughgate predominantly in the west of the country have been shown to be broadly incorrect.

THE MONEYLANDS

As we have already mentioned, the 'ounceland' was gradually replaced by the 'markland' (or merkland) in the late-thirteenth century, with the two co-existing for a time, together with 'poundlands' and 'shillinglands'. These were lands which rented for a mark, a pound or a shilling per annum respectively, and they are not to be thought of as 160, 240 or 12 pennylands respectively.[150] Rather, they give an estimate of the annual value of the land which can then be used for the calculation of rent or tax. The pennyland and ounceland terms belong to a different system, meaning lands *taxed* at a penny or an ounce of silver.[151] This was only very indirectly related to value or productivity.

These moneylands – poundlands and marklands – are to be found in the records of lands in the west, beginning in the late thirteenth century. This was just at the time the Scottish crown was starting to exercise more authority and control over the west coast and the Hebrides, which had previously been more or less semi-autonomous. Until 1266 the islands were under the jurisdiction of the King of Norway, while the mainland had returned to the Scottish crown, nominally at least when the Lord of Kintyre had acknowledged Alexander II as king, some 40 years earlier. There was no sudden break from davochs (or ouncelands) and pennylands into poundlands and marklands, but one system gradually overtook the other.

Charters of Robert I granted under the Great Seal in the years after the 1314 victory at Bannockburn are in terms of moneylands. One describes twenty 'liberates' or poundlands; another mentions 100 shilling lands, equivalent to 5 liberates; a third speaks of seven liberates and six pennylands; and another gives six marklands.[152] In the time of David II one charter describes 10 liberates of land in Cloveth in the county of Forfar, while the next charter to be listed chronologically, lists by location 16 davochs in Sutherland.[153] At the same period, on the east coast, near Edinburgh, David II granted to Henry of Niddrie, a toft (a homestead), and 2 acres of arable land with pasture for one horse and 2 cows, which shows the English influence in the use of acres.[154] McKerral cited a 1295 grant of King John Balliol of 10 marklands at Lanark.[155] But he also quoted another, of James IV and dated 1505, where the 60 marklands granted are then defined and located in the islands, with the parcels making up the 60 marklands also given in davochs and pennylands: these 60 marklands in the Hebrides were said to equal 10 davochs and 14 pennylands.[156] With the davoch being 20 pennylands in these parts, the davoch is equated to 5 to 6 marklands.

On occasion, the various units appear together in a single charter to give a glorious medley showing the diversity of terminology in use at any one time in Scotland. An example of this is a charter of 1373, already discussed in this appendix. In this charter, Robert II granted to Reginald of the Isles a number of named lands and islands off the west coast together with three ouncelands of Swynwort and Lettirlochetle, two ouncelands of Ardegoware, one ounceland of Hawlaste, and 60 marklands in Lochaber, namely 17 pennylands in Loche, half a davoch of land in Kylmalde and a davoch with half of Locharkage, extending to 420 marklands.[157]

At times we do find the word 'pennyland' being used other than in its original sense, which has been described above, namely as $\frac{1}{160}$ of a mark. For example, Gilcrist Macmorich of Beallelon granted to James Stewart the 22 shillinglands and $2\frac{2}{3}$ pennylands of Beallelon in 1513, and the 11 shillinglands and $1\frac{1}{3}$ pennyland of Achamor, lying among the lands of Berroun, in 1519. But in 1554 Stewart sold these lands to another James Stewart, who was Sheriff of Bute and Arran, and the lands of Ballelone and Achemore in Bute are given 'as of the Old Extent of 33s 4d'.[158] If we add 22s $2\frac{2}{3}$d to 11s $1\frac{1}{3}$d we get precisely 33s 4d, which is $2\frac{1}{2}$ marks. The $2\frac{1}{2}$ marks arise from a davoch of 10 marks being quartered, and the term 'quarter' frequently appears in the records of the Western Isles. Here, the use of pennies relates to the money value, and they are considered as a fraction of a mark.

Lands of 33s 4d are the commonest in the Western Isles for the rental of 1507, as Cosmo Innes showed, together with lands of 16s 8d and 8s 4d which are fractions (a half and a quarter respectively) of this.[159] Lands of 50 shillings and 25 shillings also appear frequently in these rentals, and these sizes also rest on this 33s 4d unit: one-and-a-half times the 33s 4d unit gives 50 shillings (the 'quarter' of the Old Extent), which halved gives the 25 shillings.[160] A 1547 charter of Queen Mary mentions a 50 shillingland, but here the land lies far south, near Dumfries.[161] The Register of the Great Seal shows that the money value of land was becoming increasingly used over wide areas of Scotland as time passed.

THE OLD EXTENT AND ITS RE-ESTABLISHMENT

The Old Extent was an evaluation of lands held by tenants-in-chief, during the reign of Alexander III (1248-86), when the rents generated by the lands were used as a basis for taxation by the crown.[162] The 'tenth penny' granted to Robert I by the Scottish Parliament in 1326 was to be 'according to the Old Extent of lands and revenues' of Alexander III.[163] This was a relatively stable and prosperous period, a time of peace. In the period which followed – with the Wars of Independence and the destruction in the days of David II – Scotland was devastated. It has been estimated that the crown's temporal lands in the time of Alexander III generated £50,000 but in 1366, a century later, barely half that amount; the church lands did not escape either, falling in value from £16,000 to £10,000.[164] These figures are known because a new land assessment or extent was conducted in 1357, and completed more fully in 1366; and the Old Extent and current values were then tabulated by parliament for church lands of each diocese and for the sheriffdoms, and these reveal the magnitude of the loss.[165] Rents based on the Old Extent were now about twice what they had been relative to the yield of the land. Naturally, the tenants-in-chief who paid the taxes and recovered them from their vassals sought a re-evaluation by the sheriff and were occasionally successful.

Recovery was reasonably swift. By the end of the first quarter of the fifteenth century a fair measure of prosperity was achieved, and things continued to improve with time. The ransom for James I in 1424 necessitated a new levy, greater than any previously imposed. A further valuation was required in 1474, showing the land's value according to the Old Extent together with current values.[166] This evaluation became known as the 'New Extent'. However, instead of subsequently modifying the land values continually according to current productivity, it became acceptable to consider the new value to be four times the Old Extent value. By the mid-sixteenth century the idea of a frequent change in the valuation was totally abandoned and the 4:1 ratio between the valuations of the New and Old Extents was adopted.

With the passage of time, baronies would become subdivided, with parts sold off or otherwise broken up, and it became difficult to determine what the Old Extent of a property might have been. This presented a problem, because the Old Extent still figured prominently in deed and custom, and it became apparent that the Old Extent would have to be re-established in some form.

Numerous sources have testified that land valuations and rentals were very variable. The measurement terms in legal documents often meant different things in different parts of the country. There seems to have been a delight in producing needless complications, such as using the word 'husbandland' which normally and usually meant two oxgates (or 26 acres), but could on occasion mean one oxgate of 13 acres, with these two meanings becoming confused.[167] This practice began in the sixteenth century and may stem from the davoch becoming more generally equated with two ploughgates, for then the oxgate being thought of as an eighth of a davoch, would correspond to the earlier husbandland. No less an authority than Sir John Skene in his *De Verborum Significatione* of 1597 declared that, relating to early times, a husbandland was one-eighth of a davoch and if the husbandland was worth 5 shillings, then the whole davoch was worth 40 shillings (that is, £2 or 3 marks). If the davoch was notionally 104 acres, the husbandland must be 13 acres. We have seen for the early period rentals of 4 marks and in later times 10 marks or more. The ratio of 10 marks to 3 is not quite 4:1 (it would be 3⅓:1), but is in the right neighbourhood. In addition, Skene noted that in March 1541 the Lords of Session decided that one mark of Old Extent should be 4 marks of the new and four months later that same year they 'be their decreete … esteemed and modified an Oxen gate of land, to twenty shillings in all dewties ʒeirly'.[168] This gave the then current value of an oxgate in 1541 to be £1 a year, and hence the ploughgate £8.

A series of national taxations had been levied in the 1570s, during the period when the Earl of Morton acted as regent for the young James VI, but in 1581 taxation demands reached a new peak with £40,000 to be raised, half from the Church and half from the secular estates.[169] Adam Bothwell, Bishop of Orkney and Commendator of Holyrood Abbey, was called on to pay tax on his extensive lands around Edinburgh following the enabling legislation of February 1580 and October 1581. He duly passed this tax burden on to his sub-tenants as so much 'on every poundland of old extent'.[170] The levy was 10 shillings on each poundland of Old Extent, so the individual barons were to find 6s 8d on each of their marklands. But what exactly was a poundland of Old Extent? The matter came before the Exchequer Court and in 1585 a very full judgement was recorded, in which the Lords Auditors of Exchequer:

> Find, decern and declare that 13 acres of the complainer's lands, lying within the barony of Broughtoun, extendis and sall extend to ane oxgait of land, and that four oxgait of the saids lands extendis and sall extend to ane pund land of auld extent in all tyme to cum.[171]

This ruling determined that the oxgate of Old Extent of the first half of the thirteenth century had a value of 5 shillings, one-quarter of the value current

in 1541. This defined once and for all the Old Extent, and the 4-to-1 rule
continued.

One of the earliest, if not the earliest reference to the Old Extent is in
the Register of the Great Seal, where a charter of Robert I mentions two
schmarcatas of land in Kintyre, and then goes on to define them as four
separate pennylands.[172] If we can accept McKerral's derivation of the word
'*schmarcata*' as meaning 'old markland', this could well mean 'markland of
the Old Extent'.[173] Here, then, 4 pennylands equal two marklands, which, as
we have seen earlier, gives the ounceland (davoch) of 20 pennylands the value
of 10 marks, a frequently-encountered figure, although usually at a later
period.

The Old and New Extents were replaced in 1667 by a new system of
collecting and calculating the land tax which was to be assessed not on the
real value of the land but on 'rented value'.[174] The total amount of such rent
from each county was set in 1667, and not altered thereafter, but up to 1707
it could be reallocated among the various estates within a county.[175] Thus
matters continued until 1854, when the Valuation of Lands Act introduced
annual assessments as the basis of taxation.[176]

1. William F. Skene, *Celtic Scotland*, second edition, 3 volumes (Edinburgh, 1890) III, 42-3; A. A. M. Duncan, *Scotland: The Making of the Kingdom* (Edinburgh, 1975), 47-8; Isabel Henderson, 'Pictish Territorial Divisions', in Peter G. B. McNeill and Hector L. MacQueen (eds), *Atlas of Scottish History to 1707* (Edinburgh, 1996), 9.

2. Ian A. Morrison, 'Subsistence Potential of the Land', in McNeill and MacQueen, op. cit. (1), 14-15.

3. G. A. Bergenroth, *et al.* (eds), *Calendar of Letters, Dispatches, and State Papers, relating to negotiations between England and Spain, 1485-[1558], preserved in the archives at Simanacas and elsewhere*, 13 volumes (London, 1862-1954), I, 169: Prothonotary Don Pedro de Ayla to Ferdinand and Isabella [25 July 1498]; quoted by P. Hume Brown, *Early Travellers in Scotland* (Edinburgh, 1891, reprint 1978), 43.

4. A. Gibson and T. C. Smout, 'Scottish Food and Scottish History 1500-1800', in R. W. Houston and I. D. Whyte (eds), *Scottish Society 1500-1800* (Cambridge, 1989), 59-84, esp. p. 66.

5. Ibid., 67-8.

6. Samuel Johnson, *A Dictionary of the English Language,* 2 volumes (London, 1755), s.v. *oats*.

7. Gibson and Smout op. cit. (4), 76.

8. Ibid., 72.

9. J. B. Brichan (ed), *Origines Parochiales Scotiae [OPS]*, Bannatyne Club no. 97, 2 volumes (Edinburgh, 1851-5), II, part 2, 425; T. Thomson and C. Innes (eds), *The Acts of the Parliaments of Scotland [APS]*, 13 volumes (Edinburgh, 1814-75), VIII, 295-6: Act 38, 1681.

10. John Erskine, *The Principles of the Law of Scotland*, seventh edition (Edinburgh, 1791), 212: 'When a right of pasturage is given to several neighbouring proprietors, on a moor or common belonging to the granter [*sic*], indefinite as to the number of cattle to be pastured, the extent of their several rights is to be proportioned, according to the number that each of them can fodder in winter upon his own dominant tenement, which proportions may be fixed by an action of *souming* and *rouming*, two old words signifying the form of law by which the number of cattle that each proprietor may pasture upon the moor is ascertained.'

11. Thomas Thomson, *et al.* (eds), *Registrum Magni Sigilli [RMS]*, 11 volumes (Edinburgh, 1882-1914), X, 40-3: no. 77, 17 January, 1653. This land grant also gives an example of the measurement of a salmon fishery when it records as part of the privileges granted: 'the half and ninth part of the Curroch salmon fishing on the water of Spey … extending to and comprehending one coble, half a coble, and a third of a coble of salmon fishing upon the said water …' A coble was a short flat-bottomed fishing boat, so we see the fishery defined as that which would occupy so many cobles. If $\frac{1}{2} + \frac{1}{9} = \frac{11}{18}$ of this fishery would support the activities of $1 + \frac{1}{2} + \frac{1}{3} = \frac{11}{6}$ cobles, it is evident that the entire fishery would support 3 cobles on the water. This also shows the difficulties being encountered even in the mid-seventeenth century in expressing fractions. They were often left as a sum of fractional components as above. A further example of this is given in another 1653 grant, where '… 3 eighth parts, 3 sixth parts, and 2 eighth parts of the lands', tells us that the acreage consists of fragmented parcels of land, not all in one piece (ibid., X, 56: no. 104, 28 February 1653). Professor P. Grierson has drawn our attention to the Doges of Venice who, having acquired $\frac{3}{8}$ of the territory of the Byzantine Empire after the fourth crusade of 1204, found themselves unable to express the fraction other than by saying that they were now the lords of a quarter and a half (of a quarter) of the empire of the Greeks.

12. *Spectator*, London, 17 May 1884, 642. 'Strictly speaking to *sowm* the common is to ascertain the several sowms it may hold and to *rowm* it is to portion it out amongst the dominant proprietors': W. Bell, *Dictionary of the Law of Scotland* (Edinburgh, 1838), 932.

13. F. W. L. Thomas, 'Ancient Valuation of Land in the West of Scotland', in *Proceedings of the Society of Antiquaries of Scotland*, 20 (1886), 200-13, esp. p. 211.

14. George Macleish, 'United Parishes of Saddel and Skipness … Kintyre', in Sir John Sinclair (ed), *The Statistical Account of Scotland*, 21 volumes (Edinburgh, 1791-9), XII, 475-89, see p. 477; Sir John Sinclair, *General View of the Agriculture of the Northern Counties and Islands of Scotland* (London, 1795), part I, 76 and note; James Robertson, *General View of the Agriculture of the County of Inverness* (London, 1808), 75.

15. Robertson, op. cit. (14).

16. Sir Archibald C. Lawrie, *Early Scottish Charters prior to 1153* (Glasgow, 1905): see p. 10: no. 12, 1093; and p. 12: no. 15, 1095, from the many which could have been mentioned.

17. See, for example, Joseph Robertson (ed), *Collections for a History of the Shires of Aberdeen and Banff,*

Spalding Club no. 9 (Aberdeen, 1843), 407, for a charter of the Earl of Buchan (prior to 1214), describing 3 davochs ('tres dauatas') in Aberdeen. For 'davoch', see text below.

18 For example, G. W. S. Barrow and W. W. Scott (eds), *The Acts of William I, King of Scots 1165-1214: Regesta Regum Scottorum [RRS], volume II* (Edinburgh, 1971), 248: no. 194, *c.*1178.

19 '… terram que jacet inter Forth et viam que descendit de Striueling ad naues usque ad riuum qui descendit a molendino regis in Forth per suas rectas diuisas': G. W. S. Barrow (ed), *The Acts of Malcolm IV, King of Scots 1153-1165: RRS, volume I* (Edinburgh, 1960), 161: no. 46, 1147 × 1153 (i.e. dated between 1147 and 1153).

20 As in the grant by Malcolm IV of the church of Bathgate to Holyrood Abbey with the lands which Geoffrey de Melville and Uhtred, the Sheriff of Linlithgow perambulated on the day on which the king sent them to view that land: see *RRS*, I, 235-6: no. 199, 1159 × 1163.

21 Dorothy Whitelock, *English Historical Documents Volume I, c.500-1042* (Oxford, 1955), 450 (for acre); Walter de Gray Birch, *Cartularium Saxonicum,* 3 volumes (London, 1885-93), I, 58, no. 34 (for hide).

22 R. D. Connor, *The Weights and Measures of England [WME]* (London, 1987), 35-49. In Scotland, too, the plough-team consisted in theory at least of 8 oxen: see *APS*, II, 13: Act 6 , 1426. On occasion horses could replace oxen.

23 J. M. Kemble, *The Saxons in England* (London, 1876), 96-7.

24 Dorothea Oschinsky, *Walter of Henley and Other Treatises* (Oxford, 1971), chapter 28.

25 Sir John Skene, *De Verborum Significatione* (Edinburgh, 1597), s.v. *particata.*

26 *Statutes of the Realm,* II volumes (London, 1810-28), I, 206-7; *WME*, 322.

27 Peter G. B. McNeill (ed), *The Practicks of Sir James Balfour of Pittendriech*, Stair Society nos. 21 and 22, 2 volumes (Edinburgh, 1962-3), I, 441-2.

28 Oschinsky, op. cit. (24), chapter 28; *APS*, I, 751: Act 16. The text, from the *Fragmenta*, dating from the late thirteenth and fourteenth centuries though likely describing the practice of earlier centuries, says: 'In the first tyme that the law wes maid and ordainit, thai began at the freedome of Halikirk [Holy Church] and syne [afterwards] at the mesuring of landis the plew land thai ordainet to content viij oxingang, the oxgang sall contene xiij akeris. The aker sall contene four rude, the rude xl fallis. The fall sall hald vi ellis.' In Scotland the term 'oxgang' was in use generally in the fourteenth century. By the sixteenth it had changed to 'oxgate'

though far from universally as we have seen but the meaning was the same.

29 Charles Johnson (ed), *Dialogus de Scaccario* (London and New York, 1950), chapter 17.

30 McNeill, op. cit. (27), 441-2; James Avon Clyde, Lord Clyde (ed), *Hope's Major Practicks 1608-1633,* Stair Society nos. 3 and 4, 2 volumes (Edinburgh, 1937-8), II, 119 (Book VI.21.5).

31 For example, Lawrie, op. cit. (16), 257, citing the legal historian Frederic Maitland.

32 For an early English use of 'cassatorum', see Birch, op. cit. (21), I, 108: no. 74, AD 688-90.

33 Charter of Alexander I to Scone Priory: Lawrie, op. cit. (16), 28: no. 36, 1120; and charter of David I to the Church of St John in the Castle of Roxburgh, which refers to 'unam carrucatam terre': Lawrie, op. cit. (16), 69: no. 83, 1128.

34 '… dimedia carrucata terra … cum communi pastura': Sir William Fraser (ed), *Registrum Monasterii S. Marie de Cambuskenneth AD 1147-1535,* Grampian Club no. 4 (Edinburgh, 1872), 252-3: no. 175, *c.*1174.

35 '… duas bovatas terre et duas acras prati et pasturam ad cccc oves …': *APS*, I, 389.

36 '… et de una virgata terre in Stiuecleya … et de pastura quatuor vaccarum': *RRS*, I, 240-1: no. 208, 1157 × 1165. Scottish kings over the centuries held considerable lands in England either by inheritance, conquest or enfiefment from the English king so it is not a matter for surprise that grants of English lands are made to an English institution. A virgate was a quarter ploughgate (carucate) or two oxgangs (bovates) in Scotland. In England it was a quarter of a hide. The hide was a unit essentially of land assessment varying from 40 to 160 acres (see below). On occasion the word 'virgate' meant a quarter of an acre.

37 '… duas bouates terre in Polgawin [Powgarie, Perthshire] … cum pastura viginti animalium et sex equorum': ibid., II, 255: no. 202, 1178 × 1185.

38 '… quatuor bouatis terre cum … nominatim ad quingentas oues et ad viginti vaccas et ad unam carucatam boum et ad equos et ad cetera …': ibid., II, 369: no. 373, 1189 × 1195.

39 '… unam carucatam terre in Molla, per easdem divisas per quas eis mensurata fuit et perambulata, et pasturam quingentis ovibus, et aisiamenta pasture ceteris animalibus quantum pertinet ad unam carucatam terre in eadem villa': ibid., II, 241-2: no. 184, 1173 × 1177.

40 Lawrie, op. cit. (16), 99-100, 365: no. 131, *c.*1141.

41 '… Crorstorfin cum duabus bovatis terrae et sex acris … una salina in Hereth et XXVI acris terrae … una carrucata terrae et quadraginta solidos de meo burgo de Ewinesburg ….': ibid.; this is translated

in Gordon Donaldson, *Scottish Historical Documents* (Edinburgh, 1974), 20-3.

42 '… unum toftum in Reinfry quinque particarum …' ibid., 116-9: no. 153, 1128 × 1136.

43 '… et unum manerium in burgo meo de Caraile … cum tribus rudis terre': ibid., 194: no. 242, 1152 × 1155.

44 '… unam carrucatum terrae in Hetona de quinquies viginta acris et quatuor …': Lawrie, op. cit. (16), 202: no. 251, 1152 × 1153.

45 Julian Goodare, 'The Long Hundred in Medieval and Early Modern Scotland', in *Proceedings of the Society of Antiquaries of Scotland,* 123 (1993), 395-418.

46 Joseph Bain (ed), *Calendar of Documents Relating to Scotland preserved in Her Majesty's Public Record Office,* 4 volumes (Edinburgh, 1881-8), II, 226: no. 856, 1296. See also *APS*, I, 198.

47 '12 bovatas terrarum, vulgariter tuelf oxin gang of land, de Uchiltre': *RMS*, III, 705: no. 3007, 1 April 1544.

48 For an English reference to the use of horses, see *WME*, 37.

49 *APS*, I, 198, note.

50 Joseph Stevenson (ed), *Documents Illustrative of the History of Scotland,* 2 volumes (Edinburgh, 1870), I, 257. Alternatively, a simple scribal error has given XXIV instead of XXVI.

51 *RMS*, XI, 547: no. 1098, 1667; and ibid., XI, 593: no. 1178, 1668.

52 '… vifteen hide land in Egeham': Birch, op. cit. (21), I, 58: no. 34, AD 673.

53 See *WME*, 39-43 and 54-67.

54 Ibid., 36-49 for details. See also J. H. Round, *Feudal England* (London, 1895, reprinted 1964), 41.

55 For the interpretation of much of this section we are indebted to contributors to a conference at the Centre for Advanced Historical Studies, University of St Andrews, the proceedings of which were published as L. J. MacGregor and B. E. Crawford (eds), *Ouncelands and Pennylands* (St Andrews, 1987). See also I. H. Adams, *Agrarian Landscape Terms: a Glossary for Historical Geography* (London, 1976).

56 John Bannerman, *Studies in the History of Dalriada* (Edinburgh, 1974).

57 Ibid., 134-5.

58 Ibid., 132-5 and 141-3.

59 Ibid., 48-9.

60 Ibid., 132-3. For the bench, see ibid., 153: 'The fact that bench (*sess*) was an element in the name of the naval ship in the *Senchus* makes it likely that there were two oarsmen to a bench. A warship could, in any case, hardly be rowed otherwise than by two rows of oarsmen if only for reasons of space. The Old Norse word for rowing bench was *rúm* which

was divided into two half benches (*hálf-rými*) and a ship of thirty *rúm* had sixty oars.' But W. D. Lamont, '"House" and "Pennyland" in the Highlands and Islands', in *Scottish Studies,* 25 (1981), 65-76, p. 68, disagrees, suggesting that *'sess'* means an oarsman's seat. Bannerman's suggestion is persuasive, however.

61 Ibid., 147-8.

62 A. R. Easson, 'Systems of Land Assessment in Scotland before 1400', unpublished PhD thesis, University of Edinburgh, 1986, 76.

63 K. H. Jackson, *The Gaelic Notes in the Book of Deer* (Cambridge, 1972), 31, 34.

64 G. W. S. Barrow, *The Kingdom of the Scots* (London, 1973), 273, note 60.

65 Alexis Easson, 'Medieval Land Assessment', in McNeill and MacQueen, op. cit. (1), 284-5.

66 Barrow, op. cit. (64), 269, 273.

67 Bannerman, op. cit. (56), 140-1. The earliest entries in *RMS* only date from the time of Robert I and there we find the fiscal davoch being mentioned. See, for example, *RMS*, I, Appendix I, 428: no. 9.

68 Thomas, op. cit. (13), 210. See also A. R. Easson, 'Ouncelands and Pennylands in the Western Highlands', in MacGregor and Crawford, op. cit. (55), 1-11, see p. 2.

69 Bannerman, op. cit. (56), 141.

70 Thomas, op. cit. (13); Skene, op. cit. (1); A. McKerral, 'Ancient Denominations of Agricultural Land', in *Proceedings of the Society of Antiquaries of Scotland,* 78 (1944), 39-80, esp. p. 56; continued in ibid., 82 (1947-8), 49-52; and ibid., 85 (1950-1), 52-64; MacGregor and Crawford, op. cit. (55).

71 Thomas, op. cit. (13), 201, note 4.

72 Cosmo Innes, *Lectures on Scottish Legal Antiquities* (Edinburgh, 1872), 272.

73 R. D. Oram, 'Davochs and Pennylands in South-West Scotland, a Review of the Evidence', in MacGregor and Crawford, op. cit. (55), 46-59, esp. p. 47. See also Jackson, op. cit. (63), 116; and William Elder Levie, 'The Scottish Davoch or Dauch', in J. Macdonald (ed), *Scottish Gaelic Studies,* 3 volumes (Edinburgh, 1931), III, 99-110.

74 Barrow, op. cit. (64), 269, note 36.

75 Bannerman, op. cit. (56), 141.

76 Cosmo Innes (ed), *Liber S. Thome de Aberbrothoc,* Bannatyne Club no. 86, 2 volumes (Edinburgh, 1848-56), I, 157: no. 223.

77 *RRS*, I, 227: no. 187, 1153 × 1162.

78 '… sexaginta davatas terre de Badennache cum pertinenciis … et terris ac forestis': *RMS*, I, 205: no. 558; Easson, op. cit. (62), 3.

79 Barrow, op. cit. (64), 269.

80 See, in particular, the contributions to MacGregor

and Crawford, op. cit. (55), and Easson, op. cit. (62).

81 Bruce Webster (ed), *The Acts of David II, King of Scots 1329-1371, RRS, volume VI* (Edinburgh, 1982), 114-15: no. 73, 1343.

82 *OPS*, II, part I, 351.

83 Ibid., 367.

84 Ibid., 374.

85 Ibid., 205. See also, McKerral, op. cit. (70).

86 *RMS*, I, 189: no. 520; *OPS*, II, part I, 199.

87 *OPS*, II, part II, 829.

88 Ibid., II, part I, 53. See also National Archives of Scotland (NAS) RH2/2/14/3.

89 '… sui predicti servitium unius navis viginti et sex remorum cum hominibus et victualibus pertinentibus ad eandem in exercitu nostro cum opus habuerimus et super hoc fuerint rationabiliter premoniti …': *RMS*, I, appendix I, 428: no. 9.

90 '… dimidiam davatam de Ayrsayk videlicet quinque denariatas terre de Gedeuall et quinque denariatas terre de Glenbestell et de Bethey …': ibid., appendix I, 480: no. 107.

91 Bain, op. cit. (46), II, 435: no. 1633.

92 *OPS*, II, part I, 201.

93 Thomas, op. cit. (13), 208.

94 *WME*, 55-6.

95 Easson, op. cit. (68), 4.

96 Oram, op. cit. (73), 52.

97 Easson, op. cit. (68), 3.

98 Thomas, op. cit. (13), 211.

99 George Keith Skene, *General View of the Agriculture of Aberdeenshire* (Aberdeen, 1811), 75.

100 Lamont, op. cit. (60), 65-6.

101 Skene, op. cit. (1), III, 143; Lamont, op. cit. (60), 65.

102 Lamont, op. cit. (60).

103 Duncan, op. cit. (1), 107; see also R. Andrew McDonald, *The Kingdom of the Western Isles: Scotland's Western Seaboard, c.1100-c.1336* (East Linton, 1997), and John Marsden, *Somerled and the Making of Gaelic Scotland* (East Linton, 2000).

104 Duncan, op. cit. (1), 198.

105 Jean Munro and R. W. Munro (eds), *Acts of the Lords of the Isles 1336-1493* (Edinburgh, 1986), xx.

106 Ibid., xxii, III.

107 Ibid., 311.

108 Lamont, op. cit. (60), 69; Easson, op. cit. (68), I and figs. 1 and 2.

109 McKerral, op. cit. (70), 57.

110 Lamont, op. cit. (60), 70.

111 G. Gregory Smith, *The Book of Islay* (Edinburgh, 1895), 32.

112 *APS*, I, 476, 483-4.

113 Lamont, op. cit. (60), 68.

114 Ibid., 67. See also McKerral, op. cit. (70), 73, and W. D. Lamont, 'Old Land Denominations and the "Old Extent" in Islay', in *Scottish Studies*, I (1957), 183-203 (part I) and ibid., 2 (1958), 86-106, (part 2), esp. part I, p. 191.

115 Lamont, op. cit. (60), 68-9.

116 For example, *RMS*, II, 639: no. 3001, 1506: '6 vaccatas terrarum de Progayg, in insula de Ile.'

117 Ibid., IV, 24: no. 91, 1547 and ibid., 29: no. 115, 1547.

118 Lamont, op. cit. (114), part I, 188.

119 Ibid., 190; W. D. Lamont, *The Early History of Islay* (Dundee, 1966), 82.

120 Lamont, op. cit. (114), 191.

121 Lamont, op. cit. (60), 72.

122 Malcolm Bangor-Jones, 'Pennylands and Ouncelands in Sutherland and Caithness', in MacGregor and Crawford, op. cit. (55), 13-25, p. 15.

123 'unam *lie Dawach* sive 6 denariat terrarum de Mekill Sordell ….': *RMS*, V, 35: no. 112, 19 February 1580/81.

124 'Et de xl š. de Westre Halfdavach. … Et de xl š. de Estirhalfdavach. Et eisdem, propter vastitatem unius bovate terre de Westerhalfdavoch, eciam de primo termino hujus compoti, sub eodem periculo, x š.': J. Stuart, *et al.* (eds), *Exchequer Rolls of Scotland [ER]*, 23 volumes (Edinburgh, 1878-1908), VI, 463, 468: 1458.

125 Thomas, op. cit. (13), 202-3.

126 Bangor-Jones, op. cit. (122), 17-18.

127 Barrow, op. cit. (64), 59, 267-9.

128 '… ad dimidiam Carrucatam terre scotticam in Karelis-schire': *RRS*, II, 432-3: no. 469, 1205.

129 *APS*, I, 635: *Regiam Majestatem*, book IV, no. 17. See also Thomas Mackay Cooper, Lord Cooper (ed), *Regiam Majestatem and Quoniam Attachiamenta*, Stair Society no. 11 (Edinburgh, 1947), 285-6: '*De Heryheldis*'. Lord Cooper believed that this exemption went back to the days of David I. See also Skene, op. cit. (25), s.v. *herrezelda* [heriot].

130 *APS*, IV, 548: Act 21, 1617.

131 [John Stuart (ed)], *Miscellany of the Spalding Club, Vol. IV*, Spalding Club no. 20 (Aberdeen, 1849), 261-319, esp. pp. 309 and 300; see also, Barrow, op. cit. (75), 271-2.

132 Thomas, op. cit. (13), 203.

133 Easson. op. cit. (68), 3.

134 McKerral, op. cit. (70), 52. Bangor-Jones, op. cit. (122), 16. Thomas, op. cit. (13), 208.

135 Innes, op. cit. (72), 273.

136 *Registrum Episcopatus Moraviensis*, Bannatyne Club no. 58 (Edinburgh, 1837), 433.

137 Thomas, op. cit. (13), 203.

138 Ibid.

139 Skene, op. cit. (25), s.v. *herrezelda* [heriot].

140 *APS*, IV, 548: Act 21, 1617.

141 Thomas, op. cit. (13), 205.

142 McKerral, op. cit. (70), 49.

143 Sir Frank Stenton (ed), *Documents Illustrative of the Social and Economic History of the Danelaw* (London, 1920), xxviii-xxix.

144 G. W. S. Barrow, 'Rural Settlement in Central and Eastern Scotland: the Medieval Evidence', in *Scottish Studies,* 6 (1962), 123-44.

145 John Macqueen, 'Pennyland and Davoch in South-Western Scotland', in *Scottish Studies,* 23 (1979), 69-74; Easson, op. cit. (65).

146 Basil Megaw, 'Note on "Pennyland and Davoch in South-Western Scotland"', in *Scottish Studies,* 23 (1979), 75-7. Probably the earliest extent of which we have a record which mentions 'pennylands' occurs in a Carrick charter of 1260 which states: '… xiiij denariate terre valent annuatim lxx sex marcas. Item terra de Drumfad et de Glenop scilicet decem denariate terre valent annuatim xl marcas. Item terra de Dulachoran, scilicet ii denariate terre, valent annuatim viii marcas. Item terra de Glenkenith scilicet ii denariate terre valent annuatim iii marcas et dimidiam': Isabel A. Milne, 'An Extent of Carrick in 1260', in *Scottish Historical Review,* 34 (1955), 46-9. In this it is demonstrated that the four territories named have 'pennylands' each worth 5·4, 4, 4, 1¾ marks respectively. In the west, these annual values are those associated more with lands of one davoch rather than a pennyland. McKerral suggested that the term *'denariata'* had become less precise in meaning and referred then to a much larger territory than that encountered later. He gave an example of this in Wigtownshire, where a place known as 'The Pennyland' was equated to 24 marklands: A. McKerral, 'An Extent of Carrick in 1260', in ibid., 34 (1955), 189-90.

147 See, for example, *RMS,* I, appendix 1, 435-7, 478: nos. 20 and 102.

148 Oram, op. cit. (73), 52.

149 Ibid.

150 Ibid., 53; Hugh Marwick, *Orkney Farm Names* (Kirkwall, 1952), 200-1.

151 McKerral, op. cit. (70), Part 1, 62.

152 '… in extentum viginti librataum terre': *RMS,* I, 2: no. 6, 1315 × 1321; '… in extentum centum solidataum terre': ibid., 2: no. 7, 1315 × 1321; '… septem libratis et sex denariates terre': ibid., 7: no. 24, 1315 × 1321: '… sex marcataum … terre': ibid., 20: no. 71, 1315 × 1321.

153 '… decem librates terre': ibid., I, 37: no. 131, 1362 × 1363; '… sexdecim davatas': ibid., I, 37-8: no. 132, 1362 × 1363.

154 'Henrico de Nudre, unum toftum cum duabus acris terre arabilis, et herbagio unius equi et duarum vaccarum ….': ibid., I, 41: no. 143, 1363.

155 McKerral, op. cit. (70), 62.

156 *OPS,* II, part 1, 374.

157 'Reginaldi de Insulis, … de tribus unciates terre de Swynwort et de Lettirlochetle, de duabus unciates terre de Ardegoware, de una unciata terre de Hawlaste, … et sexaginta marcatis terre in partibus de Lochabre, videlicet de decem et septem denariatis terre de Loche, de dimidia davata terre de Kylmalde, et de una davata cum dimidia de Locharkage, in extentam quadrigentarum et viginti marcataum terre.': *RMS,* I, 189: no.520, 1372/3.

158 *OPS,* II, part 1, 231.

159 Innes, op. cit. (72), 278-9.

160 Ibid.

161 *RMS,* IV, 24: no. 91, 14 April 1547.

162 This section is based on J. D. Mackie (ed), *Thomas Thomson's Memorial on the Old Extent,* Stair Society no. 10 (Edinburgh, 1946). Thomson's definitive presentation of this Memorial to the Court of Session in 1812, in the case of Cranstoun *v* Gibson is recorded in abbreviated form in *Faculty Collections, Decisions of the Court of Session, November 1815-November 1819,* XIX (Edinburgh 1821), 511: no. 164, 16 May 1818. Mackie gives the full text of Thomson's submission as well as an essay on its background and his observations on the case. Some have taken the view that the Old Extent was made in the reign of William the Lion as a levy had to be made to meet the 10,000 merks for the restoration of the independence of the kingdom: see, for example, George Chalmers, *Caledonia,* 3 volumes (Paisley, 1887), II, book IV, 745-6. However, it is stated in *ER,* I, lxxxix-xc that 'The old extent may be defined as a valuation of the whole temporal lands of the country, which in the time of Alexander III [1249-86] was preserved among the muniments of the Crown, and continued to be appealed to in subsequent reigns as the basis of territorial imposts … how old it may in some cases be is difficult to conjecture … the extent in use in the time of Alexander III, while sufficiently indicative of the relative value of the lands in the kingdom to be an equitable basis for taxation, must in all cases have fallen far below the absolute value'.

163 *APS,* I, 475, 483-4.

164 I. F. Grant, *Social and Economic Development of Scotland before 1603* (Edinburgh and London, 1930, reprinted 1971), 166.

165 *APS,* I, 499-501: 1366. See also the detailed information given county by county from the original records showing the Old Extent and the New Extent of 1366 in Chalmers, op. cit. (162), II, book IV, 816.

166 *APS,* II, 107: Act 10, 9 May 1474.

167 Robert A. Dodgshon, *Land and Society in Early Scotland* (Oxford, 1981), 75. It is here suggested that in early times the husbandland was 26 acres

but that later this term displaced the bovate as the term for 13 acres.

168 Skene, op. cit. (25), s.v. *extent*; quote from *bovata terrae*.

169 *APS*, III, 189-90: 1580, and 192: 1581.

170 Innes, op. cit. (72), 281-4, quote, p. 282.

171 Ibid., 283.

172 '… duas schanmarcatas terre in Kontyr, videlicet denariatam terre de Arydermede, denariatam terre de Ballostalfis, denariatam terre de Kyllewllane et denariatam terre de Seskamousky': *RMS*, I, 477: appendix 1, no. 99.

173 McKerral, op. cit. (70), 62.

174 *APS*, VII, 540-7: 23 January 1667.

175 Anne E. Whetstone, *Scottish County Government in the Eighteenth and Nineteenth Centuries* (Edinburgh, 1981), 74.

176 17/18 Victoria c. 91, 1854.

APPENDIX C

Land Service

TOWARDS THE END OF APPENDIX B, QUOTATIONS WERE GIVEN WHICH indicated that some element of 'service' was expected from the recipient of land when enfiefment took place. As we shall see, service in Scotland closely matched that demanded of the English nobility when land was held of the king by military tenure.

The feudalisation of Scotland may have begun with Edgar (1097-1107), and continued with Alexander I, but there can be no doubt that the degree and intensity of feudalism brought in by David I was such as to indicate that he, rather than any of his forebears, was the initiator, and the momentum generated in his reign far exceeded what might have been before.[1] This continued through the reign of his successor, Malcolm IV, with most of the Lothians feudalised and a start made on the Highlands and areas north of the Forth.[2] The process continued until the reign of Alexander III.

While some aspects of Scottish feudalism were closely akin to that in England, there were nevertheless significant and important differences. In England the new scheme of things was imposed by conquest with the victors dividing up the land as so much spoil under the aegis of the Norman conqueror and his immediate successors, with the wholesale displacement if not extirpation of the previous owners. In Scotland the Celtic land owners were not subject to sudden and total expropriation. Celtic nobility continued alongside the Anglo-Norman incomers brought in by David. For example, William the Lion granted the earldom of Fife to the native-born Duncan.[3] Subsequently, he enfiefed Duncan with West Calder in Midlothian, when the grant was confirmed by William in 1206 or 1207 to Malcolm, Duncan's son on Duncan's death. An even earlier grant of this land by David I to Duncan's father has not survived, nor has William's own original charter to Duncan.[4] The change in Scotland progressed slowly, beginning in the South with occupiers of the land finding themselves vassals of new French-speaking lords.

The matter has been put succinctly by the historian H. R. Loyn:

Scotland lacked the great clarification brought about by the brutal reality
of the Norman Conquest in England. The native Scottish nobility was
never dispossessed, and so the intrusion of an Anglo-Norman feudal
element was correspondingly more difficult. The native nobility adapted
itself slowly to the presence of the Anglo-Normans – at least in the
Lowlands – and learned new ways. To complicate matters further the
feudalisation of Scotland took place in a period of economic growth
when expectations were increased among the land-owning classes.[5]

As for the clergy, Marinell Ash noted that:

> Almost without exception all the episcopal demesne known to have
> belonged to the bishops of St Andrews in the middle ages had belonged
> to the bishops before *c*.1100. The introduction of feudalism and the
> reform of episcopal administration may have modified the bishops'
> jurisdiction over their estates, but it could not totally alter them.[6]

The king could grant a fief to a noble who in turn could sub-infeudate
parts of his fief to others who would regard the noble as their lord, and so on
down the hierarchy until the bottom level was reached, that of the actual
labourers of the land. Each level in the hierarchy swore allegiance to the
grantor 'saving the allegiance owed to the king', for ultimately, everyone was
the king's man. But this largess was not without attached strings. The king
could, and almost invariably did, demand service in the field from his
tenant-in-chief, known later in Scottish law as the 'subject-superior'.[7] More
than that, the king could on occasion demand an 'aid' in cash for a specific
purpose; for instance, the ransom of his body, the marriage of his eldest
daughter, the knighting of his eldest son, or the raising of an army. An 'aid'
was one of the feudal 'causalties' or 'incidents', of which there were not a few.
If a noble died his lands reverted to the overlord or king depending on
specific circumstances, and the heir, if of age, would be required to pay a
'relief' on taking up the lands again. If under age, the heir became the 'ward'
of the superior, the overlord holding the land 'in wardship' and retaining its
revenues until majority was reached, with the heir meantime being reasonably maintained. The lord also had the right of 'marriage', in that a male heir
under age, or an heiress of any age, might be donated in marriage to someone
of the lord's choosing, always for a pecuniary consideration.[8] It would appear
that on occasion women were simply sold off to the highest bidder.[9]
Additionally, it would seem that 'the Scots Crown, like the English, was
jealous of its rights of wardship and marriage over tenants-in-chief and
exacted reliefs from heirs to baronies'.[10] Even the lesser lords enfiefing others
could grant an estate 'for an annual rent of £6 and the incidents of ward and
relief'.[11] An example was the 1557 grant of

> … the five Oxengang of the Toun & Lands of Brey lying within the
> Shire of Inverness The new extent whereof is £100 Scotts and the old

extent £49.9.4 Which Lands hold of the King in cheif by Service of Ward and Releif.[12]

There had been a pre twelfth-century system of dues and renders from land held of the king involving not only rent in cash (or, more usually, in kind) but also military service. The 'mormaer', the chief steward of a Celtic province, all but independent of the king, could call up his levies from his tenants and sub-tenants for the protection of his province from outside attack, but more often to do battle with his neighbours to right some wrong, real or imagined. The mormaers in general become earls under feudal law, retaining their rights within the earldom to call out those liable for military service in their area. The earl led his troops, 'for each man must follow his lord'.[13]

THE KNIGHT'S FEE OR 'FEUDUM MILITIS'

David's establishment of the feudal system in Scotland on the Norman pattern, displaced the bonds of family and clan by which land had been held previously. The amount of service due was notionally based on the arable land given, whether calculated in terms of davochs, carucates (ploughgate) or arochors (in Lennox and mid-western Scotland), or on its annual value. This was often expressed in money, but in reality the service due was that specified by the charter and might or might not have any correspondence to area or worth; and, on occasion, the grant was given as a particular property defined only by name with no indication as to its extent or value. Examples of this are the granting of Coultra and Lundin, in Fife, for the service of half a knight and one knight respectively; such lands being constituted as a half or a full knight's fee, but no boundaries or areas are indicated.[14] The pattern of enfiefment in Scotland was virtually identical in all respects to that in England as the examples provided below will indicate.

Usually a grant had to be reconfirmed by successive monarchs to the baron or to his heirs with the fiefdom becoming hereditary, as in the famous reconfirmation by William I of the grant of Annandale to Robert de Brus by David I, the service being set at reconfirmation as that of ten knights: 'per seruicium x militum.'[15] This is one of the larger numbers of knights required by any grant to be found in the cartularies. David granted lands for one knight and even half a knight, while in about 1140 David's son, Earl Henry, granted lands for one-third of the service of a knight and Henry's son, William I granted part of Yester in East Lothian for the service of one-fifth of a knight.[16] Royal enfiefments rarely, if ever, went below this fraction of a knight's service, but with subsequent sub-infeudations, when a baron would grant part of his holding to another, it could go down as far as one-twentieth or even one-thirtieth of a fee.[17] In England, in 1250, there is an example of land held for one-sixtieth of a fee.[18] Knightly service was a royal prerogative; grants by the barons of a portion of their holdings to persons of lower degree frequently meant that the return was commuted into cash or rendered in kind, especially in England.[19] It is hard to see how one-thirtieth or one-sixtieth of a fee could be met otherwise. The handling of larger fractions was easier. Two nobles each holding half-fees could join and together make one knight available for the field, or every two years one or other could furnish the knight; or the service could be made into a money payment, and this must have been very common.[20] Sometimes, even for huge grants of land, the amount of service was not defined. An example of this occurred in the confirmation by William I, of the land King David had granted to the Earl of Fife, dating from about 1207. These lands were to be held *'per seruicium militum'* (by the service of knights), with no numbers specified.[21] In Fife the earl had sub-infeudated parts of his vast territory for several knights' fees. When the king issued his summons all the knights holding lands of the earl would ready themselves or a substitute for the field. On occasion the charter is even vaguer, as, for instance, in 1302-3, a half-davoch was granted 'for the service thereto appertaining'.[22] No further information is given. Perhaps all who needed to know would know full well what the requirement was.

It should not be thought that the lands enfiefed for one knight were all uniform in value or productivity, or, indeed, in area. Evidence has been advanced that knights' fees of annual value as low as £5, £10 and £13 did occur, although the later value of £20 was common in both England and Scotland.[23] The huge grant of Annandale mentioned above, of 200,000 acres to de Brus, held for ten knights, was equivalent to more than 31 square miles per knight. Yet Lundin in Fife and Inverkeilor in Angus were both rated 'for 1 knights service'.[24] Fife consists of some 500 square miles: for Lundin to consist of 31 square miles would mean this territory was some six per cent of the whole county, a highly unlikely figure when it is recalled that most of the land east of the river Leven was held by the church of St Andrews in any case, and was not available for royal giving.[25] Likewise, the small lands of Inverkeilor could not extend to four per cent of the land of Angus. A figure giving the number of acres per knight's fee can never hope to represent on a broad scale the enfiefment of Scotland. An average figure for the English fee might be four hides each of 120 acres, or 480 acres in all (although hides as low as 60 or as large as 160 acres were known, and were not rarities, and the fees of the Conqueror himself ranged from two to 14 hides).[26] As discussed in Appendix B, in some parts of the kingdom, and especially in later times, the davoch consisted of four ploughgates, each of 104 acres, which would yield 416 (Scottish) acres or 528 English acres. This is not very far from the average English fee. This might be taken to lend credence to the supposition that one davoch might correspond to the knight's fee, but such an interpretation cannot stand generally, if only in the light of a single example: the charter of about 1211 giving 'Kingower' in Gowrie and five davochs in Mearns to Ranulf, the falconer, for personal service or else the service of one archer in the army.[27] This is much less service than that of one knight for each davoch.

SERGEANTY SERVICE

This last reference to a charter indicates another form of service, namely that from an individual of lower rank than a knight, and it appears not infrequently in the twelfth and early thirteenth centuries. The following are some further examples from various charters. One mounted archer was required from two ploughgates in Roxburghshire (in a charter dating between 1153 and 1165), and the same from two ploughgates at Granton (between 1166 and 1171).[28] One foot soldier was required from a half Scottish ploughgate in Fife in a charter of 1171-4.[29] One mounted archer with haubergel (a coat of mail) was required for 13 oxgangs in the Mearns, given in 1198.[30] One sergeant on a horse with haubergel was required for various lands in Fife in 1205.[31] Finally, as we saw above, one archer in the army was required for land in Gowrie and five davochs in Mearns in 1211.[32]

Service such as that mentioned above has been called 'sergeanty' (or serjanty) service, and in the later twelfth and thirteenth centuries smaller grants, which otherwise would have been given for a fraction of a knight's service, now appear as sergeanty service, especially thirds, quarters and fifths of a knight's service. A. A. M. Duncan and G. W. S. Barrow have both drawn attention to the granting of Gilberton (or Brunstane, Edinburgh) to Michael Fleming, Sheriff of Edinburgh, for the service of a sergeant on a horse with a haubergel: when later this was reconfirmed, the same lands were to be held for a quarter knight.[33] It should not be supposed that this was a universally-held equivalent. However, sergeanty service was well-established and used in the reign of William I, and especially so for private sub-infiefments. The equipment – horse, haubergel, and so on – would be beyond the common man, but could be achieved by local squires.

On occasion, south of the border, we find a substantial land grant called a sergeanty, requiring in addition to the provision of a sergeant for the field, some portion or a whole knight's service as well. Thomas Madox gave the example of particular lands in the sergeanty of William de Paris in Ayston and Clinton, where the tenant, Richard de Crokel, was to do the service of one-thirtieth of a knight's fee, while William – for another portion of the same sergeanty – was to do the service of one knight and to provide for the king's army one sergeant or two horses for 40 days at his own expense.[34]

The men-at-arms used for the maintenance of the castles were most frequently sergeants. The Exchequer Rolls for 1264-6 tell us that Alexander III had eight of them spend six months at Invery (perhaps Blervie, Morayshire) at a cost of 104 shillings, which works out at just under a penny each per day.[35] A penny a day was the wage of a 'servien' in England also; thus, for example, in 1168 the roll of Henry II records '40 *serviens* of Blancmust for 29 weeks cost £33 16s 8d and 20 *serviens* who stayed for 23 weeks cost £13 8s 4d'.[36] A little calculation shows these payments to be at the rate of one penny per day per 'servien'.

The task of garrisoning the royal castles within their domain was a duty required of those holding land by military tenure. Occasionally this appears explicitly in the charter as, for example, one of 1160 granting Berowald the Fleming land near Elgin for the service of one knight in Elgin Castle.[37] More charters are extant giving exemption from castle ward than requiring it; an example is the charter dated 1165-73 of David I to Robert de Brus noted earlier, in which the service of ten knights was to be given *exempt from ward* of the king's castles.[38] While few charters speak of castle ward explicitly, this was something understood to be part of knight's service on enfeifment. This relationship of castle duty to the grant was made very plain when the Laird of Hadden in Roxburghshire and his heirs were required to pay 20 shillings each year for castle guard at Roxburgh Castle.[39] This duty had to be performed in person, or with a hired man. Such a sum would buy the services of a servien for 240 days or six men for 40 days. Thus, the castle duty of a fief of one-sixth of a knight's fee would be met by the payment of 40 pence which would provide one man for 40 days, and this duration of service will be discussed below.

SCOTTISH AND 'FORINSEC' SERVICE

The concept of a 'common army' (*communis exercitus*) can be traced to Dalriada in the eighth to tenth centuries. Something very similar to this pre-feudal service continued into later centuries, appearing in charters as 'Scottish service' (*servitium scoticanum*) and was intended primarily for the defence of old Scotia (north of the Forth). This was a force made up of the commonalty, quite recognisably different from knight or serjeanty service. A charter of 1240 for lands on the shores of Loch Awe required the service of half a knight and as much as pertains to one knight in aids, but in addition also 'Scottish Service, as performed by our barons and knights for their lands on the north side of the Scottish Sea', namely the Firth of Forth.[40] (It is worth noting that land could be assessed at half a knight for one type of service as well as at a full knight for another.)

'Forinsec Service' (*servitium forinsecum*) was originally a French term with the same meaning as Scottish service.[41] It represented military service expected in addition to any owing to the immediate superior, such as when the king called out his levies on a regional or national basis. It could be, and often was, in addition to the requirement of knight's service from the noble, as demonstrated in the previously-mentioned charter of 1240. The tenant had to appear in the field equipped as a knight and bring from his estates the requisite numbers of foot soldiers. Forinsec service was initially levied on lands south of the Forth, and had its origin in the force defending the old earldom of Northumbria which until the eleventh century extended from the Forth southwards. Scottish and forinsec service intermingled in later centuries and were virtually indistinguishable in the charters save for the locality of the land grant. An example of this linkage of terminology is to be found in the charter forged in about 1290, but purporting to date from 1171, in which it was stated that Morgund had made good his claim to the whole earldom of Mar, to be held as freely as any other earl, 'for the forinsec service, that is the Scottish service wont to be performed by his predecessors'.[42]

The church in England might have to provide knight service or not, according to the terms of the charter granting the lands in question. In 1156 the Abbot of Ely had to find money to pay for the hire of knights to meet his obligations.[43] While ecclesiastical institutions in Scotland could be free from knight service, although the bishop might attend the field as a tenant-in-chief, the church was not exempt from the 'common aid', the 'common army' service, or the provision of common labour. William's charter to the grandson of the Earl of Fife is a case in point: the Abbey of Abernethy in Perthshire was granted land between 1173 and 1178, free without obligation except for common aid, army service and labour service.[44] Common labour was needed, for someone had to move goods and materials and to maintain the roads and bridges. Sometimes the abbeys were free of all forinsec service except in times of national emergency when the king required a general enlistment. Thus Earl Henry, son of David I, granted freedom from army and escort service to Tynemouth Priory in 1147, unless army service was levied from the earl and from his land between Tyne and Tweed.[45] Robert the Bruce (then Earl of Carrick) agreed with Melrose Abbey in 1302 that the Abbey's Ayrshire lands were to be free from service 'except from the common army occurring for the defence of the kingdom and gathered from the whole kingdom'.[46] Where army service was imposed it could on occasion be met by donations in kind, whether foodstuffs or materials, because the army had to be fed.

In general the clergy on both sides of the border also held lands for spiritual service, masses and prayers for the repose of the souls of antecedents. Where unspecified service was required as in 'the service due from a ploughgate', the requisite number of men from the peasantry would have to be furnished or tribute in kind paid for the maintenance of the host. An example is the levy from lands in Lennox, when 'the kings common army was met', of two cheeses from each house which made cheese.[47]

THE DURATION OF THE SERVICE

Everyone called upon for service in the field was expected to serve for 40 days without cost to the king, whether they were mounted knights or footsoldiers. Attention has been drawn to a document of about 1286 calling out the army in a general summons. All who owed military service – whether free service (for example, knights' service) or Scottish service – were to be armed and ready with food for 40 days, to proceed on 24-hours' notice to the assembly point.[48] A charter of Robert I of 1329 granted a piece of land for the annual service of an armed foot soldier supplied with food for 40 days as well as forinsec service when called upon. In the same year, Robert granted eight bovates of land for a footsoldier with sword, lance and food for 40 days, as well as giving the customary forinsec service.[49] Precisely the same 40 days of service was required in England.[50] An example is the 'tallage' laid on a village of Shropshire in 1199 for the maintenance of 500 sergeants: a total of 121 marks and 3 pounds was collected, and this is almost precisely the money needed for the sergeants for 40 days at one penny each per day.[51]

Occasionally a charter will spell out the duration of service required. A charter of David II dating from 1362-3 mentions land granted for the service of a knight each year and for all the service demanded if a universal call is issued.[52] Feudal service from land holders provided a number of well armed mounted knights, squires and men-at-arms (sergeants), but their numbers would always be small. The bulk of the army came from the ordinary people as in pre-feudal days, when all men fit for service between the ages of 16 and 60 were called up, producing an inhomogeneous group equipped with a diversity of weapons. With a larger pool of man-power to call on, England could be more selective. Whereas England was always strong in cavalry, Scotland was always weak in this regard, but by means of 'wappinschaws', where everyone would appear with their arms for inspection, an attempt was made to upgrade the weaponry. The Parliament at Scone in December 1318 stipulated that a man with goods worth a cow was to have a good spear or a bow and arrows; and a subsequent act of the same year laid down the armament for those possessing ten pounds worth of goods, an 'acton' (that is, a padded jacket worn underneath a coat of mail), a basnet (a small steel headpiece, sometimes with a visor) and gloves of plate, with a spear and a sword.[53]

By the end of the thirteenth century and during the Wars of Independence, the military force which could be called up for service from land held

by military tenure could no longer meet requirements. The feudal organisation had all but collapsed and no longer supplied enough soldiers or
enough money to hire troops in sufficient numbers.[54] By the early fifteenth
century there had grown an expectation that all fit men should bear arms
with 'wapynschawings' in each sheriffdom four times a year.[55]

The type of armament expected was laid down by statute in 1429. Each
man with an income of £20, or £100 in movable goods, was to be well horsed
and fully armed 'as a gentleman ought to be'; and if he was in receipt of £10
in rent or £1 in goods he was to have plate armour. Each yeoman worth £20
in goods was to have a good doublet of 'fence' (chain mail) or a hauberger, an
iron hat with bow and arrows, shield and knife; and all yeomen with £10 in
goods were to have bow and arrows, shield and knife. Those who could not
handle a bow were to have a good stout hat for his head, a doublet of fence, a
shield, and a good axe or else a spiked staff. Each burger with £20 in goods
was to be armed as a gentleman, and similar provision was to be made by
those 'of lawer degre'.[56]

When Henry VIII, frustrated in his attempts to have a marriage arranged
between Mary of Scotland and his son Edward, sent the Earl of Hertford on
punitive raids into Scotland in 1544 and 1545, the Scottish Parliament
required the enlistment from Edinburgh of

> all manner of personis now being present in this town or salhappin
> [happening] tocum to this town And siclik all maner of man that
> dwellis outwith this toun within the sherefdoms of Lothiane be redy
> bayth on fute & horss weill abilzeit for weir cum and meit my said lord
> gouvernour at the toun of Edinburgh on monuday nixt tocum the x day
> of Nouember instant at xii houris of the day, furnist with four dayis
> wittale ...[57]

A further act in 1545 raised a thousand horsemen for the defence of the
border, parliament understanding 'that the bordouris of this realm ar almaist
uterlie distroyit brynt and hereit be our auld Inymeis of Ingland', with
£6,000 per month being provided for their upkeep, each man receiving four
shillings per day in wages.[58] This was followed immediately by an act for
raising £16,000 for this force of cavalry from the lords spiritual and temporal
and the burghs.[59] This latter sum would provide for a thousand men for
twice 40 days service. The rate of four shillings a day was certainly not excessive payment for a mounted man. At this time, Scottish money was worth one-
quarter to one-fifth of sterling of the same face value, hence four shillings was
about ten to twelve pence sterling.

Theoretically, the first 40 days of a campaign were to be at the tenant's own
expense, with the troops receiving pay from their lord. Although English
troops were paid by the English king, it certainly would not be the Scottish
king who paid the members of his army, if indeed they were paid at all. It
has been said that the element holding the Scottish host together was the

prospect of a share in the plunder that would accrue with an advance into England.[60] Moreover, the vast majority of Scottish descents upon England were of short duration, well within the 40-day period.

Although in practice knight's service became obsolete, it continued as the fictional basis of assessment for money levies, 'voluntary' gifts to the crown, and fines, until it was finally abolished in 1660 along with the other feudal incidents like marriage, and ward and relief.[61] These levies had been in use for no less than six centuries, for most of which they were merely money-raising devices, having little to do with the feudal system of which they initially formed a part. But their abolition did not mean that the monarch lost in a financial sense: the enabling legislation granted him a duty on 'beer; cider; perry; metheglin or mead; vinegar beer; aqua-vite; imported beer and on coffee, chocolate, sherbet and tea made and sold'.[62]

1   William Ferguson, *Scotland's Relations with England: A Survey to 1707* (Edinburgh, 1977), 18.

2   Ibid. See also G. W. S. Barrow, 'The Growth of Military Feudalism from about 1100 to about 1240', in Peter G. B. McNeill and Hector L. MacQueen (eds), *Atlas of Scottish History to 1707* (Edinburgh, 1996), 413.

3   G. W. S. Barrow, *The Kingdom of the Scots* (London, 1973), 283.

4   G. W. S. Barrow and W. W. Scott (eds), *The Acts of William I King of Scots 1165-1214: Regesta Regum Scottorum [RRS], volume II* (Edinburgh, 1971), 435-6: no. 472, 1206 × 1207.

5   H. R. Loyn, review of A. A. M. Duncan's *Scotland, the Making of the Kingdom*, in *Scottish Historical Review*, 55 (1976), 196.

6   Marinell Ash, 'The Diocese of St Andrews under its "Norman" Bishops', in ibid., 115.

7   G. W. S. Barrow, *The Anglo Norman Era in Scottish History* (Oxford, 1980), 133.

8   William Croft Dickinson, *Scotland from the Earliest Times to 1603* (London, 1961), 87.

9   Austin Lane Poole, *From Domesday Book to Magna Carta 1087-1216*, second edition (Oxford, 1955), 32.

10   Barrow, op. cit. (7), 142.

11   Ibid., 133.

12   J. T. Clark (ed), *Genealogical Collections concerning Families in Scotland, made by Walter Macfarlane 1750-1751, volume II*, Scottish History Society no. 34 (Edinburgh, 1900), 361.

13   Poole, op. cit. (9), 273.

14   G. W. S. Barrow (ed), *The Acts of Malcolm IV King of Scots 1153-1165 [RRS], volume I* (Edinburgh, 1960), 219-20: no. 147, 1173 × 1178; ibid., I, 270-1: no. 255, 1161 × 1164.

15   Ibid., II, 178-9: no. 80, c.1172.

16   Ibid., I, 158: no. 42, 1146 × 1153; ibid., I, 146: no. 22, 1139 × 1142; ibid., II, 182-3: no. 85, 1166 × 1170; A. C. Lawrie, *Early Scottish Charters prior to 1153* (Glasgow, 1905), 149-50: no. 186.

17   Barrow, op.cit. (3), 198; Barrow, op. cit. (7), 135.

18   Thomas Madox, *History and Antiquities of the Exchequer* (London, 1711), 453, and note (p).

19   Frederick Pollock and Frederic William Maitland, *The History of English Law before the time of Edward I*, 2 volumes (Cambridge, 1895), I, 252.

20   For example, 140 acres at Tottenham were granted for a payment of one merk of silver each Michaelmas: *RRS*, I, 239-40: no. 206, 1161 × 1164.

21   Ibid., II, 435-6: no. 472, 1206 × 1207.

22   Joseph Bain (ed), *Calendar of Documents relating to Scotland*, 4 volumes (Edinburgh, 1881-8), II, 346-7: no. 1350.

23   R. D. Connor, *The Weights and Measures of England [WME]* (London, 1987), 63; Barrow, op. cit. (3), 292-6.

24   *RRS*, I, 270-1: no. 255, 1161 × 1164; ibid., II, 242: no. 185, 1173 × 1180.

25   Ibid., I, 41.

26   William Sharp McKechnie, *Magna Carta* (Glasgow, 1905), 234-5.

27   *RRS*, II, 452: no. 497, 1209 × 1211.

28   Ibid., I, 284: no. 300, 1153 × 1165; ibid., II, 152: no. 45, 1166 × 1171.

29   Ibid., II, 207: no. 131, c.1171 × 1174.

30   Ibid., II, 389: no. 404, 1198.

31   '… unius servientis in equo cum halbergello': ibid., II, 432-4: no. 469, probably 1205.

32   Ibid., II, 452: no. 497, 1209 × 1211.

33   See A. A. M. Duncan, *Scotland: The Making of the Kingdom* (Edinburgh, 1975), 386-7; and Barrow, op. cit. (3), 301. Both refer back to the *inspeximus* of early charters given in T. Thomson and C. Innes (eds), *The Acts of the Parliaments of Scotland [APS]*, 13 volumes (Edinburgh, 1814-75), VII, 144: 1661. The original charter of 1198 × 1214 is described in *RRS*, II, 477: no. 560.

34   Madox, op. cit. (18), 452-3, and note (o) on p. 453.

35   J. Stuart and G. Burnett (eds), *The Exchequer Rolls of Scotland*, 23 volumes (Edinburgh, 1877-1908), I, xlvi, note 18.

36   '… XL servientum de Blancmust de XXIX septimanis XXXIII Li et XVI s et VIII d et XX servientilus qui remanserunt XXIII septimanas XIII Li et VIII s et IIII d': J. H. Round, *Feudal England* (London, 1964), 215 *et seq.* for this and other similar examples.

37   '… servicium unius militis in castello meo de Elgin …': *RRS*, I, 219-20: no. 175, 25 December 1160.

38   '… per seruicium · x · militum excepta custodia castellorum meorum': ibid., II, 178-9: no. 80, c.1172.

39   Duncan, op. cit. (33), 383.

40   Barrow, op. cit. (7), 165, and Duncan, op. cit. (33), 381-2, both quoting from J. R. N. Macphail (ed), *Highland Papers*, 3 volumes, Scottish History Society, second series nos. 5, 12 and 20 (Edinburgh, 1914-20), II, 121-3.

41   Duncan, op. cit. (33), 450-1; *RRS*, II, 200.

42   '… forinsecum seruicium videlicet seruicium Scoticanum sicut antecessores sui mihi et antecesssoribus meis facere consueuerunt …': *RRS*, II, 200: no. 119, c.1290.

43   Round, op. cit. (36), 232 *et seq.*

44 '… in feudo hereditate libere quiete ab omnibus seruciis · consuetudinibus excepto comuni auxilio · Comuni exercitu · Comuni operatione · ': *RRS*, II, 222-3: no. 152.

45 '… libertatem et acquietacionem de exercitu et equitatu nisi it[a euenerit quod] exercitus super me et terram meam infra Northumb' [uenerit inter Tinam] et Twedam …': ibid., I, 159: no. 43, 1147.

46 C. Innes (ed), *Liber Sancte Marie de Melros,* Bannatyne Club no. 56, 2 volumes (Edinburgh, 1837), I, 313-4: no. 351.

47 '… unum Arochor in Levenax pertinet et inveniendo in communi exercitu regis de qualibet domo in qua casei fiunt, duos caseos': James Balfour Paul (ed), *Registrum Magni Sigilli Regum Scotorum [RMS],* 11 volumes, new edition (Edinburgh, 1984), II, 41: no 187, 13 February 1431.

48 Barrow, op. cit. (7), 165-6.

49 *RMS*, I, 477-8, appendix I: nos. 100, 101.

50 *WME*, 61-4.

51 Madox, op. cit. (18), 487, and note (y).

52 'Sui servicium unius milites per annum, pro omni servicio exaccione seu demanda qualicunque …': *RMS*, I, 37-8: no. 132, 1362/3.

53 *APS*, I, 465-6 and 473-4: 1318.

54 Pollock and Maitland, op. cit. (19), 230 *et seq.*, and 256.

55 *APS*, II, 8: Act 23, 1424.

56 Ibid., II, 18: Acts 11, 12 and 14, 1429.

57 Ibid., II, 448: 8 November 1544.

58 Ibid., II, 460: Act 2, 1545. In England in 1205 the rate for the maintenance of a knight was 2 shillings: Poole, op. cit. (9), 370. Parliament had been correctly informed as to the extent of the destruction because Hertford in 1544 penetrated to Edinburgh burning that burgh and Holyrood and Leith; and the next year he boasted of having burned seven monasteries including Dryburgh, Melrose and Kelso and over 240 towns and villages: Dickinson, op. cit. (8), 317.

59 *APS*, II, 460: Act 3, 1545.

60 Ranald Nicholson, *Scotland: The Later Middle Ages* (Edinburgh, 1974), 49.

61 12 Charles II c. 24, 1660.

62 Ibid., Section 14.

APPENDIX D

ORKNEY
AND SHETLAND

ORKNEY AND SHETLAND, THE GROUPS OF ISLANDS LYING TO THE NORTH OF mainland Scotland, were overrun by Norse invaders beginning probably in the late eighth century; within a hundred years 'Caithness' had been colonised as far south as the Dornoch Firth. Subsequently, Norse influence spread throughout Caithness, the Western Isles and the Isle of Man. Orkney and Shetland remained under Norse control and were not considered part of Scotland until the second half of the fifteenth century.[1] Norse legal conventions and measurement units differed significantly from those developed in the southern part of Scotland, and this brought administrative difficulties when land assessment and rental payments were progressively integrated into Scottish practice. It is this later adjustment of Orkney and Shetland to Scottish metrological practice that is of particular relevance here, because this carries with it independent evidence of changes in the sizes of the Scottish measures discussed in Chapter 7.

When Harald Finehair became the first ruler of a unified if not always peaceful Norway, following his great victory at Hafvsfjord in the 880s, his opponents were dispossessed and fled in numbers to Orkney and Shetland. From these islands they began to harry the Norwegian coast in retaliation, as well as mounting raids on the Western Isles and parts of Scotland, Ireland and England. Harald descended on the Northern and Western Isles to put a stop to this, shortly after the battle.[2] To Rognvald, the Norse earl of Moeri, he gave Orkney and Shetland. Rognvald in turn gave them to his brother, Sigurd, to whom the title of jarl was granted. According to traditional accounts, thus was created the first earl of Orkney whose territory at his death included modern Caithness, Sutherland, Ross and Moray.[3]

By the twelfth century, the earldom was ruled jointly by, and the land was divided among, several related earls, and on the death of an earl his share could be divided among his sons or relatives. When there were two or more claimants or joint rulers, it was likely that one was soon to depart this life, as in the case of Hakon and his cousin Magnus. Miracles following the murder of Magnus in 1115 led to his canonisation but not without opposition from Hakon and his son, later to be Earl Paul II on Hakon's death. Rognvald Kolsson, a nephew of Magnus, also had a strong claim to the earldom, being heir to St Magnus, and in 1129 Rognvald was given a half interest in the earldom by King Sigard (The Crusader) of Norway. A second confirmation followed in 1134 when Rognvald sent emissaries to Paul, but without success. Alliances were created against Paul with David I of Scotland and Maddad, Earl of Atholl, nephew of Malcolm Canmore, who had married a sister of Earl Paul II. Part of the agreement was that, should Rognvald succeed, half the earldom would go to Atholl's infant son, Harald Maddadsson. An invasion of Orkney in 1135 by Rognvald failed, but another succeeded the following year. Paul was captured and was heard of no more. With Rognvald in control, Harald was brought to Orkney in 1138, becoming co-earl the next year at the age of five. Almost at once Rognvald commenced the building of the great cathedral of St Magnus at Kirkwall.

Atholl was a Scot of the Canmore line. His mother was a daughter of Earl Hakon, so Harald had a claim to Orkney in his own right. It suited David I to intrude a Scottish element and a close relative into Orkney's affairs, but Norway had not been consulted in any of this; therefore, as might have been anticipated, King Eystein made a foray to Orkney in 1151, and captured Harald at Thurso on the mainland. Harald yielded the earldom to Eystein and received it back on payment of a ransom of three gold marks (the traditional fine of an earl entering into his earldom), on condition that Harald became 'Eystein's man'. This switch of allegiance did little to commend itself to David, who in the same year granted half of Caithness to Erlend Haraldson, another grandson of Earl Hakon and a cousin of Harald. Norway, still smarting at having been ignored, granted Erlend half of Orkney. All the elements were now in place for a struggle between Erlend, Harald and Rognvald. The War of the Three Earls led to the killing of Erlend in 1154. Rognvald was assassinated in 1158. Harald alone survived.

Harald's earldom lasted from 1138 until 1206, an immense period for the times. He was undoubtedly more than a token force, for when armed men from Orkney and Shetland joined in the Norwegian civil wars in 1194 – only to be annihilated by King Sverre at the battle of Floruvoe or Florevåg – Harald thought it prudent to submit to Sverre in Norway lest he find himself the subject of a punitive expedition. As retribution, Sverre annexed Shetland in 1195, and until Scotland acquired the Northern Isles in the fifteenth century, Shetland was administered directly from Norway. This led to significant developmental differences, especially with respect to land valuations, between Orkney and Shetland.

Harald's troubles were not yet over, for in 1196 the army of William the Lion invaded Caithness, destroying Thurso. Harald now submitted to the Scottish crown, surrendering hostages, including his son Thorfinn. Caithness was removed from the earldom of Orkney and given in 1198 or 1199 to the King of Man in exchange for tribute. This goaded Harald into action. He mounted an invasion of Caithness whereupon William blinded and castrated Thorfinn, who died in Roxburgh Castle in 1201. Early in 1202 a Scottish expedition to the north brought Harald again into submission. Although the details are obscure, it appears that he was able to regain Caithness for a payment of £2,000.[4] On his death in 1206 the earldom of Orkney and Caithness passed jointly to his two sons, David (1206-14) and John (1206-31).[5] Thereafter the earldom became subject to increasing Scottish influence, and with John's death in 1231 the original native line of earls of Orkney came to an end. The earldom subsequently passed into a series of Scottish families who held Orkney from Norway, with Caithness and other lands from the Scottish crown.[6]

Earl John was succeeded by another Magnus, second son of Gilbert, Earl of Angus. The title of earl of Caithness was granted to him in 1231 by Alexander II of Scotland. Magnus's title to Orkney was probably confirmed by the Norwegian king, but from this time onwards the Norse records contain few references to Orkney or to its earls.[7]

At this time Scotland's relations with Norway were not good and they deteriorated over the next 30 years. In 1262 a major and bloody confrontation followed the invasion of Skye by the earl of Ross.[8] Hakon of Norway decided to teach the Scots a lesson and set off for the Western Isles.[9] His expedition terminated disastrously for him at the battle of Largs in 1263, with Hakon dying at Kirkwall in Orkney on his way home. He was succeeded by his son, Magnus IV, who by the subsequent Treaty of Perth in 1266 resigned the Isle of Man and all the Hebrides to Scotland, but retained Orkney and Shetland. On their side the Scots were to pay 4,000 marks of silver in four yearly instalments with an annual payment of 100 marks thereafter in perpetuity.[10] If payment was outstanding, Scotland was to pay a fine of 10,000 marks to Norway, with any dispute being referred to the Pope for adjudication.[11] Alexander III agreed to this payment for the sake of peace.[12] A contributing reason, if not the most important one, was the fear of the considerable navy Norway could muster: Hakon's 1263 fleet is estimated to have consisted of between 100 and 200 ships.[13]

These annual payments were not inconsiderable. One hundred silver marks equalled 10 gold marks (which was the exchange rate in about 1300).[14] This was the high level of 'relief' which an island ruler would be expected to pay on the accession of a new king in Norway.[15] By the terms of the Treaty of Perth this sum was to be paid annually, so financially this was a favourable arrangement for Norway, assuming the Scots lived up to their side of the bargain. The 100 marks fine was called the 'Annual of Norway'.[16] It was paid until 1291, but left outstanding in 1292 when John Balliol was on the Scottish throne, although Edward I of England was effectively ruling.[17] Edward was unlikely to exert himself on Norway's behalf, although representations were made by the Norwegians. Thereafter the Annual was often in arrears, with payment coming 'when it suited the Scots'.[18] Payment was made in 1312, and again in 1426, on the occasion of the reconfirmation of the Treaty of Perth, but there is no evidence of any subsequent instalments – indeed, representations were made in 1456 by Norway to Charles VII of France in the hope of persuading the Scots to make further payments.[19]

This matter was allowed to rest until October 1460, when ambassadors from the two countries met under Charles's aegis. James II of Scotland had died two months previously; his son James III was only eight years old; and the chief Scottish negotiator, James Kennedy, Bishop of St Andrews, was ill and not in attendance. It is not surprising therefore that the meeting fell short of resolving the difficulties, the main reason being, of course, the non-payment of the Annual.

However, a significant proposal was raised by the Scots with Charles's support. This was the annulment of the Annual as part of a marriage settlement concluded between James III and Margaret, the only daughter of Christian I of Denmark and Norway.[20] It is clear that this matter had been discussed by Charles, Christian and James II, because the Scots (if no one else) were fully prepared to state what they thought the dowry should be: first, that all claims to the Annual and any arrears be withdrawn; second, that Orkney and Shetland be transferred to Scotland; and third, that the queen bring 100,000 crowns ('scutorum') 'for her own adornment'. Not surprisingly, the Danes, overwhelmed by the enormity of these proposals, backed off.[21] Christian was perennially short of ready money, although the claim that he appealed for Papal assistance in the early 1460s to make the Scots pay up has been discounted.[22]

In 1466 the marriage question was raised in the Scottish Parliament, coupled with the matter of the Annual, followed immediately by a Statute setting the queen's 'third' or 'dowry' at one-third of the Scottish king's income from lands and customs.[23] Two years later, James agreed to give Margaret one-third of the crown's property and revenue, together with the palace of Linlithgow and the castle of Doune.[24] Although parliamentary records prior to 1466 are incomplete, we can assume that the proposed marriage had been discussed extensively in parliament because a marriage contract was agreed to by all parties on 8 September 1468. As to dowry, the Scandinavian crown was to provide remission of the Annual with all arrears, and the sum of 60,000 Rhenish florins, with 10,000 in cash. As for the remaining 50,000 florins, the lands of Orkney would be pledged to the Scottish crown, with right of redemption. No time limit was set on the possibility of redemption.

In May 1469, with Christian unable to raise the 10,000 florins in cash (largely because of a serious rebellion in his Swedish dominions), Shetland was pledged for 8,000 of the remaining 10,000 florins. There is no strong evidence to show that the remaining 2,000 florins were ever paid, although it is now believed that they were.[25] The record is only of the 10,000 florins (*'decem millium florenorum Renensium'*) and the pledging of Shetland for 8,000 florins (*'octo millium florenorum Renensium'*), with no mention of the remaining 2,000 florins. It is probable that Christian intended one of his successors to redeem the islands even if he could not, and the afterthought of pledging Shetland shows no secret arrangements to simply give the islands to Scotland *in lieu* of cash as some have suggested.[26] Twelve-year-old Margaret and 18-year-old James were married at Holyrood on 10 July 1469.[27]

It did not take the Scots long to act. In September 1470, James III bought out the incumbent earl, William Sinclair, with property in Fife, a pension of 50 marks annually (which was to come from the great customs of Edinburgh), together with other benefits.[28] The so-called 'annexation' itself followed soon after. The parliamentary act of 1472 reads in part:

> ALSUA the samyn day our souuerain lorde with deliuerance of his thre estates annext & vnijt [joined] the erledome of Orkney & the lordschip of scheteland to the crown nocht to be gevin away in tyme to cum to na personne nor persons except annerly til ane the kingis sonis ~~gott~~ of lachfull [lawful] bed.[29]

The word 'annexed' as used here should not be interpreted as seizing or absorbing another state's territory. The islands have never been annexed in the modern international sense of the word. This king simply bought an estate, which was now in his gift. The Scots crown now held the lands of the earldom and those that had belonged to the Norwegian crown. In a sense, the Scottish king was also the earl. The diocese of Orkney and that of the Isles (Sodor) had for long been within the ecclesiastical province of Nidaros (Trondheim, Norway), and this was recognised in the Treaty of Perth of 1266.[30] However, by Papal bull, both were transferred to the newly-elevated archdiocese of St Andrews in 1472.[31]

It is said that the earldom was fourteen times granted to persons other than the king's sons and was annexed to the crown no fewer than five times.[32] Even if this is something of an exaggeration, the provision of 'lawful bed' was violated when Robert Stewart, the illegitimate son of James V and Euphemia Elphinstone, acquired the earldom in 1581.[33] Robert had held the earldom in feu for an annual payment of £2,000 since 1565, when parliamentary approval was given, and although Privy Council approval was given the next year, the grant did not ever appear in the Register of the Great Seal. Because the title of earl was vested in the crown, Robert was not then the earl, although he held the earldom lands in feu. But as superior, he exercised the powers of an earl since he was sheriff of Orkney and 'foud' (chief magistrate) of Shetland. However, in 1581, he was also given the title of earl, but parliament declared this *ultra vires*. Although deprived of the title, he retained the feu (and therefore the revenues) of the estates. In December of that year he was termed 'lait Erle of Orkney', and at his last accounting in 1592 he was termed 'feuar of Orkney and Shetland'.[34] Robert died in 1593, to be succeeded by his son Patrick from 1593 to 1615. Patrick paid £6,000 for entry into the earldom lands, but his title as earl was in doubt for seven years until granted in 1600. Patrick's oppressions were spectacular, but the downfall of the Stewart line of earls did not arise from them, but from his open rebellion and that of his son Robert. The latter was hanged in January 1615, and Patrick was beheaded a month later.

At no time did Norway renounce sovereignty over the islands. In 1560, and again in 1561, Frederick II of Norway declared that *'dominium'* pertained 'to us and to our kingdom of Norway', and this was repeated in 1585.[35] It has been said by some Scottish writers that the right to redeem the islands was surrendered. This does not appear to have been so, for overtures in this regard were being made to Scotland as late as 1667.[36] On each occasion the Scots prevaricated, with the result that the long continued occupation and rule by Scotland has led to Scottish sovereignty being regarded internationally both *de facto* and *de jure.*[37]

UDAL LAW

The 1468 marriage contract between James III and Margaret made no mention of the laws under which Orkney and Shetland were to be governed, although it had been assumed that it was to be the Norse laws.[38] Having acquired the Hebrides by the Treaty of Perth, Scotland in 1425 endeavoured to ensure the establishment of Scots law in these islands by statute:

> ITEM It is ordainit be the king with the consent & deliuerance of the thre estates that all & sindry the kingis lieges of the realme leif & be gouernyt vndir the kingis lawis & statutes of this realme alanerly and vindir na particular lawis na speciale priualegis [privileges] na be lawis of vthir cuntreis nor realmis.[39]

At this time Orkney and Shetland were appendages of the Norse-Danish crown, but once these islands had been annexed the question arose as to which code of laws was to apply. An Act of March 1503 was amended in passage through parliament. It originally read:

> ITEM that all our Sovrane Lordis liegis <u>baith within Orknay Scheteland & the Ilis & other places</u> be Reulit be our sovrane lordis avne lawis & the common lawis of the Realme And be nai other lawis

but the words which are underlined were deleted and replaced by 'beand [being] under his obesance [obedience] & i speciale [especially] all the Ilis'.[40] The next day the act as amended was reintroduced and passed, with specific reference to Orkney and Shetland still deleted.[41] This Act of 1503 was in substance a re-enactment of that of 1425, to which apparently the Hebrides had not conformed. However, in 1567 the question of law in Orkney and Shetland was again raised and parliament's answer was that it should be by their own laws that the islands should be governed.[42] In 1633 it was again declared that the occupiers held their land by Danish law, the Scandinavian kingdom then being the federation of Norway and Denmark.[43] It is hardly surprising, therefore, that the editions of James Mackenzie's work on the Pundler Process (see below) identified 'the Isles' of the 1503 Act as the Hebrides exclusively, with the islands adjacent to Argyll being called 'the South Isles', while those adjacent to the counties of Ross and Inverness were 'the North Isles'.[44]

It mattered a great deal which system of laws was applicable in Orkney and Shetland. In Scotland proper at this time, land was held by 'tack', when the royal revenues from land would be given to an individual for a stated period for a set annual payment or by 'feu', a feudal term, when land would be granted, not for military service but for a return ('*reddendo*') each year of a set sum of money or its worth in grain or meat and other produce, for a (semi-) permanent period. (In modern terms a feu is a perpetual lease for a fixed rent.) Ever since the Norse occupation, land in Orkney and Shetland could be held by 'udal' or 'odal' law, in which the occupier, though having no written title, could pass his holding on to his heirs. The land was held of no one; no service was due to any superior, be he noble or monarch, but the 'udaller' paid 'skat', originally a tribute paid to the state but not feu duty and it came to be regarded as a land-tax.[45] No skat was payable on uncultivated land, moorland or wasteland. Having no superior, the udaller paid no rent for his udal lands and his occupancy was tantamount to ownership. Under these laws he could not sell to strangers without first offering the land to his relatives or obtaining their approval for the sale, and at least in Norway, if not in the Islands, a purchaser could acquire the odal rights by making a special payment.[46] When his family succeeded to the lands, each son received one share, each daughter a half share. It will be at once apparent that in the course of a few generations what may have been a substantial holding could have been sub-divided into several, perhaps many, small plots. In time these plots could become so small as to be useless, although size could be maintained by one heir buying out his siblings and, moreover, land could be acquired by marriage. In some respects, the earldom itself was considered udal property, divisible into co-earldoms, although no permanent division was ever established.[47]

The udallers were not to be left undisturbed in the enjoyment of their titleless lands. It is said that udal rights were lost in about AD 900 when, for the killing of his son Halfdan Longlegs, the islands were fined 60 gold marks by the Norse king, Harald Fairhair. The earl (Torf-Einar) paid this fine on condition that he received all the udal rights. The story goes that the udal rights could be bought back for a one-time payment of a mark of silver per 'plógsland', a unit nowhere else referred to, but perhaps a pennyland or even a ploughland. This story was probably a twelfth-century creation to justify a special taxation for building St Magnus cathedral in Kirkwall.[48] This implies, however, if true, that many of the descendants of the holders of the late tenth century still were occupying the lands two centuries later.[49]

In the period of Scottish administration, if udallers sold out, the udal nature of the land was lost in most cases and became subject to normal feudal levies. However, in 1690 a modest concession was made:

> And as to vassalls of Kirklands (churchlands) in Orknay and Zetland where their valuatione does not exceed twenty pounds Scotts, It is hereby declared that they shall bruike [*ie* have and enjoy possession of the land] by the Udall Right without necessitie of Renovatione of their Rights to infeftments.[50]

These 'vassals of kirklands' had bought their land from the church. Technically they feued their land, but this act freed petty holdings from feu duty. They would therefore have held their lands just as if they held them by udal tenure.[51]

But the battle was lost. At an early date those who held some of the bigger estates by udal tenure voluntarily surrendered their property, receiving it again by feudal charter. This gave them written rights of ownership, and therefore a greater degree of security, accompanied by the requirement of substantial annual payments. In the seventeenth century many smaller udallers in Orkney (but not in Shetland) were persuaded similarly to seek charters. Sometimes the annual payments were difficult to meet and led to repossession. Latterly, udal ownership was confined to very small properties and these petty udallers had problems in maintaining their unwritten rights. Their position was not improved by the earlier Court of Session decision of 1624 that udal lands 'behoved to be bruiked by some lawful title and that naked kindness and possession were not sufficient to possess them'.[52]

Under Scots law all land was the king's; and when he, by charter, gave lands in the islands in tack or in feu to his chosen current favourites, they often had no understanding of northern practices. Indeed, each new superior expected to be able to levy rent on the occupiers and the pressure to do so became unsupportable. The first Orkney feudal charter was issued in 1535.[53] This was when Sir James Sinclair was granted the islands of Sanday and Stronsay in feu heritage 'without division among heirs' for an annual reddendo of 200 marks.[54] As late as 1633, the udallers were declaring that for many ages they possessed land by payment of skat and teind (tithe) according to Danish law and now asked that no one come between them and the Crown 'to molest them'.[55]

The concept of udal tenure continued long after 1633. Indeed, legal cases in the twentieth century made use of it, for instance in the hearings to determine the ownership of the St Ninian's Isle treasure in 1962-3.[56] This questioned whether the sea-bed round Orkney was owned by the British Crown.[57] This was of importance in deciding whether the Commissioners of the Crown Estates could charge salmon farmers rental for use of the sea-bed. The petitioners questioned the ownership under udal law, but the Court of Session 'found that the Crown had the right of property on the territorial sea-bed from the low water mark out to the twelve mile limit'.[58] Landowners' rights to the foreshore and to salmon fishing were upheld, but not those relating to a share of whales or treasure.[59] Udal law, originating in Norse times, is still extant, helping to preserve, for some at least, their identity as islanders.

LAND DESCRIPTION AND THE ISLAND 'RENTALS'

Much of the information which we have of the land measures, and the weights and measures of Orkney and Shetland, comes from a series of documents known as the 'Rentals' which list the 'skat' (or land tax, also known as 'land male'), the rent and any other dues, property by property. The very earliest rental was called the Orkney 'Auld Parchment Rental', on which the later Orkney rentals were based, but this is now lost.[60] The first extant Orkney rental is that of Lord Henry Sinclair compiled in 1492 – only 24 years after the pawning of the islands.[61] This was followed by one of 1497 to 1503 (mistakenly referred to by early nineteenth-century writers as the 'Old' or 'First' rental), and another of 1595.[62] For Shetland, there were rentals of about 1500, 1628 and 1716.[63]

In these records we find dues paid in grain (malt, meal and barley) and butter in Orkney, and with oil, butter and cloth specifically mentioned for Shetland. These were the principal commodities in which the various taxes and rents were paid. Until a native coinage was introduced, these goods had to be measured when the agents of the Crown or the earl came to collect their dues. In communities such as these, and in the absence of a coinage, barter and exchange would be the order of trading with no need for the niceties of accurate weighing or measuring. When coin (initially foreign coin) did reach these areas, it was common to find the tax paid partly in produce and partly in money or silver.

But settlement came before taxation. The landscape determined where and to what extent a region might be made to provide a living. There had to be arable land and grazing too; and a group of families could plough and sow collectively and their grazing could be held in common.

The initial grouping of farm neighbours must have been largely pragmatic. The early island records tell of the imposition of skat on ouncelands
(sometimes 'urislands' or 'eyrislands') which almost invariably consisted of
18 'pennylands' as in Sutherland and Caithness where Norse influence was
strong.[64] In contrast, in the Hebrides we have already encountered an ounceland of 20 pennylands (see Appendix B).

It is as fiscal units that ouncelands are found in the records and the ounceland was the unit of taxation; but earlier they must have represented units of
settlement and community, with taxation coming later.[65] These territorial
settlements were in place in very early times. One study of the earliest settlements in Shetland has indicated just how far back these roots may go: when
the so-called 'spheres of influence' (represented by the defensive structures
known as brochs) of Iron Age Dunrossness were mapped, they fitted almost
exactly the much later outline of the agricultural groupings called 'scattalds',
which are described below.[66] However, Orkney has not apparently shown
the same pattern. These territorial settlement units show remarkable long-
range stability over many centuries.

Although documentary evidence is totally lacking, it might be imagined
that an ounceland was taxed at the rate of an ounce of silver annually (as in the
West of Scotland in Norse times), and a pennyland at one silver penny, perhaps
commencing in the late tenth century with subsequent development, and an
annual tax of one penny per family occurs in other Norse situations too.[67]
While Orkney lacked a native coinage, skat could be paid in pennyworths
of unminted silver, foreign coin or more commonly in produce such as grain,
oil, cloth, butter and poultry. The equivalences of these various commodities
(for example, between cloth and butter) were well understood, and apparently stable in a period when actual money values changed radically. Further,
the rentals (1492, 1497 and 1595) give the rent of an ounceland and the total
rental enables one to calculate that there were 201 ouncelands in Orkney.
The Orkney figure is equivalent to 3,618 pennylands at 18 to the ounceland
and the number of agricultural holdings, farms and crofts in the nineteenth
century was 3,373. Specifically, the Isle of Rousay is given as 6½ ouncelands,
or 117 pennylands. The number of farms there in 1885 was 117 which suggests
quite strongly that a pennyland was the farmland of a family, suitable for the
support of a family, and probably had been so for an extended period.[68]

The ounceland is not so easily equated to family holdings. Only the
largest townships were valued at one ounceland, with only a few in excess of
that value. More commonly, townships were taxed at a fraction of an ounceland, frequently between one-third to one-ninth, but neither pennylands nor
ouncelands were defined in terms of acres until the eighteenth century when
their highly variable sizes were made plain. An analysis of mid eighteenth-
century pennylands has shown their acreages to range from about 5 to 75
with a mean of about 28 acres for mainland Orkney; elsewhere they appear

nearer to 10 acres.[69] F. W. L. Thomas (writing in 1884) used figures for 1765, 1827, 1831 and 1850 for Orkney, and concluded that the pennylands at those dates ranged from 4 to 18 acres of arable land, and from 10 to 56 acres when the associated grasslands were included. However, these dates are very late, and there are indications that in Norse times that the pennyland was much smaller than the average figure of 28 acres.[70]

Ouncelands were largely determined by the features of the terrain, thus many are still identifiable on the ground, unlike pennylands; but the pennylands of a township signified the portion due to the holder. In a 9-pennyland township (half an ounceland), when land was held in common, the holder of three pennylands would be allocated every third strip of arable and one-third of all other resources of the community which included in addition to portions of land and grazing, such items as salvage from wrecks, beached whales, flotsam and other gifts from the sea. Equally one-third of all taxes and other levies would fall to the owner. It is not to be supposed that Norse rulers sent agents to count each and every household on which to levy tax of one penny. It was shown nearly two centuries ago that a lump sum was required from Orkney and from Shetland which was then divided up and levied among the inhabitants.[71]

Land measures were used not only in levying taxes and rents but they served ecclesiastical purposes as well. Tithes were assessed on the land value and there would appear to be a strong relation between the number of chapels built and the number of assessed ouncelands, in Orkney, and scattalds, in Shetland. It has been declared, with only slight exaggeration, that in Orkney each ounceland had its chapel 'for matins and vespers'.[72] This could indicate a desire on the part of the church or the people to have a place of worship on each ounceland. While the rule of one chapel per ounceland was not invariable (for two or more ouncelands could share the same chapel, and moreover they were privately owned), the frequency with which it held true encourages belief in the intent. Something similar is found in Shetland. For example, the isle of Unst in Shetland had 28 scattalds (or ouncelands) and 24 chapels.[73] The system of chapels may well go back to the pre-Norse period, perhaps to the eighth-century Northumbrian mission to Pictland.[74]

Both Orkney and Shetland had a notional currency without a coinage of their own. William Thomson has shown that for the Shetland island of Yell it was recorded that one mark of silver was equated to 12 Shetland shillings or 144 Shetland pence. Thus the ounce of silver, being one-eighth of a mark in weight, was equal to 18 Shetland pennies, the same proportion as lay between the ounceland and pennyland.[75] Naturally, a pennyland could be sub-divided into 4 farthinglands.

With the passage of time more land was brought into production from waste and scrub, but instead of generating new pennylands, the existing ones simply grew bigger, especially on Orkney with its richer soil. The potential for expansion was more limited in Shetland, thus the pennylands of Orkney presumably became larger than the corresponding units of Shetland. Nor was the increase everywhere uniform. Such became the variability of these units that they became virtually useless for purposes of assessment and levying of rent. By the thirteenth century it had become increasingly common when lands changed hands or were rented, to describe the land in terms of its purchase price in marks, a mark being £⅔ or 13s 4d in silver or produce. Initially a markland was land purchasable for a mark, although before 1400, a markland sold for more than a mark of money. Initially it was most common for a pennyland to be bought or sold for 4 marks, and generally the price was in the range between 1½ to 4 marks. This continued for the next two centuries. It is clear from the Orkney rentals that ouncelands and pennylands were used for *taxation* (skat) whereas *rent* (land male) was levied on the mark. The rent per mark would be given in 'meils' and 'settens' (or setteens) of 'cost' (sometimes barley, sometimes a mix of one-third oatmeal and two-thirds malt) and sometimes 'flesche' (meat).[76] Cost and flesche appear as though of equal value, but the weights given are not real weights of meat. Most likely rent was given originally entirely in grain with meat becoming accepted on some scale of equivalence which changed according to the market price of cattle relative to grain. With the pennyland at 1½, 2, 3 or 4 marks, the mark was most commonly rated at 10 or 12 'settens' to the mark; and since the monetary value of a setten weight of produce was around a penny in about 1500, the mark of land was rated at about 10 or 12 pence.[77]

Not every markland was rented at 10 or 12 pence. In about 1500 some were paying anywhere in the range 4 to 20 pence (although 20 pence would be a very large value in Orkney; in Shetland so large a figure as 20 is never encountered): a contemporary record for Shetland therefore might read '12 marks of land at 12 pennies the mark'. But in the mid-fifteenth century, in a period of severe economic recession, the system of valuation changed completely. After 1455, the last recorded date for a markland to be sold at a mark, marklands appeared at a variety of different prices. They were uneven both in price and extent, so that the original proportional relationship between pennylands and marklands was destroyed: this was related to the changing value of Norse money during this period of depression.

However, in Orkney the name 'pennyland' was retained. It was the rating in marks which changed with time. An example from Brough in South Ronaldsay may be given: in 1329 the rating was given as 3 marks to the pennyland, but in 1502 the same pennyland was given as 4 marks.[78] Thus the number of marks in a pennyland could and did change in the medieval period, but later the relationship between the mark and pennyland tended to solidify, with the mark becoming only slightly variable in terms of its renting value.

Considering the separation of Shetland from the control of the earldom in 1195, we would expect some differences to appear between the measures of

Shetland and Orkney, and indeed there are such differences. For example, after the introduction of the term 'marklands' into land descriptions, the older designations of 'ounceland' and 'pennyland' were retained in Orkney; however, they all but vanished, as we have already indicated, from Shetland records (although ouncelands were appearing infrequently as late as 1500). Nevertheless, an analysis of the Shetland situation can with difficulty reveal the underlying ouncelands.[79]

The word 'ure' (or ounce) in Shetland meant one-eighth of a mark or markland, and was the lowest value used in the rentals; but in Orkney it meant the ounceland of 18 pennylands or 72 marklands (there being usually 4 marks to the pennyland), and this was the largest unit used there.[80] The term 'scattald' appears in the Shetland records (and will be discussed below) but never in those relating to Orkney. Again, the large weighing device resembling a steelyard for use with heavy goods was called in Orkney a 'pundlar', the smaller one for more modest weighings was called a 'bismar'; but in Shetland the reverse was true.[81] We shall therefore consider the weights and measures of the two island groupings separately.

ORKNEY WEIGHTS AND MEASURES

F. W. L. Thomas provided examples, which are listed in Table D.1, for the 'land male' (or 'mail') in the Rental of 1497-1503, not very long after Orkney had become an appendage of the Scottish crown. This was the term used for land rent.[82] He gave a variety of marks to the pennyland, but inexplicably he omitted to give an example of the commonest rate of 4 marks to the penny-land, although he stated that 'it is plain that on an average there were four marks in a pennyland'.[83] He dealt only with a rental of 10 settens to the mark, while stating it was often 12, and omitted a value for $13\frac{1}{3}$ setteens, where $13\frac{1}{3}$ is one-twelfth of a mark of money (160 pence). At the time of the 1492 Rental, the commonest patterns show lands of $1\frac{1}{2}$, 2, 3 or 4 marks to the pennyland,

Table D.1

Examples of Orkney land-rent assessments in the rental of c.1500, from figures of F. W. L. Thomas, 1884.

PLACE	MARKS IN THE PENNYLAND	RENT IN SETTENS FOR A PENNYLAND (AT 10 SETTENS TO THE MARK)
Lynhow, North Sandwick	$1\frac{1}{2}$	15
Grymestath, Hurray Brugh	2	20
Hoy and Stromness	3	30
Orphair	5	50
Orphair (second entry)	6	60
Wallis	Several at 8	80

rented at a standard rate of either 10, 12 or 13⅓ settens to the mark.[84] Thomas's figures indicate that the rent is fairly directly related to the number of marks in a pennyland.[85] Thus the male (rent) on a markland might be 10 settens of cost, or if bere (barley) was given the male was 15 settens.[86] To illustrate the quantity of rental involved, in about 1500 the setten was 24 marks in weight, where the mark of weight was 8 ounces, so the setten was 12 pounds of produce, most probably of grain.[87]

With the pennyland frequently equated to 4 marks the corresponding ounceland would be 4 × 18 = 72 marks. With skat paid by the ounceland from at least 1502, the tax on the 72 marks no longer added up to one ounce of silver, as it may have done in the distant past. At this time the records show ouncelands paying tax of upwards of 12 ounces of silver or the equivalent and sometimes well above this, more in keeping with the current value of the yield of the land.[88] Thus an ounceland of 72 marks, paying 10 pence rent per mark would yield 720 pence or, in Norse terms, 3 marks of 240 pence in rent.[89] With the mark equal in weight to 8 ounces of silver, 3 marks amount to a payment for rent of no less than 24 ounces of silver.

An Orkney term infrequently used was 'skatland', which was a quarter of an ounceland, or 4½ pennylands.[90] (This is not to be confused with the Shetland 'scattald', discussed below.) It has been suggested that the skatland was the unit required to produce one man for the army or navy in a similar way to that in Norway and is reminiscent of the old Dalriadic naval levy.[91] However, this is a vague relationship which modern work has tended to doubt.[92]

The skatland of 4½ pennylands looks awkward and it is not impossible that this is evidence of an early merging of two systems. While some authorities view ouncelands and pennylands as appearing together, another view is that the ounceland was older and was divided into quarters with the newer pennyland being superimposed later.[93] The geographic distribution of nameable ouncelands is different from that of pennylands in the West of Scotland, and in Caithness and Sutherland there are only two references to ouncelands amidst numerous pennylands.[94] Skatlands are present in Caithness too, but there is no indication of their relationship to pennylands.[95] The skatland does not appear to have played any major role in the economic life of the Islands.

The early nineteenth-century edition of the early rentals, edited by Alexander Peterkin, gave the table of weights and measures used in determining skat.[96] The Orkney system apparently ran as follows:

FOR GRAIN: 24 marks (12 pounds) = 1 setten
              6 settens = 1 meil
              24 meils = 1 last = 144 settens

FOR BUTTER: 24 marks (12 pounds) = 1 lispund
               1 span (or spann) = 21 (or 20) pennies-worth
                                   = 5¼ (or 5) lispunds
               1 lispund = 4 pennies-worth

Incidentally, the setten is mentioned in the list of tolls and customs duties, the *Assisa de Tolloneis*, attributed to David I, which specified the duty 'for ilk mesure that is callyt a settyng a half peny'.[97] Butter in spans or lispunds was seldom weighed in units greater than one lispund in early times, although in the 1492 Rental the barrel is mentioned at about 20 lispunds or 80 (skat) pennies; however, butter was used for the payment, in part, of skat (the butter-skat), but almost never for the payment of rent.[98] (However, in Shetland, butter comprised a substantial portion of the rent.)

A span of butter was 20 or 21 pennies-worth. The lispund was worth 4 skat-pennies so there were 5 or 5¼ lispunds in the span, whichever was most convenient for the calculation then in hand. It was a very common occurrence for a pennyland to pay 7 pennies of butter-skat, one-third of a span of 21 pence. The span's use was restricted virtually entirely to assessing the butter-skat. Only one exception is known to us, which is that given by James Mackenzie's editors in 1836, quoting from another historical work:

> [25 July 1328] … the tenths due to the Pope, viz. 22 cwt. of wool, less than 16 pounds, according to the standard of Shetland, being 36 span Shetland weight of wool.[99]

It is uncertain what form the hundredweight of the early fourteenth century was measured in the islands. If it was 96 of the Bruges wool pound of 7,000 grains, which we believe was the understanding in Scotland at the end of the century (see Chapter 4), then the wool span would be about 58 pounds. Alternatively, if at this early time the lispund was 12 pounds, and the span was 5 lispunds, then the span would be 60 pounds.

In 1597 the table of weights had been expanded by John Skene.[100] His conclusions can be tabulated as:

FOR GRAIN: 24 marks = 1 setten
              6 settens = 1 meil
              24 meils = 1 last
              1 last (meal and malt) = 1 Scottish chalder
              1½ lasts = 36 meils = 1 chalder bere
FOR BUTTER: 24 marks = 1 setten
              6 settens = 1 lispund = 1 stone 2 pounds, Scots weight
              15 lispunds = 1 barrel
              12 barrels = 1 last

Here Skene has intruded some Scottish terms: the 'chalder' and 'stone'. Chalders would have been used after 1489 when Orkney rents were paid to the Scottish Exchequer. Skene had shown the grain and butter weights divided in very similar ways, but presumably he was in error in inserting a setten in the butter table since, both before and afterwards, 24 marks was one lipsund of butter. Barrels were recorded from about 1492, although by 1753 it was apparent that several different barrel types could be used:

> They take any Cask that goes under the Name of *Barrel*, sometimes Tar-barrels, sometimes Salt-barrels, sometimes Herring-barrels and that these Casks or Barrels differ considerably one from another.[101]

By the mid-eighteenth century these tables were revised somewhat, in the course of a legal discussion known as the 'Pundlar Process' (see below).[102] The agreed table was given as:

FOR GRAIN:	24 marks	=	1 setten (sixth part)
	6 settens	=	1 meil
FOR OATMEAL AND MALT:	24 meils	=	1 last or Scottish chalder
FOR BERE:	36 meils	=	1½ lasts or one chalder of bere
FOR BUTTER:	24 marks	=	1 lispund
	12 barrels	=	1 last

This lispund is the same as that of the early rental, but very different from the erroneous value given by Skene. Skat and rent were collected in Orkney's own system of weights and measures, but when barley was sent by sea to Leith it was measured in Scots units, hence the need for a table showing equivalents. It will be noticed that although the Scottish chalder was a measure of capacity, in Orkney the chalder was a weight. At this time, the chalder of bere was declared to be equivalent to a capacity of about 20 bolls Leith measure, rather than the 16 chalders that might be expected, but the difference represented an eighteenth-century equivalent to the old water measure allowance.

Rents or land male (paid by tenants, not the udallers) were levied not on ouncelands or pennylands but on marks of land.[103] Rent charged per mark of land was given in meils and settens of 'cost' and 'flesche' as if equal weights were of equal value, but the weights given in the rentals for meat are not to be taken literally. With the development of animal husbandry, meat became acceptable on an equivalent grain scale, as in 1492, when a 'mart' (a head of cattle), was counted as acceptable as 10 meils.[104] In 1597, Skene listed several types of flesh.[105] These can be tabulated as:

1 'sufficient' cow or ox	=	10 meils
1 geld ox	=	15 meils
1 'wedder' [sheep]	=	4 meils
1 goose	=	2 meils
1 capon	=	1 meil

The meil of malt was worth 6 pence, in Orkney reckoning, and a 'sufficient cow' was worth 10 meils, namely 60 pence or 2 ounces of silver in the fourteenth century, because the Norse mark was of 240 pence; and 60 pence, being worth one quarter mark, was 2 ounces.[106] In Iceland in the twelfth and thirteenth centuries a cow was worth 2 ounces of silver, which shows a common valuation running throughout Norse territory at that time.[107]

The instruments used in Orkney for weighing were called pundlars. This was a comparatively crude type of steelyard, fabricated in wood, ranging from a farmer's rough-and-ready instrument made from whatever wood was at hand to the large elegant machine now on display at the museum at Lerwick (which is not a Shetland instrument at all, but is Orcadian).[108] There were three basic instruments used in Orkney: the malt pundlar, the bere (or barley) pundlar, and the bismar. The latter device was used for relatively small weighings, and it differed from the pundlar in that it had a movable fulcrum – it was suspended by a cord which was moved along the graduated beam until a balance point was found. Strict accuracy was not required, fortunately, for these weighing instruments were notoriously inaccurate. In 1615 an order required them to be stamped with the royal initials as an indication of authenticity and standard measure.[109] However, Alexander Peterkin gave the weights recorded in 1758 on ten pundlars of a mass of 3 settens. These ranged from 6 stone 4 ounces to 6 stone 7 pounds, showing a discrepancy of about 7 per cent – a very large spread for a weighing, indicating not only the inherent imprecision of these pundlars, but perhaps also a lack of adequate adjustment.[110]

There have been two legal determinations of the weight standards of Orkney. One was established at the Court of Session in Edinburgh in an action by the Earl of Galloway against the Earl of Morton in the years between 1733 and 1759, termed in legal circles of the time, and universally from then onwards, as the 'Pundlar Process'.[111] This dealt with the status of the Orkney measures that were being used to assess rents and tithes, in which it was alleged that the weights and measures had been increased enormously in the course of the centuries, but without authority. The Earl of Morton was the superior to whom the land-skats were payable. The duration of the action – some 26 years – made it one of the more famous cases of the eighteenth century. It is little wonder that it has been called 'a stupendous litigation'.[112] The action failed when it was shown that any changes brought about by earlier superiors were in place, and had been taken into account, when the present plaintiffs entered into occupancy of their present holdings.[113]

The other legal determination of weights in Orkney was the evidence accepted, although not without dissent, by the county jury set up in 1826 to determine the conversion factors needed to bring the local measures into

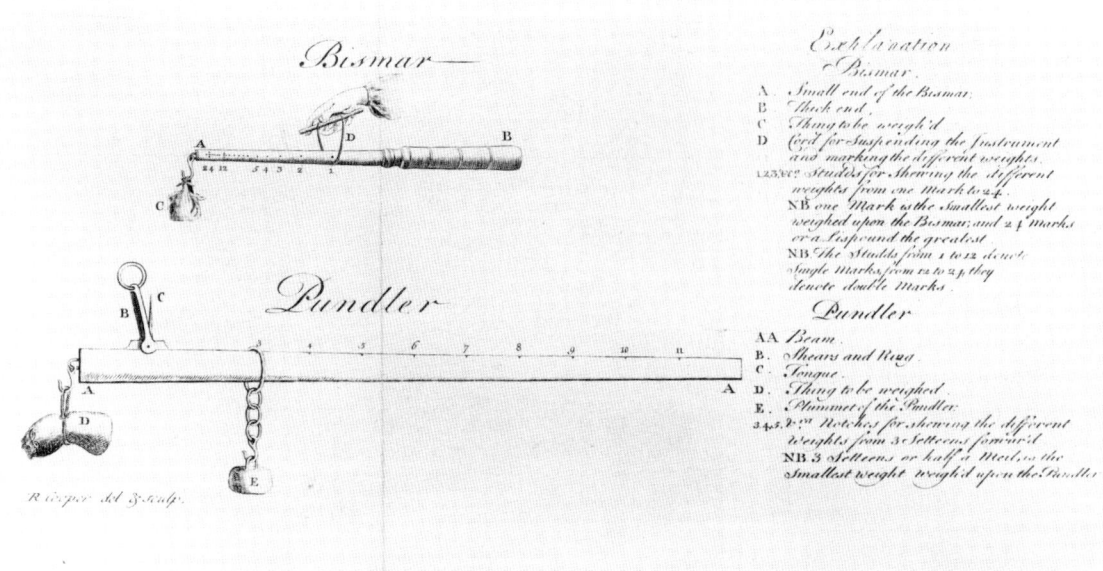

Explanation

Bismar.

A. Small end of the Bismar.
B. Thick end.
C. Thing to be weigh'd.
D. Cord for suspending the instrument and marking the different weights.
1,2,3&c. Studds for shewing the different weights from one Mark to 24.
NB one Mark is the smallest weight weighed upon the Bismar; and 24 Marks or a Lispound the greatest.
NB The Studds from 1 to 12 denote single Marks, from 12 to 24 they denote double Marks.

Pundler

AA. Beam.
B. Shears and Rigg.
C. Tongue.
D. Thing to be weighed.
E. Plummet of the Pundler.
3,4,5,&c. Notches for shewing the different weights from 3 Setteens forward.
NB 3 Setteens or half a Meil is the smallest weight weigh'd upon the Pundler.

R. Cooper del. & sculp.

Fig. D.1

Contrasted forms of the Orcadian bismar and pundlar from the *Memorial for the Proprietors of Orkney against the Earl of Morton,* drawn and engraved by Richard Cooper, 1758.

conformity with the Imperial Standards established by the Act of 1824.[114] (The role of such juries is described in Chapter **9**.)

The printed papers of the Pundlar Process provided detailed description of the pundlars and the bismar, and this has been paraphrased many times over the years by various authors. John Swinton's version of 1779 reads:

The Pundlar is a beam of wood about six feet long, and about three inches in diameter at one end, tapering gradually to the other. A hook is fixed to the greater end for suspending the goods. About six inches from that end, a tongue and shears, like those on the beam of a balance, are fixed; and, at the upper end of the shears, there is a large iron ring, through which, when the instrument is used, there is put a cross-beam for suspending the machine; and this cross-beam is generally supported by two men on their shoulders. The Pundlar is marked with notches at proper distances, corresponding to, and exhibiting the weight, from three Setteens upwards, to twelve; and the weight of the commodity is ascertained by a stone of the weight of a Setteen hung upon the Pundlar by an iron ring, which may be shifted from notch to notch, till the tongue between the shears, as in a steelyard, discovers the instrument to be in equilibrio.

There are two Pundlars; namely, the Malt-pundlar and the Bear-pundlar: the first of which is understood to be the general weight. The weight signified by both have the same denominations; but those on the Malt-pundlar are comparatively one half heavier than those on the Bear-pundlar. Thus, two Setteens on the Malt-pundlar are equal to three Setteens on the Bear-pundlar, and so on of the other pundlar-weights.

The Bysmar is a beam of wood about three feet long, whereof a little more than the half is a cylinder of about an inch in diameter. The remaining part of the beam, or but-end, is also cylindrical, but much thicker than the other, being about three inches in diameter. In the small end there is a hook, from which the goods are suspended. The small end is marked with iron studs, at unequal distances. These studs correspond to and exhibit the weight of the commodities weighed, from 1 mark to 24 marks, which make a Setteen or Lyspund. When the material to be weighed is hung upon the hook, the Bysmar is horizontally suspended in the bight or loop of a cord. The weigher holds this cord in his hand; shifting its place, until the material weighed equiponderates the but-end of the Bysmar, which serves as the counterpoise. When the instrument is thus brought to an equilibrium, the stud nearest the cord shows the weight of the commodity in marks. This instrument bears relation to the Malt-pundlar, that is, the weights on it are multiples of the Malt-pundlar.

All these instruments are very rude and imperfect, especially the Bysmar, which may be considerably affected by the manner of holding the cord. The manner of weighing, even on the Pundlar, may make a difference in the great weights, to the amount of six marks.[115]

During the Pundlar Process it was generally agreed that 24 marks made a setten, 6 settens made a meill, and 24 meills made a last. Additionally, 24 marks made a lispund and 12 barrels a last. What was in dispute was the weight of the mark and the number of lipsunds to the barrel. In early times the mark had always been 8 ounces, but by the mid-eighteenth century the standard currently in use exceeded a pound, with corresponding increases of all the other units, for it was declared that:

> The Earl of Morton, according to the showing of the heritors, in those islands, is, and, for some time past, has been in use to receive the flesh and victual payable by them, by or upon instruments which make the mark weigh about one pound four ounces, (or twenty ounces) trois, which of consequence makes the Lispund to be about 30 pounds trois weight, and so six of these fill a barrell, as now containing 180 of such pounds; whereas these gentlemen averr and lybell, that the standart of the Mark was and is eight ounces trois, the consequence of which from the agreed proportion between Mark and Lispund is, that the standart of the Lispund was and is twelve trios pound; and they also lybell and aver, that the barrell always was of the same measure or capacity that it is now, the consequence of which is, that it should take 15 true Lispunds to fill it.[116]

If the barrel was fixed in size (theoretically, at least) and if the number of lispunds to fill it was large, then the lispund was of small volume and therefore small weight. Rents were paid in lispunds not barrels, so a small lispund meant that the tenant could meet his rent more easily.

During the long drawn-out legal proceedings, the 'standards' of Orkney were examined in 1743. Table D.2 has been taken from the listing of the various items produced by the Deacon of the Wrights and Hammermen of

| DENOMINATION | DESCRIPTION OF STANDARD | WEIGHT IN SCOTS TROYE | | | |
		STONES	LB	OZ	DR
1 mark	'Boar's Tooth'	0	1	2	8
2 marks	'A Lead weight and Rope'	0	2	6	8
4 marks	'A round Freestone and rope'	0	4	10	0
7 marks	'A Freestone, Rope, and a piece of loose Lead'	0	8	1	12
9 marks	'A Freestone and a piece of loose Lead'	0	11	6	0
16 marks	'A Freestone, Boar's Tooth, and a piece of loose Lead'	1	3	4	8
1 setten	'A Freestone and a piece of Lead'	1	12	8	8
2 settens	'A Freestone and a piece of lead'	3	6	8	0
2 settens	'A Whin-stone and Boar's Tooth'	3	8	11	8
2 settens	'A Whinstone and a piece of Lead'	3	6	0	0

Table D.2

The Orkney standard weights of 1743, from the table printed by Alexander Peterkin in 1820.

Kirkwall and the weights determined from them.[117] Their production 'occasioned much derisive laughter', because nothing in this list could faintly resemble a properly-made standard.[118] Instead, it was stated that: 'The standards were stones taken from the beach, shapeless, and easily substituted by others, so that each successive collector of rents could, without any difficulty, alter his weights.'[119] The boar's tooth had been filled with lead at some unknown date, and lead had been attached to the stones, and 'there was the justifiable suspicion that extra lead had been added, either by Morton's chamberlains or their predecessors and, since one beach stone is very like another, perhaps heavier stones had been substituted for the originals'.[120]

As far as the value of the setten or lispund is concerned, it was stated that the lispund had been 12 pounds (24 marks), but that Earl Robert Stewart had increased it to 15 pounds in the 1580s. His son, Earl Patrick, advanced it to 18 pounds in about 1600. In 1691 the lispund was set at 24 pounds.[121] The rough mean of the setten from the 1743 table of 'standards' is 27¾ pounds Scots troye (or trois) or 30⅛ pounds avoirdupois. The later 1826 verdict was that the Orkney lispund was 29 pounds 10 ounces 12 drams avoirdupois, only 1½ per cent less than the 1743 value from the (rather variable) settens, and this by a majority vote only. The 1826 jury also found that:

meil of malt on the malt pundlar	=	177 lb 12 oz avoirdupois
meil on the bere pundlar	=	116 lb 7 oz avoirdupois
mark on the bismar	=	1 lb 3 oz 12½ dr avoirdupois
lispund of 24 marks	=	29 lb 10 oz 12 dr avoirdupois
barrel of butter	=	217 lb 6 377/1000 ounces, avoirdupois.[122]

The precision given for the weight of the barrel of butter is absurd, but it reveals the intent that there should be 7½ lispunds to the barrel, which, correctly calculated, would have yielded over 222½ pounds avoirdupois. Only five years earlier, the Sheriff-Clerk of Orkney had recorded it at 8 lispunds, being weighed together as a unit.[123]

It is noted, however, that frequently the butter paid into both the Earl of Morton and the Bishop's storehouses was very bad, with half a barrel weighing less than 100 pounds, 'because the same was hoved up [expanded] with whey and trash. When this filthy compound was shipped south, it sold not as butter but as grease'.[124] This was nothing new, for in the Treasurer's Accounts for 1526, and again in 1528, there is an entry for 'Orknay buttir to creische [grease] the quhelis [wheels] and extreis [axles]' of cannon, which gives an example of the use to which rancid butter was put.[125]

Although the under-tenants may have lost the Pundlar Process in the late eighteenth century, they appear to have recouped their losses somewhat, for it was subsequently revealed that only two of the nine witnesses heard in 1826 had subscribed to the above values of the meils, marks, lispunds and barrels, although they were authoritative individuals; namely, Andrew MacPherson, Deacon of the Wrights and Hammermen, and his predecessor in office, Lieutenant John Baikie. The other seven witnesses were of the opinion that currently the meil on the malt pundlar was 11¼ stone Dutch (or Scots troye) weight, or roughly 196 pounds avoirdupois weight.[126] This was some 18¼ pounds avoirdupois more than that decided upon by the jury. By the accepted verdict the superiors would lose some 10 per cent, a fact which was not long in being appreciated. The verdict is dated 1 August 1826; and on 7 August the chamberlain of the Orkney earldom wrote to his factor in Edinburgh that the results had come out much lower than he had reason to expect.[127] By this decision the superior – Henry Dundas, 3rd Viscount Melville – would lose more than a stone per meil (actually 18¼ pounds) on the malt pundlar, and almost 14 pounds on the bere pundlar. The chamberlain further pointed out the absurdity of the use of the so called 'standards', where the malt meil standard consisted of three separate stones, and the maker of the bere pundlar in adjusting it, used two of the 'standard' stones (presumably to attain the two-thirds value), but:

> he swore at the time that he did not know which two, though they were all different. The result of the verdict will affect Ld. Dundas' interest very materially … [and] the Crown's revenue will also be materially affected.[128]

Nonetheless, the 1826 verdict apparently stood, and no amendments were recorded.

It has been generally admitted that the 1826 figures showed too great a reliance on the 1743 weighing of the 'standards' with the almost complete dismissal of what was then the customary practice, which would presumably

	1743 AVERAGE (SCOTS TROYE)	1743 AVERAGE (AVOIRDUPOIS)	1826 FIGURES (AVOIRDUPOIS)
mark	1 lb 3·3 oz	1 lb 5 oz	1 lb 3·78 oz
setten/lispund (24 marks)	27 lb 10·9 oz	30 lb 1·7 oz	29 lb 10·75 oz
meil (6 settens)	166 lb 1·1 oz	180 lb 10·1 oz	177 lb 12 oz

Table D.3
Orkney units for rental payments, determined in 1743 and 1826.

have matched the chamberlain's expectation. If we look at the six weighings of the standards called '1' to '16 marks' in Table D.2, the average mark comes to 1 pound 3·3 ounces. Likewise, the average 24-mark setten (or the equivalent lispund) can be extracted from the table as 27 pounds 10·9 ounces, and hence the meil of 6 lispunds would be 166 pounds 1 ounce, all of which are in Scots troye (or trois) weight. Converting them to avoirdupois weight and comparing with the 1826 adopted values in Table D.3, it can be seen how very close the average from the standards (in avoirdupois) is to the avoirdupois figures adopted in 1826. The standards' average is in all cases a little higher: for the mark at only 3·6 per cent, for the lispund at 2·7 per cent, and for the meil at 1·5 per cent. Considering the inaccuracies inherent in even determining an average value of these three units from the data given for the standards, it is fairly obvious that the jury adopted the 1743 values and converted into avoirdupois weight as best they might.

The growth of these measures seems to have been arrested by the end of the eighteenth century, but even then the standards are so variously stated that no really precise equivalents in modern terms can be given. The descriptions have been gathered together in Table D.4. The first column contrasts the claims of the plaintiffs in the Pundlar Process with those of the Earl of Morton; the second gives John Swinton's 1779 version of the findings; the third shows the 1826 jury decision, and those expected by the chamberlain

Table D.4
Determinations of Orkney and Shetland Weight Units, in pounds. Entries in square brackets are not given explicitly, but have been obtained from other internal data.

	ORKNEY						SHETLAND		
	PUNDLAR PROCESS		SWINTON 1797		1826 JURY	CHAMBERLAIN'S EXPECTATIONS	SWINTON 1797		1826 JURY
	PLAINTIFFS (SCOTS)	EARL OF MORTON (SCOTS)	(SCOTS)	(AVOIR)	(AVOIR)	(AVOIR)	(SCOTS)	(AVOIR)	
mark	½	1¼	1¼	1·36	1·24	[1·36-1·40]	1·17	1·27	1⅓
lispund	12	30	30	32·6	29·8	32·7-35	28	30·5	32
barrel	180	180	200	217·5	217·4	–	–	–	–
meil (bere)	48	[120-122·7]	120	130·6	116·4	[130-140]	[112]	[121·9]	[128]
meil (malt)	72	180-184	180	195·8	177·8	196-210	168	182·7	[192]

(the latter in all likelihood represented the customary interpretation given to the units by the stewards of the superiors in collecting the actual payment of renders in kind in 1826 and the preceding years). For comparison, the values of the equivalent Shetland units from Swinton's 1779 account and George Buchanan's 1829 printing of the 1826 jury decision are given in the final columns. It will be seen that the situation in Shetland was not very different from that in Orkney.

The plaintiffs in the Pundlar Process identified the butter span of the early rentals through multiples of the lispund and the setten. However, from the earlier evidence of the rentals it could be deduced that that the span in Orkney was 5¼ lispunds, each of 12 pounds, for a total of 63 pounds. In Orkney, the lispund of butter was worth 4 pennies and the span 21 pennies, so the span was 5¼ lispunds. With each lispund equal to 12 pounds according to the plaintiffs, the weight of the span would be 63 pounds.[129] In 1779, on the basis of the data derived from the Pundlar Process, Swinton made the lispund equivalent to 30 pounds Scots troye (or 32·6 pounds avoirdupois), with about 6⅔ lispunds to the barrel.[130] This gives a barrel of 217·5 pounds avoirdupois, very close to the 1826 value of 217·4 pounds.[131] Another, more recent, author makes the span equal to 5 lispunds.[132] It is clear that, for an equivalent to be made, the period must be specified carefully and the printed word read with caution.

SHETLAND WEIGHTS AND MEASURES

The whole of Shetland was divided into parishes, each of which in turn was apparently sub-divided into small units called 'scattalds' or skattalds, whose very name (meaning, 'enclosed land for which tax was paid') indicates their principal role.[133] The inhabitants of this locality paid their skat jointly, and they jointly received their share of the proceeds of the community which included rights to peat cutting, thatching, taking sea-weed for manure, using the by-products of beached seals and whales. Those who did not pay tax did not participate in the privileges either. Initially the word referred to a portion of arable land with associated grazing. The scattald served as a social unit and could be considered as a unit of settlement, and does not appear to have had an ecclesiastical basis. In more recent times, and certainly by the eighteenth century, it had come to refer to grazing land only.[134] Whatever their ultimate origin, it would seem that scattalds were in roughly their final form by 1200. There would appear to have been a highly stable system of territorial organisation since antiquity, because the distribution of scattalds corresponds to the brochs of Iron Age Shetland and also to the distribution of neolithic chambered cairns; but as has been observed earlier this may be no more than the fact that settlement in Shetland throughout the ages took place in the same pockets of arable land.[135] Could it be considered possible that scattalds had neolithic origins? In Orkney these similarities have not been seen. There the unit of organisation was the parish; in Shetland it was the scattald and these were much smaller than the later parishes in Shetland. In Shetland, as in Norway, skat (taxation) and rent (land male) were closely related, with everyone paying skat whether they rented land or not, and the land measures in both Shetland and Norway were very different from those of Orkney.

The 'last of land' is one of the distinguishing features of Shetland measure. It did not appear in Orkney, where ouncelands and pennylands continued to be used, while Shetland turned to marks and lasts of land. The last was a large unit, and was rented for 144 pennies. Each mark was assessed a value, such as marks at 6 pennies the mark, and the number of pennies in the mark was usually given in the rentals. The last could therefore be made up in various ways, for example:

24 marks at 6 pennies the mark = 144 pennies,   so 24 such marks = 1 last
16 marks at 9 pennies the mark = 144 pennies,   so 16 such marks = 1 last

At 12 pence to the Shetland shilling, the last was worth 12 shillings. The monetary mark of both Orkney and Shetland, probably dating from the tenth century, was one of 12 shillings or 144 pence; and this continued at least into the sixteenth century in Shetland, although Orkney had converted to a 160-penny mark (13 shillings and 4 pence in Scots notation) long beforehand.[136] It is one of the intriguing features of these islands that, in the absence of a native coinage, they should conduct their reckoning in marks, shillings and pence from such early times. With the early mark at 12 shillings, and the mark universally equated to 8 ounces, the ounce was worth 18 pence. This is the origin of the 18 pennylands in an ounceland, as mentioned earlier.

The first record of the mark as a land unit in Shetland is in 1299, and this is also the only mention of pennylands in the Shetland records.[137] It would appear that shortly thereafter the pennyland terminology disappeared completely and marks came into general use. In this particular record there were 8 marks in a pennyland (although a 4-mark pennyland was more common in Orkney).[138] The mark was rented for 1½ meils, which in grain would amount to 9 settens, since the same grain table prevailed in Shetland as in Orkney.[139] These marks were rented for 9 settens, or 9 skat-pennies, so these marks were 9-penny marks. This is the same rental (or nearly so) as that for a mark in Orkney, rented at the end of the fifteenth century, namely 10 or 12 settens, especially if we allow a small increase for the passage of two centuries.

It is rare to find less than half a markland sold or rented. For example, in a sale of 1525, 14 separate plots of ground changed hands, ranging from half a mark to 9 marks, but no unit of less than half a mark appeared anywhere in the transaction.[140] This may be because, in Shetland, the markland was small and quite variable in extent, ranging from half an acre to 2 acres.[141]

Infrequently, we have the square fathom referred to, but land was measured in terms of value, not area, and the declaration that 'each merkland ought to contain 1600 square fathoms', as in the first *Statistical Account*, cannot be substantiated.[142] There was, however, a rather variable land unit called a 'plank' which notionally was the area of a square of side 40 fathoms. With a fathom of 2 ells, the plank was approximately one Scottish acre.[143]

In 1539, a charter gave the annual dues from a piece of land as 5 'packs' (for pack, see below) of 'wadmel' and 2 barrels of butter yearly.[144] Wadmel or wadmal (from the Norse *vadmal*) was a kind of coarse woollen cloth with which Shetlanders could pay part of their skat, and this method of payment was not used in Orkney. In 1601, the sale of 5 marks of land was recorded, which had the annual value of 10 pennies (worth) of butter and 10 pieces of wadmel.[145] Later, in 1661, a charter assessed the annual dues from a piece of land as 5 barrels of butter and 10 packs of wadmel, indicating that this land was worth about twice the land referred to in 1539.[146]

The occurrence of wadmel as an element of payment probably stems from the earlier but similar practice in Iceland and, indeed, throughout most of the Norse possessions, including Greenland and the Faroes. From the beginning of the settlement of Iceland in the ninth century, silver was the main currency, but silver became scarce in the eleventh century, particularly after about 1090 when the whole of Europe was scoured for silver to pay for the enormous costs of the Crusades – a dearth that continued until the 1160s when new mining relieved the situation.[147] Homespun cloth became the main monetary vehicle, with the ell length becoming the monetary unit when referring to cloth.[148] Lengths also became related to weights, with 6 ells equivalent to an 'ounce unit'. The shortest length of cloth tenderable in payment of a debt was 6 ells, and such a piece of cloth was called 'an ounce of wadmel' or 'one law-ounce'. The usual ounce unit of cloth was 6 ells long and 2 ells broad.[149]

It is thought very probable that in these very early times the ounce of wadmel was worth an ounce of silver, but from the time of the earliest Icelandic written sources there had been no fixed relation between the two.[150] In about 1117 or 1118 the ratio of the law-ounce to the silver ounce was 8:1, and by about 1200 it was 7½:1.[151] Silver depreciated relative to homespun in Iceland during the twelfth and thirteenth centuries as a result of the enormously increased German production of that metal, although subsequently its value recovered somewhat.

It will be recalled that a cow was normally worth two ounces of silver. In the Icelandic records we find a cow being valued in ounces of cloth as 16, 15 and 12 ounces (or 96, 90 and 72 ells respectively), so that 8, 7½ and 6 ounces of cloth were equivalent to one ounce of silver, which matches the entries in the table below.[152] It is not therefore surprising to find the Shetlanders

DATE	CLOTH TO SILVER OUNCE RATIO	NUMBER OF ELLS TO THE SILVER OUNCE
*c.*1000	3:1	18
*c.*1100	8:1	48
*c.*1200	7½:1	45
*c.*1300	6:1	36

Table D.5
The value of wadmel cloth, as derived from Icelandic sources.

tendering wadmel as skat when convenient ratios could convert ells into silver equivalent. Table D.5 shows the information that can be extracted from these Icelandic sources.[153]

Shetland ells have been mentioned so far without qualification, but it will be clear from what has been said that those ells were significantly different from either the Scottish or the English ells of 37 and 45 inches respectively. It was a short or 'cut-ell', termed variously as 'cuttels', 'cutteles', 'cutells' or 'cuttils'. The original length of the ell was 18 to 19 inches (457 to 483 mm), increasing with time to 22 to 24 inches (559 to 610 mm). A document dated 1628 (although now believed to have been written in 1640) states that the cuttel wadmel is a Shetland ell, and Brian Smith has argued that it is likely to be related to the Norwegian *'stikke'* of about 22 inches and to the Danish *'Själland'* ell of 25 inches, hence its variability.[154] The word 'cuttel' does not appear in Orkney records, but it shows up in an English statute of 1541: 'The said clothes … shall be folded either in pleights [pleats] or cuttelle as the clothes of all other countries of this realm commonly have been use.'[155] A Scottish Court of Session record of 1598 referred to 'ilk cuttill of thre quarters of an elne Scottis'; hence the cuttell is ¾ × 37 = 27¾ inches.[156] There are also two declarations of the cuttel's length from the early seventeenth century: in the first, the cuttel was 'estimat to the length off a half elne ane nail meassour'; and in the other, 'each cuttill will be half el, half quarter and ane nail in length', giving 20·8 and 25·4 inches respectively.[157] In 1786 Thomas Gifford recorded the length as follows: 'a shilling of wadmill is 6 cuttals or *curtele*, i.e., shortell, a measure containing 24 *canches* [inches] in length.'[158] It cannot be said that the cuttel was a fixed invariable unit.

In the collection known as the 'Old Country Acts' of Shetland, dating from the late seventeenth or early eighteenth century, is to be found a statement that the ell is 37 inches (probably a late Scottish intrusion), but that 'the ell called the Websters ell be 3 feet 4 inches, or 40 inches long, on which only unsco[u]red cloth is measured'.[159] This measure, however, is likely to be linked with the requirements of the cloth regulations discussed in Chapter 3.

In Orkney, neither wool nor cloth was tendered as payment for tax or rent at about 1500, but in Shetland these commodities figure prominently.[160]

Indeed, a separate table was prepared to service wadmel, for we read in the Shetland 1628 Rental:

1 cuttell of wadmel	=	1 Shetland ell of wadmel
6 cuttells of wadmel	=	1 shilling wadmel
10 shillings of wadmel	=	1 pack[161]

From this we see the Shetland shilling wadmel serving the same purpose as the law ounce of Iceland and it is tempting to consider that by this date in the seventeenth century, 10 ounce-units might equal an ounce of silver, which would be reflected in the unit of the pack.

A separate table was also produced for butter and oil (the latter extracted from fish-liver and whale blubber):

FOR BUTTER:	4 marks	=	1 penny butter
	6 penny butters	=	1 lispund
	12 lispunds	=	1 barrel
FOR OIL:	1 Scottish quart	=	1 can
	4 cans	=	1 bull
	9 bulls	=	1 barrel[162]

From this we see that there were 24 marks in a lispund, the same number as for Orkney, but only 12 lispunds to the barrel, instead of the range of between 6 and 15 lispunds encountered during the Pundlar Process. A 'penny butter' was a penny's worth Shetland reckoning. There are 72 penny butters, or pennyworths of butter, in a barrel by Shetland reckoning and a shilling butter was 2 lispunds.[163]

These Shetland 'moneysworths' were established before the devaluation of Scots money had begun and represented the value of that particular quantity of commodity. With progressive devaluation this no longer held, and although the old 'Shetland Reckoning' was retained it was common to find this accompanied by the then equivalent amount of Scots money. Thus, for example, in the rental of 1628 mentioned earlier:

1 lispund of butter	=	6 pence in Shetland payment		
	=	48 shillings Scots	=	4 shillings Sterling
1 ell of wadmel	=	2 pence in Shetland payment		
	=	4 shillings Scots	=	4 pence Sterling[164]

One of a series of complaints registered in 1576 was that if an excess of cloth was delivered to the tax-gatherer, the tenant was paid 2 pence per ell in Scots money; but if the tenant was short, the deficiency was charged at the then going-rate in the market of around 4 shillings per Scots ell.[165]

The Shetland barrel was on occasion called the Bremen barrel, reflecting that the main trade route from the islands was to Germany, Scandinavia and the Baltic generally.[166] Prior to the second half of the sixteenth century, the barrel held 48 cans, but in the 1560s Earl Robert increased the volume of the can by one-third so that the barrel now held only 36 cans of the increased size.[167] The size of the barrel was given as 27 Imperial gallons in the nineteenth century, hence the volume of the small can was 4½ Imperial pints (156·3 in³); that of the later can was 6 Imperial pints (208·4 in³) or 2 Scots pints, representing an increase of one-third.[168] These figures provide further evidence that before the mid-sixteenth century the Scots pint was small. We have shown in Chapter **6** that the volume of the Scots pint was increased from 77·8 in³ to 103·7 in³, also an increase of one-third. Although the large pint was introduced in about 1500, it is likely that the small pint was only eliminated at the Assize of 1555 or that of 1563 (see Chapter 7). Thus the changes made by the earls reflected changing legal requirements and paralleled changes in the general metrology of the Scottish burghs. For the earl to make a render to the crown, he would be expected to pay in the currently-accepted Scottish units, and consequently the earl demanded his dues in like measure. It was his good fortune that his factors could now demand payments that were one-third larger.

Over and above the requisite number of cans in the barrel, a customary allowance had been given to offset the oil which adhered to the sides of the vessel. Before the 1560s this amounted to half a mutchkin, rising thereafter to one mutchkin per can.[169] With 36 cans plus the allowance to the barrel, the allowance was 36 mutchkins, or 9 Scots pints, or 4½ cans, and so the total barrel contents would have been 40½ cans. (This equalled an allowance of 3·38 Imperial gallons for every 30·38.)[170] In 1587, a barrel of oil was worth £14 6s 8d; this is almost exactly 7 shillings per can.[171]

Marks of land were assessed in pennies, ranging from 12 to 4 pence; a mark of 12 pennies was the dearest, one of 4 pennies the cheapest. The notation is perhaps confusing because a mark of 12 pennies or a 12-penny-land is not twelve pennylands. Rental values from the Shetland 1628 Rental in the National Archives of Scotland, extracted from Thomas Gifford's account and expanded by Brian Smith, are shown in Table D.**6**.[172] The land rent (land male) payable annually to heritors and feuars used to be paid in butter and wadmel, but towards the end of the eighteenth century it was paid in butter and money. When an individual was granted a tack or lease, he was required to pay a fee on entering into these newly-acquired territories, called a 'grassum'. As tacks were usually for a three-year period, a further grassum was exacted on re-entry. The grassum began as 2 shillings per markland; later, with the tenant paying nothing at entry, it became an annual payment of 8 shillings per markland.[173]

The explanation of the table is as follows. The first line shows 1 mark of 12 pennies, where the 12 is made up of 4 pennies butter and 8 pennies wadmel, because the wadmel payment was always double that for butter (at least from about 1300 onwards). With 4 marks to the penny butter (see above), 4 penny butters is 16 marks butter. Earlier, it was shown that 1 lispund of butter was worth 48 shillings Scots, and 6 penny butters equals 1 lispund, hence 4 penny

I MARK AT	BUTTER			WADMEL		
	PENNIES BUTTER	MARKS BUTTER	SHILLINGS (SCOTS)	PENNIES WADMEL	CUTTELS WADMEL	SHILLINGS (SCOTS)
12 pennies	4	16	32s	8	4	16s
11 pennies	3⅔	14⅔	29s 4d	7⅓	3⅔	14s 8d
10 pennies	3⅓	13⅓	26s 8d	6⅔	3⅓	13s 4d
9 pennies	3	12	24s	6	3	12s
8 pennies	2⅔	10⅔	21s 4d	5⅓	2⅔	10s 8d
6 pennies	2	8	16s	4	2	8s
5 pennies	1⅔	6⅔	13s 4d	3⅓	1⅔	6s 8d
4 pennies	1⅓	5⅓	10s 8d	2⅔	1⅓	5s 4d

butters is worth 32 shillings. Likewise for wadmel, 1 penny wadmel is 2 shillings Scots, so 8 pennies equals 16 shillings Scots. One cuttel is 2 pennies wadmel so 8 pennies is 4 cuttels. The remainder of the table follows in proportion. One cuttel was sometimes called a groat (4 pence) since 1 cuttel equals 2 Shetland pennies, 4 shillings Scots, or 4 pence Sterling.[174]

To read the records of both Orkney and Shetland one cannot but be struck with the sense of timelessness in it all. We are told that there was little variation in the silver value of agricultural produce in the interval between 1137 and 1502, and none at all in the price of a cow.[175] It has been said that the way of life of Orkney crofters and fishermen of the seventeenth century was not very different from that of their neolithic predecessors.[176] If the terminology of the records appears strange to us today, it is because we see here the fusion of two (or more) cultures with very different outlooks on life than those in rural mainland Scotland, far less those of England. To find a coinless currency measured in pennies in early days is in itself a matter for study and tells us much about the manner in which assessments were paid and trade by barter conducted. Orkney supplied Shetland's need for grain and Shetland in return supplied wool which the Orcadians sold back as woven cloth. The cash received for the cloth was usually foreign coin of German or Dutch origin received in payment from foreign fishing merchants. Sea voyages are recorded in early times to the Hebrides and Ireland and, of course, trade with Norway in grain in the middle ages was important, bringing in return much-needed wood to the islands. The early land descriptions, ounceland and pennyland, once thought to have been of Norse origin, are now taken to have originated with the Scots in early Dalriada.[177] Early valuations remembered today reveal connections to other lands in the Norse sphere of influence and serve as a reminder of the rich heritage possessed by these islands.

Table D.6
Shetland rents from the 1628 Rental (with grassum throughout at 8 shillings Scots annually).

1   For the interpretation of data referring to Orkney and Shetland we are indebted to the following works: Barbara E. Crawford, *Scandinavian Scotland* (Leicester, 1987); William P. L. Thomson, *History of Orkney* (Edinburgh, 1987); Thomson, *The New History of Orkney*, (Edinburgh, 2001); Thomson, 'Ouncelands and Pennylands in Orkney and Shetland', in L. J. MacGregor and B. E. Crawford (ed), *Ouncelands and Pennylands* (St Andrews, 1987), 24-45. Similarly, communications from Brian Smith, Archivist, Shetlands Islands Council, have been of great assistance.

2   This is the traditional view. Modern scholarship suggests a date in the 880s for Hafvsfjord and indicates a confusion of Harald with Magnus Barelegs whose journeys to the islands in 1098 and 1102 are recorded in the Sagas, while those of Harald are not. The Sagas were written down some three centuries after the supposed events so may contain apocryphal elements. For details, see Per Sveaas Andersen, *Samlingen av Norge og Kristningen av Landet* (Bergen, Oslo and Tromsø, 1977), 79-84.

3   Joseph Anderson (ed), *The Orkneyinga Saga* (Edinburgh, 1873; reprinted 1990), 203-4. See also B. E. Crawford, 'Orkney and Caithness', in Peter G. B. McNeill and Hector L. MacQueen (eds), *Atlas of Scottish History to 1707* (Edinburgh, 1996), 448-9.

4   A. A. M. Duncan, *Scotland: The Making of the Kingdom* (Edinburgh, 1975), 195. See also D. D. R. Owen, *William the Lion: Kingship and Culture 1143-1214* (East Linton, 1997), esp. pp. 86-98, and P. Topping, 'Harald Maddadson, earl of Orkney and Caithness, 1139-1206', in *Scottish Historical Review*, 62 (1983), 114-5.

5   Although usually thought of as Norse, both David and John were three-quarters Scottish. See Gordon Donaldson, 'Problems of Sovereignty and Law in Orkney and Shetland', in *Miscellany, Volume II*, Stair Society no. 35 (Edinburgh, 1984), 14.

6   Ibid. For the Earls of Orkney, see articles in the *New Orkney Antiquarian Journal*, 1 (1999): Brian Smith, 'Earl Robert and Earl Patrick in Shetland: Good, Bad or Indifferent?', ibid., 6-16; Denys Pringle, 'The Houses of the Stewart Earls in Orkney and Shetland', ibid., 17-41; and Peter D. Anderson, 'Earl Patrick and his Enemies', ibid., 42-52.

7   Anderson, op. cit. (3), xlvi.

8   Arne O. Johnsen, 'The Payments from the Hebrides and Isle of Man to the Crown of Norway 1152-1263', in *Scottish Historical Review*, 48 (1969), 18-34; see p. 31; Duncan, op. cit. (4), 577-8; Thomson, *New History of Orkney*, op. cit. (1), 138-47.

9   At this time Norway held Orkney and Shetland (the Nordereys), the Inner and Outer Hebrides (the Sudreys), Caithness, and Sutherland (so called as this was the land to the south of their first mainland settlements in Caithness), see Duncan, op. cit. (4), chapter 4 and R. A. McDonald, *The Kingdom of the Isles: Scotland's Western Seaboard c.1100-c.1336* (East Linton, 1997), 21-38.

10   '… dabunt et reddent imperpetuum': Barbara E. Crawford, 'The Pawning of Orkney and Shetland', in *Scottish Historical Review*, 48 (1969), 35-53, p. 35, note 2.

11   Ibid. See also Johnsen, op. cit. (8), 28.

12   T. Thomson and C. Innes (eds), *The Acts of the Parliaments of Scotland [APS]*, 13 volumes (Edinburgh, 1814-75), I, 78.

13   Duncan, op. cit. (4), 578. See also G. W. S. Barrow, 'Scotland from about 842 to 1286', in McNeill and MacQueen, op. cit. (3), 76-7.

14   Johnsen, op. cit. (8), 30 and 30, note 4.

15   Ibid.

16   John Skene, *De Verborum Significatione* (Edinburgh, 1597), s.v. *annuell of Norway*.

17   *APS*, I, 115: 1292.

18   Crawford, op. cit. (10), 36.

19   Ibid.

20   Christian I had acceded to the Danish throne in 1448, to that of Norway in 1449, to that of Sweden in 1457 (although not without continuing opposition). He acquired title to the duchies of Schleswig-Holstein in 1460, but lost it soon after: for background see Steinar Imsen, 'Public Life in Shetland and Orkney, c.1300-1550', in *New Orkney Antiquarian Journal*, 1 (1999), 53-65.

21   Crawford, op. cit. (10), 39.

22   Kai Hørby, 'Christian I and the Pawning of Orkney', in *Scottish Historical Review*, 48 (1969), 54-63; see p. 58; corrected in Richard I. Lustig, 'The Treaty of Perth: a Re-Examination', in ibid., 58 (1979), 35-57, see p. 54, note 5.

23   *APS*, II, 85: Acts 2 and 3.

24   Crawford, op. cit. (10), 52, note 1. For Linlithgow Palace and Doune Castle, see John G. Dunbar, *Scottish Royal Palaces: the Architecture of Royal Palaces during the Late Medieval and Early Renaissance Periods* (East Linton, 1999), 5-21 and 83-7.

25 The text (from British Library Royal MS 18B VI, 13) is given in Crawford, op. cit. (10), 52-3. See also Barbara Crawford, 'The Pledging of the Islands in 1469', in Donald J. Withrington (ed), *Shetland and the Outside World 1469-1969* (Aberdeen and Oxford, 1983), 32-48, see p. 42.

26 Crawford, op. cit. (10), 38.

27 Ranald Nicholson, *Scotland: The Later Middle Ages* (Edinburgh, 1974), 416.

28 Thomas Thomson, *et al.* (eds), *Registrum Magni Sigilli [RMS]*, 11 volumes (Edinburgh, 1882-1914), II, 207-8: nos. 996-1002.

29 *APS*, II, 102: 1472.

30 Duncan, op. cit. (4), 275. See also D. E. R. Watts, 'Ecclesiastical Organisation', in McNeill and MacQueen, op. cit. (3), 336-8.

31 William Jardine Dobie, 'Udal Law', in Hector McKechnie (ed), *An Introductory Survey of the Sources and Literature of Scots Law*, Stair Society no. 1 (Edinburgh, 1936), 445-60; see p. 449.

32 Ibid. See also David Balfour, *Oppressions of the Sixteenth Century in the Islands of Orkney and Zetland*, Maitland Club no. 77, Abbotsford Club no. 31 (Edinburgh, 1859), xi.

33 Thomson, *New History of Orkney*, op. cit. (1), 204 and 262-3.

34 Ibid.; Peter D. Anderson, *Black Patie: the Life and Times of Patrick Stewart, Earl of Orkney, Lord of Shetland* (Edinburgh, 1992), 35, 137-9. For a biography of Patrick's father, Robert, see Peter D. Anderson, *Robert Stewart, Earl of Orkney, Lord of Shetland 1533-1595* (Edinburgh, 1982).

35 Donaldson, op. cit. (5), 18-9, 22; G. Goudie, *The Celtic and Scandinavian Antiquities of Shetland* (Edinburgh, 1904), 217-9.

36 Ibid., 229; Donaldson, op. cit. (5), 19.

37 Dobie, op. cit. (31), 449.

38 Thomson, *New History of Orkney*, op. cit. (1), 201-2; George Barry, *History of Orkney Islands*, second edition (London, 1808), 224.

39 *APS*, II, 9: Act 3, 1425.

40 Ibid., II, 244: Act 27, 1503.

41 Ibid., II, 252: Act 24.

42 Ibid., III, 41: Act 48, 1567.

43 Ibid., V, 55: Act 42, 1633. Sweden had separated from the Scandinavian kingdom in 1523.

44 James Mackenzie, *The General Grievances and Oppressions of the Isles of Orkney and Shetland* (Edinburgh, 1750; reprinted, 1836), xix. See also Donaldson, op. cit. (5), 26.

45 Dobie, op. cit. (31), 451. For udal law, see Crawford, op. cit. (1), 199-203.

46 Dobie, op. cit. (31), 452, note 1.

47 Thomson, *New History of Orkney*, op. cit. (1), 34-5 and 108.

48 Anderson, op. cit. (3), xxiv, 112, 205-6 and appendix 210-11. The Saga stories may have become conflated, and it is possible that the story of the 60 gold marks belongs to a later time (1136): personal communications from William P. L. Thomson, September 1993. See also Thomson, *New History of Orkney*, op. cit. (1), 218. Anderson translates plógsland as ploughland.

49 A. W. Johnston, 'Odal Law in Orkney and Shetland', in *Old-lore Miscellany of Orkney, Shetland, Caithness and Sutherland*, 8 (1920), 47-59, see pp. 49-50; Thomson, *New History of Orkney*, op. cit. (1), 35; Anderson, op. cit. (3), 112.

50 *APS*, IX, 200: Act 61, 1690.

51 Personal communication from William P. L. Thomson, September 1993.

52 Donaldson, op. cit. (5), 34-5.

53 Dobie, op. cit. (31), 449.

54 Thomson, *New History of Orkney*, op. cit. (1), 241.

55 *APS*, V, 53: Act 42, 1633.

56 Alan Small, Charles Thomas and D. M. Wilson, *St. Ninian's Isle and its Treasure*, 2 volumes (Oxford, 1973), I, 149. This gives a clear, concise description of udal law.

57 Michael R. H. Jones, 'Perceptions of Udal Law in Orkney and Shetland', in Doreen J. Waugh and Brian Smith (eds), *Shetland's Northern Links: Language and History* (Edinburgh, 1996), 186-204.

58 Ibid., 186.

59 Ibid., 200.

60 Thomson, *New History of Orkney*, op. cit. (1), 206.

61 William P. L. Thomson, *Lord Henry Sinclair's 1492 Rental of Orkney* (Kirkwall, 1996), xxi, note 128. We are greatly obliged to Mr. Thomson for providing the text of his 'Introduction' in advance of publication. The original Rental document has not survived, but there are manuscript copies in the Orkney Archives, MS D 2/7, and the Mackenzie Manuscript, MS D 8/5. A further copy is in the National Archives of Scotland (NAS): NAS GD 1/236.

62 Alexander Peterkin, *Rentals of the Ancient Earldom and Bishoprick of Orkney* (Edinburgh, 1820), 4; F. W. L. Thomas, 'What is a Pennyland?', in *Proceedings of the Society of Antiquaries of Scotland*, 18 (1884), 253. Peterkin made no mention of the 1492 Rental and perhaps Thomas was following Peterkin in failing to mention this Rental. It was known to James Mackenzie, op. cit. (44), 16, and to David Balfour, *Odal Rights and Feudal Wrongs* (Edinburgh, 1860), 94.

63 Shetland rentals are to be found in NAS GD 1/366/1, c.1500; NAS E 41/7, 1628; NAS RH 9/15/176, 1716-17.

64 Thomson, 'Ouncelands', op. cit. (1), 24.

65 Ibid., 34.

66 Neil Fojut, 'Towards a Geography of Shetland

Brochs', in *Glasgow Archaeological Journal*, 9 (1982), 38-59. However, this may reflect no more than the fact that in Shetland, arable land appears in pockets. With settlement predominantly occurring in areas of arable land, one might expect the same general pattern at different periods. See W. F. H. Nicolaisen, L. J. MacGregor and Allan Small, 'Scandinavian Place-names and Settlements', in McNeill and MacQueen, op. cit. (3), 68-70.

67 Thomson, *New History of Orkney*, op. cit. (1), 214; Alexis R. Easson, 'Ouncelands and Pennylands in the West Highlands and Islands of Scotland', in MacGregor and Crawford, op. cit. (1), 1-11, which shows the same origin for the pennylands and ouncelands in mainland Scotland. The term 'penilands' appears in a Scottish document dated 'earlier than A.D. 1200'. See National Library of Scotland MS Adv., 20.3.9, fo 287.

68 Thomson, 'Ouncelands', op. cit. (1), 25.

69 Ibid., 36-7.

70 Thomas, op. cit. (62), 277.

71 Barry, op. cit. (38), 227-8.

72 George Low, 'The United Parishes of Birsay and Harray', in John Sinclair (ed), *The Statistical Account of Scotland*, 21 volumes (Edinburgh, 1791-9), XIV, 323.

73 Thomson, 'Ouncelands', op. cit. (1), 26, 30.

74 Ibid., (1), 31. The legend of Boniface is given in W. F. Skene, *Celtic Scotland*, 3 volumes (Edinburgh, 1876-80), II, 229-30.

75 Thomson, 'Ouncelands', op. cit. (1), 28.

76 Thomson, op. cit. (61), xxi. We have adopted Thomson's preferred spelling of 'setten' rather than 'setteen', and similarly, 'span' rather than 'spann'.

77 Personal communication from Brian Smith, Archivist, Shetlands Island Council, October 1987; we are indebted to Brian Smith for his assistance with this section dealing with the value of a markland. With a meil worth 6d, there being 6 setteens to the meil, the setteen was a pennyworth: Johnston, op. cit. (49), 55-7.

78 Hugh Marwick, 'Two Orkney letters of AD 1329', in *Orkney Miscellany*, 4 (1957), 49-56.

79 Thomson, 'Ouncelands', op. cit. (1), 26.

80 Thomas, op. cit. (62), 283.

81 Personal communication from Brian Smith, October 1987; and see also Brian Smith, 'Bismars and Pundlars', in *The New Shetlander*, no. 166 (1986), 6.

82 Thomas, op. cit. (62), 270-1.

83 Ibid., 272.

84 Thomson, op. cit. (61), xxi.

85 Thomas, op. cit. (62), 271.

86 Ibid., 272.

87 Ibid., 261, 265.

88 Thomson, *New History of Orkney*, op. cit. (1), 216-17; Thomson, 'Ouncelands', op. cit. (1), 29.

89 In eleventh-century Norway, King Harald Sigurdarson had 240 pence coined from a mark of silver: Jón Jóhannesson (trans. Haraldur Bessason), *Islendinga Saga* (Winnipeg, Manitoba, 1974), 330; see also Pamela Nightingale, 'The Evolution of Weight-Standards and the Creation of New Monetary and Commercial Links in Northern Europe from the Tenth Century to the Twelfth Century', in *Economic History Review*, 38 (1985), 192-209, see p. 199 and Thomson, *New History of Orkney*, op. cit. (1), 216-9, notes that there were 30 pennies minted from the Norse ounce, in about 1030.

90 Thomson, *New History of Orkney*, op. cit. (1), 214.

91 Hugh Marwick, 'Leidang in the West', in *Proceedings of the Orkney Antiquarian Society*, 13 (1934-5), 15-30; Marwick, 'Naval Defence in Norse Scotland', in *Scottish Historical Review*, 28 (1949), 1-11. For the Dalriadic levy, see Appendix B.

92 Personal communication from William P. L. Thomson, September 1993. Per Sveaas Andersen, 'When was Regular Annual Taxation introduced in the Norse Islands of Britain? A Comparative Study of Assessment in North-western Europe', in *Scandinavian Journal of History*, 16 (1991), 73-83.

93 Personal communication, William P. L. Thomson, February 1991.

94 Easson, op. cit. (67), 1 and figure 2.

95 Malcolm Bangor-Jones, 'Pennylands and Ouncelands in Sutherland and Caithness', in MacGregor and Crawford, op. cit. (1), 14-5. See also Crawford, op. cit. (1), 86-91.

96 Thomas, op. cit (62), 261-2.

97 *APS*, I, 670: Act 12. In the Latin version: 'Pro qualibet mensura scilicet sectinge de guello dabit.'

98 Thomson, op. cit. (61), 108 and xi; rent was usually paid in meils of corn and meat.

99 Mackenzie, op. cit. (44), xxv, quoting Richard Gough (ed), *Bibliotheca Topographica Britannica*, 10 volumes (London, 1780-1800), X, 84-5.

100 Skene, op. cit. (16), s.v. *serplaith: weichtes and measures in Orknay*.

101 B. H. Hossack, *Kirkwall in the Orkneys* (Kirkwall, 1900), 411.

102 H. Marwick, 'Old-Time Weights and Measures', in *Proceedings of the Orkney Antiquarian Society*, 15 (1939), 9-13.

103 Thomson, op. cit. (61), xxi.

104 Ibid., xxvi.

105 Skene, op. cit. (16), s.v. *serplaith: weichtes and measures in Orknay'*. See also Habakkuk Bisset (P. J. Hamilton-Grierson [ed]), *Rolment of Courtis*, 3 volumes (Edinburgh, 1922), II, 210; Peterkin, op. cit. (62), glossary, appendix, 135.

It is statute and ordainit that na baxter baik na mayne breid to sell fra hine furthwart [henceforward], saiffing allenarly [only] at Whitsounday, Sanct Geillis messe, Yule, and Pasche; and that the said breid sall nocht be sauld at nane of the said festivall tymes bot endurand aucht days, that is to say begynnand at the evin of ilk [each] ane of the said feists and endurand quhill that day awcht [eight] dayes, and gif any mayne breid be sauld any vther tyme it sall be chete [escheated], and the said baxteris sall nocht bake the said mayne forowtyn pase [without permission] to be given thame be the baillies.[9]

In times of scarcity, manchet (and presumably simnel) was prohibited. Thus, at Edinburgh in October 1595, in a time of dearth, it was declared that the use of mayne-flour and mayne-bread was not economical and led to loss. No baker was to have, bake or sell mayne-flour or bread, and this prohibition was repeated in March 1600 and again in April of that year.[10]

The four early English assizes appear more generous to the bakers, for the baker was allowed four servants and two boys. For each baking of a quarter of wheat he was allowed 3d or 4d, the bran and two loaves as well; his servants were normally paid ½d each as in Scotland, and the boys were paid ⅛d in earlier times and ¼d each later. The English baker was also allowed a total of 4¾d for salt, yeast, wood (fuel), candles and the use of his sieve, none of which was mentioned in the Scottish assizes.[11] However, the English baker had to bake an entire quarter to get his allowance. It is not known accurately what the volume was of an English quarter in the thirteenth century. It will be suggested below that it might have been between 4 and 4½ times that of the contemporary Scottish boll. If so, the baker's money payments were about equal, but the Scottish helpers fared much better than their English counterparts, and there was still no sign of a payment for the Scottish materials apart from the grain. At this time, the penny was worth the same amount in the two countries. Only in 1373 did the Scottish penny slip to being ¾d sterling.[12]

THE ASSIZE OF BREAD

The pivot of the assize table is the weight of wastel (best white) bread given for ¼d (in England) or ½d (in Scotland), which varied in accordance with the price of wheat. As the price of grain rose, the weight of bread for a fixed payment fell. In Scotland the weight of the loaf changed for every 2-penny rise in grain price per boll. In England it changed for every 6-penny rise in the price of grain per quarter (8 bushels).

The first entry in David's Assize of Bread is for a boll of wheat priced at 10d, when the ½d wastel loaf, of white and well-sieved flour, was to weigh 6 marks and 4 shillings (that is, 84 shillings); when well-baked and new it was to weigh 6 marks (that is, 80 shillings); and when well-baked and dry it was to weigh 5 marks, 9 shillings and 4 pence (that is, 76 shillings). This was to apply also when the boll cost 9d and 11d. With wheat at 12d per boll, the ½d wastel loaf of white and well-sieved flour was to weigh 5 marks 40d (that is, 70 shillings); when well-baked, 5 marks (that is, 66⅔ shillings); and when well-baked and dry, 63 shillings and 4 pence (that is, 63⅓ shillings). This was to apply also when the boll cost 11d, and 13d. This pattern was repeated down the table, until wheat is shown at 2 shillings (or 24d) per boll.[13]

Other types of bread in David's Assize were related to the wastel loaf and the pattern closely followed the thirteenth-century English assizes. It may be instructive to compare (see Table E.1) the English assizes for 1266 and 1290, which were identical, with that (probably contemporaneously) attributed to David, for wheat at 12d a quarter and 12d a boll respectively, remembering that the penny had the same value in both kingdoms at this time.[14] The Latin and Scottish texts of David's Assize differ, with the former being the more extensive: the Latin text is therefore the one used for most points here in translation.

It can be seen that there were marked similarities between the Scottish and the English assizes, the only differences being that the one used the boll and the ½d loaf, the other the quarter and the ¼d loaf. Table E.1 shows that for wastel well-baked and dry, probably the most frequently traded kind, the English buyer purchased 4.29 times the weight of bread that the Scottish buyer would get for the same money (½d).[15] It is not known precisely the size of the English quarter of this period, but it is not likely to be far from that of Henry VII of 1497, which was 8 bushels each of 2,145 in³ (see Chapter 6), namely 17,160 in³. With David's boll of 12 gallons, each of 331 in³, the boll was about 3,970 in³, so the ratio of the sizes of the vessels was 4.32 : 1. This is close enough to 4.29 : 1 to suggest that the weight of bread given for a fixed sum was

Table E.1

Comparison between David's Assize of Bread and the English assizes of 1266 and 1290.

DAVID'S ASSIZE OF BREAD	ENGLISH ASSIZES OF 1266 AND 1290
When the Boll of wheat costs 12d:	When the Quarter of wheat costs 12d:

½d wastel loaf of white well-sieved flour weighs
    5 marks 40d (70s),
    well-baked 5 marks (66⅔s),
    well-baked and dry 63s 4d (63⅓s).
	¼d wasted loaf weighs £6 16s (136s)

Cocket loaf weighs more than wastel by 2s	Cocket loaf weighs more than wastel by 2s
Simnel weighs less than wastel by 2s	Simnel weighs less than wastel by 2s
Mixed grains shall weigh 1½ cockets	
Wholemeal shall weigh 1½ cockets	Wholemeal shall weigh 1½ cockets
Treet shall weigh 2 wastels	Treet shall weigh 2 wastels
Bread of all grains shall weigh 2 cockets	'Common wheat' shall weigh 2 wastels

strictly in the inverse proportion to the price of wheat at this time in both kingdoms. This may be further illustrated from David's Assize in Table E.2, where the weight of bread given for ½d is listed for wheat, in the price range 10d to 2 shillings a boll, against the weight of ½d wastel bread ranging from 84 shillings to 35 shillings. Since the ounce of 20d was 450 grains and the shilling therefore 270 grains, these loaf weights are 22,680 and 9,450 grains, amounting to 48 and 20 ounces of Paris weight exactly (ounce of 472½ grains). All other entries in Table E.2 give either a whole number of French ounces, or a whole number plus a recognisable fraction. That these ounces should come to precisely whole numbers, illustrating the use of Paris weight in David's day (or perhaps a century later), supports the contention that David's gallon was based on Paris weight, as discussed in Chapter 4. And, the weight of bread given for ½d can be seen to vary strictly in inverse proportion to the price of the boll of wheat.

The first indication of a change in the policy of controlling bread prices came as late as 1555, when the Privy Council intervened in what had previously been largely a matter for the burghs. The Privy Council concluded that when the boll of wheat sold for 4 marks (4 × 160d), bakers should sell 14 ounces of bread for 4 pennies. Their minute of January 1556 continues:

> And if any man delivers wheat to a Baxter to be baked to him, that he receive seven score [140] pound weight of baked bread good and sufficient stuff for each boll.[16]

It can be seen, therefore, that the cost of the boll was paid for by the sale of 160 loaves of 14 ounces or, alternatively, 140 sixteen-ounce pounds of bread. Fixing this limit of 140 pounds of bread, as a means of linking the price of the raw material to that of the finished product, remained an important part of the bread assize for a considerable time. We can obtain some idea of the change from the baxter's perspective by attempting to recast the David Assize of Bread in the units current in 1556. If, for example, we choose wheat at 10d per boll from Table E.2 (where all the entries are equivalent), the sale of

PRICE OF BOLL OF WHEAT	WEIGHT OF ½d WASTEL BREAD
10d	84s
12d	70s
14d	60s
16d	52½s
18d	46⅔s
20d	42s
22d	38⅛s [to nearest penny]
24d	35s

Table E.2

Weight of wastel bread against price of wheat.

20 half-penny loaves was required to pay for the grain. Each loaf weighed 84 shillings, so 20 loaves weighed 1,680 shillings. David's pound was 25 shillings or 6,750 grains (see Chapter 4), so this weight of bread was 67·2 pounds. But what weight of bread would be required to pay for a grain boll of the size used in 1556? We first need to make some assumptions about the size of boll in the thirteenth to fourteenth-century period to which the figures in the assize related. This might be David's boll of 12 gallons, with the gallon at about 331 in³, which we have argued at the end of Chapter 4 incorporated an initial allowance of a sixteenth; so the boll was about 3,970 in³. In 1556 the customary boll was four times 19¹⁄₁₆ pints (see Chapter 7 and Fig. **7.9**), which would be about 7,570 in³ of grain, or about 1·90 of David's boll. So the number of David pounds required to pay for the grain in 1·90 David bolls was 67·2 × 1·90 = 128 pounds. In terms of the 7,560-grain pounds of 1556, this is about 115 pounds rather than the 140 pounds stipulated in 1556.

Although it is doubtful how far such a comparison can be taken, it is apparent that the earlier profits of the baxters were high (and must have easily met production costs). For comparison, the parallel David Assize of Ale is discussed below; and here concerns for the quality of the ale required a more direct expression of the profit level than was given in the Assize of Bread. It will be seen that the expenses and profit were equivalent to two gallons in twelve, or more specifically that the chalder of malt brewed into 192 gallons, of which 160 paid for the malt, leaving 32 gallons profit and expenses. This profit margin of 20 per cent (also found in the William I version) is closely comparable with that suggested for the early Assize of Bread (assuming the equivalent of the later 140 pounds was actually baked). By the mid-sixteenth century, however, the adjustment of prices implied by the 140-pound limit had squeezed the baxters' profits.

A recent survey of prices, food and wages in early modern Scotland has demonstrated that 'in the case of Edinburgh, it is possible … to derive a series of wheat prices from the known quotations of bread prices … to allow us to chart the relationship between the two from 1495 to 1681'.[17] This appears to be possible only for Edinburgh, amongst the burghs, and the Edinburgh magistrates recorded both the price of wheat and the weight of the 4-penny loaf on four occasions in the mid-sixteenth century, which enables the local weight of bread to be determined necessary to be sold in order to meet the cost of the grain. These were:

7 December 1548:	wheat 36s per boll, the 4d loaf to weigh 20 ounces
11 September 1550:	wheat 36s per boll, the 4d loaf to weigh 20 ounces
11 January 1556:	wheat 53¹⁄₃s per boll, the 4d loaf to weigh 14 ounces
24 February 1556:	wheat 43¹⁄₃s per boll, the 4d loaf to weigh 17 ounces.[18]

From this it follows that the numbers of 4-penny loaves required to be sold to pay for the boll of grain were respectively 108, 108 (each of of 20 ounces),

160 (of 14 ounces) and 130 (of 17 ounces), yielding total weights of bread of 135, 135, 140 and 138⅛ pounds, all of 16 ounces. All were at, or very near, the figure of 140 pounds decreed by the Privy Council. A recent study of medieval Aberdeen bread assizes has concluded that over a certain grain price, 'the Aberdeen authorities seem to have expected a heavier penny loaf than that demanded in Edinburgh under the Privy Council system'.[19]

Alexander Hunter's quasi-official *A Treatise of Weights, Mets and Measures of Scotland,* printed in 1624, reported an assize of bread based on a trial by the Edinburgh magistrates in 1556, and he reproduced a tabulation of the results for wheaten bread, with wheat prices per boll ranging from £4 to £20 for loaves costing 2 shillings, and 18, 16, 12, 8 and 6 pennies. (Hunter's weights are in pounds, ounces, drops and grains, not money weight. The grains used by him were in terms of Scots weight, where 36 grains equals 1 drop, 16 drops equals 1 ounce, and 16 ounces equals 1 pound of 9,216 Scots grains.) The table is strictly proportional in all respects. For a doubling of the price of wheat, the weights of all the loaves were halved; and for a given price of wheat the weights of the various loaves were strictly inversely proportional to the price. A few excerpts (Table E.3) will illustrate this.

PRICE OF WHEAT PER BOLL	2s BREAD	12d BREAD	6d BREAD
£4	3 lb 8 oz	1 lb 12 oz	14 oz
£8	1 lb 12 oz	14 oz	7 oz
£12	1 lb 2 oz 10 dr 24 gr	9 oz 5 dr 12 gr	4 oz 10 dr 24 gr
£16	14 oz	7 oz	3 oz 8 dr
£20	11 oz 3 dr 7 gr	5 oz 9 dr 21 gr	2 oz 12 dr 28 gr

Here again it is demonstrated that the sale of 140 pounds of bread was necessary to meet the cost of the grain. Taking any entry, say wheat at £16 per boll, and the 2-shilling loaf weighing 14 ounces, meeting the wheat price required the sale of 160 loaves, each 14 ounces, totalling 140 sixteen-ounce pounds.

In all of the above, the sale of 140 pounds of baked bread appears to generate precisely the 'cost of a boll of wheat'. Were the boll to *bake* into only 140 pounds of bread the baker would be in an impossible position, for the Scottish assizes stated no pecuniary allowances for the baker for salaries, fuel, materials such as salt and yeast, nor any 'free' bread as in the English assizes.[20] So it must be asked: did the boll in fact only yield 140 pounds of bread?

Alexander Hunter indicated that the boll of wheat yielded a greater quantity of bread:

Table E.3
Sample wheat prices and bread weights, extracted from Alexander Hunter's 'Table to finde out the weight of Wheat Bread', 1624.

But now the said Counsell [of Edinburgh] finding that albeit some of
the Bakers makes better bread than the rest: yet the best bread is not of
that finenesse, that was ordained by that ordinance: and therefore are of
intention to make new trialls: … which trialls being made and reported,
I thinke that they will finde that the Bowe of wheate, may render a greater
quantitie of bread, then is set downe in the said ordinance.[21]

As in England, the assize was set to give a modest excess and, again as in
the early English assizes, the sale of the prescribed amount of bread more
than paid for the grain. It also paid for all the costs of turning the grain into
bread which are mentioned above. Even in the early *Leges Quatuor Burgorum*
and in the *Iter Camerarii,* the bakers' allowances were given, so it would be
expected that something similar would occur in the later Scottish legislation.

The text of the Privy Council minute of 1556 stated that in exchange for a
boll of wheat, the donor would be *given* 140 pounds of bread, in exchange for
the wheat, *not* that the boll bakes into 140 pounds. Thus the retail value of
140 pounds of bread bought the grain, leaving the excess as the baker's profit.
This is reminiscent of bullion being delivered to the English mint and 240
pennies of alloyed silver being given for each pound of pure silver brought
in.[22] The second point is that when the assize tables refer to the 'price of the
boll of wheat', the meaning is the 'price of the boll of wheat plus all production costs'.

Hunter, in stating that his bread table had been drawn up in 1597 for the
bailies of Edinburgh, based on their 1556 trial, wrote:

… the counsell of the said Burgh in Anno 1555: who (after good consideration of the labour and all charges needefull to bee allowed and deduced
to the Bakers:) concluded that there should bee made 140 poundes weight
of very fine wheate bread out of everie Bow of wheate.[23]

This means that after meeting all the costs of production, the milling of the
grain, wages, fuel, and so on, there remained still 140 pounds of bread to pay
for the grain. This presents a much more realistic picture of the functioning
of the industry.

After the Union of 1707, statutes enacted by the Westminster Parliament
affecting Scotland were acts passed by 'the Commons of Great Britain', or
'the Parliament of Great Britain'. The first assize of this new era was that of
1710, whose greatest change was to record bread weight in avoirdupois
units.[24] By this time avoirdupois weights had been imposed on Scotland, but
they were far from being universal. Consignments of standards for these
weights, for English bushels and for other units, were sent to Scotland subsequent to the Union, and their distribution to the burghs is described in
Chapter **9**.

A later assize of 1757 was more explicit, referring to 'the Price of the Bushel
of Wheat *and Baking'*, while the preamble stated that the price in the table
was that of a Winchester bushel of wheat, ranging from 2s 9d to 14s 6d per

bushel, with the allowance of the magistrates or justices to the baxter for baking being included. A worked example accompanying a table in the legislation described how, if the price of wheat in the market was 5s a bushel and 1s 6d was allowed for baking, 6s 6d should be found in the table and used to get the weights of the various loaves.

This assize put its finger on a difficulty. It was the price of grain, not the price of flour, that was used to construct the table. A bushel of wheat does not generate a bushel of flour, and the assize stated that

> ... no provision being made how they should know what price the respective sorts of meal and flour should be esteemed to bear in proportion to the price of wheat, they are therefore to take notice that every sack of meal or flour is to weigh 2 cwt 2 qrs net [namely, 280 pounds avoirdupois weight] and from every sack of meal or flour on average 20 such peck loaves.[25]

Peck loaves, half-peck and quarter-peck (quartern) loaves were new to legislation for Scotland. These were the loaves baked from these measures of flour, and were sold at prices set by the price of flour, together with the baking costs and so on. The loaves were called 'Prized Bread'; it was the price of the loaf of fixed weight which varied, whereas the old 'Bread of Assize' gave a loaf of variable weight for a fixed sum, such as a half-penny or a penny. The peck loaf was baked from a peck of flour; and since there were, by definition, 4 pecks to the bushel, it follows that the 280-pound avoirdupois sack of flour constituted 5 bushels, and the bushel of flour was 56 pounds. The 1757 Assize stipulated that each peck loaf weighed 17 pounds 6 ounces, so the bushel of flour baked into four times this weight, or 69½ pounds of wheaten bread. Thus, 1 pound of flour produced 1·24 pounds of bread.

Taking the worked example of the statute, with wheat at 5 shillings a bushel and 1s 6d baking costs, where the table states 6s 6d, we find the peck (wheaten) loaf is to be sold for 2s 6d. The bushel of flour baked into 4 peck loaves, and sold for 10 shillings. The baker would recover the cost of grain and his 1s 6d allowance from the sale of 2·6 peck loaves and still have 3s 6d worth of bread (1·4 peck loaves) extra. Grain and baking allowances together (6s 6d) accounted for 65 per cent of the final worth of the bread (10 shillings). This percentage was close to that recorded in England for the period 1600 to 1684 by John Powell, Clerk of the Market of the Royal Household under Elizabeth I and James I, who declared that when wheat cost 12 shillings per quarter, the baker was to bake at 18 shillings.[26] That is, the grain and baking cost is 66⅔ per cent of the money yield of the quarter.

The year 1757 was a period of dearth, with bread being made from a variety of other grains besides wheat. The second part of this assize dealt with breads made from rye, barley, oats, beans and 'maslin' (defined as two-thirds wheat and one-third rye), and all the produce of the grain was to be used except the bran and hull (or husks). A later act of 1801 (a time of desperate

Fig. E.1
Official control of bread
prices continued well into the
nineteenth century. These
two weights, made into the
shape of loaves of bread, are
standards of the Dundee
Guildry of 1826, giving the
weights of the quartern and
half-quartern loaf in terms of
avoirdupois weight.

scarcity), raised to 100 per cent the extraction rate which had previously been over 90 per cent, and it permitted bread of bran only, or bran and 'pollards' (bran flour or bran mixed with flour).[27]

Matters improved after 1757, for in 1779 John Swinton gave an abbreviated assize table, in which he considered one kind of (unspecified) bread made from wheat of a medium weight of 57 pounds avoirdupois per bushel.[28] Taking the bushel, he declared that the 57 pounds of wheat generated 42 pounds of flour, and 52 pounds 2 ounces of bread. This again shows that 1 pound of flour baked into 1·24 pounds of bread, with a more modest extraction rate of 74 per cent. Instead of the four peck loaves, or 16 quartern loaves made from a bushel in 1757, in a time of scarcity, Swinton specified 12 quartern loaves, indicating that by 1779 the yield of flour from a bushel of wheat was being restricted to 74 to 75 per cent, while that of 1757 was over 90 per cent.

Unfortunately Swinton gave no price for the loaf, but he did show a baker's allowance of 1s 6d per bushel, as in 1757. He stated that a boll of wheat weighing 232 pounds 15 ounces (avoirdupois) milled into 171 pounds 10 ounces of flour, which in turn baked into 213 pounds of bread. The baker's allowance which Swinton gave was now 6s 1½d per boll (which is the same as 1s 6d per bushel, because the table shows the conversion of 4·09 Winchester bushels to one 'standard' boll).[29] Thus, once again, 1 pound of flour baked into 1·24 pounds of bread.[30]

It will be clear that those assizes which give a strictly inverse proportional weight of bread to the cost of grain or grain plus baking costs are to some extent flawed, for the baking costs do not follow the price of grain. Indeed, the evidence is that as grain prices fluctuated, especially in the short term, the baking allowance remained fixed. The weights given in the assize table can only be approximate, even although the baker was liable to a fine or worse if he was caught breaking the assize. It would be very hard, even with modern technology, to create loaves of precisely the required weight, especially when these changed so frequently. It would seem that shortly after Hunter published his *Treatise*, the myth of proportionality was laid to rest. In the records of the Burgh of Stirling for June 1638 there is a table (Table E.4) for oat bread weights and prices inserted in the minutes.[31]

PRICE OF OATMEAL PER BOLL	12d BAPS SHOULD WEIGH*	10d BAPS SHOULD WEIGH*	8d BREAD SHOULD WEIGH	6d BREAD SHOULD WEIGH
£5	16 oz trois	14 oz trois	11¼ oz trois	10 oz trois
£6	14	12	10	9¼**
£7	11½	9¼	8	7½
£8	9½	7 [*recte* 7½]***	7½ [*recte* 7]***	6
£9	8¼	6 [*recte* 6½]***	6½ [*recte* 6]***	5¼
£10	7½	5½ [*recte* 6½]****	5¼	4¾

   * a bap roll is a bread roll
  ** clearly this should be 9 or 8½
 *** it is obvious that these entries in the two columns have been transposed
**** this entry should probably be 6½

Table E.4
Oat bread prices, Stirling, 1638.

The first point to make is that this table lacks any proportionality. When meal doubled in price the loaf was less than half the previous weight, and the ratio was not even constant for the loaves of different prices. When meal cost £10 the boll, loaves selling at both 8d and 10d weighed the same, according to the table: so there must be at least some scribal errors, possibly those suggested in the table. However, in addition, for meal at a fixed price, the weights of the loaves are not proportional to their prices as might be expected, other things being equal. The cheaper loaves weigh relatively more. A little calculation shows that, although there is a fair amount of variation across the table, an average of about 102 pounds of bread sold would pay for the meal, and not the 140 pounds of the bread regulations. The table would appear to have been based on the notion that a boll of meal *baked into* 140 pounds of bread, as this would yield a margin of profit to the baker which is met with elsewhere. If there were 140 pounds of bread generated, and around 102 pounds paid for the grain as set out in the table, then the cost of the meal was about 75 per cent of the return on all the bread when sold, but it must be borne in mind that the tentative nature of the corrections to the table reduces the reliance that can be placed on this data.

It is probable that it was this Stirling data that the economic historians Edgar Lythe and John Butt were referring to when they wrote that, 'according to an analysis of baking costs published in 1638, [grain] represented about three-quarters of the final cost of the loaf, [so] it would be reasonable to expect that bread prices would be highly sensitive to the price of grain'.[32] The non-proportionality in even those parts of the table where proportionality is to be expected was probably due to the intervention of the magistrates; and Lythe and Butt have pointed out the effectiveness of this control in Stirling between 1600 and 1625 when, although wheat fluctuated between £6 and £12 per boll, the weight of the 12d loaf neither rose above 16 ounces nor fell below 12 ounces.[33] A very similar pattern is revealed in the graph (Fig. E.2) of the Edinburgh prices for bread. This stabilising of the price of bread in the face of market trends meant, of course, that the bakers' profits would fluctuate accordingly.

Oat cakes or oat bread was also referred to at Paisley in the early seventeenth century. In October 1603 it was ordained:

> … that the quarter kaik of gude and sufficient aetmeil whereof there sall be onlie fyve kaiks in the peck and ilk [each] kaik three quarters allenarlie [only] be sauld for vi pennies the quarter.[34]

The Paisley report may be taken to mean that the peck baked into five cakes; each cake, of course, consisted of four quarters, as always, with three quarters being sold at 6d each to defray the meal cost. This preserved the figure of 75 per cent of the cost of the finished product being due to the meal. The whole peck sold for 10 shillings. At the same time in Aberdeen, in December 1602, it was recorded that 10 (presumably smaller) cakes were to be baked 'in pect meill' (from a peck of meal), each cake to contain 15 ounces. In July 1603, the cake now being required to weigh 16 ounces, ten bakewives were each fined 6s 8d for selling cake dearer than allowed.[35] Presumably, the bakewives had continued to give only 15 ounces in each cake.

The Aberdeen record shows the peck baked into 150 or 160 ounces; and the peck sold for 10 shillings using the Paisley price. So, approximately 10 pounds of oatcake sold for 10 shillings; thus one pound sold for 12d. From the Stirling oatbread table (Table E.4), a 12d unit is not an unreasonable price for oatmeal at £5 the boll, although we have to qualify this by noting that the Stirling table is some 28 years later than the Aberdeen and Paisley data. Such burgh records for the price of oats that survive shows the price to be very variable.[36] For instance, on 10 November 1592, Edinburgh's records show a peck costing 3 shillings, or 48 shillings a boll, and on 13 October 1598 at Stirling the boll cost 60 shillings.[37] However, recent work in economic history has determined that '[Rosalind] Mitchison's conclusion that anything resembling a national oatmeal market did not develop until the 1730s must be amended … it is quite feasible that much important headway in the integration of local markets had been made prior to 1630'.[38]

The bakers were not universally happy with these regulations imposed upon them, which made their profit depend on how the assize was set relative to grain prices, and with these prices controlling the retail price of bread through the apparent convention that the cost of the grain should be about 75 per cent of the total money yield. However, this was not always adequate. A dispute at Dundee came to a head in May 1707 when John Taylor, deacon of the bakers, demanded that the assize be altered, the bakers threatening to go on strike and bake no bread until this was done. The town council at Dundee promptly fined Taylor £10 Scots and ordered him to pay, or stay in prison until he had paid the fine. The council also ordered the bakers to serve the lieges with bread, or otherwise they would enable others to do so.[39] This revolt failed to raise the assize, but the bakers returned in May 1734, requesting the assize to be set at £9 for the boll of wheat, and this time the council conceded an increase to £8 10s.[40] Then, in October 1741, the price of wheat being £8 and the bakers' allowance £2, the council agreed for a period of three months to permit bread at £10 the boll and all bread was to conform to that assize.[41] Here, the component of the grain cost in the final price had risen to 80 per cent.

In his 1624 book Hunter declared that 'if the Meale were solde also by weighte, it might prove proffitable to the Leiges',[42] but he would have had to wait some 50 years to see changes in the way meal was to be measured. Not until December 1679 did the Register of the Privy Council state that

> … wee … have ordained that in time coming the constant rule of selling meall shall be by weight and not by measure, and that in place of a boll measure the same be sold and received at the weigh of eight stone troas weight. And the saids magistrats of Linlithgow are to make a standard conforme …[43]

We have reason to believe this was implemented fairly rapidly, although we can find no 8-stone weights and the confirming parliamentary act came as much as 17 years later. This act of September 1696 reads in part:

> … considering the many abuses committed in the measuring of meal in several parts of this Kingdom and that these abuses may be obviat by selling of meal by weight do for remeid thereof statute and ordain that from and after the first day of January nixt to come, all sorts of meal bought and sold within this Kingdom shall be sold and delivered by weight at eight stone Trois weight in place of the boll of Linlithgow measure …[44]

Eight Scots trois or troye stone is 128 troye pounds, equivalent to 139·2 avoirdupois pounds, often rounded up to 140 avoirdupois pounds (see Chapter **8**). It will be clear from the statute that this weight of meal was intended to be a good match for a Linlithgow boll for meal. Although meal had been measured previously by the wheat firlot, oats had been measured by the larger barley firlot, as discussed in Chapter **6**. The advocate John

Swinton reported two instances of counties claiming that the boll of oats would generate a boll of meal, but with the oat boll size depending on the quality of grain.[45] This is a rather different definition from that of the Privy Council and the 1696 Act and it may perhaps have reflected a convention of the following century.

We can get a feel for the accuracy of such statements from a table published by Swinton of the weights of various qualities of oats and the meal that could be ground from it, which was said to have been 'founded upon twenty years of experience of a gentleman in Aberdeenshire'.[46] This tabulated the weights of Winchester bushels of oats against avoirdupois pounds of meal, but also the Scots troye equivalent of bolls of oats against troye pounds of meal. Taking a comparatively good quality of grain that generated the required 8 troye stone of meal (139 avoirdupois pounds) indicated that the large boll would contain 219 troye pounds of oats, which converts to 239 avoirdupois pounds. Whereas Swinton's standard wheat boll, for meal, was about 8,790 in³, the larger standard barley boll, for oats, was about 12,820 in³.[47] This allows us to calculate the bulk density for oats as 32·2 pounds per cubic foot, and for meal as 27·3 pounds/ft³. These appear to compare well with modern values of about 33 pounds/ft³ for oats and between 21 and 29 pounds/ft³ (depending on the size of the flake) for meal, and they bear out the approximate truth of both the 1696 Act and the eighteenth-century convention.[48] It is remarkable how little meal was generated from a given weight of oats and it is not easy to specify the weight of a given volume with any accuracy, for everything depended on the coarseness (or fineness) of the meal.

Swinton listed the measures used in the various Scottish counties for meal, and most of his returns reported 'a boll of 8 stone Dutch weight', or simply, 'a boll by Dutch weight'. (Dutch weight in Scotland is discussed in Chapter **8**.) A few examples measured 8½ or 9 stone, but in Nairn the boll was shown to be 9½ stone. A few counties still recorded their measure by boll. As for oats measured in firlots, Swinton recorded all counties as having their measures as little larger than what he termed the Scottish standard. Seventeen were less than 10 per cent in excess, although those of three counties in south-west Scotland were so much larger that they were clearly based on different multiples or heaping traditions.[49] As Gibson and Smout comment, 'the national standards were hardly immune from either re-interpretation or mis-interpretation, and the same processes must have been working on the presumably much less closely managed local standards'.[50] However, we have argued at the end of Chapter 7 that the sizes of the firlots across the east central belt of Scotland shared remarkable stability and uniformity.

In 1829 George Buchanan recorded 13 Scottish counties trading meal by the boll of 8 stone Dutch troye weight, and only three – Inverness, Nairn and Ross – using a boll of 9 stones.[51] This may perhaps represent yet another instance of dealers taking an extra eighth over and above the regulatory figure (see Chapter **6**). The Poor Law Enquiry (Scotland) of 1844 recorded an enormous amount of information connected with the measures and prices of meal, potatoes, coal and other fuel. About half of the parishes responding stated that the meal boll was 140 pounds avoirdupois, and the other half used twice this amount, called a 'load' of meal of 280 pounds, sometimes referred to as 20 stone avoirdupois (there being 14 pounds to the avoirdupois stone).[52]

BAKERS AND BREAD PRICES

To be a burgess brought certain privileges, not the least of which was a favourable assize of bread, which set bread prices that did not reflect commercial pressure from the bakers. Those who were from outside the burgh, called 'strangers' or 'unfreemen', had to provide a heavier loaf than was required of the baker within the burgh for a given amount of money. For instance, in Edinburgh, in October 1518, 'the iiij d. [4d] laif of qhheitt breid sawld be the said unfremen sall wey iiij unces mair than the iiij d. laif baikin within this burgh'.[53] In October 1536 the two-penny loaf was to weigh 24 ounces, but bakers based outside the burgh had to provide 26 ounces for the same money.[54] The excess to be provided by strangers often amounted to one ounce for each pennyworth of bread, as was demanded by the burgh in June 1547 when the 4d loaf was to weigh 20 and 24 ounces respectively, and again a year later in June 1548, when a loaf of the same price was to weigh 22 and 26 ounces.[55] This simple rule broke down in the 1590s during times of great privation. For example, in August 1599, the 12d loaf of a burgess was to weigh 13 ounces, and that of a stranger 16 ounces; and in November 1599, the 16d loaf of the burgh was to weigh 20 ounces, while the stranger had to provide one of 24 ounces.[56] By May 1600, the 16d loaf of town bread weighed 18 ounces while 'outland' bread had to weigh 22 ounces; and by December 1602, 16d town bread weighed 14 ounces, and outland bread 17 ounces.[57] There seems to have been no simple rule being observed; but for loaves larger than the 2d loaf, the weight differential was usually either 3 ounces, or 4 ounces in the loaf priced anywhere between 4d and 16d.

The few examples given above for town bread show that the sixteenth century was a time of spiralling prices. Depreciation of the silver content of the coinage was one factor which contributed in large measure to this. A second was the continually increasing size of the firlot itself, which is discussed below. The coinage was debased at least six times between 1540 and 1600, its value falling relative to the English penny from a quarter to a twelfth, and the price of silver doubled between 1544 and 1580.[58] From the figures given below it would appear that the English penny was taken as a standard in Scotland even though this was itself far from being rock steady.[59]

Every few years the new number of Scottish pennies equal to a sterling penny grew: the two were the same until 1355, but by 1601 there were 12 Scots pennies to the sterling.[60] These perturbations led to marked oscillations in prices from year to year and even from month to month, and were aggravated by crop failures and scarcities.

The trend of bread prices in Edinburgh from 1493 until 1653, from the published extracts from the burgh records, has been examined. Initially loaves were priced at 1d or 2d (Scots), then at 4d for a long period until these became inconveniently small. Thereafter, we have 6d, 8d and 12d loaves. The violent short-term fluctuations in the price of bread tends to obscure the upward trend in prices, but this can be made clearer if we graph the price (in Scots money) of 16 ounces of best wheaten bread for the period in question on a semi-logarithmic scale (see Fig E.2).

From 1510-94 there was an approximately exponential nine-fold increase in bread prices, followed by a sudden jump in price from 9 to 14 pence for the one-pound loaf. This reflects the effects of the 1595 famine. The years of scarcity 1550-53 are also indicated on the graph by the violent price

Fig. E.2

Edinburgh bread prices 1493-1653. Periods of dearth or bad harvests are shown.

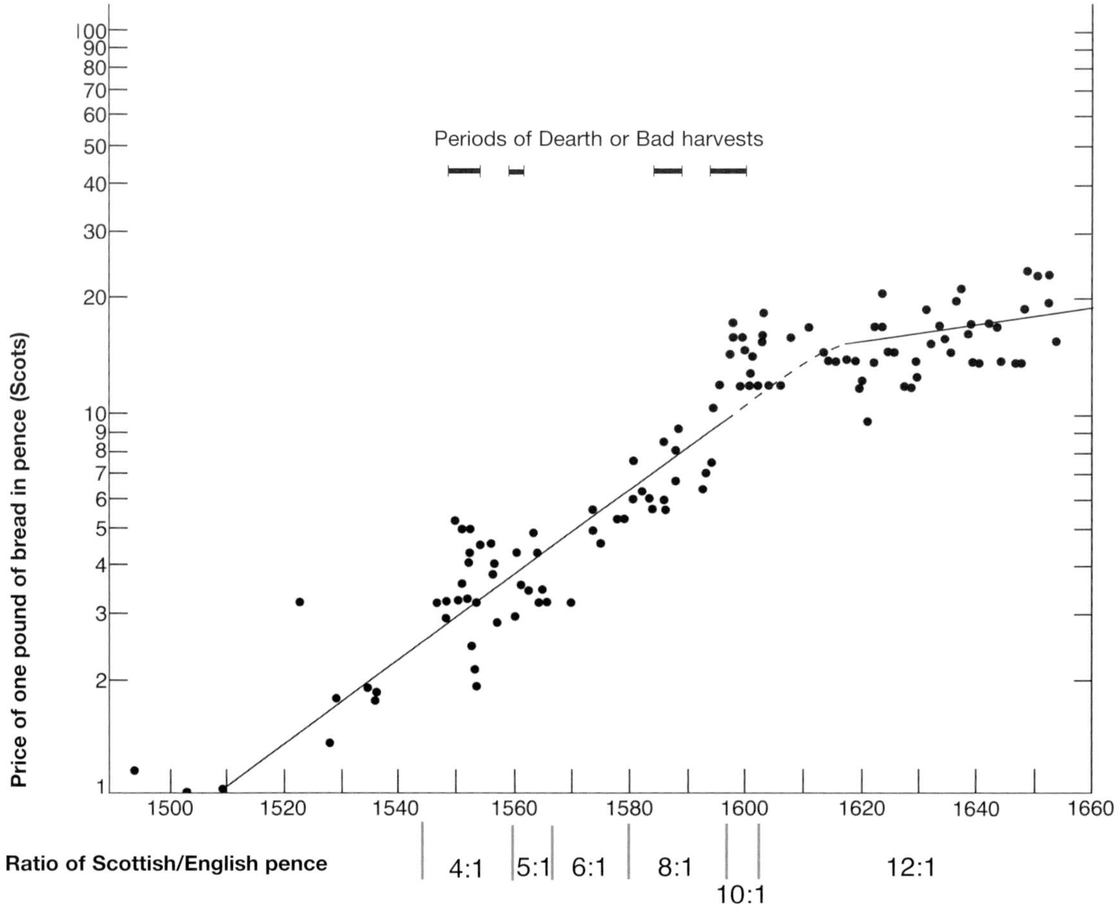

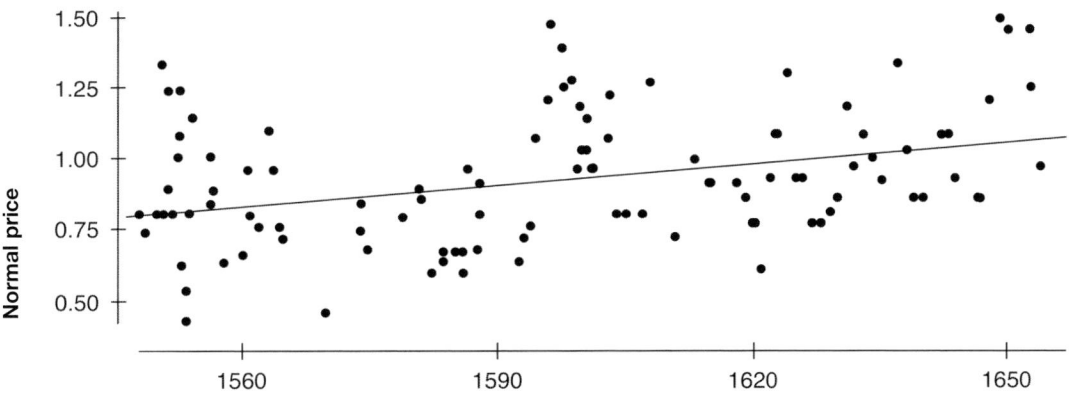

Fig E.3
Statistical analysis of the
log$_{10}$ of Edinburgh bread
prices 1547-1653 in English
sterling, normalised to a
constant firlot as at 1547.
The regression line is given
by price = 0·794 + 0·0025
(Year − 1547).

fluctuations which greatly exceed anything encountered until about 1595. From 1600 the overall price rise is modest, the increase being on average from about 14d to 18d over the succeeding 60 years and continuing to the end of the century.

The conclusions given above for the prices of bread are in accord with the views of modern Scottish historians.[61] This period has been summarised by Christopher Smout as follows:

> The second half of the sixteenth century was a terrible time: twenty-four years out of fifty were marked by abnormally high food prices. Though not every one of these amounted to a famine, there were many seasons in the 1570s and 1590s that saw men dying of starvation. The first half of the seventeenth century was rather better – there were seventeen dearth years before 1660, many of them concentrated in the period between 1630 and 1650. Then apparently quite suddenly there was an improvement. Between 1660 and 1695 only four seasons were marked by high prices, and the rest were distinguished by remarkably low grain prices ... The peasants in time of scarcity attempted first to fall back on other goods. The Highlanders tried to live off cheese and the Lowlanders off herring.[62]

A totally different picture arises if the price of a pound of bread is given in terms of English sterling normalised to a constant size of firlot (that is, to the size of the firlot at the beginning of the period in question), for throughout this period the value of Scots money depreciated relative to sterling while the firlot increased in size and so would cost more to buy as time went on. A statistical analysis was undertaken of the data from 1547 to 1653 since the earlier information was sparse (see Fig. E.3).[63] Looking at the logarithmic graph, it can be seen that when prices are given in sterling for a constant firlot size, only a gentle rise is indicated from about 0·80 in 1547 to 1·06 in 1655. The enormous price rise shown in Scottish currency indicates that it was largely due to the debasement of the coinage, and only in a very secondary way to the change in the size of the measures, which amounted only to about 20 per cent overall.

The sixteenth-century records of the Scottish burghs contain many declarations concerning the price of bread. These were frequently accompanied by
penalties for breaking the terms of the assize, which usually ranged from
seizure of the bread (this would be given to the poor), to a simple fine. So, for
instance, at Edinburgh in July 1531 David Gillaspye was fined £10 Scots for
baking bread found to be insufficient both in weight and in fineness of the
flour.[64] The Town Council enacted in October 1551 that

> … the iiij. d. laif wey xx uns [20 ounces], vnder the pane of xl s. [40s] for
> the first falt, v li. [£5] for the secund falt, and the third spanyng of thame
> fra the occupatioun [the baker shall be barred from the occupation], and
> that al baxtaris [bakers] within this burgh haue thair mark vpoun thair
> awin breid, togidder with sa mony hoillis [indentations, holes] besyde the
> samyn as the breid will cost pennyis …[65]

At Stirling the penalties described for Edinburgh were those normally
enacted, but occasionally the magistrates' patience became exhausted. The
entry for 20 March 1607 in the Stirling record states that

> The baillies and counsall, convenit, declaires thair will on Laurence
> Thomsoun, baxter and his complices as followis, viz., that on ane mercatt
> day the said Laurence salbe cariet [carried] in ane sled throw the toun,
> beir futtit and beir heidit [bare-footed and bare-headed], with ane quhite
> sark [white shirt] on him and ane paper on his heid beiring the caus of his
> punishment, and the rest of his complices to gang beir futtit and beir
> heidit throw the toun, and thaireftir to be brocht and bund [tied] at the
> croce, to stand thair during the baillies will.[66]

This has a familiar ring to it, for it is the punishment meted out to bakers
in fourteenth- and fifteenth-century London for giving short weight.[67]

It would appear that the last sentence taken to the hurdle in London was
in October 1438, so it seems that the Scots kept this penalty for a further two
centuries.[68] The Stirling record shows yet again that enforcement of the assize
was very much in the hands of the bailies, with only intermittent interven-
tions by Parliament or Privy Council in times of difficulty.

THE ASSIZE OF ALE

That ale was the subject of an assize in early times is clear from the *Leges
Quatuor Burgorum*, the *Iter Camerarii* and from the assizes attributed to
King David I and William the Lion.[69] In the *Leges* two clauses describe the
need to ensure the quality and availability of supply:

Of the maner of ale brewing be assise

Quhat woman that wil brew ale to sell sall brew al the yhere [year] thruch eftir the custume of the toune And gif scho [she] dois nocht scho sal be suspendyt of hir office be the space of a yhere and a day And scho sall mak gud ale and approbabill as the tym askeis [acceptable according to the times] And gif scho makis ivil ale and dois agane the custume of the towne and be convykkyt of it scho sall gif till hir mercyment viii s. or than thole the lauch of the toune that is to say be put on the kukstule and the ale sall be geyffin to the pure folk the tua part and the thryd part send to the brethyr of the hospitale And rycht sic dome sal be done of meide [mead] as of ale And ilke broustare sal put hir alewande ututh hir house at hir window or abone hir dur that it may be seabill communly [generally seen] til al men the quhilk gif scho dois nocht scho sal pay for hir defalt iiij d.

Of sellaris of met [meat] and drynk

All broustaris the quhilkis sellis ale and thai that sellis brede or flesche or fysche and all hukstaris [middle-men] the quhilkis byis and sellis communly sal sell til al men als well gangand as command [leaving and entering the burgh] quhat somevir and thai sall halde na mare in thair house to the oyse [use] of thair hushalde gif that ony man wil by it bot to the valur of iiii d. oure nycht and al the layss [remainder] sal be common til al maner of man passand and cummand for thair payment And quha dois the contrare of this and tharof be convyct he sal pay to his forfalt viii s.[70]

The *Iter Camerarii* is quite specific. An assize was to be held by the bailies of the burgh weekly. Ale inspectors (also known as cunsters or conners) had been appointed under oath and were to perform their duties without fear or favour.[71]

Cosmo Innes included a short version of an ale assize attributed to King David in the Record Edition of the Acts, with a brief Scots text, accompanied by a more extended text in Latin. In the first part, up to the chalder costing 20 shillings, the Latin and Scots affirm each other, after which they diverge with the Latin being the more reliable. The amended Scots text reads:

Off the assise of aill eftir the malt ·

Qwhen the chalder of malt is sauld for half a merk thane sall the galloun of aill be sald for a halfpeny · Quhen the chalder is sald for · x · s · the galloun salbe sald for three ferdingis · Quhen the chalder is sald for a merk the galloun salbe at a peny · [Quhen] the chalder sald for · xvj · s · [and 8d] than the galloun salbe at a peny and a ferding · Quhen the chalder is sauld for · xx · s · the galloun salbe at iij halpenijs · quhen the chalder is sauld at · xxiij · s · [and 4d?] the galloun salbe at a peny and · iij · ferdingis · quhen it is sauld for · xxvj · s · the galloun salbe at · ij · penijs · And sua ay vp ascendand eftir the custum of the burcht ·

After this, the Latin text continues (in translation):

> And it should be known that the said Lord, King David, used to receive at his buttery from his brewers who brewed the grain of his own lands at least ten gallons of ale for every boll of good malt and at least one hundred and sixty gallons of ale from every chalder of good malt. Of which two parts of the said ale were best ale which was called knights' ale and the third part of the said ale was of smaller worth and was called attendants' ale.
>
> It is stipulated by the said Lord, King David that every male or female brewer making ale for sale should get from each boll of good malt no less than twelve gallons of tavern ale. In this way each of the male or female brewers who brew ale for sale would be able to have two gallons of ale (or its value) in payment for his or her expenses and profit.[72]

This addition tells us that the chalder was 16 bolls in early times, just as it was later, and since the boll itself was twelve gallons, the brewing of a boll of malt produced the same volume of ale, with ten gallons going to the king (the land owner), and two gallons (that is 16⅔%) going to the brewers for their wages and materials.

Innes followed this by an assize of ale, which he attributed to William the Lion (1165-1214):

> The assise of aill maide be kyng Willȝiam
>
> Qwhen the chalder of malt is sauld for · x · s · thane sall twa gallounis of aill be saulde for a peny · Qwhen the chalder of malt is saulde for · xv · s · the galloun and a halfe salbe sauld for a peny · Qwhen the chalder is sauld for · xx · s · the galloun salbe at a peny · Qwhen the chalder is sald for · xxx · s · the galloun salbe for three halfpenijs · Qwhen the chalder is sauld for · xl · s · the galloun salbe at · ij · penijs · And sua ay up ascendand · [73]

For comparison purposes aspects of the ale assizes of David and William have been placed in Table E.5.

From this it appears that ale was cheaper in William's day relative to that of David's for malt at a given price. In fact, the price of David's ale was approximately 50 per cent dearer and the steps in the price of malt are in units of 3s 4d (¼ mark). The price of ale was strictly proportional to the price of malt. In William's Assize the steps are in terms of 5 or 10 shillings; the price of ale is proportional to that of malt except for the erroneous entry at 15s, where instead of ⅔d it should have read ¾d. It is clear that these assizes relate to the buying of malt in the marketplace and the retailing of ale.

In the time of David, one boll of malt of 12 gallons brewed into 12 gallons of ale, so the chalder brewed into 16 × 12 = 192 gallons, of which 160 gallons paid for the malt (see Table E.5). For William, one boll of 12 gallons brewed into 18 gallons, the chalder into 288 gallons (otherwise, had it brewed into 192

| DAVID | | | WILLIAM | | |
MALT, PRICE OF CHALDER	ALE, PRICE OF GALLON	PRICE RATIO	MALT, PRICE OF CHALDER	ALE, PRICE OF GALLON	PRICE RATIO
6s 8d	½d	160			
10s	¾d	160	10s	½d	240
13s 4d	1d	160	15s	⅔d [*recte* ¾d]	270 [*recte* 240]
16s 8d	1¼d	160			
20s	1½d	160	20s	1d	240
23s 4d*	1¾d	160			
26s 8d**	2d	160	30s	1½d	240
40s**	2d	160	40s	2d	240

  * Apparently incomplete entry in Scots text of David Assize.

** Entry in Latin in text of David Assize.

Table E.5

Table of malt and ale prices, David I and William the Lion.

gallons, it would not have met even the price of the malt, which required 240 gallons). William's ale was only two-thirds the strength of David's ale, which accounts for its being cheaper by that fraction for the same price of malt in the two assizes. (On the English scene, the near contemporary Account Book of Beaulieu Abbey of 1270 states that 4 quarters (or 256 gallons) of barley make 240 gallons of ale.)[74]

We get another view of ale strength from a much later statute of 1644, when the boll of 4 firlots of malt made 15 gallons of ale.[75] By this time, the gallon was 8 pints and the firlot of bere (or rough barley) was 31 pints. Hence, 124 pints of malt made 120 pints of ale, which again was very nearly a given volume of malt creating the same volume of ale, just as in David's Assize. There are glimpses of other ale assizes in the intervening period, of which unfortunately no details survive. For example, there is mention of an assize of ale and fish at Aberdeen in the early fourteenth century.[76] At about the same time, there was an inspection between 1318 and 1327 mentioned in the charter of David I to Holyrood Abbey, which refers to an existing assize of bread and ale in the control of the burgesses.[77]

The earliest assizes recorded malt as sold by the chalder; but it was also sold by the boll, as in 1618 (and again in 1695), when a tax of 2 marks was laid on every boll of malt brewed.[78] It was also commonly sold by the 'load' of 9 barley firlots, and this is apparently another example of a unit being augmented by one-eighth for use in the market-place. The Edinburgh records show that the load of malt was sold for 20 shillings in January 1492; and when the load was first defined as 9 firlots in October 1532, the price had risen to 32 shillings.[79]

An analysis of Edinburgh malt prices shows little movement from 1520 until 1545, followed by 15 years of fluctuating prices. Overall, between 1500 and 1590 there was a price rise of a factor of seven, followed by a sharp peak, so that, for the sixteenth century as a whole, malt prices escalated by a factor of ten. Average prices for both malt and ale in this century show an approximate exponential increase, indicating rampant inflation fuelled by repeated periods of dearth accompanied by frequent debasement of the coinage.

One would expect to find a good correlation between the price of malt and the price of ale, and semi-logarithmic graphs (Fig. E.4) for the Edinburgh prices of a load of malt (9 firlots) and a gallon of ale demonstrates both trends to be very nearly the same, and the straight lines drawn are almost but not quite parallel.

Had these trends been parallel, that would have revealed a constant ratio between the two prices. As it appears, ale prices rose slightly more steeply than those of malt. For most purposes the price of ale can be regarded as bearing a fixed ratio to the price of malt. From 1510 to about 1590 the price increase for ale is a factor of about 1 : 6·4. The sudden jump in ale prices in the 1590s is a reflection of the period of dearth at this time, just as there was for bread. Once the price had jumped from about 70d to about 120 to 125d per gallon, it remained reasonably static to the mid-seventeenth century, followed by a significant increase.

The Privy Council in 1666 intervened by imposing an assize for ale, with rough bear ranging from £6 to £10 per boll, with ale required to be sold in the corresponding range 12d to 24d per pint. Subsequently, in 1680 the

Fig. E.4
Edinburgh malt and ale prices from 1493 and 1504 respectively.

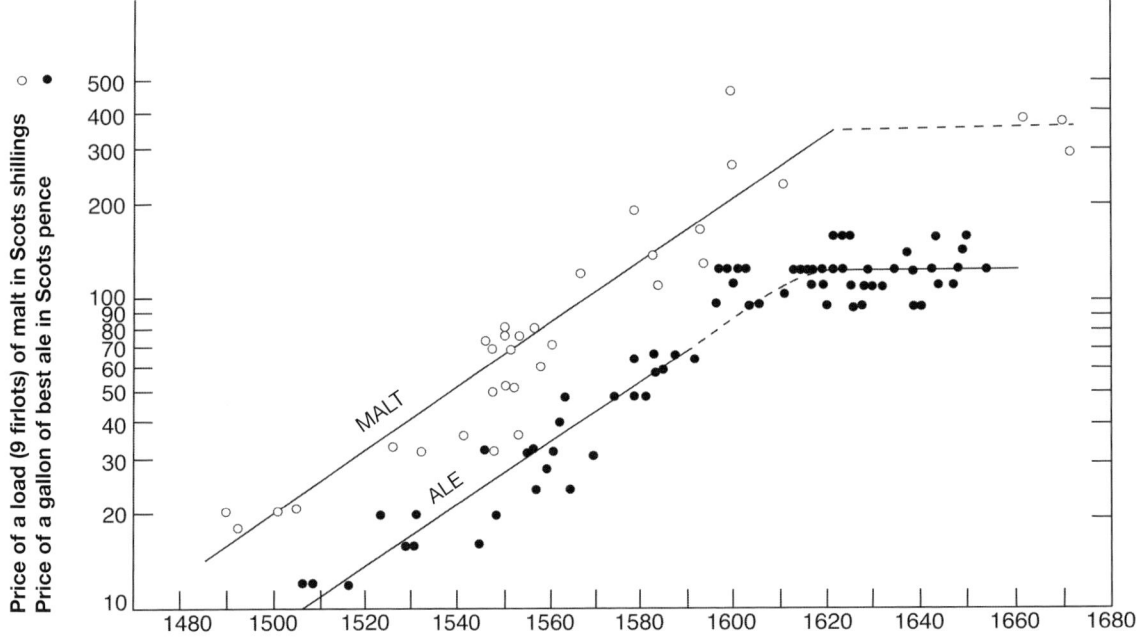

Privy Council again fixed prices, with rough bear in the range 6 marks to 12 marks per boll; a pint of ale was to range in cost from 12d to 28d per pint.[80] The statutes for July 1695 record that the lowest price for ale was to be 28d per pint, or 224d per gallon, which is almost twice what it had been during the first half of the seventeenth century.[81] A similar statistical analysis was performed on these data yielding the logarithmic graphs below, normalising to the price in English sterling and to a constant firlot (see Fig. E.5).

As in the case for bread discussed earlier, so these graphs demonstrate that when the price is given in English sterling normalised to a constant size of firlot, the trends in the costs of both malt and ale reveal a very gentle price rise, indicating that the massive price rises for both commodities was largely due to the debasement of the Scottish currency, and only in a minor way to the continuing increase in the size of the firlot.

On occasion, the Edinburgh records give simultaneously the price of a boll of malt and the price for which ale is to be sold. Some of these costs for the period 1546 to 1610 are given in Table E.6.[82] No strict proportionality can be expected with the first two entries, which have the same price of ale with differing prices of malt; but on average, about 12¾ gallons of ale will pay for the malt. If it is supposed that the declaration of 1644 (stating that a boll of malt made 15 gallons of ale) held also for the previous century, it can be deduced that the brewer's profit would be 2¼ gallons, from which he would have to pay for his materials, generate his income and pay his workers. As the gallon became more expensive, so in like proportion his allowance would be worth more. It can be seen that the ratio of the two prices is not quite constant, but becomes somewhat less as time progresses, indicating again that ale prices were rising a little more rapidly than those for malt. The brewer's profit margin on 2¼ gallons in 15 gallons, or 15 per cent, is a little less than that allowed in the post-David period when 2 gallons out of 12 (17%) were allowed the brewer.

THE MALTMAN'S PROFIT

At Aberdeen, maltmen in 1483 were permitted to take one boll from each chalder of bear barley, converted into malt, namely one-sixteenth (6¼%), which would go towards their expenses and profit.[83] When the differential in price is given in money, it is often difficult to determine whether the gain was calculated on the price of the bear or on the worth of the malt. With highly-variable grain prices the difference was hardly significant, but we can get a feel for the fractional profit being aimed at by the authorities, from a parliamentary act of 1552 which gave the price allowed the maltster for converting into malt a boll of barley (variously referred to as 'bere', 'beir' or 'bear'):

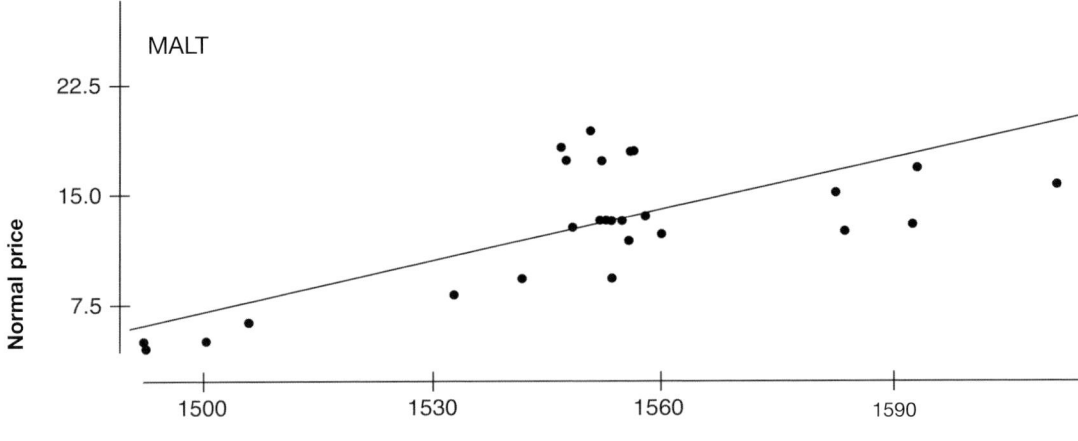

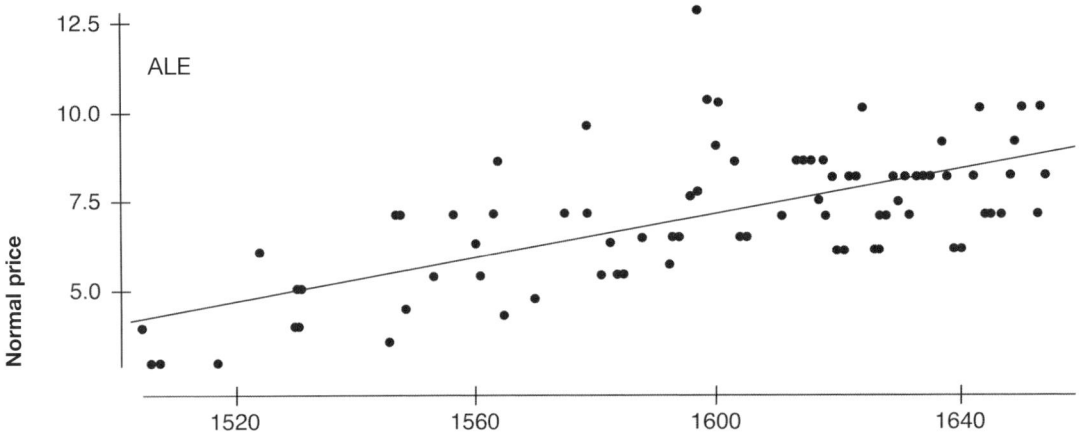

Fig. E.5
Malt and ale prices from 1492 and 1504 respectively. In English sterling pence, normalised to a constant firlot. The regression lines are respectively: for malt, price = 6·12 + 0·114 (Year − 1492); for ale, price = 4·428 + 0·027 (Year − 1504).

… That because of the greit derth of fewall [fuel] presentlie occurrand euerie malt man sall haue of euerie boll of malt maid be him for his labouris and fewall · iiij · s · of ilk boll of malt mair nor [than] the boll of beir is commounlie sauld …[84]

This does not appear to have been acted upon immediately, for it was not until 1555 that this 4-shilling allowance showed in the pricing.

In the interval, the maltmen appear to have been granted 2 shillings, as is seen from the Edinburgh record. On 10 February 1552 (just nine days later than the statute noted above), the Edinburgh records declared that the maximum price of a 9-firlot load of malt was to be £3 8 shillings (that is, 68 shillings) when bear barley had reached a price of 28 shillings to 30 shillings per boll.[85] The boll of 4 firlots of malt was to cost 30⅔ shillings (68 × ⁴⁄₉); so the maltman only obtained 2⅔ shillings down to ⅔ shillings per boll, or a gain of about 7½% (at 28 shillings to the boll) down to ¾% (at 30 shillings). Only when malt was at 28 shillings could the maltman get as much as 2 shillings, let alone the 4 shillings that parliament had allowed.

DATE	PRICE OF BOLL OF MALT (SHILLINGS)	PRICE OF GALLON OF ALE (PENCE)	GALLONS SOLD TO PAY FOR MALT
12 October 1546	32	32	12
June 1547	30·2	32	11·3
1 February 1548	22·2	20	13·3
24 February 1556	35·6	32	13·4
20 April 1556	35·5	32	13·3
18 September 1557	26·7	24	13·3
14 October 1560	54	48	13·5
15 June 1582	59·25	56	12·7
16 October 1583	48·9	48	12·2
7 June 1592	56·9	56	12·2
10 November 1592	74·1	64	13·9
7 December 1610	102·9	104	11·9

Table E.6

Relative prices of malt and ale in Edinburgh, 1546-1610.

By February 1556 the Edinburgh records proclaimed that when 'the boll [of] beir is commonlie sauld for xxxvj s. [36s.] or thairby, … the nyne furlettis [9 firlots] grundin malt *of the auld mesour* [our emphasis] is to be sauld for iiij li. [£4 or 80 shillings] …'.[86] It is clear that the bere was not being sold here by the same measure as malt, for then the boll of malt would be sold for 35⁵⁄₉ shillings (equivalent to 80 × ⁴⁄₉), which was less than the cost of the bere. The interpretation is that bere was sold by the new measure, and malt still sold by the old measure. (This action, prompted by the Privy Council, was only weeks before the 1555 Assize commissioners formally modified the measures.) The old legal firlot at this time was 15 pints and the new was 16⅞ pints, as outlined in Chapter 7 and Fig. 7.9. (The price scatter shown in Fig. E.4 at this time was sufficiently large to obscure this 13% increase.) Bringing these commodities to the new measures, the new boll of malt would be priced at 40 shillings (equivalent to 35⁵⁄₉ × 16⅞ ÷ 15), using the legal values. With bere at 36 shillings, the difference between the bere cost and the malt price was 4 shillings (a gain of 11%), matching the 4 shillings allowed to the maltster in the 1552 parliamentary act. (Although we would expect the legal sizes of the measures to be used in this instance, the equivalent customary boll sizes of 16⅞ and 19¹⁄₁₆ pints give almost the same result.)

More than a century later, in 1679, the Privy Council intervened to set prices, having been duly authorised so to do by statute in 1669.[87] We have already described how the Privy Council reformed the sale of meal in 1679; and the same change from a volume measurement to a weight measurement was applied at the same time to bere barley. After weighing different samples of barley, a compromise weight was recommended for the boll's content:

a meane proportion betuixt best and worst is that new bear be sold by the weigh of fifteen stone troas weight till the fifteenth of Aprile inclusive nixt after its grouth [growth], and at fourteen stone after the said day, each stone consisting of sixteen pounds, and each pound of sixteen ounces, all troas weight …[88]

The 14-stone transitional 'weigh' was eliminated in 1680, and the larger weigh's equivalence to the boll was made clear in its description as the 'weigh of barley consisting of fifteen stone trois weight (which is now in place of the boll) …'.[89]

When the boll of bere sold for £6, £8 or £10, the maltmen were to receive 10 shillings for the making of a 'weigh' of malt, and the pint of ale and best drinking beer was to be priced at 12d, 20d and 2 shillings respectively.[90] These were the same prices as had previously been given by the Privy Council in 1666 and repeated in 1679.[91] For the middle price of £8, the 10-shilling profit on the price of bere was 6¼ per cent.

The average price of malt between 1660 and 1670 was 300 to 380 shillings per load, or 133⅓ to 169 shillings the boll. Removing the 10-shilling malting charge, the average price for bere was 123⅓ to 159 shillings the boll. Ten shillings on a boll of bere costing 123⅓ shillings was a gain of about 8 per cent; on a boll of bere priced at 159 shillings it was 6¼ per cent. A study of the figures indicates that the percentage gain was being calculated on the price of bere, which was the maltman's outlay; and that an endeavour was made to keep the profit between 6¼ per cent and 8½ per cent, with only the 1556 figure exceeding this range with a value of about 11 per cent.

EXCISE AND THE BREWER'S PROFIT

In 1644 a new element entered the picture – the introduction of an excise tax on malt liquors. Whereas wine had been taxed for centuries, ale and beer only became liable to excise duty during the Interregnum when Oliver Cromwell was in power and money needed to be raised to support the army. Scottish acts of 1644 followed closely on the heels of their English counterpart of 1643.[92]

The first Scottish act was passed in January 1644 and this was followed in July of the same year by a much longer statute which laid down – among other things – what records were to be kept, who was to pay what and when, who was to collect the excise, and so on.[93] The Roll of Excisable Commodities and the amounts of tax levied was the same in both statutes, which read more like a customs list than an excise roll, for, in addition to items involving beer, ale, wines and spirits, there was enumerated a very mixed bag of goods, including thread, cloth, beaver hats, gold and silver lace, ox and sheep carcases. The items involving malt liquors may be summarised as follows: first, on every pint of ale or small beer the brewer must pay 4d, as must each house-holder who brewed for his own use; second, that foreign imported beer would be taxed at 12d per pint; and third, that on every pint of strong beer the brewer or householder must pay 6d. These levies were on small quantities (pints) and were imposed at the retail level. It was the brewer who had to pay the excise on ale or beer, and the acts declare that the levy is 'to be allowed to him in the price'. The purchaser paid the cost of the ale plus the excise; and this tax was remitted by the brewer to the collector. For malt liquors, a boll of malt made 15 gallons of ale. This was reiterated in the July act:

> That the excise of the Ale and beir shall be payed by the bruer either for sale or private Use According to the rate of Four pennies for ilke [each] pynt of Ale and beir strong and small overheid Counting [without exception] and allowing fyfteine Gallounes for ilke boll of Malt brued in Ale And Tuentie Gallounes for ilke boll of Malt brewed in beir strong and small overheid And the bruer for sale for his releefe shall exact from the buyer for ilke pynt of Ale or strong beir four pennies And for ilke pynt of small beir tuo pennies.[94]

This confirms the expectation that a boll of malt brewed into 15 gallons of ale, with the additional information that the boll brewed into 20 gallons of beer.

The provisions of these two acts can be summarized for our purposes as shown in Table E.7, from which we see that the rather high duty on beer was reduced after six months by 2d, with that on ale unchanged.

It appears from the burgh records that there was no sudden jump in the price of a pint through the critical date of 1644 or for another decade, as can be seen in Table E.8. Certainly an increase of 4d or 6d in the price of a pint would have been noticed, and the implication must be that the Edinburgh magistrates specified the price on top of which the excise tax was to be levied. Gibson and Smout have, however, detected a rise in the prices in Stirling.[95] Although the excise duty was decreased to 2d per pint in 1647, Gibson and Smout have also pointed to the additional complication in Edinburgh's figures from 1650 caused by a special imposition intended to repay the council's debts; initially this was 4d per pint, rising to 12d in 1658, but dropping to a constant 2d by 1666.[96] The Privy Council's malt and ale prices of

	JANUARY 1644	JULY 1644
Ale	4 pence	4 pence
Small beer	4 pence	2 pence
Strong beer	6 pence	4 pence
	No duty on malt liquors for the provisioning of ships	Ships now pay duty

Table E.7
Excise tax on malt liquors, 1644.

DATE	ALE		BEER	
	STRONG	SMALL	STRONG	SMALL
26 October 1626	12d	6d	18d	9d
9 November 1627	14	7	20	10
14 November 1628	16	8	20	10
16 October 1629	14	7	18	9
1 December 1630	16	–	–	–
26 October 1629	14	–	–	–
2 November 1632	16	–	20	–
1 November 1633	16	–	20	–
12 November 1634	16	–	20	–
23 November 1636	18	–	20	–
1 November 1637	16	–	20	–
26 October 1638	12	–	16	–
22 November 1639	12	–	16	–
3 December 1641	16	–	20	–
18 November 1642	20	–	24	–
27 October 1643	14	–	18	9
6 December 1644	14	–	18	9
4 December 1646	14	7	20	10
12 November 1647	16	–	20	10
22 November 1648	18	12	20	10
16 November 1649	20	10	–	–
2 April 1652	24	–	24	–
27 October 1652	20	10	20	10
21 October 1653	16	–	20	–

Table E.8
Edinburgh prices of ale
and beer, per pint.

1666 and 1669, already mentioned, required the following: first, that when bere was £6 per boll, ale should be sold at 12d per pint with, in addition, an excise tax of 2d per pint; second, that when bere reached the price of £8 a boll, ale should be sold at 20d, with the excise as before; third, when bere cost £10 a boll, ale should be sold at 24d, with the excise as before; and finally, if bere was less than £6 or more than £10, the price of ale was to be in proportion, with the sheriff checking that the act was being adhered to.[97]

This would be a rather stiff task, for the three statements are not themselves in proportion. Taken at face value, the ale sold to pay for the grain would be: first, 15 gallons; second, 12 gallons; and third, 12½ gallons. The 15 gallons from a boll of malt (when it is £6 a boll), was impossible, as this would mean that all 15 gallons from brewing a boll would go to pay for the grain, with nothing for expenses or wages. Possibly the objective was 12 gallons, but this would mean amending the 12d pint to 15d and the 24d pint to 25d. Tentatively, if there was to be a constant profit level, and with 15 gallons brewed from the boll, the brewer made three gallons gross profit, so that the gain was

20 per cet of the end product; but he still had to pay for the malting of the bere and meet his other expenses.

In 1680, the Privy Council laid down the following assize for the brewing of a weigh (or boll):

> When barley is 7 marks, rough bear 6 marks, the pint shall sell for 14d.
> When barley is 8 marks, rough bear 7 marks, the pint shall sell for 16d.
> When barley is 9 marks, rough bear 8 marks, the pint shall sell for 18d.
> When barley is 10 marks, rough bear 9 marks, the pint shall sell for 2s.
> When barley is 11 marks, rough bear 10 marks, the pint shall sell for 2s 2d.
> When barley is 12 marks, rough bear 11 marks, the pint shall sell for 2s 4d.
> When barley is 13 marks, rough bear 12 marks, the pint shall sell for 2s 6d.
> And for every mark of price more, 2d shall be added to the pint of drink; *and the prices foresaids are to be for both drink and Excise* [our emphasis].[98]

It should be noted that up to and including the Assize of 1679, the excise was shown separately, but in 1680 it was built into the price of the pint. The addition of 2d for every mark increase in the price of bere or malt implied a proportional table. However, there is a discontinuity after ale at 18d: instead of a 2d rise to 20d, the price jumps by 6d to 2 shillings (or 24d). Thereafter, the increase is a uniform 2d per mark increase. The Privy Council's arithmetic often appears to be in error. Probably a scribal error was made in which 2 shillings was written for 20d, and so on.

If these changes are made and it is remembered that it is rough bere that was malted and made into ale, with the 10 shillings malting charge included in the cost of the boll (thus, when the text reads 'rough bear' it really means 'malt'), then a new table (Table E.9) can be produced in which the excise contribution of 2d per pint is removed. (Amending the first three entries to conform with the last three, does not lead to a constant ratio.)

Throughout, the boll makes 15 gallons of ale. Eighty pints sold will pay for the malt, so the brewers had 5 gallons or its value from each boll; thus they made 33⅓ per cent of the final worth of the brew including fuel, wages and income. This high figure did not go unnoticed by the Privy Council.[99] The

PRICE OF MALT (BOLL)	PRICE OF PINT OF ALE	RATIO: PRICE OF MALT TO PRICE OF ALE
6 marks (960d)	14 − 2 = 12d	80
7 marks (1120d)	16 − 2 = 14d	80
8 marks (1280d)	18 − 2 = 16d	80
9 marks (1440d)	20 − 2 = 18d	80
10 marks (1600d)	22 − 2 = 20d	80
11 marks (1760d)	24 − 2 = 22d	80
12 marks (1920d)	26 − 2 = 24d	80

Table E.9
Privy Council assize of ale, 1680.

brewers' gain throughout this period can be summarised thus: King David 17%; 1546-1610, 15%; 1666 and 1679, 20%; and 1680, 33⅓%. This is only a factor of two, first to last, but during this long period all prices rose considerably, although the actual cash amount of the gain rose by a much greater amount, probably in keeping with inflation. The maltmen did not fare nearly so well.

How closely did the weigh of barley (bere) of 15 stone trois (later 'troye') match the boll by measure? There is no reliable value for the density of barley at this time, but we may take the modern value of about 605 kg/m³, which is 47 avoirdupois pounds per Winchester bushel (and this may not be strictly applicable to the late seventeenth century).[100] In comparison, Swinton, writing in the late eighteenth century, gave for barley at this density a boll weight of 16 stone 1 pound 10¼ ounces troye.[101] So 15 stones was less than the boll by measure by something like 7 per cent. These uncertain figures may explain the cautionary note sounded by the bailies of Glasgow in June 1682 (only two and a half years after the ruling of the Privy Council) and their subsequent action:

> … no person who are resting any victwall [victual] bowght be them in the mercatis of Glasgow, be the new measouris, aught to pay the same untill as much of the pryce be rebaitted be the magistratis as the new measouris are made les nor [than] the old measouris were.[102]

But the bailies went on to declare Linlithgow measure to be the standard, statute or no statute.

It was probably the above discrepancies which proved fatal to the plan to have barley sales by weight, because the boll of 15 stone does not seem to have continued for any length of time and certainly not into the eighteenth century. The figures imply that there was no charity, so the declaration was seen to be a 'legal' one. We have not located any 14- or 15-stone weights (or 8-stone weights for meal, for that matter), but there are extant examples of firlots of the eighteenth century and Swinton, writing in 1779, makes no reference to barley measured by a boll of 15 stone. Further, no county in 1826-7 recorded a barley boll of 15 or 14 stone troye.[103] By Swinton's time, the influence of the Act of Union was strong and while he worked in bolls, firlots and troye weight he also gave their equivalence in English measure, quarters, bushels and avoirdupois weight. Linlithgow measure was prominent everywhere in the records.

1   For a description of the medieval bread assize, see Nicholas Mayhew, 'Medieval Bread in Scotland', in *Review of Scottish Culture,* 8 (1993), 91-5; a discussion of the effects of the bread assizes between 1555 and 1773 is to be found in Alexander Gibson, 'The Bread Assize in Early-Modern Scotland', ibid., 10 (1997), 22-32. We have, however, arrived at a different interpretation at some points to those in this paper.

2   R. D. Connor, *The Weights and Measures of England [WME]* (London, 1987), 194-231.

3   For the relevant sections of the *Leges Quatuor Burgorum,* see T. Thomson and C. Innes (eds), *The Acts of the Parliaments of Scotland [APS],* 13 volumes (Edinburgh, 1814-75), I, 327-56; for the *Iter Camerarii,* see ibid., I, 680-2 693-702; and for the Assize of Bread, Ale and Wine attributed to David I, and, in part, to William I, see ibid., I, 675-9. Cosmo Innes, who was responsible for editing this section on the Assize of Bread, Ale and Wine, was baffled by the version found in the early fourteenth-century Ayr MS (which he reproduced in a footnote), and he followed the more coherent text in the later fourteenth-century Cromartie MS: ibid., 50, 56. He clearly appreciated that the weights and prices had been modified to bring them more up to date, and errors in doing so must account for some of the variation.

4   Gibson, op. cit. (1), 23.

5   *APS,* I, 339, Act. 36; ibid., 344, Act. 59; ibid., 344, Act. 60, which reads: 'Et pistor habeat ad lucrum de qualibet celdra secundum quod videatur probis hominibus ville' [And each baker shall gain from each chalder according to the view of the goodmen of the town]. The Scots version adds, just to make sure, 'and nocht eftir his awne discrecione'.

6   Ibid., I, 336, no. 19.

7   Ibid., I, 695; *Iter Camerarii,* Art. 4.

8   Ibid., I, 697; *Iter Camerarii,* Art. 9. The Scots text is not a faithful translation of the Latin.

9   J. D. Marwick, *et al.* (eds), *Extracts from the Records of the Burgh of Edinburgh 1403-1718 [Edinburgh Extracts],* 14 volumes (Edinburgh, 1869-1967), I, 7: 9 April 1443.

10   Ibid., V, 142: 30 October 1595. Also ibid., 267: 26 March 1600 and ibid., 268: 11 April 1600.

11   *WME,* 197-8.

12   R. W. Cochran-Patrick, *Records of the Coinage of Scotland,* 2 volumes (Edinburgh, 1876), I, lxxvi and 10.

13   *APS,* I, 676-8. Some figures are only in the Latin or the parallel Scots texts.

14   *WME,* 195-6.

15   $2 \times 136 \div 63\frac{1}{3} = 4\cdot29$.

16   J. D. Marwick, *et al.* (eds), *Records of the Convention of the Royal Burghs of Scotland (1295-1779) [RCRB],* 7 volumes (Edinburgh, 1866-1918), I, 556: 11 January 1555/6; *Edinburgh Extracts,* II, 230-2.

17   A. J. S. Gibson and T. C. Smout, *Prices, Food and Wages in Scotland 1550-1780* (Cambridge, 1995), 29.

18   *Edinburgh Extracts,* II, 144, 149, 230, 237.

19   Elizabeth Gemmill and Nicholas Mayhew, *Changing Values in Medieval Scotland: A Study of Prices, Money, and Weights and Measures* (Cambridge, 1995), 'The assize of bread', 30-42, p. 35.

20   *WME,* chapter 11, especially p. 211.

21   Alexander Huntar [*sic*], *A Treatise of Weights, Mets and Measures of Scotland* (Edinburgh, 1624), 68.

22   See *WME,* 110-12; also Philip Grierson, 'Weight and Coinage', in *Numismatic Chronicle,* 4 (1964), ix.

23   Hunter, op. cit. (21), 63.

24   8 Anne c. 19, 1710; given as c. 18 of 1709 in the 'Common Printed Edition'. See also Gibson, op. cit. (1), 27-9 for the 'Post-Union Assize of Bread'.

25   31 George II, c. 29, 1757; and see *WME,* 210.

26   John Powell, *Assize of Bread* (London, 1684). See also *WME,* 205-7.

27   41 George III, c. 12, 1801.

28   [John Swinton], *A Proposal for Uniformity of Weights and Measures in Scotland* (Edinburgh, 1779), 50.

29   Ibid. The Winchester bushel was defined by the Act 8 and 9 William III c.22, s. of 1696-7 as $2,150\cdot42$ in^3, which makes the 'standard' boll $8,795\cdot2$ in^3, or near enough 8,800 in^3. This is simply four times the wheat bushel size that emerged from the experiments of the 1750s: see Chapter **9**.

30   Swinton's values of 57 pounds of grain yielding 42 pounds of flour gives a ratio of $0\cdot737$. For a modern equivalent, the Canadian Grain Institute stated that 1 tonne of wheat yielded $0\cdot722$ tonnes of flour and $0\cdot278$ tonnes of bran: *Selkirk Journal,* 10 September 1991. Thus, one unit of weight of flour made $1\cdot2$ units of bread; Swinton shows a value of $1\cdot24$. From this, it can be seen that

milling yields and baking practices of today are hardly different at all from those near the end of the eighteenth century.

31 R. Renwick (ed), *Extracts from the Record of the Royal Burgh of Stirling 1519-1752 [Stirling Extracts]*, 2 volumes (Glasgow, 1887-9), I, 180: also discussed by Gibson and Smout, op. cit. (17), 35-6, who comment that the 'relationship … between the price of oatmeal and the price of oatbread is far from straightforward'.

32 S. G. E. Lythe and J. Butt, *An Economic History of Scotland 1100-1939* (Glasgow and London, 1975), 37.

33 Ibid., 38.

34 W. M. Metcalfe, *A History of Paisley 600-1918* (Paisley, 1909), 336.

35 A. M. Munro (ed), *Records of Old Aberdeen 1157-1891, vol. I*, Spalding Club, new series no. 21 (Aberdeen, 1899), 32-3.

36 See the price series of oatmeal fiars from 1556 to 1814 in Gibson and Smout, op. cit. (17), chapter 3, and between the 1550s and 1710 in A. Gibson, 'Prices and Wages', in Peter G. B. McNeill and Hector L. MacQueen (eds), *Atlas of Scottish History to 1707* (Edinburgh, 1996), 326; also Gemmill and Mayhew, op. cit. (19), 41-2.

37 *Edinburgh Extracts*, I, 75: 10 November 1592; *Stirling Extracts*, I, 89: 13 October 1598.

38 A. J. S. Gibson and T. C. Smout, 'Regional Prices and Market Regions: the Evolution of the Early Modern Scottish Grain Market', in *Economic History Review*, 48 (1995), 258-82; quotation p. 281, referring to Rosalind Mitchison, 'The Movements of Scottish Corn Prices in the Seventeenth and Eighteenth Centuries', in ibid., 18 (1965), 278-91.

39 William Hay (ed), *Charters, Writs, and Public Documents of the Royal Burgh of Dundee: the Hospital and Johnston Bequest 1292-1880* (Dundee, 1880), 157.

40 Ibid., 158.

41 Ibid.

42 Hunter, op. cit. (21), 3.

43 J. Hill Burton, *et al.* (eds), *The Register of the Privy Council [RPCS]*, three series (Edinburgh, 1877-1970), third series, VI, 365-8: 23 December 1679.

44 The proposal to replace meal measure by weight, initially proposed in 1695 (*APS*, IX, 362), was allowed to 'lie on the table' on 15 September, was agreed to on 19 September, 'touched by the sceptre' on 25 September, promulgated as 8 stone trois instead of the Linlithgow boll, then approved on 12 October, all in less than two months in 1696. See *APS*, X, 13, 18, 25, 34 and 75.

45 Swinton, op. cit. (28), 78, 88.

46 Ibid., 52.

47 Ibid., 32.

48 We are obliged to the Canadian International Grains Institute, Winnipeg, Canada, and Can-Oat Milling, Portage la Prairie, Manitoba, for supplying the density figures of the present day.

49 Swinton, op. cit. (28), *passim*.

50 Gibson and Smout, op. cit. (17), 375.

51 George Buchanan, *Tables for Converting the Weights and Measures hitherto in use in Great Britain into those of the Imperial System* (Edinburgh, 1829), 221, 239, 255. Inverness and Ross used a boll of 8 stone in determining ministers' stipends.

52 *Poor Law Inquiry (Scotland)*, Parliamentary Papers, Session 1 February to 5 September 1844 (Reports from Commissioners), volume XXIII, appendix, part IV, 459. Discussed by Ian Levitt and Christopher Smout, 'Some Weights and Measures in Scotland, 1843', in *Scottish Historical Review*, 56 (1977), 146-52.

53 *Edinburgh Extracts*, I, 178: 8 October 1518.

54 Ibid., II, 80: 7 October 1536.

55 Ibid., 130: 17 June 1547; ibid., 134: 20 June 1548; ibid., 230: 11 January 1556: '… utheris baxteris duelland in landwart and unfremen sall eik ane unce weycht for ilk penny to the quantite foirsaid …' [Outland bakers give an ounce weight for each penny to the quantity aforesaid.]

56 Ibid., V, 252, 258.

57 Ibid., V, 268, 315.

58 Rosalind Mitchison, *A History of Scotland,* second edition (London and New York, 1982), 145.

59 In the early sixteenth century, the sterling weighed 15 grains, in 1543 it was down to 10 g, in 1549 to only 6⅔ grains, but under Elizabeth it rose to 8 grains before falling to 7¾ grains in 1601, which would have roughly kept pace with the rising price of silver had the percentage of silver in the coinage remained at 92½% but by 1546 it had fallen to 33⅓%; it rose to 50% in 1549 and was only finally restored to sterling quality in 1601.

60 Cochran-Patrick, op. cit. (12), I, lxxvi.

61 For example, see Michael Lynch, *Scotland, A New History* (London, 1991), 183.

62 T. C. Smout, *A History of the Scottish People 1560-1830* (London, 1969), 143-4.

63 We are very grateful to Dr. Brian D. Macpherson of the Department of Statistics, University of Manitoba, for programming and conducting this analysis as well as those for Malt and Ale.

64 *Edinburgh Extracts*, II, 50: 1 July 1531.

65 Ibid., II, 157: 10 October 1551.

66 *Stirling Extracts*, I, 116: 20 March 1607.

67 *WME*, 201. Henry Thomas Riley (ed), *Munimenta*

Gildhallae Londoniensis: Liber Albus, 3 volumes (London, 1859-62), III, 82-3. See also ibid., I, 264-5.

68 *WME*, 202.

69 The assize of ale from these early dates is discussed by Gemmill and Mayhew, op. cit. (19), 48-59; Gibson and Smout, op. cit. (17), 26-8, and Nicholas Mayhew, 'The Status of Women and the Brewing of Ale in Medieval Aberdeen', in *Review of Scottish Culture,* 10 (1997), 16-21.

70 *APS*, I, 345, 346: nos. 63 and 67.

71 Ibid., I, 683: *Juramenta Officiariorum.*

72 Ibid., I, 675. The additional Latin text is as follows: 'Et sciendum est quod quia dictus dominus Rex Dauid solebat recipere ad Boteleriam suam de brasiatoribus suis brasiantibus braseum firamarum suarum · videlicet de qualibet bolla boni brasei ordiacii nisi · x · lagenas seruisie · et de qualibet Celdra boni brasei ordiacei nisi octies viginti lagenas seruisie · De quibus due partes dicte seruisie erant optima seruisia que tunc vocabatur seruisia militum · Et tertia pars dicte seruisie erat minoris precii que vocabatur seruisia armigerorum · uel seruientum ·

Statutum est per dictum dominum Regem Dauid quod quilibet brasiator uel brasiatrix faciens seruifiam venalem reciperet de qualibet bolla boni brasei ordiacii nisi · xij · lagenas seruisie taberne · Et sic poterit quilibet brasiatorum uel brasiatricum brasians seruisiam venalem duas lagenas seruisie uel pretium earundem pro expensis suis ad lucrum habere ·'

73 Ibid., I, 676.

74 S. H. Hockley (ed), *Account Book of Beaulieu Abbey*, Camden Society, fourth series, no. 16 (London, 1975).

75 *APS*, VI, part I, 240: 1644.

76 A. A. M. Duncan (ed), *The Acts of Robert I, King of Scots, 1306-1329, Regesta Regum Scottorum [RRS],* volume V (Edinburgh, 1988), 514: no. 247, 25 September 1323. Regulation of fish in medieval Aberdeen is discussed by Gemmill and Mayhew, op. cit. (19), 47-8.

77 *RRS*, V, 654: no. 410.

78 *APS*, VIII, 247: 6 September 1681; ibid., IX, 451: 16 July 1695.

79 *Edinburgh Extracts*, I, 61: 17 January 1491/2; ibid., II, 59: 5 October 1532. The variable relationship between the price of ale and malt in Edinburgh

(and elsewhere) is discussed by Gibson and Smout, op. cit. (17), 37-41.

80 *RPCS*, third series, VI, 365, 530-1: 1666 and 1680.

81 *APS*, IX, 451-2: Act 52, 16 July 1695.

82 Prices of malt and ale in Edinburgh between 1546 and 1610 are to be found in *Edinburgh Extracts*, II, 124: 12 October 1546; ibid., 130: 17 June 1547; ibid., 132: 1 February 1548; ibid., 237: 24 February 1556; ibid., 240-1: 20 April 1556; ibid., III, 11: 18 September 1557; ibid., 84: 14 October 1560; ibid., IV, 237: 15 June 1582; ibid., 300: 16 October 1583; ibid., V, 65: 7 June 1592; ibid., V, 75: 10 November 1592; ibid., VI, 68: 7 December 1610.

83 J. Cooper (ed), *Cartularium Ecclesiae Sancti Nicholai Aberdonensis*, Spalding Club, new series no. 2, 2 volumes (Aberdeen, 1888-92), I, 336.

84 *APS*, II, 486: Act 13, 1 February 1551/2.

85 *Edinburgh Extracts*, II, 162-3: 10 February 1552.

86 Ibid., II, 237-8: 24 February 1556.

87 *APS*, VII, 574: Act 36, 1669.

88 *RPCS*, third series, VI, 365-8: 23 December 1679; quotation p. 366.

89 Ibid., 530: 10 August 1680.

90 Ibid., 368 and 367 respectively: 23 December 1679.

91 Ibid., third series, II, 130-2: 18 January 1666.

92 Excise Ordinance, 22 July 1643.

93 *APS*, VI, part I, 76, 237-45: January and July 1644.

94 Ibid., 243.

95 Gibson and Smout, op. cit. (17), 39.

96 Ibid.

97 *RPCS*, third series, II, 130-2: 18 January 1666; ibid., VI, 365-8: 23 December 1679.

98 Ibid., VI, 530: 10 August 1680.

99 Ibid., 365, 1679: 'Forasmuch as the Lords of our Privy Councill, in consideration of the exorbitant prices taken by brewars and ventners for ale and drinking beer, at their pleasure, without observing any just proportion betuixt the same and the boll of bear …'

100 Herbert W. Ockerman, *Source Book for Food Scientists,* 2 volumes (Connecticut, 1978) I, 26; L. C. Buchanan, *et al., Principles and Practices of Commercial Farming*, fifth edition (Winnipeg, 1977), 509.

101 Swinton, op. cit. (28), 51.

102 J. D. Marwick (ed), *Extracts from the Records of the Burgh of Glasgow, 1663-1690* (Glasgow, 1905), 316-7: 19 June 1682.

103 Buchanan, op. cit. (51), *passim.*

APPENDIX F

A SYNOPSIS OF THE
SCOTTISH SYSTEM OF
WEIGHTS AND MEASURES,
INCLUDING PARTICULAR MEASURES

IN THE MAIN TEXT WE HAVE PURSUED A LINE OF ARGUMENT WHICH IS designed to explain the structure and development of the system of Scottish weights and measures. Inevitably we have concentrated on particular aspects which have allowed that structure to be revealed. We have not, therefore, covered the great range of Scots terms covering all types of measurement and customary usage, nor indeed regional variations or dialect terms. In this appendix we have tried to provide an account of some of the special terms used in the principal reference books which have formed the basis for this study. We are conscious that this brief compilation is in no sense comprehensive in scope, and would refer interested readers to entries in the *Dictionary of the Older Scottish Tongue* or the *Scottish National Dictionary*. For the early period, up to about 1500, Elizabeth Gemmill and Nicholas Mayhew, *Changing Values in Medieval Scotland: A Study of Prices, Money, Weights and Measures* (Cambridge, 1995) have provided an excellent account of the terms often encountered in charters and early documents. However, we have serious reservations about the frequently-cited work by R. E. Zupko, 'The Weights and Measures of Scotland before the Union', in *Scottish Historical Review*, 56 (1977), 119-45, and are inclined to agree with the comments on this made by Ian Levitt and Christopher Smout: 'Any list giving local or county standards, however comprehensive and carefully compiled, needs to be used with caution, because slips are easily made and because such standards could evidently vary in a disconcerting way depending on the period of history and on the district even within counties. Weights and measures are a bramble bush full of good fruit, but no one can come away completely unscratched' ('Some Weights and Measures in Scotland, 1843', ibid., 152).

In common with other mercantile nations, Scotland had well-defined systems of weights and measures for external and internal trade and the legal and administrative structures for enforcing their use. As well as forming the basis of trade of all sorts in the Scottish burgh markets, specifically Scottish units were used in land contracts, in levying taxation, and in rental and tenancy agreements which involved the payment of dues in kind.

The Scottish system may well have been imported from England by David I and his predominantly Anglo-Norman followers in the twelfth century, but the two systems evolved separately and subsequent Scottish units were strongly influenced by the metrologies of principal external trading partners, notably the Low Countries (particularly Flanders) and France. A number of the Scots measurement terms have familiar names which are also found in English metrology, including inches, pounds, stones, pints and gallons. However, others have distinctive Scottish names, such as the grain 'firlot' and 'boll'. Only the slightest traces survive of the earlier Celtic units (except for more widely encountered land assessment terms), and we are concerned here largely with the practices of the Scots-speaking royal burghs, with their monopoly of local markets and their foreign trading privileges, to which metrological legislation was addressed.

The size of certain units, and the numerical relationships between them, were specified periodically by the Scots parliament or Privy Council in the form of 'assizes' of weights and measures. The earliest such assize is traditionally associated with David I, but details survive of a succession of later assizes between 1426 and 1618 at which the permitted sizes of measures were changed or redefined.

As in England, linear measure was based on an inch – in fact on the same inch, and identically defined in terms of the size of conventional grains of barley. Larger units used, for example, in laying out plots of land, were the ell of 37 inches and the fall of 6 ells. Weights were specified in ounces, pounds and stones and, as in England, we can determine that they were initially related by the same multipliers and were based on the same Cologne bullion ounce (although again specified only in the weights of conventional grains). Liquid was measured by the pint and gallon, defined by the weight of their contents; and dry goods were measured in large coopered vessels by the firlot and boll (the approximate analogues of the English bushel and quarter) which in turn were defined by their dimensions and capacities in pints.

The apparent simplicity of this, however, disguises the fact that there were often several separate variants of each unit. Thus, for example, ale gallons differed from wine gallons, and different pounds were appropriate for different goods. In addition, market practices allowed the measures to be used with various types of traditional allowance, or 'charity', which had the effect of increasing their sizes within carefully controlled limits. The amount of charity depended on the measurement circumstances, but also changed over time. Thus, for most practical purposes, the wet and dry measures used were larger than the sizes defined in legislation. For all these reasons, it may not be possible to give a clear equivalent in modern units for a quantity of goods described in any particular reference.

Each burgh was required to hold physical standards of the measurement units so that the weights and measures used in trade, and particularly in the markets, could be checked by burgh officials and destroyed if found inaccurate. A number of early burgh standards exist in museum collections and elsewhere and these are described in the Inventory. These standards were checked periodically against national standards by the Chamberlain, in the course of his ayre, or progress round the royal burghs. Cases of using false weights were tried in the burgh courts but appealed in the Chamberlain's Court of the Four Burghs (or subsequently heard at the Convention of Royal Burghs). The construction of national standards (none of which are now definitely known to survive) is associated with specialist workers at the Mint and at the royal artillery workshops; and at times of major assizes when changes were introduced, authenticated copies of new standards were issued to a limited number of principal burghs. In due course, Edinburgh, Stirling, Linlithgow and Lanark (the four burghs of the Chamberlain's former court) were given the right to issue copies of their retained standards of the ell, pint, firlot and stone respectively. In addition, Aberdeen held the standard of the salmon barrel; Perth came to be associated with a subsidiary linear standard for thread (the reel) and Edinburgh with the later white-fish barrel.

Scottish units were officially displaced (in theory, at least) by those of England under the provisions of the Act of Union of the parliaments of the two kingdoms in 1707; and new sets of English standards were issued to the main Scottish burghs. However, this had very little effect on the internal markets of the burghs, and the traditional units remained in almost exclusive use for ordinary trade until the mid-nineteenth century, when they were progressively replaced by those of the new British 'Imperial' system. Before this, the largely self-contained nature of these local and regional trading areas, and the lack of centralised control after 1707, meant that some geographical variation became established in the later sizes of the traditional Scots units.

Much of the received view of Scots metrology is based on somewhat naïve comparisons made in the 1820s between the new Imperial units and existing local standards, and on earlier antiquarian enquiries which aimed to settle the legal size of the units of the last major assize, of 1618. Although the nineteenth-century analyses appear to be accurate and definitive, they are based on incorrect assumptions about medieval measurement practices and they show no appreciation of the existence or effect of trading allowances.

However, these results have become firmly embedded in the literature, and are reproduced, for example, in the table 'Scottish Currency, Weights and Measures' in volume 10 of the *Scottish National Dictionary* (Edinburgh, 1976), 316-7, which has subsequently been reprinted in a number of other works. For a brief corrective to this, see A. D. C. Simpson and R. D. Connor, 'Measures: Interpreting Scots Measurement Units', in Glen L. Pryde, *Dictionary of Scottish Building* (Edinburgh, 1996), 104-5.

WEIGHT

The David Assize, in its surviving fourteenth-century form, defined the weight series as:

> 1 pound = 15 ounces
> 1 stone = 15 pounds,

where the ounce weighs 20 pence, matching the contemporary English bullion ounce (the Cologne or tower ounce) of 29·14 g, so the stone was 6·55 kg. By about the mid-fourteenth century there was a merchant weight (called 'Scottis' weights in 1426) of 16 such ounces, and a 16-pound (7·46 kg) stone. This may also mark the introduction of the trone series – a goods pound and stone for the internal market, maintained at 1¼ times the merchant weights (hence a trone stone of 9·33 kg).

New 'trois' weights appear first in the definitions of the 1426 Assize, identical with the English troy bullion weight (ounce of 31·08 g), with:

> 1 trois pound = 16 ounces
> 1 stone = 15 trois pounds = 16 Scottis pounds (of 16 Cologne ounces).

The system was re-orientated from the English troy to the French troyes bullion ounce (30·60 g) in about 1510. However, merchant trade was often conducted with weights based on the marginally heavier Flemish troois ounce (operating in Scotland at about 30·72 g), sometimes known as trose or *fleur-de-lys* weight, which was apparently adopted officially in 1587. These, and later Scots weights, are generalised as 'troye' weights in this study.

By the mid-sixteenth century, troye weight had become the main merchant weight series for fine goods, and in 1563 the troye stone was officially increased from 15 pounds to 16 pounds (7·84 kg), with the trone pound and stone increasing in proportion (trone stone of 9·80 kg). The relationship of the various parts of the weight series was now:

> 1 ounce = 16 drops (or 24 deniers, each of 24 grains, for bullion weight)
> 1 pound = 16 ounces
> 1 stone = 16 pounds.

The weight standard of the burghs was the Lanark trone stone until 1618, when a revised troye (or later 'Dutch') stone of about 7·89 kg was substituted. From the 1680s, oatmeal was measured by weight rather than volume, reckoned at 8 troye stones per boll of meal, or 63 kg.

After the passing of the 1824 Act which introduced the Imperial system, the troye and trone stones were determined at equivalents to 7·89 and 9·98 kg respectively, based on the Edinburgh standards.

LIQUID CAPACITY

The standard unit of liquid capacity was the pint defined by the weight of water it contained.

> 1 pint = 2 chopins
> = 4 mutchkins (each of 4 'jowcats', subsequently gills).

The David Assize defined only the gallon and not the pint, recording it as 12 pounds of (mixed) waters. This was a 6-pint gallon (described here as 6 'old' pints) which can be identified as the ale gallon. A similarly early wine gallon, of 6 wine pints, was one-and-a-half times the size of the ale gallon. However, ale and wine were both sold with an allowance of ¹⁄₁₆ added, so that the retail vessels were enhanced by this amount. A dimensioned gallon of 5·42 litres described in the assize was a practical vessel which included the allowance. The basic sizes of the fourteenth-century ale and wine pints were 0·85 and 1·27 litres, and 0·90 and 1·36 litres with the allowance.

Dry fills of the pint (fills of grain) defined the dry measures from the 1426 Assize, and for burghs this may have been the principal purpose for holding standards of this size. In 1426 the size of the statutory pint was increased by a half (described here as the 'small' pint) to match the basic wine pint, and by a further third in about 1500 (the 'large' pint, of 1·70 litres), remaining subsequently at this size.

The pint was defined as containing 41 trois ounces in 1426, and after the increase of c.1500 as 55 French ounces, neither of which are exact figures. Changes of official ounce type led to small variations in the capacity of newly-issued pint standards. From at least 1550, the enhanced versions of the large pint (about 1·80 litres) and its chopin were the usual vessels for the sale of ale. The 1426 gallon was defined as an 8-pint gallon. It is not clear whether this immediately affected the size of the ale gallon, but the wine gallon was considered to be 8 pints in the late 1450s. Indirect evidence suggests that different pint sizes long remained in parallel use.

The Stirling Jug came to be considered as the definitive standard, and it was determined to be 1·71 litres after the passing of the 1824 Act.

DRY CAPACITY

The standard of dry capacity, for grains, oatmeal, and so on, was the firlot (which was the Scots equivalent of the English bushel). The firlot was a shallow cylindrical wooden coopered vessel, which was defined in the legislation by its dimensions and also as the volume of a set number of pints of grain. Its volume was progressively increased over several assizes (by a process of engrossing allowances in the use of the measure), until it reached 21¼ dry pints (34·6 litres) at the 1618 Assize. Barley and oatmeal, which were previously measured as 3 fills of the wheat firlot for 2 firlots of barley or malt, were from 1618 measured in a separate series of larger measures, where the barley firlot was 31 dry pints (49·5 litres). The theoretical relationships between the volume sizes is:

> 1 firlot = 4 pecks, each of 4 lippies or forpets
> 1 boll = 4 firlots
> 1 chalder = 16 bolls.

In practice, however, the relationships were different, because heaping allowances were provided for the merchants conducting transactions and the permitted trade measures were enlarged to incorporate these allowances.

The boll of the David Assize contained 12 ale gallons; and a dimensioned vessel, in two versions of the assize, is 60 or 63 litres. By the 1426 Assize the customary boll was 92 litres, considerably larger than four 16-pint firlots (each of 19·7 litres). An allowance of ¹⁄₁₆ in the firlot was first officially admitted in the 1426 Assize (although it clearly pre-dated this), and after about 1500 an additional charity of ¹⁄₁₆ for trading in bolls (the 'peck to the boll') was also being taken at the firlot level. The combined increase of ⅛ was the amount by which the firlot was increased at the major assizes until 1618, the legal size of the firlot rising in stages from 15 to 21¼ pints in a century. Only ¹⁄₁₆ seems to have been sanctioned for the 1618 firlots, implying trading volumes of about 36·8 and 52·6 litres, but quantities were still liable to increases from charities allowed on trading in bolls and chalders, as well as equivalent amounts for all dry produce imported by sea (and measured by the larger 'water metts').

By the late eighteenth century, the Linlithgow firlots were described (incorrectly) as by water fill, from the 1618 specifications, which should have resulted in capacities of 36·0 and 52·5 litres, approximately matching the earlier correct usage. However, they were typically found in the burghs at about 37 and 54 litres, and in the 1820s the standards were officially determined to be 36·3 and 52·9 litres.

LENGTH AND AREA

The Scottish and English inches were identical, and in Scottish use:

> 1 foot = 12 inches
> 1 ell = 37 inches (0·940 m), not divided in inches, but in ½, ¼, ⅛, and so on.

A longer ell of 37·2 inches (0·945 m), and therefore equivalent to 37 over-sized inches, is found in the eighteenth century. It appears to arise from a late seventeenth-century re-measurement of the yarn reel, but instead of being restricted to textile use it was also inappropriately applied from the late eighteenth century in land measurement. In markets for coarse and un-bleached cloth, 'plaiding' ells and yards were used, which incorporated a shrinkage allowance of approximately 1/16, and were therefore usually of about 39-40 inches and 38¼-38½ inches (about 1·00 and 0·98 m) respectively.

For land measurement:

> 1 [lineal] fall, raip [rope] or rod = 6 ells or 18½ feet (5·64 m),
> but initially the rod for measuring plots in burghs was 20 feet.

> 1 [Scots] chain (late 16th century) = 4 falls or 24 ells (74 feet or 22·5 m),
> with 100 links to the chain. (The link was also the same as the small foot,
> of about 9 inches or 0·22 m long, used by tradesmen and for measuring
> material such as glass.)

> 1 [Scots] furlong = 40 falls, also considered as the length that could be
> ploughed by a team of oxen without a rest.

> 1 [Scots measured] mile (from sixteenth century) = 8 furlongs (1·12
> English statute miles, 1·80 km), but the older Scots common mile of
> 1,500 paces (each of 5 feet) was 1·42 English miles or 2·29 km.

Land area:

> 1 rood (earlier, 'particate') of land = 40 square falls.

> 1 [Scots] acre = 4 roods, 160 square falls or 10 [Scots] square chains (0·509
> hectares), and an acre strip of arable land was a furlong in extent and 4
> falls wide (the 'acre's breadth').

> 1 ploughgate (earlier 'carrucate') = 8 oxgangs, taken as 104 acres (53
> hectares), being the area that could be ploughed by an 8-ox team in a
> ploughing season.

Work area:

> 1 square fall = 36 square ells (31·8 square m): the 'rood of work' or 'small
> rood' for measuring the work of craftsmen such as mason, slaters or
> wrights.

PARTICULAR MEASURES

Acre (Akker): Notionally and usually, a rectangular area of land 40 falls (q.v.) by 4 falls or an equivalent number (160) of square falls (5,760 square ells) following the English practice of defining the acre as 40 rods by 4 rods. First found in Southern Scotland where Anglian influence was strong. One Scottish acre equalled 1·26 English acres with 13 acres equal to one oxgang (q.v.) in use from early twelfth century onwards.[1] But there is evidence of an oxgang being divided into 12 acres each of 6,240 square ells.[2] This seems to have been rare and died out in the fourteenth century. (See Appendix B.)

Anker: A liquid measure of 20 Scottish pints (1812).[3] In Orkney, a firlot, dry measure, often for potatoes and fish.[4]

Arachor (Arachar): An area of land measurement, from the Gaelic corruption of Latin *aratrum*, a plough; hence an arachor is a ploughgate or ploughland (q.v.). Principally used in Lennox.[5]

Avoirdupois weight: A system of weights based on a pound of 7,000 English troy grains, probably originating in the wool pound of Bruges which was equivalent to 7,000 troy grains associated here with the 6,989-grain *grosso* pound of Florence. (See Chapter 4.)

Barrel:
of apples is 8 gallons (1779);[6]
of ashes, beef, bacon, butter and honey is 200 pounds (1779);[7]
of beef and pork is 8 gallons at least, to contain not less than 200 pounds of beef or pork (1705);[8]
of brass is 10 stone (c.1500);[9]
of butter (English or Dutch) is 15 lispunds (1579);[10]
of butter (English or Danish) 12 stone (1612);[11]
of coal is 1¼ cwt (140 pounds) (post-1707);[12]
of flax is 18 pecks (post-1707);[13]
of general goods is 10 stone (1624);[14]
of gunpowder is 10 stone (c.1500);[15]
of herring is 9 gallons by the Stirling pint (1573);[16]
of lead is Vᶜ (long hundred, hence 600) pounds (c.1500);[17]
of lead ore is 5 cwt (40 stone) (1612);[18]
of lemons is 10 stone (1661);[19]
of limestone is 32 gallons English measure (post-1707);[20]
of oil (Shetland) held 48 cans (q.v.) prior to mid-1500s.
  In the 1560s the Earl of Orkney increased the volume of the can by one-third, so the barrel held 36 large cans: the small can was 4½ Imperial pints (156.3 in³), the large was 6 Imperial pints (208 in³). (See **Can**, below, and Appendix D);
of salmon was 14 gallons, that is gallons each of 8 pints

(1487),[21] later 12 gallons by the Stirling jug (1573),[22] and subsequently, 10 gallons by the Stirling Jug (seventeenth century).[23] (Although these salmon barrels are apparently different, it was the gallon size which was changing and the declared volumes were only approximate. See discussion in Chapter 6);
of salt (small) is 2 bolls (c.1500);[24]
of wine is ⅛ tun (q.v.) (1624).[25]

Bind: An Aberdeen capacity measure for salmon; another name for the salmon barrel (q.v.);[26] of skins of schorling (recently shorn sheep) is 24;[27] a standard measure for wine (sixteenth century).[28]

Boll (Bow): A capacity measure for dry goods, especially grain, consisting originally of 4 firlots (q.v.) but frequently augmented by one-sixteenth by giving a peck to the boll as a charity. The boll increased with time as the firlot increased.
of Teviotdale is 5 firlots (1649);[29]
of Dumfries and Annandale for bear and meal is 2½ bolls Linlithgow measure (1649);[30]
of Wigtown and Kirkcudbright for bear and meal is 1½ bolls Linlithgow measure and their double boll is three Linlithgow bolls (1649);[31]
of marl (for manure) is a 2 foot cube containing 8 cubic feet (1791);[32]
of meal, Linlithgow measure, is to be 8 stone trois weight (1679);[33]
of bark, unbeaten mallow bark, is 22 gallons; and of beaten bark the boll is to be that of the Linlithgow barley measure (1686).[34]

Bolt: A length of cloth of 5⅓ ells (1328).[35]

Bovate: (See **Oxgang**.)

Cade (Caid): of fish (red herring) is Vᶜ (or 600) fish.[36]

Can: A Shetland measure for oil, being 4½ pints before the mid-sixteenth century, thereafter 6 pints. (See **Barrel**, and Appendix D.)

Cap: A variable measure of capacity for local use. Thus, in 1779:
Annan: 1 cap = 7 pints 1 mutchkin 1¾ gills Scots = 765·3 in³;[37]
Forfar: 1 cap = 1⅜ pints = 143 in³;[38]
Kelso: 1 cap = 1/12 of the local wheat firlot of 2,274·9 in³ = 189·6 in³.[39]

Carrowcat: A measure of land, normally of 4 husband-lands (q.v.).[40]

Carucate: From the thirteenth century onwards in Scotland, a ploughgate (q.v.).[40]

Caslamos: A weight of cheese or meal worth 7 pence (1326).[42]

Castlelaw (Castellaw): A weight of meal or cheese, similar to a Caslamos (q.v.) but worth 8 pence.[43]

Cathan (Caithness) weight: A weight mentioned by Skene in his first manuscript of 1601 and attributed to David II but without description or definition. It has been shown by A. D. C. Simpson that it is most likely the trone stone of the period.[44]

Chain, Scots: A surveyor's measuring chain equal in length to the acre's breadth (as in England) but here it comprised 4 linear falls (q.v.), or 74 English feet, and was divided into 100 links. (See Chapter **3**.)

Chalder: A dry measure of 14 heaped or 16 stricken bolls (q.v.).[45]

Chopin: A liquid measure of half a pint (see Chapter **6**); two mutchkins (q.v.).

Chudreme, Cudreme, Cudermose: An ancient measure of weight for cheese, equal to a cogall (q.v.).[46]

Clavinos: An unidentified grain measure only described as a 'bag'.[47]

Codrum: (West Coast) An early Gaelic trone weight of cheese, worth 7, 7½ or 8 pence, but usually 7 pence (1326).[48] (See **Cogall**.)

Cogall (Cowgall, Tonegall): A weight, often of cheese given as payment in kind, of 6 stone, worth 3 shillings in the thirteenth century.[49] The stone was therefore worth 6 pence. It is possible that the cogall and codrum (q.v.) were related in the same way as were merchant and trone weights since $6 \times 1\frac{1}{4} = 7\frac{1}{2}$. Alternatively known as a chudreme (q.v.).

Cowland: The land needed for the support of 7 cows and which paid rent of one cow per year (used in early Celtic times).[50]

Cran: A measure for herrings recognised officially by the Herring Fishery (Scotland) Act of 1815,[51] but in use long before as 45 English wine gallons (each of 231 in³).

Crear: 2 pecks.

Cumal: The land sufficient to maintain 3 cows and their calves.[52]

Cutt: (See **Reel**.)

Cuttell (Cuttel): The short Shetland ell of 27¾ inches (sixteenth century);[53] but also recorded as 21, 25·6 and 24 inches in the seventeenth and eighteenth centuries.[54] One cuttell was equivalent to 2 Shetland pennies of wadmel, or 4 shillings Scots, or 4 pence Sterling. Because of this last equivalence, the cuttell was sometimes termed a 'groat' (that is, 4 pence) after 1600.

Dacre (Dicker, Dacer, Daiker): of hides is 10 hides (*c.*1400 and 1597).[55]

Darge: The amount of meadow that can be mown in a day.[56] Often simply a day's work, and similar to the notion that the tilling of an acre is a day's work.[57]
of marl: as much as can be cast up with one spade in one day, often amounting to 200 bolls;[58]
of peats: as many as a man can cut in one day (1803);[59]
of coals, 28 hutches (q.v.).[60]

Davoch, Davach (Dauch): (Gaelic, *Tirunga*) Anciently, a 20 'house unit' in Dalriada: 'The amount of land necessary to produce (or require for sowing it) a fixed amount of grain, enough to fill a large vat of fixed size; this perhaps not being the total yield of grain but only the proportion of it due to a fixed render.'[61] Originally one ploughgate (q.v.), later two, and still later four ploughgates.[62]

Dozen, Dusane: A unit of length for Scots cloth and plaiding is 12 ells.[63]

Ell (Eln, Elne): A unit of length of 37 inches (there being no difference between the Scottish and English inches), equivalent to 939·8 mm.

The '**linen**' ell of 37·2 inches (late seventeenth century) arose from a failure to take into account the length of thread passing over the four pegs of the reel, with the distances between the centres of the pegs being made ⅝ of an ell (of 37 inches): this made the circumference 93 inches, but still described as 2½ ells. (See Chapter **3** and Fig **3**.6.)

The **ell of 37·06 inches** is the length of the ell bed of Edinburgh's sixteenth-century ell bed as measured accurately (as 37·0598 inches) by James Jardine and Patrick Copland in 1811 and incorrectly assumed to be the true length of the ell.[64] The additional 0·06 inches is now understood to allow an ell wand of exactly 37 inches to be inserted in the bed with some play. (See Chapter **2**.)

The **plaiding ell** for coarse twilled woollen cloth allowed for shrinkage in the cloth and was somewhat variable within small limits but was usually 38·4 to 38·5 inches.[65] (See Chapter 3.)

Eyrisland: (Orkney) A fiscal ounceland always of 18 pennylands (q.v.), sometimes (though inaccurately) called a urisland (q.v.).[66]

Fall: A linear measure of 6 ells.[67] Sometimes used to denote a square fall.

Farthingland: One-quarter of a pennyland (q.v.).

Fathom: of peats, 'originally a cube of 6ft square containing 216 solid ft … now increased from 216 to 1,008 solid ft. The present fathom is no less than 12 ft square and 7 ft high.'[68]

Fidder (Fother, Fodder): A measure of weight for lead, variously recorded as:
of 120 stones (1474);[69]
nearly 128 stone (1597);[70]
2,000 pounds weight (1612);[71]
126 stone or 2,000 pounds weight (1624);[72]
2,000 pounds weight where 112 pounds make the
   hundredweight (1655);[73]
two thousandth weight (1661);[74]
19½ cwt or 2,148 pounds avoirdupois (nineteenth-
   century English measure);[75]
20 cwt or 2,240 pounds (nineteenth-century English
   measure).[76]

Firlot (Firlat): A quarter of a boll (q.v.), legally 16 pints but 17 allowed in the market (1426). With allowances, this grew to be (legally) 21¼ pints for wheat and 31 pints for bear (barley) in 1618. Subsequently, an additional charity of a sixteenth was often added.

Foot: A linear measure of 12 inches. The 'glaziers' or 'glass' foot was a hundredth part of the Scots chain, so 1 link or 8·88 inches (used as 9 inches) divided again into 10 'inches' of 0·888 inches. (The corresponding English unit was also used in Scotland, namely a hundredth part of the English chain or 7·92 inches, used as 8 inches). (See Chapter 3.)

Forpit (Forpet): One-fourth of a peck (q.v.); also known as a lippie (or lippy).

Fotmal (Fotmel, Formel): The thirtieth part of a fidder (q.v.) or load of lead ore.[77] Principally an English measure.

French weight: The pound of 16 ounces each of 8 gros, the gros being divided into 3 deniers each of 24 French grains. Adopted in Scotland in the sixteenth century. The ounce of 576 French grains equalled 472½ English troy grains; and the pound of 9,216 French grains equalled 7,560 English troy grains. Hence 1 English troy grain is equal to 1·219 French grains.

Furlong: In Scotland, a distance of 40 falls (q.v.); an eighth of a Scots mile (q.v.).

Gallon: As defined in the Assize of David I, the ale gallon from dimensions was 330 in³ or 12 pounds of divers waters; but as derived from the 1426 Assize the gallon was 311 in³. The 330 in³ vessel was the merchants' standard, being a sixteenth greater than the legal 311 in³. The gallon contained 6 pints. The 1426 Assize doubled the size of the gallon to a weight of water of 20 pounds 8 ounces trois and was equal to 8 new pints each of 41 ounces. After 1426 the gallon is not mentioned in the statutes, the boll and firlot being given as so many pints.

Gill: A measure of liquid capacity, being ¹⁄₁₆ of a pint, ¼ of a mutchkin (q.v.).

Grain: The English Imperial troy grain is 64·8 mg. For the French grain see French weight (q.v.).

Hamburg barrel: 14 gallons (1597).[78]

Hesp: (See **Reel**.)

Horsegang: A land area, where two horsegangs equalled three cowlands (q.v.); a quarter of a ploughgate (q.v.).

House: (Dalriada) An ancient rental unit, the basis of levying tribute and producing men for military or naval service. Having a 'house' was the property qualification of a freeman in the eighth century; the houses were grouped under a leader and the most common grouping was one of 20 houses. Before the twelfth century, the 'house' had 3 cowlands. Two 'houses' or 6 cowlands made the small ploughland of Islay. (See Appendix B.)

Husbandland: A land area normally of 26 acres or 2 oxgates (q.v.), but also 'commonly 6 [correctly 26] aikers, quhair pleuch and syith may gang … Swa I find na certaine rule preceived anent the quantitie and valour of ane husband land'.[79]

Hutch: of peat is one-quarter of an ordinary cart load; of coal is one-fifth of a cartload.[80]

Inch: A linear measure; traditionally, the inch is the breadth of the thumb of an average-sized man measured at the root of the nail; alternatively, it is the length of three good barley corns end to end without tails. Now 25·4 mm.

Jowcat: A gill (q.v.) (1587).[81]

Kip: of hart horns weighs 10 stone (*c.*1500).[82]

Ladleful: A tariff of a quarter of a peck exacted by certain burghs (Glasgow, Dundee and Rutherglen) on each load of victuals entering the burgh (1592, 1633, 1662).[83] At Rutherglen this was short-lived, but Glasgow was still exacting the ladleful in 1793 though it had been commuted to a cash payment equal to the worth of half a peck.

Landmale: (See **Rent**.)

Last: of copper is 14 schippund (q.v.) (1597);[84]
of wool is 10 sacks (1597);[85]
of bere (in Orkney) is 1½ Scots chalders (1597);[86]
of cost or coist (a mixture of malt and bere or malt and barley), is 1 Scots chalder (Orkney) (1597);[87]
of goods generally is 12 barrels (q.v.) (1597);[88]
is 2 packs or 126 stone trois (1597);[89]
of rye is 18, sometimes 19 bolls (1597);[90] later 18? bolls (1624);[91]
of hides is 20 dacres each of 10 hides, making 200 hides (1597);[92]
of pitch or tar is 12 Hamburg barrels (q.v.), each of 14 gallons (1597);[93]
of goods from Danzig or of light goods is half a serplath (q.v.), commonly 12 or 14 barrels (1597);[94]
of salmon is 12 barrels[95] (for a discussion of salmon barrels, see Chapter **6**);
of wax is 14 schippund (q.v.) (1597);[96]
of red herrings is 10,000 fish (sixteenth century);[97]
of herrings is 20 loads each of 500 fish (sixteenth century);[98]
of herrings and fish generally, is 12 barrels (*c.*1500) (1624);[99]
of goods is 120 stone or 1920 pounds (1624);[100]
of corn is 24 herring barrels full (sixteenth century);[101]
of woolfels is 126 stone (1597);[102]
in Orkney is 24 meils.[103]

Last of land: (Shetland) A large unit of land or a group of marklands (q.v.), rented for 144 Shetland pennies, which equated to a mark of silver or 12 Shetland shillings.[104] Each mark was assessed a value, such as 6 pennies to the mark, so a last of such land would amount to 24 marks. (See Appendix D.)

Leat, Leet: A measure for peat being 'a stack 12 feet long, 12 feet broad and high in proportion'.[105]

Liberate: (See **Poundland**.)

Lippie (or **Lippy**): A forpit (q.v.).

Lispund: ⅟₂₀ of a shippund (q.v.) (*c.*1580),[106] or 18 pounds Scottish weight (1597).[107]
In Orkney a lispund is a weight for butter, equal to a setten (q.v.): during the eighteenth century 'Pundlar Process' it was agreed at 24 marks, with the mark given at 8 ounces, but as 1·25 pounds by the Earl of Morton; 24 marks of 8 ounces is 12 pounds, 24 marks of 1·25 pounds is 30 pounds. In 1826 the lispund was declared equal to 29 pounds 10¾ ounces avoirdupois. (See Appendix D.)

Load: of malt is 9 firlots (see Appendix E);
of lead is 24 fotmals (q.v.);[108]
of meal is 280 pounds avoirdupois, that is, 20 stone or 2 bolls (1844);[109]
of herring is 500 fish (sixteenth century).[110]

Male: is half a boll of 5 firlots (Orkney and Western Isles) (1587-8).[111]

Mark: Usually a weight of eight ounces, and in money £⅔ or 13s. 4d. The mark of silver in Shetland equalled 12 Shetland shillings or 144 Shetland pennies. The Shetland ounce (one-eighth of a mark) was 18 Shetland pennies, just as the ounceland was 18 pennylands.

Mark of land (**Markland, Merkland**): A mark of land was land yielding a mark of money or its equivalent in produce annually. In Orkney and Shetland it was originally the land purchasable for one mark of money;[112] but later the markland was quite variable in price.

Meil: (Orkney and Shetland) A weight of 6 settens or ⅟₂₄ of a last (q.v.).[113]

Mela: A twelfth-century weight for cheese or measure for grain.[114] The term is undefined in the texts.

Mese (**Maze, Maise**): of herrings is 500 fish (1597).[115]

Mett: In general a measure, usually of capacity.[116] The 200-herring mett of 42 pints requested by Crail from the Convention of Royal Burghs in 1707 was granted in 1722 and used in the Firth of Forth: five fills passed as 1,000 herring.[117] The mett of 66 pints for herring was used in the North and West.[118] For Scots coals at Dundee the mett or burden was 10½ stones Scots troye [for this to match the 1830 declaration, these stones must be trone weight, not troye]; for English coals, 54 Scottish pints, stricken measure (1829).[119] For Scots coals the mett was one hundredweight 5 stone 1¾ Imperial pounds avoirdupois (1830).[120] By English measure, English coals were two Imperial heaped bushels to the mett (1830).[121]

Mile: As in England, the Scots mile was 8 furlongs, but each furlong was of 40 falls, so the mile was 1,973⅓ yards or 1·12 English miles.[122] The 'common mile' was of 1,500 paces, each of 5 feet, hence was 2,500 yards or 1·42 English miles. (See Chapter **3**.)

Mutchkin: One-quarter of a Scots pint.

Nail (Nallis): One-sixteenth of several different sorts of units; for instance, as a weight, the nail for Scottish wool was 6 pounds (*c*.1400),[123] the pound being that of 7,000 English troy grains, namely the wool pound of Bruges. Six of these pounds made the clove (¹⁄₁₆ of the Bruges hundredweight of 96 pounds);
the nail for wool was later 7 pounds (*c*.1500),[124] being ¹⁄₁₆ of the hundredweight of 112 pounds;
as a unit of length, the nail of the yard was 2¼ inches, being ¹⁄₁₆ of 36 inches.[125]

Neiff: A handful. An exaction by the burgh of Glasgow of a handful from every weigh of wool or fleece that entered the burgh (1592).[126]

Old Extent: A survey of the worth of lands said to have been made in the reign of Alexander III, 1249-86 ('in time of peace'). Following the Wars of Independence, lands had been damaged (roughly by the mid-fourteenth century) and so ('in time of war') were worth less. This was only about half of what had previously been recorded, but marked recovery took place and the worth of land was latterly taken to be four times that recorded on the Auld Extent. This was termed the New Extent. (See **Oxgate** and **Ploughgate** (q.v.), and Appendix **B**.)[127]

Ounce (Once): Originally, in the Assize attributed to King David, a mass (weight) of 450 English troy grains and ¹⁄₁₅ of the then pound of 6,750 grains.
Trois ounce: (See **Trois Weight** below);
Trone ounce: an ounce 1¼ times larger than the current merchant's ounce. (See **Trone Weight** below);
English troy ounce: 480 grains;
Tower (England) and Cologne ounce: 450 grains;
Shetland ounce: 18 Shetland pennies.[128]

Ounceland: Notionally, this was land taxed at an ounce of silver annually, equal to 20 pennylands but in the Northern Counties and in Orkney equal to 18 penny-lands.

Oxgang (Oxgate, Oxgait) (Latin, *Bovate*): On the Old Extent (q.v.) this was land worth 5 shillings annually, so determined in 1585.[129] Four oxgates on the Old Extent

was a ploughland (q.v.) worth £1 with the oxgate equal to 52 acres. In 1541 it was worth 20 shillings in all duties annually, and amounted to 13 acres of land, in agreement with the ploughland of 8 oxgates being 104 acres.[130]

Pack: of woolskins was half a sack, containing 36 Prussian stones in weight, each stone being 28 pounds trois, totalling 1,008 pounds trois.[131]
of wool is 12 stone each of 24 pounds of white wool or 25½ pounds of 'laid' wool (wool to which there has been a light application of tar) (1797);[132]
of wadmel (q.v.) (Shetland) a length of 60 cuttells of this cloth.[133]

Peck: A quarter of a firlot (q.v.).[134]

Pennyland: Notionally, land taxed at a penny per year.

Piece: (See **Roule**.)

Pint: The 'old pint', the ale pint; before 1426, ⅛ of a gallon, which was legally 51·9 in³ but customarily ¹⁄₁₆ larger, 55·1 in³.
The 'small pint', one eighth of a gallon (1426), which was legally 77·8 in³ but customarily ¹⁄₁₆ larger, 82·7 in³.
the 'large' pint (*c*.1510), which was legally 103·7 in³ but customarily ¹⁄₁₆ larger, 110·2 in³.
the measure of the Stirling stoup or jug (*c*.1510), 103·7 in³.
the wine pint, which was legally 77·8 in³ but customarily ¹⁄₁₆ larger, 82·7 in³, with 704 pints to the tun (*c*.1400).[135] After 1426 both ale and wine pints were of the same size. (For pints see Chapter **6**.)

Ploughgate, ploughland (Latin, *Carucate*): Notionally, the ploughland of a team of 8 oxen; hence equal to 8 oxgangs (q.v.);[136] or 104 acres (twelfth to thirteenth century);[137] dropping to 52 acres then later raised to 104 acres. (See **Oxgang** and Appendix **B**.)
The small 'ploughgate' of Islay rented for £1 per year. (In early Islay there was no mention of davochs, ploughgates or pennylands, probably because the island was the demesne lands of the Lords of the Isles and was never subdivided.) (See **House**.)

Pound (Pund): A unit of mass (weight).
Early Scots pound of 15 ounces, each of 450 grains (thirteenth century);
Scots pound of 16 ounces, each of 450 grains (1426),[138] subsequently reduced to an ounce of about 433 grains. (See Chapters **5** and **8**.)
For trois pound, and trone pound, see below.

Poundland: of Old Extent (q.v.) was 52 acres or 4 oxgaits (mid to late thirteenth century).[139]

Quadrica: (See **Wayne**.)

Quart: A quarter of a gallon (q.v.).

Quarter Land: In ancient times, a quarter of a davach (q.v.) or 6 cowlands (q.v.).

Rath: (Of Irish origin, originating in seventh-century Dalriada.) In North Scotland, a husbandland;[140] alternatively, a homestead.[141]

Reel (Reill): A large winding frame for linen thread with a circumference of 2½ ells of 37 inches, 92.5 inches (usually termed ten-quarters of an ell), hence the 'ten quarters reel', one turn of which took up a 'thread' and 120 threads made a 'cutt'. Twelve cutts made a 'hesp', 'hank' or 'skein' by which the yarn could be sold by weight.[142]

Rent (Landmale): The annual due to the superior of the land. In Scotland this was usually stated in money terms, but invariably accepted in agreed amounts of produce. In Orkney rent was levied by the mark and was given as so many meils and settens (q.v.) per mark. In Shetland, the marks were assessed at so many pennies each partly paid in butter, partly in wadmel (q.v.). (See Appendix D.)

Rod (Rude): (See **Rood**.)

Rood (Rude): of land, a strip 40 falls long by 1 fall broad, or otherwise 40 square falls; a quarter acre.[143]

Rood of work: A square fall, 36 square ells. (In England, a square rod, 30¼ square yards.)

Roule or Piece: of sackcloth; 15 ells (1612).[144]
of missellanes (muslin?): 30 ells (1612);[145]
of broadcloth; ¾ ell broad and 18 ells at least in length; 1 ell broad and 24 ells at least in length (1693).[146]

Sack: of wool is 24 stone, each stone of 15 pounds, so equals 360 pounds Bruges wool weight, thus each pound is 7,000 English troy grains. (See Chapter 4.)[147]
Otherwise, a sack is 2 tuns (q.v.);[148]
of wool is 2 weighs (q.v.);[149]
of sheepskins is 500 skins;[150]
of goatskins is 680 skins;[151]
of woolfels, is for tax purposes, equal to 240 woolfels.[152]

Scattald: (Shetland) Originally, the arable land and hill grazing belonging to a township or group of townships; latterly, meaning only the township's common hill grazing. It was a fundamental unit of settlement.[153]

Scots weight: (See Scots pound under **Pound**.)

Seam (Sowme): of nails is 10,000 nails (c.1500).[154]

Sek: A measure of wool weighing about 680 pounds, probably intended to be a sarplar (q.v.) of 2 sacks net weight.[155]

Serplath (Serplaith, Sarpler): of merchandise, commonly 80 stone or 1,280 pounds (1527);[156]
a serplath is 2 sacks, that is, 2 tuns (1597);[157]
of lamb skins is 8,000 skins (1597);[158]
of 'cunning' (rabbit) skins is 16,000 skins;[159]
of 'futfelles' (skins of newborn lambs) is 4,000 skins.[160]

Setten (Setteen, Settin): (Orkney and Shetland) A weight for grain of 24 marks, 6 of which made a meil.[161] (See Appendix D.)

Sheaf: of arrows is 24 arrows (1318);[162]
of iron, 16 'gaddys' (bars);[163]
of steel, 30 'gaddys' (bars).[164]

Sheafland: A parcel of land of annual value of 28 shillings 1 pence; 7 sheaflands are 1½ oxgangs (q.v.) (1635).[165]

Shillingland: Land whose produce was worth a shilling annually under the 'Auld Extent' (q.v.).[166] Later, ⅟₄₀ of a ploughgate or 2·6 acres, the ploughgate having been established well before to be 104 acres.[167]

Shippund (Shippound): Was 16½ stone Scots trois [correctly, trone] weight,[168] or 20 lispunds (q.v.).[169]

Shok: of cloth is 28 ells (c.1500);[170]
of corn (shock), is 12 sheaves (1811).[171]

Skat: (Norse) A land tax levied in Orkney on ouncelands and pennylands (q.v.).

Skatland: (Orkney) A quarter of an ounceland or 4½ pennylands.

Sleek: A measure for fruit, 20 Scots pints (1816);
for onions, 16 Scots pints (1816);[172]
for fruit, 21½ pints of 103·7 in³, or 2,230 in³, or 43 pounds avoirdupois (1824);
for onions, 1,785 in³, or 39 pounds 5 ounces avoirdupois (1824);[173]
for potatoes, ⅛ boll (1829).[174]
for fruit, 21¼ pints (equal to 1618 wheat firlot); for onions, 17 pints (1832).[175]

Soum: the grassland which will support an ox or a cow. Two soums were required for a horse.

Span (Spann): (Orkney) A unit of weight for butter equal to 5 lispunds (q.v.).

Stone (Stane): a unit of mass (weight) for goods in bulk comprising a set number of pounds (latterly 16). First defined in the Assize attributed to David I as 15 pounds, where the pound was 15 Cologne or tower ounces, with the stone of wax at 8 pounds. The stone, of which 12 made the waw (wey) of wool, hence 24 to the woolsack in the same assize, was of 15 Bruges wool pounds. The stone of the 1426 Assize was of 16 merchant or 'Scots' pounds of 16 Cologne ounces (although defined as 15 trois pounds) and, from 1563, of 16 French or Flemish 16 ounce troye pounds.
Trone stone (Lanark stone) of 16 trone pounds was the weight unit for internal trade, equivalent to 20 merchant or 'Scots' pounds up to 1563 and subsequently 20 troye pounds. The trone stone was thus equivalent to 1¼ troye stone.
The Prussian stone was 28 pounds troye (1597).[176]

Stook: Ten or twelve sheaves of corn set up in a field to dry.[177]

Stoop (Stope): A liquid measure, sometimes specifically 2 pints;[178]
The Stirling stoop or jug was 103·7 in³. (See Chapter **6**.)

Terung: Gaelic equivalent of ounceland (q.v.).

Thrave: A number of sheaves of corn:
1 thrave is four sheaves (1523);[179]
1 thrave wheat is 28 sheaves and a thrave of barley or peas is 24 sheaves (1812);[180]
1 thrave is 24 sheaves (nineteenth-century).[181]

Thread: (See **Reel**.)

Timber (Timmer): of furs is 40 pelts (1597).[182]

Tonegall: (See **Cogall**.)

Trois (Troyis and Troye [Troas]) weight: Trois weights were first mentioned by statute in 1426 (in modern English): 'the stone shall weigh 15 true trois pounds and the stone shall be divided into 16 true Scots pounds.'[183]
The ounce of the trois pound is 480 grains; that of the 'Scots' pound is 450 grains. In the 1426 Assize, it was recorded that 'the stone [was] to contain 15 pounds trois and each trois pound to contain 16 ounces'.[184] Here the ounce is of 480 grains. (Equally, this could be described as a stone of 16 pounds, each of 16 ounces of 450 grains.) Trois weight gradually displaced 'Scots' weight in terms of this last definition.
The designation 'troy' or 'troye' was used for a

number of slightly different weights as time progressed. In about 1511 the official ounce of Scots troy weight was changed from 480 grains to the French ounce of 472½ grains. The ounce of Scots weight of 450 grains dropped correspondingly to about 443 grains, giving a 16-ounce pound of about 7,090 grains (equivalent to 15 Paris ounces of 472½ grains). Officially the ounce was 472½ grains Paris weight, but the merchants were using Flemish weight (trose), whose pound had an ounce of about 474·5 grains (while in the Low Countries the ounce was about 474·9 grains).
The Craigengelt Weight shows that at the time of its creation the stone had not yet risen to 16 pounds troye; however, the Assize of 1563 had as one of its aims the raising of the stone to 16 pounds, each pound of 16 ounces, each ounce of 472½ grains.
Ultimately in 1578 the administration appeared to capitulate to merchant usage and adopted Flemish weight. (See Chapter **8**.)

Tron: A large equal arm balance, or the place where the scale is located in the burgh market-place for weighing local produce or a similar device at ports for weighing imported and exported goods, thus enabling duties to be assessed.

Trone weight: The weight standard for locally produced commodities, the stone, pound and ounce of which was 1¼ times the merchants' stone and pound. (See Chapter **8**.)

Trose weight: (See **Trois Weight**.)

Tun: A French barrel standard for wine containing 704 Scots pints (c.1400);[185]
of merchandise, half a sack or a quarter of a serplath;[186]
a sack of 600 pounds troye weight (1597).[187]

Udal (Odal) law for land: Broadly, this refers to the whole system of Norse law based mainly on the Magnus code of 1274. In Orkney and Shetland it refers to land held without written title and held with an entail on the family. The udaller held his land of no one and owed no service, feudal or otherwise, to any superior. The udaller paid skat (q.v.) but no feu duty. Occupancy was tantamount to ownership especially as occupancy could pass down to the udaller's heirs. The land had to be offered for sale first to the family and if to be sold to others the family's consent was required. In certain cases, it still applies today.[188]

Urisland: (Orkney) A social-ecclesiastical district akin to an ounceland (eyrsland) but not necessarily containing 18 pennylands.[189]

Wadmel, Wadmal (Norse, **Vadmal**): (Shetland) A kind of coarse woollen cloth with which Shetlanders could pay part of their skat (q.v.) or rent.

Water measure: The capacity rate for bulk goods landed at ports, as much as 1¼ times the basic landward amount in both Scotland and England, which existed from early times. (See Chapters **6** and **7**.)

Wayne (Latin, *Quadrica*): A load of lead of 24 fotmals (q.v.) (fourteenth-century).[190]

Wey (Waw, Weigh): of wax is 12 stone each of 8 pounds (each of 7,000 grains) (twelfth to fourteenth centuries);[191]
of wax is 30 nails each of 6 pounds (*c*.1400);[192]
of wool is 30 nails each of 6 pounds (*c*.1400);[193]
of wool is half a sack (*c*.1400);[194]
of wool is 12 stone (each of 15 pounds) (*c*.1500);[195]
of salt is 40 bushels (1707);[196]
of barley is the measure of a boll and weighs 15 stone, or 14 stone if of last year's crop.[197]

1   Sir Archibald C. Lawrie, *Early Scottish Charters prior to 1153 [ESC]* (Glasgow, 1905), 116-9: no. 153, 1128-36; Sir William Craigie, *et al.* (eds), *A Dictionary of the Older Scottish Tongue from the Twelfth century to the End of the Seventeenth [DOST]*, 9 volumes (Chicago, London and Aberdeen, 1937-2002), I, 18.

2   T. Thomson and C. Innes (eds), *The Acts of the Parliaments of Scotland [APS]*, 13 volumes (Edinburgh, 1814-75), I, 198 (fourteenth century); Joseph Stevenson, *Documents Illustrative of the History of Scotland,* 2 volumes (Edinburgh, 1870), I, 275; Peter J. B. McNeill (ed), *The Practicks of Sir James Balfour of Pittendreich*, Stair Society nos. 21 and 22, 2 volumes (Edinburgh, 1962), I, 441.

3   John Henderson, *A General View of the Agriculture of Caithness* (London, 1812), 242.

4   G. Smellie, 'Parish of St. Andrew, Orkney', in *The New Statistical Account of Scotland,* 15 volumes (Edinburgh, 1840-5), XV, 182.

5   James Johnston, *Place Names of Scotland* (London, 1892, reprinted 1972), 89.

6   [John Swinton], *A Proposal for Uniformity of Weights and Measures in Scotland* [Swinton] (Edinburgh, 1779), 29.

7   Ibid. For various other barrels, see *DOST*, I, 195.

8   *APS*, XI, 295: 1705.

9   C. Innes (ed), *Ledger of David Haliburton, Conservator of the Privileges of the Scottish Nation in the Netherlands, 1492-1503 [Haliburton]* (Edinburgh, 1867), 336.

10 John Skene, *De Verborum Significatione* [Skene] (Edinburgh, 1597), s.v. *serplaith: weichtes and measures in Orknay.*

11 *Haliburton*, 293.

12 William Grant (ed), *Scottish National Dictionary [SND]*, 10 volumes (Edinburgh, 1931-76), s.v. *barrel.*

13 Ibid.

14 Alexander Huntar [*sic*], *A Treatise of Weights, Mets and Measures of Scotland* (Edinburgh, 1624), 3.

15 *Haliburton*, 339.

16 *APS*, III, 82-3: Act 4, 1573.

17 *Haliburton*, 338.

18 C. Innes, 'Book of Rates 1612', in *Haliburton*, 340.

19 *APS*, VII, 252: 1661.

20 *SND*, s.v. *barrel.*

21 *APS*, II, 178-9: Act 16.

22 *APS*, III, 82-3: Act 4, 1573 and J. D. Marwick, *et al.* (eds), *Records of the Convention of the Royal Burghs of Scotland, 1275-1779 [RCRB]*, 7 volumes (Edinburgh, 1866-1918), I, 482: 1 July 1596.

23 J. H. Burton and D. Masson (eds), *The Register of the Privy Council of Scotland [RPCS]*, 14 volumes (Edinburgh, 1877-87), first series, XII, 16-7: 15 July 1619, and *APS*, VII, 230: Act 245, 1661.

24 *Haliburton*, 339.

25 Hunter, op. cit. (14), 4.

26 *APS*, II, 119: Act 9, 1478; ibid., 178: Act 16, 1487; ibid., 213: Act 3: 1488; ibid., 237: Act 3, 1493.

27 *APS*, I, 668: *Assisa de Tolloneis.*

28 *DOST*, I, 260.

29 *APS*, VI, part II, 524: 4 August 1649.

30 Ibid.

31 Ibid.

32 John Sinclair (ed), *The Statistical Account of Scotland [OSA]*, 21 volumes (Edinburgh, 1790-9), III, 689: Eckford, Roxburghshire.

33 *RPCS*, third series, VI, 365-8: 1679.

34 *APS*, VIII, 609: Act 41, 1686.

35 John Stuart, *et al.* (eds), *The Exchequer Rolls of Scotland [ER]*, 23 volumes (Edinburgh, 1877-1908), I, 117: 1328: '… centum triginta vlnarum de karde, pro viginti quinque boltis', so 25 bolts equals 130 ells, and thus 1 bolt equals 5·2 ells. See also *DOST*, I, 298-9.

36 *Haliburton*, 304; see also *DOST*, I, 413.

37 Swinton, op. cit. (6), 72.

38 Ibid., 82.

39 Ibid., 117-8.

40 *APS*, VIII, 433: 1681.

41 *ESC*, 202: no. 251, 1152-3 and *APS*, I, 751: Act 16. See also *DOST*, I, 443.

42 *ER*, I, 55-6: 1326.

43 *ER*, XII, 698: 1505. See also *DOST*, I, 457.

44 Allen D. C. Simpson, 'Scots 'Trone' Weight: Preliminary Observations on the Origins of Scotland's Early Market Weights', in *Northern Studies*, 29 (1992), 62-81.

45 *ER*, I, 42 and 48: 1288-9. See also *DOST*, I, 477-8.

46 *ESC*, 10: no. XI, before 1093; ibid., 141: no. 179, 1147.

47 G. W. S. Barrow (ed), *The Acts of Malcolm IV King of Scots 1153-1165: Regesta Regum Scotorum [RRS]*, *volume I* (Edinburgh, 1960) 264: no. 243, 1163-4; *ESC*, 287, note to p. 27.

48 *ER*, I, 55-7: 1326.

49 *ER*, I, 49-50: 1290. An earlier reference is *ER*, I, 6: 1264.

50 William F. Skene, *Celtic Scotland*, second edition, 3 volumes (Edinburgh, 1876-80), III, 143; W. D. Lamont, '"House" and "Pennyland" in the Highlands and Islands', in *Scottish Studies*, 25 (1981), 65-76, see p. 65.

51 55 George III c.94 1815; John Alfred O'Keefe, *The Law of Weights and Measures* (London, 1966), 629; R. D. Connor, *The Weights and Measures of England [WME]* (London, 1987), 174-5.

52 Lamont, *op. cit.* (50), 65.

53 National Archives of Scotland (NAS) CS 7/175, ff.247-8: 24 June 1598. See also *DOST*, I, 797.

54 Brian Smith, 'The Humble Cuttell', in *The New Shetlander*, no. 165 (1988), 10; G. Goudie, *The Celtic and Scandinavian Antiquities of Shetland* (Edinburgh, 1904), 178.

55 Alison Hanham, 'A Medieval Scots Merchant's Handbook', in *Scottish Historical Review*, 50 (1971), 117; Skene, s.v. *serplaith*.

56 J. M. Thomson and J. H. Stevenson (eds), *Registrum Magni Sigilli Regum Scotorum [RMS]*, 11 volumes (Edinburgh, 1912), XI, 507: no. 1016, 1667.

57 *WME*, 37-8.

58 John Jamieson, *An Etymological Dictionary of the Scottish Language* [Jamieson], 4 volumes (Edinburgh, 1818), s.v. *darge*.

59 Jamieson and *SND*, s.v. *darge*.

60 *SND*, s.v. *darge*.

61 K. Jackson, *The Gaelic Notes in the Book of Deer* (Cambridge, 1972), 116.

62 A. McKerral, 'Ancient Denominations of Agricultural Land in Eastern Scotland', in *Proceedings of the Society of Antiquaries of Scotland*, 78 (1944) part I, 52; F. L. W. Thomas, 'Ancient Valuation of Land in the West of Scotland', in ibid., 20 (1886), 208; C. Innes, *Lectures on Scottish Legal Antiquities* (Edinburgh, 1872), 272-3. See also *DOST*, II, 13.

63 *Haliburton*, Appendix, cxiv.

64 George Buchanan, *Tables for Converting the Weights and Measures hitherto in use in Great Britain into those of the Imperial System* (Edinburgh, 1829), 198-9.

65 *SND*, s.v. *ell*.

66 W. P. L. Thomson, 'Ouncelands and Pennylands in Orkney and Shetland', in L. J. MacGregor and B. E. Crawford (eds), *Ouncelands and Pennylands* (St Andrews, 1987), 24-45, see pp. 31-2.

67 *APS*, I, 751: Act 16. See also *DOST*, II, 398.

68 Sinclair, *OSA*, XIX, 413: 1797, Ophir in Orkney.

69 John Stuart (ed), *Extracts from the Council Register of Aberdeen 1398-1510*, Spalding Club no. 12 (Aberdeen, 1844), 32: 3 November 1474. Also given by E. Gemmill and N. Mayhew, *Changing Values in Medieval Scotland: A Study of Prices, Money, Weights and Measures* (Cambridge, 1995), 393: '3 fothers lead each containing 6 score (120) stones', from the manuscript Aberdeen Council Registers, VI, 316.

70 Skene, s.v. *serplaith*. See also *DOST*, II, 466.

71 Innes, 'Book of Rates', in *Haliburton*, 338.

72 Hunter, op. cit. (14), 3.

73 *APS*, VI, part II, 82: 22 May 1655.

74 *APS*, VII, 252: 1661.

75 J. E. Thorold Rogers, *A History of Agriculture and Prices in England from the Year after the Oxford Parliament (1259) to the Commencement of the Continental War (1793)*, 7 volumes (Oxford, 1866-1902), I, 168.

76 *Second Report of Commissioners to Consider the Subject of Weights and Measures, 13 July 1820*, Parliamentary Papers 1820 (HC314), VII, appendix A.

77 Samuel Jeake, *Logisticelogia, or Arithmetic Surweighed and Reviewed*, second edition (London, 1696), 80. See also *DOST*, II, 546.

78 Skene, s.v. *serplaith*. See also *DOST*, III, 31.

79 Skene, s.v. *husbandland*. See also *DOST*, III, 187.

80 Sinclair, *OSA*, IV, 277: 1792, Kilmacolm and Renfrew; Jamieson gives this as 2 Winchester bushels, s.v. *hutch*.

81 *APS*, III, 521: Act 136, 1587. See also *DOST*, III, 358.

82 *Haliburton*, 335.

83 *RCRB*, I, 381: 12 June 1592; ibid., I, 398: 12 June 1593; *APS*, V, 48-9: Act 34, 1633; David Ure, *The History of the Rutherglen and East Kilbride* (Glasgow, 1793, reprinted 1981), 46-8. See also *DOST*, III, 503.

84 *Haliburton*, appendix, cxv: 22 May 1597; see also Skene, s.v. *bullion*.

85 *APS*, I, 668: *Assisa de Tolloneis*; Skene, s.v. *serplaith*.

86 Skene, s.v. *serplaith: weichtes and measures in Orknay*.

87 Ibid.

88 Ibid., s.v. *serplaith*.

89 Ibid.

90 Ibid.

91 Hunter, op. cit. (14), 5.

92 Skene, s.v. *serplaith*.

93 Ibid.

94 Ibid.

95 Ibid.

96 *Haliburton*, cxv: 22 May 1597; see also Skene, s.v. *bullion*.

97 McNeill, op. cit. (2), I, 87.

98 Ibid., 88.

99 *Haliburton*, 353; Hunter, op. cit. (14), 4.

100 Hunter, op. cit. (14), 3.

101 McNeill, op. cit. (2), I, 87.

102 Skene, s.v. *serplaith*.

103 William P. L. Thomson, *New History of Orkney* (Edinburgh, 2001), 451; see also, Skene, s.v. *serplaith: weichtes and measures in Orknay*.

104 Thomson, op. cit. (66), 28; and communication from Brian Smith, Archivist, Shetland Islands Council, Lerwick, 22 March 1991. See also *DOST*, III, 565.

105 Sinclair, *OSA*, XIII, note 71: 1794, Kennethmont, Aberdeenshire.

106 McNeill, op. cit. (2), I, 88.

107 Skene, s.v. *serplaith: weichtes and measures in Orknay*. See also *DOST*, III, 810.

108 *APS*, I, 669: *Assisa de Tolloneis*.

109 Poor Law (Scotland) Enquiry 1844: Session 1 February-5 September 1844 (Reports from Commissioners), XXIII, appendix, part IV. Also given in *Parliamentary Papers* XXIII, appendix, part IV, 459.

110 McNeill, op. cit. (2), I, 88.

111 Donald Gregory and William F. Skene (eds), *Collectanea de Rebus Albanicis, consisting of Original Papers and Documents relating to the History of the Highlands and Islands of Scotland* (Edinburgh, 1847), 172-3, notes 5 and 6.

112 Thomson, op. cit. (103), 259, 261. See also *DOST*, IV, 108-9.

113 Thomson, op. cit. (103), 259, 261; see also Skene, s.v. *serplaith: weichtes and measures in Orknay*; Swinton, op. cit. (6), 106-7; and *DOST*, IV, 168.

114 *ESC*, 210: no. CCLXIII, *c.*1153; *RRS*, I, 264: no. 243, *c.*1164; ibid., II, 348: no. 347, 1189 × 1195.

115 Skene, s.v. *mese*.

116 *DOST*, IV, 226-9.

117 *RCRB*, IV, 406: 5 July 1707; ibid., V, 314: 6 July 1722.

118 Ibid., V, 314: 6 July 1722.

119 Buchanan, op. cit. (64), 214 and 218.

120 William Shiress, *Tables for Converting the Weights and Measures hitherto in use in Forfar and Kincardineshire into the Imperial Standards* (Brechin, 1830), 181, 188, 215.

121 Ibid.

122 Hunter, op. cit. (14), 9-10; also *WME*, 68-78.

123 Hanham, op. cit. (55), 115; *DOST*, IV, 456.

124 *Haliburton*, 354.

125 *WME*, 84; *DOST*, IV, 456.

126 *RCRB*, I, 381: 15 June 1592; ibid., 398: 12 June 1593.

127 J. D. Mackie (ed), *Thomas Thomson's Memorial on the Old Extent*, Stair Society no. 10 (Edinburgh, 1946). See also Skene, s.v. *extent*, and *APS*, I, 121: 1327.

128 Thomson, op. cit. (66), 28

129 Skene, s.v. *bovata terrae* (citing Lords of the Exchequer, 11 March 1585).

130 *APS*, I, 751: fourteenth century; Innes, op. cit. (62), 272. The full judgement of 11 March 1585 is given in Mackie, op. cit. (127), 10. See also *DOST*, V, 240-1.

131 Skene, s.v. *serplaith*.

132 Sinclair, *OSA*, II, 304 (Galashiels) and VIII, 528 (Hawick).

133 Smith, op. cit. (54), 11.

134 *DOST*, V, 381-2.

135 Hanham, op. cit. (55), 119.

136 *APS*, I, 751: *Fragmenta*.

137 Ibid*.; ESC*, 202: no. 251, 1152 × 1153; Joseph Bain, *et al.* (eds), *Calendar of Documents Relating to Scotland preserved in Her Majesty's Public Record Office, London, AD 1272-1307*, 4 volumes (Edinburgh, 1881-1988), II, 226: no. 856, 1296.

138 *APS*, II, 10: Act 14, 1426; see also *DOST*, VI, 82.

139 Innes, op. cit. (62), 283; Mackie, op. cit. (127), 312; *DOST*, IV, 83.

140 Skene, op. cit. (50), III, 243.

141 Ibid.

142 *APS*, IX, 311: Act 48, 1693.

143 *APS*, I, 751: *Fragmenta*, late thirteenth century. See also *DOST*, VII, 549.

144 *Haliburton*, 326.

145 *Haliburton*, 322: Customs 1612.

146 *APS*, IX, 312: 1693.

147 *Haliburton*, 360; Skene, s.v. *serplaith*.

148 Skene, s.v. *bullion*.

149 *APS*, I, 668: *Assisa de Tolloneis*, section 5; *APS*, VI, part II, 829: 1655.

150 Skene, s.v. *bullion*; also *Haliburton*, appendix, cxvi.

151 Ibid.

152 Iris Origo, *The Merchant of Prato* (Harmondsworth, 1963), 72.

153 Brian Smith, 'What is a Scattald?: Rural Communities in Shetland 1400-1900', in B. E. Crawford (ed), *Essays in Shetland History* (Lerwick, 1984), 99-124. See also *DOST*, VIII, 230.

154 *Haliburton*, 322.

155 *Haliburton*, 358. See also *DOST*, VIII, 501.

156 Skene, s.v. *serplaith*.

157 *Haliburton*, cxv: Customs list of 22 May 1597. Skene, s.v. *serplaith*.

158 Skene, s.v. *bullion*.

159 Ibid.

160 Ibid.

161 Thomson, op. cit. (103), 259.

162 *APS*, I, 474: 1318.

163 *APS*, I, 670: *Assisa de Tolloneis*.

164 Ibid.

165 *APS*, II, 470: 1635.

166 *SND*, s.v. *shillingland*.

167 *APS*, I, 751: *Fragmenta*.

168 Skene, s.v. *serplaith*.

169 McNeill, op. cit. (2), I, 88.

170 *Haliburton*, Customs, 258: 1612; ibid., glossary, 358: 1612.

171 James Trotter, *A General View of the Agriculture of West Lothian* (Edinburgh, 1811), 96.

172 James Cleland, 'Inventory of Weights and Measures belonging to the City of Glasgow, taken by James Cleland, 22 November 1816': City of Glasgow Archives, Mitchell Library, Glasgow, MS volume A2.1.3 (Inv. XIII (5) 3). See also Agnes Baird,

A Review of the Historical and Topographical Works of James Cleland LL.D. (Glasgow, 1830).

173 John Sinclair, *Analysis of the Statistical Account* (Edinburgh, 1825), Appendix, 29.

174 Buchanan, op. cit. (64), 210.

175 Baird, op. cit. (175).

176 Skene, s.v. *serplaith*.

177 *SND*, s.v. *stook*.

178 Hanham, op. cit. (55), 119.

179 Sir James Balfour Paul, *et al.* (eds), *Accounts of the Lord High Treasurer 1473-1580*, 13 volumes (Edinburgh, 1877-1978), V, 230: 27 October 1523.

180 Sir John Sinclair, *An Account of the Systems of Husbandry adopted in the more Improved Districts of Scotland*, second edition, 2 volumes (Edinburgh, 1813), I, 330.

181 *SND*, s.v. *thrave*; Jamieson, s.v. *thrave*; *Second Report of the Commissioners on Weights and Measures, 13 July 1820*, op. cit. (76), 35.

182 *Haliburton*, 359 and 305-7; Skene, s.v. *timbria*.

183 *APS*, II, 10: Act 14, 1426.

184 *APS*, II, 12: Act 22, 1426.

185 Hanham, op. cit. (55), 107-20; see p. 119, note 5.

186 *Haliburton*, cxv.

187 Skene, s.v. *serplaith*.

188 William Jardine Dobie, 'Udal Law', in Hector McKechnie (ed) *An Introductory Survey of the Sources and Literature of Scots Law*, Stair Society no. 1 (Edinburgh, 1936), 445-60; see also Michael R. H. Jones, 'Perceptions of Udal Law in Orkney and Shetland', in Doreen J. Waugh and Brian Smith (eds), *Shetland's Northern Links: Language and History* (Edinburgh, 1996), 186-204, and D. J. Cusine, 'Udal Law', in *Northern Studies*, 32 (1997), 33-45.

189 Thomson, op. cit. (66), 31-2.

190 *APS*, I, 669: *Assisa de Tolloneis*.

191 *APS*, I, 673: Assize of David.

192 Hanham, op. cit. (55), 119.

193 Ibid.

194 Ibid.

195 *Haliburton*, 360.

196 *APS*, XI, 407, 448: Exemplification of the Act of Union, Article 8, 15 January and 19 March 1707.

197 *RPCS*, third series, VI, 530: 10 August 1680.

JOHN WHITE & SON'S

ILLUSTRATED

PRICE LIST

OF

SCALE BEAMS, COUNTER BALANCES,

AND

PLATFORM WEIGHING MACHINES.

MANUFACTORY ESTABLISHED 1760.

JOHN WHITE & SON,

Scale Works,

AUCHTERMUCHTY.

APPENDIX G

DIRECTORY OF SCOTTISH SCALE, WEIGHT AND MEASURE MAKERS TO 1900

THE MAKERS OF SCALES, WEIGHTS AND MEASURES WERE METALWORKERS. In the medieval period those who can be identified – such as Hans Cochran, maker of the 'Craigengelt' weight of 1553 (Item **31** in the Inventory), or David Rowan, who constructed the standard half-pint measure for Edinburgh in 1555 (Item **109**) – were involved with important crafts in the Scottish Mint, or with the construction of heavy artillery. They were not specialist scale or weight makers, and often they were not Scots. Many of the contractors of the early standards remain anonymous, while other precision weights and measures were purchased by the royal burghs from European centres which specialised in the production of such items, in particular from Nuremberg (see Items **33**, **38**, **39** and **88**). The continental influence on Scots metrology is shown in the shapes and sizes of eighteenth-century weights and measures, while the artefacts were increasingly produced in Scotland. However, the makers and retailers whose names we find on surviving weights and measures are only occasionally found to be also makers of precision instruments, and are much more likely to be those involved in heavy smithwork and the embryonic engineering trades.

The most obvious source available which can be used to uncover the names and numbers of tradesmen and firms connected with the manufacture and retail of weights and measures – not just the standards described in the Inventory – has been the local street directories, which started to appear as a direct result of increasing commerce both within Scotland and with the rest of the United Kingdom from the mid-eighteenth century onwards. Despite the knowledge that there are a number of inherent problems in using directories, the information found in them has often been taken at face value. These problems concern the original compilation of the directories, a lack of information concerning the motives of the contributors, the topicality and veracity of the information, and whether entries had to be paid for. Directories, produced intermittently, are a flawed source, yet remain the most important starting point for uncovering the names and probable longevity of

Opposite:
John White & Son's factory from a late nineteenth-century catalogue in the firm's archive.

businesses, and have been used as such here. Edinburgh directories were produced almost on an annual basis from 1773, Glasgow intermittently from 1783, Dundee from 1809 and Aberdeen from 1824. The local history libraries of each centre have runs of these volumes, and many are to be found in the National Library of Scotland.

Other smaller towns, such as Paisley and Greenock, occasionally produced directories, and there were also larger compendiums produced in the English Midlands which listed the names and occupations of tradesmen in towns across Scotland, undertaken first by Bailey in 1781, and subsequently by Holden, Pigot, Slater, Kelly and ultimately, by the Post Office (PO).[1] All these have been used to uncover the extent of various related trades such as 'beam-maker' 'scale-maker' 'smith' and 'weighing machine maker'.

We have also included the specialist scientific instrument makers known to have made balances or specialised metrological apparatus: for these, we have relied on work in this area, in particular the published work of the National Museums of Scotland.[2] Over the years, we have exchanged information with the late Michael Crawforth, and his wife, Diana Crawforth-Hitchins, and we have contributed to and admired the journal of the International Society of Antique Scale Collectors, *Equilibrium*. We are grateful to the Crawforths for their help and encouragement over the years, particularly in the complicated area of identifying individual craftsmen.[3] Another major source has been the objects themselves. Where we have found makers' names, we have investigated them. Often a name has led us to further research, and to check local newspapers and archives. For access to these, we again acknowledge the assistance of the curators and collectors of these items, and the librarians and archivists who have helped us compile this directory. We would particularly like to thank Edwin White of John White & Son, Auchtermuchty, who helped us with the history of his family's firm.

Each directory entry follows the same pattern, based on that used by Gloria Clifton, successor to Michael Crawforth at Project SIMON (Scientific Instrument Makers, Objects and Notes), whom we also would like to acknowledge for assistance.[4] First, there is the trading name, together with the town and dates, if known. This is followed by listed addresses with dates (if at all possible) in chronological order, followed by descriptions as they appear in the directories. Finally, there is a small section indicating precursors or successors, or any useful literature. This includes examples of items which are marked with the manufacturers' name, and the current location. For makers who pre-date the advent of the street directories, but whose names are known from objects or literature, the sources for these are given.

Although the contract for the manufacture of the official standards – in particular, those for 1707 and 1824 – was won by London manufacturers, the necessity of having local standards produced by indigenous makers soon arose. We have therefore discussed the role of the London makers – John Snart, John Savidge and Robert Brettell Bate[5] – in the main text, and in the Inventory where the particular items they made for a Scottish customer are described.

NOTES AND REFERENCES

1   The nature of directories, their history and listings, are given in C. W. F. Goss, *The London Directories 1677-1855* (London, 1932); J. E. Norton, *Guide to the National and Provincial Directories of England and Wales, excluding London, published before 1856* (London, 1950) and Gareth Shaw and Allison Tipper, *British Directories: A Bibliography and Guide to Directories published in England and Wales (1850-1950) and Scotland (1773-1950)* (Leicester, 1988). See also Gareth Shaw, 'British Directories as Sources in Historical Geography', in *Historical Geography Research Series*, no. 8 (1982).

2   Much of this is covered in T. N. Clarke, A. D. Morrison-Low and A. D. C. Simpson, *Brass & Glass: Scientific Instrument Making Workshops in Scotland* (Edinburgh, 1989).

3   Another specialist area is that of coin-scales, and here we acknowledge the work of Paul and Bente Withers, especially their book *British Coin-Weights: A Corpus of the Coin-Weights made for use in England, Scotland and Ireland* (Llanfyllin, 1993); and also the assistance of our colleague in the National Museums of Scotland, Nick Holmes.

4   Gloria Clifton, *Directory of British Scientific Instrument Makers 1550-1851* (London, 1995).

5   For a business biography of Bate, see Anita McConnell, *R. B. Bate of the Poultry 1782-1847: the Life and Times of a Scientific Instrument Maker* (London, 1993), esp. pp. 19-28 for his role in constructing the 1824 Imperial standards.

SCOTTISH SCALE, WEIGHT
AND MEASURE MAKERS AND RETAILERS

ADAMS, C. & R., Glasgow, 1844-50

1. 181 Trongate, Glasgow, 1844-9
2. 49 Trongate, Glasgow, 1850

- Smiths and scale beam makers: PO Glasgow Directory 1844-50

John Malloch (q.v.) previously at first address; succeeded by C. Adams (q.v.).

ADAMS, C., Glasgow, 1851-64

49 Trongate, Glasgow, 1851-64

- Smith and scale beam maker: PO Glasgow Directory 1851-54
- Scale beam maker: PO Glasgow Directory 1855-64

Previously C. & R. Adams (q.v.).

ADAM, James, Glasgow, 1816-27

1. 48 Rutherglen Loan, Glasgow, 1817-20
2. 47 Rutherglen Loan, Glasgow, 1821-3
3. 46 Rutherglen Loan, Glasgow, 1824-5
4. 71 Rutherglen Loan, Glasgow, 1826
5. 77 Rutherglen Loan, Glasgow, 1827

- Scale beam maker: McFeat's Glasgow Directory 1817
- Scale and beam maker: Glasgow Directory 1818-27

Admitted member of the Incorporation of Hammermen of Glasgow 1816 as a smith stranger, with a 'small weighing beam' as essay: Harry Lumsden and P. Henderson Aitken, *History of the Hammermen of Glasgow* (Paisley, 1912), 307.

ADAM, James, Kilmarnock, 1820-3

Lower Glencairn Street

- Beam and scale maker: Pigot's Directory 1820-3

ADIE, Alexander, Edinburgh, 1822-34

1. 15 Nicholson Street, Edinburgh, 1822-9
2. 58 Princes Street, Edinburgh, 1830-4

- Mathematical, optical and philosophical instrument maker: PO Edinburgh Directory 1823-34

Previously in business with his uncle, John Miller *junior* (son of John Miller *senior* [q.v.]), Alexander Adie (1775-1858) was involved in metrological adjustment work on the Edinburgh weights in 1817, after the death of John Robison, with James Jardine (Item **106**); with making a copy of the standard ell (Item **27**), and with adjusting surveyors' chains from Scots measure to Imperial standard. Advertisement in the *Scotsman*, 14 October 1826: 'Notice is hereby given, that a STANDARD of the IMPERIAL CHAIN of 66 feet has lately been marked off with great accuracy, on the parapet wall in front of the College Buildings, by Mr Adie, optician, under the direction of Mr Wallace, Professor of Mathematics in the University … Those situated at a distance may have their chains adjusted, by transmitting them to Mr Adie, who has kindly undertaken to do so at a moderate charge for each.' See Adie & Wedderburn below.

Succeeded by Adie & Son (q.v.).

ADIE (Alexander) & Son, Edinburgh, 1835-80

1. 58 Princes Street, Edinburgh, 1835-43
2. 50 Princes Street, Edinburgh, 1844-76

3. 37 Hanover Street, Edinburgh, 1877-80

- Mathematical, optical and philosophical instrument makers: PO Edinburgh Directory 1835-80.

Alexander Adie was in business with his eldest son, John Adie (1805-57), and the firm produced a wide range of products, including a chondrometer, or device for measuring corn (now NMS T.1965.38), and a Robinson-type balance (now NMS T.1968.25). See J. T. Stock and D. J. Bryden, 'A Robinson Balance by Adie & Son of Edinburgh', in *Technology and Culture*, 13 (1972), 44-54. See Adie & Wedderburn below.

Succeeded by Adie & Wedderburn (q.v.).

ADIE & WEDDERBURN, Edinburgh, 1881-1900+

1. 37 Hanover Street, Edinburgh, 1881-2
2. 17 Hanover Street, Edinburgh, 1883-1902
3. 33 Hanover Street, Edinburgh, 1903-8
4. 52 George Street, Edinburgh, 1909-13

- Mathematical, optical and philosophical instrument makers: PO Edinburgh Directory 1881-1913.

Known to have sold a Walker balance.

After the death of both proprietors by 1858, the firm had been run by Alexander Adie's third son, Richard (1810-81); after his death the business was run by the foreman, Thomas Wedderburn, under whose name it traded until 1913, although Wedderburn had died in 1886. The firm continued, first as Richardson, Adie & Co., and subsequently from 1918 as a limited company, until 1933. For the history of the firm at each stage, see the chapter 'The Adie Business' in T. N. Clarke, A. D. Morrison-Low and A. D. C. Simpson, *Brass & Glass: Scientific Instrument Making Workshops in Scotland* (Edinburgh, 1989), 25-74.

AITKEN, Thomas, Kirkwall, 1730-54

As Assayer of the Orkney weights, Aitken was called before the Dean of Guild in 1743 to explain where he had obtained them: James Mackenzie, *The General Grievances and Oppression of the Isles of Orkney and Shetland* (Edinburgh, 1750, reprinted 1836), 66. Described as 'Wright, Maker and Adjuster of the Pundars and Bysmars in Kirkwall', Aitken was a witness in a court action brought by Alexander Earl of Galloway, and others against James, Earl of Morton, the so-called 'Pundlar Process', 1733-59 (see Appendix D); he had succeeded Thomas Foubister (q.v.) as Assayer in 1730; Thomas Aitken 'was Apprentice five years to the said Thomas Foubister (his predecessor in Office)': *Memorial and Abstract of the Proof for the Earl of Galloway and Others, Udalmen, and Proprietors of land in Orkney, Pursuers, against James Earl of Morton, Defender* (Edinburgh, 1758), 30, 33, 89-90; *Memorial for James, Earl of Morton, Defender, against Alexander, Earl of Galloway … Pursuers* (Edinburgh, 1758), 17, 23 and 42.

ANDERSON, David, Glasgow, 1855-8

1. 84 & 86 Stockwell Street, Glasgow, 1855-7
2. Workshop 29 Turner's Court, Glasgow; warehouse 84 & 86 Stockwell Street, Glasgow, 1858

- Ironmonger and tinsmith: PO Glasgow Directory 1854
- Ironmonger, tinsmith and gas-fitter: PO Glasgow Directory 1856
- Ironmonger, tinsmith and grocers' shopfitting manufacturers: PO Glasgow Directory 1857
- Ironmonger, standard beam and weighing machine maker, tinsmith, gas-fitter and bell-hanger: PO Glasgow Directory 1858

Succeeded by Anderson Brothers (q.v.).

ANDERSON Brothers, Glasgow, 1859-1900+

1. 93 Stockwell Street, Glasgow, 1859-85
2. Sale shop 35 Stockwell Street, Glasgow; works 81 & 91 Stockwell Street, Glasgow, 1886
3. 35 Stockwell Street, Glasgow, 1887-1900+

- General ironmongers, standard beam and weighing machine makers, tinsmiths, gas-fitters and coppersmiths: PO Glasgow Directory 1859
- Ironmongers, tinsmiths, general grocers' outfitters, standard beam and weighing machine makers, japanners, bellhangers, gas-fitters and coppersmiths: PO Glasgow Directory 1860-1
- Ironmongers, tinsmiths and gas-fitters, grocers' canisters, and standard beam and weighing machine makers: PO Glasgow Directory 1862-7
- [variations on the above]: PO Glasgow Directory 1868-1900+

Beam scales by Anderson Brothers of Glasgow: NMS T.1981.96.

ANDERSON, James *younger*, Glasgow, *fl.*1712
Admitted member of the Incorporation of Hammermen of Glasgow, 1712, with an essay which included 'ane tinn pint stoup': Harry Lumsden and P. Henderson Aitken, *History of the Hammermen of Glasgow* (Paisley, 1912), 291.

ANDERSON, James, Glasgow, 1876-81

1. 126 Renfield Street, Glasgow, 1876-8
2. 69 Stockwell Street, Glasgow; works 140 Stockwell Street, Glasgow, 1879
3. 69, 117 & 140 Stockwell Street, Glasgow, 1881

- Scale beam and weighing machine maker, etc.: PO Glasgow Directory 1876-8
- Shopfitter and weighing machine maker: PO Glasgow Directory 1879
- Weighing machine maker and grocers' outfitter: PO Glasgow Directory 1880

Succeeded by Anderson & Co. (q.v.) at 140 Stockwell Street, Glasgow; and by D. Spence (q.v.) at 69 Stockwell Street, Glasgow.

ANDERSON & Co., Glasgow, 1881
140 Stockwell Street, Glasgow, 1881

- Weighing machine makers: PO Glasgow Directory 1881

Previously James Anderson (q.v.); succeeded by James Anderson & Son (q.v.).

ANDERSON, James & Son, Glasgow, 1882
62, 63, 69 & 140 Stockwell Street, Glasgow, 1882

- Ironmongers and scale makers: PO Glasgow Directory 1882

Previously Anderson & Co. (q.v.); succeeded by D. Spence (q.v.) at 69 Stockwell Street, Glasgow.

ANDERSON, James, Glasgow, 1872-4

1. 21 Macfarlane Street, Glasgow, 1872
2. 27 Graeme Street, Glasgow, 1873-4

- Grate, fender, fire-iron, shovel, ashpan and beam maker and grinder: PO Glasgow Directory 1872
- Ironmonger, smith and grinder: PO Glasgow Directory 1873-4

Succeeded by Anderson & Rankin (q.v.).

ANDERSON (James) & RANKIN (James), Glasgow, 1875-8

1. 123 & 129 Graeme Street, Glasgow, 1875

2. Gallowsmuir Ironworks, 70 & 72 Port Dundas Road, Glasgow, 1876-8

• Smiths, grate, fender and beam manufacturers, steam power grinders, and ironmongers: PO Glasgow Directory 1875
• Manufacturing ironmongers and wholesale hardware merchants, smiths, steam power grinders and polishers … manufacturers of … beams and standard iron bedsteads: PO Glasgow Directory 1876-8

Previously James Anderson (q.v.).

ANDERSON, James M., Glasgow, 1852-60
1. 15 Main Street, Gorbals, Glasgow, 1852-4
2. 42 East Howard Street, Glasgow, 1855
3. 41 & 43 East Howard Street, Glasgow, 1856
4. 32 Great Clyde Street, Glasgow 1857-9
5. 13½ East Howard Street, Glasgow, 1860

• Copper and tinsmith, gas-fitter, bellhanger and ironmonger: PO Glasgow Directory 1852-3
• Copper and tinsmith, bellhanger and ironmonger: PO Glasgow Directory 1854
• Tinsmith, and gas-fitter, bellhanger, coppersmith. Beams, gasaliers and grocers' furnishings made to order: PO Glasgow Directory 1855
• Tinsmith, bellhanger and gas-fitter: PO Glasgow Directory 1856
• Tinsmith, gas-fitter, ironmonger, beam maker, japanner and grocers' outfitters: PO Glasgow Directory 1857-60

ANDERSON, John, Glasgow, fl.1726
Admitted member of the Incorporation of Hammermen of Glasgow 1726, with an essay which included 'a pint stoup of pewter': Harry Lumsden and P. Henderson Aitken, *History of the Hammermen of Glasgow* (Paisley, 1912), 293.

ARMSTRONG, Henry, Edinburgh, 1806-33
1. Niddry Street, Edinburgh, 1806-10
2. 14 Niddry Street, Edinburgh, 1811-25
3. 12 Niddry Street, Edinburgh, 1826-7, 1831-3
4. 13 Niddry Street, Edinburgh, 1828-30

• Coppersmith: Edinburgh Directory 1806-33

Previously William Armstrong, coppersmith, West Bow, 1773-86; Thomas Armstrong, coppersmith, West Bow, 1788; Thomas Armstrong, coppersmith, Niddry Street, 1793-1803; T. & H. Armstrong, coppersmiths, Niddry Street, 1804-5; maker of barley and wheat firlots for Aberdeen in 1810 (see Item **222**).

ARMSTRONG, William & George, Edinburgh, 1790-4
1. Grassmarket, Edinburgh, 1790
2. Niddry Street, Edinburgh, 1794

• Founders and ironmongers: Williamson's Edinburgh Directory 1790
• Ironmongers and brass founders: Williamson's Edinburgh Directory 1794

Stamp 'W&GA EDIN' on a set of English avoirdupois weights at Dundee City Council, Trading Standards office, Dundee (Dundee City Council, Environmental and Consumer Protection), and a set of cup weights at North Berwick.

Previously Mrs Armstrong, founder, Castle Wynd, 1774-80; William Armstrong, founder, Castle Wynd, 1782-6. Subsequently Armstrong & Co., brass founders, Niddry Street, 1796-1800; William Armstrong, founder, Niddry Street, 1805-10; W. & R. Armstrong, brass founders, 24 Niddry Street, 1820.

AULD, Patrick, Glasgow, 1828-33
1. 34 Queen's Street, Glasgow, 1828-9

2. 34 & 36 Queen Street, Glasgow, 1830-2
3. 36 Queen Street, Glasgow, 1832-3

- Coppersmith: PO Glasgow Directory 1828-9
- Coppersmith, pewterer, tinplate worker and brass founder: PO Glasgow Directory 1830
- Coppersmith, etc.: PO Glasgow Directory 1831-3

Patrick Auld, merchant, admitted member of the Incorporation of Hammermen of Glasgow 1815, with an essay of 'a sauce pan': Harry Lumsden and P. Henderson Aitken, *History of the Hammermen of Glasgow* (Paisley, 1912), 306. He first appears in the Glasgow Directory for 1818, at 57 Queen Street. In 1827, still listed as a merchant, he is associated with J. & H. Wardrop (q.v.). From 1831-3, he was associated with Patrick C. Auld, coppersmith, etc., at the same address.

AUTOMATIC WEIGHING MACHINE Co. Ltd, Edinburgh, 1887-93
1. 8 York Buildings, Edinburgh, 1887
2. 15 Queen Street, Edinburgh, 1888-93

- Scales and steelyard makers: PO Edinburgh Directory 1887-93.

AVERY, W. & T., Ltd, Aberdeen, 1900+
46A Queen Street, Aberdeen, 1900

- Weighing machine and scale makers: PO Aberdeen Directory 1900

AVERY, W. & T., Ltd (Agent A. Wood & Sons [Glasgow]), Edinburgh, 1897-1900+
1. 68 Lady Lawson Street, Edinburgh, 1897
2. 23 Bread Street, Edinburgh, 1898-9
3. St James' Court (repair depot), 501 Lawnmarket, Edinburgh, 1900+

- Manufacturers of all sizes of weighing apparatus for railways, collieries, etc., shop and store outfitters for grocers, butchers, etc., estimates given for repair and upkeep of scales and weighing machines: PO Edinburgh Directory 1897-1900+ (and see A. Wood & Sons, Glasgow, below)

AVERY, W. & T., Glasgow (see Alex Craig, 1872-8)

AVERY, W. & T., Glasgow, 1889-1900+
1. 38 & 40 Robertson Street, Glasgow, 1889-90
2. 28, 38 & 40 Robertson Street, Glasgow, 1891-3
3. Partick Weighbridge Works, 179 to 185 Dumbarton Road, Partick, Glasgow, 1894-6
4. 8 & 10 Stockwell Street and 23 Bread Street, Edinburgh; works and offices, Dumbarton Road, Partick, Glasgow, 1899-1900+

- Weighing machine apparatus manufacturers: PO Glasgow Directory 1889-96
- (see Alexander Wood & Sons, Glasgow): PO Glasgow Directory 1898
- [Vast description of weighing apparatus]: PO Glasgow Directory 1899-1900+

Became W. & T. Avery Ltd in 1891, with Edwin John Turner as manager. Did not appear in the directory for 1897. Acquired the business of Alexander Wood & Son (Glasgow) in 1897: L. H. Broadhurst, *The Avery Business, 1730-1918* (Birmingham, 1949), 83.

Dutch-end weigh beam for the Burgh of Paisley made by Avery, Glasgow, *c.*1890: NMS T.1989.40.

BAIRD (Hugh Harper) & TATLOCK (John), Glasgow & Edinburgh, 1881-1900+
1. 100 Sauchiehall Street, Glasgow, 1881-8
2. 40 Renfrew Street, Glasgow, 1889-1900+
3. 50 Renfrew Street, Glasgow, 1898-1900+
4. 10 Drummond Street, Edinburgh, 1897-1900+

- Laboratory furnishers, chemical, mathematical and philosophical instrument makers: PO Glasgow Directory 1881-1900+

Exhibited chemical balances at the 1886 International Exhibition, Edinburgh. Branches established in London in 1889 (in 1903 this became the separate firm of Baird & Tatlock (London) Ltd), Edinburgh in 1897, Liverpool in 1904, Manchester 1911. Merged in 1925 with John J. Griffin & Sons Ltd. See the chapter 'Baird & Tatlock' in T. N. Clarke, A. D. Morrison-Low and A. D. C. Simpson, *Brass & Glass: Scientific Instrument Making Workshops in Scotland* (Edinburgh, 1989), 288-91.

Makers of the set of standard capacity measures for the Burgh of Govan, 1894 (Item **255**). Hydrostatic balance retailed by Baird & Tatlock, *c.*1890: NMS T.1982.104.

BANNERMAN, James, Edinburgh, 1883-96
1. 17 Burgess Street, Edinburgh, 1883-5
2. 22 Broad Wynd, Edinburgh, 1886-91
3. 109 Giles Street, Edinburgh, 1892-6

- Smith, beam and scale maker: PO Edinburgh Directory 1883-96

BEATON, Charles Macdonald, Glasgow, 1870-1
80 Oswald Street, Glasgow, 1870-1

- Canister, scale, weighing machine, japanner, and beer engine manufacturer, pewter measure maker and general ironmonger: PO Glasgow Directory 1870
- Beam, weighing, canister and beer engine manufacturer, coppersmith, and pewterer, grocers' and spirit dealers' outfitter: PO Glasgow Directory 1871

Succeeded by Beaton & McCrindle (q.v.).

BEATON (Charles M.) & McCRINDLE (John), Glasgow, 1872-4
6 Clyde Place, Glasgow, 1872-4

- Beam, scale and weighing machine makers, beer engine manufacturers, coppersmiths and pewterers, grocers' and spirit dealers' outfitters, japanners and general ironmongers: PO Glasgow Directory 1872-4

Previously Charles Macdonald Beaton (q.v.); succeeded by C. M. Beaton & Co. (q.v.).

BEATON, C. M. & Co., Glasgow, 1875-8
1. 6 Clyde Place, Glasgow, 1875
2. 60 West Howard Street, Glasgow, 1877-8

- Grocers' and spirit dealers' furnishers, beam, scale and weighing machine makers, coppersmiths, pewter measure makers, beer pump manufacturers, ironmongers, etc.: PO Glasgow Directory 1875, 1877-8

BENNIE, John, Glasgow, 1877-1900+
1. Star Engine Works, 39 Macfarlane Street, Glasgow, 1877-80
2. Star Engine Works, 149-55 Moncur Street, Glasgow; former works 39 Macfarlane Street, Glasgow, 1881-1900+

- Hydraulic engineer and machine maker and maker of all kinds of hydraulic steam and hand hoists; maker of the patent water engine for pumping presses, hand pumps, hydraulic presses and tobacco machinery; in the trades directory as a weighing machine maker: PO Glasgow Directory 1877-1900+

BIGGAR (William F.) & HENDRY (William B.), Glasgow, 1887-1900+

1. 72 West Howard Street, Glasgow, 1887-90
2. 84 Maxwell Street, Glasgow, 1891-9
3. 84 Maxwell Street, Glasgow; works Great Wellington Street, 1900+

- Engineers' ironmongers, etc. (in trades directory as weighing machine makers): PO Glasgow Directory 1887-93
- Brass founders and engineers' factors (no longer in trades directory): PO Glasgow Directory 1894-1900+

BLACK, John, Paisley, late nineteenth century
Scale maker, 33 Moss Street, Paisley

Bushel adjusting table: NMS T.1988.41.

BLACKIE, John, Selkirk, *fl.*1823
Carpenter; one of the makers of the 1823 Selkirk measures (Items **228** and **229**).

BLYTH, Colin, Glasgow, 1783-90

1. Saltmarket, Glasgow, 1783-4
2. Wilson's Close, near the foot of the Saltmarket, Glasgow, 1787
3. Buchanan's Court, back of Virginia Street, Glasgow, 1789-90

- Founder: Tait's, and Jones's Glasgow Directory 1783-4, 1787
- Brass founder: Jones's Glasgow Directory 1789-90

Paisley trone weights dated 1786 (Item **91**) and Paisley avoirdupois bell weights dated 1786 (Item **104**).

BOOTH, George, Aberdeen, 1800-50
Watchmaker and jeweller: Gordon's Aberdeen Directory 1824-35

Became George Booth & Son in 1827, but the son died in 1839; the father carried on until retiring in 1850: see I. E. James, *The Goldsmiths of Aberdeen* (Aberdeen, 1981), 103-5, who gives his dates as 1800-50. A quadrant balance, signed 'G. Booth, Aberdeen, 1844', is in the Natural Philosophy Museum, University of Aberdeen, ABDNP 200031a.

BORTHWICK, Robert, Edinburgh, *fl.*1511-32
Possibly the maker of the Stirling Jug (Item **108**). (See Chapter 7.)

BOYLE, Andrew, Perth, *fl.*1826
Maker of dated Imperial bushels in Perth (Items **257** and **258**) and Stirlingshire (Items **259** and **243**).

BRASH, David, Glasgow, 1880
8 Sauchiehall Street and 261 Buchanan Street, Glasgow, 1880

- Smith, scale, beam and weighing machine maker: PO Glasgow Directory 1880

BRASH, David & Sons (Robert and James), Glasgow, 1881-1900+
167 & 169 Renfield Street, Glasgow, 1881-1900+

- Scale, beam, and weighing machine makers, also chain makers: PO Glasgow Directory 1881-1900+

BROADFOOT, Robert, Glasgow, 1852-4
Works 38 Miller's Place, Glasgow; warehouse 45 & 47 Miller's Place, Glasgow, 1852-4

- Smith, scale, beam and weighing machine maker: PO Glasgow Directory 1852-4

BROTHERSTON, A. & J., Edinburgh 1890-1900+
27 & 46 Candlemaker Row, Edinburgh, 1890-1900+

- Grocers' ironmongers and tinsmiths: PO Edinburgh Directory 1880
- Grocers' ironmongers, weighing machine makers and tinsmiths: PO Edinburgh Directory 1891-6
- Grocers' ironmongers, weighing machine makers and dairy utensil manufacturers: PO Edinburgh Directory 1897-1900+

BROWN, William, Edinburgh, 1834-57
1. 26 High Calton, Edinburgh, 1834-44, 1856-7
2. 26 Calton, Edinburgh, 1845-55

- Smith and jack maker: PO Edinburgh Directory 1834-43
- Smith and scale beam maker: PO Edinburgh Directory 1844-57

BRUCE, John & Co., Edinburgh, 1844
Paul's Works, Edinburgh, 1844

- Scale beam and steelyard maker: PO Edinburgh Directory 1844

Successors of W. & J. Bruce (q.v.); succeeded by John Bruce (q.v.).

BRUCE, John, Edinburgh, 1846-50
1. Chalmer's Close, Edinburgh, 1846-8
2. 47 High Street, Edinburgh, 1849-50

- Scale beam and steelyard maker: PO Edinburgh Directory 1845-50

Successors of John Bruce & Co. (q.v.).

BRUCE, W. & J., Edinburgh, 1832-44
1. Paul's Work, Edinburgh, 1832-5
2. 7 Old Physic Gardens, Edinburgh, 1836-44

- Scale beam and steelyard makers: PO Edinburgh Directory 1832-54

Succeeded by John Bruce & Co. (q.v.).

CALLAM, Thomas, Leith, 1822-59
1. 54 Shore, Leith, 1822-34
2. 64 Shore, Leith, 1835-40
3. 59 Shore, Leith, 1841-6
4. 56 Shore, Leith, 1847-59

- Tin and coppersmith: PO Edinburgh Directory 1822-5
- Tin brazier and imperial measure manufacturer: PO Edinburgh Directory 1826-9
- Copper and tinplate worker: PO Edinburgh Directory 1830
- Gas-fitter, copper and tinplate worker: PO Edinburgh Directory 1831-59

Maker of sets of Imperial capacity measures at Anstruther Wester and in Edinburgh City collection (Items **243** and **244**).

Succeeded by Thos. Callam & Son, gas-fitters, braziers and tinplate workers in 1860.

CAMPBELL (Colin) & McMILLAN (James), Glasgow, 1857-8
21 Kirk Street, Gorbals, Glasgow, 1857-8

• Smiths, scale, beam and weighing machine makers: PO Glasgow Directory 1857-8

CARRON COMPANY, Carron, Stirlingshire, 1759-1900+
Manufactory, Carron, Stirlingshire, 1759-1900+

• Carron & Co., ironmasters and founders [Glasgow Warehouse]: McFeat's Glasgow Directory, PO Glasgow Directory, 1801-1900+

Listed in McFeat's Glasgow Directory for 1801, on the west side of Queen Street; by 1810 this was numbered 23 Queen Street, but by 1820 had become 20 Buchanan Street. By 1845 this had been renumbered 123 Buchanan Street; by 1875 it was 127 Buchanan Street, reverting in 1884 to 125 Buchanan Street.

Carron Company [*Trade catalogue*], 3 volumes (Carron, Stirlingshire, *c.*1902), II, 368: illustrations of five different forms of weights. Business histories include: *Carron Company from the Reign of George II to the Reign of George VI, 1759-1938* (Carron, Stirlingshire, 1938); Roy H. Campbell, *Carron Company* (Edinburgh, 1961), and Brian Watters, *Where Iron Runs Like Water! A New History of the Carron Iron Works 1759-1982* (Edinburgh, 1998).

CHALMERS, Alexander, Dundee, 1853
16 New Inn Entry, Dundee, 1853

• Smith, weighing machine and mangle maker: Dundee Directory 1853

CHALMERS, Thomas, Glasgow, 1809-12
Hospital Street, Hutcheston, Glasgow

• Smith: McFeat's Glasgow Directory 1809-12

Admitted member of the Incorporation of Hammermen of Glasgow 1810, with an essay of 'a small waste beam': Harry Lumsden and P. Henderson Aitken, *History of the Hammermen of Glasgow* (Paisley, 1912), 304.

Possibly previously Chalmers & Lindores (q.v.); succeeded by Chalmers & Renton (q.v.).

CHALMERS & LINDORES, Glasgow, 1805-7
Gorbals, Glasgow

• Smiths: McFeats Glasgow Directory 1805-7

Possibly succeeded by Thomas Chalmers (q.v).

CHALMERS (Thomas) & RENTON (William), Glasgow, 1813-14
Broad Street, Hucheston, Glasgow, 1813-14

• Smiths: McFeat's Glasgow Directory 1813-14

See William Renton and Renton & Ritchie (q.v.); succeeded by Thomas Chalmers (q.v.).

CHALMERS, Thomas, Glasgow, 1815-29
1. Hospital Street, Hutcheson Town, Glasgow, 1815
2. 7 Hospital Street, Hutcheson Town, Glasgow, 1816-27
3. 7 & 18 Dunlop Street, Glasgow, 1828
4. 7 Hospital Street, Glasgow, 1829

• Smith: McFeat's Glasgow Directory 1815-25
• Smith and bellhanger: McFeat's Glasgow Directory, 1826-7; PO Glasgow Directory 1828-9

Succeeded by Thomas Chalmers & Son (q.v.).

CHALMERS, Thomas & Son, Glasgow, 1830-2
75 Maxwell Street, Glasgow, 1830-2

- Smiths and bellhangers: PO Glasgow Directory 1830-2

Succeeded by Thomas Chalmers *senior* (q.v.).

CHALMERS, Thomas *senior*, Glasgow, 1833
Howard Street, Glasgow, 1833

- Smith and bellhanger: PO Glasgow Directory 1833

Previously Thomas Chalmers & Son (q.v.).

COATS, A. (Archibald) & Co., Glasgow, 1787-1807
1. Warehouse above no. 1, head of the Gallowgate, Glasgow, 1787-90
2. Saracen's Lane, Gallowgate, Glasgow, 1801-5, 1807
3. Saracen's Lane, Gallowgate; warehouse, 598 Argyll Street, Glasgow, 1806

- Hardware merchants: Jones's Glasgow Directory 1787
- Wholesale hardware merchants: Jones's Glasgow Directory 1789-90
- Brass founders: McFeat's Glasgow Directory 1801
- Manufacturers of copper, pewter, brass, tin, wrought iron, and cast iron goods: McFeat's Glasgow Directory 1803-5
- Brass and iron founder: McFeat's Glasgow Directory 1806-9

Listed in 1787 and 1789 as William and Archibald

Coats & Co., and as William & Archibald Coats from 1790-1804; Archibald Coats from 1805-7.

Stamp on base of trone weights at Dumbarton, *c.*1800 (Item **102**).

COCHRAN, Archibald, & Son, Falkirk, 1826, 1835
Distributor (and maker?) of Imperial weights, Stirling, 1826 and 1835

Advertisement in the *Stirling Journal and Advertiser*, 12 January 1826: 'New Weights and Measures – The general adoption of the Imperial Standard Weights and Measures has not been so rapid throughout the country as might have been anticipated. We understand that at Falkirk the Weights were used in the market at the Tron on Thursday Last, and that today the butcher-meat and butter will also be sold by them. Mr Cochran disposed of a great quantity, but a considerable time must elapse, however, ere they can be in use in remote districts; and, we believe, an erroneous notion prevails that no penalty is attached to selling by the old weights ...' In 1835 there was a dispute between Archibald Cochran & Son and J. & W. Grant (q.v.), as to which firm was supposed to be authorised to do the stamping in Falkirk: *Stirling Journal and Advertiser*, 6 and 13 February 1835.

COCHRAN, Hans, Edinburgh, *fl.*1553-8
Maker of the Craigengelt Weight, 1553 (Item **31**). (See Chapters 7 and **8**.)

COCHRANE, H., Edinburgh, 1835
Advertisement in the *Scotsman*, 26 September 1835, for a new weighing machine for 'carriages, carts, hay, straw, turnips &c.'. Henry Cochrane, Haymarket was a spirit dealer, according to PO Edinburgh Directory 1835.

CORBET, Walter, Glasgow, 1803-9
49 Saltmarket, Glasgow, 1803-9

- Smith: McFeat's Glasgow Directory 1803-9 (but not in 1806)

Admitted member of the Incorporation of Hammermen of Glasgow 1803, with an essay of 'a beam': Harry Lumsden and P. Henderson Aitken, *History of the Hammermen of Glasgow* (Paisley, 1912), 302.

COUPER, Willie, Edinburgh, *fl.*1504
Measure maker: J. D. Marwick, *et al.* (eds), *Extracts from the Records of the Burgh of Edinburgh*, 14 volumes (Edinburgh, 1869-1967), I, 98: 2 March 1503/4.

COWIE, Thomas, Edinburgh, 1849
13 McDowall Street, Edinburgh, 1849

- Scale beam and gas-burner maker: PO Edinburgh Directory 1849

CRAIG, Alex., Glasgow, 1871-8
4 Woodlands Road, Glasgow, 1871-8

- Agent for J. A. Gilbert & Co., grocers' fitters, London: PO Glasgow Directory 1871
- Agent for W. & T. Avery's patent agate brass beams and scales: PO Glasgow Directory 1872-4
- Agent for W. & T. Avery's agate beams and Gilbert's tea canisters and caddies: PO Glasgow Directory 1875
- Avery's Patent agate brass gilt beams and scales and weights: agent Alex. Craig: PO Glasgow Directory 1876-8

CRAIG, John, Glasgow, 1864-9
108 Dumbarton Road, Glasgow, 1864-9

- Furnishing ironmonger: PO Glasgow Directory 1864
- Furnishing ironmonger and beam and scale manufacturer: PO Glasgow Directory 1865-9

CRAIGIE, David, of Over-Sanday, Orkney, *fl.*1663
In 1663 David Craigie was paid '20 marks Scots for the sixteen Pound Weight of Brass, which is in the custody of George Mouat our juster's hand, for regulating the weights of the country': James Mackenzie, *The General Grievances and Oppression of the Isles of Orkney and Shetland* (Edinburgh, 1750, reprinted 1836), 54. This transaction is also mentioned in *Memorial and Abstract of the Proof for the Earl of Galloway and Others, Udalmen, and Proprietors of land in Orkney, Pursuers, against James Earl of Morton, Defender* (Edinburgh, 1758), 77.

CRAIGIE, George, Kirkwall, *fl.*1686
'At that time [1686] it [the Standard steelyard] was consigned over to George Craigie, an intrant Assayer, in room of George Mouat, the former Assayer ...: *Memorial and Abstract of the Proof for the Earl of Galloway and Others, Udalmen, and Proprietors of land in Orkney, Pursuers, against James Earl of Morton, Defender* (Edinburgh, 1758), 79. Subsequently, a witness for the Earl of Morton affirmed that 'his Pundar ... was made by George Craigie, the Maker and Adjuster of those Instruments in Kirkwall [before 1708] ... and that it was adjusted by Thomas Aitken (q.v.), the present Maker and Adjuster': *Memorial for James, Earl of Morton, Defender, against Alexander, Earl of Galloway ... Pursuers* (Edinburgh, 1758), 19.

CRAWFOORD, Stephen, Glasgow, *fl.*1708
Coppersmith contracted by Stirling to construct measures, 1708: R. Renwick (ed), *Extracts from the Records of the Royal Burgh of Stirling*, 2 volumes

(Glasgow, 1887-9), II, 123: 7 October 1710. (See Items **179-98.**)

CRICHTON, A., Glasgow, 1801-15
Old Vennal [*sic*], Glasgow, 1801-15

- Smith: McFeat's Glasgow Directory 1801-15

Probably becomes A. M. Crichton (q.v).

CRICHTON, A. M., Glasgow, 1817-27
1. 20 Saltmarket, Glasgow, 1817-18
2. 33 Gallowgate, Glasgow, 1819
3. 20 Saltmarket, Glasgow, 1820-5
4. 7 London Street, Glasgow; works 19 Saltmarket, Glasgow, 1826-7

- Beam, scale, etc., maker: McFeat's Glasgow Directory 1817-27

Probably becomes Andrew Crichton, hatter and smith, 1828 only, whose business divides, with hatters at 54 Gallowgate until 1834; for **Andrew Crichton**, smith, see below.

CRICHTON, Andrew, Glasgow, 1828-38
1. 54 Gallowgate, Glasgow; smithworks 20 Saltmarket, Glasgow, 1828
2. 19 London Street, Glasgow; works London Lane, 1830-2
3. 32 London Street, Glasgow; works London Lane, 1833-4
4. 32 London Street, Glasgow, 1835
5. 18 London Street, Glasgow, 1836-8

- Hatter and smith: PO Glasgow Directory 1828
- Scale and beam maker: PO Glasgow Directory 1830-4
- Smith, scale and beam maker: PO Glasgow Directory 1835-6

- Smith, beam maker and gas-fitter: PO Glasgow Directory 1837
- Smith, beam maker, gas-fitter, manufacturer of iron gas tubes and gas burners: PO Glasgow Directory 1838

Succeeded by A. Crichton & Son (q.v.).

CRICHTON, A. & Son, Glasgow, 1839
18 London Street, Glasgow; works London Lane, Glasgow, 1839

- Beam makers and gas-fitters, manufacturers of gas tubes and gas burners: PO Glasgow Directory 1839

Succeeded by John Crichton (q.v.) at this address.

CRICHTON, John, Glasgow, 1840-3
1. 18 London Street, Glasgow, 1840-1
2. 26 London Street, Glasgow; works London Lane, Glasgow, 1842-3

- Beam maker, gas-fitter, etc.: PO Glasgow Directory 1840
- Manufacturer of iron gas tubes, beam maker and gas-fitter: PO Glasgow Directory 1841-3

Previously A. Crichton & Son (q.v.); succeeded by Crichton & Eadie (q.v.).

CRICHTON (John) & EADIE, Glasgow, 1844-56
1. 26 London Street, Glasgow; works London Lane, Glasgow , 1844
2. 44 & 46 London Street, Glasgow; works London Lane, Glasgow, 1845-50
3. 44 & 46 London Street, Glasgow; works London Lane and Dalmarnock Bridge, Glasgow, 1851-6

- Smiths, beam makers, gas-fitters and manu-facturers of malleable iron gas pipes: PO Glasgow Directory 1844-51
- Manufacturers of Cutler's patent lap-welded tubes for locomotive and marine boilers: PO Glasgow Directory 1852
- Smiths, beam makers, gas-fitters and manu-facturers of wrought-iron tubes, etc.: PO Glasgow Directory 1853
- Patent iron tube makers, smiths and ironfounders, etc.: PO Glasgow Directory 1854-6

Previously John Crichton (q.v.).

CRICHTON [Chrichton], James, Glasgow, 1775-90

1. Saltmarket, Glasgow, 1783
2. House above 129 Gallowgate, Glasgow, 1789
3. East side Charlotte Street, Glasgow, 1790

- Hammerman: Tait's Glasgow Directory 1783
- Smith: Jones's Glasgow Directory 1789
- Lockmaker: Jones's Glasgow Directory 1790

Does not appear in the directory for 1787. Admitted member of the Incorporation of Hammermen of Glasgow 1775, with a 'small black beam' submitted as essay: Harry Lumsden and P. Henderson Aitken, *History of the Hammermen of Glasgow* (Paisley, 1912), 294.

CRICHTON [sometimes Crighton], J. (James), Glasgow, 1801-35

1. Charlotte Street, Glasgow, 1801-11
2. 5 Charlotte Street, Glasgow, 1812
3. 2 Charlotte Street, Glasgow, 1813-18, 1820-5
[4. 9 Charlotte Street, Glasgow, 1819, 'John Crichton']
5. 5 Charlotte Street, Glasgow, 1826-35

- Chemical and philosophical instrument maker: McFeat's Glasgow Directory 1801-27; PO Glasgow Directory 1828-35

This may be the same man who was listed 1775-90. Helped verify the Imperial standard capacity measures for the City of Glasgow 1826 (Item **241**). The McLean Museum, Greenock, has a balance signed 'Crichton, Glasgow', presented to the Greenock Philosophical Society in 1863 as having belonged to James Watt.

CRICHTON, Robert, Glasgow, 1807-33

1. 297 High Street, Glasgow, 1807-25
2. 60 High Street, Glasgow, 1826-33

- Smith: McFeat's Glasgow Directory 1807-20
- Smith, beam and scale maker: McFeat's Glasgow Directory 1821-7
- Beam and scale maker: PO Glasgow Directory 1828-33

Succeeded by Walter Crichton & Co. (q.v.).

CRICHTON, Walter, Glasgow, 1819-32

1. 36 Gallowgate, Glasgow, 1819-25
2. 57 Gallowgate, Glasgow, 1826-32

- Smith: McFeat's Glasgow Directory, 1819-22
- Smith, beam and scale maker: McFeat's Glasgow Directory 1823-7; PO Glasgow Directory 1828-32

Presumably became Walter Crichton & Co. (q.v.) at Robert Crichton's address in 1834.

CRICHTON, Walter & Co., Glasgow, 1834

60 High Street, Glasgow, 1834

- Beam and scale makers: PO Glasgow Directory 1834

Reverts to Walter Crichton (q.v.).

CRICHTON, Walter, Glasgow, 1835-41

60 High Street, Glasgow, 1835-6
66 High Street, Glasgow, 1837-41

- Smith, scale and beam maker: PO Glasgow Directory 1835-41

CUNNINGHAM, Murdoch, Edinburgh, 1830-43

14 Royal Exchange, Edinburgh, 1830-43

- (Of Imperial Standard Weights and Measures Office): PO Edinburgh Directory 1830
- Weights and Measures Office: PO Edinburgh Directory 1831-6

- (Goldsmith) Weights and Measures Office: PO Edinburgh Directory 1837-43

Stamp 'IMPERIAL / CUNNINGHAM EDIN.' on a weight with the initials of James Hill, Edinburgh Dean of Guild 1827 and 1828, at Lothian Regional Council Trading Standards Department, Edinburgh.

DAVIDSON, Jonathan, Edinburgh, 1825-76
(and see **Jonathan Davidson & Co.**)
1. 123 High Street, Edinburgh, 1825-30
2. Paul's Work, Edinburgh, 1837-42
3. 24 Barony Street, Edinburgh, 1857-74
4. 37 Broughton Street, Edinburgh, 1876

- Ironmonger: PO Edinburgh Directory 1825-9
- Furnishing warehouse: PO Edinburgh Directory 1830
- Not in 1831 directory
- Smith: PO Edinburgh Directory 1837-42
- Steelyard maker: PO Edinburgh Directory 1857-65
- Steelyard maker (patentee and manufacturer of

self-reefing sails): PO Edinburgh Directory 1866-8
- Steelyard maker: PO Edinburgh Directory 1869-74 and 1876
- No trade [but see Charles Robertson (q.v.) at same address 1875]

Advertisement in the *Scotsman*, 26 April 1826, as an ironmonger at 123 High Street: 'J. D. has just received a supply of New Weights and Copper Measures well worthy the attention of the public.'

Intermittently this firm became Jonathan Davidson & Co. (q.v.). Succeeded at Barony Street by Charles Robertson (q.v.).

DAVIDSON, Jonathan & Co., Edinburgh, 1832-1900+ (and see **Jonathan Davidson**)

1. 123 High Street, Edinburgh, 1832-6
2. 24 Barony Street, Edinburgh, 1843-56
3. East London Street, Edinburgh, 1877-1900+

- Ironmongers: PO Edinburgh Directory 1825-9
- Ironmongers and smiths: PO Edinburgh Directory 1833-6
- Steelyard makers: PO Edinburgh Directory 1844-56; 1877-8
- Inventors and manufacturers of the Gart steelyard and weighing machines: PO Edinburgh Directory 1879-84
- Inventors and manufacturers of the celebrated Prize Gart steelyards and weighing machines in all sizes for general use, established in 1825: PO Edinburgh Directory 1885-93
- Inventors and manufacturers of the celebrated Prize 'Gart' steelyard and weighing machines, beams and scales as supplied to Her Majesty's Customs, established in 1825: PO Edinburgh Directory 1894-1900+

Inventors and manufacturers of weighing machines, exhibited at the Great Exhibition, 1851: see *Great Exhibition of the Works of All Nations, 1851, Official Descriptive and Illustrated Catalogue,* 4 volumes

(London, 1851), I, 252: item 774. Advertisement for (1851) prize gart steelyard in PO Dundee Directory 1880. Glasgow Directories carry advertisements for Davidson's Edinburgh business in the 1880s.

Examples of Davidson's gart balance: NMS T.1982.152; T.1983.59, T.1990.100 and T.2002.136. Beam at Cupar: CUPMS 1984.171

Intermittently this firm became Jonathan Davidson (q.v.).

DAVIES, Thomas O., Glasgow, 1895
80 Elcho Street, Glasgow, 1895

- Weighing machine maker and butchers' outfitter: PO Glasgow Directory 1895

DAVIES & NISBET (A. W.), Glasgow, 1896-7
80 Elcho Street, Glasgow, 1896-7

- Weighing machine makers and shop outfitters: PO Glasgow Directory 1896-7

DAVIES, T., & Co., Glasgow, 1898-1900+
80 Elcho Street, Glasgow, 1898-1900+

- Weighing machine makers and shop outfitters: PO Glasgow Directory 1898
- Weighing machine makers: PO Glasgow Directory 1899-1900+

DEMPSTER (John), MOORE (W. P.) & Co., Glasgow, 1873-1900+
1. 49 Robertson Street, Glasgow, 1873
2. 49 Robertson Street, Glasgow; Iron warehouse, 28 York Street, Glasgow, 1874-83
3. 49 Robertson Street, Glasgow; Iron warehouse, 41 York Street, Glasgow, 1884-5
4. 49 Robertson Street, Glasgow; Iron warehouse, 41 York Street, Glasgow; London warehouse, 60 Queen Victoria Street, E. C., 1886-9
5. 41 York Street, Glasgow; London warehouse, 60 Queen Victoria Street, E. C. 1890-5
6. Head office and general warehouse 49 Robertson Street, Glasgow; Iron warehouse, 41 York Street, Glasgow, 1896-1900+

- Iron and hardware merchants and engineers' ironmongers: PO Glasgow Directory 1873
- Iron, steel, hardware indiarubber merchants, machinery agents, engineers' ironmongers: PO Glasgow Directory 1874
- Iron and hardware merchants and engineers' ironmongers: PO Glasgow Directory 1875-7
- Iron and steel merchants and engineers' ironmongers, bolt, nut, rivet and chain makers: PO Glasgow Directory 1878-89
- [Large description, 1890-1900+] In trades directory as weighing machine makers from 1879-1900+

DICK, Alexander, Glasgow, 1826-76
1. 101 Bridgegate, Glasgow, 1826-32
2. Buchanan Court, 105 Stockwell, Glasgow, 1833-76

- Wireworker: McFeat's Glasgow Directory 1826
- Wireworker and land measure chain maker: McFeat's Glasgow Directory 1827
- Land measuring, chain and safety lamp maker: PO Glasgow Directory 1828-43
- Land surveying chain maker: PO Glasgow Directory 1844-76

Surveyor's brass land chain with brass handles and tallies, 'ALEX^R DICK GLASGOW', offered for sale by David Stanley Auctions, Osgasthorpe, Leicestershire, 27 November 1993, Lot 755.

From 1875-6 the firm became Alexander Dick & Co., at the same address.

DICKESON, Charles, Edinburgh, 1614-42
Charles Dickson or Dickesoun, goldsmith and sinker of the irons at the Scottish Mint, was authorised in 1619 to make weights. Quoted by R. W. Cochran Patrick, *Records of the Coinage of Scotland*, 2 volumes (Edinburgh, 1876), I, 292, 25 November 1619: A proclamation for reforming 'sindrie inconveniences touching the Coynes of his M. Realmes … We … have given command, warrand, & direction to Charles Dickeson sinker of our irons, to prepare & make readdie sufficient numbers of upright & true weghts, …; which wechts, … the said Charles Dickeson shall bee holden to sell to our leiges at the pryce of twelfe shillings usuall moneye of this kingdom for everie stand …'. Also mentioned by T. Sheppard and J. F. Musham, *Money Scales and Weights* (London, 1923), 26; and Paul and Bente Withers, *British Coin-Weights: A Corpus of the Coin-Weights made for use in England, Scotland and Ireland* (Llanfyllin, 1993), 327.

DINNESTOUNE, Andro, Glasgow, *fl.*1691
Admitted member of the Incorporation of Hammermen of Glasgow in 1691 as a white iron smith stranger, with an essay which included 'ane pynt stoup': Harry Lumsden and P. Henderson Aitken, *History of the Hammermen of Glasgow* (Paisley, 1912), 290.

DONALD, Robert, Elderslie, 1831
73 & 74 High Street, Elderslie [nr. Johnstone], 1831-2

* Engineer, machine maker, ironfounder: Fowler's Directory 1831-2

Adjuster of weights and measures.

DONALDSON, Alexander, Edinburgh, 1818-30
South Niddry Street, Edinburgh, 1818-30

* Tool maker, turner and globe manufacturer: PO Edinburgh Directory 1818-30

Maker of standard wooden yard for Kinghorn (Item **26**). For more about the business, see A. D. C. Simpson, 'Globe Production in Scotland in the Period 1770-1830', in *Der Globusfreund*, nos. 35-7 (1987), 21-32: Proceedings of the VIth International Coronelli Symposium, Amsterdam, 1986.

DREW, John, Glasgow, 1890-1900+
78 York Street, Glasgow, 1890-1900+

* Japanner, bronzer and clock-dial manufacturer: PO Glasgow Directory 1890-1900+

R. G. Wood, manager (in trades directory as a beam and scale maker). This firm appears to have had its beginnings in 1819 with J. Drew, japanner and clock-dial maker, 297 High Street. Not listed in the directory for 1899.

DRUMMOND, Robert, Glasgow, *fl.*1825
Admitted member of the Incorporation of Hammermen of Glasgow in 1825 as a coppersmith stranger, with 'a copper half-gallon measure' as essay: Harry Lumsden and P. Henderson Aitken, *History of the Hammermen of Glasgow* (Paisley, 1912), 310.

DRYSDALE, Andrew, Glasgow, 1827-32
1. Bishop Street, Glasgow, 1827
2. Union Place, Anderston and 218 Broomielaw, Glasgow, 1828
3. Union Place, Anderston, Glasgow, 1829-32

* Coppersmith, brass founder, etc.: PO Glasgow Directory 1827

- Coppersmith, brass founder, tinplate, plumber, etc.: PO Glasgow Directory 1828-30
- Coppersmith: PO Glasgow Directory 1829-32

Admitted member of the Incorporation of Hammermen of Glasgow in 1828 as a coppersmith stranger, with 'a half-gallon measure' as essay: Harry Lumsden and P. Henderson Aitken, *History of the Hammermen of Glasgow* (Paisley, 1912), 313.

DUNCAN, William, Glasgow, *fl.*1722

Admitted member of the Incorporation of Hammermen of Glasgow 1722, with an essay which included 'ane tin pint stoup': Harry Lumsden and P. Henderson Aitken, *History of the Hammermen of Glasgow* (Paisley, 1912), 293.

DUNN, John, Edinburgh, 1824-42

1. 7 West Bow, Edinburgh, 1824
2. 25 Thistle Street, Edinburgh, 1825-7
3. 52 Hanover Street, Edinburgh, 1828-31
4. 50 Hanover Street, Edinburgh, 1832-42

- Mathematical, optical and philosophical instrument maker: PO Edinburgh Directory 1824-42

Advertisement in the *Scotsman*, 18 February and 6 June, 1835 for '… Balances, … Manufactured by himself, and which he can confidently recommend as being of the best quality.' For information about his business, see 'John and Thomas Dunn' in T. N. Clarke, A. D. Morrison-Low and A. D. C. Simpson, *Brass & Glass: Scientific Instrument Making Workshops in Scotland* (Edinburgh, 1989), 88-95.

Succeeded by his brother, Thomas Dunn.

DURWARD, John, Aberdeen, *c.*1820-36

1. Ferguson's Court, 30 Gallowgate, Aberdeen, 1829
2. 23 Frederick Street, Aberdeen, 1831
3. 42 Frederick Street, Aberdeen, 1832-6

- Beam and scale maker: Gordon's Aberdeen Directory 1829

Constructed a precision beam for G. S. Keith's experiments on the densities of grain: see George Skene Keith, 'On the Specific Gravity of Barley and Scotch Bigg, with the description of a New Instrument for Measuring it', in *Edinburgh Philosophical Journal*, 5 (1821), 173-7.

EDMISTON, George, Greenock, 1894

4 Dalrymple Street, Greenock, 1894

- Scale beam manufacturer: Kelly's Metal Trades 1894

EDMISTON, Thomas, Glasgow, 1881

61 Surrey Street, Glasgow, 1881

- Beam, scale and weighing machine maker: PO Glasgow Directory 1881

EGLIN (William) & GARDNER, Glasgow, 1879-82

70 York Street, Glasgow, 1879-82

- American factors and merchants: PO Glasgow Directory 1879
- American factors and importers of joinery, veneers, perforated seating, furniture, spring hinges, weighing machines, barb wire fencing, hardware, machinery, etc.: PO Glasgow Directory 1880-2

Succeeded by Eglin & Gilchrist (q.v.).

EGLIN (William) & GILCHRIST, Glasgow, 1883
70 York Street, Glasgow, 1883

- American factors and importers of joinery, veneers, perforated seating, furniture, spring hinges, weighing machines, barb wire fencing, hardware, machinery, and sole agents for the Patent Air and Steam Blower and Patent Boiler Tube Cleaner: PO Glasgow Directory 1883

Succeeded by William Eglin & Co. (q.v.).

EGLIN, William & Co., Glasgow, 1884-8
1. 70 York Street, Glasgow, 1884
2. 124 Queen Street, Glasgow; New York [import] Office, 29 Chambers Street, 1885-8

- American factors and importers of joinery, veneers, perforated seating, furniture, spring hinges, weighing machines, barb wire fencing, hardware, machinery, and sole agents for the Patent Air and Steam Blower and Patent Boiler Tube Cleaner: PO Glasgow Directory 1884-8

EVANS, William, Selkirk, *fl*.1823
Cooper; one of the makers of the 1823 Selkirkshire measures (Items **228** and **229**).

EWEN, John, Aberdeen, 1770-d.1821
Retailer of weights, scales and balances: Castle Street, Aberdeen

A set of brass weights, steel scales with brass pans in a wooden box, with a table inside the lid, describing 1772 guineas, also reads: 'Complete sets of Gold Weights and Scales, also, Hydrostatical, Bal-lances, sold by J. Ewen, at his Ware-room, Castle-street, Aberdeen': in Marischal Museum, University of Aberdeen, ABDUA 18280. I. E. James, *The Goldsmiths of Aberdeen* (Aberdeen, 1981), 86-7, describes him

as being at various times a jeweller, watchmaker, goldsmith, and gives his dates as 1770, dying in 1821.

FALKIRK IRON WORKS, Falkirk, *c*.1840-1900+
Manufactory, Falkirk

Falkirk Iron Company [*Trade Catalogue*] (Falkirk, *c*.1900), plate 344: illustrations of three different forms of weights. Also gives addresses of warehouses in London, Liverpool, Edinburgh and Glasgow. (See below.)

FALKIRK IRON COMPANY, Glasgow, 1840-1900+
1. 118 Argyll Street, Glasgow, 1840-5
2. 118 Argyll Street and 10 Buchanan Street, Glasgow, 1846-54
3. Warehouse, St Enoch Square and Dixon Street, Glasgow, 1854
4. 22 Dixon Street, Glasgow, 1855-65
5. 18 & 22 Dixon Street, Glasgow, 1866-70
6. 77 & 81 Union Street, Glasgow, and 56 & 70 Alston Street, Glasgow, 1871-6
7. 16 Bothwell Street, and 39 & 45 St Vincent Lane, Glasgow, 1877-83
8. 16 Bothwell Street, and 39 St Vincent Lane, Glasgow, 1884
9. 16 Bothwell Street, and 39 & 41 St Vincent Lane, Glasgow, 1885
10. 16 Bothwell Street, Glasgow, 1886-92
11. 32-4 Bothwell Street, Glasgow, 1893-1900+

- No description: PO Glasgow Directory 1840-54
- Ironfounders, etc.: PO Glasgow Directory 1855-86
- Ironfounders, enamellers, etc.: 1887-1900+

FERGUSON (John) & DAWSON, Glasgow, 1885

31 North Street, Glasgow, 1885

- Beam and weighing machine makers: PO Glasgow Directory 1885

Succeeded by John Ferguson & Co. (q.v.).

FERGUSON, John & Co., Glasgow, 1886

31B North Street, Glasgow, 1886

- Beam and weighing machine makers and general blacksmith: PO Glasgow Directory 1886

FERGUSON, J. & R., Glasgow, 1819-28

1. 151 Stockwell, Glasgow, 1819-23
2. 34 Stockwell, Glasgow, 1824-5
3. 69 Stockwell, Glasgow, 1826-7
4. 39 Stockwell, Glasgow, 1828

- Tinplate worker: McFeat's Glasgow Directory 1819-22
- Tinplate workers, smiths and brass founders: McFeat's Glasgow Directory 1823-7
- Brass founders and tinplate workers, etc.: PO Glasgow Directory 1828

Advertisement in 1825 McFeat's Directory includes 'Copper, Pewter and Tin Measures; Beam Scales & Stands, Meal Scales …; Iron weights, English Dutch and Tron; Brass weights of all kinds'. John Ferguson previously at 151 Stockwell, described as 'tinplate worker'.

FERGUSON, J. & R., Glasgow 1851-83

1. 11 Argyll Street, Glasgow, 1851
2. 11 Argyll Street, Glasgow; workshop 71 Stockwell Street, 1852-65

3. 11 Argyll Street, Glasgow; works Tontine Back Buildings, 1866-70
4. 11 Argyll Street, Glasgow; works 34 Trongate, Glasgow, 1871-82
5. 34 Trongate, Glasgow, 1883

- General and furnishing ironmongers, tinplate workers, brass founders and gas-fitters: PO Glasgow Directory 1851-3
- General and furnishing ironmongers, tinplate workers, gas-fitters and blacksmiths: PO Glasgow Directory 1854
- Ironmongers, scale beam and weighing machine makers, blacksmiths, tinsmiths and gas-fitters: PO Glasgow Directory 1855-8
- Ironmongers, scale beam and weighing machine makers, blacksmiths, tinsmiths, gas-fitters and grocers' outfitters: PO Glasgow Directory 1859-79
- Ironmongers, scale beam and weighing machine makers, blacksmiths, tinsmiths, gas-fitters and grocers' outfitters, model engine and steamship builders: PO Glasgow Directory 1880-3

Succeeded by J. & R. Ferguson & Co. (q.v.).

FERGUSON, J. & R. & Co., Glasgow, 1884-92

34 Trongate, Glasgow, 1884-92

- Ironmongers, scale beam and weighing machine makers, blacksmiths, tinsmiths, gas-fitters and grocers' outfitters, model engine and steamship builders: PO Glasgow Directory 1884-92

FISHER, Robert, Glasgow, 1806-34

1. 113 John Street, Glasgow, 1820-5
2. 42 John Street, Glasgow, 1827
3. 40 John Street, Glasgow, 1828-30
4. Market Lane, John Street, Glasgow, 1831-3
5. 6 Frederick Lane, Glasgow, 1834

- Smith: McFeat's Glasgow Directory 1820-7 (but not in 1826); PO Glasgow Directory 1828-33
- Smith and bellhanger: PO Glasgow Directory 1834

Admitted member of the Incorporation of Hammermen of Glasgow 1806 as a smith stranger, with a 'small beam' as essay: Harry Lumsden and P. Henderson Aitken, *History of the Hammermen of Glasgow* (Paisley, 1912), 302.

FLEMING, Peter & Co., Glasgow, 1811-24

1. 7 Argyll Street, Glasgow, 1811-20
2. 8 Argyll Street, Glasgow, 1820-4

- Ironmongers: McFeats Glasgow Directory 1811-24

Succeeded by P. & R. Fleming (q.v).

FLEMING, P. & R., Glasgow, 1825-69

1. 8 Argyll Street, Glasgow, 1825
2. 17 Argyll Street, Glasgow, 1826-9
3. 17 Argyll Street, Glasgow; Iron warehouse, Moodie's Court, Glasgow, 1830-6
4. 17 Argyll Street, Glasgow; warehouse 18 Stockwell Street, Glasgow, 1837-9
5. 29 Argyll Street, Glasgow; 18 Stockwell Street, Glasgow, 1840-53
6. 29 Argyll Street; 18 Stockwell Street, Glasgow; and Moodie's Court, Glasgow, 1854
7. 29 Argyll Street, Glasgow, 1855-62
8. 29 Argyll Street, Glasgow; Iron warehouse, 18 Stockwell Street, Glasgow, 1863-8
9. 29 Argyll Street, Glasgow; Iron warehouse, 18 Stockwell Street, Glasgow; branch 1 Dowanhill Place, Partick, 1869

- Ironmongers: McFeat's Glasgow Directory 1825-7; PO Glasgow Directory 1828-35
- Ironmongers and iron merchants: PO Glasgow Directory 1837-53

- Ironmongers and iron merchants, smiths and weighing machine makers: PO Glasgow Directory 1854-6
- Ironmongers and agricultural implement warehouse, smiths and weighing machine manufacturers: PO Glasgow Directory 1857
- Ironmongers and iron merchants (in trades directory as scale beam makers): PO Glasgow Directory 1858-62
- Ironmongers, iron merchants, smiths, bell-hangers, wire fencers, and agricultural implement warehouse: (in trades directory as scale beam makers): PO Glasgow Directory 1863-9

Previously Peter Fleming & Co. (q.v.); succeeded by P. & R. Fleming & Co. (q.v.).

FLEMING, P. & R. & Co., Glasgow, 1870-1900+

1. 29 Argyll Street, Glasgow; Iron warehouse, 18 Stockwell Street, Glasgow; branch, 1 Dowanhill Place, Partick, 1870-2
2. 29 Argyll Street, Glasgow; Iron warehouse, 18 Stockwell Street, Glasgow; branch, 1 Dowanhill Place, Partick; Vulcan Smithworks, Cranstonhill 1873
3. 29 Argyll Street, Glasgow; warehouse 18 Stockwell Street, Glasgow; 1 Dowanhill Place, Partick, 1874-5
4. 29 Argyll Street, Glasgow; warehouse 18 Stockwell Street, Glasgow; 1 Dowanhill Place, Partick; works, Kelvin Street, Partick, 1876-8
5. 29 Argyll Street, Glasgow; 18 and 24 Stockwell Street, Glasgow; 1 Dowanhill Place, Partick; works Kelvin Street, Partick, 1878-91
6. 29 Argyll Street, Glasgow; 16 Graham Square; 18 and 24 Stockwell Street, Glasgow; 1 Dowanhill Place, Partick; works Kelvin Street, Partick, 1892-1900+

- Ironmongers, iron merchants, smiths, gas-fitters, bellhangers, wire fence and gate manufacturers, and agricultural implement warehouse: (in trades directory as scale beam makers): PO Glasgow Directory 1870-1900+

**FORREST & DAVIDSON, Glasgow and
Stirling, 1874-5**

1. 15A Brown Street, Glasgow; works Stirling, 1874
2. 15 Brown Street, Glasgow; works, Stirling, 1875

- Ironfounders and weighing machine makers:
  PO Glasgow Directory 1874-5

FORSYTH & SMITH, Glasgow, 1871-2

Agent for Cox & Luckman, Birmingham
45 Union Street, Glasgow, 1871-2

- Manufacturers of fire-irons, carpenters' planes
  of every description, counter weighing
  machines, etc., sole agents for Scotland:
  PO Glasgow Directory 1871-2

FORSYTH & Son, Glasgow, 1873-6

Agent for Cox & Luckman & Sons, Birmingham
40 Union Street, Glasgow, 1873-6

- Manufacturers of fire-irons, carpenters' planes
  of every description, counter weighing
  machines, etc., sole agents for Scotland:
  PO Glasgow Directory 1873-4
- Agents for Cox Luckman & Sons, Birmingham:
  PO Glasgow Directory 1875-6

FOUBISTER, Thomas, Kirkwall, 1712-30

Thomas Foubister was appointed Assayer of the
Orkney Weights and Measures by the Earl of
Morton in 1712 and 'held that Office till the Year
1730, and then the present Assayer [Thomas
Aitken (q.v.)] came in his Place': *Memorial and
Abstract of the Proof for the Earl of Galloway and
Others, Udalmen, and Proprietors of land in Orkney,
Pursuers, against James Earl of Morton, Defender*
(Edinburgh, 1758), 33. See also ibid., 89-90;
Foubister's role was also discussed in *Memorial for
James, Earl of Morton, Defender, against Alexander,
Earl of Galloway … Pursuers* (Edinburgh, 1758), 17,
18, 23 and 42.

FRASER, Alex., Glasgow, 1887-8

100 Clarence Street, off Paisley Road, Glasgow,
1887-8

- Scale, beam and weighing machine maker and
  blacksmith: PO Glasgow Directory 1887-8

FRASER, Alexander, Glasgow, 1894-1900+

1. 18 Portland Street, Glasgow, 1894-6
2. 18 North Portland Street, Glasgow, 1897
3. 3 Canning Street, Calton, Glasgow, 1899
4. 42 Craignessock Street, Glasgow, 1900+

- Practical scale, beam and weighing machine
  maker: PO Glasgow Directory 1994-7
- Scale, beam and weighing machine maker:
  PO Glasgow Directory 1899-1900+

FRASER, John, Glasgow, 1844-52

8 Surrey Lane (house, 10 Salisbury Place), Glasgow,
1844-6

- Smith and weighing machine maker:
  PO Glasgow Directory 1844-6

Succeeded by John Fraser & Co. (q.v.).

FRASER, John & Co., Glasgow, 1847-8

Salisbury Street (house, 8 Surrey Lane), Glasgow,
1847-8

- Weighing machine makers and general smiths:
  PO Glasgow Directory 1847-8

Reverts to John Fraser (q.v.).

FRASER, John, Glasgow, 1849-52

Salisbury Street (house, 8 Surrey Lane to 1850),
Glasgow, 1849-52

- Weighing machine maker and general smith: PO Glasgow Directory 1849-52

FULTON, David, Glasgow, 1898-1900+
21 Stockwell Street, Glasgow, 1898-1900+

- Weighing machine maker: PO Glasgow Directory 1898-1900+

GALBRAITH, James, Glasgow, 1826-39
53 Bridgegate, Glasgow, 1826-39

- Coppersmith: Glasgow Directory 1826-7
- Brazier: PO Glasgow Directory 1828-36
- Brazier and tinplate worker: PO Glasgow Directory 1837-9

Advertised with Robert Galbraith (q.v.) in 1835 as manufacturing milk measures as approved by James Cleland, in the *Glasgow Herald*, 5 January 1835: 'MILK MEASURES. The Subscribers having, by the kindness of the Inspector-General of Weights and Measures for the County of Lanark, been favoured with the dimensions of the Standard Measures for SWEET and BUTTERMILK, to come into operation on 1st February next, have to announce that they Manufacture, and have on hand for Sale, a supply of such Measures, properly marked and stamped. The Trade in Town and Country supplied with Standard sets. JAMES GALBRAITH, 53 Bridgegate. ROBERT GALBRAITH, 118 King Street, Opposite the Flesh Market …'

GALBRAITH, Robert, Glasgow, 1829-71
1. 118 King Street, Glasgow, 1829-44 and 1852-71
2. 118 King Street, Glasgow; works 105 King Street, Glasgow, 1845-6 and 1848-52
3. 118 King Street, Glasgow; 59 Gallowgate, Glasgow, 1847

- Brazier and tinplate worker: PO Glasgow Directory 1829, 1835-6
- Tinplate worker: PO Glasgow Directory 1830-4
- Brazier, pewterer and tinplate worker: PO Glasgow Directory 1837, 1839, 1858-65
- Brazier, pewterer, measure manufacturer: PO Glasgow Directory 1838
- Brazier, tinplate worker and gas-fitter: PO Glasgow Directory 1840-1, 1845-53
- Brazier, tinplate worker, etc.: PO Glasgow Directory 1842-3
- Brazier and pewterer: PO Glasgow Directory 1855-7
- Pewterer and tinsmith: PO Glasgow Directory 1866-71

Advertised with James Galbraith (q.v.) in 1835 (see above).

GARDNER, John, Edinburgh, 1774-1804
1. West Bow, Edinburgh, 1774
2. Nether Bow, Edinburgh, 1775-88, 1799-1804
3. Head of Fountain Close, Edinburgh, 1793, 1795-6
4. High Street, South side, Edinburgh, 1794
5. Opposite Fountain Close, Edinburgh, 1797

- Pewterer: Williamson's Edinburgh Directory 1774-86, 1793-1804
- Tinplate worker: Williamson's Edinburgh Directory 1788

Brass coin weight for £3. 12s. coin, marked 'J. Gardner Edin!', recorded in a private collection. Boxed set of coin scales with seven brass weights, in NMS K.1998.726. Trade label inside the lid reads: 'John Gardner, pewterer, at the back of the Fountain Well, Edinburgh, also all sorts of copper and brass work.' See also Paul and Bente Withers, *British Coin-Weights: A Corpus of the Coin-Weights made for use in England, Scotland and Ireland* (Llanfyllin, 1993), 327.

GARDNER, John, Glasgow, *c.*1769-1792
Crawford's Land, Bell's Wynd, Glasgow, 1773-90

- Mathematical, optical and philosophical instrument maker: Tait's Glasgow Directory 1787; Jones's Glasgow Directory 1789-90

Author of *The Description and Use of an Instrument for Weighing and Detecting Frauds in Light or Counterfeit Gold Coin: Improved by John Gardner … *(Glasgow, 1773). An example sold at Sotheby's 16 June 1975, Lot 57. See D. J. Bryden, *Scottish Scientific Instrument Makers 1600-1900* (Edinburgh, 1972), 20, note 91 for advertisement in the *Glasgow Herald*, 9 September and 11 November 1773; also D. J. Bryden, 'John Gardner's Coin-Scale: A Scottish Gold Coin Scale of 1773', in *Equilibrium*, no. 1 (1990), 1303-11; Roger Davis, 'A What's It Identified', in *Tools and Trades History Society Newsletter*, no. 30 (Summer 1990), 46-8. For John Gardner, see the chapter 'The Gardners of Glasgow' in T. N. Clarke, A. D. Morrison-Low and A. D. C. Simpson, *Brass & Glass: Scientific Instrument Making Workshops in Scotland* (Edinburgh, 1989), 88-95.

John Gardner was James Watt's senior journeyman and successor to his instrument business. He remained in business as an instrument maker and retailer in several family and other partnerships, until his death in 1822 when he was succeeded by his son's widow. She in turn was followed by her two sons, trading as Gardner & Co. (q.v.) from 1837.

GARDNER & Co., Glasgow 1837-83
1. 44 Glassford Street, Glasgow, 1837-8
2. 21 Buchanan Street, Glasgow, 1839-59
3. 53 Buchanan Street, Glasgow, 1860-82
4. 53 St. Vincent Street, Glasgow, 1883

- Mathematical, optical and philosophical instrument makers: PO Glasgow Directory 1833-83

An advertisement for the business in PO Glasgow Directory 1842, 166-7, includes 'improved MEASURING CHAINS … [and] GARDNER'S SELF-WINDING MEASURE, for the Vest Pocket'. For the Gardner business, see the chapter 'The Gardners of Glasgow' in T. N. Clarke, A. D. Morrison-Low and A. D. C. Simpson, *Brass & Glass: Scientific Instrument Making Workshops in Scotland* (Edinburgh, 1989), 88-95.

GARDNER, John, Glasgow, 1810
44 Gallowgate, Glasgow, 1810

- Ironmonger: McFeat's Glasgow Directory 1810

Successor to John Gardner & Co. (q.v.), ironmongers of Candleriggs.

GARDNER, John & Co., Glasgow, 1803-9
99 Candleriggs, Glasgow, 1803-9

- Ironmongers: McFeat's Glasgow Directory 1803-9

Succeeded by John Gardner (q.v), of the Gallowgate.

GENTLE, C. & Co., Glasgow 1880
52 Saltmarket, Glasgow, 1880

- Tinsmith, plumber and gas-fitter (in trades directory as weighing machine maker): PO Glasgow Directory 1880

GERARD, William, Edinburgh 1832-46
1. 73 North back of Canongate, Edinburgh, 1832-6
2. 10 New Street, Edinburgh, 1839-44

3. Moffat's Close, Edinburgh, 1845-6

- Smith: PO Edinburgh Directory 1832-6 and 1846
- Not in 1837-8 directory
- Smith and bellhanger: PO Edinburgh Directory 1839-45

Advertisement in the *Scotsman*, 11 March 1835: 'To farmers, victual dealers, &c., WILLIAM GERARD, SMITH, most respectfully intimates, that he has now on hand a few WEIGHING MACHINES, constructed on the most improved plan, weighing down from 1lib. to 30 stone, which we can recommend as being the most useful Weighing Machines for Farmers, Bakers and Victual Dealers, being so constructed as to suit for the Weighing of Potatoes, wholesale or retail, or any other Commodity, with as much dispatch and correctness as the common beam. The Machine will be seen, and every information given, by calling at the Shop, 73, North Back of Canongate, opposite foot of New Street. Smith Work done in all its Branches.'

GIBB & HOGG, Airdrie, 1870
Victoria engine works, Airdrie, Lanarkshire, 1870

- Weighing machine manufacturers: Kelly's Metal Trades 1870

GILBERT, E., Glasgow, 1892-3
117 Hydepark Street, Glasgow, 1892-3

- Weighing machine maker: PO Glasgow Directory 1892
- Smith and weighing machine maker: PO Glasgow Directory 1893

C. W. Ingram (q.v.) at this address from 1894.

GILBERT, John, Glasgow, 1858-90
29 Tureen Street, Glasgow, 1858-89

- Smith and weighing machine maker: PO Glasgow Directory 1858-71
- Weighing machine maker: PO Glasgow Directory 1872-3
- Beam and weighing machine maker: PO Glasgow Directory 1874-85
- Scale, beam and weighing machine maker: PO Glasgow Directory 1886-9
- (Trades directory only, as beam and scale maker): PO Glasgow Directory 1890

GLASGOW WEIGHING MACHINE MANUFACTORY, Glasgow, 1881-97

1. Works, 49 William Street, Glasgow; showrooms, 14 & 18 Stockwell Street, Glasgow, 1881-2
2. Works, 49 William Street, Glasgow; showrooms, 8 & 10 Stockwell Street, Glasgow, 1883
3. Works, Baltic Street, Glasgow; showrooms 8 & 10 Stockwell Street, Glasgow, 1884-5
4. Works, 130 Baltic Street, Glasgow; showrooms 8 & 10 Stockwell Street, Glasgow, 1886-97

- Makers of the famed counter beam and scales suitable for grocers, butchers etc.; also bucket-ended beams to weigh to 50 cwt. All sizes and adapted for every purpose: the oldest Glasgow house, Alex. Wood & Sons (trades directory only): PO Glasgow Directory 1881-97

See Alex. Wood & Sons (q.v.).

GLASS, James & Co., Glasgow, 1882-6
1. Thistle Engine Works, 46 Washington Street, Glasgow, 1882-3
2. Thistle Tool Works, 46 Washington Street, Glasgow, 1884-6

- Engineers and machine makers (in trades section as weighing machine makers from 1883): PO Glasgow Directory 1882-6

GORDON Brothers, Glasgow, 1880-2

64 West Howard Street, Glasgow, 1880-2

- Agents for Parnall & Sons, Bristol, 'engineers' and grocers' outfitters, inventors and patentees of Parnall's patent scales and balances, with agate bearings': PO Glasgow Directory 1880-2

GRAHAM & WARDROP, Glasgow, *c.*1830

In James Cleland, *An Historical Account of the Local and Imperial Weights and Measures of Lanarkshire, and an Inventory of those belonging to the Corporation of Glasgow* (Glasgow, 1832), 32: one set of Imperial brass spherical avoirdupois weights: 'one 56 lb., one 28 lb., one 14 lb., one 7 lb., one 4 lb., one 2 lb., one 1 lb., one 8 oz., and one 4 oz. Each Weight has its denomination and the Letters A V stamped on it. This set was made by Graham & Wardrop, Glasgow'; also ibid., 33: one set of Imperial brass spherical troy weights: 'one of 1 lb., one of 6 oz., one of 3 oz., one of 2 oz., and one of 1 oz. This set was made by Graham & Wardrop, Glasgow.' Probably a misidentification of J. & H. Wardrop (q.v.).

GRANDY (William) & SCOTT (Thomas), Dundee, 1861-4

193 Overgate, Dundee, 1861-4

- Ironmongers, beam makers, japanners and tinsmiths, and advertising 'beams repaired': PO Dundee Directory 1861-4

Previously McLaren & Grandy (q.v.). Subsequently Mrs Henry Grandy, ironmonger, tinsmith and gas-fitter, 193 Overgate, 1867.

GRANT, James and William, Stirling, *c.*1835

Advertisement in the *Stirling Journal and Advertiser*, 14 November 1834: 'BURGH OF STIRLING – WEIGHTS AND MEASURES The magistrates of STIRLING do hereby intimate that they have appointed Messrs. JAMES and WILLIAM GRANT, Baker Street, Stirling, INSPECTORS of WEIGHTS and MEASURES, and KEEPERS of the MODELS or COPIES verified in Exchequer, of the IMPERIAL STANDARD WEIGHTS and MEASURES provided for the BURGH, in terms of the Statute 4 and 5 William IV, Chapter 49 …' Article in the *Stirling Journal and Advertiser*, 6 February 1835: 'WEIGHTS AND MEASURES – Considerable uneasiness had prevailed in Falkirk and its neighbourhood as to the proper quarter to be applied to, in order to have the regular adjustment made of weights and measures, in consequence of the Magistrates of the Burgh giving authority to an ironmonger in the town to stamp the same. Our readers know that the Messrs. Grant are regularly appointed for that purpose in the county, by the Justices in the last Quarter Session …'. In 1835 there was a dispute between Archibald Cochran & Son (q.v.) and J. & W. Grant as to which firm was supposed to be authorised to do the stamping in Falkirk: *Stirling Journal and Advertiser*, 6 and 13 February 1835. Advertisement Article in the *Stirling Journal and Advertiser*, 13 February 1835: 'STIRLINGSHIRE WEIGHTS AND MEASURES – The Inspectors for the County, J. & W. GRANT, hereby intimate that they intend making a Circuit, next week in the County …' on specified dates for 'Denny, Kilsyth, Lennoxtown, Balfron and Kippen'.

GRANT, James, Stirling, 1860-8

28 Murray Place, Stirling, 1860-8

- Inspector of weights and measures: Slater's Directory of Scotland 1860-7
- Smith and Inspector of weights and measures: Duncan and Jamieson's Directory for Perth, etc., 1868.

HUTCHISON, D., [Dalkeith?], *c.*1765

Maker of the 1765 Aberdeen Jug (Item **200**). A possible identification is that of David Hutchison, nephew of James Gray, who inherited the Iron Mill at Dalkeith after the death of his uncle in 1761. He may have been involved with Gray's earlier metrological work in the 1740s and 1750s (see Chapter **9**).

HUTCHISON, William, Perth, *fl.*1835

Described himself in an advertisement in the *Stirling Journal and Advertiser* for 16 January 1835: 'PERTHSHIRE WEIGHTS AND MEASURES WILLIAM HUTCHISON, Smith, Inspector and Stamper of WEIGHTS and MEASURES for the County and Burgh of Perth, hereby intimates that he will be in attendance at his shop, Kirkside, Perth, every lawful day … to Verify, and, if found correct, to Stamp all Imperial Weights and Measures submitted to him …', and that he would travel on specified dates to Dunblane, Doune, Thornhill, Callander and Kincardine. On 6 February he advertised specified dates for Yetts of Muckhart, Doune and Auchterarder. Dispute in the *Stirling Journal and Advertiser*, 17 and 26 April 1835, over Hutchison's methods of verification by William Stirling (q.v), a cooper of Kippen, Stirlingshire: 'My mode of proceeding was originally approved of by Dr Anderson, the distinguished Rector of the Perth Academy.'

INGRAM, James, Glasgow, 1861-80

1. 39 Stockwell Street, Glasgow, 1861-8
2. 22 Argyll Street, Glasgow, 1868-74
3. 13 Back Wynd, Glasgow, 1875-80
4. 109 Trongate, Glasgow, 1875-9

*   Smith, scale beam and weighing machine maker: PO Glasgow Directory 1861 (and locksmith from 1862, and gas-fitter 1862-5)
*   Smith, scale beam and weighing machine maker and locksmith: PO Glasgow Directory 1866-9
*   Scale beam and weighing machine maker and locksmith: PO Glasgow Directory 1870-80

INGRAM, Charles W., Glasgow, 1883-7

1. 60 Wellington Street, Glasgow, 1883-6
2. 19 Robertson Street, Glasgow, 1887

*   Smith, scale, beam and weighing machine maker: PO Glasgow Directory 1883-7

Succeeded by James Ingram (q.v.).

INGRAM, James, Glasgow, 1888

19 Robertson Street, Glasgow, 1888

*   Weighing machine maker: PO Glasgow Directory 1888

INGRAM, C. (Charles) W., Glasgow, 1894-1900+

117 Hydepark Street, Glasgow, 1894-1900+

*   Smith, scale, beam and weighing machine maker: PO Glasgow Directory 1894-8

E. Gilbert (q.v.) previously at this address.

JAMIESON, Joseph, Dundee, 1826-9

Overgate, Dundee, 1829

*   Brass founder: PO Dundee Directory 1829

Maker of Item **242** in the Inventory, dated 1826.

JOHNSTON, Daniel, Greenock, 1853-61

1. 20 Cathcart Street, and 19 Shaw Street, Greenock, 1853
2. 20 Cathcart Street, Greenock, 1861

*   Smith: PO Greenock Directory 1853
*   Blacksmith and beam maker: PO Greenock Directory 1861

KEMP & Co., Edinburgh, 1837-87
1. 7 South College Street, Edinburgh 1835-8
2. 53 South Bridge, Edinburgh 1839-44
3. 12 & 13 Infirmary Street, Edinburgh 1845-87

- Chemical suppliers and philosophical
instrument makers: PO Edinburgh Directory
1837-87

Hand-held apothecary's balance: NMS T.1971.15; see
A. D. Morrison-Low, 'Kemp & Co., Laboratory
Suppliers', in J. T. Stock and M. V. Orna (eds), *The
History and Preservation of Chemical Instrumentation* (Dordrecht, 1986), 163-86.

KENNEDY, BENNET & Co., Glasgow, 1872-6
28 St Enoch's Square, Glasgow, 1872-5

Agents (1872-6 only) for F. E. Duckham, London,
inventor and maker of the patent hydrostatic
weighing machine. The manufacturer was
subsequently J. Cowdy of London, and Cowdy's
Glasgow agency was held by James Houston (q.v.).

KERR, Alexander, Glasgow, *fl.*1818
Admitted member of the Incorporation of
Hammermen of Glasgow 1818 as a tinsmith, by
right of his father, with 'a choppin jug' as essay:
Harry Lumsden and P. Henderson Aitken, *History
of the Hammermen of Glasgow* (Paisley, 1912), 308.

KERR, Robert, Glasgow, *fl.*1825
Admitted member of the Incorporation of
Hammermen of Glasgow in 1825 as a stranger
coppersmith, with 'a gallon measure' as essay:
Harry Lumsden and P. Henderson Aitken, *History
of the Hammermen of Glasgow* (Paisley, 1912), 311.

LANDALE & TOD, Edinburgh, 1824-9
1. 6 Hunter Square, Edinburgh, 1824-5
2. 8 West Register Street, Edinburgh, 1828-9

- General ironmongers: PO Edinburgh
Directory 1824-5
- Wholesale ironmongers: PO Edinburgh
Directory 1828-9

See J. Katz, 'Contemporary Comment, 1825', in
Equilibrium (1998), 2250-2, about R. B. Bate and
the sale of standard measures through local agents
such as Landale & Tod.

LAWRENCE, John, Aberdeen, *c.*1800-1838
1. Milner's Court, 21 Guestrow, Aberdeen,
1824-34
2. 29 Guestrow, Aberdeen, 1835-8

- Brass founder: Gordon's Aberdeen Directory
1824 (first issue of an Aberdeen Directory)

John Lawrence, Brazier, was admitted to the
Aberdeen Incorporation of Hammermen 5 July
1800, and admitted a burgess of his own trade by
Aberdeen Town Council, 22 September 1801.

Stamp 'LAURENCE ABD' on nesting weights in
Aberdeen City Museum, ABDN 16636. Two sets of
5 bronze nesting weights from 4 ounces, stamped
'LAURENCE, ABERDEEN', are held in Marischal
Museum, University of Aberdeen, ABDUA 18230
and 36837.

LEITH, CARDLE & Co., Glasgow, 1894-1900+
Plantation Engine Works, 29 Eaglesham Street,
Glasgow, 1894-1900+

- Makers of … suspended lever weighing
machines, chain makers and general engineers
(appear in previous directories, but first in 1894
trade directory as weighing machine makers):
PO Glasgow Directory 1894-1900+

LIDDEL, James & Co., Glasgow, 1812-24

1. 156 Stockwell, Glasgow, 1812-14
2. 155 Stockwell, Glasgow, 1815-22
3. 157 Stockwell, Glasgow, 1823
4. 180 Saltmarket, Glasgow, 1824

- Smiths and bellhangers: McFeat's Glasgow Directory 1812-23
- Warehousemen: McFeat's Glasgow Directory 1824

Stamp 'IL&Co' recorded on Glasgow tron standards, in James Cleland, *An Historical Account of the Local and Imperial Weights and Measures of Lanarkshire, and an Inventory of those belonging to the Corporation of Glasgow* (Glasgow, 1832), 31: 'Seven Brass Weights, with cross handles, viz. one 16 lb., one 8 lb., one 4 lb., one 2 lb., one 1 lb., one ½ lb., and one ¼ lb. Each Weight has the following inscription on it: – STANDARD TRON FOR BUTCHER MEAT, I.L. & Co.'

Previously Wilsone & Liddel (q.v.). James Liddell jun. [*sic*], smith, became a member of the Incorporation of Hammermen of Glasgow in 1806, by right of his father (admitted 1795), with 'a small beam' as essay: Harry Lumsden and P. Henderson Aitken, *History of the Hammermen of Glasgow* (Paisley, 1912), 303. It was presumably his father who is recorded separately as a smith at Boyd's Land, Stockwell, 1801-15. Alexander Wood (q.v.) is listed as at James Liddel & Co. from 1812 to 1823; he subsequently took over the premises when Liddel & Co. moved to Saltmarket.

LOCKHART, James, Glasgow, 1757-76

James Lockhart, merchant, became a member of the Incorporation of Hammermen of Glasgow in 1757, with 'a pair of silver piercers' as essay: Harry Lumsden and P. Henderson Aitken, *History of the Hammermen of Glasgow* (Paisley, 1912), 294.

Coin scales, NMS T.1996.146; see A. D. C. Simpson, 'Coins Weighed by the Drop', in *Equilibrium* (1996), 2068-2.

LOUDEN, John, Glasgow, 1886

192 New Dalmarnock Road, Glasgow, 1886

- Beam and weighing machine maker: PO Glasgow Directory 1886

LOW, P., Perthshire[?], *c.*1830

Steelyard at the Highland Folk Museum, Kingussie.

LOW, William, Perthshire[?], *c.*1830

Another of the same pattern at the Highland Folk Museum, Kingussie, both of a similar pattern to that by A. Thomson (q.v.).

LOWE, Charles, Glasgow, 1857-70

1. 151 George Street, Glasgow, 1857-62
2. 46 & 50 Bath Street and 86 Renfield Street, Glasgow, 1863-6 and 1869
3. 86 Renfield Street, Glasgow, 1868 and 1870

- Grocers' japanner, tinsmith, scale, weighing machine and mill manufacturer: PO Glasgow Directory 1857-8
- Scottish Canister and Scale Works: PO Glasgow Directory 1859
- Canister maker: PO Glasgow Directory 1860-1
- Canister, scale beam and weighing machine manufacturer: PO Glasgow Directory 1862
- Japanner and scale maker: PO Glasgow Directory 1863-6
- Japanner: PO Glasgow Directory 1868
- Canister and scale maker: PO Glasgow Directory 1869-70

Succeeded by Lowe & Co. (q.v.).

LOWE (Charles) & Co., Glasgow, 1872-92

1. 146 Renfield Street, Glasgow, 1872-6

2. 147 & 149 Renfield Street, Glasgow, 1877-8
3. 158 Renfield Street, Glasgow, 1879-92

- Canister and scale makers: PO Glasgow Directory 1872-3
- Grocers' outfitters (in trades directory as scale makers) : PO Glasgow Directory 1874-92

Previously Charles Lowe (q.v.).

LOWDON, George, Dundee, 1850-1900+
1. 25 Union Street, Dundee, 1850-61
2. 1 Union Street, Dundee, 1864-74
3. 23 Nethergate, Dundee, 1876-80
4. 60 Reform Street, Dundee, 1882-1900+

Optician and philosophical instrument maker. Recorded as having sold a Walker-type steelyard 'made by G Lowdon': see *Equilibrium* (2002), 2718. George Lowdon (1825-1912) wrote an auto-biography: see 'George Lowdon of Dundee' in T. N. Clarke, A. D. Morrison-Low and A. D. C. Simpson, *Brass & Glass: Scientific Instrument Making Workshops in Scotland* (Edinburgh, 1989), 146-51.

LUMSDEN, Benjamin, Aberdeen, 1765-82
Goldsmith, Aberdeen.

Portable folding brass coin scale, 1773: example in NMS T.1978.90; another in a private collection; see A. D. C. Simpson, 'Coins Weighed by the Drop', in *Equilibrium* (1996), 2068-72. See also I. E. James, *The Goldsmiths of Aberdeen* (Aberdeen, 1981), 70, who gives his dates as 1765-82.

LUNAN, Charles, Aberdeen, 1771-1809
Watch and clockmaker, associated with Patrick Copland, professor of natural philosophy, Marischal College, Aberdeen.

Maker of the Aberdeen standard measure for oats

and bear under the superintendence of Patrick Copland, George Skene Keith and George Hogarth; tested by them 1811: see John S. Reid, 'Patrick Copland (1748-1822)', unpublished University of Aberdeen MLitt thesis, 1983, appendix 2, p. 10. Maker of the standard pint and fractions (Item **207**). For biographical details about Lunan, see R. M. Lawrance, 'Notes on the Old Clockmakers, no. VI', *Aberdeen Journal*, 20 April 1921 and I. E. James, *The Goldsmiths of Aberdeen* (Aberdeen, 1981), 80.

LYON, William & George, Glasgow, 1819-29
In the local directories from 1819

Advertised in the *Glasgow Herald*, 20 March 1826: 'IMPERIAL MEASURES. WILLIAM & GEORGE LYON, Braziers, Coppersmiths and Tin Plate Workers, 96 CANDLERIGGS, Makers of IMPERIAL COPPER and TIN MEASURES, respectfully intimate, that having received from the Authorities an Adjusted Set of Measures, they can now furnish their Friends and the Public with Measures warranted correct and, equal in strength and workmanship to any in the Trade, Orders punctually attended to.'

Subsequently George Lyon *junior* operated as coppersmith and tinplate worker at 16 Candleriggs (1830) and William Lyon & Son as tinplate workers at 124 Buchanan Street (1830).

McADAM (Alexander) & DAWSON, Glasgow, 1873-80
19 Landressy Street, Bridgton Cross, Glasgow, 1873-80

- Smiths, tinsmiths, gas-fitters, bellhangers, scale, beam and weighing machine makers: PO Glasgow Directory 1873-9

Subsequently Alexander McAdam, tinsmith and gas-fitter, listed at this address to 1883.

McALLISTER, J. B., Glasgow, 1881
64 Howard Street, Glasgow, 1881

- Agent for P. Rogers & Co., Birmingham, scale beam makers: PO Glasgow Directory 1881 (trades only)

McALPINE, Colin, Glasgow, 1848-59
5 Brown Street, Glasgow, 1848-59

- Ship and anchor smith, chain cable manu-facturer: PO Glasgow Directory 1848-51
- Ship and anchor smith, chain cable manu-facturer, and beam maker, and inspector of weights and measures for the Western Districts: PO Glasgow Directory, 1852-9

Previously Colin McAlpine (1837-44) and Colin McAlpine & Co. (1845-7), at the same address.

McCLURE, John, Glasgow, 1820-7
Engraver at various addresses, 1820-7

Advertisement in the *Glasgow Herald* for 28 April 1826: 'OLD AND NEW MEASURES. JOHN McLURE, ENGRAVER, having obtained through the politeness of Professor Wallace, accurate proportionals between the OLD and NEW WINE, SPIRIT, and ALE MEASURES, begs leave to inform the Public that he is at present employed in Engraving on the <u>Old Measures</u> the proportion which they bear to the New. J.McC. understands that all dealers subject themselves to the penalties of the act (which will no doubt be levied), if they continue to use the Old Measures without being correctly marked; and that if they are so marked, they may be legally employed as long as they last, although not stamped <u>anew</u>.'

McCULLOCH, Andrew, Glasgow, *fl.*1786
Admitted member of the Incorporation of Hammermen of Glasgow 1786 as a smith, with a 'twelve inch beam' as essay: Harry Lumsden and P. Henderson Aitken, *History of the Hammermen of Glasgow* (Paisley, 1912), 296.

14-pound trading weight, marked McCulloch & Co., Glasgow, 1826, is possibly by the same firm, NMS T.1988.58

McDONALD (Robert) & OMNET (John), Glasgow, 1877-8
13 Norfolk Street, Glasgow, 1877-8

- Smiths, scale, beam and weighing machine makers: PO Glasgow Directory 1877

Succeeded by Robert McDonald (q.v.) and John Omnet (q.v.).

McDONALD, Robert, Glasgow, 1879-92
1. 148 Trongate and 13 Brunswick Lane, Glasgow, 1879-84, and 1886-7
2. 148 Trongate and 9 Brunswick Lane, Glasgow, 1888-92

- Smith and scale beam maker: PO Glasgow Directory 1879-82 and 1887
- Jobbing smith and machine maker: PO Glasgow Directory 1883 and 1884
- Jobbing smith, scale beam and weighing machine maker: PO Glasgow Directory 1886
- Smith and machine maker (in trades directory as beam scale maker): PO Glasgow Directory 1888-92

McFEE, Robert, Glasgow, 1870-2
103 Great Hamilton Street, Glasgow, 1870-2

- Scale beam maker: PO Glasgow Directory 1870-2

McGREGOR, Daniel, Glasgow, 1851
44 Bridge Street, Glasgow, 1851

- Smith, scale, beam and weighing machine maker: PO Glasgow Directory 1851

Subsequently A. & D. McGregor (q.v.).

McGREGOR, A. & D., Glasgow, 1852
27 Stockwell Street, Glasgow, 1852

- Smiths, scale beam and weighing machine makers and gas-fitters: PO Glasgow Directory 1852

McINTYRE, John & Co., Glasgow, 1855-8
1. 38 Great Dovehill, Glasgow, 1855-6
2. 27 Burnside Street, Glasgow, 1857-8

- Scale, beam and weighing machine makers: PO Glasgow Directory 1855-7
- Smiths', scale, beam and weighing machine makers: PO Glasgow Directory 1858

McINTYRE, John, Glasgow, 1859-61
1. 161 London Street, Glasgow, 1859
2. 20 St Mungo Street south, Glasgow, 1860-1

- Smiths', scale, beam and weighing machine makers: PO Glasgow Directory 1859-61

McKAY, Mrs E., Glasgow, 1850-61
8 Frederick Lane, Glasgow, 1850-61

- Broker: PO Glasgow Directory 1850-1
- Dealer in ironmongery, scales, beams, weights etc.: PO Glasgow Directory 1852-9
- Dealer in ironmongery (in trades as scale maker): PO Glasgow Directory 1860-1

McKEY, Archibald, Glasgow, *fl.*1702
Admitted member of the Incorporation of Hammermen of Glasgow 1702, by right of his father, with 'ane beam and skeals' as essay: Harry Lumsden and P. Henderson Aitken, *History of the Hammermen of Glasgow* (Paisley, 1912), 291.

McKINLAY, William, Glasgow, 1852-72
1. 17 Laigh-kirk Close, Trongate, Glasgow, 1852-5
2. 35 Gallowgate and Trongate, Glasgow, 1856-9
3. 47 King Street, Glasgow, 1860-72

- Tinsmith, mounting, holdfast, heel and toe plate and nail manufacturer, fender and shovel maker: PO Glasgow Directory 1852-5
- Tinsmith, mounting, holdfast, heel and toe plate and nail manufacturer, fender, shovel maker and grinder: PO Glasgow Directory 1856-60
- Smith, grinder and manufacturer of fenders, fire-irons, tinsmiths' mountings, holdfasts, bolts, nuts, beams and stands etc.: PO Glasgow Directory 1861-70
- Smith and grinder (in trades as a beam maker): PO Glasgow Directory 1871-2

McLAREN (William) & GRANDY (William A.), Dundee, 1846-58
1. 111 Overgate, Dundee, 1846
2. 171 Overgate, Dundee, 1850-8

- Brokers and ironmongers: PO Dundee Directory 1846
- Blacksmiths and ironmongers, advertising 'standards, beams, and scales; counter weighing machines; weights': PO Dundee Directory 1850
- Ironmongers, gas-fitters and beam makers: PO Dundee Directory 1856
- Ironmongers: PO Dundee Directory 1858

Previously William Maclaren [*sic*], smith, Gellatly's Buildings, Dock Street North, 1837; East Harbour,

1840-5; William A. Grandy, broker, auctioneer and appraiser, Wellgate, 1840-2; auctioneer, 35 Long Wynd, 1845. Subsequently Grandy & Scott (q.v.).

McLAREN & MEIKLE, Glasgow, 1890-1900+
25 Gordon Street, Glasgow, 1890-1900+

- Diesinkers and engravers (in trades directory as brass yard-measure makers): PO Glasgow Directory 1890-1900+

McLINTOCK, Walter, Glasgow, 1878-87
1. 41 College Street, Glasgow, 1878-81
2. 30 North Albion Street, Glasgow, 1882-7

- Tinsmith, gas-fitter, grocers' outfitter for canisters, scales, etc.: PO Glasgow Directory 1878
- Plumber, tinsmith, bellhanger and gas-fitter: PO Glasgow Directory 1879-81
- Plumber, tinsmith and gas-fitter: PO Glasgow Directory 1882-7

Does not appear in the trades directory, apart from 1878.

McMILLAN, EDMONDSTON & CRAWFORD, Glasgow, 1874
36½ Thistle Street, Glasgow, 1874

- Scale, beam and weighing machine makers: PO Glasgow Directory 1874

MACPHERSON, Daniel, Edinburgh, 1851
7 Salisbury Street, Edinburgh

Inventor and manufacturer of a weighing machine, exhibited at the Great Exhibition, 1851: see *Great Exhibition of the Works of All Nations, 1851, Official Descriptive and Illustrated Catalogue*, 4 volumes (London, 1851), I, 470: item 684A.

MAITLAND, James, Glasgow, *fl.*1718
Admitted member of the Incorporation of Hammermen of Glasgow 1718, with an essay which included 'ane pint stoup': Harry Lumsden and P. Henderson Aitken, *History of the Hammermen of Glasgow* (Paisley, 1912), 292.

MALLOCH, James, Glasgow, 1841-9
24 Jackson Street, Glasgow, 1841-9

- Smith and scale beam maker: PO Glasgow Directory 1841
- Scale beam and weighing machine maker: PO Glasgow Directory 1845

Succeeded by J. & A. Malloch (q.v.).

MALLOCH, J. & A., Glasgow, 1850-1
48 Stockwell Street, Glasgow, 1850-1

- Scale, beam and weighing machine manufacturers: PO Glasgow Directory 1850-1

Succeeded by Malloch & Williamson (q.v.).

MALLOCH & WILLIAMSON, Glasgow, 1852
14 Jackson Street, Glasgow, 1852

- Scale, beam and weighing machine makers: PO Glasgow Directory 1852

MALLOCH, James, Kilmarnock, 1860
Regent Street, Kilmarnock, 1860

- Scale beam and weighing machine maker: Slater's Directory of Scotland 1860

MALLOCH, John, Glasgow, 1821-43
1. 120 John Street, Glasgow, 1821
2. 113 John Street, Glasgow, 1822-6
3. 40 John Street, Glasgow, 1827-8
4. 187 Trongate, Glasgow, 1829-38
5. 181 Trongate, Glasgow, 1839-43

- Smith: McFeat's Glasgow Directory 1821
- Smith and scale beam maker: PO Glasgow Directory 1834

Admitted member of the Incorporation of Hammermen of Glasgow 1827, as a blacksmith stranger, with 'a beam' as essay: Harry Lumsden and P. Henderson Aitken, *History of the Hammermen of Glasgow* (Paisley, 1912), 312.

Succeeded by C. & R. Adams (q.v.) at this address.

MAXWELL, Robert, Edinburgh, 1700-43
Founder and Hammerman: see C. E. Whitelaw, *Scottish Arms Makers*, (London, 1977), 133.

Maker's mark on 1707/8 Culross weights (Item 87).

MELDRUM, William & Sons, Colinsburgh, Fife, *c.*1835
William Meldrum died 15 March 1852 aged 72: A. J. Campbell, 'Fife Deaths 1822-54', typescript in Department of Special Collections, St Andrews University Library. Campbell's 'Fife Trades and Professions 1820-70' lists no Colinsburgh or Kilconquhar records.

Imperial bushel for the County of Ross, *c.*1835, now NMS T.1994.5; Imperial bushel, NMS W.1999.76; Imperial half-bushel recorded *c.*1960.

MELDRUM, James & Co., Colinsburgh, Fife, 1860
Colinsburgh, 1860

- Coopers: Slater's Directory of Scotland 1860

MILLAR or MILLER, Stephen & Co., Glasgow, 1811-47
1. Saracen's Lane, Glasgow, 1811-25
2. 48 Saracen's Lane, Glasgow, 1826-47

- Founders: McFeat's Glasgow Directory 1811-29
- Coppersmiths, brass and ironfounders: PO Directory 1830-47

In James Cleland, *An Historical Account of the Local and Imperial Weights and Measures of Lanarkshire, and an Inventory of those belonging to the Corporation of Glasgow* (Glasgow, 1832), 32-3: 'Imperial Weights … One set of Brass Weights, with cross handles, viz. one 14 lb., one 7 lb., one 4 lb., one 2 lb., one 1 lb., one 8 oz., and one 4 oz. Each weight has the City Arms and the letters I.D.G.G. stamped on it. This set by which the magistrates occasionally prove the Weights used in the market, was made by Stephen Millar & Co., Glasgow and is kept in the Hall of the Corporation of Fleshers … One set Brass Weights, flat, viz., one 8 oz., one 4 oz., one 3 oz., one 2 oz., one 1 oz., one 8 dr., one 2 dr., and two of 1 dr. This set was made by Stephen Millar & Co., Glasgow.' Ibid., 34: 'Imperial Measures liquid … One set of Duplicates, conical shape, made of strong copper, with brass necks seven-eights of an inch diameter, and perforated brass stoppers, for adjustment, viz.:- One two Gallons, One one Gallon, One Half Gallon, One Quart, One Pint, One Half Pint, One Gill, One Half Gill. Each Measure has the City Arms and the following words engraved on it:- "Duplicate of Imperial" (the denomination) "for the City of Glasgow adjusted by Dr MEIKLEHAM, Dr THOMSON, Dr URE, JAMES CLELAND, JAMES CRICHTON, 1826." These Duplicates were made by Stephen Millar & Co., Glasgow … One set Copper [Common swelled] shape, made by Stephen Millar and Co., Glasgow, viz., – One two Gallons, One one Gallon, One Half Gallon, One Quart, One Pint, One Half Pint, One Gill, One Half Gill.' Makers of Item **241**.

MILLER, John *senior*, **Edinburgh, c.1754**
Turner; one of makers of 1754 Stirlingshire firlots (Items **216** and **217**); see T. N. Clarke, A. D. Morrison-Low and A. D. C. Simpson, *Brass & Glass: Scientific Instrument Making Workshops in Scotland* (Edinburgh, 1989), 26; also A. D. C. Simpson, '"Handle with Care": Handling Warnings on Early Scientific Instruments and Two Early Scottish Grain Measures', in *Bulletin of the Scientific Instrument Society*, no. 30 (1991), 3-4.

MILNE, James, Edinburgh, 1808-36
Chalmer's Close, Edinburgh, 1808-36

- Brass founder: PO Edinburgh Directory 1808-36
- Successor to John Milne & Sons and succeeded by James Milne & Son (q.v.).

Successor, with John Milne II (q.v.), to John Milne & Son (q.v.).

MILNE, James & Son, Edinburgh, 1837-1900+
1. Chalmer's Close, Edinburgh, 1837-40
2. Chalmer's Close, 81 High Street, Edinburgh, 1841-53
3. Milton House, 90 Canongate, Edinburgh, 1854-80
4. Milton House, and Arcade 105 Princes Street, Edinburgh, 1881-4
5. Milton House Works Abbeyhill and 12 St Andrews Square, Edinburgh, 1888-96
6. Milton House Works, Milton Street, Edinburgh, 1897-1900+

- Brass founders and gas meter manufacturers: PO Edinburgh Directory 1837-56

- Gas engineers, brass founders, gas meter manufacturers, etc.: PO Edinburgh Directory 1857-60
- Engineers, brass founders, gas meter manufacturers, etc.: PO Edinburgh Directory 1861-84
- Millwrights, engineers, brass founders and gas-meter manufacturers: PO Edinburgh Directory 1885-94
- Electrical engineers, millwrights, brass founders and gas-meter makers: PO Edinburgh Directory 1895-1900+

On John Milne (son of James), see his obituary in D. Bruce Peebles, 'Address by the President', *Transactions of the Royal Scottish Society of Arts,* 11 (1887), 391-3. See also [Anon], 'British Engineers no. 12. Messrs. James Milne & Son, Engineers, Brass-founders, &c. Milton House Works, Edinburgh,' in *The Mercantile Age,* 12 (1887), 354. 'Revolving Ballance, or Weighing Machine', in the *Catalogue of the Exhibition of Models and Manufactures, &c. at the Tenth Meeting of the British Association for the Advancement of Science* (Glasgow, 1840), 20, item 136.

Hand-held scales by Milne & Son: NMS T.1981.114 and T.1987.333: for similar, see Paul and Bente Withers, *British Coin-Weights: A Corpus of the Coin-Weights made for use in England, Scotland and Ireland* (Llanfyllin, 1993), 328.

MILNE, John (I), Edinburgh, 1742-67
Founder and ironmonger 1742; continued in partnership with his son John and already trading as Milne & Son in 1767; described as brass founder in 1808; dead by 1810.

Succeeded by John Milne & Son (John II) (q.v.).

MILNE, John & Son (John II), Edinburgh, 1767-1807

1. The Golden Church, High Street; branch Bishop's Land, High Street, Edinburgh, 1767
2. Chalmer's Close, Edinburgh, 1773
3. Bishop's Land, Edinburgh, 1774-82; 1788-94; 1795-1804
4. Head of Gray's Close, Edinburgh, 1784-6
5. Opposite the Head of Blackfriair's Wynd, Edinburgh, 1794
6. High Street, 1805-7

- Founders: Williamson's Edinburgh Directory 1773
- Ironmongers: Williamson's Edinburgh Directory 1780-2
- Founders and ironmongers: Williamson's Edinburgh Directory 1784-1803; 1806-7
- Founders: Williamson's Edinburgh Directory 1804-5

John II became a burgess by patrimony 1797, and his brother James a burgess by patrimony in 1808; John succeeded by James in c.1825. John II was maker of weights to John Robison's design, including Items **95, 98** and **99**.

Succeeded by John Milne (John II), and James Milne (q.v.).

MILNE, John (II), Edinburgh, 1808-24

1. Bishop's Land, High Street, Edinburgh, 1808-10
2. 129 High Street, Edinburgh, 1811-12; 1814-20
3. 41 North Bridge, Edinburgh, 1813
4. 15 Greenside Place, Edinburgh, 1821
5. 21 Greenside Street, Edinburgh, 1822-3
6. Milne's Court, Edinburgh, 1824

- Brass founder and ironmonger: PO Edinburgh Directory 1808-23
- Late ironmonger: PO Edinburgh Directory 1824

Successor to John Milne & Son (John II) (q.v.).

MIRRLEES, Charles, Glasgow, 1828-32
105 Stockwell, Glasgow, 1828-32

- Superintendent of weighing machines: PO Glasgow Directory 1828-32

MITCHELL, John, Alloa, *fl.*1738
Maker of the beam, dated 1738, supplied to Stirling Guildry, but otherwise unrecorded: STIGM 19817/15/10.3.

MONTROSE FOUNDRY COMPANY, Montrose, 1850
Advertisement in Dundee Directory 1850 as 'manufacturers of cast-iron goods, … weighing machines …'.

MOUWAT, George, Kirkwall, 1659-86
Appointed keeper and adjuster of Kirkwall standard weights in 1659: James Mackenzie, *The General Grievances and Oppression of the Isles of Orkney and Shetland* (Edinburgh, 1750, reprinted 1836), 51-4. Mentioned also as the authorised maker and adjuster of pundlars in 1662; succeeded by George Craigie in 1686: *Memorial and Abstract of the Proof for the Earl of Galloway and Others, Udalmen, and Proprietors of land in Orkney, Pursuers, against James Earl of Morton, Defender* (Edinburgh, 1758), 76-7, 79; *Memorial for James, Earl of Morton, Defender, against Alexander, Earl of Galloway … Pursuers* (Edinburgh, 1758), 38.

MUIR, William, Kelso, *fl.*1826
Advertisement in the *Kelso Mail*, 30 January 1826: 'WEIGHTS AND MEASURES. WILLIAM MUIR, Coppersmith in KELSO, intimates to the Public, that having been appointed by His Majesty's Justices of the Peace for the District of KELSO, to adjust WEIGHTS

and MEASURES according to the NEW IMPERIAL STANDARD, and to authenticate the same by a proper Stamp, he is ready to adjust, verify and stamp any such Weights and Measures as may be brought to him for the purpose.

WILLIAM MUIR has on hand for Sale, a considerable number of IMPERIAL BUSHELS, HALF BUSHELS, and CAST METAL WEIGHTS; and in the course of a few days, he will receive a large supply of LIQUID MEASURES and BRASS WEIGHTS, of the New Standard. OLD WEIGHTS and MEASURES taken in EXCHANGE.' The same advertisement with the following additional paragraph appeared on 6 February and 3 April 1826: 'Persons employing WILLIAM MUIR in adjusting Weights and Measures, may expect every attention being paid to accuracy of adjustment, as he has got erected an Apparatus for adjusting the Bushels by Water, according to the Act of Parliament; and for adjusting the Weights, Beams and Scales of the best description. He also alters Copper Measures to the Imperial Gallon ...'.

MUNRO, Donald, Glasgow, 1887-1900+
1. 13 Robertson Street, Glasgow, 1887-91 and 1899-1900+
2. Excelsior Tinplate Works, 13 Robertson Street, Glasgow, 1892-8

*   Wholesale ironmonger ... (brass yard-measure maker in trades directory): PO Glasgow Directory 1887-1900+

MURRAY, David, Glasgow, 1825-35
1. [At] J. & H. Wardrop's (q.v.), Glasgow, 1826-7
2. Union Street, Glasgow, 1828
3. 37 Carrick Street, Glasgow, 1829-32
4. 5 Carrick Street, Glasgow, 1833

*   Coppersmith: PO Glasgow Directory 1826-35

Admitted member of the Incorporation of Hammermen of Glasgow 1825 as a coppersmith stranger, with 'a gallon measure' as essay: Harry

Lumsden and P. Henderson Aitken, *History of the Hammermen of Glasgow* (Paisley, 1912), 310. The firm became D. & W. Murray in 1834-5, subsequently William Murray (q.v).

MURRAY, James, Dundee, 1829-53
1. 11 High Street South, Dundee, 1829
2. 12 High Street South, Dundee, 1834-7
3. 11 High Street, Dundee, 1840-2
4. 9 High Street and Victoria Dock, Dundee, 1845
5. 11 High Street and South Side Victoria Dock, Dundee, 1846
6. 9 & 11 High Street and South Side Victoria Dock, Dundee, 1850
7. 9 & 11 High Street, Dundee, 1853

*   Ironmonger: Dundee Directory and Register 1829
*   Ironmonger and blacksmith: Dundee Directory and General Register 1834
*   Ironmonger: Dundee Directory 1840
*   Ironmonger and blacksmith: Dundee Directory 1842
*   Ironmonger, shipsmith, etc.: PO Dundee Directory 1845
*   Ironmonger, advertising 'elegant scale beams' in 1850: PO Dundee Directory 1846, 1850

Previously James Murray & Co., smiths and ironmonger, High Street, 1824 (and perhaps preceded by William & James Murray, smiths and ironmongers, Nethergate North, 1818).

MURRAY, William, Dundee, 1829-53
1. 66 Murraygate, Dundee, 1829-37
2. 148 Murraygate, Dundee, 1840-5
3. 143 Murraygate, Dundee, 1846
4. 92 Murraygate, Dundee, 1850-3

*   Ironmonger: Dundee Directory and Register 1829
*   Ironmonger, smith and scale beam maker: Dundee Directory and General Register 1834

- Ironmonger and blacksmith: Dundee Directory 1842
- Ironmonger: PO Dundee Directory 1845

Previously William Murray & Co., smiths and ironmongers, 16 Murraygate, 1818-24 (and perhaps James Murray & Son, smiths and ironmongers, opposite head of Horse Wynd, Murraygate).

MURRAY, William, Glasgow, 1834-6
57 Carrick Street, Glasgow, 1834-6

- Coppersmiths: PO Glasgow Directory, 1834-6

Admitted member of the Incorporation of Hammermen of Glasgow 1834 as a coppersmith stranger, with 'a copper scale' as essay: Harry Lumsden and P. Henderson Aitken, *History of the Hammermen of Glasgow* (Paisley, 1912), 315. Joined David Murray to form D. & W. Murray, coppersmiths, 1834-5, trading as William Murray for 1836 only.

NEAVE, David, Dundee, 1853-64
King William's Dock, Dundee, 1853-64

- Shipsmith: Dundee Directory 1853
- Shipsmith and beam maker: Dundee Directory 1856
- Shipsmith, scale and beam maker: PO Dundee Directory 1858
- Shipsmith, harpoon gun, and beam maker: PO Dundee Directory 1861
- Shipsmith, harpoon gun, rivet, and beam maker: PO Dundee Directory 1864

Neave was previously listed as grocer of 51 Bucklemakers Wynd in 1846, and 95 Seagate (listed as his home address when he traded at King William's Dock) in 1850. Subsequently in partnership with his son Thomas as David Neave & Co., shipsmiths, etc. (but not beam makers), East Side King William Dock, 1867-80.

NICOL, James, Dundee, 1808-19
1. High Street north side, Dundee, 1809
2. St Clement's Lane and foot of Tindal's Wynd, Dundee, 1818

- Smith: Dundee Directory 1809

Maker of the brass yard-measure (Item **25**), stamped 'J. NICOL' and dated 1819 and 1826. Nicol is not listed in the 1824 Dundee Register and Directory, and there are no directories between 1818 and 1824.

NICOLSON, W. B., Glasgow, 1899-1900+
Hill Street, Garnethill, Glasgow, 1899-1900+

- Philosophical instrument maker: PO Glasgow Directory 1899-1900+

Butchart balance by Nicolson, Glasgow: NMS T.1986.113.

OMNET, George, Glasgow, 1899-1900+
13 Norfolk Street, Glasgow, 1899-1900+

- Smith and scale beam maker: PO Glasgow Directory 1899-1900+

OMNET, John, Glasgow, 1887-98
1. 13 Norfolk Street, Glasgow, 1887-95, 1897-8
2. 33 High Cart Craigs, Pollokshaws, Glasgow, 1896

- Smith and scale beam maker: PO Glasgow Directory 1887-98

Previously McDonald & Omnet (q.v.); subsequently George Omnet (q.v.).

PATERSON, James, Glasgow, 1836-56

1. 28 London Street, Glasgow, 1836-7
2. 16 Dunlop Street, Glasgow, 1838-54
3. 12 Great Clyde Street, Glasgow, 1855-6

- City inspector of weights and measures: PO Glasgow Directory 1836-56

PATERSON, William, Glasgow, 1810-39

1. Near Canal, Old Bason [*sic*], Glasgow, 1810-13
2. Jamaica Street, Glasgow, 1814-19
3. Ann Street, Jamaica Street, Glasgow, 1820-1
4. 4 Ann Street, Glasgow, 1822-6 and 1830-3
5. 17 Ann Street, Glasgow, 1827
6. 4 and 5 Ann Street, Glasgow, 1828-9
7. 19 Ann Street, Glasgow, 1835-9

- Smith: McFeat's Glasgow Directory 1810-21
- Smith and bellhanger: McFeat's Glasgow Directory 1822-7; PO Glasgow Directory 1828-9
- Smith, bellhanger and scale beam maker: PO Glasgow Directory 1830-3
- Smith, bellhanger and cart and wagon axle maker: PO Glasgow Directory 1834-9

Succeeded by William Paterson & Son (q.v.).

PATERSON, William & Son (George), Glasgow, 1840-50

19 Ann Street, Glasgow, 1840-50; 1852

- Smith and bellhangers, scale, beam, coach, cart, wagon axle, lock and hinge makers: PO Glasgow Directory 1840-50; 1852

PATERSON, William & Co., Glasgow, 1851

19 Ann Street, Glasgow, 1851

- Smith and bellhangers, scale, beam, coach, cart, wagon axle, lock and hinge makers: PO Glasgow Directory 1851

PATERSON, William & Sons, Glasgow, 1853

19 Ann Street, Glasgow, 1853

- Smith and bellhangers, scale, beam, coach, cart, wagon axle, lock and hinge makers: PO Glasgow Directory 1853

PATERSON, William & Son, Glasgow, 1854-1900+

1. 19 Ann Street, and 1 Windmill croft, Glasgow, 1854-6
2. 19 Ann Street, and 10 Springfield Place, Broomielaw, Glasgow, 1857
3. 19 Ann Street, and 16 Springfield Place, Broomielaw, Glasgow, 1858-77
4. 83 Pitt Street, Glasgow; 68 Bishop Street, and 10 Springfield Quay, Glasgow, 1878-83
5. 83 Pitt Street, Glasgow; 68 Bishop Street, and Finnieston Quay, Queen's dock, Glasgow, 1884-96
6. Head office 65 Pitt Street, 68 Bishop Street, Glasgow; branch 169 Finneston Street, Queen's dock, Glasgow, 1897-1900+

- Smiths, chain makers, gas-fitters, bellhangers, scale, beam, cart, axle and lock and hinge makers: PO Glasgow Directory 1854-5
- Smiths, shipsmiths, chain makers, gas-fitters, bellhangers, scale, beam, cart and coach axle, and lock and hinge makers: PO Glasgow Directory 1856-1900+

An account of the firm – 'they manufacture … large scale beams (of various sizes) …' – is to be found in *Glasgow of Today* (London, 1888), 217.

POOLEY, Henry & Son, Aberdeen, 1894-1900+

37 Guild Street, Aberdeen, 1894-1900+

- Weighing machine makers: PO Aberdeen Directory 1894-1900+

A branch of Henry Pooley & Son of Birmingham (subsequent addresses, as for Henry Pooley, Dundee).

POOLEY, Henry & Son, Dundee, 1894-1900+
Caledonian Railway Yard, Dundee, 1894-1900+

Weighing machine manufacturers: PO Dundee Directory 1894-1900+

A branch of Henry Pooley & Son of Birmingham, in business at the Caledonian Railway Yard, then Bell Street, and subsequently King Street, until 1954.

POOLEY, Henry & Son, Edinburgh, 1894-1900+
15 Crighton Place, Leith Walk, Edinburgh, 1894-1900+

A branch of Henry Pooley & Son of Birmingham. Eastern district foreman W. Inglis.

POOLEY, Henry & Son, Glasgow, 1875-1900+
1. 113 West Nile Street, Glasgow, 1875
2. 113 West Nile Street, Glasgow; workshop and stores 169 & 171 Renfield Street, Glasgow, 1876-8
3. 41 Hope Street, Glasgow, 1879
4. 41 Hope Street, Glasgow ; works Albion Foundry, Liverpool, 1880-2
5. McAlpine Street, Glasgow, 1883
6. Albion Works, 69 & 71 McAlpine Street, off Argyll Street, Glasgow, 1884-91
7. Albion Works, 69 McAlpine Street, Glasgow, 1892-3
8. Office and showrooms 22 Queen Street; Albion Works, 25 South Kinning Place, Paisley Road, Glasgow, 1894-1900+

- Manufacturers of patent weighing machines and weighbridges: PO Glasgow Directory 1875

- Patentees and manufacturers of every description of weighing apparatus for railways, ironworks, collieries, mills, warehouses etc., self-acting machines for colliers, granaries etc.: PO Glasgow Directory 1876-1900+

James Hines manager of Glasgow branch from 1876 and Scottish Head Office from 1884.

The business of Henry Pooley & Son Ltd – a total of 52 branches, plus 40 depots and 107 mobile repair shops exclusively devoted to railway work – was acquired by W. & T. Avery of Birmingham in 1914: L. H. Broadbent, *The Avery Business, 1730-1918* (Birmingham, 1949), 65.

PRIMROSE (Adam), & ROSS (John), Glasgow, 1837-57
1. 57 Carrick Street, Glasgow, 1837-43
2. 61 McAlpine Street and 57 Carrick Street, Glasgow, 1844
3. 61 McAlpine Street, Glasgow, 1845-52
4. 61 & 65 McAlpine Street, Glasgow, 1854-7

- Coppersmiths: PO Glasgow Directory 1837-48
- Coppersmiths and brass founders: PO Glasgow Directory 1849-57

Adam Primrose, coppersmith, 8 Commerce Street, Glasgow listed in the directory for 1836, and admitted member of the Incorporation of Hammermen of Glasgow 1837 as a coppersmith stranger, with 'a gallon measure' as essay: Harry Lumsden and P. Henderson Aitken, *History of the Hammermen of Glasgow* (Paisley, 1912), 316. John Ross also admitted member in 1837, having served an apprenticeship as a coppersmith, with 'a half-gallon measure' as essay: ibid., 316. Ross appears in the directory on his own account at 63 McAlpine Street, Glasgow, in 1858, and at 218 Dumbarton Road in 1859.

PRIMROSE & Co., Glasgow, 1858-77

1. Clyde Copper Works, 31 Washington Street, Glasgow, 1858-60
2. Clyde Copper Works, 29 & 31 Washington Street, Glasgow, 1861-5
3. Clyde Copper Works, 31 & 33 Washington Street, Glasgow, 1866-70

- Coppersmiths and brass founders: PO Glasgow Directory 1858-70
- Makers of patent and common distilling apparatus, vacuum pans and other sugar refining utensils, marine and general copper and brass work: PO Glasgow Directory 1871-7

Adam Primrose was joined in the business by Edward Primrose, who ran it alone from 1863.

RAE, Adam, Glasgow, *fl.*1704

Admitted member of the Incorporation of Hammermen of Glasgow as a stranger in 1704, with an essay which included 'ane pynt stoup': Harry Lumsden and P. Henderson Aitken, *History of the Hammermen of Glasgow* (Paisley, 1912), 291.

RAE, Neil, Glasgow, 1869

19 Carlton Court, Bridge Street, Glasgow, 1869

- Scale, beam and weighing machine maker: PO Glasgow Directory 1869

RAMSAY, Alexander, Glasgow, 1865-99

1. 228 Buchanan Street, Glasgow, 1865-7
2. 22 Waterloo Street, Glasgow, 1868-99

- Coppersmith and brazier: PO Glasgow Directory 1865-6; 1871; 1879-85
- Coppersmith, brazier and pewterer: PO Glasgow Directory 1867-70
- Coppersmith: PO Glasgow Directory 1872-8

- Coppersmith and measure maker: PO Glasgow Directory 1886-7
- Coppersmith, brazier and measure maker: PO Glasgow Directory 1888
- Coppersmith, brazier and tinner: PO Glasgow Directory 1889-99

Pair of conical Imperial measures, *c.*1880, marked 'ALEX^R RAMSAY / MAKER / GLASGOW' on base: NMS T.1988.35. Additional ¼ gill, added to set of seven original Imperial capacity measures adjusted by James Clelland, 1835, marked 'ALEX^R RAMSAY / MAKER / GLASGOW' on base; made for County of Lanark, Item **249**. Also similar item for Burgh of Calton, Item **251**. A graduated part-set of four copper and brass conical measures (2 gallon, 1 quart, 4 gill and 2 gill) by 'ALEX^R RAMSAY / MAKER / GLASGOW', offered at Christie's Scotland, 11 February 1998, Lot 4.

RAMSAY, William & Co., Glasgow, 1845-1900+

1. 254 Argyll Street, Glasgow; works 281 Argyll Street, Glasgow, 1889-91
2. 254 Argyll Street, Glasgow; works 285 Argyll Street, Glasgow, 1892-4
3. 195½ Argyll Street and 11 Jamaica Street, Glasgow; works 285 Argyll Street, Glasgow, 1895-6
4. 281 Argyll Street and 11 Jamaica Street, Glasgow; works 285 Argyll Street, Glasgow, 1897
5. 195½ Argyll Street and 11 Jamaica Street, Glasgow; works Robertson Street, Glasgow, 1898
6. 195½ Argyll Street and 11 Jamaica Street, Glasgow; works Ann Street, Glasgow, 1899-1900+

- Ironmongers, lampmakers, gas-fitters … tinsmiths, coppersmiths … : PO Glasgow Directory 1889-1900+

By 1892 in trades directory as beam and scale makers, although not previously; began in 1845 as William Ramsay, general and furnishing ironmonger, smith, bellhanger and gas-fitter; became William Ramsay *junior* & Co. from 1879-89. An account of the firm is to be found in *Glasgow of Today* (London, 1888), 113.

REID, B. & Co., Aberdeen, late nineteenth century
Chain maker, Aberdeen.

Surveyor's chain, NMS W.1959.1012, VH.74.

REID, William, Edinburgh, 1773-7
Opposite the Guard, Edinburgh, 1773-7

- Hard-ware merchant: Williamson's Edinburgh Directory, 1773
- Ironmonger: Williamson's Edinburgh Directory, 1774-7

Succeeded by (presumably) his widow, Mrs Reid, who continued as an ironmonger opposite the Guard until 1784; then at the head of Lyon's Close between 1786 and 1793, and her final appearance in 1794 at the foot of Lyon's Close.

Coin weights and a trade label, which reads 'Weights and Scales made … By WILLIAM REID Iron-Monger; opposite to … City Guard Edinburgh', illustrated in Paul and Bente Withers, *British Coin-Weights: A Corpus of the Coin-Weights made for use in England, Scotland and Ireland* (Llanfyllin, 1993), 329. A half-guinea coin weight is NMS K.1999.1441 (W.2852g va.).

REID & HANNAY, Paisley, 1831-8
1. Orr Street, Paisley, 1831
2. 12 High Street, Paisley, 1838

- Adjuster of weights and measures, blacksmiths and bellhangers: Fowler's Directory 1831-2

- Smith and iron boat builders: Fowler's Directory 1838-9

RENTON, William, Glasgow, 1819-41
1. 12 Hospital Street, Glasgow, 1819
2. 108 Commerce Street, Glasgow, 1820-6
3. 54 Commerce Street, Tradeston, Glasgow, 1828-35
4. 76 Commerce Street, Glasgow, 1836-41

- Smith: McFeat's Glasgow Directory 1819-26 (does not appear in 1824-5 or 1827); PO Glasgow Directory 1828
- Smith and bellhanger: PO Glasgow Directory 1829-41

Firm appears to have begun as William Renton, smith, in Campbell Street, Glasgow, 1810-12; subsequently Chalmers & Renton (q.v.) and then Renton & Ritchie (q.v.).

RENTON (William) & RITCHIE, Glasgow, 1817-18
12 Hospital Street, Glasgow, 1817-18

- Smiths: McFeat's Glasgow Directory 1817-18

Subsequently reverted to William Renton (q.v.).

ROBERTSON, Charles *junior*, Edinburgh, 1876-1900+
1. 24 Barony Street Lane, Edinburgh, 1876-88
2. 81A Pleasance, Edinburgh, 1889-92
3. 93 Pleasance, Edinburgh, 1893-1900+

- Steelyard maker: PO Edinburgh Directory 1875-1900+

Successor to Jonathan Davidson (q.v.).

ROBERTSON, David, Edinburgh, *c.*1754

Foot of Kinloch's Close, south, Edinbugh, 1752

- Smith: James Gilhooley, *A Directory of Edinburgh in 1752* (Edinburgh, 1988), 70.

One of the makers of the 1754 Stirlingshire firlots (Item **216** and **217**): see A. D. C. Simpson, '"Handle with Care": Handling Warnings on Early Scientific Instruments and Two Early Scottish Grain Measures', in *Bulletin of the Scientific Instrument Society*, no. 30 (1991), 3-4.

ROBINSON, James, Glasgow, 1880-1900+

1. 13 and 21 Norfolk Street, Glasgow, 1880
2. 114 Main Street, Glasgow; 30 Rutherglen Loan, Glasgow, 1881
3. 114 Main Street, Glasgow, 1882-1900+

- Beam and weighing machine maker: PO Glasgow Directory 1880

- Scale, beam and weighing machine maker; blacksmith, coppersmith, tinsmith, gas-fitter and japanner; grocers' and spirit merchants' outfitter: PO Glasgow Directory 1881
- Smith, scale, beam and weighing machine maker: PO Glasgow Directory 1882-7
- Scale, beam and weighing machine maker: PO Glasgow Directory 1888-1900+

ROMANIS, John, Kelso, *fl.*1826

Advertisement in the *Kelso Mail*, 16 January 1826: 'JOHN ROMANIS IRON MERCHANT, KELSO, Returns his thanks for former favours, and begs to say that he has on hand a large Assortment of IRON [WEIGHTS] of all kinds and sizes; and on Wednesday first, he will have NEW METAL WEIGHTS, adjusted by the IMPERIAL STANDARD.'

ROSS, Peter, Glasgow, 1891-1900+

1. 141 Gallowgate, Glasgow, 1891-5
2. 9½ Hunter Street, Glasgow, 1898-9
3. 9 Hunter Street, Glasgow, 1900+

- Beam and scales manufacturer: PO Glasgow Directory 1891-5
- Beam and scale maker: PO Glasgow Directory 1898-9
- Weighing machine, beam and scale maker: PO Glasgow Directory 1900+

ROWAN, David, Edinburgh, *fl.*1543-93

Maker of late sixteenth century measures for Edinburgh, St Andrews and Dundee (Items **109**, **111** and **112**). See Chapter 7.

RUSSELL, Alexander, Kirkcaldy, *fl.c.*1797-1833

Alexander Russell, ironfounder, of the Kirkcaldy Foundry, was the originator of the business of Alexander Russell & Son, which flourished from about 1833-50. Russell was initially in business with his brother Robert, with a retail shop in the High Street and foundry premises at the top of Kirk Wynd. Robert left the partnership to join his sons in another business. After 36 years in the business, Alexander's son Thomas was made partner in May 1833, and shortly after this Alexander died: for further details of the firm, see A. J. Campbell, 'Fife Trades and Professions 1820-70', 2 volumes [November 1989], unpaginated, typescript in Fife Public Library and Department of Special Collections, St Andrews University Library.

Gallon and half-gallon standards at Fife Council, Trading Standards (Item **240**), and iron bushel at Kirkcaldy Museum (Item **256**).

RUSSELL, Robert, Glasgow, 1834-5
60 High Street, Glasgow, 1834-5

- Scale and beam maker: PO Glasgow Directory (trades section) 1834

RUSSELL (Thomas) & BELL (Peter), Edinburgh, 1822-31
337 High Street, Edinburgh, 1822-31

- Ironmongers or wholesale and retail ironmongers: PO Edinburgh Directory 1822-31

Advertisement in the *Kelso Mail*, 3 July 1826, and the *Scotsman*, 28 June, 1 July and 28 October 1826: 'IMPERIAL MEASURES. RUSSELL and BELL have received on consignment a considerable quantity of COPPER MEASURES, of superior workmanship, embracing every size, from one half gill to five gallons, which will be sold on terms advantageous to the purchaser. R. & B's Stock of IMPERIAL WEIGHTS is also extensive and complete. No 337, HIGH STREET. EDINBURGH.'

Thomas Russell and Peter Bell were wholesale ironmongers, succeeded by Russell & Anderson in 1832 at 154 High Street, and by Russell & Clark in 1833, who moved to Hunter Square in 1837. Thomas Russell ran the firm on his own account from 1842 until 1855.

RUTHVEN, John, Edinburgh, *c.*1820-40
Steelyard in [Highland and Agricultural Society], *Catalogue of Museum Models of Agricultural Implements and Machinery*, etc. ([Edinburgh], *c.*1840), 40, no. B/54. For Ruthven see 'Retailing Instrument Makers', in T. N. Clarke, A. D. Morrison-Low and A. D. C. Simpson, *Brass & Glass: Scientific Instrument Making Workshops in Scotland* (Edinburgh, 1989), 96-8.

SALTER, George & Co., Glasgow, 1883-1900+
Branch of George Salter & Co., West Bromwich; agents, Hendry Brothers, 79, 80, Great Clyde Street, Glasgow, 1885-1900+

- Manufacturers of spring balances, steel springs, Bourdon's pressure and vacuum gauges: PO Glasgow Directory 1885-1900+

SCOTT, Robert, Edinburgh, 1894-1900+
1. 353 High Street, Edinburgh, 1894-5
2. 357 High Street, Edinburgh, 1896-1900+

- Brass finisher, scale and beammaker: PO Edinburgh Directory 1894-7
- Brass founder: PO Edinburgh Directory 1898-1900+

SHUTTLEWORTH, Joseph P., Dundee, 1891-1900+
1. 21 North Tay Street, Dundee, 1891-3
2. 9 North Lindsay Street, Dundee, 1893-1900+

- Scale, beam and weighing machine manufacturer: PO Dundee Directory 1891-1900+

Shuttleworth advertised in PO Dundee Directory 1894 as 'late with H. Pooley & Son'. In business at this address until 1968; as J. P. Shuttleworth & Son from 1922. Obituary of Joseph Percival Shuttleworth in *Dundee Courier & Advertiser*, October 1958.

SIMPSON, Robert D., Edinburgh, 1854-82
1. 13 Clerk Street, Edinburgh, 1854-7
2. 2 Crosscauseway, Edinburgh, 1854-82
3. 14 Clerk Street, Edinburgh, 1858-79

- Scale and beam maker: PO Edinburgh Directory 1854-74
- Scale, beam and steelyard maker: PO Edinburgh Directory 1875-82

Subsequently R. D. Simpson & Son (q.v.). *The Mercantile Age*, 3 August 1887, 323, claims the firm was established 1850 in Cross Causeway.

Pharmaceutical balance: NMS T.1968.84.

SIMPSON, R. D. *junior*, Edinburgh, 1887-1900+

1. 15 Sciennes, Edinburgh, 1887-90
2. 10 St Leonards Street, Edinburgh, 1891-1900+
3. 150 High Street, Edinburgh, 1896-9
4. 30 Victoria Street, Edinburgh, 1900+

- Scale, beam and steelyard maker: PO Edinburgh Directory 1888-1900+

Account in *The Mercantile Age*, 3 August 1887, 323, records the firm's range of products and medals at the Glasgow Industrial Exhibition and the Edinburgh International Exhibition.

SIMPSON, R. D. & Son, Edinburgh, 1883-7

15 Sciennes, Edinburgh, 1883-7

- Scale, beam and steelyard maker: PO Edinburgh Directory 1883-7

SIMPSON, Thomas, Edinburgh, 1773-84

1. Head of Halkerston's Wynd, Edinburgh, 1773-6
2. Head of Bridge Street, Edinburgh, 1776-81
3. Head of the Fleshmarket Close, Edinburgh, 1782-3
4. Fleshmarket Close, Edinburgh, 1784

- Pewterer: Williamson's Edinburgh Directory 1773-83

Pewterer who adjusted the burgh weights in 1779 (see Chapter **9** and Item **37**); and probably made the water check weights for John Robison (Items **93** and **105**). In the 1784 directory he appears as 'late con-

veener'; there is no directory for 1785; between 1786 and 1788 his widow is noted at Fleshmarket Close.

SLATER, Robert, Glasgow, 1847-87

1. 11 Centre Street, Glasgow, 1847-82
2. 29 Centre Street, Glasgow, 1883-7

- Shipsmith: PO Glasgow Directory 1847-61
- Blacksmith: PO Glasgow Directory 1862-6
- Shipsmith and beam scale maker: PO Glasgow Directory 1867-72
- Shipsmith, beam and scale maker: PO Glasgow Directory 1873-87

SMITH, Andrew, Mauchline, 1857-62

Maker of decorative woodware, with manufactories in Ayrshire and Birmingham.

Various letter balances; Registered design, 22 April 1857; British Patent for postal balance, No. 1456 of 1862.

See A. D. C. Simpson, 'Brewster's Society of Arts and the Pantograph Dispute', in *Book of the Old Edinburgh Club*, new series, 1 (1991), 47-73.

SMITH, A. (Alex.) & W. (William) & Co. (W. Robertson Smith and Hugh Smith) Glasgow, 1872-1900+

Eglinton Engine Works, 57 Cook Street, Glasgow, 1872-1900+

- Engineers, millwrights, boilermakers and sugar-mill manufacturers: PO Glasgow Directory 1872-4

- Engineers, millwrights, boilermakers, sugar-mill manufacturers and weighing machine makers: PO Glasgow Directory 1876-1900+

Becomes A. & W. Smith & Co. Ltd from 1899.

SMITH Brothers (Hugh Wilson, James and Osbourne) & Co., Glasgow, 1855-1900+

1. 112 Stirling's Road, Glasgow, 1855
2. Kingston Engine Works, 32 Kinning Street, Glasgow, 1856-65
3. Park Street, near Paisley Road toll bar, Glasgow, 1866-74
4. 71 Park Street, Kinning Park, Glasgow, 1875-1900+

- Engineers, millwrights, and agricultural machinery makers: PO Glasgow Directory 1855-7
- Engineers, millwrights, agricultural machine makers, founders, boiler makers, smiths and weighing machine makers: PO Glasgow Directory 1858-65
- Engineers, millwrights, founders, weighing machine makers, boiler makers and smiths: PO Glasgow Directory 1866-71
- Engineers, millwrights, founders and boiler-makers: PO Glasgow Directory 1872-4
- Engineers, boilermakers, millwrights, ironfounders, shipbuilders and weighing machine makers: PO Glasgow Directory 1875-7
- Engineers, boilermakers, millwrights, iron-founders, shipbuilders and weighing machine makers, machinists and railway plant merchants: PO Glasgow Directory 1877-86
- Engineers, boilermakers, millwrights, ironfounders, shipbuilders, weighing machine makers and machinists: PO Glasgow Directory 1887-1900+

SMITH, Hugh & Co., Glasgow, 1875-91, 1895, 1900+

1. 16 St Enoch Square, Glasgow, 1875
2. Possil Engine Works, Possil Road, Glasgow, 1876-91, 1895, 1900+

- Engineers, millwrights, founders and boiler-makers (in trades directory as weighing machine makers): PO Glasgow Directory 1875-6
- Engineers: (in trades directory as weighing machine makers): PO Glasgow Directory 1877-83
- Engineers, hydraulic machinery makers, boiler makers and weighing machine manufacturers: PO Glasgow Directory 1884-91
- Engineers, hydraulic machinery makers, ship-builders' machinery and weighing machine manufacturers: PO Glasgow Directory 1895 and 1900+

Succeeded by Hugh Smith (q.v.).

SMITH, Hugh, Glasgow, 1892-4, 1896-9

Possil Engine Works, Possil Road, Glasgow, 1892-4, 1896-9

- Engineer (in trades directory as weighing machine maker): PO Glasgow Directory 1892-4 and 1896-9

Reverts intermittently to Hugh Smith & Co. (q.v.).

SMITH, James, Edinburgh, 1835-40

6 Beaumont Place, Edinburgh, 1835-40

- Engineer: PO Edinburgh Directory 1835-40

Probably the maker of Imperial bushel marked 'J. SMITH EDIN^R': NMS W.1974.212.

SMITH, John, Glasgow, *fl.*1794

Member of the Incorporation of Hammermen of Glasgow 1794, admitted as an apprentice smith, with a 'black beam' as essay: Harry Lumsden and P. Henderson Aitken, *History of the Hammermen of Glasgow* (Paisley, 1912), 299. He may perhaps be identified with either John Smith, smith of Frederick Street (1801) then 11 Frederick Lane (1803-24), or John Smith, tinsmith of 75 Bridgegate (1801-6) then 261 High Street (1807).

Perhaps the maker of a coin balance signed 'I. SMITH' carrying lion and thistle markings, sold at Christie's South Kensington, 16 November 1995, lot 121.

An account of this firm, linking it with that of James Anderson (q.v.), is to be found in *Glasgow of Today* (London, 1888), 230. Succeeded by David Spence & Co. (q.v.).

SMITH, William & Sons, Glasgow, 1881
Partick Engine Works, Glasgow, 1881

- Engineers, boiler makers, sugar mill manufacturers and weighing machine makers: PO Glasgow Directory 1881

SMITH, William, Edinburgh, 1884-96
10 New Broughton, Edinburgh, 1884-96

- Engineer and weighing machine maker: PO Edinburgh Directory 1884-96

SMITH, William & Co., Edinburgh, 1897-1900+
10 New Broughton, Edinburgh, 1897-1900+

- Edinburgh Steelyard Works: PO Edinburgh Directory 1897-1900+

SPENCE, David, Glasgow, 1882-97
1. 75 Robertson Street, Glasgow, 1882
2. 69 Stockwell Street, Glasgow, 1883
3. 83 Stockwell Street, Glasgow, 1884-8
4. 83 Stockwell Street and 41 Robertson Lane, Glasgow, 1889-93, 1896-7
5. 41 Robertson Lane, Glasgow, 1894

- Scale beam maker (trades directory only): PO Glasgow Directory 1882
- Ironmonger (in trades directory as a scale beam maker): PO Glasgow Directory 1883-8
- Ironmonger and shopfitter (in trades directory as a scale beam maker): PO Glasgow Directory 1889-97

SPENCE, David & Co., Glasgow, 1898-1900+
75 Robertson Street, Glasgow; works, 41 Robertson Lane, Glasgow, 1898-1900+

- Scale beam and weighing machine makers, grocers', spirit merchants' and general shop outfitters, cycle enamellers and nickel platers, fire extinguisher patentees: PO Glasgow Directory 1898-1900+

SPRING, William, Aberdeen, 1804-27
Instrument maker. Maker of the brass bushel measure for G. S. Keith, 1804, and the large set of brass grains from 4,000 to 0·04 troy grains used by Keith in his work on the densities of grain. These weights were based on a medium troy pound derived from Aberdeen's 1707 Act of Union set of troy cup-weights. Spring, described by Keith as 'this ingenious man', worked with Gordon, Barron & Co., listed in the first Aberdeen Directory in 1824 as cotton spinners and manufacturers, 20 Belmont Street, Aberdeen, also at Woodside from 1827: see George Skene Keith, 'On the Specific Gravity of Barley and Scotch Bigg, with the description of a New Instrument for Measuring it', in *Edinburgh Philosophical Journal*, 5 (1821), 173-7.

STARK, James, Edinburgh, 1733-3
Joiner of 'Bristow nr Edin'; one of makers of 1754 Stirlingshire firlots (Items **216** and **217**): see A. D. C. Simpson, '"Handle with Care": Handling Warnings on Early Scientific Instruments and Two Early Scottish Grain Measures', in *Bulletin of the Scientific Instrument Society*, no. 30 (1991), 3-4.

Stark clearly made a range of items in wood.
For example, the Saltoun papers in the National
Library of Scotland demonstrate that he could
provide the estate with a range of joinery services,
including the role of furniture maker: 'May 1 1758
Thertein [thirteen] Chineas Chairs of Mahogine
with Leather Bottoms Mattresed & Brass nailes at
27 sh. Each … £17 – 11', MS 16884/99. Our thanks
to David Jones, University of St Andrews, for this
reference. Francis Bamford, 'A Dictionary of
Edinburgh Wrights and Furniture Makers 1660-
1840', in *Furniture History*, 19 (1983), 110, lists Stark
as a wright and undertaker, became a Burgess,
20 June 1733, and appearing in the Edinburgh
directory at Bristow Street in 1773-4.

STEEDMAN, William, Edinburgh, 1872-9
1. 12 Blackfriars Street and 103 South Back of
    Canongate, Edinburgh, 1872
2. 12 Blackfriars Street, Edinburgh, 1873
3. 18 High Street, Edinburgh, 1874-5
4. 18 South Back of Canongate, Edinburgh,
    1876-9

- Smith and weighing machine maker:
  PO Edinburgh Directory 1872-3
- Jobbing smith and ironmonger:
  PO Edinburgh Directory 1874-5
- Smith and steelyard maker: PO Edinburgh
  Directory 1876-9

Does not appear in 1880 Directory. Thereafter
moves to various addresses, and is listed as a
blacksmith, ironmonger, dry salter and wringing
machine maker.

STIRLING, William, Kippen, *fl.*1835
A cooper who disputed, in the *Stirling Journal and
Advertiser*, 17 and 26 April 1835, the methods of
verification used by the Inspector of Weights and
Measures for Perthshire, William Hutchison (q.v).

STRANG, John, Glasgow, *fl.*1726
Admitted member of the Incorporation of
Hammermen of Glasgow 1726, with an essay
which included a 'tin pint stoup': Harry Lumsden
and P. Henderson Aitken, *History of the Hammermen of Glasgow* (Paisley, 1912), 293.

STRUTHERS, Gavin, Glasgow, *fl.*1718
Admitted member of the Incorporation of
Hammermen of Glasgow 1718, with 'ane shipman's
weighting beam' as essay: Harry Lumsden and
P. Henderson Aitken, *History of the Hammermen
of Glasgow* (Paisley, 1912), 292.

STURROCK (James), Dundee[?], 1826
A 4-pound brass weight of 1826 held by Dundee
City Council, Trading Standards office, Dundee
(Dundee City Council, Environmental and Consumer Protection), is stamped 'I. STURROCK', and
this may be a manufacturer's name. If so, the maker
is likely to have been James Sturrock, millwright,
Hawkhill north, Dundee, 1829; machine manufacturer, Tannage Court, 1834-7; perhaps the same
as James Sturrock, manufacturer, 22 Ann Street,
Maxwelltown, Dundee, 1845-6.

TAIT, William, Kirkwall, 1710
A witness for the Earl of Morton (Thomas Aitken,
q.v.) recalled 'that he was Apprentice five years to
the said Thomas Foubister (his Predecessor in
Office), and during the Time of his Apprenticeship, the said Thomas Foubister (q.v.) had in his
Custody the very self same Standards, and none else
which he [Aitken] had always made and adjusted
Pundars and Bysmars by … And whenever his
Master Thomas Foubister had occasion to make or
adjust Pundars and Bysmars, he saw his Manner of
doing it, and that it was by the very same Standards
now mentioned. That he knows one William Tait
a Wright, and also Deacon of the Hammermen in

Kirkwall, to have been the said Thomas Foubister's immediate predecessor in his Office of making and adjusting of Pundars and Bysmars, and that he [Aitken], served him a Year, and saw these very Standards': *Memorial for James, Earl of Morton, Defender, against Alexander, Earl of Galloway … Pursuers* (Edinburgh, 1758), 17-8.

TAYLOR, James, Glasgow, 1855-63
1. 111 Trongate, Glasgow, 1855-6
2. 122 Trongate, Glasgow, 1856-63

- Ironmonger, scale and beam maker, tinsmith and gas-fitter: PO Glasgow Directory 1855-8
- Scale and beam maker, tinsmith and gas-fitter: PO Glasgow Directory 1858-61
- Wholesale ironmonger: PO Glasgow Directory 1862
- Wholesale ironmonger and hardware merchant: PO Glasgow Directory 1863

Succeeded by James Taylor *junior* (q.v.).

TAYLOR, James *junior*, Glasgow, 1864-1900+
1. 122 Trongate, Glasgow, 1864
2. 122, 126 & 128 Trongate, Glasgow, 1865-8
3. 124, 126 & 128 Trongate, Glasgow, 1869-74
4. 124, 126 & 128 Trongate, Glasgow, and Birmingham warehouse, 115 Great Charles Street, 1875
5. 124, 126 & 128 Trongate, Glasgow, and Birmingham warehouse, 116 Great Charles Street, 1876
6. 61, 63 & 65 Mitchell Street, Glasgow, and Birmingham warehouse, 116 Great Charles Street, 1877-85
7. 61, 63 & 65 Mitchell Street, Glasgow, 1886-1900+

- Wholesale ironmonger and hardware merchant: PO Glasgow Directory 1864-5, 1868-72
- Wholesale ironmonger: PO Glasgow Directory 1866-7

- Large description, including '… beams and scales': PO Glasgow Directory 1876-85
- Wholesale ironmonger and hardware merchant (in trades directory as a scale beam maker): PO Glasgow Directory 1886-1900+

Previously James Taylor (q.v.).

TAYLOR, John, Glasgow, 1851-63
84 Bridgegate Street, Glasgow, 1851-63

- Smith and beam maker: PO Glasgow Directory 1851
- Smith and scale beam maker: PO Glasgow Directory 1852-63

Perhaps the same as John Taylor below.

TAYLOR, John, Glasgow, 1866-76
1. 43 Miller's Place, Glasgow, 1866
2. Workshop 103 Saltmarket, Glasgow; sale shop 43 Miller's Place, Glasgow, 1867-76

- Smith: PO Glasgow Directory 1866
- Smith, scale, beam and weighing machine maker: PO Glasgow Directory 1867
- Wholesale and retail smith, scale, beam and weighing machine maker: PO Glasgow Directory 1869-76

Succeeded by John Taylor & Son (q.v).

TAYLOR, John & Son, Glasgow, 1877-83
1. Workshop 65 Bridgegate, Glasgow; sale shop 43 Miller's Place, Glasgow, 1877-8
2. Workshop 48 Market Street, Glasgow; sale shop 43 Miller's Place, Glasgow, 1879-80
3. Workshop Saltmarket, Glasgow; sale shop 49 Miller's Place and 79 King Street, Glasgow, 1881
4. Workshop Saltmarket, Glasgow; sale shop 12 Miller's Place, Glasgow, 1882-3

- Wholesale and retail smith, scale, beam and weighing machine maker: PO Glasgow Directory 1877-80
- Wholesale and retail scale, beam and weighing machine maker: PO Glasgow Directory 1881-2
- Scale, beam and weighing machine maker: PO Glasgow Directory 1883

TAYLOR, John *junior*, Glasgow, 1884-7

1. 102, 104 & 106 Saltmarket, Glasgow, 1884-6
2. 102 & 104 Saltmarket, Glasgow; workshop, 2 Osborne Street, Glasgow, 1887

- Scale, beam and weighing machine maker: PO Glasgow Directory 1884
- Wholesale and retail smith, scale, beam and weighing machine maker and cartwright: PO Glasgow Directory 1885
- Scale beam and weighing machine maker, grocers; outfittings, wholesale and retail: PO Glasgow Directory 1886
- Wholesale scale beam and weighing machine maker and cartwright: 1887

Succeeded by John Taylor (q.v.).

TAYLOR, John, Glasgow, 1888-92
2 Osborne Street, Glasgow, 1888-92

- Scale, beam and weighing machine maker, smith and cartwright: PO Glasgow Directory 1888
- General smith and cartwright (in trades as a beam scalemaker): PO Glasgow Directory 1889-92

Previously John Taylor *junior* (q.v.); succeeded by John Taylor & Son, general smith and cartwrights at 52 Market Street, Glasgow in business until 1896.

TAYLOR, John, Glasgow, 1877-9
151 Stockwell Street, Glasgow, 1877-9

- Shop furnisher (in trades directory as a weighing machine maker): PO Glasgow Directory 1877-9

THE WEIGHING MACHINE MAKERS (Limited), Glasgow, 1894-6
43 Slatefield Street, off Gallowgate, Glasgow, 1894-6

- Scale, beam and weighing machine makers: PO Glasgow Directory 1894-6

THOMSON, A., Snaigow, Perthshire, *c.*1830
Steelyard weighing scale: NMS T.1996.144

THOMSON, David & John, Edinburgh, 1855-1900+

1. Broughton Market, Edinburgh, 1855-64
2. 9 Stead's Place, Edinburgh, 1864-71
3. 144 Leith Walk, Edinburgh, 1872-96
4. 136 Leith Walk, Edinburgh, 1896-1900+

- Steelyard makers: PO Edinburgh Directory 1855-8
- Steelyard makers and smiths: PO Edinburgh Directory 1859-63
- Weighing machine makers and smiths: PO Edinburgh Directory 1864-82
- Weighing machine makers and steelyard makers: PO Edinburgh Directory 1883-1900+

Platform steelyard: NMS T.1983.225.

THOMSON, Robert, Stirling, *fl.*1827
Advertisement in the *Stirling Journal and Advertiser*, 5 July 1827: 'WEIGHTS AND MEASURES County of Stirling By Order of the Justices of the Peace of the County of Stirling whereas, by Act of Parliament – … Notice is hereby given that the Justices having procured sets of the said Standard Weights and Measures have placed the same in the

hands of Mr Robert Thomson, Wellington Place, Stirling, so as the Lieges may have an opportunity of having their Weights and Measures adjusted thereby …'. Advertisement in the *Stirling Journal and Advertiser*, 30 August 1827: 'IMPERIAL WEIGHTS AND MEASURES Town and County of Stirling From the numerous applications made for adjusting the Weights and Measures agreeable to the New Standards, the Subscriber finds it absolutely necessary, from and after 1st day of September next to make the following Regulations for the convenience of the public, and the correctness of his own duty viz: That Saturday, Monday, and Tuesday, from 10 till 3 o'clock to be solely appropriated for the adjustment of Weights and Liquid Measures, and Wednesday, Thursday, and Friday, for the Dry Measures, from 10 till 3 o'clock. ROBERT THOMSON Stirling, Wellington Place.'

Imperial bushel in Perth Museum: PERGM 1979.213.

THOMSON, William & Co., Glasgow, 1875-88, 1896-1900+

1. Portman Engine and Boiler Works, Smith Street and Park Street, Kinning Park, Glasgow, 1875-7
2. Smith Street and Park Street, Kinning Park, Glasgow, 1878-82
3. Smith Street, Kinning Park, Glasgow, 1883-6
4. 57 Smith Street, Kinning Park, Glasgow, 1887-8, 1896-1900+

- Engineers: PO Glasgow Directory 1875-82
- Engineers and patentees (in trades directory as weighing machine makers from 1884): PO Glasgow Directory 1883-5
- Engineers, chain makers and patentees (in trades directory as weighing machine makers): PO Glasgow Directory 1886
- Sole makers of Thomson's patent pulley blockes, tube expanders and weighing machines; also chain makers: PO Glasgow Directory 1887-8, 1896-1900+

Succeeded by William Thomson (q.v.).

THOMSON, William, Glasgow, 1889-95

57 Smith Street, Kinning Park, Glasgow, 1889-95

- Sole maker of Thomson's patent pulley blocks, tube expanders and weighing machines; also chain maker: PO Glasgow Directory 1889-95

Previously William Thomson & Co. (q.v.); and subsequently 1896-1900+

TOD (or TODD), John, Glasgow, 1819-22

1. 10 Thistle Street, Glasgow, 1819-21
2. 83 King Street, Glasgow, 1822

- Smith and beam maker: McFeat's Glasgow Directory 1819-21
- Bellhanger and gas-fitter [J. Todd]: McFeat's Glasgow Directory 1822

Admitted member of the Incorporation of Hammermen of Glasgow 1821, as a beam maker stranger, with 'a beam' as essay: Harry Lumsden and P. Henderson Aitken, *History of the Hammermen of Glasgow* (Paisley, 1912), 308.

Succeeded by J. Todd & Co. (q.v.).

TODD, J. (John) & Co., Glasgow, 1823-7

1. 449 Gallowgate, Glasgow, 1823-4
2. 17 Old Wynd, Glasgow, 1825-6
3. 33 Blackfriars Street, Glasgow, 1827

- Smiths and bellhangers: McFeat's Glasgow Directory 1823-7

Previously John Tod (or Todd), (q.v.).

TODD, John, Glasgow, 1838-9

1. 147 Nile Street and South Sauchiehall Street, Glasgow, 1838
2. 147 Nile Street, Glasgow, 1839

- Smith, scale and beam maker: PO Glasgow Directory 1838
- Smith, scale beam maker and bellhanger: PO Glasgow Directory 1839

Succeeded by John Todd & Son (q.v.).

TODD, John & Son, Glasgow, 1840-9

1. 147 Nile Street, Glasgow, 1840
2. 52 Sauchiehall Street, Glasgow, 1841-3 and 1845-9
3. 52 Sauchiehall Street and 52 Renfrew Street, Glasgow, 1844

- Smiths, scale and beam makers, bellhangers, gas-fitters, etc.: PO Glasgow Directory 1840-9

Previously John Todd (q.v.), subsequently Todd, Sons & Co. (q.v.).

TODD, Sons & Co., Glasgow, 1850-1

14 Turner's Court, 87 Argyll Street, Glasgow, 1850

- Ironmongers, wholesale iron merchants, hardware and machine merchants, auctioneers, appraisers and agents: PO Glasgow Directory 1850
- (In trades directory only as scale beam maker): PO Glasgow Directory 1851

TODD & GARDNER, Glasgow, 1859-61

1. 12 Melville Lane, off Gordon Street, Glasgow, 1859
2. 53 Cadogan Street, Glasgow, 1861

- Smiths, bellhangers and gas-fitters: PO Glasgow Directory 1859
- Smiths, scale beam makers, bellhangers, gas-fitters: PO Glasgow Directory 1860
- (In trades directory only, as scale beam makers): PO Glasgow Directory 1861

VEITCH, James, Inchbonny, 1826-38

Local inspector as well as maker of Imperial weights and measures: see 'James Veitch of Inchbonny', in T. N. Clarke, A. D. Morrison-Low and A. D. C. Simpson, *Brass & Glass: Scientific Instrument Making Workshops in Scotland* (Edinburgh, 1989), 16-24.

VEITCH, Thomas, Edinburgh, 1840-62

33 St Andrews Square, Edinburgh

- Recorded as a stationer and mapseller: PO Edinburgh Directory 1840-62

Printed trade label for Veitch in folding coin scale by Stephen Houghton & Son recorded in a private collection.

'I W', Glasgow[?], 1767

Mark on Paisley standard weight of 1767, probably Glasgow-made (Item **90**).

WALE, James, Edinburgh, 1830-46

1. Paul's Work, Edinburgh, 1830-2
2. 90 Leith Wynd, Edinburgh, 1833-45
3. McLaren Place, Edinburgh, 1846

- Beam and coffee mill manufacturer: PO Edinburgh Directory 1830-5
- Beam, coffee mill and scale manufacturer: PO Edinburgh Directory 1836-43
- Scale, beam, coffee mill and steelyard manufacturer: PO Edinburgh Directory 1844-6

Beam weighing scale by James Wale, Edinburgh: NMS T.1994.41.

WARDROP, J. (John McKenzie) & H. (Henry), Glasgow, 1815-28

1. 44 Argyll Street, Glasgow; works 58 Queen Street, Glasgow 1815-17
2. 57 Queen Street, Glasgow, 1818-24
3. 59 Queen Street, Glasgow, 1825-6
4. 34 Queen Street, Glasgow, 1827-8

- Coppersmiths, etc.: McFeat's Glasgow Directory 1815
- Copper and tinsmiths, pewterers and brass founders, etc.: McFeat's Glasgow Directory 1825

Henry Wardrop *junior* was admitted a member of the Incorporation of Hammermen of Glasgow, with 'a pair of copper scales' as essay, on the same day as John McKinnon Wardrop in 1806, both by right of their father Henry, admitted 1785: Harry Lumsden and P. Henderson Aitken, *History of the Hammermen of Glasgow* (Paisley, 1912), 302-3. Henry *senior* was perhaps the Henry Wardrop listed as japanner in Maxwell Street, 1801-13. Subsequently in partnership as J. & H. Wardrop & Patrick Auld, coppersmiths, at 34 then 34 & 36 Queen Street, 1829-33. See Patrick Auld, 'An Account of the Imperial Standard Measure, made by and for Messrs. J. & H. Wardrop, no. 59, Queen-Street, Glasgow, under the direction of Andrew Ure, MD, FRS, Professor in the Andersonian Institution, Glasgow', in *Glasgow Mechanics' Magazine* 4 (1826), 357-8.

Probably the makers of the weights identified by James Cleland as being by 'Graham & Wardrop' (q.v.).

WARDROP, J. & H. and Patrick AULD, Glasgow, 1829-33

34 and 36 Queen Street, Glasgow, 1829-33

- Coppersmiths, etc.: PO Glasgow Directory 1829-33

Successors to J. & H. Wardrop (q.v.) and Patrick Auld (q.v).

WATERSTON, George & Son, Edinburgh, 1832-1900+

1. 14 Hanover Street, Edinburgh, 1832-9
2. 29 Hanover Street, Edinburgh, 1840-6
3. 56 Hanover Street, Edinburgh, 1865-1900+
4. 60 Hanover Street, Edinburgh, 1872-1900+

- Stationers: PO Edinburgh Directory 1832-1900+

Postal scale by Avery also marked with name of the retailer 'G. Waterston & Son, Edinburgh' recorded in a private collection.

WATSON, James, Glasgow, *fl.*1815

Admitted member of the Incorporation of Hammermen of Glasgow 1815, as a blacksmith stranger, with 'a swan-necked beam' as essay: Harry Lumsden and P. Henderson Aitken, *History of the Hammermen of Glasgow* (Paisley, 1912), 306.

WATSON, W. & D., Glasgow, 1873-7

1. 13 Watt Street, Paisley Road, Glasgow, 1873-4
2. 13 & 15 Watt Street, Paisley Road, Glasgow, 1875
3. 204 & 206 Paisley Road, Glasgow, 1876-7

- Tinsmiths, gas-fitters, bellhangers, etc.: PO Glasgow Directory 1873
- Tinsmiths, gas-fitters, bellhangers, weighing machine, canister, and beam and scale makers, japanners, and bakers' outfitters: PO Glasgow Directory 1874-7

WATT, Charles, Glasgow, 1837-49

Of John Watt & Son, Glasgow, 1837-49

Admitted member of the Incorporation of Hammermen of Glasgow 1837 as a coppersmith, by right of his father, with 'a half-gallon measure' as essay: Harry Lumsden and P. Henderson Aitken, *History of the Hammermen of Glasgow* (Paisley, 1912), 316. Appears to have joined his father as John Watt & Son (q.v.).

WATT, George, Edinburgh, 1750-76

House, West Bow Head, Edinburgh, 1752: James Gilhooley, *A Directory of Edinburgh in 1752* (Edinburgh, 1988), 58.

George Watt, founder, was married to Lilias Johnstone, 8 April 1750: H. Paton (ed), *The Register of Marriages for the Parish of Edinburgh 1701-1750* (Edinburgh, 1908), 571; and 'George Watt, founder, burgess and guild brother, … 18 December 1776 for good services' [taking down large bell in high steeple, hanging new bell in its place, and hanging bell taken from high steeple in steeple of Tron Church]: C. B. Boog Watson, *Roll of Edinburgh Burgesses and Guild Brethren 1761-1841* (Edinburgh, 1933), 166.

Maker of the Stirling ell, 1755 (Item **16**).

WATT, John & Son, Glasgow, 1837-49

1. Saracen Lane, Glasgow, 1837-48
2. Saracen's Lane Foundry, Glasgow, 1849

- Coppersmiths, brass and ironfounders: PO Glasgow Directory 1837-49

John Watt first appears in the Glasgow directory as 'of Stephen Miller & Co.' (q.v.) in 1825, and between 1837 and 1848 appears as 'of Stephen Miller & Co., and John Watt & Son'. See Charles Watt (q.v.).

WHARTON, George, Edinburgh, 1849-86

1. 59 Leith Wynd, Edinburgh, 1849-51
2. 23 Leith Wynd, Edinburgh, 1852-64
3. 37 Leith Wynd, Edinburgh, 1865-7
4. 150 High Street, Edinburgh, 1868-86

- Stamp maker: PO Edinburgh Directory 1846-8
- Stamp and beam maker: PO Edinburgh Directory 1849-95

WHARTON, George & Co., Edinburgh, 1887-95

1. 150 High Street, Edinburgh, 1887-95
2. 37 Niddry Street, Edinburgh, 1891-5

- Stamp and beam maker: PO Edinburgh Directory 1887-95

WHISTON, Robert & Sons, Glasgow, 1871

135 Main Street, Bridgeton, Glasgow, 1871

- Jobbing smiths and beam and scale makers: PO Glasgow Directory 1871

WHITE, John & Son, Auchtermuchty, Fife, 1715-1900+

1. High Street, Auchtermuchty 1715 to present
2. Westfield Foundry, Edinburgh 1907-20
3. 28-30 Victoria Street, Edinburgh 1921-74
4. Kinleith Industrial Estate, Currie, Edinburgh 1975-8
5. 63 Brown Street, Glasgow 1920-4
6. 34 McIntyre Street, Glasgow 1925-48
7. Queenslie Industrial Estate, Glasgow 1949-present
8. 87 North Street, Forfar 1929-
9. 122 East High Street, Forfar
10. Kirkhill Industrial Estate, Dyce, Aberdeen, 1975

1. John White, born *c*.1680, assumed to be the maker of large iron beam marked 'John White 1715'
2. Thomas, b.1714
3. John, b.1739
4. Andrew, b.1768
5. John, b.1798
6. Andrew, b.1830 (these last two, assumed to be 'John White & Son'). His brother Henry, b.1838, had three sons who went into the business:
7. John, b.1873; Andrew, b.1885 and Edwin, b.1889. Andrew was the father of Henry, b.1917; Edwin the father of John, b.1935 and Edwin, b.1944.

John White & Son, beam and scale makers, pre-1715 to date; firm founded sometime before 1715; during the 1850s a partnership of John and Andrew White ran the business. This partnership was dissolved in 1854, when John White took Henry G. White as his new co-partner. John White died at Auchtermuchty on 23 May 1873 aged 74. The firm celebrated its 250th anniversary in November 1965, and continues to be run by the family: A. J. Campbell, 'Fife Trades and Professions 1820-70', 2 volumes [November 1989], unpaginated, typescript in Fife Public Library and Department of Special Collections, St Andrews University Library. See also Edwin J. White, 'Whites of Auchtermuchty', *Auchtermuchty Festival Programme 1984* (1984), 12-15; Leslie Fraser, '200 Years in the Weighing-Machine Business', in the *Scotsman*, 18 February 1975; and 'Muchty Firm Weigh up the Past, Present and Future', in *Dundee Tayside*, 19 (1976), 43-6. Also 'Tayport Foundry to Close', in the *Fife News*, 18 February 1967; ''Muchty Foundry to Close', in the *Fife Herald News*, 23 March 1977; 'Electronics Weigh in behind Whites', in the *Fife Herald News*, 1 October 1980.

WHITE BROTHERS & Co., Auchtermuchty and Glasgow, 1871-1900+

1. High Road, Auchtermuchty, 1871-1964
2. 41 King Street, Glasgow, 1900+

John Brown White and William Beveridge White, sons of Andrew White, left their father's business and set up in competition. This business subsequently passed to J. B. White's son, Andrew, who eventually sold out to John White & Son in 1964. See Edwin White, 'Whites of Auchtermuchty', *Auchtermuchty Festival Programme 1984* (1984).

WILLIAMS, James & Co., Glasgow, 1887-1900+

1. 5 Salkeld Street, off Cook Street, Glasgow, 1887-95
2. 156 Eglinton Street, Glasgow; works 5 Salkeld Street, off Cook Street, Glasgow, 1896-1900+

- Engineers, makers of shop window fittings (brass yard-measure makers in trades directory): PO Glasgow Directory 1887-9
- Engineers, bicycle makers and repairers, etc. (brass yard-measure makers in trades directory): PO Glasgow Directory 1890-2
- Bicyclemakers and repairers (brass yard-measure makers in trades directory): PO Glasgow Directory 1893-1900+

WILSON & SINGER, Auchterless, Aberdeenshire, *fl.*1874

3-cwt platform weighing machine dated 1874, with weights: NMS W.1962.1217.

WILSON, William and Alexander, Glasgow, 1876-1900+

1. 156 James' Street, Bridgeton Cross, Glasgow, 1876-8
2. 156 & 158 James' Street, Bridgeton Cross, Glasgow, 1879-81
3. 151 & 153 Canning Street, Bridgeton Cross, Glasgow, 1882-91
4. 151 Canning Street, Bridgeton Cross, Glasgow, 1892-6
5. 24, 26 & 28 Broad Street, near Bridgeton Cross, Glasgow, 1897-1900+

- Iron, steel, india rubber and gutta percha merchants, engineers, general and furnishing ironmongers, colliery and mill furnishers (and variations): PO Glasgow Directory 1876-83
- Large description (in trades directory as weighing machine makers): PO Glasgow Directory 1884-1900+

WILSONE (John) & LIDDEL (James), Glasgow, 1801-11

1. Stockwell, Glasgow, 1801
2. 154 Stockwell, Glasgow, 1803-9
3. 155 Stockwell, Glasgow, 1810-11

- Smiths and ironmongers: McFeat's Glasgow Directory 1801

Wilsone & Dickson (or Dixon), ironmongers, listed at 'head Stockwell' in 1789 and 1790. James Liddle [*sic*] admitted a member of the Incorporation of Hammermen of Glasgow, having served an apprenticeship, in 1795: Harry Lumsden and P. Henderson Aitken, *History of the Hammermen of Glasgow* (Paisley, 1912), 299. James Liddle [*sic*], smith, listed at Boyd's Land, Stockwell, 1801-15. Alexander Wood (q.v.) listed as at Wilsone & Liddel from 1805. Subsequently (1812-24) traded as James Liddel & Co. (q.v.).

WITHERSPOND, James, Glasgow, *fl.*1729

Admitted member of the Incorporation of Hammermen of Glasgow 1729, with an essay which included 'two pounds of brass weights': Harry Lumsden and P. Henderson Aitken, *History of the Hammermen of Glasgow* (Paisley, 1912), 293.

WOOD, Alexander, Glasgow, 1805-32

1. 'at Wilson & Liddell's' (q.v.), Stockwell, Glasgow, 1805-11
2. 'at James Liddel & Co's Stockwell' (q.v.), Glasgow, 1812-23
3. 157 Stockwell, Glasgow, 1824-6
4. 18 Stockwell, Glasgow, 1827
5. 16 Stockwell, Glasgow, 1828-32

- [Wilson & Liddel, smiths and ironmongers]
- [James Liddel & Co., smiths and bellhangers]
- Smith and bellhanger: McFeat's Glasgow Directory 1824
- Smith and beam maker: PO Glasgow Directory 1828
- Smith, beam and coffee mill maker: PO Glasgow Directory 1830

Alex Wood, smith, admitted as a stranger to the Incorporation of Hammermen of Glasgow 1810,

with 'a small beam' as essay: Harry Lumsden and P. Henderson Aitken, *History of the Hammermen of Glasgow* (Paisley, 1912), 304. Took over James Liddel & Co.'s premises at 157 Stockwell in 1824 when Liddel & Co. move to the Saltmarket.

'ALEXANDER WOOD, Smith, Stockwell Street, Adjuster of Weights and Measures': James Cleland, *Exemplification of the Weights and Measures of the City of Glasgow* (Glasgow, 1822), 74; 'Six very fine Balances with Chains and Copper Scales, made by Wood, Glasgow: – One 33 inches, One 24 inches, One 18 inches, One 16 inches, One 12 inches, One 10 inches ... the Balances are so very delicate, that four of them turn with the tenth part of a grain, and one with less': James Cleland, *An Historical Account of the Local and Imperial Weights and Measures of Lanarkshire, and an Inventory of those belonging to the Corporation of Glasgow* (Glasgow, 1832), 35; and 'Balances MADE BY WOOD, GLASGOW. One four feet ten inches long, One four feet six inches long, One four feet long, One three feet long. Wooden Scales are attached to these balances with Ropes': ibid., 36.

WOOD, Alex. & Son (Alexander), Glasgow, 1833-40

1. 16 Stockwell, Glasgow, 1833-6
2. 7 and 9 Stockwell, Glasgow, 1837-40

- Smiths, beam and coffee-mill makers: PO Glasgow Directory 1833-40
- (and sugarmill manufacturers from 1836, and furnishing ironmongers from 1838)

Alexander Wood *junior*, blacksmith, admitted member of the Incorporation of Hammermen of Glasgow by right of his father 1830, with a 'small polished beam' as essay: Harry Lumsden and P. Henderson Aitken, *History of the Hammermen of Glasgow* (Paisley, 1912), 313.

7-pound weight by Wood & Son: People's Palace Museum, Glasgow, GLAMG. T. 9896; 14-pound weight by Woods: People's Palace Museum, Glasgow, GLAMG. T. 9855.

WOOD, Alex. & Sons (Alex. *junior*, Archibald Colquhoun and Archibald Moncrieff), Glasgow, 1841-97

1. 7 & 9 Stockwell Street, Glasgow, 1841-2
2. 14 & 18 Stockwell Street, Glasgow, 1843-78
3. 14 & 18 Stockwell Street, Glasgow; works 49 William Street, Glasgow, 1879-82
4. 8 & 10 Stockwell Street, Glasgow; works 49 William Street, Glasgow, 1883
5. 8 & 10 Stockwell Street, Glasgow; works Baltic Street, Bridgeton, Glasgow, 1884-5
6. 8 & 10 Stockwell Street, Glasgow; works 130 Baltic Street, Bridgeton, Glasgow, 1886-90
7. 8 & 10 Stockwell Street, Glasgow, 1891-6

- Smiths, beam, coffee and sugar mill manufacturers and furnishing ironmongers: PO Glasgow Directory 1841-5
- Smiths, scale beam, coffee and sugar mill manufacturers, weighing machine makers and furnishing ironmongers: PO Glasgow Directory 1845-76 (addition of gas-fitting, 1857)
- Smiths, scale beam, coffee and pepper-mill manufacturers, gas-fitters, weighing machine makers and general furnishing ironmongers; agents for W. & T. Avery's patent agate brass beams, stands and scales: PO Glasgow Directory 1877-9
- Shop and store outfitters, manufacturers of beams, scales, platform, cart, lorry, railway wagon and coal hutch weighing machines, grocers' canisters and showcases, butchers' fittings and mincing machines, etc., tinsmiths, coppersmiths, Japanners and smiths: PO Glasgow Directory 1880
- Glasgow weighing machine manufactory, shop and store outfitters, manufacturers of bucket-ended beams and scales, platform, cart, lorry, railway wagon and coal hutch weighing

machines, grocers' beams, scales and canisters and show cases, … spirit merchants' measures … repairs done at our works on all kinds of weighing apparatus: PO Glasgow Directory 1881-96
- (Now incorporated with W. & T. Avery Ltd), 8 and 10 Stockwell Street: PO Glasgow Directory 1897-1900+.

Archibald Colquhoun Wood, blacksmith, admitted member of the Incorporation of Hammermen of Glasgow 1836, by right of his father: Harry Lumsden and P. Henderson Aitken, *History of the Hammermen of Glasgow* (Paisley, 1912), 316.

The business of Alexander Wood & Son (Glasgow) was acquired by W. & T. Avery Ltd, Birmingham, in 1897: L. H. Broadbent, *The Avery Business, 1730-1918* (Birmingham, 1949), 83. See also the entry for the Glasgow Weighing Machine Manufactory (q.v).

WRIGHT, Robert E., Glasgow, 1883-93
11 Argyll Street, Glasgow, 1883-93

- Cutler, ironmonger, tool, beam, scale and mathematical instrument maker: PO Glasgow Directory 1883
- Ironmonger, cutler, optician, saw, plane, edgetool, mathematical instrument, scale, beam and weighing machine maker, shop and window fitter: PO Glasgow Directory 1884
- Ironmonger, cutler, scale, beam and weighing machine maker and japanner: PO Glasgow Directory 1885
- Ironmonger, cutler, scale and beam maker: PO Glasgow Directory 1886-7
- Ironmonger, cutler, scale and beam maker and japanner: PO Glasgow Directory 1888-9
- Ironmonger, cutler, scalemaker, japanner, tinsmith and shop outfitter: PO Glasgow Directory 1890-3

Succeeded by R. E. Wright & Co. (q.v.).

WRIGHT, Robert E., Glasgow, 1894-1900+
1 and 3 Trongate and 2, 4 and 6 Saltmarket, Glasgow, 1894-1900+

- Ironmongers, cutlers, scale makers, japanners, tinsmiths and shop outfitters: PO Glasgow Directory 1894-1900+

WYLIE, James, Glasgow, 1801-46
1. Bridgegate, Glasgow, 1801
2. 53 Wilson Street, Glasgow, 1803-24
3. 52 Wilson Street, Glasgow, 1825-46
4. 67 Hutcheson Street, Glasgow, 1836 (adjusting office)

- Tinsmith: McFeat's Glasgow Directory 1801
- Brazier and tinsmith, and office for adjusting weights and measures: PO Glasgow Directory 1828
- Brazier, pewterer and tinsmith, and office for adjusting weights and measures: PO Glasgow Directory 1835
- Brazier, pewterer and tinsmith: PO Glasgow Directory 1837

Probably the James Wylie coppersmith who was admitted member of the Incorporation of Hammermen of Glasgow in 1825, having previously served an apprenticeship, with 'a copper gallon measure' as essay: Harry Lumsden and P. Henderson Aitken, *History of the Hammermen of Glasgow* (Paisley, 1912), 310. Succeeded as inspector of weights and measures by David Wylie (q.v.) at the Hutcheson Street address.

WYLIE, David, Glasgow, 1837-45
1. 67 Hutcheson Street, Glasgow, 1837-45
2. Sheriff Chambers, 60 Stockwell Street, Glasgow, 1843 only

- Inspector of weights and measures: PO Glasgow Directory 1837
- Inspector of weights and measures for Lanarkshire: PO Glasgow Directory 1840

Probably the David Wyllie [*sic*] admitted in 1840 with his brother Alexander, both tinsmiths, as members of the Incorporation of Hammermen of Glasgow, by right of their father, James Wylie: Harry Lumsden and P. Henderson Aitken, *History of the Hammermen of Glasgow* (Paisley, 1912), 317.

YOUNG, [Adam?], Glasgow *c*.1830
In James Cleland, *An Historical Account of the Local and Imperial Weights and Measures of Lanarkshire, and an Inventory of those belonging to the Corporation of Glasgow* (Glasgow, 1832), 34: 'A steel model of the Imperial Standard Yard, in a mahogany case. This model was made by Young, Glasgow.' Probably Adam Young, engineer, Glasgow Foundry, who appears in the PO Glasgow Directory for 1829 only.

INDEX